Advances in Non-Destructive Testing Methods

Advances in Non-Destructive Testing Methods

Editors

Grzegorz Peruń
Tangbin Xia
Bogusław Łazarz

MDPI

Basel • Beijing • Wuhan • Barcelona • Belgrade • Novi Sad • Cluj • Manchester

Editors
Grzegorz Peruń
Faculty of Transport and Aviation Engineering, Department of Road Transport, Silesian University of Technology
Katowice
Poland

Tangbin Xia
Department of Industrial Engineering & Management, Shanghai Jiao Tong University
Shanghai
China

Bogusław Łazarz
Faculty of Transport and Aviation Engineering, Department of Road Transport, Silesian University of Technology
Katowice
Poland

Editorial Office
MDPI
St. Alban-Anlage 66
4052 Basel, Switzerland

This is a reprint of articles from the Topic published online in the open access journals *Materials* (ISSN 1996-1944), *Sensors* (ISSN 1424-8220), and *Applied Sciences* (ISSN 2076-3417) (available at: https://www.mdpi.com/topics/non_destructive).

For citation purposes, cite each article independently as indicated on the article page online and as indicated below:

Lastname, A.A.; Lastname, B.B. Article Title. *Journal Name* **Year**, *Volume Number*, Page Range.

ISBN 978-3-7258-1081-9 (Hbk)
ISBN 978-3-7258-1082-6 (PDF)
doi.org/10.3390/books978-3-7258-1082-6

Contents

About the Editors

Grzegorz Peruń

Grzegorz Peruń obtained his M.Sc. degree from the Silesian University of Technology, Katowice, Poland, in 2004, followed by his Ph.D. degree in 2010. His doctoral thesis was defended with honors at the Silesian University of Technology, and also for which he received an award from the Fiat Group (Gliwice - Turyn 2011). Habilitation "Modeling dynamic of power transmission system with planetary gear in computer-aided design and diagnostics" was obtained from Air Force Institute of Technology, Warsaw, Poland, in 2018.

He is an author of over 230 published scientific works—articles, conference papers, and other scientific publications. He is a co-author of two patents, author of the monograph, academic textbook "Engineering design with software HiCAD" (2014), and co-author of two academic textbooks: "Aircraft Airframe" (2014) and "Aircraft Systems" (2015). His research interests include mechanical engineering, construction of machines, especially modeling and simulation of toothed gears, vibroacoustics, gearbox diagnostic, signal processing, transport, GIS, and for several years also non-destructive testing techniques.

Since 2019, he has been a professor SUT in the Faculty of Transport and Aviation Engineering, Department of Road Transport. He is a certified specialist in NDT testing, author of, among others, software for resonance defectoscope. He is a Member of The Polish Association of Technical Diagnostics (PTDT) and the Vice-Chairman of PTDT Audit Committee. He is a contributor to research grants and an editor of Special Issues of many journals with national and international reach. He is proficient in several programming languages, which he uses effectively in his academic work and research. He has given lectures as a part of the Erasmus program at FH Köln, Fakultät für Fahrzeugsysteme und Produktion, and Technical University of Košice, Faculty of Aeronautics.

Tangbin Xia

Tangbin Xia received a Ph.D. degree in Mechanical Engineering (Industrial Engineering) from Shanghai Jiao Tong University in 2014. He was a Visiting Scholar with WuMRC, University of Michigan, and ISyE, Georgia Institute of Technology. Currently, he is a Tenured Associate Professor and serves as Vice Dean of School of Mechanical Engineering in SJTU, Deputy Secretary General of The Shanghai Society for Modern Design Theory and Methodology Research, and the Review Expert for the Government Quality Award of Shanghai Municipality. He was awarded Young Chang Jiang Scholar of the Ministry of Education in 2022. His major research interests are Intelligent Manufacturing Systems, Quality and Reliability Engineering, and Big Data and Artificial Intelligence. His research group has published more than 100 SCI papers.

Bogusław Łazarz

Bogusław Łazarz is a graduate of the Institute of Transport, Silesian University of Technology, holding a Ph.D. in technical sciences from the Faculty of Materials Engineering, Metallurgy, and Transport, Silesian University of Technology, and a D.Sc. obtained in 2002 from the Faculty of Automotive and Construction Machinery Engineering, Warsaw University of Technology. In 2014, he was awarded the academic title of Professor at the Faculty of Transport, Warsaw University of Technology.

With extensive experience in research, teaching, and organizational and administrative roles, he is currently employed at the Silesian University of Technology in the Department of Automotive

Vehicle Construction within the Faculty of Transport. He has previously served as Deputy Dean for Organization and Development and later as Dean of the Faculty of Transportation. Presently, he holds the position of Vice-Rector for General Affairs at the Silesian University of Technology.

He is the supervisor of many engineering and master's theses as well as awarded and distinguished doctoral theses and lecturer of the following subjects: basics of machine construction, power transmission systems, engineering graphics, and methodology of scientific research. He has been honored with numerous awards from the Rector of the Silesian University of Technology for his individual and team scientific, didactic, and organizational contributions.

Additionally, he is a member of the Presidium of the Polish Academy of Sciences, 4th Faculty of Technical Sciences, Transport Committee, and serves on the Regional Council for a Just Transition of the Silesian Voivodship.

He is an author of over 380 scientific publications, and in the years 2015–2024, he has authored 121 publications, including over 50 articles in high-scoring scientific journals, monograph, conference papers, and chapters in collective works (in Polish and English) and two patents.

Preface

Non-destructive testing methods (NDT) are a group of tests, allowing to detect external (surface) as well as internal defects of the structure. They allow the determination of the state of micro- and macro-structures without interfering with the structure of the tested object and provide information about the functional properties of the tested object. This basic feature and the variety of non-destructive testing methods make them applicable in many industries. Despite the numerous advantages, NDT methods also have limitations, one of which may be the economic factor of the cost of purchasing testing equipment. NDT can be carried out at various stages of production, operation, or repair. They are used to carry out the quality control of production and to determine the technical condition of an object.

Unlike the well-known destructive testing methods that have been in use for a long time, a large number of the testing methods classified as non-destructive have been in use relatively recently. At present, a wide variety of non-destructive testing methods, operating on the basis of different phenomena, can be specified. A detailed classification and description can be found in the EN-ISO/ISO standards, among others. The most common groups of non-destructive tests include visual methods, penetrant methods, magnetic powder methods, methods using eddy currents, thermographic methods, vibration methods, acoustic methods, ultrasonic methods, and radiographic methods. Visual, penetrant, and electromagnetic methods, including eddy current, are ideal for detecting surface discontinuities. The detection of subsurface defects is made possible by eddy currents, magnetic particles, and thermographic methods. Among the methods of non-destructive volumetric testing, acoustic, ultrasonic, and radiographic methods should be listed first.

Grzegorz Peruń, Tangbin Xia, and Bogusław Łazarz
Editors

materials

Editorial

Advances in Non-Destructive Testing Methods

Grzegorz Peruń

Department of Road Transport, Faculty of Transport and Aviation Engineering, Silesian University of Technology, 40-019 Katowice, Poland; grzegorz.perun@polsl.pl

1. Introduction

Non-destructive testing (NDT) methods are a group of tests allowing one to detect external (surface) as well as internal defects of a structure. They allow one to determine the state of micro- and macro-structures without interfering with the structure of the tested object, providing information about the functional properties of the tested object. This basic feature and the variety of non-destructive testing methods make them applicable in many industries. They are notable in the diagnosis of infrastructure defects and means of air transportation and rail transportation, but are also applicable in the automotive industry [1–3]. The most obvious field of application of NDT is the quality control of final products, as well as the verification of materials and semi-finished products used in manufacturing processes. Theoretically, non-destructive testing, unlike destructive testing, can be applied to all manufactured pieces of a product; moreover, these methods are used during operation to assess the technical conditions of objects. A limitation of the number of tests conducted may be the cost of implementation and the time required to conduct the tests [4,5]. Despite these numerous advantages, NDT methods also have limitations, where one of which is the economic factor of the cost of purchasing testing equipment [6]. NDT methods can be carried out at various stages of production, operation, or repair. They are used to carry out quality control in production and to determine the technical conditions of an object, respectively. With NDT, it is also possible to detect operational defects, including corrosion, erosion, or cracks.

Unlike the well-known destructive testing methods that have been in use for a long time, a large number of testing methods classified as non-destructive have been in use relatively recently. At present, a wide variety of non-destructive testing methods, operating on the basis of different phenomena, can be specified. A detailed classification and description can be found in the work of [7–9] and in the EN-ISO/ISO standards, among others. The most common non-destructive tests include the following:

- Visual methods;
- Penetrant methods;
- Magnetic powder methods;
- Methods using eddy currents;
- Thermographic methods;
- Vibration methods;
- Acoustic methods;
- Ultrasonic methods;
- Radiographic methods.

Visual, penetrant, and electromagnetic methods, including methods using eddy currents, are ideal for detecting surface discontinuities. The detection of subsurface defects is made possible by eddy current, magnetic particle, and thermographic methods. Among the methods of non-destructive volumetric testing, acoustic, ultrasonic, and radiographic methods stand out.

Citation: Peruń, G. Advances in Non-Destructive Testing Methods. *Materials* **2024**, *17*, 554. https://doi.org/10.3390/ma17030554

Received: 20 December 2023
Accepted: 11 January 2024
Published: 24 January 2024

2. A Statistical Look at the Research Presented in this Topic

The purpose of this Topic was to present both the development of various non-destructive testing methods and examples of applications of this group of tests in various industries. Thanks to the fact that the scope of this topic was not strictly limited to defined methods, this study includes the presentation of a very wide spectrum of research and applied methods. The presented research was carried out by research centers located all over the world, and a brief description of selected studies is presented in the next chapter. Thirteen papers were published in *Applied Sciences*, eleven were published in *Sensors*, and four were published in *Materials*. The published papers include 142 authors from many institutions. These simple statistics, presented in Figure 1, show that the topic has received considerable interest and thus confirm the universality and popularity of NDT research.

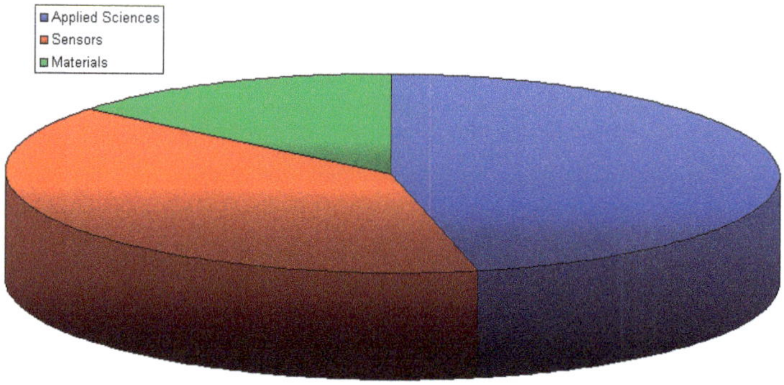

Figure 1. Participation of individual journals in the topic "Advances in Non-Destructive Testing Methods".

3. Summary of the Contributions

In one article [10], the authors propose a convolutional-neural-network-model-based through-focus scanning optical microscopy (TSOM) method for measuring individual high-aspect-ratio (HAR) grooves on silicon up to 30 μm wide and 440 μm deep. The standard deviation and error were about one hundred nanometers or less in value.

Grazing-incidence small-angle X-ray scattering (GISAXS) is an effective method for the structural analysis of ordered arrays of quantum dots, nanoparticles, or similar objects. In [11], the authors demonstrate a software platform to analyze and simulate 2D GISAXS intensity distributions from ordered networks of various nanoparticles.

The possibilities of using non-destructive testing methods in transportation are presented in [12] using air transportation as an example. High-energy radiation sources are used to provide strong safety inspection capabilities using a non-invasive imaging system. The authors present an air cargo inspection system using multiple radiation sources, a D-T neutron generator, and an accelerator-based gamma-ray source to classify plastics, metals, and organics among various sample materials. Radiation can penetrate materials tens of centimeters thick in an air cargo container.

Similarly, aerospace transportation is an area of interest for the authors of [13]. The purpose of this paper is to conduct a reliable non-destructive evaluation (NDE) of glued aerospace components by developing new multidimensional data fusion techniques that combine information obtained using ultrasonic and X-ray NDE methods.

NDT methods are also being applied to marine transportation. The marine composite market is driven by growing demand for lightweight, corrosion-resistant, and impact-resistant boats. Polymer matrix composites are currently the most popular composite materials in marine applications. Ref. [14] presents a numerical analysis of the feasibility of using vibrothermography to detect defects in fiberglass-reinforced laminates in marine applications.

Pipelines are also related to transportation, which are also of interest to those involved in non-destructive testing. Problems related to the classification and identification of defects in polyethylene (PE) gas pipelines are presented in reference [15]. The paper proposes a framework for classifying and identifying defects of underground PE gas transmission pipes. The authors prove the possibility of fast and accurate detection of defects in PE pipelines.

Ref. [16] presents a less expensive and simpler way to apply the time delay estimation method to locate leaks in underground pipelines. The accuracy of the TDE depends on the acoustic velocity and attenuation of the leak signal propagating along the pipeline. The analytical predictive model is the typical approach for obtaining the propagation velocity and attenuation of leakage waves. To determine the effective design variables in the algorithm based on the embedment parameters of the pipe, this study analyzed the sensitivity of soil parameters to the wave number.

Three-dimensional localization of buried polyethylene pipes is a problem addressed in the work of [17]. The authors developed a multi-sensor method based on the principle of pipe vibration, which allows estimating the depth, is non-invasive and non-destructive, and does not require prior information about the propagation medium.

In order to solve the problem of quantifying magnetic flux leakage (MFL) detection signals damaged in the operation of oil and gas pipelines, an inhomogeneous magnetic charge model based on magnetic effects was developed. Reference [18] discusses the force–magnet coupling relationship in the inhomogeneous magnetic charge model, develops a theoretical model of complex magnetic charge, and analyzes the pattern of changes in the characteristic values of MFL signals under elastic stress and plastic deformation.

In reference [19], the authors address the problems associated with measuring the moisture content of building materials. This is important from the point of view of both the process of building construction and the subsequent operation of said buildings. In engineering practice, indirect moisture measurement techniques, mainly resistive and capacitive, are the most popular. The main objective of this research work was to compare the classical TDR measurement technique with non-invasive surface TDR sensors.

Ref. [20] proposes an inversion algorithm for predicting dynamic moduli using CNNs which is trained using samples related to the dynamic response of a multilayer concrete pavement structure obtained using SEM. The proposed method acts as a reliable test tool for the falling weight deflection gauge technique of pavement structures.

The contribution of the study of [21] is the use of broadband ultrasonic attenuation techniques in concrete-filled steel pipe samples to locate air voids. The results show a linear relationship between the average BUA in the window of concrete-filled steel pipes and the percentage of air voids: the higher the percentage of air voids, the higher the BUA values.

The meso-scale failure properties of concrete and mortar were investigated experimentally in [22] via Brazilian disc split tensile tests combining X-ray computed tomography (CT) and digital image correlation (DIC) technology. Taking into account water/cement ratios and loading rates, the effects of meso-components on the macro tensile properties and failure modes of concrete were studied.

In reference [23], the longitudinal nonlinear ultrasonic parameters of carbonatized concrete were detected using an embedded composite piezoelectric transducer (ECPT) as a transducer, and the effect of temperature on the nonlinear parameters of carbonatized concrete was investigated. Higher variation in nonlinear parameters at the same time of carbonation was observed for samples with a high water–cement ratio than for samples with a low water–cement ratio.

The purpose of the work of [24] was to develop a technique for measuring the thickness of non-conductive coatings on non-magnetic conductive base metals. The authors used the amplitude-phase eddy current method and algorithms for processing measurement information, taking into account the effect of the base metals' specific electrical conductivity. They found two-dimensional thickness gauge graduation characteristics using several base metals with different values of specific electrical conductivity.

An innovative analytical protocol based on external reflection FTIR spectroscopy (ER-FTIR), combined with spectral deconvolution and multivariate data analysis, was used in [25]. The ER-FTIR technique is fast, portable, and widely used in the field of cultural heritage, but rarely applied to the study of textiles. This is the first time the ER-FTIR band assignment for silk has been discussed.

A completely different topic was addressed in reference [26]. A tracking model called Road-TransTrack based on transformer optimization was proposed to track and count the number of damages in road videos. The tracking and counting model was improved with a transformer and a self-observation mechanism to improve the accuracy of damage detection and counting in road videos.

The research topic of [27] is the generation of high-quality phase-contrast computed tomography images with given incomplete projections. The authors present the potential for expanding applications of PCCT techniques in the fields of composite and biomedical imaging and describe a two-domain (i.e., projection sinogram domain and image domain) deep-learning-based enhancement framework for PCCT with sparse projections. It consists of two convolutional neural networks (CNNs) in two domains and a phase-contrast Radon inversion layer (PCRIL) to connect them.

Ref. [28] presents an approach to efficiently detect local defect resonances in solids with localized defects and a method for determining the frequency of local defect resonances. The algorithm is based on the observation that vibration amplitudes are significantly higher for the modal shape of a local defect than the average value for a whole structure.

The work of [29] investigates surface roughness measurements using piezoelectric acoustic discs and corresponding signal processing. Surface roughness is one of the characteristics of surface texture, which can have various irregularities related to manufacturing methods. The study concludes that acoustic friction measurement shows promising results as a new method for measuring the surface roughness states of some materials.

In [30], the authors study various contact states of a high-voltage SF6 circuit breaker considering different currents in power system applications. This paper aims to present a novel method for detecting the state of HVCB contacts based on a vibration signal. The experimental and theoretical analyses used can provide a reference for future fault monitoring and diagnosis in HVCBs.

Conflicts of Interest: The author declares no conflicts of interest.

References

1. Żurek, Z.; Idziak, P.; Kowalski, K.; Peruń, G. Miniature resonant defectoscope for pre-diagnosis in rail transportation Pomiary magnetyczne i magnetoindukcyjne. *Napędy I Sterow.* **2018**, *20*, 11.
2. Artagan, S.; Ciampoli, L.; D'Amico, F.; Calvi, A.; Tosti, F. Non-destructive Assessment and Health Monitoring of Railway Infrastructures. *Surv. Geophys.* **2020**, *41*, 447–483. [CrossRef]
3. Hovanec, M.; Korba, P.; Jencova, E.; Koscak, P.; Pil'a, J.; Yagnik, P. Non Destructive Testing in the Case of Aircraft Wheel Overhaul. In Proceedings of the Transport Means—International Conference, Trakai, Lithuania, 3–5 October 2018.
4. Jolly, M.R.; Prabhakar, A.; Sturzu, B.; Hollstein, K.; Singh, R.; Thomas, S.; Foote, P.; Shaw, A. Review of Non-destructive Testing (NDT) Techniques and their applicability to thick walled composites. The Fourth International Conference on Through-life Engineering Services. *Procedia CIRP* **2015**, *38*, 129–136. [CrossRef]
5. Mix, E.P. *Introduction to Nondestructive Testing: A Training Guide*, 2nd ed.; Willey-Interscience: Hoboken, NJ, USA, 2005.
6. Electromagnetic Testing. In *NDT Handbook*, 2nd ed.; McIntyre, P.; Mester, M. (Eds.) ASNT: Columbus, OH, USA, 1986; Volume 4.
7. Peruń, G.; Krawiec, S. Badania nieniszczące. Charakterystyka i klasyfikacja. Wprowadzenie do metody prądów wirowych. In *Utrzymanie Ruchu*; Wydawnictwo Elamed: Katowice, Poland, 2017; p. 44.
8. Peruń, G. Badania nieniszczące—Charakterystyka badań wizualnych. In *Utrzymanie Ruchu*; Wydawnictwo Elamed: Katowice, Poland, 2022; pp. 48–52.
9. Peruń, G. Badania nieniszczące magnetyczno-proszkowe MT. In *Utrzymanie Ruchu*; Wydawnictwo Elamed: Katowice, Poland, 2023; pp. 1–5.
10. Li, G.; Shi, J.; Gao, C.; Jiang, X.; Huo, S.; Cui, C.; Chen, X.; Zhou, W. MEMS High Aspect Ratio Trench Three-Dimensional Measurement Using Through-Focus Scanning Optical Microscopy and Deep Learning Method. *Appl. Sci.* **2022**, *12*, 8396. [CrossRef]
11. Mekterović, I.; Svalina, G.; Isaković, S.; Mičetić, M. GisaxStudio—An Open Platform for Analysis and Simulation of GISAXS from 3D Nanoparticle Lattices. *Appl. Sci.* **2022**, *12*, 9773. [CrossRef]

12. Park, J.Y.; Mun, J.; Lee, J.H.; Yeon, Y.-H.; Chae, M.; Lee, M.; Lee, N.-H. Development of a Dual-Modality Gamma-ray/Fast Neutron Imaging System for Air Cargo Inspection. *Appl. Sci.* **2022**, *12*, 9775. [CrossRef]
13. Jasiuniene, E.; Yilmaz, B.; Smagulova, D.; Bhat, G.A.; Cicenas, V.; Žukauskas, E.; Mažeika, L. Non-Destructive Evaluation of the Quality of Adhesive Joints Using Ultrasound, X-ray, and Feature-Based Data Fusion. *Appl. Sci.* **2022**, *12*, 12930. [CrossRef]
14. Swiderski, W.; Strag, M. Possibilities of Detecting Damage Due to Osmosis of GFRP Composites Used in Marine Applications. *Appl. Sci.* **2023**, *13*, 4171. [CrossRef]
15. Wang, Y.; Fu, Q.; Lin, N.; Lan, H.; Zhang, H.; Ergesh, T. Identification and Classification of Defects in PE Gas Pipelines Based on VGG16. *Appl. Sci.* **2022**, *12*, 11697. [CrossRef]
16. Lu, P.; Chen, S.; Sheng, X.; Gao, Y. Application of the Differential Evolutionary Algorithm to the Estimation of Pipe Embedding Parameters. *Sensors* **2022**, *22*, 3942. [CrossRef]
17. Xerri, W.; Saracco, G.; Ribodetti, A.; Zomero, L.; Picon, P. Three-Dimensional Localization of Buried Polyethylene Pipes Using Acoustic Method. *Sensors* **2022**, *22*, 9433. [CrossRef]
18. Liu, B.; Ge, Q.; Wu, Z.; Lian, Z.; Yang, L.; Geng, H. The Signal Characteristics of Oil and Gas Pipeline Leakage Detection Based on Magneto-Mechanical Effects. *Sensors* **2023**, *23*, 1857. [CrossRef]
19. Pasnikowska-Łukaszuk, M.; Wlazło-C' wiklin' ska, M.; Zubrzycki, J.; Suchorab, Z. Comparison of Measurement Possibilities by Non-Invasive Reflectometric Sensors and Invasive Probes. *Appl. Sci.* **2023**, *13*, 665. [CrossRef]
20. Chen, G.; Chen, X.; Yang, L.; Han, Z.; Bassir, D. An Inversion Algorithm for the Dynamic Modulus of Concrete Pavement Structures Based on a Convolutional Neural Network. *Appl. Sci.* **2023**, *13*, 1192. [CrossRef]
21. Callejas, A.; Palma, R.; Hernández-Figueirido, D.; Rus, G. Damage Detection Using Ultrasonic Techniques in Concrete-Filled Steel Tubes (CFSTs) Columns. *Sensors* **2022**, *22*, 4400. [CrossRef]
22. Zhang, H.; Pan, Q.; Zheng, K.; Jin, C.; Pan, L. Mesoscale Study on Splitting Tensile Damage Characteristics of Concrete Based on X-ray Computed Tomography and Digital Image Correlation Technology. *Materials* **2022**, *15*, 4416. [CrossRef]
23. Zhao, J.; Wu, J.; Chen, X.; Zeng, R. Effect of Temperature on Ultrasonic Nonlinear Parameters of Carbonated Concrete. *Materials* **2022**, *15*, 8797. [CrossRef]
24. Syasko, M.; Solomenchuk, P.; Soloviev, I.; Ampilova, N. A Technique for Multi-Parameter Signal Processing of an Eddy-Current Probe for Measuring the Thickness of Non-Conductive Coatings on Non-Magnetic Electrically Conductive Base Metals. *Appl. Sci.* **2023**, *13*, 5144. [CrossRef]
25. Geminiani, L.; Campione, F.P.; Canevali, C.; Corti, C.; Giussani, B.; Gorla, G.; Luraschi, M.; Recchia, S.; Rampazzi, L. Historical Silk: A Novel Method to Evaluate Degumming with Non-Invasive Infrared Spectroscopy and Spectral Deconvolution. *Materials* **2023**, *16*, 1819. [CrossRef]
26. Wang, N.; Shang, L.; Song, X. A Transformer-Optimized Deep Learning Network for Road Damage Detection and Tracking. *Sensors* **2023**, *23*, 7395. [CrossRef]
27. Zhang, C.; Fu, J.; Zhao, G. Learning fromProjection to Reconstruction: A Deep Learning Reconstruction Framework for Sparse-view Phase Contrast Computed Tomography via Dual-domain Enhancement. *Appl. Sci.* **2023**, *1*, 6051. [CrossRef]
28. Zdziebko, P.; Krzemiński, M.; Okon, M.; Loi, G.; Aymerich, F.; Pieczonka, Ł.; Klepka, A. An Approach to the Automated Characterization of Out-of-Plane and In-Plane Local Defect Resonances. *Materials* **2023**, *16*, 3084. [CrossRef]
29. Kursun, K.; Güven, F.; Ersoy, H. Utilizing Piezo Acoustic Sensors for the Identification of Surface Roughness and Textures. *Sensors* **2022**, *22*, 4381. [CrossRef]
30. Guo, Z.; Li, L.; Han, W.; Guo, Z. SF6 High-Voltage Circuit Breaker Contact Status Detection at Different Currents. *Sensors* **2022**, *22*, 8490. [CrossRef]

sensors

MDPI

Technical Note

Application of the Differential Evolutionary Algorithm to the Estimation of Pipe Embedding Parameters

Ping Lu [1,*], Shuang Chen [2,*], Xiaozhen Sheng [3] and Yan Gao [4]

1 State Key Laboratory of Traction Power, Southwest Jiaotong University, Chengdu 610031, China
2 Rail Transit College, Chengdu Vocational & Technical College of Industry, Chengdu 610213, China
3 School of Urban Railway Transportation, Shanghai University of Engineering Science, Shanghai 201620, China; shengxiaozhen@hotmail.com
4 Key Laboratory of Noise and Vibration Research, Institute of Acoustics, Chinese Academy of Sciences, Beijing 100190, China; gaoyan@mail.ioa.ac.cn
* Correspondence: lupingbean@163.com (P.L.); cscdgzy@163.com (S.C.)

Abstract: The time-delay estimation (TDE) method is the primary method for predicting leakage locations in buried water distribution pipelines. The accuracy of TDE depends on the acoustic speed and attenuation of the leakage signal propagating along the pipeline. The analytical prediction model is the typical approach for obtaining the propagation speed and attenuation of leakage waves. However, the embedding parameters of the buried pipe in this model must be measured using soil tests, which are very difficult, costly, and time-consuming. These factors restrict the application of the TDE method in pinpointing pipeline leakage. A method for inverse identification of pipe embedding parameters using discrete wavenumbers obtained in field testing is presented in this paper, and the differential evolution algorithm is introduced as an optimization solution. A field experiment is conducted to validate the method, and the test wavenumbers are measured in a cast-iron pipeline. The estimated sensitive parameters in the analytical model using the method are soil elastic modulus, Poisson's ratio, and pipe–soil contact coefficient, while the conventional soil test is used to measure the soil density due to the character of the optimization algorithm and the soil properties. The application effects show that the estimated parameters are close to those measured from a conventional soil test. The wave speed based on the estimated parameters was an excellent match for the on-site test in the engineering application. This work provides a less costly and more straightforward way to apply the TDE method for leak localization in buried pipelines.

Keywords: buried water pipe; pipe embedding parameters estimation; differential evolutionary algorithm; wavenumber estimation

Citation: Lu, P.; Chen, S.; Sheng, X.; Gao, Y. Application of the Differential Evolutionary Algorithm to the Estimation of Pipe Embedding Parameters. *Sensors* **2022**, *22*, 3942. https://doi.org/10.3390/s22103942

Academic Editor: Grzegorz Peruń

Received: 25 April 2022
Accepted: 20 May 2022
Published: 23 May 2022

1. Introduction

Water leakage, especially in buried pipelines, is a subject of increasing concern across the world because of the potential public health danger, economic constraints, environmental damage, and wastage of energy. Detecting and pinpointing leakage provides a key means to solving the issue. The leak detection methods in pipes can be classified into three categories [1]: methods based on signal processing, methods based on state estimation, and methods based on knowledge. Most methods based on signal processing focus on utilizing measurements collected from different sensors and applying different analytical techniques to detect and localize faults. The state estimation methods thus far focus on developing and using models based on fundamental principles to detect and localize leaks. Most of the methods based on knowledge were initially designed to detect leaks in systems with single flow. Among these methods, acoustic-based technology is more suitable for leak detection in water distribution pipelines.

Acoustic-based leak detection techniques have been in common use in water-distribution networks over the past 30 years [2,3]. They usually use the cross-correlation function between leak noise signals measured using hydrophones or accelerometers placed on both sides of the leak. The efficacy of a correlator depends upon knowledge of the speed at which the leak noise propagates along the pipe as well as how much it attenuates with distance. As is known, the fluid-borne wave in the pipe is the predominant energy-carrying mode in the pipe leakage detection field, and its propagating characteristics are profoundly influenced by the pipe and soil properties, especially in the pipe–soil strong coupling pipeline systems, e.g., buried plastic pipe systems. Muggleton [4] and Gao [5,6] observed that the medium outside the pipe acts on the pipe with additional mass and additional stiffness through the interface shear effect, affecting the acoustic speed and attenuation. Based on the analytical and finite element models, Brennan [7] further studied the effects of the soil properties surrounding the pipe on the propagation speed of the leakage acoustic wave, pointing out that the shear properties of the soil mainly affected the propagation speed of the acoustic wave, and the compression properties mainly affected the attenuation of the acoustic wave. The models were subsequently applied to two different types of soil—one sandy, the other clay—to validate the above study.

The theoretical model plays an important role in predicting leak wave speed and attenuation along the pipe. Although the pipe property parameters in this model can be determined relatively easily, estimation of the soil property parameters surrounding the pipe is more problematic. Representative soil samples—used for parameter testing—are difficult to obtain due to the large covering scale and the complex embedding conditions of the on-site pipeline. The subsequent time-consuming soil tests of the soil samples make them costly for practical applications. Meanwhile, the presumption of the theoretical model generally reduces agreement between predicted and actual wave propagating speed. Accordingly, in leakage-location engineering practice, a more accurate estimation method for soil properties is needed, along with a more convenient and precise method of determining propagating speed and attenuation of the leak wave.

Currently, in many industrial applications, the inversion identification of model parameters using field test data is an emerging approach to obtaining the soil parameters in a model [8,9]. Jesenik et al. [10] tested different soil models on measured data and used different metaheuristics to determine soil parameters. These inversion methods of soil parameters are worthy of reference, but the required test methods are not suitable for buried pipe conditions. Scussel et al. [11] introduced the idea of inversion into the buried liquid-filled pipe and used the cost-function algorithm to estimate the bulk modulus and shear modulus of the soil. However, the two above estimated parameters are not independent, which easily leads to multiple solutions. Meanwhile, it is easy to be trapped in local optima when solved by the cost-function algorithm for optimization problems with unclear gradient information.

An embedding parameters identification method for water-filled pipes based on the differential evolutionary algorithm is put forward using field test data. With this method, the pipe embedding parameters can be quickly inverted based on a few discrete leakage test data. The continuous speed in the full frequency band of interest can be obtained easily through the amended analytical model.

2. Method for Wavenumber Prediction

The leakage acoustic wave propagation has multiple modes and dispersion characteristics for the liquid-filled pipe system with a high coupling of pipe/soil. To predict the wave speed more accurately, Fuller [12] and Pinnington [13] proposed a wave-speed prediction analytical model for a liquid-filled pipe in a vacuum, which considered the dispersion characteristics for the modes of sound propagation along with the fluid and pipe. Muggleton [4] and Gao [5] proposed a theoretical model of wavenumber prediction considering the effect of external pipe medium on sound-wave propagation under the assumed extreme condition of lubricated contact and compact contact between pipe and soil. They then proposed the method for calculating the propagation speed of buried liquid-filled pipe, meanwhile observing that the fluid-dominated axisymmetric wave is the primary mode in leak detection.

Lu et al. [14] introduced the pipe–soil contact state variable into the fluid-dominated axisymmetric wave equations for the first time, which broke through the assumption of the extreme buried conditions, and the general wave equation of the axisymmetric wave of the buried liquid-filled pipe is obtained as follows:

In the above equations, the parameters of the pipe and fluid are easily available. The factors needing identification in the system are usually the soil parameters, which include Young's modulus, Poisson's ratio, density, and contact coefficient between pipe and soil. The first three parameters are generally obtained through conventional soil tests. The main on-site test methods for soil density are the ring knife method and the irrigation method, which are both mature technology [15]. The Poisson's ratio is obtained by measuring the volume change, and is difficult to test accurately. The determination of the elastic modulus requires a high production level of instrument and equipment [16], which are not easy to operate in practice. Meanwhile, the determination method for the deformation modulus and shear modulus is easier to realize. The actual elastic modulus of the soil is often converted from the bulk modulus and shear modulus, and the conventional test method for both is the three-axis test [17]. At present, there is no suitable test technology for the contact coefficient of pipe and soil interface, which can only be gained by experience.

$$k_1^2 = k_f^2\left(1 + \frac{\beta}{1 - \Omega^2 + \alpha}\right), \tag{1}$$

where:

$$\begin{cases} \alpha = -SL_{22} - \frac{[v_p + \mathrm{i}SL_{12}/k_1 a][v_p - \mathrm{i}\xi SL_{21}/k_1 a]}{1 - \xi SL_{11}/k_1 a} \\ \beta = 2\frac{a}{h}\left(\frac{1 - v_p^2}{E_p}\right)B_f \end{cases}, \tag{2}$$

where α stands for the surrounding medium loading and pipe parameters, which can be used to evaluate the influence of soil load on the pipe wall displacement, β refers to fluid and pipe parameters that can be used to evaluate the influence of fluid load on the pipe wall displacement, Ω is the non-dimensional frequency, $\Omega = \omega a/c_L = k_L a$, $k_f = \omega/c_f$ is the fluid wavenumber, $c_f = \sqrt{B_f/\rho_f}$ is the free-field fluid wave speed, c_L is the shell compressional wave speed, k_L is the shell compressional wavenumber, and k^r_{fs} is the internal fluid radial wavenumber, which can be expressed as $(k^r_{fs})^2 = {k_f}^2 - {k_s}^2$.

$$\begin{cases} SL_{11} = -\mu_m \frac{(1 - v_p^2)}{E_p} \frac{a}{h} \frac{k^r_{ds} a k^r_{rs} a^2}{k^r_{rs} a k^r_{ds} a\left[\mathrm{H}_0(k^r_{rs} a)/\mathrm{H}_0'(k^r_{rs} a)\right] + k_s^2 a^2\left[\mathrm{H}_0(k^r_{ds} a)/\mathrm{H}_0'(k^r_{ds} a)\right]} \\ SL_{12} = i\mu_m \frac{(1 - v_p^2)}{E_p} \frac{a}{h} k_s a \left\{ 2 - \frac{k_r^2 a^2 \mathrm{H}_0(k^r_{ds} a)/\mathrm{H}_0'(k^r_{ds} a)}{k^r_{rs} a k^r_{ds} a\left[\mathrm{H}_0(k^r_{rs} a)/\mathrm{H}_0'(k^r_{rs} a)\right] + k_s^2 a^2\left[\mathrm{H}_0(k^r_{ds} a)/\mathrm{H}_0'(k^r_{ds} a)\right]} \right\} \\ SL_{21} = SL_{12} \\ SL_{22} = -\mu_m \frac{(1 - v_p^2)}{E_p} \frac{a}{h} \left\{ 2 + \frac{k^r_{rs} a k^2_{rs} a^2 \left[\mathrm{H}_0(k^r_{rs} a)/\mathrm{H}_0'(k^r_{rs} a)\right]\left[\mathrm{H}_0(k^r_{ds} a)/\mathrm{H}_0'(k^r_{ds} a)\right]}{k^r_{rs} a k^r_{ds} a\left[\mathrm{H}_0(k^r_{rs} a)/\mathrm{H}_0'(k^r_{rs} a)\right] + k_s^2 a^2\left[\mathrm{H}_0(k^r_{ds} a)/\mathrm{H}_0'(k^r_{ds} a)\right]} \right\} \end{cases}, \tag{3}$$

k^r_{ds}, k^r_{rs} are the compression and shear wavenumbers of soil in the radial direction respectively, which can be expressed by compressed wavenumber k_d, shear wavenumber k_r, and wavenumber in the axil direction k_s as follows:

$$\begin{cases} (k^r_{ds})^2 = {k_d}^2 - {k_s}^2 \\ (k^r_{rs})^2 = {k_r}^2 - {k_s}^2 \end{cases}, \tag{4}$$

$$\begin{cases} {k_d}^2 = \rho_m \omega^2/(\lambda_m + 2\mu_m) \\ {k_r}^2 = \rho_m \omega^2/\mu_m \end{cases}, \tag{5}$$

λ_m, μ_m are Lame coefficients, and ρ_m is the density of the medium.

$$k_L = \omega^2 \rho_p \left(1 - {v_p}^2\right) / E_p, \tag{6}$$

$$\begin{cases} \lambda_m = E_m v_m / (1 + v_m)(1 - 2v_m) \\ \mu_m = E_m / 2(1 + v_m) \end{cases}, \tag{7}$$

$\xi \in [0,1]$ represents the contact coefficients to the actual boundary conditions at the pipe–soil interface, among which $\xi = 1$ represents compact contact, and $\xi = 0$ represents lubrication contact. The other parameter details in Equations (1)–(7) are given in reference [14].

In the above equations, the parameters of the pipe and fluid are easily available. The factors needing identification in the system are usually the soil parameters which include Young's modulus, Poisson's ratio, density, and contact coefficient between pipe and soil. The first three parameters are generally obtained through conventional soil tests. The main on-site test methods for soil density are the ring knife method and the irrigation method, which are both mature technology [15]. The Poisson's ratio is obtained by measuring the volume change, and is difficult to test accurately. The determination of the elastic modulus requires a high production level of instrument and equipment [16], which are not easy to operate in practice. Meanwhile, the determination method for the deformation modulus and shear modulus is easier to realize. The actual elastic modulus of the soil is often converted from the bulk modulus and shear modulus, and the conventional test method for both is the three-axis test [17]. At present, there is no suitable test technology for the contact coefficient of pipe and soil interface, which can only be gained by experience.

By Equation (2), β can be obtained directly, but α, related to the unknown wavenumber k_1, cannot be solved directly. As it is difficult to obtain the closed analytical solution of Equation (1), the numerical method is adopted by transforming the differential equation solving problems into optimization problems.

The above theoretical model can be solved using the Nelder–Mead method [18] to obtain the wavenumber information. To pick out effective design variables in the algorithm in Section 3, the sensitivity of the wavenumber was analyzed for each parameter, as shown in Figures 1–3. The parameters of the pipe system are shown in Table 1.

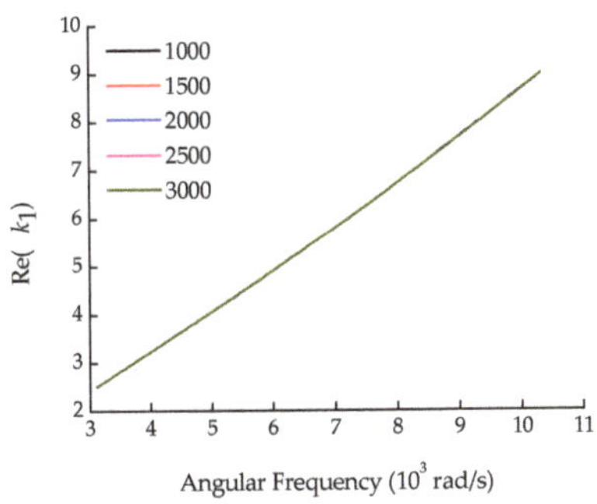

(a)

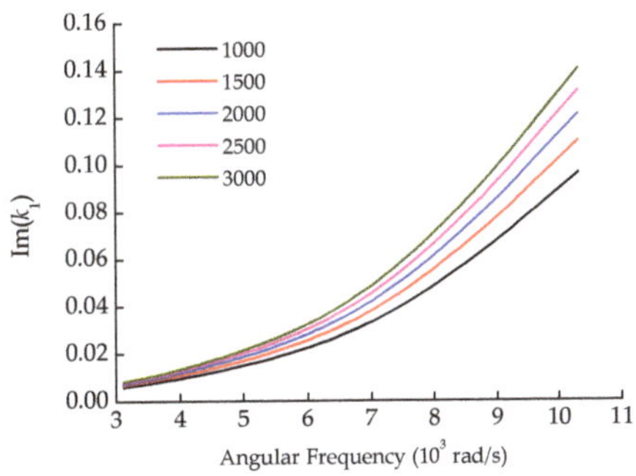

(b)

Figure 1. The influence of soil density on wavenumber. (**a**) Real part; (**b**) Imaginary part.

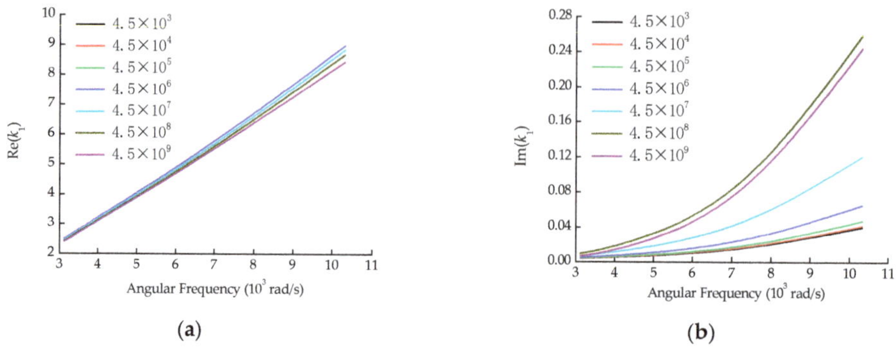

Figure 2. The influence of soil elastic modulus on wavenumber. (**a**) Real part; (**b**) Imaginary part.

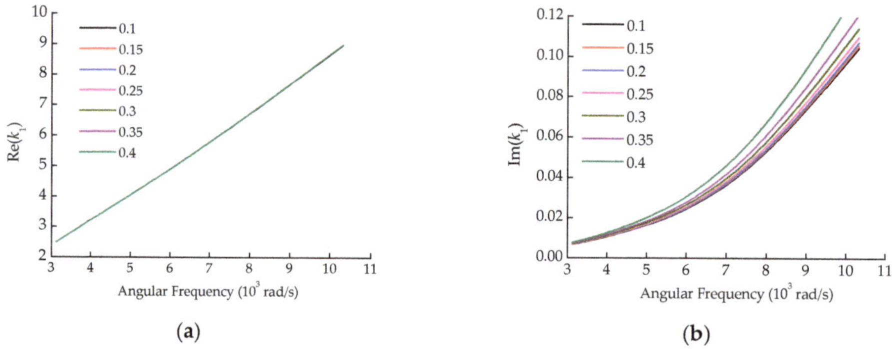

Figure 3. The influence of Poisson's ratio on wavenumber. (**a**) Real part; (**b**) Imaginary part.

Table 1. Properties of the theoretical model for the cast iron pipeline.

Properties	Fluid	Pipe	Surrounding Medium
Density (kg/m^3)	1000	7800	1999
Young's modulus (N/m^2)	-	1.22×10^{11}	4.5×10^7
Bulk modulus (N/m^2)	2.25×10^9	-	5.0×10^7
Shear modulus (N/m^2)	-	-	1.67×10^7
Poisson's ratio	-	0.25	0.35
Material loss factor	-	0.01	-

It can be seen from Figures 1–3 that the effects of soil density and Poisson's ratio on the wavenumber are similar, both having a considerable influence on the imaginary part, but little influence on the real part. The elastic modulus affects both the real and imaginary parts of the wavenumber, especially when the elastic modulus increases above 4.5×10^7 N/m^2. From the optimization algorithm viewpoint, the design variables with similar influences on the objective function are likely to produce thematical multiple solution problems. Thus, soil density and Poisson's ratio could not be used simultaneously as design variables for the optimization algorithm. Density is relatively stable and easy to measure, while Poisson's ratio is the physical parameter that is not easy to measure accurately by the current test means. Therefore, the Poisson's ratio was taken as the identification target parameter of the optimization algorithm, while the density was still determined by the conventional soil test method.

3. The Estimation Method of Pipe Embedding Parameters

According to the theory of wavenumber prediction, there is a mapping relationship between pipeline embedding parameters and wavenumbers. When the wavenumbers corresponding to some frequencies are obtained by field testing, theoretically, the pipe embedding parameters approaching the test wavenumber can be obtained through mathematical optimization.

Traditional mathematical optimization methods rely on the derivative or gradient matrix of each iteration step to determine the next step's search direction and step length. However, the above theoretical wavenumber is not directly expressed by the wave equation but obtained by numerical methods, which leads to the difficulty in obtaining a gradient matrix, and even to singularity. Consequently, the methods dependent on gradient information are difficult to use directly for pipeline embedding parameter identification.

Due to the particularity of the engineering problems, optimization algorithms that do not require a gradient matrix are widely adopted. The EM algorithm is a commonly used tool for estimating the parameters for a mixture model but is more dependent on the initial values [19]. Bio-inspired optimization that does not need to iterate with a gradient matrix and does not depend on the initial values is a growing research topic to solve large-scale complex optimization problems. Jesenik, et al. [20] employed bio-inspired methods to determine a DC motor and drive parameters. Due to its huge computational amount, complex structures and many parameters are needed to control the bio-inspired algorithm in the application. LaTorre et al. [21] proposed methodological guidelines to prepare a successful proposal through many surveys. Liu et al. [22] introduced a parameter control approach utilized as feedback to control evolution processes adaptively.

Genetic algorithm (GA), particle swarm optimization (PSO), and differential evolution algorithm (DE) are very excellent bio-inspired optimization processes; however, each would need to be adapted according to the actual engineering problem. DE was chosen mainly because the algorithm does not have patent protection, which means it is possible to popularize the soil parameter estimation method. DE was put forward by Storn and Price [23] in 1995 and has gained wide applications [24–26].

DE is a parallel direct search method that utilizes NP dimensional parameter vectors as population $\mathbf{X}$ for each generation. In searching the optimal solution, two-parent vectors were selected and subtracted to obtain the differential vector. DE generates new parameter vectors by adding the weighted difference between two population vectors to a third vector, called the mutation operation. Then, the mutated vectors are mixed with the parameters of another predetermined vector to yield the trial vector. If the trial vector yields a lower cost function value than the target vector, the trial vector replaces the target vector in the following generation, which is called the selection operation. Through the several-generation evolution of mutation, crossover, and selection operation, the optimal individuals are retained, inferior individuals are eliminated, and the population is guided to approach the optimal value gradually.

According to the analysis in Section 2, the parameters to be identified are the Elastic modulus E, Poisson's ratio v of the medium surrounding the pipe, and the contact coefficient of pipe and soil ξ, which constitute the vectors of the optimized design variables $\mathbf{x}$:

$$\mathbf{x} = (E, v, \xi). \tag{8}$$

The parameters based on their physical meaning are set as:

$$\begin{cases} \mathbf{x}_{\min} = [1 \times 10^3, 0, 0] \\ \mathbf{x}_{\max} = [1 \times 10^{10}, 0.5, 1] \end{cases}. \tag{9}$$

NP sets of initial values are randomly generated within the constraints of the vector, which form a population $\mathbf{X}$:

$$\mathbf{X} = [\mathbf{x}_1, \mathbf{x}_2, \cdots, \mathbf{x}_{NP}], \tag{10}$$

The approximation of the wavenumber corresponding to the target value of each group of the individual vector was calculated using the function f:

$$f(E, v, \xi) = \sqrt{(w_{\text{re}} f_{\text{re}})^2 + (w_{\text{im}} f_{\text{im}})^2} \to \text{min}. \tag{11}$$

Since the wavenumber is complex, it is commonly difficult to simultaneously achieve the same degree of approximation in real and imaginary parts. Therefore, the weight coefficient w_{re}, w_{im} is introduced to consider the relative importance of the real part f_{re} and the imaginary part f_{im}:

$$\begin{cases} f_{\text{re}}(E, v, \xi) = \sum_{i=1}^{M} \frac{\left| k_{\text{re}}{}^{tar}(\omega_i) - k_{\text{re}}{}^{calc}(\omega_i) \right|}{M k_{\text{re}}{}^{tar}(\omega_i)} \\ f_{\text{im}}(E, v, \xi) = \sum_{i=1}^{M} \frac{\left| k_{\text{im}}{}^{tar}(\omega_i) - k_{\text{im}}{}^{calc}(\omega_i) \right|}{M k_{\text{im}}{}^{tar}(\omega_i)} \end{cases} \tag{12}$$

where M is the number of frequencies requiring calculation, k^{tar} is the target wavenumber, and k^{calc} is the calculated wavenumber in the current step.

After several generation populations of mutation, crossover, and selected operation, the function f approximates to a minimum. When the smallest function value f in the population no longer decreases in the subsequent t cycles, the loop will exit, and the exit condition can be expressed as:

$$\left| \min f_g - \min f_{g+t} \right| < \varepsilon, \tag{13}$$

where g is the current number of cycles, termed the generation number, and ε is the tolerance error.

The final identified pipe embedding parameters are the vector corresponding to the individual minimizing the function f in the last population. The solution process is shown in Algorithm 1.

In Algorithm 1, j is the individual sequent number, and $j \in [1, NP]$; $\mathbf{b}_j$ is the mutated individual, $\mathbf{u}_j$ is the crossovered individual, and $\mathbf{x}_j$ is the individual in current population, respectively; F is the amplification factor of the differential variation; $rand(m)$ indicates generating a random value from [0, 1] in each dimension; $\mathbf{q}$ is a dimensional sequent of the individual vector; and CR is the crossover constant that influences computational efficiency and accuracy—in this paper, $CR = 0.5$.

Choosing a reasonable amplification factor F is essential for the algorithm to weigh the global convergence difficulty and computational efficiency. Larger values of F can help the function jump from the local optimum to the global optimum but deteriorate the algorithm's convergence. Smaller values of F are in favor of the convergence; however, against the computational efficiency and prone to a local optimum. The elastic modulus as the design variable varies widely, and its effect on the objective function is non-monotonic. It is improper to use a traditionally constant value for F. A variable F is employed with the increasing generation number, termed $F = 2 - g/300$, and 300 is the maximum number of cycles in this paper. At the beginning of the algorithm's operation, a particularly high value should be set to ensure the population's approximation to the global optimum. In the following stages of the algorithm, the value of F becomes smaller to ensure convergence. F is a real and constant factor in a particular generation and varies with each generation. If a mutant individual exceeds the boundary values shown in Equation (9), the mutant individual will be replaced by the boundary values.

Algorithm 1 Pseudo-Code for the DE Process

set control parameters NP, CR, ε.
randomly generate the initial population of vectors in 3-dimensional search space
$g = 1$
repeat
for each individual j in the population **do**
    // begin mutation operation
    $F = 2 - g/300$
    select three mutually exclusive random individuals $\mathbf{x}_{r1}, \mathbf{x}_{r2}, \mathbf{x}_{r3}$
    where, $r1 \neq r2 \neq r3$
    generate a donor individual by Equation (14)

$$\mathbf{b}_j = \mathbf{x}_{r1} + F(\mathbf{x}_{r2} - \mathbf{x}_{r3}) \tag{14}$$

    // end mutation operation
    do
        while (g $\leq$ 300 **AND** Equation (13) is false)
        // begin crossover operation
        m = a random integer in the range of [1, 3]
        randD = [1, 2, 3]
        generate a trial individual $\mathbf{u}_j$ employing crossover by Equation (15)

$$u_{jq} = \begin{cases} b_{jq}, \text{ if rand}(m) \leq CR \text{ or randD}(q) = m,\ q \in [1,\ 2,\ 3] \\ x_{jq}, \text{ for all other dimensions} \end{cases} \tag{15}$$

        // end crossover operation
        // begin selection operation
        evaluate the candidate individual $\mathbf{u}_j$ using Equation (11)
        replace $\mathbf{x}_j$ with $\mathbf{u}_j$, if fitness of $\mathbf{u}_j$ is better than fitness of $\mathbf{x}_j$
        // end selection operation
        end while
    $g = g + 1$
end for

4. Application of the Method

To validate the feasibility of the method, the cast-iron pipe embedding parameters were estimated using a set of wavenumbers obtained from the on-site test. Figure 4 shows a schematic diagram of the test of the acoustic wave propagation in a buried water pipe. The test platform is shown in Figure 5. Sensor 0 was installed on a hydrant to collect the vibration response in the normal direction caused by water discharge from the hydrant. Sensors 1 and 2 were installed on the pipe wall in the inspection wells on both sides of the hydrant to collect the leakage propagation signals.

The time delay between the two-way leak signals collected by Sensor 1 and Sensor 2 can be acquired using the cross-correlation analysis method. The corresponding wave speed and attenuation are easily available through the acquisition of the interval distance from the hydrant to either test point. Accordingly, the test can obtain wavenumbers at the characteristic frequencies, as shown in Table 2, which shows the target wavenumber k^{tar} in Equation (12).

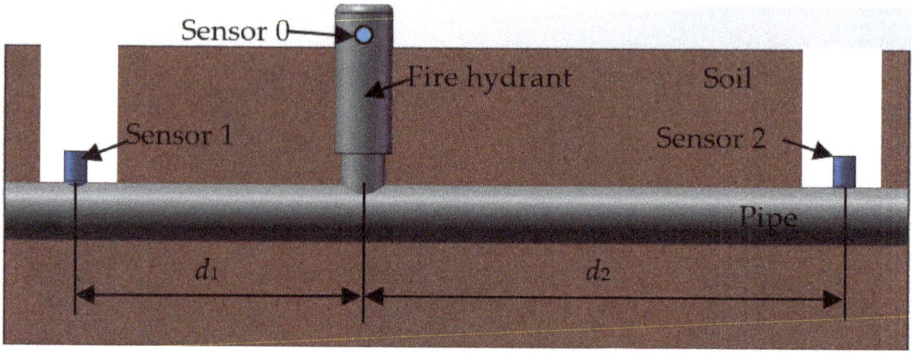

Figure 4. Schematic of the set-up for acoustic wave propagation estimation.

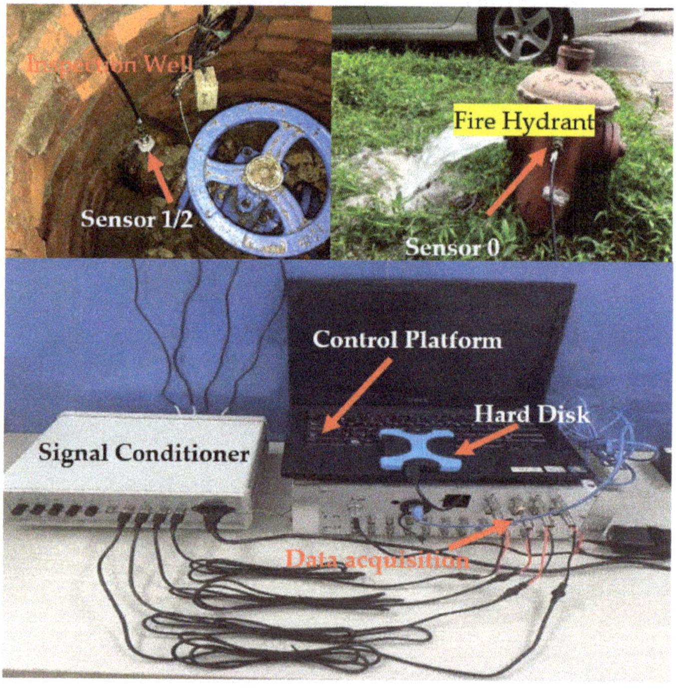

Figure 5. Test platform.

Table 2. The real and imaginary parts of the test wavenumber.

No.	Angular Frequency (rad/s)	Real Part of Wavenumber	Imaginary Part of Wavenumber
1	3121.9580	2.5080	0.00544
2	4167.0400	3.3276	0.00882
3	7221.7360	5.9966	0.02740
4	10,314.4800	8.1600	0.05960

In DE calculation, each population vector serves once as the target vector, so that *NP* competitions occur in one generation. The larger the *NP*, the more computation in a single cycle; however, there are fewer total cycles with larger *NP*. Because the *NP* competitions within the same generation can be performed in parallel, *NP* should be set as a multiple of the number of computer threads. Moreover, it is also undesirable to significantly increase *NP* to reduce the cycles, as this would result in a longer overall calculation time. In this calculation, the number of computer threads is 10, setting *NP* as 50, and $w_{re} = 1$, $w_{im} = 5$ according to experience and convenience. Figure 6 shows the trend of the approximation function *f* with the elastic modulus and the Poisson's ratio. The approximation function has a local minimum on each side of the elastic modulus of 1×10^9 N/m^2. The proposed algorithm randomly generates the initial population in the whole domain, which can effectively jump from the local optimum to the global optimum. The final estimate is the parameters corresponding to the minimum value of *f* in the figure.

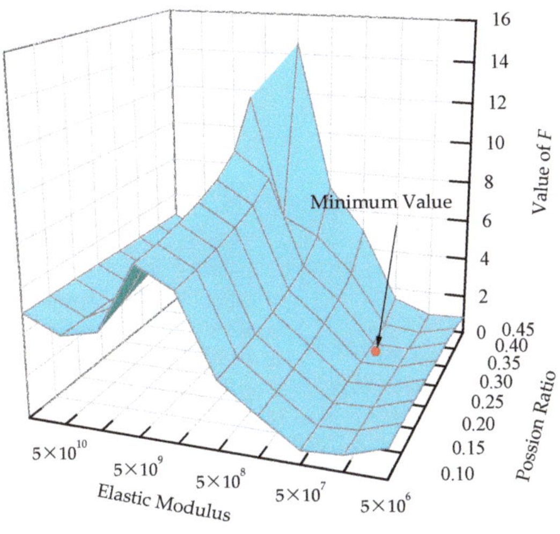

Figure 6. The variety of approximate functions with the identified soil parameters.

Table 3 shows the respective values of the pipe embedding parameters obtained according to the conventional soil test and estimated method. As shown in the table, the values estimated by the inversion identification approach are very close to the soil test measurement. The deviation rate of the elastic modulus is about 6.44%, and the deviation rate of Poisson's ratio is about 4.97%, meeting the accuracy requirements for soil engineering applications.

Table 3. Comparison between the identified values and the test values for soil parameter.

Items	Elastic Modulus N/m^2	Poisson Ratio	Contact Coefficient
Test Values	4.5×10^7	0.35	-
Estimated Values	4.79×10^7	0.3326	0

The parameter estimation aims to provide a quick and accurate way to obtain the continuous wave speed in the full interest frequency domain through the wavenumber theoretical model. The soil test and estimated parameter values in Table 3 were each substituted into the theoretical formula in Section 2. Theoretical Wavenumbers 1 and 2 were obtained, as shown in Figure 7, and compared with the test wavenumber. The Wavenumber 1 curve represents the results calculated by soil test values, and the Wavenumber 2 curve represents those calculated by the estimated values. As is generally believed, the test wavenumber is the most accurate for water leakage pinpoint localization. There was a good consistency for the real part; however, there was a notable difference for the imaginary part between Wavenumber 1 and Wavenumber 2. Wavenumber 2 was closer to the test value for both the real and imaginary parts.

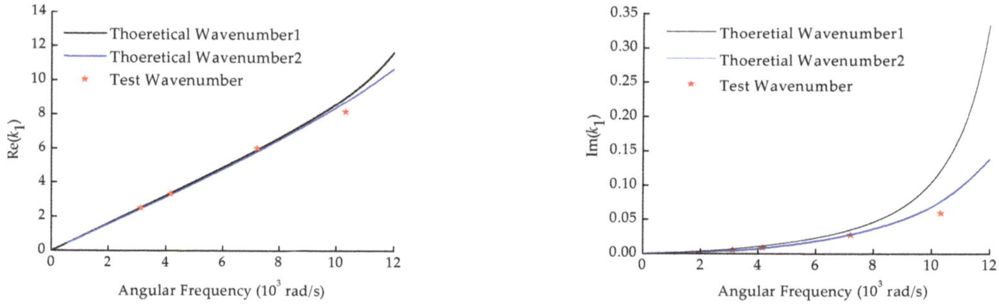

(a) (b)

Figure 7. The comparison between the theoretical wavenumbers and the field test wavenumbers. (**a**) Real part; (**b**) Imaginary part.

The deviation rate of the two theoretical wavenumbers relative to the test wavenumber is shown in Table 4. The theoretical wavenumber deviation rate of the real part calculated using the estimated values of pipe embedding parameters was slightly higher than that calculated using the soil test values generally. The deviation rates were within 5.68%, meeting the engineering requirements. Meanwhile, regarding the deviation rate of the imaginary part, Theoretical Wavenumber 2 had a better performance. For the soil samples used in the soil test, it was difficult to fully represent embedding soil properties along the pipeline due to the inhomogeneity of the site soil. This may be one of the reasons for the large deviation of the imaginary part, as well as the real part at high-frequency in Theoretical Wavenumber 1. Since they were derived from on-site buried conditions, the estimated values of pipe embedding parameters adequately matched the on site pipe buried situation in the engineering application.

Table 4. The deviation rate relative to the test wavenumber.

No.	Angular Frequency (Hz)	Real Part of Wavenumber 1	Real Part of Wavenumber 2	Imaginary Part of Wavenumber 1	Imaginary Part of Wavenumber 2
1	3121.9580	0.41%	2.60%	34.38%	19.12%
2	4167.0400	1.16%	1.85%	34.01%	18.37%
3	7221.7360	1.32%	4.36%	29.45%	13.47%
4	10,314.4800	10.17%	5.68%	103.29%	9.06%

5. Conclusions

An estimation method for pipe embedding parameters was put forward to improve the applicability of the wavenumber prediction theory model for leakage location. This method can quickly identify the sensitive pipeline embedding parameters according to the test wavenumber based on the differential evolutionary algorithm.

To ascertain the effective design variables in the algorithm from the pipe embedding parameters, an analysis of the sensitivity of the soil parameters to wavenumber was carried out in this work. Since the conventional soil test can easily measure the soil density, it was reasonable and feasible to choose the soil elastic modulus, Poisson's ratio, and the pipe/soil contact coefficient as the design variables to be estimated.

The DE algorithm was briefly introduced in Section 3 of this paper, and the procedure of the DE algorithm was given considering the specific situation of the soil parameter identified. Control methods for population mutation and crossover in soil parameter estimation were given, and the feasibility of the method was validated based on the cast-iron pipeline field test. The outputs showed that the estimated parameters were very close to those obtained from the soil test, the deviation rate of the elastic modulus was about 6.44%, the Poisson's ratio is about 4.97%, and the estimated pipe–soil contact coefficient is 0, which is consistent with traditional cast-iron pipes. Compared with the tested wavenumbers, the maximum deviation rate of the theoretical wavenumbers calculated with the estimated parameters was 5.68% for the real part and 19.12% for the imaginary part, which satisfies the leakage localization requirement in engineering applications.

According to the acoustic wave propagation theory of the buried liquid-filled pipe, the coupling effect of soil on the plastic pipe is enhanced, and wave propagating speed and attenuation are seriously affected by the surrounding medium. Future work will explore methods for estimating soil parameters for plastic pipes. Meanwhile, research on the application of this method to on-site pipe leak localization engineering will also be conducted.

Author Contributions: Conceptualization, P.L. and X.S.; methodology, P.L and S.C.; validation, Y.G.; investigation, P.L.; resources, P.L.; data curation, P.L. and X.S.; writing—original draft preparation, P.L.; writing—review and editing, X.S.; visualization, X.S.; supervision, S.C.; project administration, S.C.; funding acquisition, Y.G. All authors have read and agreed to the published version of the manuscript.

Funding: This research was funded by the National Natural Science Foundation of China (No. 11774378) and the Science and Technology Project of Huaneng Group Headquarters (No. HNKJ21-HF197).

Institutional Review Board Statement: Not applicable.

Informed Consent Statement: Informed consent was obtained from all subjects involved in the study.

Data Availability Statement: Not applicable.

Conflicts of Interest: The authors declare no conflict of interest.

References

1. Behari, N.; Sheriff, M.Z.; Rahman, M.A.; Nounou, M.; Hassan, I.; Nounou, H. Chronic leak detection for single and multiphase flow: A critical review on onshore and offshore subsea and arctic conditions. *J. Nat. Gas. Sci. Eng.* **2020**, *81*, 103460. [CrossRef]
2. Fuchs, H.V.; Riehle, R. Ten years of experience with leak detection by acoustic signal analysis. *Appl. Acoust.* **1991**, *33*, 1–19. [CrossRef]
3. Hunaidi, O.; Chu, W.T. Acoustical characteristics of leak signals in plastic water distribution pipes. *Appl. Acoust.* **1999**, *58*, 235–254. [CrossRef]
4. Muggleton, J.M.; Yan, J. Wavenumber prediction and measurement of axisymmetric waves in buried fluid-filled pipes: Inclusion of shear coupling at a lubricated pipe/soil interface. *J. Sound. Vib.* **2013**, *332*, 1216–1230. [CrossRef]
5. Gao, Y.; Sui, F.; Muggleton, J.M.; Yang, J. Simplified dispersion relationships for fluid-dominated axisymmetric wave motion in buried fluid-filled pipes. *J. Sound. Vib.* **2016**, *375*, 386–402. [CrossRef]
6. Gao, Y.; Liu, Y.; Muggleton, J.M. Axisymmetric fluid-dominated wave in fluid-filled plastic pipe effects of surrounding elastic medium. *Appl. Acoust.* **2017**, *116*, 43–49. [CrossRef]
7. Brennan, M.J.; Karimi, M.; Muggleton, J.M.; Almeida, F.C.L.; Lima, F.K.; Ayala, P.C.; Obata, D.; Paschoalini, A.T.; Kessissoglou, N. On the effect of soil properties on leak noise propagation in plastic water distribution pipes. *J. Sound. Vib.* **2018**, *427*, 120–133. [CrossRef]
8. Rao, S. The Inversion of Frequency Domain Soil Parameters of Horizontally Multilayered Earth Based on Global Optimization Algorithms. Master's Thesis, China Electric Power Research Institute, Beijing, China, 2019.
9. Sun, J. On-Line Deformable Terrain Parameter Identification Based on Wheel-Terrain Interaction Mechanics. Master's Thesis, Shanghai University, Shanghai, China, 2015.
10. Jesenik, M.; Mernik, M.; Črepinšek, M.; Ravber, M.; Trlep, M. Searching for soil models' parameters using metaheuristics. *Appl. Soft Comput.* **2018**, *69*, 131–148. [CrossRef]
11. Scussel, O.; Brennan, M.J.; Muggleton, J.M.; Almeida, F.C.L.; Paschoalini, A.T. Estimation of the bulk and shear moduli of soil surrounding a plastic water pipe using measurement of the predominantly fluid wave in the pipe. *J. Appl. Geophys.* **2019**, *164*, 237–246. [CrossRef]

12. Fuller, C.R.; Fahy, F.J. Characteristics of wave-propagation and energy distributions in cylindrical elastic shells filled with fluid. *J. Sound. Vib.* **1982**, *81*, 501–518. [CrossRef]
13. Pinnington, R.J.; Briscoe, A.R. Externally applied sensor for axisymmetric waves in fluid-filled pipe. *J. Sound. Vib.* **1994**, *173*, 503–516. [CrossRef]
14. Lu, P.; Sheng, X.; Gao, Y.; Wang, R. Influence of shear effects on the characteristics of axisymmetric wave propagation in a buried fluid-filled pipe. *Chin. J. Mech. Eng.* **2022**, *35*. [CrossRef]
15. Nanjing Hydraulic Research Institute. *Specification of Soil Test—Density Test: SL237-04-1999*; China Water & Power Press: Beijing, China, 1999.
16. Nanjing Hydraulic Research Institute. *Specification of Soil Test—Elastic Modulus Test: SL237-029-1999*; China Water & Power Press: Beijing, China, 1999.
17. Nanjing Hydraulic Research Institute. *Specification of Soil Test—Soil Deformation Parameter Test: SL237-030-1999*; China Water & Power Press: Beijing, China, 1999.
18. Xue, B.; Cui, D.; Li, L.; Du, X.; Wen, Z.; Jin, X. Parallel inverse design method of wheel profile. *J. Mech. Eng.* **2013**, *49*, 8–16. [CrossRef]
19. Panić, B.; Klemenc, J.; Nagode, M. Improved initialization of the EM algorithm for mixture model parameter estimation. *Mathematics* **2020**, *8*, 373. [CrossRef]
20. Jesenik, M.; Hamler, A.; Trbušić, M.; Trlep, M. The use of evolutionary methods for the determination of a DC motor and drive parameters based on the current and angular speed response. *Mathematics* **2020**, *8*, 1269. [CrossRef]
21. LaTorre, A.; Molina, D.; Osaba, E.; Poyatos, J.; Del Ser, J.; Herrera, F. A prescription of methodological guidelines for comparing bio-inspired optimization algorithms. *Swarm Evol. Comput.* **2021**, *67*, 100973. [CrossRef]
22. Liu, S.-H.; Mernik, M.; Hrnčič, D.; Črepinšek, M. A parameter control method of evolutionary algorithms using exploration and exploitation measures with a practical application for fitting Sovova's mass transfer model. *Appl. Soft Comput.* **2013**, *13*, 3792–3805. [CrossRef]
23. Storn, R.; Price, K. Differential evolution—A simple and efficient heuristic for global optimization over continuous spaces. *J. Glob. Opt.* **1997**, *11*, 341–359. [CrossRef]
24. Das, S.; Suganthan, P.N. Differential evolution: A survey of the state-of-the-art. *IEEE Trans. Evol. Comput.* **2011**, *15*, 4–31. [CrossRef]
25. Tanabe, R.; Fukunaga, A. Success-history based parameter adaptation for differential evolution. In Proceedings of the 2013 IEEE Congress on Evolutionary Computation, Cancún, México, 20–23 June 2013. [CrossRef]
26. Jesenik, M.; Mernik, M.; Trlep, M. Determination of a hysteresis model parameters with the use of different evolutionary methods for an innovative hysteresis model. *Mathematics* **2020**, *8*, 201. [CrossRef]

sensors

MDPI

Article

Utilizing Piezo Acoustic Sensors for the Identification of Surface Roughness and Textures

Kayra Kurşun [1], Fatih Güven [2] and Hakan Ersoy [1,*]

1 Department of Mechanical Engineering, Akdeniz University, Antalya 07070, Turkey; kayrakursun@akdeniz.edu.tr
2 Department of Machinery and Metal Technologies, Başkent OSB Vocational School of Technical Sciences, Hacettepe University, Ankara 06909, Turkey; fatihguven@hacettepe.edu.tr
* Correspondence: hakanersoy@akdeniz.edu.tr

Abstract: This study examines surface roughness measurements via piezo acoustic disks and appropriate signal processing. Surface roughness is one characteristic of surface texture that can have various irregularities inherent to manufacturing methods. The surface roughness parameters and corresponding surface profiles are acquired by a stylus profilometer. Simultaneously, elastic waves propagated along metal surfaces caused by the friction of a diamond tip are obtained in the form of raw sound via piezo acoustic disks. Frequency spectrum analysis showed apparent correlations between the traditionally obtained measurement parameters and the piezo acoustic measurement data. Thus, it is concluded that acoustic friction measurement shows promising results as a novel measurement method for the surface roughness states of certain materials.

Keywords: roughness; machining; acoustic; measurement

Citation: Kurşun, K.; Güven, F.; Ersoy, H. Utilizing Piezo Acoustic Sensors for the Identification of Surface Roughness and Textures. *Sensors* **2022**, *22*, 4381. https://doi.org/10.3390/s22124381

Academic Editors: Bogusław Łazarz, Grzegorz Peruń and Tangbin Xia

Received: 4 May 2022
Accepted: 7 June 2022
Published: 9 June 2022

1. Introduction

The surfaces of machine parts often contain irregularities that result from manufacturing methods. Surface roughness, one such irregularity, is one of the important parameters used to describe surface texture. Surface roughness can cause nonlinear problems in fluid systems, especially in machine parts that contact each other. Studies have shown that surface roughness affects contact pressure in interference fit [1,2], preload loss in bolted connections [3,4], surface wear properties [5,6], surface coating properties [7], the fatigue strength of machine parts against dynamic stresses [8,9], the strength of adhesive-bonded joints [10,11], and frictional losses in fluid systems [12].

Surface roughness measurement is defined in ISO 21920-2:2021 [13] and has different parameters. *Ra* is the most commonly used parameter in determining the surface texture state, as it provides the most general and adequate information about the surface roughness. However, when the fluctuations are sudden and large in amplitude, the sensitivity decreases. Therefore, it is a better option to use the *Rz* parameter when sensitivity becomes important. Because the mathematical formulations depend on the five highest peaks and the five smallest valleys in the surface profile, they capture sudden spikes more accurately [14]. In this study, the main goal is to establish a correlation between acoustic measurements and surface roughness measurements. Therefore, the comparison with *Ra* in terms of the general condition of the surface roughness and the comparison with *Rz* in terms of sensitivity have gained importance in terms of determining which parameter is more proportional to the acoustic measurement values. Therefore, both parameters are considered within the scope of this study.

Surface roughness measurements are generally divided into contact and noncontact methods. Stylus profilometry, a contact measurement method, is generally performed by moving a diamond tip over the surface. Noncontact methods generally comprise measurement methods that optically monitor the surface. However, in these methods, the surface to be measured must be exposed and accessible. Laser profilometry is a highly preferred noncontact method. In this method, the laser beam intensity is dependent on the material surface brightness. In this way, the surface roughness profile can be mapped. The roughness mapping process plays an important role in health and pharmaceutical sciences, where two-dimensional data acquisition is critical [15]. In addition, modulalography is a special-case surface mapping method that is used for reconstructing Young's modulus of coronary arteries from radial-strain ultrasound images [16].

Surface roughness studies remain one of the most interesting topics in mechanical engineering. In addition to traditional methods, new surface roughness measurement methods continue to be developed [17]. These studies also note how important it is to measure surface roughness values accurately, quickly, and effectively. Although acoustic wave sensors and technologies have been used in certain engineering system measurements, their use to define surface roughness has only recently attracted attention [18,19].

Through the acquisition of acoustic or elastic wave propagation within solids, various physical data can be obtained and analyzed accordingly. Related studies can be seen in the literature as follows: the analysis of cutting parameters of machine tools with acoustic measurements [20,21]; tool life prediction via sound frequency analysis [22–24], fault detection in roller bearings by wavelet packet transform [25], in train bearings by improved harmonic product spectrum [26], and in low-speed bearings using acoustic emission sensors [27]; plate fatigue evaluation using high-mode Lamb waves [28]; investigation of the effects of elastic waves on fractures [29]; and surface stress measurement with Rayleigh wave detection [30]. In addition, while low-cost lead zirconate titanate (PZT) transducers were previously used as buzzers for random sound generation, there are also studies where they are used as acoustic sensors [31] for structural health monitoring [32]. Moreover, PZT transducers and acoustic emission (AE) sensors are also used together to measure the surface quality in the grinding process, and it was determined that the AE sensor and PZT transducer gave similar responses to the stimuli from the grinding process [33]. In a similar study in which AE sensors and PZT transducers were compared by applying the PLB-(pencil lead break) test on alumina ceramic material, it was determined that the obtained frequency values showed that PZT could be used as an alternative to AE sensors [34].

In this study, the measurement and calculation of the *Ra* and *Rz* surface roughness parameters are first investigated. Subsequently, the surface wave propagation equations are defined. The *Ra* and *Rz* values of the prepared test specimens are then measured. Finally, all obtained data are transferred to a computer environment to be compared and analyzed.

2. Materials and Methods

2.1. Surface Roughness Equations

Surface roughness parameters are calculated according to ISO 21920-2:2021 [13]. The average absolute value (*Ra*) is shown in Figure 1 and calculated according to Equation (1).

$$Ra = \frac{1}{l_n}\int_0^{l_n} |z(x)|dx \tag{1}$$

where l_n and $z(x)$ are the relative length and amplitude function of the profile, respectively.

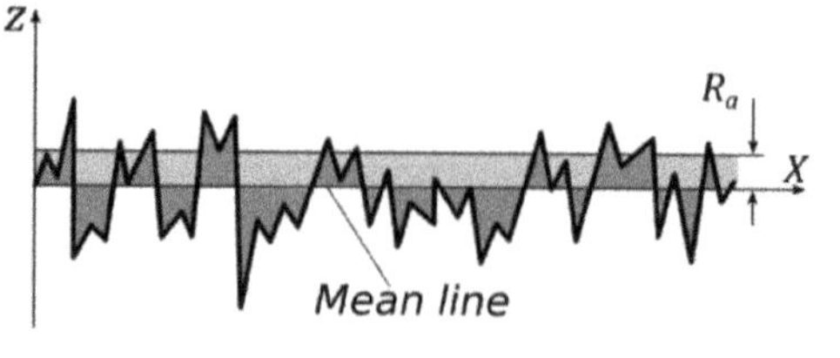

Figure 1. Average absolute value (*Ra*).

The average profile height (Rz) is the average of 5 heights, as shown in Figure 2, and calculated according to Equation (2).

$$Rz = \frac{1}{5}(Rz_1 + Rz_2 + Rz_3 + Rz_4 + Rz_5) \tag{2}$$

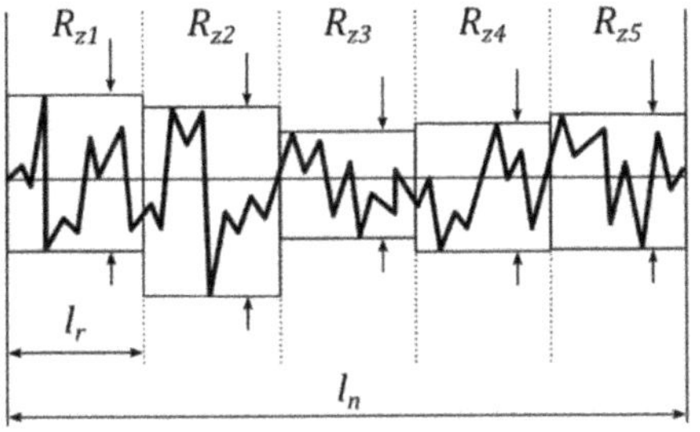

Figure 2. Average profile height (Rz).

2.2. *Governing Equations for Sound Propagation on Solid Surfaces*

Surface Sound propagation equations for solids were investigated thoroughly in [35]. Based on this study, the equations of interest are addressed as follows.

The wave equation in solids can be written for all coordinates as:

$$\rho \frac{\partial^2 \boldsymbol{u}}{\partial t^2} = (\lambda + \mu) grad\ div \boldsymbol{u} + \mu \nabla^2 \boldsymbol{u} \tag{3}$$

In Equation (3), t is time, ρ is density, ∇ is the Laplace operator, $\boldsymbol{u}$ is the displacement vector, ν is the Poisson ratio, E is Young's modulus, and G is the shear modulus; the Lamé constants are $\lambda = \nu E/(1+\nu)(1-2\nu)$ and $\mu = G = E/2(1+\nu)$.

It is possible to prove that two types of waves propagate at different speeds in an infinite solid medium. Accordingly, it is possible that any vector field can be presented from the vector analysis as a summation of two vectors—one with a scalar potential and the other with a vector potential:

$$\boldsymbol{u} = \boldsymbol{u}_l + \boldsymbol{u}_t = grad\varphi + rot\boldsymbol{\psi} \tag{4}$$

where φ is the vibration velocity potential, $\boldsymbol{u}_t$ is the transversal displacement vector, and $\boldsymbol{u}_l$ is the longitudinal displacement vector. It can be noted that $rot\boldsymbol{u}_l = div\boldsymbol{u}_t = 0$. Substituting Equation (4) into Equation (3) and applying the rotation (curl) and divergence operations, the equation converts to:

$$\partial^2 \boldsymbol{u}_l/\partial t^2 - c_l^2 \nabla^2 \boldsymbol{u}_l = 0; \quad c_l = \sqrt{(\lambda + 2\mu)/\rho} \tag{5}$$

and

$$\partial^2 \boldsymbol{u}_t/\partial t^2 - c_t^2 \nabla^2 \boldsymbol{u}_t = 0; \quad c_t = \sqrt{\mu/\rho} \tag{6}$$

where c_t is the transversal wave speed and c_l is the longitudinal wave speed.

The abovementioned equations describe the propagation of elastic waves for an unbounded solid. However, if the propagation of elastic waves on the solid surface is considered, the equations should be expanded. These are called Rayleigh equations [36] and are obtained by substituting Equation (7) into Equation (5) and Equation (8) into Equation (6):

$$\boldsymbol{u}_l = Ae^{-\sqrt{k_s^2 - k_l^2}y}e^{-jk_s x} \tag{7}$$

$$\boldsymbol{u}_t = Be^{-\sqrt{k_s^2 - k_t^2}y}e^{-jk_s x} \tag{8}$$

where $k_s = \omega/c_s$ and $k_t = \omega/c_t$. The elastic properties of the materials used in this study are shown in Table 1:

Table 1. Elastic properties of specimen materials [35].

Material	Density ρ, 10^3kg/m^3	Velocity of Wave Propagation c, 10^3m/s		
		Longitudinal	Transversal	Surface
Aluminum	2.7	6.35	3.08	2.80
Stainless Steel	8.03	5.73	3.12	2.90
AISI 1040	7.80	5.92	3.28	3.01

2.3. Significant Equations for the Principle of Piezoelectric Sensors

In this study, the piezoelectric principle is used for sensing acoustic waves. In [37], the constitutive and governing equations of this principle for detection and actuation are discussed extensively. Here, we will focus on the equations significant to a piezoelectric sensor. The direct piezoelectric effect, which makes the sensing applications of dynamic signals (including acoustic wave signals) possible, is the ability of some crystalline materials to generate an electrical charge in proportion to the applied mechanical disturbance.

As shown below, the constitutive equations for a sensor made of a one-dimensional piezoelectric material are [37]:

$$S = s^E T + d_{33} E \tag{9}$$

$$D = d_{33} T + \varepsilon^T E \tag{10}$$

where S and D (Coulomb/m^2) are the strain and electric displacement, respectively; s^E is the compliance; T (N/m^2) is the stress; E (V/m) is the electric field; d_{33} (m/V or Coulomb/N) is the piezoelectric constant; and ε^T is the dielectric constant.

Parameters T and E are independent variables, while S and D are dependent variables in Equations (9) and (10). If the equations are rearranged such that E and S are independent and T and D are dependent, they take the form of:

$$T = -\frac{d_{33}}{s^E} E + \frac{1}{s^E} S \tag{11}$$

$$D = \left(1 - \frac{d_{33}^2}{s^E \varepsilon^T}\right) \varepsilon^T E + \frac{d_{33}}{s^E} S \tag{12}$$

$$T = -e_{33} E + c^E S \tag{13}$$

$$D = \left(1 - k^2\right) \varepsilon^T E + e_{33} S \tag{14}$$

In these equations, $e_{33} = d_{33}/s^E$ (C/m^2) is the product of d_{33} by the Young modulus, $c^E = 1/s^E$ (N/m^2). The parameter k is called the electromechanical coupling factor and can be defined as:

$$k^2 = \frac{d_{33}^2}{s^E \varepsilon^T} = \frac{e_{33}^2}{c^E \varepsilon^T} \tag{15}$$

The conversion of mechanical disturbance to electrical current or applied electric current to mechanical response is evaluated by this parameter. From Equations (13) and (14), the dielectric constant under zero strain can be seen as $\varepsilon^T(1 - k^2)$. If a sensor is stacked, it is possible to obtain the constitutive equations by integrating Equations (13) and (14) over the volume.

$$\Delta = \frac{1}{K_a} f + n d_{33} V \tag{16}$$

$$Q = CV + n d_{33} f \tag{17}$$

$$f = -n d_{33} K_a V + K_a \Delta \tag{18}$$

$$Q = C\left(1 - k^2\right)V + nd_{33}K_a\Delta \quad (19)$$

The following applies to Equations (17) and (19). f is the total force (mechanical disturbance), which is equal to $A \cdot T$. Q is the total electric charge on the electrodes of the sensor and is equal to $nA \cdot D$—where n is the number of elements, and A is the cross-sectional area. $\Delta = Sl$ is the total extension, where $l = nt$ is the sensor length. C is the capacitance of the sensor ($C = \varepsilon^T \cdot A \cdot n^2/l$), and K_a is the stiffness ($K_a = A/s^E l$). V is the voltage applied between the electrodes of the sensor. The electromechanical coupling factor can be written with the n, K_a, and C parameters as:

$$k^2 = \frac{d_{33}^2}{s^E \varepsilon^T} = \frac{n^2 d_{33}^2 K_a}{C} \quad (20)$$

Subsequently, Equations (18) and (19) can be inverted and rewritten as:

$$V = \frac{1}{C(1-k^2)}(Q - K_a n d_{33}\Delta) \quad (21)$$

$$f = \frac{K_a}{1-k^2}\left(-\frac{nd_{33}}{C}Q + \Delta\right) \quad (22)$$

It can be seen from Equations (21) and (22) that the stiffness with open electrodes ($Q = 0$) is $K_a/1 - k^2$, and the capacitance for a fixed geometry ($\Delta = 0$) is $C(1 - k^2)$. For large k values, the stiffness changes significantly with the electrical boundary conditions, while the capacitance depends on the mechanical boundary conditions.

If a discrete sensor is subjected to a mechanical disturbance, such as a force with an amplitude of f, the total power transmitted to the sensor is the sum of the mechanical ($f\dot{\Delta}$) and electric (Vi) power. Thus, the network can be written in the differential form as:

$$dW = Vidt + f\dot{\Delta}dt = VdQ + fd\Delta \quad (23)$$

For a conservative element, Equation (23) is converted into a stored energy equation. Equation (24) is obtained by integrating this equation from the reference state to the final state and by substituting the equivalents of f and V in the corresponding Equations (21) and (22) to obtain the analytical expression of stored electromechanical energy for a piezo sensor.

$$W_e(\Delta, Q) = \frac{Q^2}{2C(1-k^2)} - \frac{nd_{33}K_a}{C(1-k^2)}Q\Delta + \frac{K_a}{1-k^2}\frac{\Delta^2}{2} \quad (24)$$

The first term of Equation (24) is the electrical energy stored in the capacitance $C(1 - k^2)$. The second term is the piezoelectric energy, and the last term is the elastic strain energy stored in spring stiffness as $K_a/(1 - k^2)$.

Up to this point, the equations are defined for a one-dimensional sensor. For more complicated applications, generalized equations must be addressed. Consequently, the constitutive equations of the general piezoelectric material are [37]:

$$T_{ij} = c_{ijkl}^E S_{kl} - e_{kij}E_k \quad (25)$$

$$D_i = e_{ikl}S_{kl} + \varepsilon_{ik}^S E_k \quad (26)$$

In Equations (25) and (26), ε_{ik}^S is the dielectric constant under constant strain, e_{ikl} (Coulomb/m^2) is the piezoelectric constant, c_{ijkl}^E is Hooke's tensor, T_{ij} is the stress tensor, and S_{kl} is the strain tensor.

In these equations, classic tensor notations are used, in which all indices (i, j, k, l = 1, 2, 3, 4) and all repeated indices are summarized. Equations (25) and (26) are the generalized forms of Equations (13) and (14). Similar to the transition from Equations (9) and (10) to

Equations (13) and (14), the dependency of the parameters can be shifted, and Equations (25) and (26) can be rewritten as:

$$S_{ij} = s^E_{ijkl} T_{kl} + d_{kij} E_k \tag{27}$$

$$D_i = d_{ikl} T_{kl} + \varepsilon^T_{ik} E_k \tag{28}$$

where d_{ikl} is the piezoelectric constant (Coulomb/N), ε^T_{ik} is the dielectric constant under constant stress, and s^E_{ijkl} is the tensor of compliance under a constant electric field.

The required engineering vector notations can be used instead of tensor notations:

$$T = \begin{Bmatrix} T_{11} \\ T_{22} \\ T_{33} \\ T_{23} \\ T_{31} \\ T_{12} \end{Bmatrix} \quad S = \begin{Bmatrix} S_{11} \\ S_{22} \\ S_{33} \\ 2S_{23} \\ 2S_{31} \\ 2S_{12} \end{Bmatrix} \tag{29}$$

With these changes, Equations (25)–(28) can be rewritten in matrix form:

$$\{T\} = [c]\{S\} - [e]\{E\} \tag{30}$$

$$\{D\} = [e]^T\{S\} + [\varepsilon]\{E\} \tag{31}$$

$$\{S\} = [s]\{T\} + [d]\{E\} \tag{32}$$

$$\{D\} = [d]^T\{T\} + [\varepsilon]\{E\} \tag{33}$$

In Equations (32) and (34), the superscript T indicates the transpose of the matrix, and for the sensing applications, the explicit form of Equation (33) is:

$$\begin{Bmatrix} D_1 \\ D_2 \\ D_3 \end{Bmatrix} = \begin{bmatrix} 0 & 0 & 0 & 0 & d_{15} & 0 \\ 0 & 0 & 0 & d_{24} & 0 & 0 \\ d_{31} & d_{32} & d_{33} & 0 & 0 & 0 \end{bmatrix} \begin{Bmatrix} T_{11} \\ T_{22} \\ T_{33} \\ T_{23} \\ T_{31} \\ T_{12} \end{Bmatrix} + \begin{bmatrix} \varepsilon_{11} & 0 & 0 \\ 0 & \varepsilon_{22} & 0 \\ 0 & 0 & \varepsilon_{33} \end{bmatrix} \begin{Bmatrix} E_1 \\ E_2 \\ E_3 \end{Bmatrix} \tag{34}$$

2.4. Experimental Setup

A schematic of the experimental setup is shown in Figure 3a. The experiment is performed on six specimens comprising three materials and two surface treatments for each material: one milled and one ground aluminum specimen; one milled and one ground AISI 1040 specimen; and one lathed and one milled stainless-steel specimen. The geometric forms of the test specimens are rectangular prismatic, and their dimensions are 60 × 60 × 10 mm. The surface roughness measurement device used in the experiment is the Mahr Perthometer M2, and a piezo disk with a diameter of 35 mm is placed on each sample, as shown in Figure 3b. The radius of the profilometer stylus tip used for surface roughness measurements is r_{tip} = 2 μm. The measurement distance (λ_t) of the diamond tip of the Mahr M2 device is set to 17.5 mm; the cutoff distance (λ_c) is set to 2.5 mm. By subtracting the cutoff length (λ_c) from the start and end of the total measurement distance (λ_t), the resulting final distance provides the measurement result. The obtained final distance is the one used in the roughness measurement result parameters. For this reason, cutoffs are made proportionally to the first and last parts of the raw sound signal vectors.

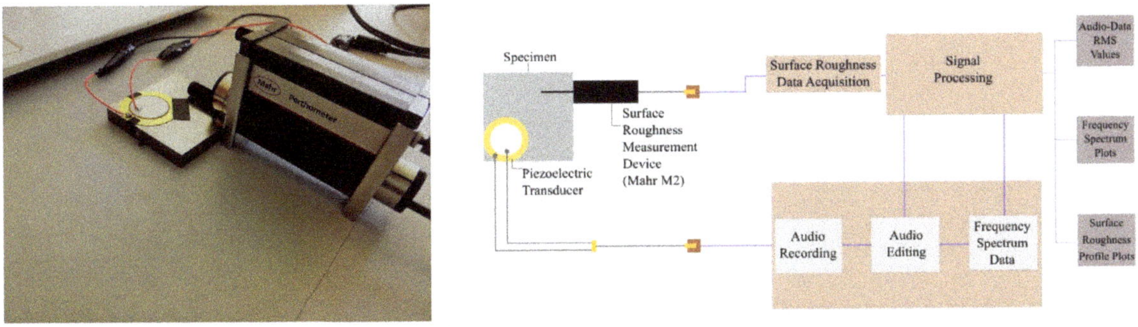

(a) (b)

Figure 3. Photograph (**a**) and schematic (**b**) of the experimental setup.

The piezo disk is connected to the computer through an AUX cable, and surface roughness profile graphics are digitized. Using Audacity software, the piezo disk is detected as a microphone. While the diamond tip of the Mahr M2 is moved on the specimen to measure the *Ra* and *Rz* parameters, a sound recording was simultaneously initiated, and the friction sound of the diamond tip moving on the specimen was recorded. The main purpose here is to make the sounds of friction and crackling—which are barely perceptible to the human ear—into interpretable signals using the ability of the piezo transducers to detect very low-pitched signals or low amplitude vibrations as certain voltage values.

The Mahr Perthometer M2 surface roughness measuring device performs the roughness measurement by moving the diamond tip forward and backward on the sample. While the forward motion speed is constant, the backward return speed can be adjusted in three stages. For the sound data to be evaluated accurately, the sound occurring at all operating speeds of the device is recorded and analyzed. The sampling frequency for all audio data is 44,100 Hz. Frequency spectrum data of each vector are created separately. Subsequently, the obtained surface roughness profile data, sound recording vector data, and voice recording spectrum data are all imported into the software.

3. Results

In the experiments, the *Ra* and *Rz* parameters and the surface roughness profiles are obtained from the Mahr M2 device, while the original audio raw signal (A_{rs}), audio power spectrum (A_{ps}) parameters, and power spectrum plots are created using the recorded sound data from the piezo acoustic sensor. A_{rs} and A_{ps} parameters are calculated from the raw sound data and power spectrum data based on the *Rz* Equation (2). The parameter values are presented in Table 2:

Table 2. *RMS*, A_{rs}, A_{ps}, values of acoustic measurements; *Ra* and *Rz* values of surface roughness measurements.

	Aluminum (Milled)	Aluminum (Ground)	Stainless Steel (Milled)	Stainless Steel (Lathed)	AISI 1040 (Milled)	AISI 1040 (Ground)
Sound *RMS*	3.6027	1.0279	1.2262	1.3868	0.8783	0.8085
Ra (µm)	1.106	0.20	0.984	3.404	3.359	1.903
Rz (µm)	6.50	2.56	5.26	14.30	20.10	14.80
A_{rs}	28.3730	8.0338	27.5645	33.7739	19.8591	18.3818
A_{ps}	11.2109	9.4554	30.3371	39.7898	32.8009	24.7656

An investigation of the roughness profiles for the aluminum specimens (Figures 4 and 5) showed that there are no sudden fluctuations on the milled surface. However, on the ground surface, although the smoothness of the surface profile seemed to be preserved throughout the measurement length, sudden discontinuities were found in the two small regions. Accordingly, when a parameter comparison is made between the milled surface and the ground surface, the increase in milled surface values is found to be 5.53 times in the *Ra* parameter and 2.56 times in the *Rz* parameter. Therefore, it is confirmed that a more sensitive value is obtained with the *Rz* parameter, and the results obtained in acoustic measurements are compared with this parameter.

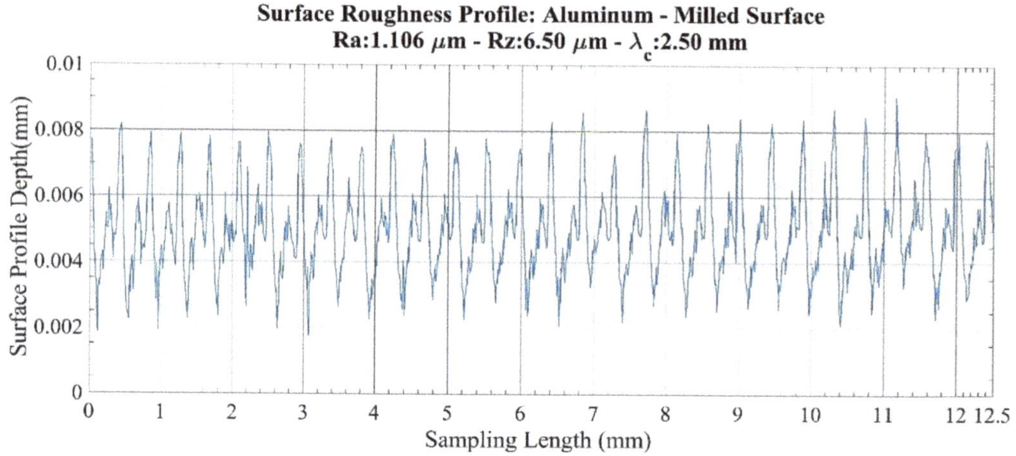

Figure 4. Surface roughness profile: aluminum–milled surface.

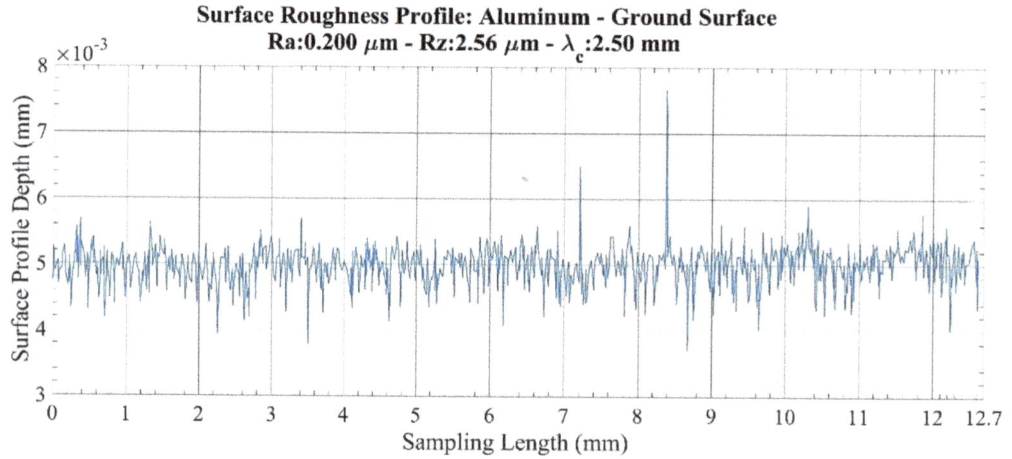

Figure 5. Surface roughness profile: Aluminum-ground surface.

The audio power spectrum of the aluminum specimen is provided in Figure 6. From this point on, for all audio recordings, the sampling frequency is 44,100 Hz. The local maxima values for this specimen are observed between 800 Hz and 5000 Hz. Legend abbreviations and corresponding sampling time values are provided in Table 3. If the shortest sampling times are considered during the return stroke of the device tip, the frequency response is largely similar in form for both the milled and ground surfaces, but the amplitudes are different. Additionally, it is determined that the amplitude values of the audio are inversely proportional to the length of the sampling time for all measurements.

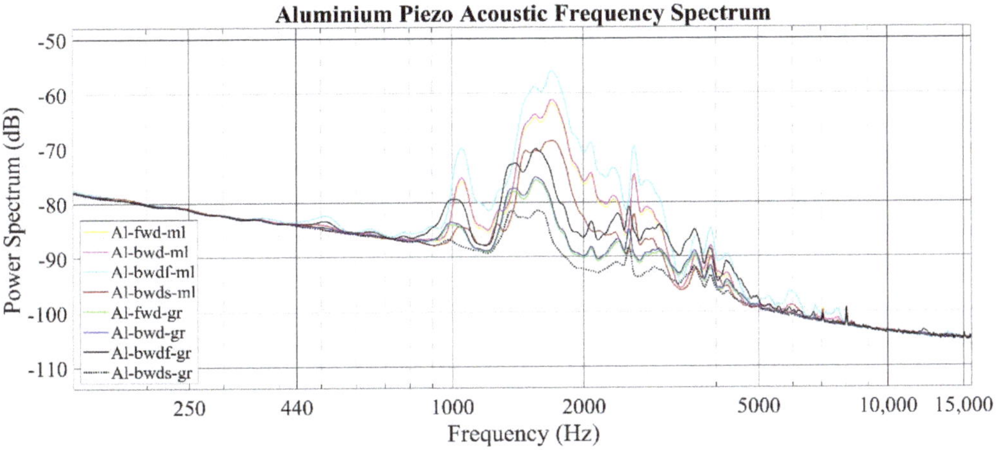

Figure 6. Piezo acoustic measurement frequency spectrum of the aluminum specimen (frequency resolution: 50 Hz).

Table 3. Power spectrum plot legend abbreviations and corresponding sampling time values.

Legend	Specimen	Stroke-Speed	Machining Operation	Sampling Time, Sec
Al-fwd-ml	Aluminum	Forward–Standard	Milling	55
Al-bwd-ml	Aluminum	Backward–Standard	Milling	48
Al-bwdf-ml	Aluminum	Backward–Fast	Milling	12.3
Al-bwds-ml	Aluminum	Backward–Slow	Milling	35.3
Al-fwd-gr	Aluminum	Forward–Standard	Grinding	55
Al-bwd-gr	Aluminum	Backward–Standard	Grinding	47.5
Al-bwdf-gr	Aluminum	Backward–Fast	Grinding	12
Al-bwds-gr	Aluminum	Backward–Slow	Grinding	35.4
SStl-fwd-ml	Stainless Steel	Forward–Standard	Milling	56.5
SStl-bwd-ml	Stainless Steel	Backward–Standard	Milling	48
SStl-bwdf-ml	Stainless Steel	Backward–Fast	Milling	12.3
SStl-bwds-ml	Stainless Steel	Backward–Slow	Milling	35.7
SStl-fwd-lt	Stainless Steel	Forward–Standard	Lathing	52.5
SStl-bwd-lt	Stainless Steel	Backward–Standard	Lathing	46
SStl-bwdf-lt	Stainless Steel	Backward–Fast	Lathing	12
SStl-bwds-lt	Stainless Steel	Backward–Slow	Lathing	34
1040-fwd-ml	AISI 1040	Forward–Standard	Milling	60
1040-bwd-ml	AISI 1040	Backward–Standard	Milling	50
1040-bwdf-ml	AISI 1040	Backward–Fast	Milling	14.5
1040-bwds-ml	AISI 1040	Backward–Slow	Milling	35.5
1040-fwd-gr	AISI 1040	Forward–Standard	Grinding	62
1040-bwd-gr	AISI 1040	Backward–Standard	Grinding	48
1040-bwdf-gr	AISI 1040	Backward–Fast	Grinding	12.5
1040-bwds-gr	AISI 1040	Backward–Slow	Grinding	35.5

An investigation of roughness profiles for the stainless-steel specimens (Figures 7 and 8) shows that no significant discontinuity is detected on either the milled or lathed surfaces. However, a small deviation and an increase in profile amplitude are observed from the 4 mm point on the milled surface. Thus, compared to the 3.459 times increase in *Ra* values, 4.201 times increase in the value of *Rz* is determined. Therefore, it can be deduced that the sensitivity of the parameters does not create a serious issue for these specimens.

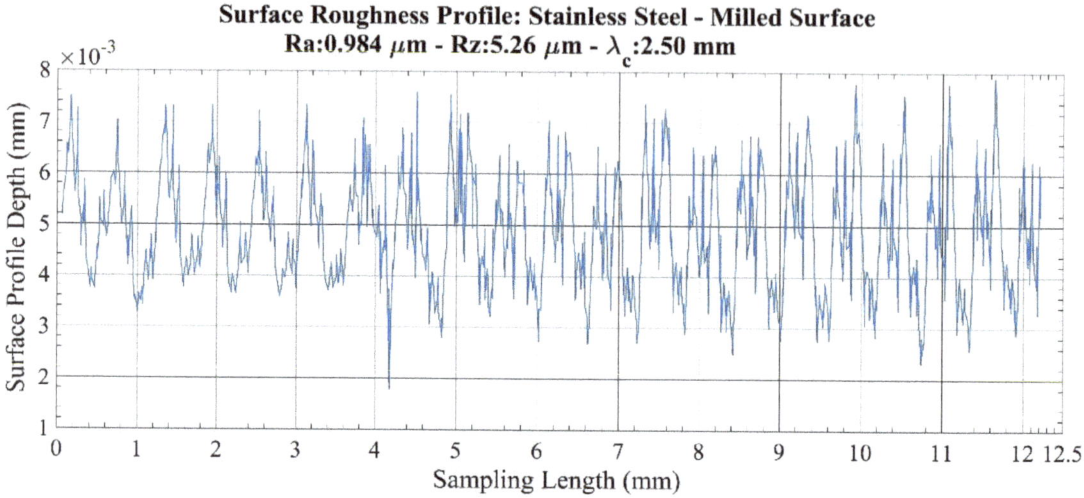

Figure 7. Surface roughness profile: Stainless steel-milled surface.

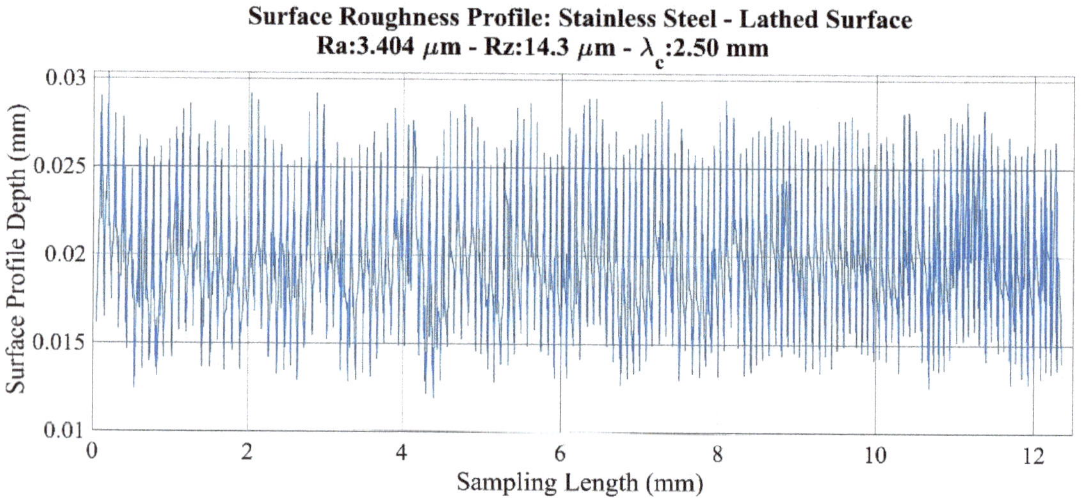

Figure 8. Surface roughness profile: Stainless steel–lathed surface.

The audio power spectrum of the stainless-steel specimen is provided in Figure 9. The variation between the milled surface and lathed surface values appears as a shift in the frequency spectrum. The frequency response for the milled surface shows two steps between 700–2000 Hz and 2000–3200 Hz; for the lathed surfaces, there are three steps between 700–1250 Hz, 1250–2000 Hz, and 2000–3200 Hz. The local maxima, as the signal strength for the lathed surfaces, are between 700–1250 Hz.

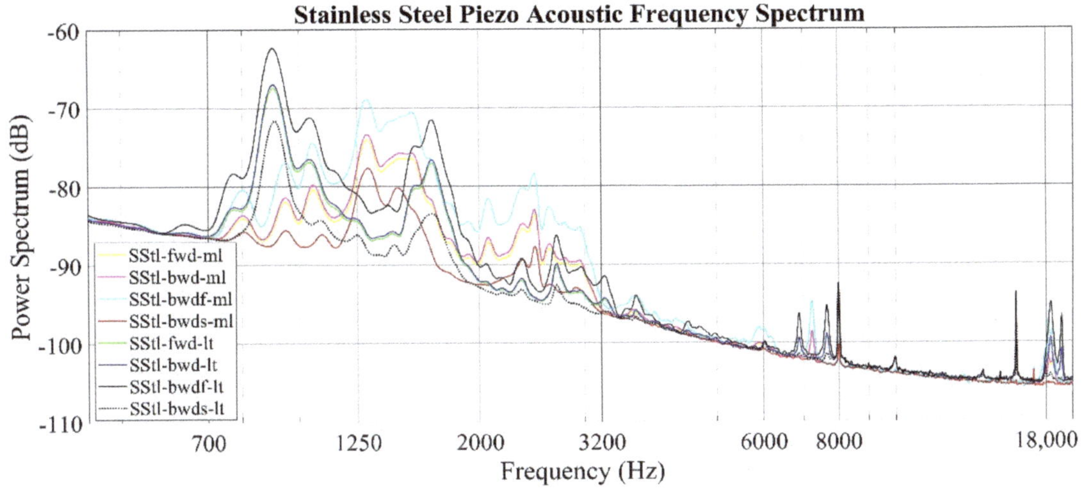

Figure 9. Piezo acoustic measurement frequency spectrum of the stainless steel specimen (frequency resolution: 50 Hz).

An investigation of roughness profiles for the AISI 1040 specimens (Figures 10 and 11) shows that considerable discontinuities are detected on both the milled surface and the ground surfaces in terms of surface quality. This situation creates large variations between *Ra* and *Rz* values, especially for the same machining operation. When considering these differences, it was determined that the *Ra* parameter, in particular, does not provide sensitive values for this specimen.

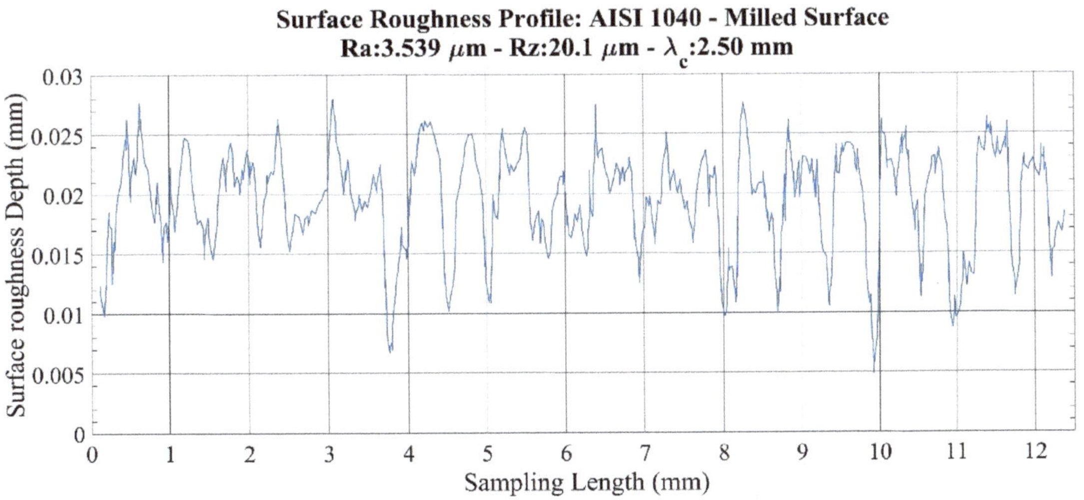

Figure 10. Surface roughness profile: AISI 1040–milled surface.

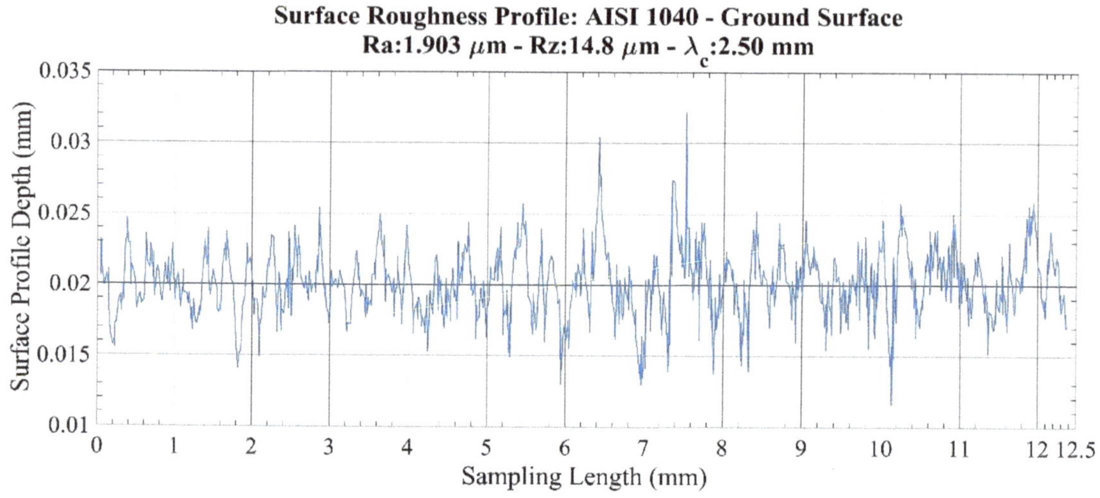

Figure 11. Surface roughness profile: AISI 1040–ground surface.

The audio power spectrum of the AISI 1040 specimen is provided in Figure 12. The frequency response for the ground surface spans a wide bandwidth and is observed between 700–1200 Hz, 1200–2000 Hz, and 2000–3000 Hz. The frequency response for the milled surface shows a unique amplitude spike, unlike other specimens, at approximately 7000–8000 Hz.

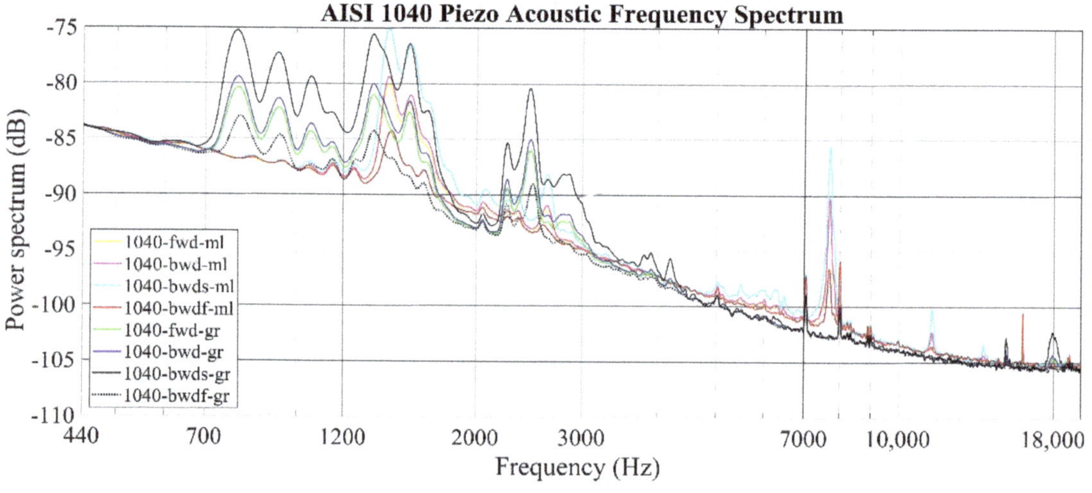

Figure 12. Piezo acoustic measurement frequency spectrum of the AISI 1040 specimen (600–10,000 Hz) (Frequency Resolution: 50 Hz).

4. Discussion

Piezo acoustic data obtained for all samples and production operations were evaluated in three stages: a parametric analysis, a power spectrum analysis, and finally, a comparative analysis with standard parameters. The RMS and A_{rs} acoustic surface roughness parameters obtained within the scope of the first stage provided accurate information, especially about the surface quality obtained with different production methods of the same material. These parameters guided the other stages by taking lower values on the smoother surface for all materials. Then, sound power spectrum graphs, including all sampling rates, materials, and production methods, were produced. The power spectrum of the aluminum sample showed a frequency response compatible with the surface roughness profile obtained by standard measurements of the same material. In this graph, it is determined that the smoother surface creates a low-intensity signal strength and the signal strength increases when the surface roughness increases. When stainless-steel samples are examined, it is seen that the greatest difference between the lathed and milled surfaces is the difference in frequency responses. It is clear that this situation is due to the very tight formation of the hills and pits, especially on the lathed surface. Due to the aforementioned tight formation, the timbre and the frequency of the sound were increased in the sound recordings taken from that surface, and this was determined as the frequency shift in the power spectrum graph. This result was shown in the power spectrum graph together with the roughness profile graphs, where the surface quality of the AISI 1040 material is problematic. Unlike other materials, the surface condition showed sudden "spikes", even around 7000 Hz. When considering all these results, it can be seen that an examination of the power-dependent power spectrum graphs will provide as much information as the profiles obtained by the traditional method, including the information related to the maximum and minimum values of the peak and trough values, as well as the frequency axes in the power spectrum graphs. It has been observed that it can be obtained by working on it. The A_{ps} value obtained from the sound power data of the system also obtained a numerical value comparable with the Rz value among all materials depending on the correction parameters, the material density, and the surface speeds, especially for the horizontal material on the horizontal axis, whether compact compression for stainless steel or irregular compression for AISI 1040. A surface roughness value with high sensitivity, including spreading information, has been reached.

Author Contributions: Conceptualization, K.K. and H.E.; methodology, F.G. and H.E.; Experimental setup, K.K. and F.G.; validation, K.K. and H.E.; formal analysis, K.K.; investigation, K.K. and F.G.; resources, K.K. and H.E.; data curation, H.E.; writing—original draft preparation, F.G. and K.K.; writing—review and editing, K.K. and F.G.; visualization, K.K. and F.G.; supervision, H.E.; project administration, K.K. and H.E.; funding acquisition, H.E. All authors have read and agreed to the published version of the manuscript.

Funding: This research received no external funding.

Data Availability Statement: The data presented in this study are available on request from the corresponding author.

Conflicts of Interest: The authors declare no conflict of interest.

References

1. Seifi, R.; Abbasi, K.; Asayesh, M. Effects of contact surface roughness of interference shaft/bush joints on its characteristics. *Iran. J. Sci. Technol. Trans. Mech. Eng.* **2017**, *42*, 279–292. [CrossRef]
2. Yang, G.-M.; Coquille, J.; Fontaine, J.F.; Lambertin, M. Influence of roughness on characteristics of tight interference fit of a shaft and a hub. *Int. J. Solids Struct.* **2001**, *38*, 7691–7701. [CrossRef]
3. Decker, K.-H.; Kabus, K.; Rieg, F.; Hackenschmidt, R.; Engelken, G.; Weidermann, F. *Maschinenelemente-Funktion, Gestaltung und Berechnung: Mit 871 Bildern, 164 Berechnungsbeispielen und einem Tabellenband mit 334 Tabellen und Diagrammen. 19*; Hanser: München, Germany, 2014.
4. Steinhilper, W.; Röper, R. *Maschinen-und Konstruktionselemente 3: Elastische Elemente, Federn Achsen und Wellen Dichtungstechnik Reibung, Schmierung, Lagerungen*; Springer: Berlin/Heidelberg, Germany, 2013.
5. Bayer, R.; Sirico, J. The influence of surface roughness on wear. *Wear* **1975**, *35*, 251–260. [CrossRef]
6. Svahn, F.; Kassman-Rudolphi, Å.; Wallén, E. The influence of surface roughness on friction and wear of machine element coatings. *Wear* **2003**, *254*, 1092–1098. [CrossRef]
7. Podgornik, B.; Hogmark, S.; Sandberg, O. Influence of surface roughness and coating type on the galling properties of coated forming tool steel. *Surf. Coat. Technol.* **2004**, *184*, 338–348. [CrossRef]
8. Bayoumi, M.R.; Abdellatif, A. Effect of surface finish on fatigue strength. *Eng. Fract. Mech.* **1995**, *51*, 861–870. [CrossRef]
9. Murakami, Y. *Metal Fatigue: Effects of Small Defects and Nonmetallic Inclusions*; Academic Press: Cambridge, MA, USA, 2019.
10. Sekercioglu, T.; Gulsoz, A.; Rende, H. The effects of bonding clearance and interference fit on the strength of adhesively bonded cylindrical components. *Mater. Des.* **2005**, *26*, 377–381. [CrossRef]

11. Uehara, K.; Sakurai, M. Bonding strength of adhesives and surface roughness of joined parts. *J. Mater. Process. Technol.* **2002**, *127*, 178–181. [CrossRef]
12. Munson, B.R.; Okiishi, T.H.; Huebsch, W.W.; Rothmayer, A.P. *Fluid Mechanics*; John Wiley & Sons Singapore: Singapore, 2013.
13. *ISO 21920-2:2021*; Geometrical Product Specifications (GPS)—Surface Texture: Profile—Part 2: Terms, Definitions and Surface Texture Parameters. International Organization for Standardization (ISO): Geneva, Switzerland, 2021; p. 78.
14. Gadelmawla, E.; Koura, M.M.; Maksoud, T.M.; Elewa, I.M.; Soliman, H. Roughness parameters. *J. Mater. Process. Technol.* **2002**, *123*, 133–145. [CrossRef]
15. Seitavuopio, P.; Rantanen, J.; Yliruusi, J. Use of roughness maps in visualisation of surfaces. *Eur. J. Pharm. Biopharm.* **2005**, *59*, 351–358. [CrossRef]
16. Baldewsing, R. *Modulography: Elasticity Imaging of Atherosclerotic Plaques*; Erasmus University Rotterdam: Rotterdam, The Netherlands, 2006.
17. Windecker, R.; Tiziani, H.J. Optical roughness measurements using extended white-light interferometry. *Opt. Eng.* **1999**, *38*, 1081–1087. [CrossRef]
18. Frigieri, E.P.; Campos, P.H.S.; Paiva, A.P.; Balestrassi, P.P.; Ferreira, J.R.; Ynoguti, C.A. A mel-frequency cepstral coefficient-based approach for surface roughness diagnosis in hard turning using acoustic signals and gaussian mixture models. *Appl. Acoust.* **2016**, *113*, 230–237. [CrossRef]
19. Vellekoop, M.J. Acoustic wave sensors and their technology. *Ultrasonics* **1998**, *36*, 7–14. [CrossRef]
20. Tekıner, Z.; Yeşılyurt, S. Investigation of the cutting parameters depending on process sound during turning of AISI 304 austenitic stainless steel. *Mater. Des.* **2004**, *25*, 507–513. [CrossRef]
21. Washabaugh, P.D.; Knauss, W. A reconciliation of dynamic crack velocity and Rayleigh wave speed in isotropic brittle solids. *Int. J. Fract.* **1994**, *65*, 97–114. [CrossRef]
22. Ai, C.S.; Sun, Y.J.; He, G.W.; Ze, X.B.; Li, W.; Mao, K. The milling tool wear monitoring using the acoustic spectrum. *Int. J. Adv. Manuf. Technol.* **2011**, *61*, 457–463. [CrossRef]
23. Downey, J.; O'Leary, P.; Raghavendra, R. Comparison and analysis of audible sound energy emissions during single point machining of HSTS with PVD TiCN cutter insert across full tool life. *Wear* **2014**, *313*, 53–62. [CrossRef]
24. Lu, M.-C.; Wan, B.-S. Study of high-frequency sound signals for tool wear monitoring in micromilling. *Int. J. Adv. Manuf. Technol.* **2012**, *66*, 1785–1792. [CrossRef]
25. Hemmati, F.; Orfali, W.; Gadala, M.S. Roller bearing acoustic signature extraction by wavelet packet transform, applications in fault detection and size estimation. *Appl. Acoust.* **2016**, *104*, 101–118. [CrossRef]
26. Zhao, M.; Lin, J.; Miao, Y.; Xu, X. Detection and recovery of fault impulses via improved harmonic product spectrum and its application in defect size estimation of train bearings. *Measurement* **2016**, *91*, 421–439. [CrossRef]
27. Van Hecke, B.; Yoon, J.; He, D. Low speed bearing fault diagnosis using acoustic emission sensors. *Appl. Acoust.* **2016**, *105*, 35–44. [CrossRef]
28. Lin, W.; Deng, X.H.; Li, X. High sensitive evaluation fatigue of plate using high mode Lamb wave. *Appl. Acoust.* **2013**, *74*, 1018–1021. [CrossRef]
29. Hentati, H.; Kriaa, Y.; Haugou, G.; Chaari, F.; Wali, M.; Zouari, B.; Dammak, F. Influence of elastic wave on crack nucleation—Experimental and computational investigation of brittle fracture. *Appl. Acoust.* **2017**, *128*, 45–54. [CrossRef]
30. Hu, E.; He, Y.; Chen, Y. Experimental study on the surface stress measurement with Rayleigh wave detection technique. *Appl. Acoust.* **2009**, *70*, 356–360. [CrossRef]
31. Castro, B.; Clerice, G.; Ramos, C.; Andreoli, A.; Baptista, F.; Campos, F.; Ulson, J. Partial discharge monitoring in power transformers using low-cost piezoelectric sensors. *Sensors* **2016**, *16*, 1266. [CrossRef]
32. Budoya, D.E.; Baptista, F.G. A comparative study of impedance measurement techniques for structural health monitoring applications. *IEEE Trans. Instrum. Meas.* **2018**, *67*, 912–924. [CrossRef]
33. Aulestia Viera, M.A.; Aguiar, P.R.; Oliveira Junior, P.; Alexandre, F.A.; Lopes, W.N.; Bianchi, E.C.; da Silva, R.B.; D'Addona, D.; Andreoli, A. A time-frequency acoustic emission-based technique to assess workpiece surface quality in ceramic grinding with PZT transducer. *Sensors* **2019**, *19*, 3913. [CrossRef]
34. Viera, M.A.A.; Gotz, R.; de Aguiar, P.R.; Alexandre, F.A.; Fernandez, B.O.; Oliveira Junior, P. A low-cost acoustic emission sensor based on piezoelectric diaphragm. *IEEE Sens. J.* **2020**, *20*, 9377–9384. [CrossRef]
35. Nazarchuk, Z.; Skalskyi, V.; Serhiyenko, O. Propagation of elastic waves in solids. In *Acoustic Emission*; Springer: Cham, Switzerland, 2017; pp. 29–73.
36. Rayleigh, L. On waves propagated along the plane surface of an elastic solid. *Proc. Lond. Math. Soc.* **1885**, *1*, 4–11. [CrossRef]
37. Preumont, A. *Vibration Control of Active Structures: An Introduction*; Springer: Berlin/Heidelberg, Germany, 2018; Volume 246.

sensors

MDPI

Article

Damage Detection Using Ultrasonic Techniques in Concrete-Filled Steel Tubes (CFSTs) Columns

Antonio Callejas [1,2,3,*], Roberto Palma [1], David Hernández-Figueirido [4] and Guillermo Rus [1,2,3]

1 Department of Structural Mechanics, University of Granada, 18071 Granada, Spain; rpalgue@ugr.es (R.P.); grus@ugr.es (G.R.)
2 Instituto de Investigación Biosanitaria, ibs.GRANADA, 18012 Granada, Spain
3 Excellence Research Unit "ModelingNature" (MNat), 18071 Granada, Spain
4 Department of Mechanical Engineering and Construction, Universitat Jaume I, 12071 Castellón de la Plana, Spain; hernandd@uji.es
* Correspondence: acallejas@ugr.es

Abstract: Concrete-filled steel tubes (CFSTs) are structural elements that, as a consequence of an incorrect elaboration, can exhibit internal defects that cannot be visualized, being usually air voids. In this work, the detection of internal damage in CFST samples elaborated with a percentage of contained air voids in concrete, was carried out by performing a complete ultrasound scan using an immersion tank. The analysis of the ultrasound signals shows the differences presented in the amplitude of the fundamental frequency of the signal, and in the Broadband Ultrasound Attenuation (BUA), in comparison with a sample without defects. The main contribution of this study is the application of the BUA technique in CFST samples for the location of air voids. The results present a linear relationship between BUA averages over the window of the CFSTs and the percentage of air voids contained (Pearson's correlation coefficient r = 0.9873), the higher percentage of air voids, the higher values of BUA. The BUA algorithm could be applied effectively to distinguish areas with defects inside the CFSTs. Similar to the BUA results, the analysis in the frequency domain using the FFT and the STFT was sensitive in the detection of internal damage (Pearson's correlation coefficient r = −0.9799, and r = −0.9672, respectively). The results establish an improvement in the evaluation of CFST elements for the detection of internal defects.

Keywords: nondestructive evaluation; ultrasound; Broadband Ultrasound Attenuation; concrete-filled steel tubes

Citation: Callejas, A.; Palma, R.; Hernández-Figueirido, D.; Rus, G. Damage Detection Using Ultrasonic Techniques in Concrete-Filled Steel Tubes (CFSTs) Columns. *Sensors* **2022**, 22, 4400. https://doi.org/10.3390/s22124400

Academic Editors: Grzegorz Peruń, Tangbin Xia and Bogusław Łazarz

Received: 25 May 2022
Accepted: 9 June 2022
Published: 10 June 2022

1. Introduction

The use of Concrete-Filled Steel Tubular (CFST) columns is a popular solution for high rise buildings, in modern bridges, sports stadia, towers, and offshore structures [1]. A CFST column is formed by infilling concrete into a hollow steel tube. The steel tube confines the concrete core and provides a permanent formwork, and the concrete core avoids or delays the local buckling of the steel tube. This combination enhances the structural behaviour of both materials increasing the strength capacity, ductility, and fire resistance of the column which leads to reduced sections and more usable floor area and thus the economic benefit of the building [2]. In addition, CFSTs are fast to erect, and it is possible to work in different levels at the same time due to concrete can be infilled pumping from the bottom. In addition, B.K. Oh et al. [3] demonstrated that CFSTs are more sustainable, in terms of CO_2 emissions than other traditional solutions. Many studies were carried out on CFST columns subjected to concentric load, eccentric load, impact load [4], stiffened with V-shaped grooves [5], with internal stiffeners [6], or with outer annular stiffener [7], and the behavior of this type of structural elements are well-known. In contrast, a reduced number of studies were carried out to measure the importance of the imperfections in CFST columns.

Two materials are used in CFST columns, and both are prone to imperfections. Imperfections in steel tubes due to the manufacturing process affect the local and global buckling of the columns. In CFST columns, the influence of the concrete core reduces these effects [2]. In concrete, the main problems are related to the influence of concrete compaction during execution, corrosion of steel, working loads, and changes in temperature. Two recent works studied the detection of cracks due to steel corrosion in reinforced cement mortar, one by means of a correlation of the nonlinear elastic features of ultrasonic waves with the critical events of the corrosion [8], and another through intermodulation generation of ultrasonic waves [9]. Some of the typical defects in CFSTs are internal voids and separations between steel and concrete that can be regional or complete. Main defects in the form of spherical gaps between concrete core and steel tube make a negative effect on the confinement contribution causing a decrease of compressive strength of CFSTs by almost 2–14% [10–13] compared with a healthy specimen, creating a partially confined effect between concrete and steel, that the compressive strength of the CFST is similar to the one of concrete without steel. Air voids located through the matrix of concrete in CFSTs minimize the mechanical contribution of concrete and affect it similarly as the percentage of porosity does, making a decrease of compressive strength when the percentage of porosity increases [12]. Experiments in CFSTs with simulated air voids [14] show an increase in buckling deformation compared to a healthy specimen, and a reduction in its ultimate compressive load and ductility. For this reason, the prevention of failures through the detection of defects is of great importance in the elaboration of CFST elements, and existing structures using CFSTs. A new ultrasonic index were proposed by Chao G. et al. [15,16] to determine the influence of air-void on the final strength of the arch bridge rib of CFST; however, the work did not study the presence of air voids inside the concrete.

CFSTs have the inconvenient that internal defects in concrete cannot be visualized. An approach to detect internal defects is the analysis of ultrasound signals in the time-frequency domain, using several methods such as Fast Fourier Transform [17–19], Short-Time Fourier Transform [20,21] or Wavelet Transform [22,23]. Analyzing the spectra of frequencies of ultrasound signals helps in the detection of internal flaws and composition of materials since it is difficult to make a good comparison and a characterization of the ultrasound signals in the time domain in highly attenuated materials, such as concrete. The evaluation of ultrasound signals in the frequency domain to detect damage in concrete is mainly applied since there is a more sensitive change in amplitudes of the signals, attenuation, and energy of harmonics than in the ultrasonic pulse velocity method (UPV), when internal damage (small inclusions, internal voids and cracks) is presented in concrete [22,24]. Evaluation in the frequency domain using the Fast Fourier Transform of ultrasound signals in concrete with internal cracks made by induced damage shows in [19] that a decrease in the energy of the fundamental harmonics appears to be compared to the analysis of ultrasound signals in locations of the same specimen without damage. In addition, in [18] showed that the energy spectra of the fundamental frequency of signals in damaged concrete induced by compression tests decreased significantly compared to the healthy elements, although UPV remained with almost the same values. In experimentations evaluating concrete with induced porosity by freeze-thaw cycling and salt-scaling in [17], UPV decreased just 3–9% compared to a healthy specimen, and values of the amplitude of the fundamental harmonic, evaluating the spectrum of the Fourier Transform, resulted in a more sensitive parameter than a change of UPV, presenting higher values of energy of the fundamental frequency of ultrasound signals through healthy concrete specimens than in damaged ones.

Analysis of ultrasound signals through the Wavelet Transform decomposes the signal into a specific wave-like oscillation (wavelet), different from the Fourier Transform that makes a decomposition of the signal into sine waves of different frequencies. Short-Time Fourier Transform and Wavelet Transform are usually applied in pulse-echo technique, since the detection of echoes caused by internal flaws, when ultrasound signals are submerged in noise, can be done evaluating the signals in the time-frequency domain [25]. Even though, an investigation of interest [23] of a concrete structure with holes located in different depths, demonstrated that the UPV, using the through-transmission technique, does not have a linear relation of the size of the internal hole, since the time of flight of the ultrasonic pulse of the signals traveling through an internal hole were similar to the ones in the intact concrete, but there was a clear difference in the location in time and amplitude of the energy of the fundamental frequency evaluating the received signal in the time-frequency domain using a combination of the Short-Time Fourier Transform and the Wavelet Transform, called S Transform. Similar results were obtained in a simulation study carried out by Nadom K. et al. [26]. The detection of damage and its size is subject to the change in the colour scheme of the STFT spectrogram. The resultant images show that the increase in the frequency of the excitation signal gives better results.

Broadband Ultrasound Attenuation (BUA) is a physically meaningful way of obtaining attenuation of material as the slope of the linear regression over a certain range of frequencies. BUA was introduced initially by Langton et al [27], applied in medical experimentations to determine the effects of attenuation, scattering, and porosity of cancellous bones. The BUA is obtained by measuring the difference of the spectra of an ultrasonic wave transmitted through reference material, such as water, and through the material to be analyzed [28]. The difference of both spectra results in an attenuation versus frequency curve, being the BUA the slope of this curve in a certain range of frequency. As R. Strelitzki et al. [29] mentioned, a certain range of frequencies could be selected for the BUA regression slope, usually a range between 200 kHz to 1MHz, depending on the fundamental resonant frequency of the transducer used, being in most of the case the selected range of frequencies the one that best fits a linear regression. Even though it is possible that different materials have similar BUA values, this parameter can be used to do a comparison between healthy and unhealthy samples.

BUA parameter is commonly used in experimentations in the field of bioengineering applied to identify properties of bones [30] as differences between healthy bones and cancellous bones [31–33], and relationships between BUA parameter and porosity of bones [34], but there are some references of experimentations done in other materials. J.B Hull et al. [35] made an important approach of BUA applied in the identification of polymer materials and porosity of ceramics. In this experimentation, they performed a relation of the attenuation versus frequency regression slope (BUA) with the time of flight of the ultrasound signal through the material to obtain Hull/Langton index (HL), used to make a comparison with porosity in ceramics measured with hydration test, resulting in a linear relationship of HL index increasing as the percentage of porosity in ceramics does. BUA is shown in [36] that can be applied in the same way for detection of defects (as drill holes) in aluminum bars making an interesting approach on how BUA has higher values when a material is analyzed in a fully drilled area than in an area with smaller drill holes, applied with both pulse-echo and through-transmission methods. A comparative result [37] in mortars with different degrees of porosity demonstrated that mortars with a higher percentage of porosity have higher regression slope values (BUA) in the attenuation versus frequency curve for a certain range of frequencies (between 1.5 to 1.7 MHz) than mortars with less percentage of porosity. Although the degree of porosity cannot be obtained from BUA results, there is an approach about how BUA values allow differentiating between an element that has a percentage of damage.

Based on the studies presented, there is a knowledge gap about the effectiveness of the different analysis methods that can be applied to the existing ultrasound techniques for the detection of damage, specifically air voids inside the concrete matrix of CFSTs. The objective of this study is to measure possible internal defects in the concrete. In this line, CFST stub columns with internal defects were manufactured including random inclusions.

The remainder of this paper is organized as follows. Section 2 describes the preparation of samples, the experimental setup, and the algorithms employed to detect internal damage in CFSTs. Section 3 describes the results and the discussion. Finally, Section 4 offers the conclusions of this research and suggestions for the future.

2. Materials and Methods

2.1. Preparation of Samples

In this work, a total of 5 CFST stub columns were tested to detect internal imperfections. All these columns were manufactured in the laboratory of the Department of Mechanical Engineering and Construction of the Universitat Jaume I in Castellón, Spain. The columns were 300 mm in length with a nominal cross-section of the tubes (height (h) × width (b) × steel thickness (t)) 100 × 100 × 4 mm (see Figure 1). The steel tubes were cold-formed carbon steel and supplied by the same manufacturer. The nominal yield strength of the tubes was S275JR.

Figure 1. CFST stub columns.

Air voids were simulated using a certain amount of expanded polystyrene beads with a volume of 25 mm in diameter. This material is conformed to 98% of air, for this reason, expanded polystyrene pieces in the concrete can be simulated as internal air voids. The percentage of voids for each CFST is 0%, 1%, 2%, 3% and 4%. The approximate number of air voids can be estimated with the previous information, as seen in Table 1.

Table 1. Approximate number of added polystyrene beads.

Percentage of Air Voids	Polystyrene Beads
0%	0
1%	3
2%	6
3%	9
4%	12

Elaboration of concrete, fully described in [38] (proportions, Table 2), was done using a planetary mixer to prepare the concrete mix. Concrete and polystyrene pieces were poured in steel tubes by stages using a vibrator rod to compact concrete correctly. A concrete specimen was used to obtain the values of the characteristic resistance. Concrete and CFSTs were standardly cured for 28 days covered with wet clothes, then external surfaces were treated to have a proportional dimension.

Table 2. Concrete mix proportions.

Infill	Proportions
Cement (kg/m^3)	348
Water (l/m^3)	220
Sand (kg/m^3)	1065
Gravel (kg/m^3)	666

2.2. Experimental Setup

The through-transmission technique was used to perform a 2D scan of the specimens in an immersion tank. Figure 2 shows a controlled arm used to correct alignment between the transducers and the specimens and to scan them without changing the distance emitter-receiver. Taking into account the compromise of the attenuation of waves with the frequency, immersion transducers with a central frequency of 1 MHz (0.5″-V303) were selected. The transducer with a smaller diameter was used to increase the resolution in the 2D scan. The maximum voltage registered by the receiver was obtained by rotating the transducers on their vertical and horizontal axis. The distance from the emitter transducer to the CFST samples was set to avoid near-field effects. In this case, the separation was 30 mm. The receiver was positioned at 50 mm from the CFST sample to prevent reflections from interfering with the recorded signal. The edge of the transducer was aligned to the lateral edge of the CFST samples, as shown in Figure 3. The ultrasonic device was programmed to perform measurements every 2 mm in the two dimensions of the movement.

An ultrasonic testing device with integrated pulser and receiver was employed (OP-BOX 2.1). A maximum voltage of 360 V was set and the gain pre-amplifier of +24 dB was used. An analog filter of 0.5–25 MHz was applied. Signals were acquired with a sampling frequency of 100 MHz, and an averaging of 64 samples was set.

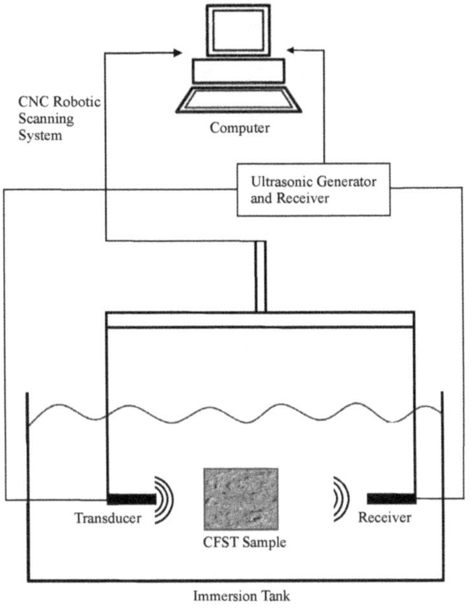

Figure 2. Experimental setup.

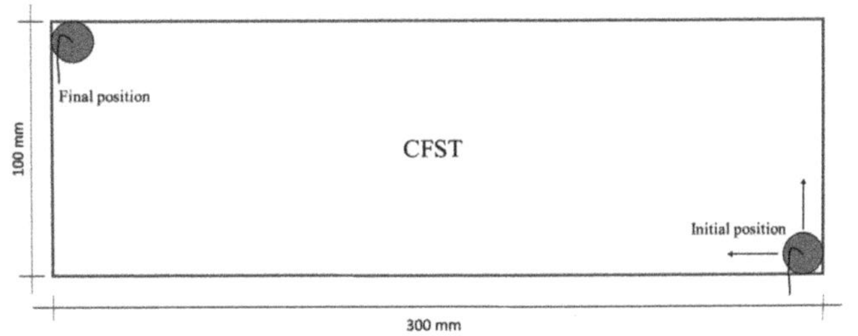

Figure 3. Schematic path of transducers during the scanning process. The gray circles represent the diameter of the transducers. The arrows show the directions of movement of the same.

2.3. Computational Algorithms

2.3.1. Fast Fourier Transform and Short-Time Fourier Transform

Transformation of the signals in the frequency domain was done using the Fast Fourier Transform (FFT) algorithm in Matlab (R2018b, The MathWorks Inc., Natick, MA, USA). To prevent leakage and aliasing, 8192 signal points were employed.

Short-time Fourier transform (STFT) is a sequence of Fourier transforms of a windowed signal. STFT provides the time-localized frequency information for situations in which frequency components of a signal vary over time. STFT algorithm of an ultrasound wave signal s(t) can be written as,

$$S(\tau,\omega) = \int_{-\infty}^{+\infty} s(t)\theta(t-\tau)e^{-i\omega t}\,dt \tag{1}$$

where $\theta(t-\tau)$ is the window function with time and duration, τ is the time resolution, and ω is the radial frequency. Here, a 512-point Hamming window with 50% overlap was employed. The spectrogram allows to represent the signal in the time-frequency domain and evaluate the evolution of the frequencies of the signal over time to detect, in a more precise way, differences between ultrasound signals as its phase velocity and the energy spectrum of the fundamental frequency emitted by the transducer.

In STFT analysis, there exists a tradeoff between time and frequency resolution when determining the window size. In other words, although a narrow-width window results in a better resolution in the time domain, it generates a poor resolution in the frequency domain, and vice versa.

2.3.2. Broadband Ultrasound Attenuation Analysis

The attenuation parameter to compare intact and damaged structures was obtained by comparing the frequency spectrum of the transmitted signal through a reference material that does not contain defects or has low attenuation and the spectrum of the analyzed specimen to be characterized (Figure 4). In this work, degassed water was used for reference signal.

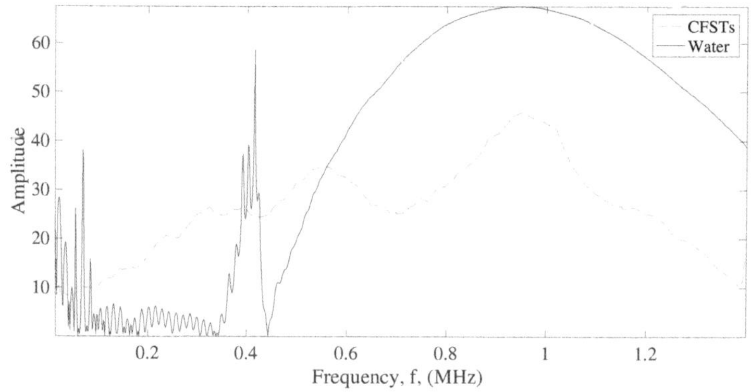

Figure 4. Representation of frequency spectra of reference material (degassed water) and CFST sample (dashed line).

Attenuation at a certain frequency is the difference between the amplitudes of both spectra, the reference material, and the characterized material [29]. Therefore, to obtain the attenuation parameter, the subtraction of the absolute values of the spectra of the reference material (VR), obtained from the FFT, and the absolute values of the spectra of the material to be characterized (VS) should be obtained as,

$$\alpha = 20 \cdot Log_{10}|VR| - 20 \cdot Log_{10}|VS| \tag{2}$$

where α is the attenuation.

The BUA parameter was determined by fitting the attenuation values within a selected frequency range to a straight line (Figure 5) as follows,

$$\alpha = a + BUA \cdot f \tag{3}$$

where f is the frequency in MHz and a is the intersection value of the curve with the vertical axis. The range of frequencies to obtain BUA regression slope highly depends on the fundamental resonant frequency of the transducers used, in this case, the transducers used were 1 MHz. The range of frequencies to obtain BUA rate was from 0.95 MHz to 1.1 MHz (see Figure 5) because it is the range of frequencies around the fundamental frequency of the transducer and that the variation of the attenuation as a function of frequency has a linear behavior.

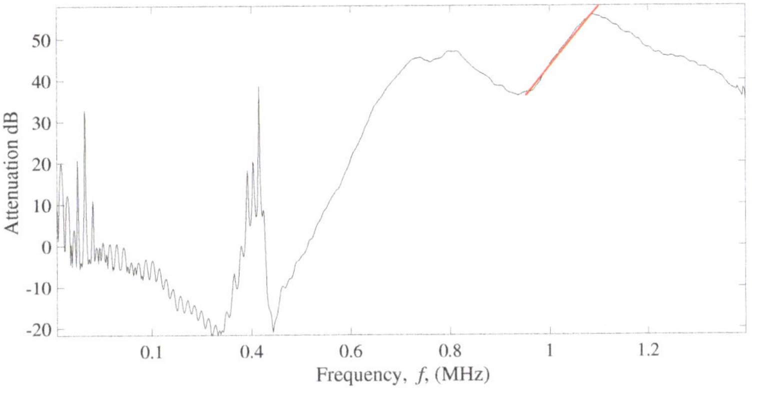

Figure 5. Example of the attenuation versus frequency curve (black solid line) with regression slope (red solid line).

3. Results and Discussion

3.1. FFT and STFT C-Scan Results

The FFT algorithm was applied to the received signals of the 5 CFST samples, and a C-Scan was obtained evaluating the amplitude of the fundamental harmonic of every signal. Results in Figure 6 present a clear difference of the energy spectrum of the signals in the CFST samples according to the percentage of air voids contained.

As shown in Figure 6, a decrease in the amplitude of the fundamental frequency is exhibited on the sides of the CFST samples, compared with values at the center of the CFST. This drop in energy is given due to the high attenuation of the signals in this area since part of the energy from the transducer travels through the water. As the transducer moves away from the edges this effect disappears.

Average of the fundamental harmonic amplitudes, excluding the lateral part of the CFST sample (30 mm on each side) (see Figure 7) shows a decrease of the amplitude of the fundamental harmonic energy as the percentage of air voids contained in the CFSTs increments.

From the obtained result of the fundamental harmonic amplitude averages over the window of the CFST samples, it is shown that the average energy value decreases as the percentage of polystyrene beads increases.

Similarly, the algorithm of STFT was applied to the signals of the 5 CFST samples using the parameters described previously. The C-Scan was performed evaluating the highest value of the energy spectrum, close to 1 MHz.

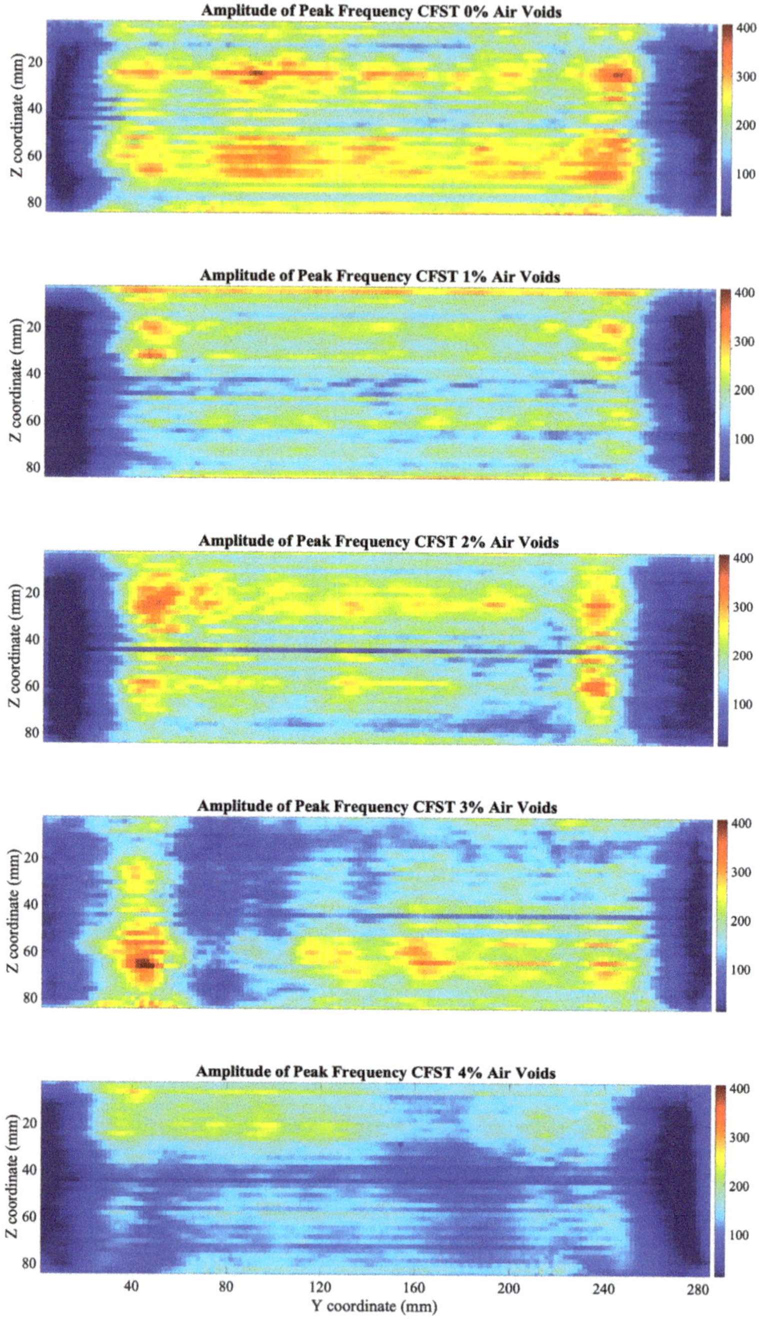

Figure 6. C-Scan of fundamental harmonic amplitudes using FFT for the five CFST samples.

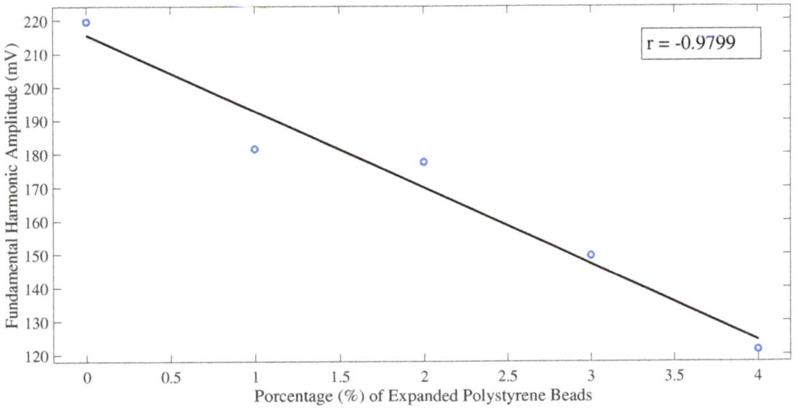

Figure 7. Fundamental harmonic amplitude averages over the window of the CFST samples using the FFT algorithm. The solid black line is a linear fit of the data. Pearson correlation coefficient r = −0.9799.

Although similar results are presented in Figure 8, in comparison with the use of the FFT algorithm, they present some differences. As seen in the C-Scan, in the top and bottom of the CFST, there is an increment in the amplitude of the fundamental frequency due to the signals transmitted through steel. In addition, the range of the energy spectrum is considerably less than the obtained using the FFT algorithm.

The results of the STFT analysis show, as with the FFT algorithm, a reduction in the average value over the analyzed window as the percentage of gaps increases (Figure 9), motivated by the fact that the ultrasounds when traveling through the polystyrene beads reduce the amplitude considerably due to the difference in impedance between two media.

Assumptions of the areas of damaged concrete, due to air voids contained, could be done. The air voids that are contained in the lateral parts of the CFST cannot be visualized, since at these positions the wave travels through the CFTS sample and part through the water.

The analysis in the frequency domain using the FFT and the STFT was sensitive in the detection of internal damage. When transforming the signals in the frequency domain using the FFT a decrease in the amplitude of the fundamental frequency was presented if the signal travels through an air void. Similar results were displayed evaluating the time-frequency energy spectrum using the STFT where a reduction of the level of energy of the fundamental frequency was presented. C-Scan using previous algorithms allowed to identify damaged areas in the CFST samples. Results were validated since the average values of both C-Scan presented a decrease in the energy of the fundamental frequency as the percentage of air voids increased. Several works presented in the literature to detect internal defects in the time-frequency domain showed similar results, such as Fast Fourier Transform [17–19], and Short-Time Fourier Transform [20,21]. Analyzing the frequency spectrum helps in the detection of internal flaws and composition of materials since it is difficult to make a good comparison and a characterization of the ultrasound signals in the time domain in highly attenuated materials, such as concrete.

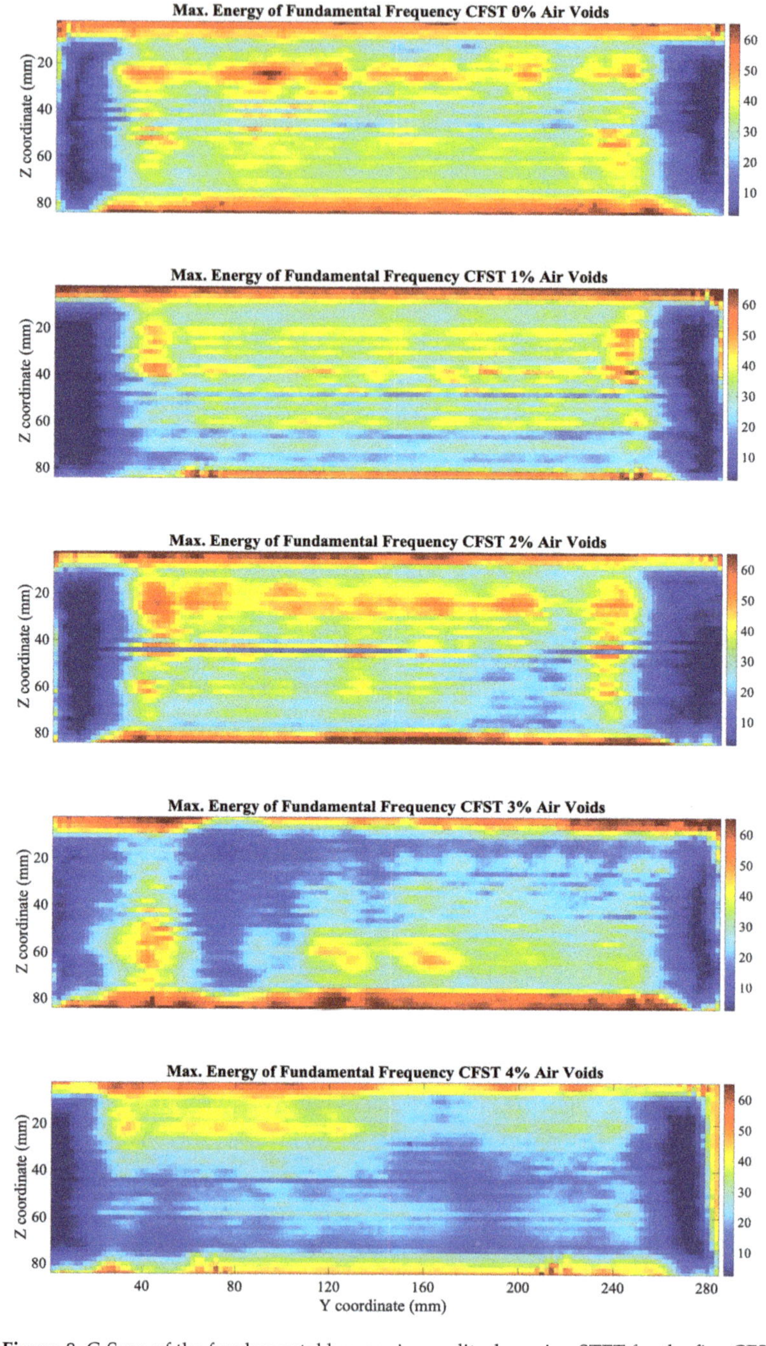

Figure 8. C-Scan of the fundamental harmonic amplitudes using STFT for the five CFST samples.

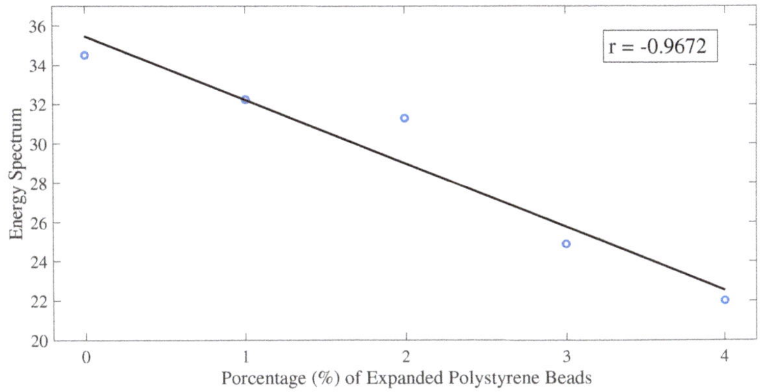

Figure 9. Fundamental harmonic amplitude averages over the window of the CFST samples using the STFT algorithm. The solid black line is a linear fit of the data. Pearson correlation coefficient r = −0.9672.

3.2. BUA C-Scan Results

First, from the obtained signals in the five different CFST samples, a comparison between the attenuation versus frequency curves obtained from signals taken at the center of the CFST sample was done. As shown in Figure 10, the linear regression can be applied in the same range of frequencies (950 kHz to 1.1 MHz) to obtain a difference in the BUA values that allow making a distinction between signals that pass through an air void, and from signals passing through an intact area of the CFST.

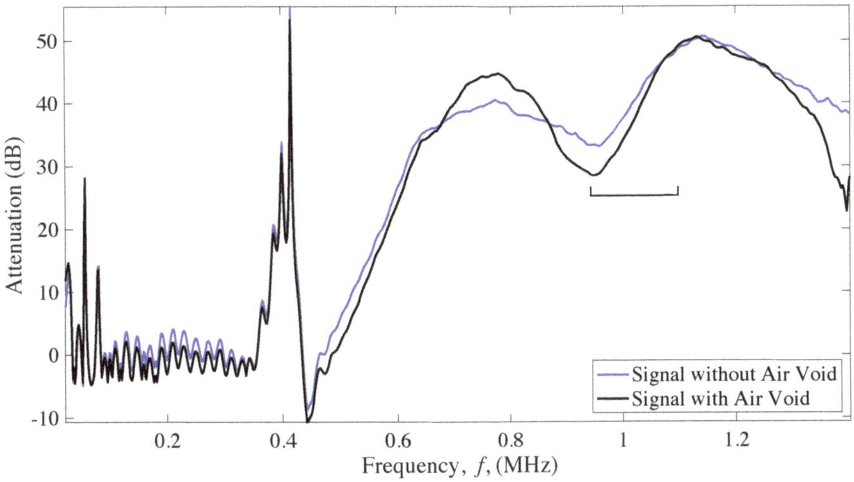

Figure 10. Attenuation versus frequency curve from ultrasound signals measured in the 5 CFST samples. The square bracket indicates the frequency range in which the linear regression is performed.

From the previous result, it can be distinguished that, in the attenuation versus frequency curves a higher slope (BUA) is presented in signals that travel through air voids than in those that travel through the concrete. Therefore, a C-Scan of the BUA values was performed to detect the damaged areas.

Results from the BUA C-Scan are presented in Figure 11. Quantification of results was obtained with averages of the BUA parameter over the window of the CFST samples, and results were plotted against the percentage of air voids, as shown in Figure 12. Results, although affected by the lateral parts of the CFSTs, present a linear relationship between BUA and the percentage of air voids contained, the higher percentage of air voids, the higher values of BUA.

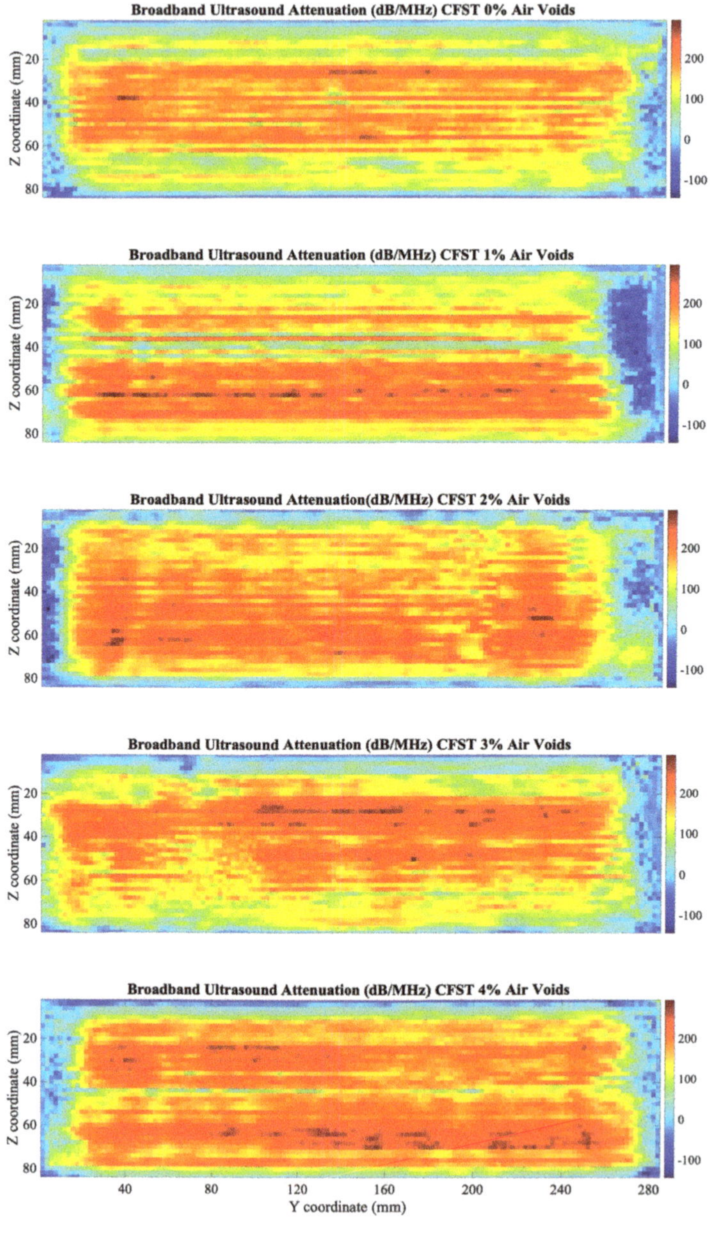

Figure 11. C-Scan of BUA values in dB/MHz for the five CFST samples.

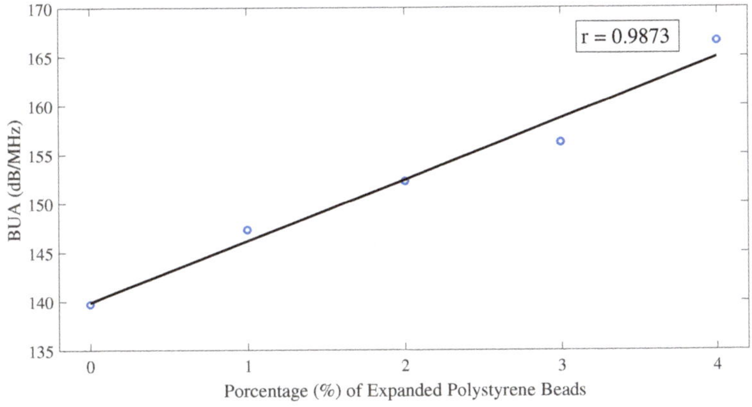

Figure 12. BUA averages over the window of the CFST samples. The solid black line is a linear fit of the data. Pearson correlation coefficient r = 0.9873.

Similar to frequency domain analysis results from the lateral parts of the CFST are different from the values at the center of the CFST samples. For a better comparison of the C-Scan, results excluding the lateral and upperparts that affect the range of BUA values were analyzed (15 mm on each side, and 10 mm top and bottom were excluded), as seen in Figure 13.

Damage can be detected in a more accurate way area in the CFSTs samples (Figure 13), allowing to do an estimation of the location of the expanded polystyrene beads included in the elaboration of the CFST samples. For that purpose, a damage detection algorithm was programmed in Matlab environment. The algorithm performs a sweep over the entire C-scan. In each of the positions, a circle of the diameter of the polystyrene beads is positioned. A BUA threshold is established and if the average of the values inside the circle is greater than the determined threshold, that zone is identified with an inclusion. The set threshold was 240 dB/MHz, obtained through iteration to capture the estimated approximate number of voids in each CFST sample (see Table 1). Figure 13 shows the reconstruction of the air voids, which, in general, are aligned with the longitudinal direction of the samples, which makes sense due to the introduction of a vibrator rod for sample compaction.

Results of analyzing the signal through BUA indicate that this method could be used to detect defects inside CFSTs. The selected range of frequencies to analyze BUA was from 950 kHz to 1.1 MHz, which has a relation with the central frequency of the transducers used. As mentioned by R. Strelitzki et al. [29] there is certain range of frequencies that could be selected for the BUA regression slope, depending on the fundamental resonant frequency of the transducer used, being in most of the case the selected range of frequencies the one that best fits a linear regression. This frequency range was analyzed since more differences were presented in the regression slope of the attenuation versus frequency curve if an ultrasound signal travels through an area with defects. BUA averages over the window of the CFST samples presented a linear increase with the percentage of air voids contained in the sample. Similar results were found in mortars with different degrees of porosity [37], mortars with a higher percentage of porosity have higher regression slope values (BUA) in the attenuation versus frequency curve for a certain range of frequencies than mortars with less percentage of porosity. Although the degree of porosity cannot be obtained from BUA results, there is an approach about how BUA values allow differentiating between elements with different percentage of damage. In addition, the work performed by Rosalba et al. [36] showed that BUA has higher values when aluminum bars are analyzed in a fully drilled area than in an area with smaller drill holes, applied with through-transmission methods.

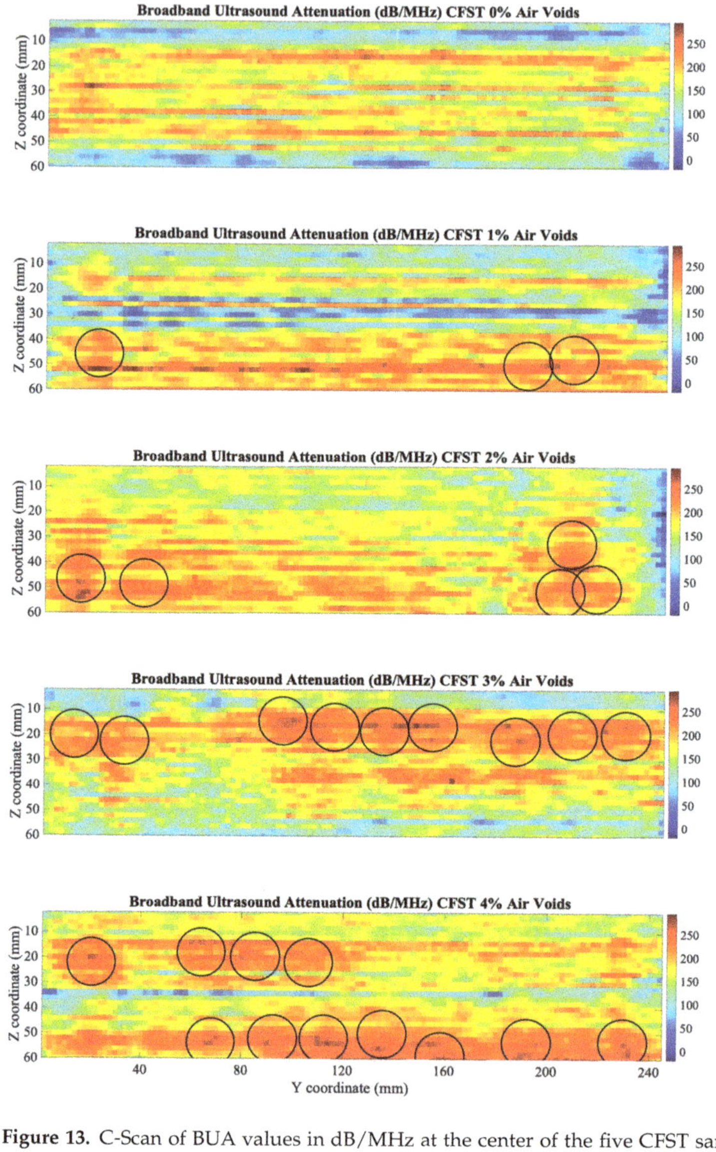

Figure 13. C-Scan of BUA values in dB/MHz at the center of the five CFST samples. The circles represent the position of the air voids calculated by the proposed algorithm.

Results from Pearson's correlation study were shown in Figures 7, 9, and 12. Pearson's correlation coefficient r = −0.9799 for the FFT study, r = −0.9672 for the STFT study, and r = 0.9873 for the BUA study. Despite that the three studies show high correlation coefficients, the BUA study is the one that shows the highest value (r = 0.9873). Despite this higher correlation, the edge effects of the C-scans (left and right areas) in both the FFT and STFT methods (Figures 6 and 8) mask information related to possible concrete imperfections; however, with the BUA method (Figure 11) these edge effects are reduced, obtaining more information in the mentioned areas.

4. Conclusions

Concrete-filled steel tubes (CFSTs), formed by infilling concrete into a hollow steel tube, are structural elements that can present internal defects such as air voids. In this work, the nondestructive evaluation using the analysis of ultrasound signals in the frequency domain was presented. The Fast Fourier Transform, Short-Time Fourier Transform, and Broadband Ultrasound Frequency techniques were employed. The detection of internal damage was carried out by doing a complete ultrasound C-Scan in the CFST samples using an immersion tank. Five CFST stub columns were tested, which have a certain percentage of failure being this 0%, 1%, 2%, 3% and 4%, based on the percentage of expanded polystyrene beads added to the concrete specimen, simulating air voids, in comparison with the volume of concrete. For the three proposed techniques, fundamental harmonic amplitude (FFT and STFT techniques) or BUA averages over the window of the CFST samples were performed. The results present a linear relationship between BUA averages over the window of the CFSTs and the percentage of air voids contained (Pearson's correlation coefficient $r = 0.9873$), the higher percentage of air voids, the higher values of BUA. The BUA algorithm could be applied effectively to distinguish areas with defects inside the CFSTs. Similar to the BUA results, the analysis in the frequency domain using the FFT and the STFT was sensitive in the detection of internal damage (Pearson's correlation coefficient $r = -0.9799$, and $r = -0.9672$, respectively). When transforming the signals in the frequency domain using the FFT a decrease in the amplitude of the fundamental frequency, corresponding to the central frequency of the transducer (1 MHz), was presented if the signal travels through an air void. Similar results were displayed evaluating the time-frequency energy spectrum using the STFT where a reduction of the level of energy of the fundamental frequency was presented. In summary, the results of this investigation establish an improvement in the evaluation of CFST elements for the detection of internal defects. As future research works, based on the limitations of this work, a study of the samples is proposed using the pulse-echo technique, to solve the inspection difficulties by requiring two accessible faces; as well as 2D scanning with smaller diameter transducers to achieve higher resolution and to be able to inspect defects at the edges between the concrete and the steel.

Author Contributions: A.C.: conceptualization, methodology, software, validation, writing—original draft. R.P.: conceptualization, methodology, writing—review and editing. D.H.-F.: methodology, validation, writing—review and editing. G.R.: conceptualization, methodology, validation, writing—review and editing. All authors have read and agreed to the published version of the manuscript.

Funding: This research was funded by Ministerio de Educación, Cultura y Deporte grant numbers DPI2017-83859-R, DPI2014-51870-R, UNGR15-CE-3664 and EQC2018-004508-P; Ministerio de Sanidad, Servicios Sociales e Igualdad grant numbers DTS15/00093 and PI16/00339; Instituto de Salud Carlos III y Fondos Feder; Junta de Andalucía grant numbers PI-0107-2017, PIN-0030- 2017 and IE2017-5537; MCIN/AEI 10.13039/501100011033 grant number PRE2018-086085 (Co-funded by European Social Fund "Investing in your future"); Consejería de economía, conocimiento, empresas y universidad and European Regional Development Fund (ERDF) SOMM17/6109/UGR, B-TEP-026- IE2017-5537 and P18-RT-1653.

Institutional Review Board Statement: Not applicable.

Informed Consent Statement: Not applicable.

Acknowledgments: The authors acknowledge David Sánchez for his participation in the experimental tests.

Conflicts of Interest: The authors declare that they have no known competing financial interests or personal relationships that could have appeared to influence the work reported in this paper.

Abbreviations

The following abbreviations are used in this manuscript:

CFSTs	Concrete-filled steel tubes
BUA	Broadband Ultrasound Attenuation
CFST	Concrete-Filled Steel Tubular
UPV	Ultrasonic pulse velocity method
HL	Hull/Langton index
FFT	Fast Fourier Transform
STFT	Short-time Fourier transform

References

1. Han, L.H.; Li, W.; Bjorhovde, R. Developments and advanced applications of concrete-filled steel tubular (CFST) structures: Members. *J. Constr. Steel Res.* **2014**, *100*, 211–228. [CrossRef]
2. Tao, Z.; Uy, B.; Han, L.H.; Wang, Z.B. Analysis and design of concrete-filled stiffened thin-walled steel tubular columns under axial compression. *Thin Walled Struct.* **2009**, *47*, 1544–1556. [CrossRef]
3. Oh, B.K.; Park, J.S.; Choi, S.W.; Park, H.S. Design model for analysis of relationships among CO_2 emissions, cost, and structural parameters in green building construction with composite columns. *Energy Build.* **2016**, *118*, 301–315. [CrossRef]
4. Figueirido, D.H. Estudio Experimental del Pandeo de Perfiles Tubulares Rectangulares de Acero, Rellenos de Hormigón de Alta Resistencia, Bajo Carga Axial y Diagrama de Momentos Variables. Ph.D. Thesis, Universitat Politècnica de València, Valencia, Spain, 2012.
5. Al Zand, A.W.; Badaruzzaman, W.H.W.; Al-Shaikhli, M.S.; Ali, M.M. Flexural performance of square concrete-filled steel tube beams stiffened with V-shaped grooves. *J. Constr. Steel Res.* **2020**, *166*, 105930. [CrossRef]
6. Al-Zand, A.W.; Badaruzzaman, W.W.; Ali, M.M.; Hasan, Q.A.; Al-Shaikhli, M.S. Flexural performance of cold-formed square CFST beams strengthened with internal stiffeners. *Steel Compos. Struct. Int. J.* **2020**, *34*, 123–139.
7. Xu, P.; Wang, Z.; Mou, B.; Gao, D. Seismic performance of CFST column to steel beam joint with outer annular stiffener. *J. Build. Eng.* **2022**, *54*, 104679. [CrossRef]
8. Climent, M.Á.; Miró, M.; Eiras, J.N.; Poveda, P.; de Vera, G.; Segovia, E.G.; Ramis, J. Early Detection of Corrosion-Induced Concrete Micro-cracking by Using Nonlinear Ultrasonic Techniques: Possible Influence of Mass Transport Processes. *Corros. Mater. Degrad.* **2022**, *3*, 235–257. [CrossRef]
9. Miró, M.; Eiras, J.N.; Poveda, P.; Climent, M.Á.; Ramis, J. Detecting cracks due to steel corrosion in reinforced cement mortar using intermodulation generation of ultrasonic waves. *Constr. Build. Mater.* **2021**, *286*, 122915. [CrossRef]
10. Liao, F.Y.; Han, L.H.; Tao, Z. Behaviour of CFST stub columns with initial concrete imperfection: analysis and calculations. *Thin Walled Struct.* **2013**, *70*, 57–69. [CrossRef]
11. Liao, F.Y.; Han, L.H.; He, S.H. Behavior of CFST short column and beam with initial concrete imperfection: Experiments. *J. Constr. Steel Res.* **2011**, *67*, 1922–1935. [CrossRef]
12. Lu, Z.; Guo, C.; Li, G. Air void and ring gap effect on CFST arch bridges dynamic performance. *J. Constr. Steel Res.* **2021**, *177*, 106418. [CrossRef]
13. Han, L.H. The influence of concrete compaction on the strength of concrete filled steel tubes. *Adv. Struct. Eng.* **2000**, *3*, 131–137. [CrossRef]
14. Han, L.H.; Yao, G.H. Influence of concrete compaction on the strength of concrete-filled steel RHS columns. *J. Constr. Steel Res.* **2003**, *59*, 751–767. [CrossRef]
15. Guo, C.; Lu, Z. A Comprehensive Research of the Air-void Defect of Concrete-filled Steel Tube. *Period. Polytech. Civ. Eng.* **2021**, *65*, 474–485. [CrossRef]
16. Guo, C.; Lu, Z. CFST Rib with Circumferential Gap and SWS Composite Defects. *Iran. J. Sci. Technol. Trans. Civ. Eng.* **2021**, *45*, 1441–1456. [CrossRef]
17. Selleck, S.F.; Landis, E.N.; Peterson, M.L.; Shah, S.P.; Achenbach, J.D. Ultrasonic investigation of concrete with distributed damage. *Mater. J.* **1998**, *95*, 27–36.
18. Shah, A.; Ribakov, Y. Non-destructive evaluation of concrete in damaged and undamaged states. *Mater. Des.* **2009**, *30*, 3504–3511. [CrossRef]
19. Zhao, M.; Nie, Z.; Wang, K.; Liu, P.; Zhang, X. Nonlinear ultrasonic test of concrete cubes with induced crack. *Ultrasonics* **2019**, *97*, 1–10. [CrossRef]
20. Le, M.; Kim, J.; Kim, S.; Lee, J. B-scan ultrasonic testing of rivets in multilayer structures based on short-time Fourier transform analysis. *Measurement* **2018**, *128*, 495–503. [CrossRef]
21. Petro, J.T., Jr.; Kim, J. Detection of delamination in concrete using ultrasonic pulse velocity test. *Constr. Build. Mater.* **2012**, *26*, 574–582. [CrossRef]
22. Shiotani, T.; Aggelis, D.G. Wave propagation in cementitious material containing artificial distributed damage. *Mater. Struct.* **2009**, *42*, 377–384. [CrossRef]
23. Xu, J.; Wei, H. Ultrasonic testing analysis of concrete structure based on S transform. *Shock Vib.* **2019**, *2019*, 2693141. [CrossRef]

24. Chai, H.; Momoki, S.; Kobayashi, Y.; Aggelis, D.; Shiotani, T. Tomographic reconstruction for concrete using attenuation of ultrasound. *NDT E Int.* **2011**, *44*, 206–215. [CrossRef]
25. Sharma, G.K.; Kumar, A.; Rao, C.B.; Jayakumar, T.; Raj, B. Short time Fourier transform analysis for understanding frequency dependent attenuation in austenitic stainless steel. *NDT E Int.* **2013**, *53*, 1–7. [CrossRef]
26. Mutlib, N.K.; Ismael, M.N.; Baharom, S. Damage Detection in CFST Column by Simulation of Ultrasonic Waves Using STFT-Based Spectrogram and Welch Power Spectral Density Estimate. *Struct. Durab. Health Monit.* **2021**, *15*, 227. [CrossRef]
27. Langton, C.; Palmer, S.; Porter, R. The measurement of broadband ultrasonic attenuation in cancellous bone. *Eng. Med.* **1984**, *13*, 89–91. [CrossRef]
28. Maia, J.M.; Costa, E.T.; Neto, J.F.M.; Button, V.L. Broadband ultrasound attenuation in the calcaneal region: A comparative study of single-position versus scanning systems. *IEEE Trans. Ultrason. Ferroelectr. Freq. Control.* **2008**, *55*, 64–73. [CrossRef]
29. Strelitzki, R.; Evans, J. An investigation of the measurement of broadband ultrasonic attenuation in trabecular bone. *Ultrasonics* **1996**, *34*, 785–791. [CrossRef]
30. Töyräs, J.; Nieminen, M.; Kröger, H.; Jurvelin, J. Bone mineral density, ultrasound velocity, and broadband attenuation predict mechanical properties of trabecular bone differently. *Bone* **2002**, *31*, 503–507. [CrossRef]
31. Tasinkevych, Y.; Falińska, K.; Lewin, P.A.; Litniewski, J. Improving broadband ultrasound attenuation assessment in cancellous bone by mitigating the influence of cortical bone: Phantom and in-vitro study. *Ultrasonics* **2019**, *94*, 382–390. [CrossRef] [PubMed]
32. Chappard, C.; Laugier, P.; Fournier, B.; Roux, C.; Berger, G. Assessment of the relationship between broadband ultrasound attenuation and bone mineral density at the calcaneus using BUA imaging and DXA. *Osteoporos. Int.* **1997**, *7*, 316–322. [CrossRef]
33. Langton, C.; Ali, A.; Riggs, C.; Evans, G.; Bonfield, W. A contact method for the assessment of ultrasonic velocity and broadband attenuation in cortical and cancellous bone. *Clin. Phys. Physiol. Meas.* **1990**, *11*, 243. [CrossRef]
34. Hodgskinson, R.; Njeh, C.; Whitehead, M.; Langton, C. The non-linear relationship between BUA and porosity in cancellous bone. *Phys. Med. Biol.* **1996**, *41*, 2411. [CrossRef]
35. Hull, J.; Langton, C.; Barker, S.; Jones, A. Identification and characterisation of materials by broadband ultrasonic attenuation analysis. *J. Mater. Process. Technol.* **1996**, *56*, 148–157. [CrossRef]
36. Da Costa, R.; Maia, J.M.; Assef, A.A.; Pichorim, S.F.; Gewehr, P.M.; Costa, E.T. Defect Detection in Aluminum Bars Using Impedance and Ultrasonic Attenuation. *IEEE Sens. J.* **2020**, *20*, 7400–7413. [CrossRef]
37. Vergara, L.; Miralles, R.; Gosálbez, J.; Juanes, F.; Ullate, L.; Anaya, J.; Hernández, M.; Izquierdo, M. NDE ultrasonic methods to characterise the porosity of mortar. *NDT E Int.* **2001**, *34*, 557–562. [CrossRef]
38. Iba nez, C.; Hernández-Figueirido, D.; Piquer, A. Shape effect on axially loaded high strength CFST stub columns. *J. Constr. Steel Res.* **2018**, *147*, 247–256. [CrossRef]

materials

MDPI

Article

Mesoscale Study on Splitting Tensile Damage Characteristics of Concrete Based on X-ray Computed Tomography and Digital Image Correlation Technology

Hua Zhang *, Qi Pan, Kai Zheng, Chuanjun Jin and Luoyu Pan

College of Civil and Transportation Engineering, Hohai University, Nanjing 210009, China; 191304020005@hhu.edu.cn (Q.P.); zhengkai1993@163.com (K.Z.); 201304020009@hhu.edu.cn (C.J.); 201304020031@hhu.edu.cn (L.P.)
* Correspondence: 20020018@hhu.edu.cn

Abstract: In this paper, the mesoscale damage properties of concrete and mortar were studied experimentally under Brazilian disc splitting tensile tests combining X-ray computed tomography (CT) and digital image correlation (DIC) technology. Considering the factors of water/cement ratios and loading rates, the influence of meso components on the macro tensile properties and failure modes of concrete were studied. The experimental results and analysis indicate that the following: (1) the existence of coarse aggregate makes the tensile strength of concrete lower than that of mortar and reduces the sensitivity of tensile strength to the loading rates; (2) the failure modes of mortar and concrete Brazilian discs differ in the crack initiation positions and localization phenomena. Under high loading rates, the local failure plays a critical role in the strength improvement of concrete; (3) for concrete, interface failure and mortar failure are the main failure modes under low loading rates, whereas aggregate failure gradually becomes the main failure mode with increasing loading rates. The decrease in water/cement ratios improves the strength of the mortar matrix and interfacial bonding performance, leading to more serious aggregate damage and higher strength.

Keywords: concrete; splitting tensile strength; X-ray CT technology; digital image correlation technology; mesoscale damage

Citation: Zhang, H.; Pan, Q.; Zheng, K.; Jin, C.; Pan, L. Mesoscale Study on Splitting Tensile Damage Characteristics of Concrete Based on X-ray Computed Tomography and Digital Image Correlation Technology. *Materials* **2022**, *15*, 4416. https://doi.org/10.3390/ma15134416

Academic Editor: Bogusław Łazarz

Received: 27 April 2022
Accepted: 20 June 2022
Published: 22 June 2022

1. Introduction

Concrete, with a complex heterogeneous structure, is considered as a three-phase composition of cement mortar, aggregates, and interfacial transition zones at the mesoscale [1,2]. Unlike macroscopic analysis assuming a homogeneous material, the internal stress distribution of concrete is extremely intricate, resulting in the complexity and difference of mesoscale damage distribution in the failure progress, which would considerably influence the macro tensile performance [3].

On the macroscopic scale, splitting tensile tests (Brazilian disc tests) are the most popular methods to determine the uniaxial tensile strength of concrete due to simple preparation and easy operation [4,5]. In this method, the tensile stress is indirectly generated by applying the radial compressive pressure perpendicular to the tensile direction [6,7]. Therefore, the compressive stress concentrations near the loading strip usually disturb the distribution of horizontal normal tensile stresses. To ensure the validity of the experiments, the initial splitting position and failure modes are of great concern, as the initial cracks should appear in the center of the Brazilian disc specimens and propagate along the diameter [8].

However, with the increase in loading rates, the macro tensile behavior of concrete specimens usually changes significantly [9]. As generally recognized, there will be an apparent increase in the tensile strength of concrete under higher loading rates [10]. Previous studies usually proposed that the energy accumulation caused by insufficient time for internal crack development is the key reason [11], while abundant numerical simulation results concluded that the viscosity effect of free water under high loading rates, as well as the inertial effect derived from the dynamic response, cannot be ignored [12,13]. For failure modes, the crack initiation of specimens tends to be well predicted under static or quasi-static loading [14]. Conversely, it is difficult for specimens to reach the internal stress balance under high loading rates, and the main failure modes are hard to ensure. The first crack may appear at any position, followed by numerous multiple secondary cracks. Finally, multiple parallel main cracks or fragment failure modes can be observed [15,16]. A change in the failure pattern is usually attributed to the complexity of the stress distribution or the stress concentration at the loading end in macro analysis [17], which ignores the localization characteristics and stress redistribution in the fracture process [18]. In summary, it is very difficult to accurately reveal the damage characteristics and truly understand the behavior of concrete under dynamic tension on the basis of macroscopic damage analysis. Thus, it is necessary to analyze the tensile failure process from a mesoscopic perspective and explore the characteristics of mesoscopic damage to highlight the effect of different mesoscopic phases on the strengthening tensile strength.

On the mesoscopic scale, the distribution and change in internal stress are closely associated with the volume fraction, distribution, and interaction of the three phases (i.e., mortar, aggregates, and the interface) of concrete [19,20]. Currently, with the development of the nondestructive testing technology, the damage characteristics of the three phases can be extracted using various methods such as X-ray computed tomography (X-CT), scanning electron microscopy (SEM), ultrasonic methods, and digital image correlation technology (DIC). Among them, indirect methods such as nonlinear ultrasonic technology have been proven to be able to identify and quantify the damage evolution in the failure process, which can meet the actual industrial needs [21–23]. However, it cannot directly provide information regarding defects as the information is deduced on the basis of parameters related to the defect [24]. In direct methods, SEM can monitor crack morphology with extremely high resolution and quantitatively analyze the parameters at the fracture boundary to compare the fracture mechanism [25], but the sample size is limited to about 10 mm. For specimens of tens of centimeters in size, X-CT and DIC technology are more commonly used to detect mesoscopic damage [26].

DIC technology can establish the displacement field and strain field of the specimen surface by using a camera and receiving surface images of the specimen. Numerous experiments on specimens of other shapes have demonstrated the capacity of DIC for the study of the deformation characteristics and the stress variation [27,28]. For Brazilian disc specimens, Jin et al. [14] captured the surface of concrete specimens using a DIC camera to evaluate the failure modes. Martínez-López et al. [8] identified the fracture properties and elastic properties of mortar specimens by combining the DIC method and the Brazilian disc test. Nonetheless, as a two-dimensional system, it lacks the ability to describe the internal strain distribution for thicker specimens. To investigate internal damage propagation, X-CT is commonly used for further analysis [29]. The images of each layer are obtained intuitively by slicing the specimens under the X-ray, in which multiphase materials are distinguished by density [30]. Suchorzewski et al. [5] simulated quasi-static splitting tensile tests on concrete specimens and evaluated the effect of the different loading types on the fracture process on the basis of CT images. Skaryński and Tejchman [31] applied X-ray CT technology to obtain a quantitative description of the size and distribution of pores and cracks in the fracture process. Zhu et al. [20] used the porosity obtained from CT images to calculate the degree of mesoscale damage under tensile tests. In addition, researchers have also explored more accurate threshold methods to separate phases before analysis, as the gray values of solid phases in CT images are relatively close [32]. Loeffler et al. [33] presented a quasi-3D geometric segmentation technique to accurately identify and isolate both void clusters and large cracks, which improved the deficiency of previous segmentation methods in identifying small holes. Hao et al. [34] classified CT images under different deformation stages to make a connection between the cavity distribution and the stress change in the failure progress of concrete. Thus, the damage parameters of concrete could be analyzed quantitatively. However, previous research on mesoscopic damage of concrete with the application of CT or DIC technology mainly focused on the internal pores and cracks. The specific influence of the damage evolution of other phases on the failure progress, especially under splitting tensile tests, remains to be further analyzed.

The aim of this paper is to investigate the influence of mesoscopic damage characteristics on the macro splitting tensile properties of concrete. For this purpose, splitting tensile tests of concrete and mortar Brazilian disc specimens under different loading rates were carried out to compare the different effects of mesoscopic materials. Then, on the basis of the tensile strength and failure modes obtained from the macroscale experiments, DIC and X-CT technology were employed for the subsequent analysis. The distribution characteristics of each phase of concrete, their roles in the failure process, and their effects on the macrocrack morphology were discussed. Lastly, a three-dimensional model of crack distribution was established for statistical research on the volume proportion of each phase and the analysis of the three-dimensional morphology of cracks.

2. Experimental Program

2.1. Specimens Preparation

The specimens were categorized into a concrete group and a mortar control group for comparative study. The mix proportions for concrete strength are summarized in Table 1. In preparation, the cement was of grade 32.5. The size of the fine aggregates was controlled between 0.4 and 2.5 mm, and the coarse aggregate size was within the range of 5–10 mm.

Table 1. The mix proportions for concrete and cement mortar.

Materials	Type	Water	Cement	Fine Aggregate	Coarse Aggregate	Design Strength
Concrete	CA	0.64	1.00	2.12	3.22	C25
	CB	0.56	1.00	1.85	2.86	C35
	CC	0.41	1.00	1.02	2.09	C45
Cement mortar	MA	0.64	1.00	2.12	/	/
	MB	0.56	1.00	1.85	/	/
	MC	0.41	1.00	1.02	/	/

In the design of the size of specimens, it is necessary to consider the obvious size effect of concrete materials due to the nonuniformity of the meso composition in concrete [35,36]. When the size of specimens increases, the energy absorbed in the local failure zone is smaller than that released in residual unloading regions under decreasing load, causing the lower splitting tensile strength [5,36]. It also affects the transformation of fracture modes, as a larger specimen is more brittle [37]. After comprehensive consideration, the size of Brazilian disc specimens in this paper was designed as Φ 74 mm $\times$ H 37 mm.

2.2. Brazilian Disc Splitting Tensile Test

The Brazilian disc test was based on the Chinese standard GB/T 50081-2002 (standard test method for mechanical properties of ordinal concrete) and was carried out on an Instron 8802 servo-hydraulic testing machine, as shown in Figure 1. The Instron 8802 servo-hydraulic testing machine can be controlled by displacement or pressure. In order to control the loading rates, the tests in this paper were controlled by displacement, i.e., the falling displacement of the upper plate of the machine.

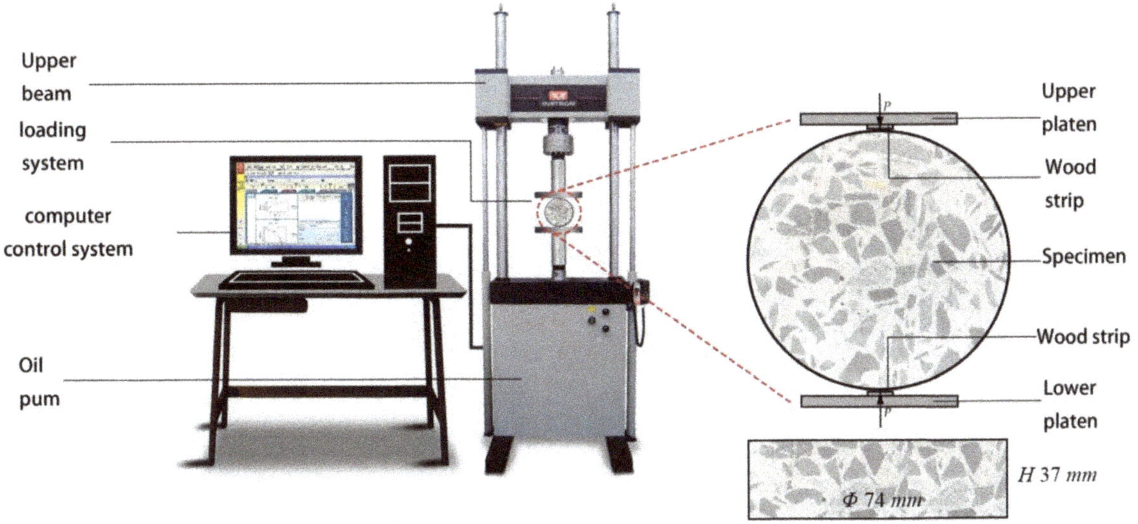

Figure 1. Schematic diagram of Brazilian disc splitting tensile test.

Six groups of specimens were tested under four different loading rates. The naming rules of the specimens were as follows: type of specimens-loading rates. For the convenience of recording, 02 m, 002 s, 02 s, and 2 s were used as loading rates corresponding to loading rates of 0.2 mm/min, 0.02 mm/s, 0.20 mm/s, and 2.00 mm/s, respectively. There were five specimens under each working condition. Moreover, two 10 mm wide and 2 mm thick wood cushions were glued to both ends of the specimen as loading strips.

2.3. DIC Test System

According to the literature, the DIC test system can obtain the deformation information by matching the corresponding subregions of the image before and after deformation. A square reference subregion is taken out from the image before deformation, and then the differences between the deformed and reference images are measured by the standard covariance cross-correlation function Equation (1). The coordinates of one point in the reference subregion are denoted as (x, y), and the coordinates of this point after deformation are (x', y'). When the correlation coefficient $C(u, v)$ reaches the maximum, the subregion after deformation is most similar to the reference subregion, and the displacement value of each point in the calculation area can be obtained using an iterative method.

$$C(u,v) = \frac{\sum_{x=-M}^{M} \sum_{y=-M}^{M} [f(x,y) - f_m][g(x',y') - g_m]}{\sqrt{\sum_{x=-M}^{M} \sum_{y=-M}^{M} [f(x,y) - f_m]^2} \sqrt{\sum_{x=-M}^{M} \sum_{y=-M}^{M} [g(x',y') - g_m]^2}}, \quad (1)$$

where M is half the side length of the reference subarea, $f(x, y)$ is the gray value of the point (x, y), $g(x', y')$ is the gray value of the point (x', y'), f_m is the average gray value of the reference subarea, g_m is the average gray value of the deformed area, u is the displacement of the point in the X-direction, and v is the displacement of the point in the Y-direction.

During this procedure, the surface preparation of the specimens is a key factor affecting the experimental quality, as obvious and random distribution characteristic points are needed for correlation matching. Moreover, the experimental accuracy can be affected by inconspicuous spraying points or light intensity fluctuations. Therefore, black speckles were evenly and randomly sprayed on the specimen surface with a spacing of 0.1–1 mm before experiments. In addition, a constant light source was set near the test piece to eliminate the influence of light changes on the image.

In the system shown in Figure 2a, the shutter speed of the lens was 1/800 s, the aperture size was f/5, and the ISO was 3200. The exposure interval of the camera was set as 1 s. Due to the relatively small size of the specimens, the measuring field range was set over the entire surface. As shown in Figure 2b, the collected images were set as 8 bit grayscale images with a resolution of 1440 × 1440 pixels, and the diameter of the specimens occupied 1020 pixels. After calibration, the object plane resolution of the system was 0.0725 mm/pixel.

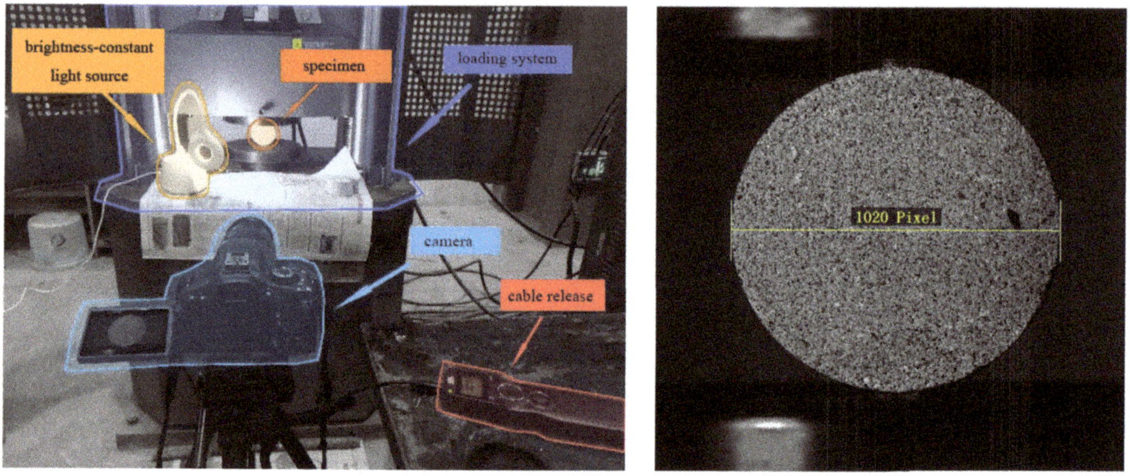

(a) **(b)**

Figure 2. Digital image correlation (DIC) test system: (**a**) details of test setup, (**b**) images captured by camera.

2.4. CT Scanning Test

In this paper, CT scanning test was carried out according to Chinese standard GB/T 37166-2018 (nondestructive testing method of composite materials for industrial computed tomography). In the CT scanning system, the X-ray from the radiation source reaches the detector after scanning the sample. After data processing, slice files in all directions are received, and the three-dimensional shape of the sample is reconstructed. There are two main factors affecting the accuracy of industrial CT systems [38]. One is the scanning resolution of the machine. Generally, a smaller voxel size results in higher resolution and greater accuracy. Another is the segmentation method of the images. In CT images, materials with different densities can be distinguished by different gray levels. Therefore, different phases of concrete can be divided after the threshold procedure. Appropriate threshold segmentation is crucial to obtain clear and accurate mesoscale structures.

Herein, the YXLON microfocus X-ray computed tomography system was used in the CT scanning test. In the CT scanning system shown in Figure 3, the number of detector pixels was 1024 × 1024, and the size of the pixels was 200 microns. In images, the resolution of the *X*-axis and *Y*-axis was 901 × 901 pixels, and the *Z*-axis resolution was approximately 450 pixels. The pixel size was calibrated to 0.0812 mm/pixels. In addition, the scanning time of each projection was 2.5 s, and approximately 450 layers were sliced along the thickness direction of each specimen.

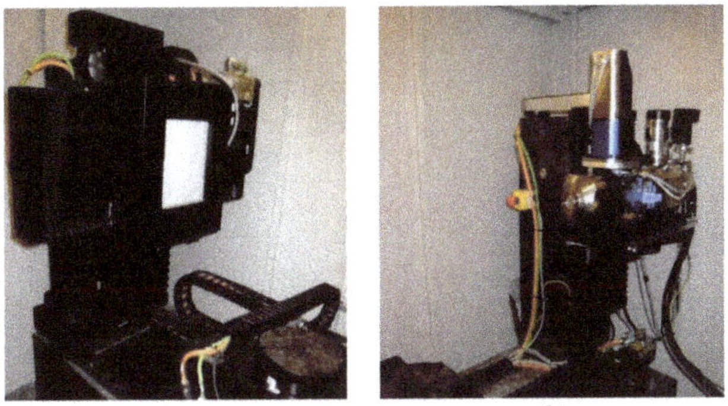

(**a**) (**b**)

Figure 3. CT scanning system: (**a**) flat panel detector, (**b**) X-ray source.

3. Results and Discussion

3.1. Splitting Tensile Strength Properties

As depicted in Figure 4, the load–displacement curves of concrete and mortar were obtained under different loading rates.

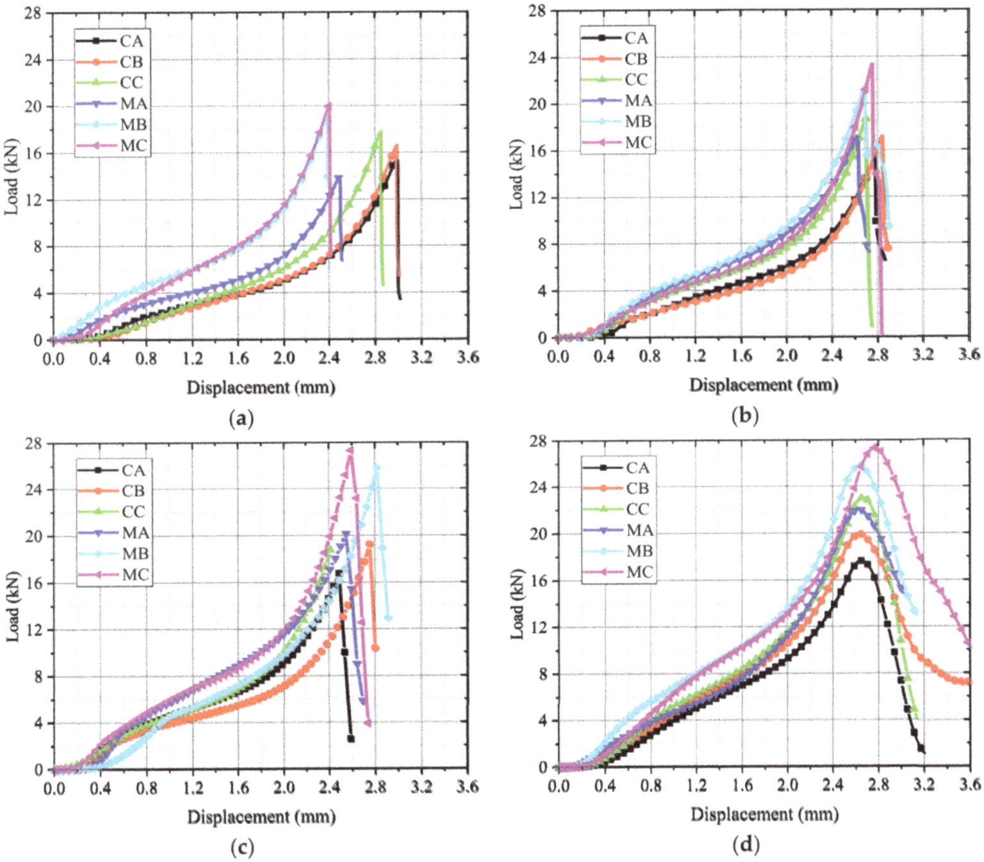

Figure 4. Load–displacement curves of splitting tensile test under different loading rates: (**a**) 0.2 mm/min; (**b**) 0.02 mm/s; (**c**) 0.2 mm/s; (**d**) 2 mm/s.

At the initial stage of loading, the load increased slowly with increasing displacement due to the deformation of the gasket. After the gasket was compacted, the stress increased sharply and declined soon after reaching the peak. Comparing the curves under different loading rates, it can be seen that the specimens presented obvious brittle failure under low loading rates, while the specimens presented ductile facture at a high loading rate of 2 mm/s. Additionally, it can be found that the failure displacements of mortar had an obvious increasing trend when the loading rate increased, gradually increasing from 2.38 mm (loading rate = 0.2 mm/min) to 2.77 mm (loading rate = 2 mm/s), which indicated that the failure displacement of mortar was sensitive to the loading rates. However, concrete has no obvious trend of the displacements at failure, and it varied with an average value of 2.72 mm. The aggregate may play a critical role in reducing the sensitivity of concrete.

According to ASTM D3967-16, the theoretical formula of tensile strength obtained from Brazilian disc splitting test is shown in Equation (2).

$$f_t = \frac{2P_{max}}{\pi DH}, \quad (2)$$

where P_{max} is the peak load, D is the diameter of the Brazilian disc, and H is the height of the Brazilian disc.

After calculation, Figure 5 presents the tensile strength and standard deviation of each group of specimens; the tensile strength of mortar was generally higher than that of concrete under the same water/cement ratio. The average tensile strength of mortar under loading rates of 0.2 mm/min, 0.02 mm/s, 0.2 mm/s, and 2 mm/s was 10.3%, 17.8%, 21.3%, and 21.8% higher than that of concrete, respectively. Due to the presence of the interface around aggregates, the damage of concrete developed faster at higher loading rates, which made the strength of concrete lower than that of mortar. At water/cement ratios of 0.64, 0.56, and 0.41, the average tensile strength of the mortar specimens was 13.6%, 21.4%, and 22.6% higher than that of the concrete, respectively. All specimens with lower water/cement ratios had higher tensile strength because the interfacial strength increased with decreasing water/cement ratio, making specimens more difficult to crack.

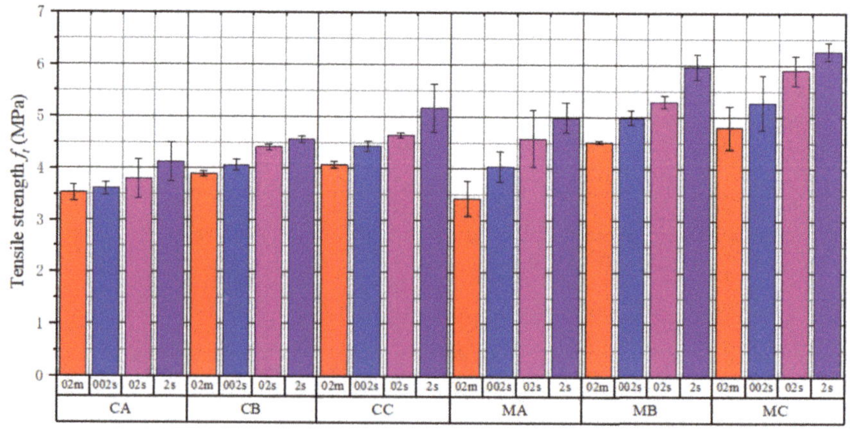

Figure 5. Statistical chart of tensile strength of specimens.

3.2. *Failure Modes*

3.2.1. Analysis of the Displacement Field and Strain Field Based on DIC Technology

In this paper, CA (concrete, water/cement ratio = 0.64) and MA specimens (mortar, water/cement ratio = 0.64) under the loading rate of 0.2 mm/min were selected for a comparative study of DIC images.

The characteristic points were marked on the test piece, as shown in Figure 6, and the strain–time curves of the characteristic points were extracted, as presented in Figure 7. The loading time of the failure progress was divided into 22 moments, and the peak load was at approximately the 17th moment.

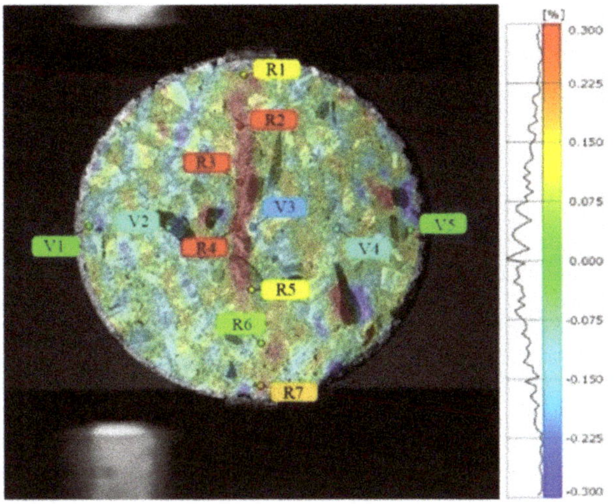

Figure 6. Characteristic points.

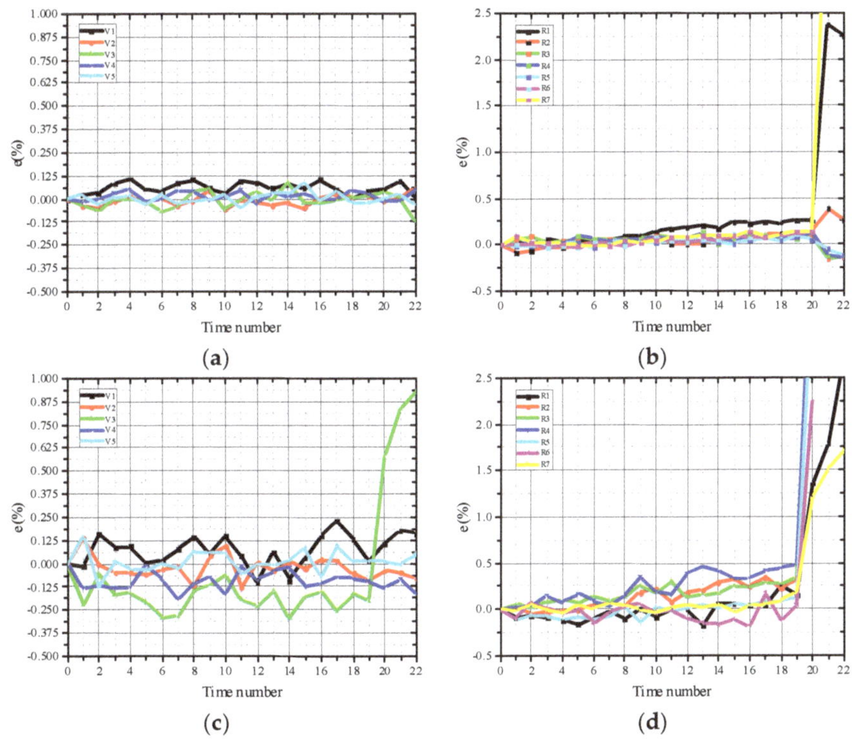

Figure 7. Strain—time curves of characteristic points: (**a**) vertical loading direction of mortar specimens; (**b**) loading direction of mortar specimens; (**c**) vertical loading direction of concrete specimens; (**d**) loading direction of concrete specimens.

In general, the strain level of the concrete characteristic points was higher than that of the mortar, which indicated that the aggregates improved the deformation ability of concrete. After reaching the peak load, the strain–time curves of the points in the vertical loading direction fluctuated slightly while the strain of the characteristic points in the load direction increased with time, which indicated the nonuniformity of strain at the characteristic points of the specimens. In the horizontal direction, the strain of point V3 nearest to the center was relatively large compared with the points at the edge of the specimens (V1 and V5). Thus, the strain level decreased from the center to the end, and the center of the specimen was the first to fail. However, along the loading direction, the strain levels at the edge of the mortar specimen were always higher than those at the center, which meant that macro cracks first occurred at the end of the specimens. For the concrete specimens, the strains at the center of the specimens were always at the highest level, indicating that the cracks first appeared at the center of the concrete specimens, and the crack propagated to the end of the specimen along the radial direction.

Figure 8 presents the strain maps of mortar. It can be noted that the localization phenomenon at the loading end was obvious. Near the loading strip, the cracks often deviated from the center of the specimen, resulting in secondary cracks. The largest tensile strain on mortar specimens was 0.0237 along the diametral loading direction, which meant the mortar specimens began to crack at this moment. This is very close to the value in the literature [14]. For the strain maps of concrete specimens shown in Figure 9, the localization phenomenon at the end was improved successfully. There was no deviation from the center or secondary cracks. Therefore, the existence of aggregate improved the crack development process and made the test results more reliable. After the peak time, the high strain zones extended from the periphery of the aggregate to the end. The largest tensile strain was 0.0836 along the diametral loading direction. According to the strain maps and strain–time curves, the failure process could be divided into three stages. In the first stage, the internal stress level of the specimens was small, and the specimens were in the stage of linear elastic deformation under tension. At this time, the number of microcrack initiation was small, and stress increased relatively slowly. In the second stage, the internal stress level of the specimen increased, and the stress localization was obvious. Although macro cracks did not yet occur, a large number of internal microcracks were already initiated. The third stage came after reaching the peak stress, the microcracks propagated and converged rapidly, and a macro crack occurred.

Before the macro cracks occurred, the size of the local stress concentration area also affected the failure process of specimens. In DIC images, the uniform width of a localized zone of concrete could be described following Equations (3) and (4). The particle displacements were fitted by the error function *ERF*, as shown in Equation (3). Then, the error function evaluated at $\frac{x}{s\sqrt{2}}(x > 0)$ gave the probability that the measurement under the influence of normally distributed errors with the standard deviation s had a distance smaller than x from the mean value. Then, the width of localized zone W_{lz} could be calculated according to the fitting function parameter s, as shown in Equation (4). In this way, 95% of the values of the normal distribution function area were within the distance of two standard deviations from the average value [39].

$$ERF(x) = \frac{2}{\sqrt{x}} \int_0^x e^{-t^2} dt, \quad (3)$$

$$W_{lz} = 4 \cdot s, \quad (4)$$

where x is the measured distance, s is the fitting function parameter, and W_{lz} is the width of the localized zone.

Lastly, the width of the localized zone in concrete was 3.78 mm at the peak load, which was little larger than that of mortar (3.45 mm). The localized zone of mortar appeared earlier, but did not spread significantly before failure. The existence of aggregate made the high strain region distribute around the aggregate in the strain map of concrete, and the different elastic moduli of aggregate and mortar led to uneven deformation, which gradually expanded the localization zone from center to both ends. The width of the localized zone in concrete was also little larger than the width recorded in other studies such as 3.41 mm [5] and 2.80 mm [40]. This is because the width of strain localization is also affected by the specimen size and test methods.

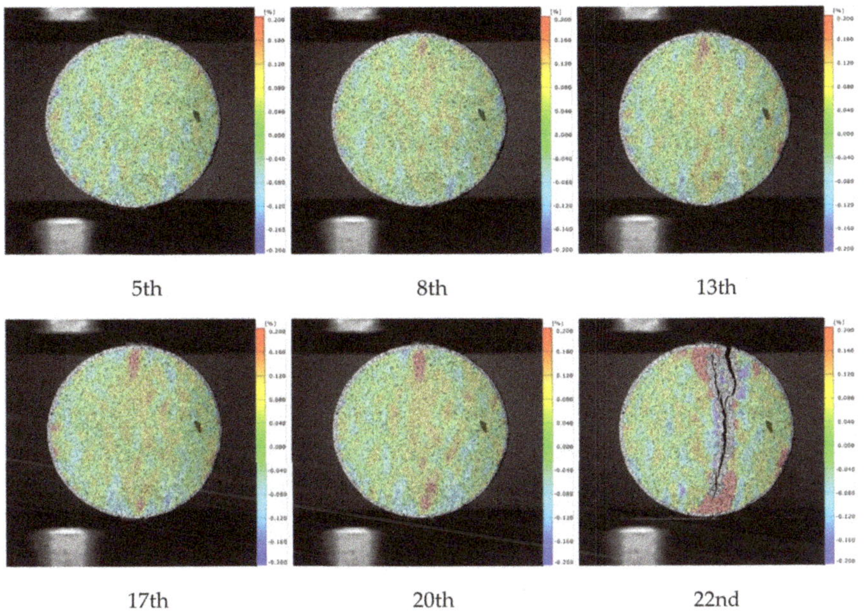

Figure 8. Strain maps of mortar.

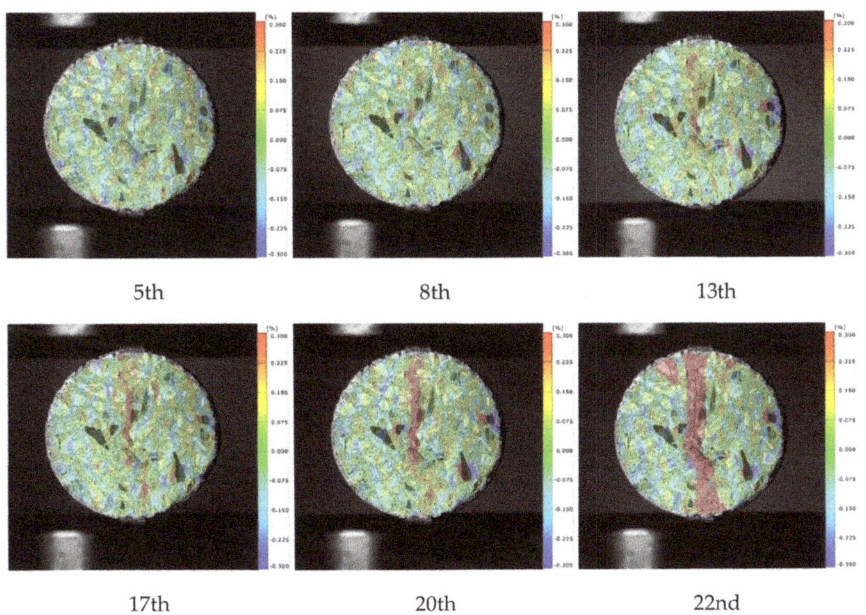

Figure 9. Strain maps of concrete.

3.2.2. Analysis of Crack Propagation Based on CT Images

CA specimens (concrete, water/cement ratio = 0.64), CC specimens (concrete, water/cement ratio = 0.41), and MA specimens (mortar) under loading rates of 2 mm/s and 0.2 mm/min were selected for the comparative study of CT slice images. Approximately 450 slice images were derived from each specimen along the thickness direction, and the typical slice images from the 250 images of the middle layer were extracted for analysis. Specifically, the 100th, 170th, 240th, and 310th slices of each specimen are presented in Figure 8. On a whole, all specimens had a main crack along the loading diameter direction when they were destroyed, and the distribution of cracks changed along the specimen thickness direction.

According to the crack distribution of the MA specimens shown in Figure 10a,b, the cracks were fine at low loading rates but relatively coarse with secondary cracks at high loading rates. The width of the central crack increased from 0.81 mm to 2.84 mm when the loading rates increased from 0.2 mm/min to 2 mm/s. Furthermore, the distance between the primary crack and the secondary crack was about 10 mm, which was equal to the width of the gasket. It could be noted that the gaskets under the high loading rate resulted in local failure at the loading end, forming a wedge-shaped area.

According to the typical section images of the group CA specimens in Figure 10c,d, the crack morphology of each section was tortuous along the thickness direction, as the aggregate around the diameter affected the crack evolution. At low loading rates, the cracks were mainly distributed along the interface between the aggregates and mortar. With increasing loading rates, the cracks passed through the large aggregates and extended along the periphery of the small aggregates, forming local fracture zones around small aggregates. The crack usually expanded along the weak surface between the aggregate and mortar under low loading rates. However, under high loading rates, the stress at the crack increased to the extent that the stone could be broken before the crack could expand along the weak interface; thus, the crack passed directly through the coarse aggregate by looking for the shortest path. Furthermore, the width of the central crack increased from 1.31 mm to 3.75 mm when the loading rates increased. The shear failure occurred at the loading end of the specimens, but there were wedge-shaped fragmentation zones instead of complete areas left behind. This further demonstrated that concrete had more initial defects than the mortar, which made the stress conditions at the load-bearing ends more complex.

As illustrated in Figure 10e,f, cracks also penetrated the aggregate at low loading rates but were mainly distributed along the interface. At high loading rates, there was a fan-shaped local failure close to 60° that occurred at the loading end instead of local punching failure. Similarly, the fixed ends of the specimens also produced relatively small areas of local crushing. The width of the central crack increased from 1.65 mm to 2.53 mm. Under the higher loading rate, the width of the main crack was smaller than that of CA, which meant more energy was absorbed by the local failure at the loading end under lower water/cement ratios, resulting in slower energy release during crack propagation.

On a whole, aggregates affected the distribution of cracks along the radial and thickness directions, making the crack morphology complex. The strength of concrete improved with decreasing water/cement ratios and aggregate volume as the strength of hardening cement mortar increased, which made crack expansion more difficult. Meanwhile, the bonding performance of the interface between aggregate and mortar was also improved, making the interface strength higher. In terms of the loading rates, the tensile strength increased, and more secondary cracks appeared when the loading rates increased. The failure modes also turned to local impact failure instead of a single main crack. It could be inferred that multiple points of one specimen could reach the peak tensile strain at the same time under high loading rates; thus, several local failures appeared simultaneously.

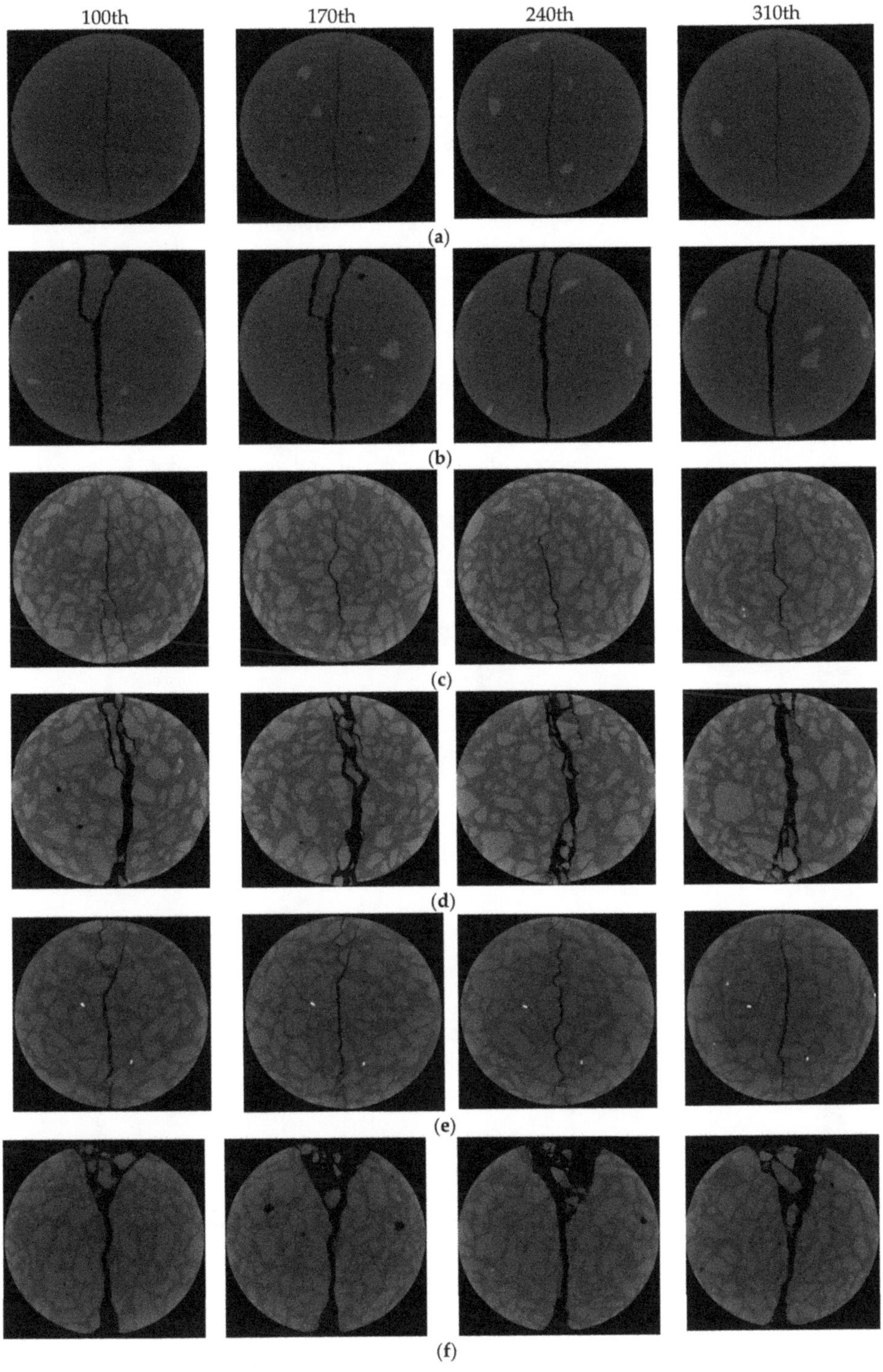

Figure 10. Typical slice images: (**a**) MA-02 m; (**b**) MA-2 s; (**c**) CA-02 m; (**d**) CA-2 s; (**e**) CC-02 m; (**f**) CC-2 s.

In addition, local failure became more obvious under high loading rates, especially for the concrete specimens, as the stress increased rapidly at the loading end. When the stress diffused to the center of the specimen, cracks at the weak interface developed with the main crack penetrating, forming band or block crushing zones, including aggregate fracture and mortar cracks. However, the main crack still grew slowly when the load level increased sharply. It can be inferred that local failure was the main reason for the enhanced strength. The cracks at the loading end expanded and penetrated rapidly under high-level confining pressure, causing damage to the aggregates, mortar, and interface, finally leading to an increase in strength. Therefore, the distribution of cracks and the failure modes of specimens also had a certain influence on the splitting tensile strength.

3.3. CT Image Segmentation and Three-Dimensional Reconstruction

3.3.1. Analysis of Gray Histogram Characteristics

A histogram can display the data frequency of each group in a graphical form, which represents the distribution probability of the CT number of materials with different densities and expresses the change in the content of each phase through the change in CT numbers [41]. As different CT numbers corresponded to different material brightness, the range of CT numbers of the voids, mortar, and aggregate could be determined through the CT image histograms [18]. Additionally, when cracks were formed inside the specimens, the gray value decreased as the average density decreased [42]. Statistical analysis of gray histograms was carried out on the characteristic images of 100–300 layers of specimens under various working conditions. The results are shown in Figures 11–13.

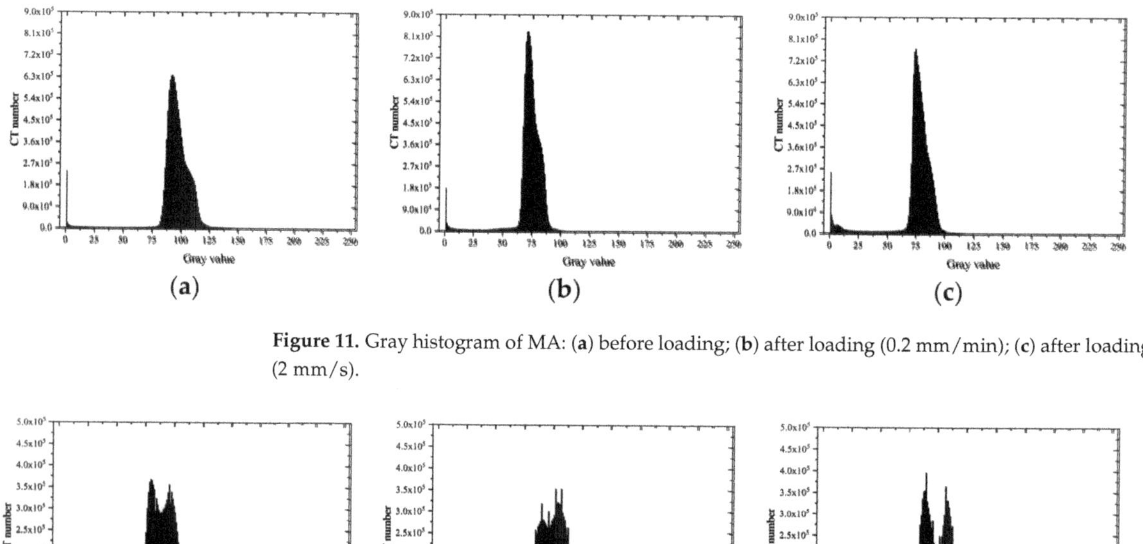

Figure 11. Gray histogram of MA: (**a**) before loading; (**b**) after loading (0.2 mm/min); (**c**) after loading (2 mm/s).

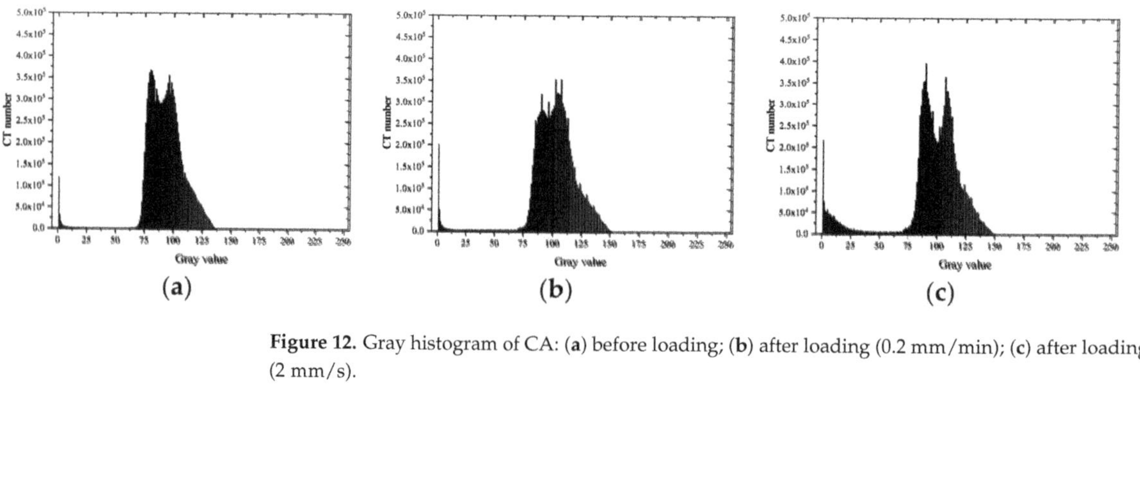

Figure 12. Gray histogram of CA: (**a**) before loading; (**b**) after loading (0.2 mm/min); (**c**) after loading (2 mm/s).

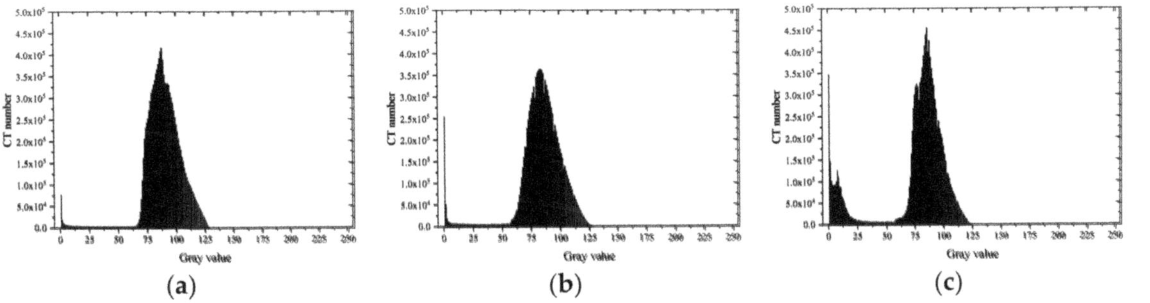

Figure 13. Gray histogram of CC: (**a**) before loading; (**b**) after loading (0.2 mm/min); (**c**) after loading (2 mm/s).

In general, the mortar had a relatively simple microstructure, presenting a single peak shape with a peak, while the concrete histogram had an obvious bimodal shape. It can be roughly inferred that a CT number in the range of 0–50 grayscale represented the pores and cracks, in the range of 50–100 represented the mortar matrix in the concrete, and in the range of 100–125 represented the aggregate by quantitative statistical analysis of CT numbers.

As shown in Figure 11, the gray histograms of MA specimens presented a single peak shape. The single peak of the histogram moved forward from [75, 125] to [60, 90] and became more concentrated after loading, indicating that the dispersion of mortar density decreased after failure. In addition, the CT number in the [0, 25] interval increased as the loading rates increased, indicating that the cracks developed better at higher loading rates.

The gray histograms of the type CA specimens in Figure 12 presented obvious bimodal properties. At low loading rates, the CT numbers of mortar decreased, but those of aggregate remained almost unchanged after failure. Therefore, interface failure and mortar failure were the main failure modes. At high loading rates, the bimodal property became more obvious with the CT numbers of mortar almost unchanged, but the CT numbers of aggregate decreased, indicating that aggregate failure was the main failure mode at this time.

As shown in Figure 13, the gray histograms of the CC concrete specimens under the two loading rates presented a single peak. The peak gray value was almost equal to the average value of the gray value of aggregate and mortar. It can be inferred that the intensity of the X-ray wave was attenuated during the CT scanning process, making the images present a ring-shaped gray value reduction phenomenon.

3.3.2. CT Image Segmentation Method

As mentioned before, appropriate threshold segmentation is crucial to obtain clear and accurate mesoscale structures. Since the bimodal feature of the histogram was not obvious, the Otsu algorithm was adopted. However, the Otsu method applied to the global specimen made the inner aggregate nebulous. Thus, a block multi-threshold segmentation method was purposed to explore a more accurate threshold segmentation method, as shown in Figure 14. According to the gray value of each pixel, the image could be divided into three blocks along the radial direction, namely, the circular block within the radius of 220 pixels, the block between 220 pixels and 340 pixels, and theblock between 340 pixels and 420 pixels. Then, the gray histogram of the CT number of each block presented the bimodal characteristics in Figure 14b.

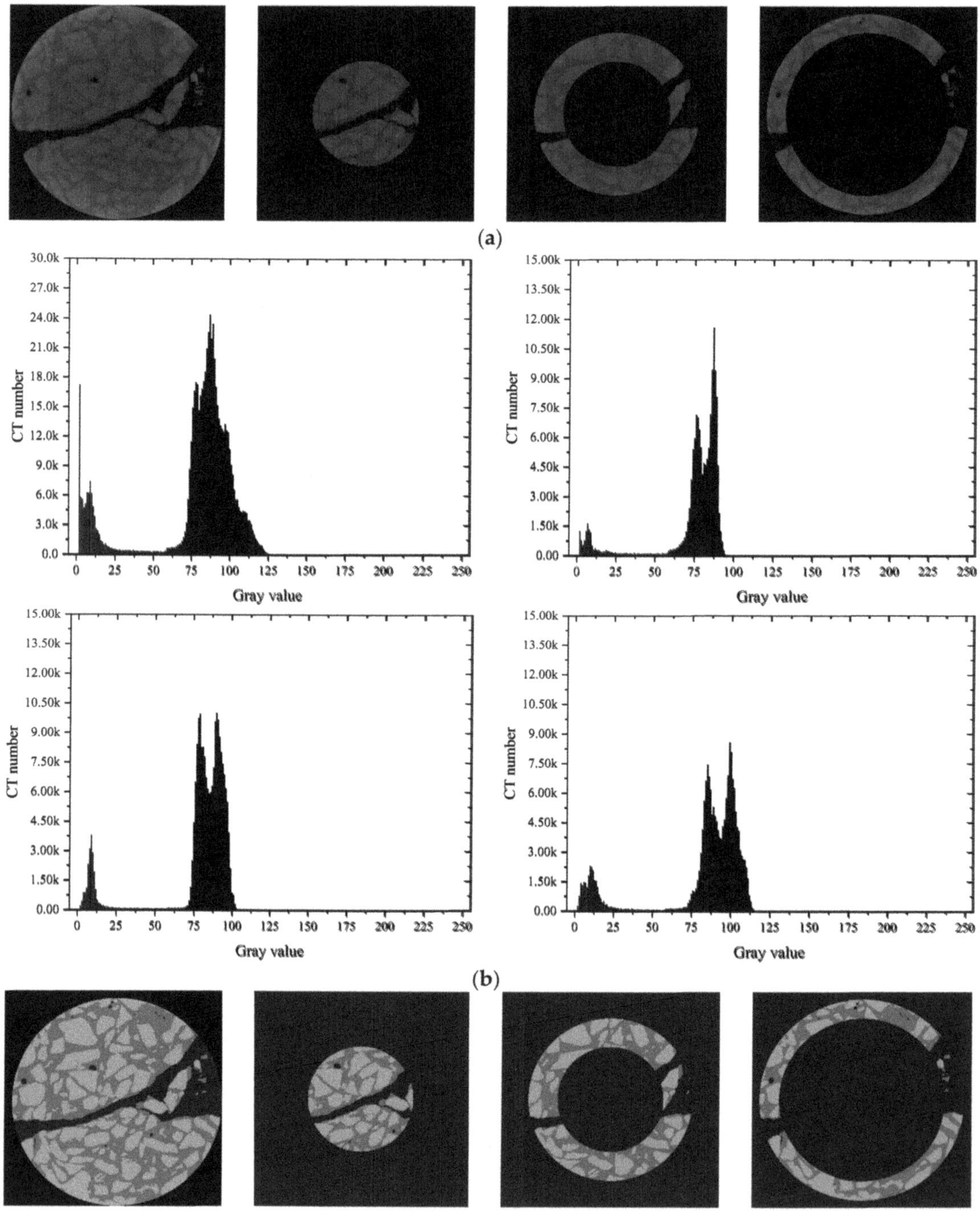

Figure 14. Block multi-threshold segmentation method: (**a**) original image and block processed images; (**b**) CT histograms of original images and segmented images; (**c**) segmented block image after multi-threshold segmentation.

Thus, multiple threshold segmentation was conducted to form a trivalued image $g^i_{block}(x,y)$ from the extracted image $f^i_{block}(x,y)(i = 1,2,3)$, as shown in Equation (5).

$$g^i_{block}(x,y) = \begin{cases} 0.2 & f^i_{block}(x,y) < T_1 \\ 0.5 & T_1 \le f^i_{block}(x,y) < T_2, \\ 0.7 & T_2 \le f^i_{block}(x,y) \end{cases} \quad (5)$$

where T_1 and T_2 are the segmentation threshold of the pore and mortar, respectively, and block number $i = 1, 2, 3$.

Finally, the threshold segmentation image was obtained by the combination calculation shown in Equation (6).

$$g(x,y) = \begin{cases} 0 & (x,y) \notin f^i_{block}(x,y) \\ \sum_i g^i_{block}(x,y) & (x,y) \in f^i_{block}(x,y) \end{cases}. \quad (6)$$

The segmented operation realized the decoupling of the gray value of each phase material and was used to segment and derive mortar matrix, aggregate, cracks, and holes in the image at one time. It can be seen from Figure 14b that the image could clearly reflect the mesoscale characteristics of concrete after threshold segmentation.

3.3.3. Mesoscale Research on the Three-Dimensional Characteristics of Concrete Damage

The three-dimensional internal damage of concrete was reconstructed according to the segmentation method mentioned above. Then, the volume proportion of each phase of specimens under different working conditions was arranged, as presented in Table 2.

Table 2. Volume proportion of each phase.

Type	CA-02 m	CA-2 s	CC-02 m	CC-2 s	MA-02 m	MA-2 s
Pore	0.05%	0.14%	0.07%	0.37%	0.24%	0.25%
Crack	1.36%	7.69%	1.51%	12.17%	1.18%	5.57%
Mortar	54.70%	50.61%	45.67%	36.13%	98.58%	94.18%
Aggregate	43.89%	41.56%	52.75%	51.33%	-	-

For concrete specimens, the volume ratio of aggregate was approximately 50%, which was slightly larger than that of mortar. Under a high water/cement ratio, there were more pores and cracks in the concrete specimens, and the volume of mortar decreased. With increasing loading rates, the proportion of mortar and aggregate decreased while the volume proportion of pores and cracks increased. When the loading rate increased from 0.2 mm/min to 2 mm/s, the crack volume ratio of CA specimens increased from 1.36% to 7.69%, and that of CC specimens increased from 1.51% to 12.17%. This indicated that specimens with higher strength cracked more seriously under the influence of the loading rates. However, the proportion of pores was always less than 0.5%, and it was actually smaller than the actual value, considering that the micropores along the crack propagation path were included in the crack volume ratio by mistake. Furthermore, CT images also missed some smaller pores. The pore proportion of the mortar specimen was larger than that of the concrete specimen, but the crack volume was approximately 2% less than that of the concrete, which also shows that the existence of aggregates could make the crack develop more maturely.

The single porosity parameter could not completely reflect the relationship between the tensile strength and the pore structure, as the distribution of pore size also had a critical influence on strength properties [43]. Herein, specimens MA-02 m and CC-02 m were selected, and the equivalent diameters d_{eq} were defined according to Equation (7) for the calculation of the distributions of pore size and aggregate size. The separation results are shown in Figure 15.

$$d_{eq} = \sqrt[3]{\frac{6V_{3d}}{\pi}}. \tag{7}$$

Figure 15. Separation results: (**a**) results of mortar pore separation; (**b**) results of aggregate separation.

For specimen MA-02 m, after removing the largest volume elements (cracks), the pores were numbered from large to small to draw the distribution curves of pore volume and cumulative volume, as shown in Figure 16a. According to the calculated equivalent aperture, 13 intervals were divided at a resolution of 0.2 mm, forming the probability distribution diagram shown in Figure 16b.

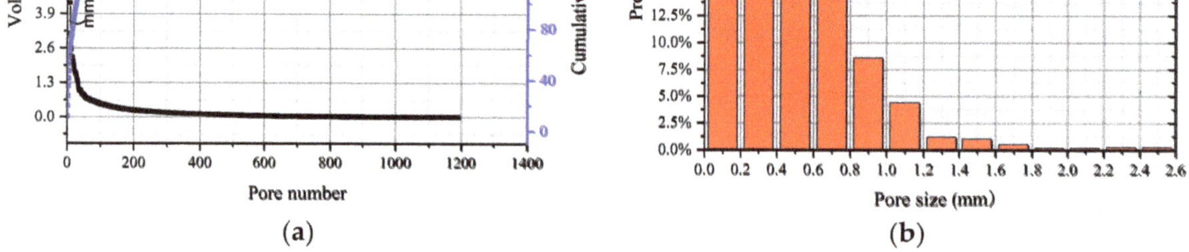

Figure 16. Statistical curves of pore distribution: (**a**) pore volume and cumulative volume distribution curves; (**b**) proportion of the pore size.

The pore volume curve was overall smooth, as presented in Figure 16a, which indicated that the variation in the internal pore size of the mortar was continuous. Moreover, the cumulative volume of pores was 247.41 mm^3. According to Figure 16b, the number of pores had an approximately normal distribution. In addition, the pore size was mainly distributed in the range of 0.2–0.4 mm, accounting for 27.5% of the total pores, followed by the sizes of 0–0.2 mm and 0.4–0.6 mm, both accounting for about 20% of the total pores. However, the size of almost all pores was below 1 mm. This also indicated that the matrix was compacted; hence, the strength was relatively high.

For specimen CC-02 m, the aggregate volume curve and cumulative volume distribution curve are shown in Figure 17a. The original curves were smooth with longer platform segments, indicating that a large number of small volume components were contained in the data, which is inconsistent with the aggregate separation results in Figure 15b. It could be inferred that the broken aggregate caused errors. The volume of broken aggregate could be roughly calculated by combining the CT images and the aggregate separation results, and it was 12.08×10^3 mm^3 for CC specimens under a loading rate of 2 mm/s. Therefore, data extraction was required before analyzing particle size distribution statistics. After calculation, data points with slopes of less than 1 per tangent were eliminated. Then, the equivalent particle sizes of the aggregates were calculated, and the interval was divided according to the 2 mm resolution, as illustrated in Figure 17b. It can be seen that the number of aggregates with particle sizes ranging from 3 mm to 6 mm accounted for more than 50%. The presence of small aggregates made more interfaces, and the interfaces were weak during failure, which further explained that the mortar specimens had higher tensile strength than the concrete specimens.

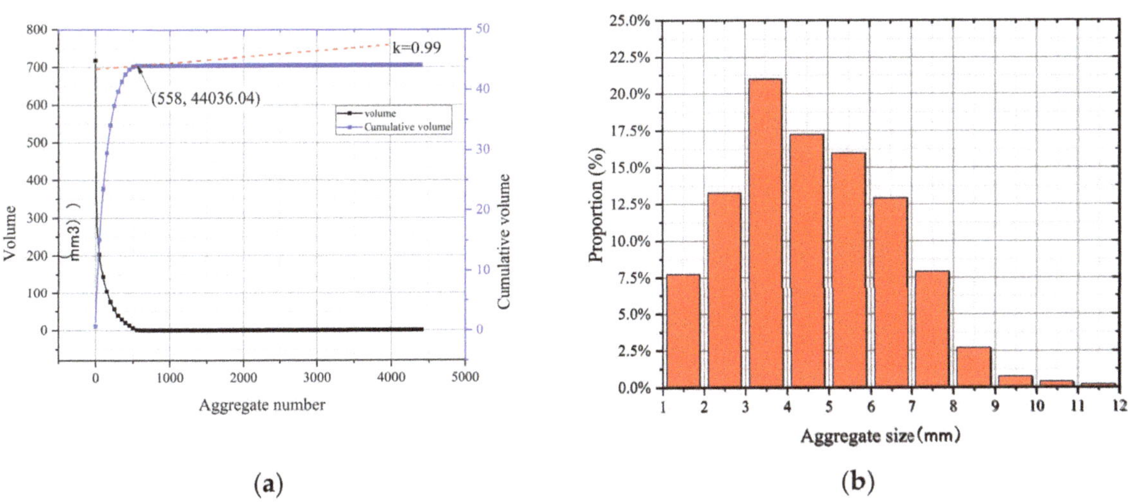

Figure 17. Statistical curves of aggregate distribution: (**a**) aggregate volume and cumulative volume distribution curves; (**b**) proportion of the aggregate size.

A change in the stress–strain state of the specimens at the macro level leads to a change in the deformation in the local material region [44]. To further study the influence of various factors on the crack morphology in the failure process, the 3D crack morphology of specimens under six working conditions was extracted, as shown in Figure 18. The width of the crack is shown at the top right of each picture, facilitating interpretation of the shape. The cracks of the mortar specimens were straighter and smoother than those of the concrete specimens, and the crack texture mainly developed along with the original defects in the specimens. With an increase in the loading rates, the cracks became coarser and bifurcated at the loading end of the specimen, but the overall trend of the cracks changed little along the thickness.

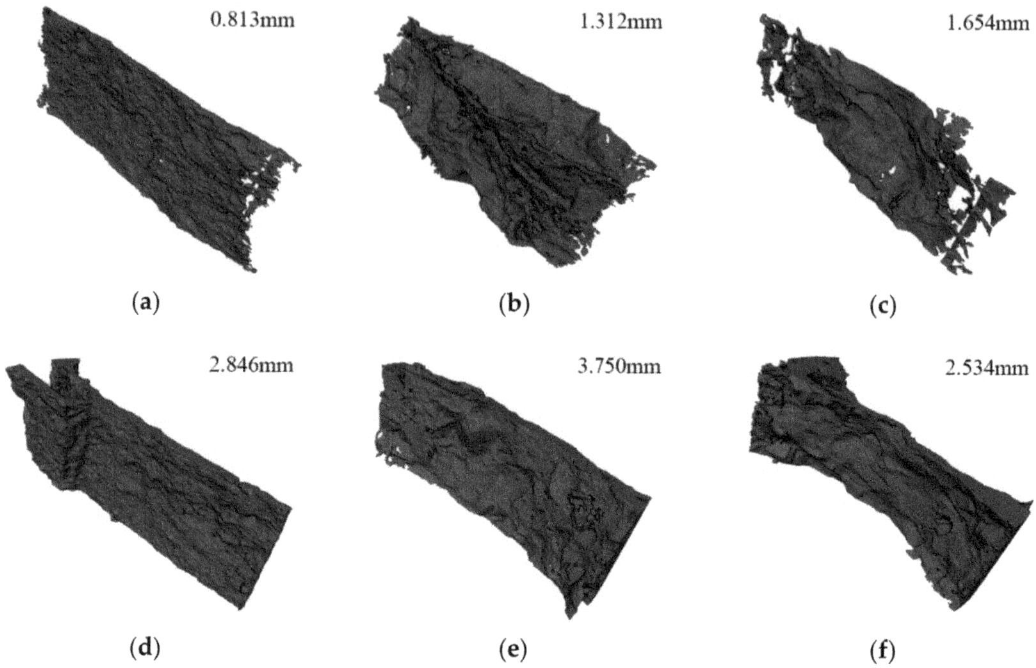

Figure 18. The 3D morphology of reconstructed cracks: (**a**) MA-02 m; (**b**) CA-02 m; (**c**) CC-02 m; (**d**) MA-2 s; (**e**) CA-2 s; (**f**) CC-2 s.

Compared with mortar, the secondary cracks of concrete specimens appeared more frequently. At low loading rates, there were many concavities on the crack surface due to the aggregates on the crack propagation path, for which the shape was the same as the aggregate shape at the corresponding position. When the cracks extended to the surface of the larger aggregates, they cracked along the surface of the aggregates, dividing the cracks into two parts. This meant that the cracks in concrete were mainly distributed around the aggregates at low loading rates, and that aggregates were an important factor influencing the crack surface distribution. At high loading rates, the crack surfaces became relatively smooth but were still affected by aggregates with larger inner concave shapes. In addition, specimens with higher strength also had smoother crack surfaces. With the decrease in water/cement ratios and the increase in loading rates, fragmentation zones appeared near small aggregates, forcing small aggregates to break away from the matrix, and this process also improved the strength of the concrete. On a whole, the aggregate distribution influenced the distribution of cracks, and aggregate failure was an important factor to enhance the strength of concrete.

4. Conclusions

In this paper, Brazilian disc splitting tensile tests were carried out, and X-ray CT technology and DIC technology were introduced to study the mesoscale damage characteristics and failure process of the concrete and mortar specimens. The main conclusions could be drawn as follows:

1. For the splitting tensile tests, the tensile strength of concrete was 11–30% lower than that of mortar under different loading rates. Compared with concrete, the failure displacement of mortar was more sensitive to the loading rates and increased by 16.38% when the loading rate increased from 0.2 mm/min to 2 mm/s.

2. The stress field obtained by the DIC test showed that the cracking modes of the mortar and concrete specimens were different. Concrete specimens cracked from the center, and mortar specimens cracked from the edge, which could affect the reliability of the tensile strength calculation of mortar specimens. Macro cracks gradually formed when the tensile strain of mortar and concrete reached 0.0237 and 0.0836, respectively.
3. According to the images obtained from the CT test, the failure mode of the concrete and mortar specimens changed from a single main crack to local punching failure at the loading end with the increase in the loading rates and the decrease in the water/cement ratio. For concrete, interface failure and mortar failure were the main failures under low loading rates, whereas aggregate failure became the main failure mode under a loading rate of 2 mm/s.
4. According to the three-dimensional reconstruction results, the internal pores of mortar were small and dense. The size and distribution of aggregates affected the morphology of cracks and made the cracks concave. With the decrease in the water/cement ratios and the increase in the loading rates, the crack surfaces became relatively smooth, as the small aggregates were forced to break away from the matrix.

Author Contributions: Conceptualization, H.Z. and K.Z.; Data curation, C.J.; Formal analysis, L.P.; Writing—original draft, Q.P. and K.Z.; Writing—review & editing, Q.P. All authors have read and agreed to the published version of the manuscript.

Funding: This research received no external funding.

Conflicts of Interest: The authors declare no conflict of interest.

References

1. Liu, T.; Qin, S.; Zou, D.; Song, W.; Teng, J. Mesoscopic modeling method of concrete based on statistical analysis of CT images. *Constr. Build. Mater.* **2018**, *192*, 429–441. [CrossRef]
2. Tian, W.; Han, N. Analysis on meso-damage processes in concrete by X-ray computed tomographic scanning techniques based on divisional zones. *Measurement* **2019**, *140*, 382–387. [CrossRef]
3. Ren, W.; Yang, Z.; Sharma, R.; Zhang, C.; Withers, P. Two-dimensional X-ray CT image based meso-scale fracture modelling of concrete. *Eng. Fract. Mech.* **2015**, *133*, 24–39. [CrossRef]
4. Hou, C.; Wang, Z.; Liang, W.; Li, J.; Wang, Z. Determination of fracture parameters in center cracked circular discs of concrete under diametral loading: A numerical analysis and experimental results. *Theor. Appl. Fract. Mech.* **2016**, *85*, 355–366. [CrossRef]
5. Suchorzewski, J.; Tejchman, J.; Nitka, M. Experimental and numerical investigations of concrete behaviour at meso-level during quasi-static splitting tension. *Theor. Appl. Fract. Mech.* **2018**, *96*, 720–739. [CrossRef]
6. Li, D.; Wong, L. The Brazilian disc test for rock mechanics applications: Review and new insights. *Rock Mech. Rock Eng.* **2013**, *46*, 269–287. [CrossRef]
7. Carmona, S.; Aguado, A. New model for the indirect determination of the tensile stress–strain curve of concrete by means of the Brazilian test. *Mater. Struct.* **2012**, *45*, 1473–1485. [CrossRef]
8. Martínez-López, M.; Martínez-Barrera, G.; Nunes, L.; Reis, J.; Da Costa Mattosa, H.S. Mixed mode fracture analysis in a polymer mortar using the Brazilian disk test. *Eng. Fract. Mech.* **2016**, *154*, 140–151. [CrossRef]
9. Hevroni, D.; Kochavi, E.; Kofman, B.; Gruntmana, S.; Sadot, O. Experimental and numerical investigation on the dynamic increase factor of tensile strength in concrete. *Int. J. Impact Eng.* **2018**, *114*, 93–104. [CrossRef]
10. Malvar, L.J.; Ross, C.A. A Review of Strain Rate Effects for Concrete in Tension. *ACI Mater. J.* **1998**, *95*, 735–739.
11. Wu, S.; Zhou, J.; Chen, H. Mechanism and unified model of concrete dynamic tensile strength improvement based on microstructure characteristics. *J. Water Conserv.* **2010**, *41*, 419–428. (In Chinese) [CrossRef]
12. Květoňa, J.; EliáŠa, J. Influence of inertia and material properties on discrete simulation of dynamic fracture of concrete. *Procedia Struct. Integr.* **2018**, *13*, 1367–1372. [CrossRef]
13. Zhou, R.; Chen, H.M.; Lu, Y. Mesoscale modelling of concrete under high strain rate tension with a rate-dependent cohesive interface approach. *Int. J. Impact Eng.* **2020**, *139*, 103500. [CrossRef]
14. Jin, X.; Hou, C.; Fan, X.; Lu, C.; Yang, H.; Shu, X.; Wang, Z. Quasi-static and dynamic experimental studies on the tensile strength and failure pattern of concrete and mortar discs. *Sci. Rep.* **2017**, *7*, 15305. [CrossRef]
15. Zhou, Z.; Zhou, Y.; Li, X.; Jiang, Y. Stress evolution and failure process of Brazilian disc under impact. *J. Cent. South Univ.* **2013**, *20*, 172–177. [CrossRef]
16. Xudong, C.; Limei, G.; Jikai, Z.; Sheng, X.W. Experimental Study on Split Hopkinson Pressure Bar Pulse-Shaping Techniques for Concrete. *J. Mater. Civ. Eng.* **2016**, *28*, 04015196. [CrossRef]

17. Xudong, C.; Limei, G.; Jikai, Z.; Shengxing, W. Dynamic Brazilian Test of Concrete Using Split Hopkinson Pressure Bar. *Mater. Struct.* **2017**, *50*, 1. [CrossRef]
18. Yang, W. Finite element model of concrete material based on CT image processing technology. *J. Vis. Commun. Image Represent.* **2019**, *64*, 102631. [CrossRef]
19. Lang, L.; Zhu, Z.; Zhang, X.; Qiu, H.; Zhou, C. Investigation of crack dynamic parameters and crack arresting technique in concrete under impacts. *Constr. Build. Mater.* **2019**, *199*, 321–334. [CrossRef]
20. Zhu, L.; Dang, F.; Xue, Y.; Ding, W.; Zhang, L. Comparative study on the meso-scale damage evolution of concrete under static and dynamic tensile loading using X-ray computed tomography and digital image analysis. *Constr. Build. Mater.* **2020**, *250*, 118848. [CrossRef]
21. Castellano, A.; Fraddosio, A.; Piccioni, M.; Kundu, T. Linear and Nonlinear Ultrasonic Techniques for Monitoring Stress-Induced Damages in Concrete. *J. Nondestruct. Eval. Diagn. Progn. Eng. Syst.* **2021**, *4*, 041001. [CrossRef]
22. Payan, C.; Garnier, V.; Moysan, J. Potential of Nonlinear Ultrasonic Indicators for Nondestructive Testing of Concrete. *Adv. Civ. Eng.* **2014**, *2010*, 238472. [CrossRef]
23. Antonaci, P.; Bruno, C.L.E.; Gliozzi, A.S.; Scalerandi, M. Monitoring evolution of compressive damage in concrete with linear and nonlinear ultrasonic methods. *Cem. Concr. Res.* **2010**, *40*, 1106–1113. [CrossRef]
24. Shen, J.; Xu, Q.; Liu, M. Statistical analysis of defects within concrete under elevated temperatures based on SEM image. *Constr. Build. Mater.* **2021**, *293*, 123503. [CrossRef]
25. Maruschak, P.O.; Chausov, M.G.; Konovalenko, I.V.; Yasnii, O.P.; Vlasov, I.V. Effect of Shock and Vibration Loading on the Fracture Mechanisms of a VT23 Titanium Alloy. *Strength Mater.* **2020**, *52*, 252–261. [CrossRef]
26. Wang, X.; Jin, Z.; Liu, J.; Chen, F.; Feng, P.; Tang, J. Research on internal monitoring of reinforced concrete under accelerated corrosion, using X-CT and DIC technology. *Constr. Build. Mater.* **2021**, *266*, 121018. [CrossRef]
27. Li, D.; Huang, P.; Chen, Z.; Yao, G.; Guo, X.; Zheng, X.; Yang, Y. Experimental study on fracture and fatigue crack propagation processes in concrete based on DIC technology. *Eng. Fract. Mech.* **2020**, *235*, 107166. [CrossRef]
28. Mamand, H.; Chen, J. Extended digital image correlation method for mapping multiscale damage in concrete. *J. Mater. Civ. Eng.* **2017**, *29*, 04017179. [CrossRef]
29. Lawler, J.; Keane, D.; Shah, S. Measuring three-dimensional damage in concrete under compression. *Mater. J.* **2001**, *98*, 465–475.
30. Kim, J.; Chung, S.; Han, T.; Stephan, D.; Elrahman, M. Correlation between microstructural characteristics from micro-CT of foamed concrete and mechanical behaviors evaluated by experiments and simulations. *Cem. Concr. Compos.* **2020**, *112*, 103657. [CrossRef]
31. Skaryński, L.; Tejchman, J. Investigations on fracture in reinforced concrete beams in 3-point bending using continuous micro-CT scanning. *Constr. Build. Mater.* **2021**, *284*, 122796. [CrossRef]
32. Chung, S.; Kim, J.; Stephan, D.; Han, T. Overview of the use of micro-computed tomography (micro-CT) to investigate the relation between the material characteristics and properties of cement-based materials. *Constr. Build. Mater.* **2019**, *229*, 116843. [CrossRef]
33. Loeffler, C.; Qiu, Y.; Martin, B.; Heard, W.; Williams, B.; Nie, X. Detection and segmentation of mechanical damage in concrete with X-Ray microtomography. *Mater. Charact.* **2018**, *142*, 515–522. [CrossRef]
34. Hao, S.; Dang, F.; Chen, H.; Liang, X. Three dimension reconstruction of concrete meso-structure based on CT images by ANSYS. *Concrete* **2009**, *3*, 13–15. (In Chinese)
35. Jin, L.; Yu, W.; Du, X.; Yang, W. Mesoscopic numerical simulation of dynamic size effect on the splitting-tensile strength of concrete. *Eng. Fract. Mech.* **2019**, *209*, 317–332. [CrossRef]
36. Kadlеček, S.; Modrý, S.; Kadlеček, J. Size effect of test specimens on tensile splitting strength of concrete: General relation. *Mater. Struct.* **2002**, *35*, 28–34. [CrossRef]
37. Huang, Y. *Study on the Size Effect of Concrete Brittleness and Mechanical Parameters and Their Relationship*; Chongqing University: Chongqing, China, 2002.
38. Wang, J.; Xu, L.; Lu, Y. Application of edge detection in industrial CT image. *Nondestruct. Test.* **2007**, *29*, 61–65. (In Chinese) [CrossRef]
39. Skaryński, L.; Kozicki, J.; Tejchman, J. Application of DIC Technique to Concrete—Study on Objectivity of Measured Surface Displacements. *Exp. Mech.* **2013**, *53*, 1545–1559. [CrossRef]
40. Wu, Z.M.; Hua, R.; Zheng, J.J.; Feng, X.; Wei, D. An experimental investigation on the FPZ properties in concrete using digital image correlation technique. *Eng. Fract. Mech.* **2011**, *78*, 2978–2990. [CrossRef]
41. Zheng, K. *Study on Mesostructure and Damage Characteristics of Coal Gangue Particles Based on X-ray CT*; China University of Mining and Technology: Beijing, China, 2016.
42. Mao, L.; Sun, Q.; Yuan, Z.; Liu, Y.L.; Hb, L. Crack and strain field analysis in concrete under uniaxial compression based on CT images. *J. Build. Mater.* **2016**, *19*, 449–455. (In Chinese)
43. Hao, J.; Jiang, Y.; Mei, S.; Zhu, L. Identification of crack in concrete interior based on X-ray CT. *Concrete* **2010**, *10*, 44–47. [CrossRef]
44. Maruschak, P.O.; Konovalenko, I.V.; Bishchak, R.T. Effect of thermal fatigue cracks on brittle-ductile deformation and failure of cbcm roller surface layers. *Metallurgist* **2012**, *56*, 30–36. [CrossRef]

applied sciences

MDPI

Article

The Restored Premolars Biomechanical Behavior: FEM and Experimental Moiré Analyses

Jose Luis Valin Rivera [1,*], Edison Gonçalves [2], Paulo Vinicius Soares [3], Giovana Milito [3], Jorge Octavio Ricardo Perez [4], Guillermo Francisco Palacios Roque [4], Meylí Valin Fernández [5], Henry Figueredo Losada [6], Fabrícia Araújo Pereira [7], Gilberto Garcia del Pino [8] and Alexander Rodriguez Soto [1]

1 Escuela de Ingeniería Mecánica, Pontificia Universidad Católica de Valparaíso, Quilpué 2430000, Chile; angel.rodriguez@pucv.cl
2 Polytechnic School, University of Sao Paulo, São Paulo 05508-010, Brazil; edison@usp.br
3 Faculty of Dentistry, University of Uberlandia, Uberlândia 38405-266, Brazil; paulovsoares@foufu.ufu.br (P.V.S.); giovanamilito@gmail.com (G.M.)
4 Faculty of Natural Science, University of Oriente, Santiago de Cuba 90100, Cuba; jrperez@cnt.uo.edu.cu (J.O.R.P.); palaciosg226@gmail.com (G.F.P.R.)
5 Department of Mechanical Engineering (DIM), Faculty of Engineering (FI), University of Concepción, Concepción 4030000, Chile; mvalin@udec.cl
6 Faculty of Engineering, University of the Republic, Montevideo 11200, Uruguay; henryf@fing.edu.uy
7 Department of Dentistry, University of Brasilia, Brasília 70910-900, Brazil; fabricia_pereira@hotmail.com
8 College of Technology, State University of Amazonas, Manaus 69850-000, Brazil; gpino@uea.edu.br
* Correspondence: jose.valin@pucv.cl

Abstract: This study applied the finite element method (FEM) and the moiré strip projection method to evaluate the biomechanical behavior of healthy and endodontic-treated premolar teeth. The finite element method and the moiré strip projection method were applied to evaluate the influence of restored materials in association with cervical lesions and were considered as strain estimates for a tooth sample with 21 units, under loads of 25, 50, 75, and 100 N, frontal and oblique applied. The focused cases were: tooth H healthy; tooth A-MOD amalgam; tooth AL-MOD amalgam + lesion; tooth ALR-MOD amalgam + injury restored; tooth R-MOD resin; tooth RL-MOD resin + lesion; tooth RLR-MOD resin + injury restored. The results obtained by FEM simulation can be considered perfectly validated by the results presented by the experimental moiré projection analysis, demonstrating that the FEM numerical analysis can be used to evaluate the biomechanical behavior of healthy and endodontically treated teeth. Developing an alternative protocol to generate FEM three-dimensional models will lead to a ready and inexpensive tool since there is no need for costly equipment for tooth extraction prognosis.

Keywords: fringe projection; premolars teeth; interferometry; FEM

Citation: Valin Rivera, J.L.; Gonçalves, E.; Vinicius Soares, P.; Milito, G.; Ricardo Perez, J.O.; Palacios Roque, G.F.; Valin Fernández, M.; Figueredo Losada, H.; Araújo Pereira, F.; Garcia del Pino, G.; et al. The Restored Premolars Biomechanical Behavior: FEM and Experimental Moiré Analyses. *Appl. Sci.* **2022**, *12*, 6768. https://doi.org/10.3390/app12136768

Academic Editors: Bogusław Łazarz, Grzegorz Peruń and Tangbin Xia

Received: 26 May 2022
Accepted: 21 June 2022
Published: 4 July 2022

1. Introduction

The loss of dental structure due to caries, trauma, cavity preparation, and endodontic treatment has a negative impact on tooth fracture resistance [1–3]. The commonly proposed functional reconstructions for premolars with extensive structure reduction have been the direct adhesive techniques, using composite resins due to the high adhesion capacity to the dental structures [4] and to seek minimal dental damage during the repair preparations [5]. Other types of treatments, generally less conservative, are available to treat posterior teeth using indirect treatments, nowadays generally performed with CAD/CAM technology [6,7]. Endodontic treatments are not considered complete and successful until definitive coronary restoration is performed [8–10].

Endodontic repair that restores the lost tooth structure's biomechanical behavior would allow for a good lifetime tooth prognosis under masticatory loads. Some authors recommend the use of amalgam fillings and other resinous materials fillings [11,12]. Dental amalgams, due to their favorable mechanical properties, are widely used as restorative materials, since they use a simple restorative technique and have a good predictable clinical performance. However, they present some disadvantages, such as low dental structure adhesion and a reduced strain resistance of the remaining dental structure. Amalgam-restored teeth tend to fracture due to the multiplication of micro cracks under alternated fatigue loads. The composites, as restoring materials, have mechanical properties, such as elastic modulus similar to the tooth structure, that allow better distributions of stresses and strains throughout the material and the remaining tooth structure. Even so, they have not been successful as an alternative to amalgams to restore non-vital teeth [13]. The loss of tooth structure in the cervical region is very common [14,15]. Cervical tooth structure wear occurs near the cement enamel junction and affects mainly premolars, mainly as a result of the stress concentration in the cusps of the occlusal region.

This study evaluates the influence of different types of restorative material in premolar cavities, associated with the presence of cervical lesions under the action of axial and oblique loads. The biomechanical behaviors of endodontic-treated premolars are analyzed by using numerical, computational, and experimental non-destructive techniques. In order to fulfill this purpose, frontal and oblique loads were applied in the values of 25, 50, 75, and 100 N, in both experimental and numerical-computational biomechanical behavior evaluations. The aforementioned loads were selected in order to compare the results obtained with previous work developed by the authors of this work, in which the extensometry technique was applied.

2. Materials and Methods

2.1. Selection and Preparation of Tooth Samples

From the Faculty of Dentistry Bank of Teeth, University of Sao Paulo, twenty-one (21) healthy human pre-upper molars (1st and 2nd premolars), with an indication for extraction, were selected. This selection was approved by the Research Ethics Committee: Final Analysis no. 520/11 of the Research Ethics Committee for the Protocol Registration CEP/UFU No. 171/11 and Final Analysis 372/11 of the Research Ethics Committee for the Protocol Registration CEP/UFU 065/11 the Research Ethics Committee of the Federal University of Uberlandia.

The teeth, after the coronary surfaces were measured, were divided into seven groups with three teeth each, which received the following treatments: (1) endodontic treatment; (2) mesial-distal-occlusal cavity (MOD) preparation; (3) mod resin composite/amalgam restoration to first amalgam group and restoration in the second group; (4) cervical ripening; (5) cervical resin composite/amalgam restoration. Thus, the teeth were classified according to the following nomenclature, which will be used in development work: tooth H healthy; tooth A-MOD amalgam; tooth AL-MOD amalgam + lesion; tooth ALR-MOD amalgam + injury restored; tooth R-MOD resin; tooth RL-MOD resin + lesion; tooth RLR-MOD resin + injury restored [16,17].

The teeth were prepared with endodontic access and different restorative materials. The cavities made in the teeth were the mesial-distal occlusion-type (MOD) and cervical preparation (Figure 1). The cavities were standardized following the standard thickness of the diamond bur 1151 (KG Sorensen, Barueri, SP, Brazil). The wear amount considered was occlusal box with an opening of 3.65 mm and the proximal box with a depth of 2.0 mm. The inclination of the occlusal and proximal boxes was also defined by the diamond stylus configuration and therefore parallel (Figure 2). Figure 3 shows the prepared proximal boxes, with special attention to the detail of the rounding of angles. Figure 4 shows restoration MOD models with cervical cavity and restoration MOD and cervical restoration, both trying to imitate the shape of the coronary healthy tooth.

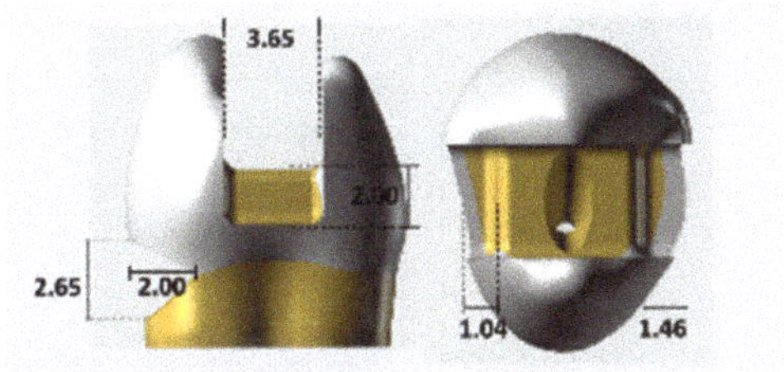

Figure 1. Cavities made in the models of the mesial-distal occlusion-type (MOD).

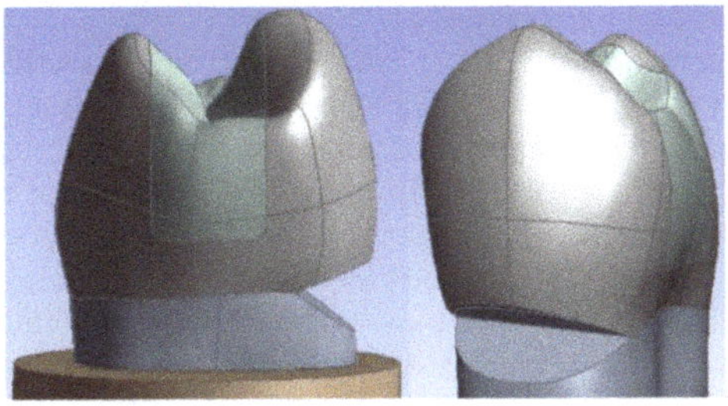

Figure 2. Cavities made in models of type cervical preparation.

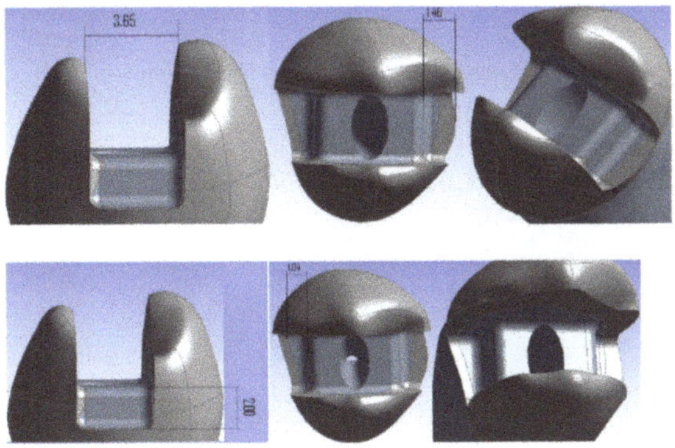

Figure 3. Making the proximal boxes, observe the detail of the rounding of angles. Amount of wear: occlusal box with a 3.65 mm box opening, a proximal box with a depth of 2.0 mm.

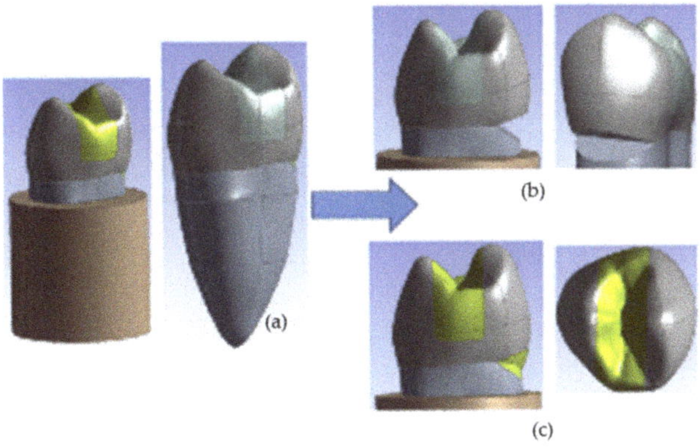

Figure 4. Restoration: (**a**) MOD, (**b**) models with a cervical cavity, and (**c**) restoration MOD and cervical restoration.

2.2. Finite Element Analysis

Finite element analysis has been used in other studies to determine stress and strain distribution. In [18] the hypothesis that restoration of class II mesio-occlusal-distal (MOD) cavities can be strengthened through judicious choice of restoration geometry and material properties was proven. Research by Shah et al. 2021 [19] used computational analysis of prepared crowned teeth to differentiate the possibility of using porcelain material for typical clinical conditions.

In the present work, the FEM analysis was developed using the computational code ANSYS Academic. The establishment of the physical model was performed using the mean dimensions of the premolars selected set and favorable coronary anatomy as a standard for the healthy and the repaired cases. Thereafter, with the help of a scanner, the outer tooth surface was generated, using a 0.2 mm contact tip trace calibration. The external geometry data was stored in the type "*.STL" computer files (Stereolithographic), generated by the scanner system. These files were exported to Bio-CAD modeling software (Computer Assisted Design; Rhino3D 4.0, rhinoceros USA) to produce the standard three-dimensional model of the healthy tooth. The repaired main dental regions were generated by using the software called NURBS (Non-Uniform Rational Basis Spline) suitable for modeling complex geometries. Figure 5 shows: (a) Model STL computer connected to the scanner system; (b) Software Bio-modeling CAD (Computer Assisted Design; Rhino3D 4.0, Rhinoceros, New York, NY, USA) NURBS surfaces (Non-Uniform Rational Basis Spline); (c) three-dimensional model generation that serves as a standard healthy tooth. Figure 6 shows the model generation of healthy teeth based on the main anatomical landmarks and with the help of a specific tool program (NURBS surfaces). Finally, Figure 7 shows the volumes of the internal and external structures of the reference teeth.

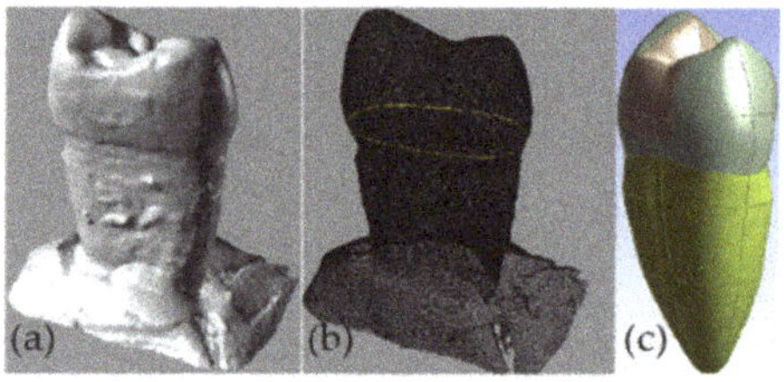

Figure 5. (**a**) Model STL, (**b**) Bio-CAD modeling, and (**c**) three-dimensional model generation.

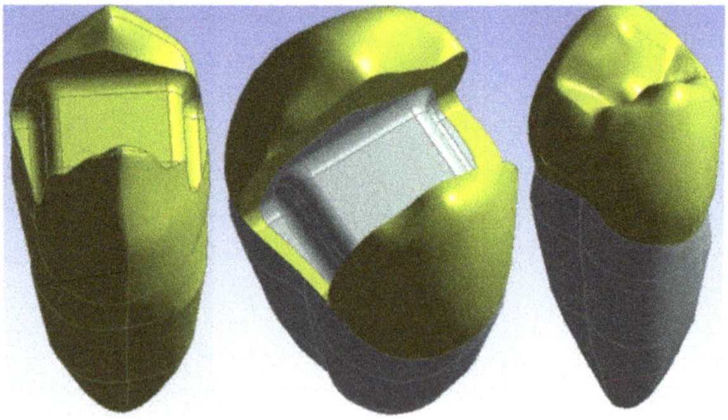

Figure 6. Model generation is healthy, based on the main anatomical landmarks and with the help of a specific tool program (NURBS surfaces).

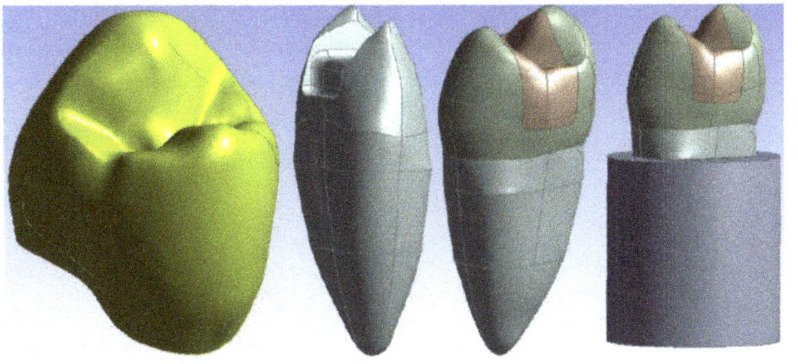

Figure 7. Generation of volumes of internal and external structures of the reference tooth.

The geometric models were exported to ANSYS Academic preprocessing software and, for each model, meshes were generated using tetrahedral elements of the linear type, with a total of 68,860 nodes and 40,358 elements. The generated meshes on the models with MOD cavity [20], and endodontic access plus restoration on models with cervical [21], and corresponding restoration cavity are shown, respectively, in Figures 8 and 9. As shown in Figure 10, the boundary conditions were defined for perfectly simulating contacts between the fixing structure (gray cylinder) and models with MOD cavities restored with composite resin.

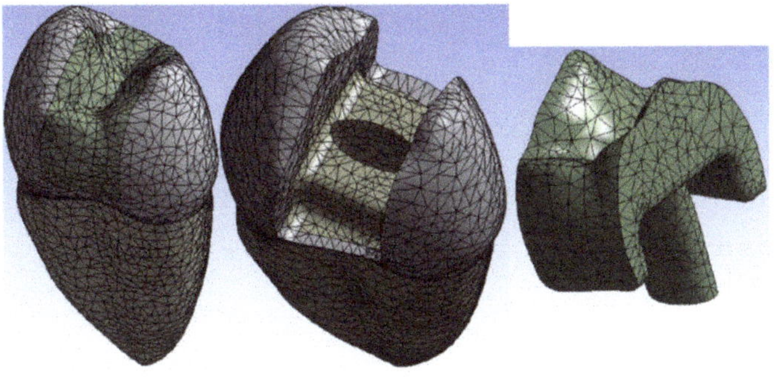

Figure 8. Mesh generated on models with MOD cavity and endodontic access with the corresponding restoration.

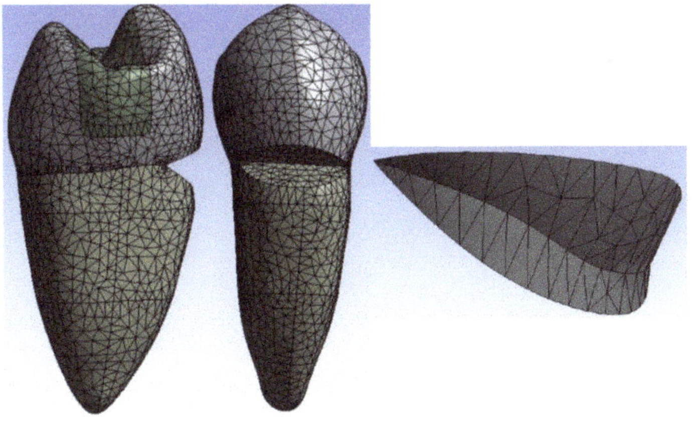

Figure 9. Mesh generated on models with cervical and corresponding restoration cavity.

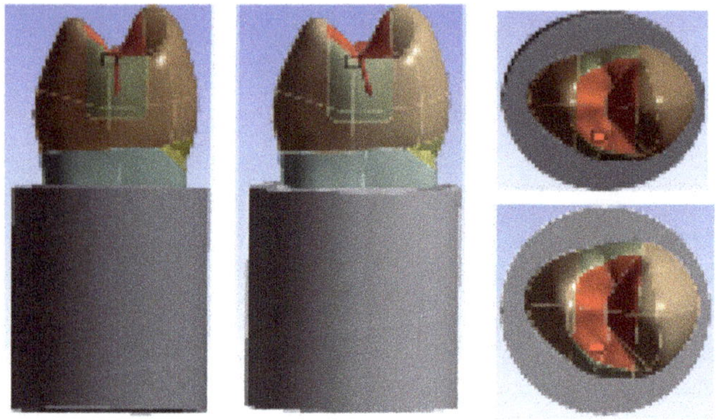

Figure 10. Contact area simulation of applied loads.

The tetrahedral model consists of 68,860 nodes and 40,358 elements. The analysis employed was a structural, linear, and elastic type for all such structures were isotropic, linear, and homogeneous. For this, it was necessary to obtain the elastic moduli and Poisson's ratios of the structures which characterize each model, which are described below in Table 1.

Table 1. Mechanical properties applied models.

Material	Modulus of Elasticity E (Mpa)	Poisson's Ratio (υ)
Enamel	46.8×10^3	0.30
Dentine	18.0×10^3	0.31
Polystyrene resin	13.5×10^3	0.30
Composite resin	16.6×10^3	0.24
Amalgam	15.0×10^3	0.30

For the teeth modeled with FEM, the periodontal tissue PDL was not considered to be maintained under the same conditions as in the moiré experiments.

Axial and oblique loads of 25 N, 50 N, 75 N, and 100 N were applied in the red area shown in Figure 10. The models were exported to the software processing core (Ansys Academic). For analysis of the results, criteria for normal direction deformation were used. The materials were considered homogeneous, isotropic, and with linear-elastic behavior.

The FEM numerical simulation estimated the normal elastic strain (for *X*-, *Y*-, and *Z*-axis) for all the specified cases mentioned before. Figure 11 shows, as an example, only for loading 100 N, a typical 3D color strain representation on the Z-axis direction.

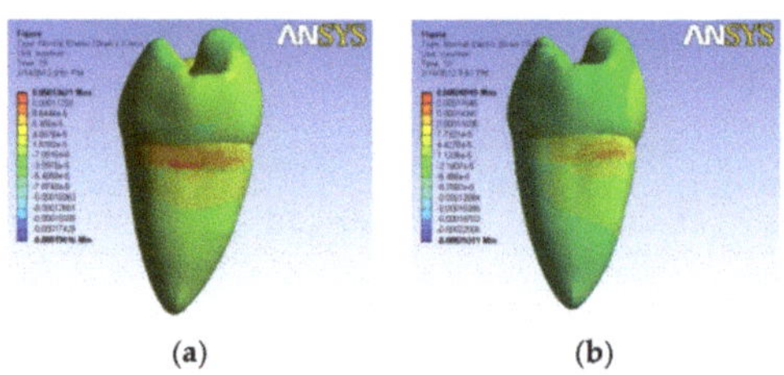

Figure 11. *Cont.*

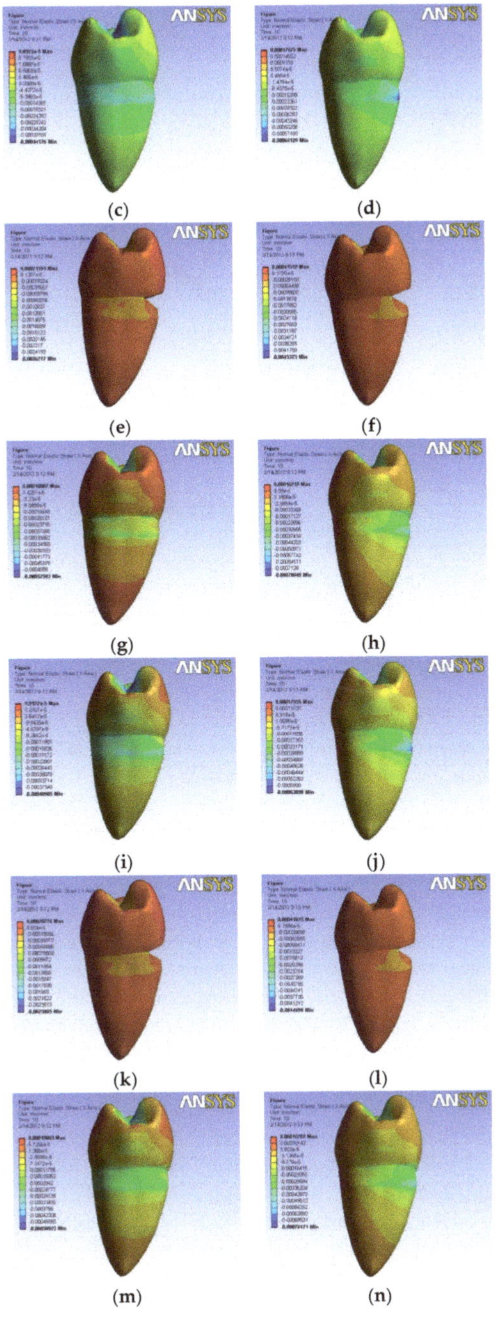

Figure 11. Typical 3D color strain representation on the Z-axis direction for 100 N loading: (**a**) tooth H healthy; (**b**) tooth H healthy load oblique 15°; (**c**) tooth A-MOD amalgam; (**d**) tooth A-MOD amalgam load oblique 15°; (**e**) tooth AL-MOD amalgam + lesion; (**f**) tooth AL-MOD amalgam + lesion oblique load 15°; (**g**) tooth ALR-MOD amalgam + restore lesion; (**h**) tooth ALR-MOD amalgam + restore lesion load oblique 15°; (**i**) tooth R-MOD resin; (**j**) tooth RMOD resin load oblique 15°; (**k**) tooth RL-MOD resin + lesion; (**l**) tooth RL-MOD resin + lesion oblique load 15°; (**m**) tooth RLR-MOD resin + restore lesion; (**n**) tooth RLR-MOD resin + lesion oblique load 15°.

The FEM technique strain average results, in the Z-axis direction and for all cases mentioned, are plotted in the graph presented in Figure 12.

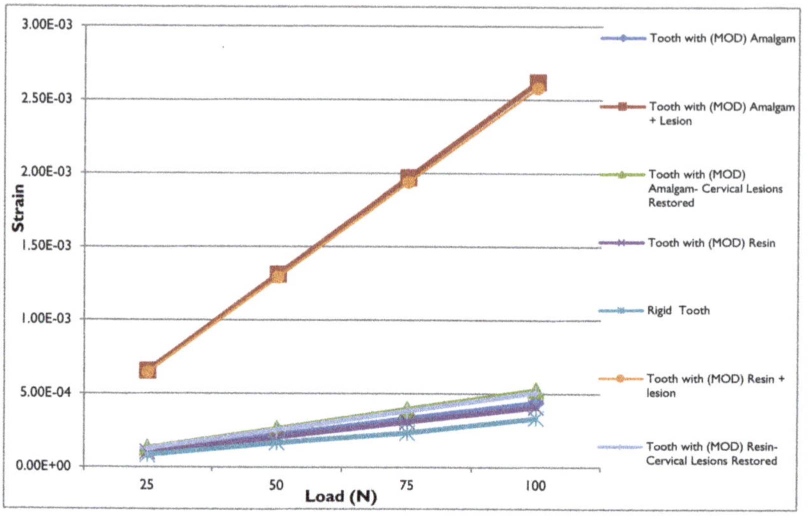

Figure 12. Graph of the average results of the tension of the FEM technique in the direction of the Z-axis for all the mentioned cases.

2.3. Moiré Projection Analysis

The moiré projection technique is a useful tool in structural engineering for measuring and controlling complex body geometries [17]. The method was applied to strain experimental evaluation of premolar tooth samples under the same conditions mentioned for the FEM analysis. The moiré projection technique is applied according to the system diagram shown in Figure 13a, where the projection system (SP) directs light beams (RP) over the study object (PR-D: tooth), which reflects a network image to the observation system (OS), which captures and superimposes with another network (RO), for the evaluation purpose.

In the implementation of the moiré technique, the same loads used for the simulation of FEM, frontal and oblique loads, were applied in the values of 25, 50, 75, and 100 N.

This technique allows the 3D shape determination by the projection of a regular pattern of straight fringes in an equally spaced pattern after the resulting image processing. The distance between camera and object is 500 mm, the projector lens and camera lens are 110 mm [17]. The best experimental results can be achieved with a small lens and standard projection fringes of 140 lines. This configuration considers the tooth object as the system origin, where the teeth fixing apparatus is called the Originator, which provides good flexibility to define different positions for the object. The equipment used can be considered very common; a Mini Projector Led Benq Gp1 and a CCD camera.

The 3D shape was determined by a technique called "phase shift" which assigns a phase angle to each pixel of the image. There is high uncertainty in the calculation of the phase angle, mainly because the light intensity profile is nearly sinusoidal. This can be mitigated by the acquisition and processing of three or four image fringes.

The software used for data acquisition and processing was FRINGEP, developed specifically for this research and based on the academic version of MATLAB. Considering the pixel size of the camera (CCD) and the pixels on the image plane, the FRINGEP software processes the digital reconstruction of the moiré image, including reconstructions of image intensities, interferogram intensities, phase interferogram, demodulation, and finally, integration to get a 3D representation of the shape and displacement of the deformed tooth.

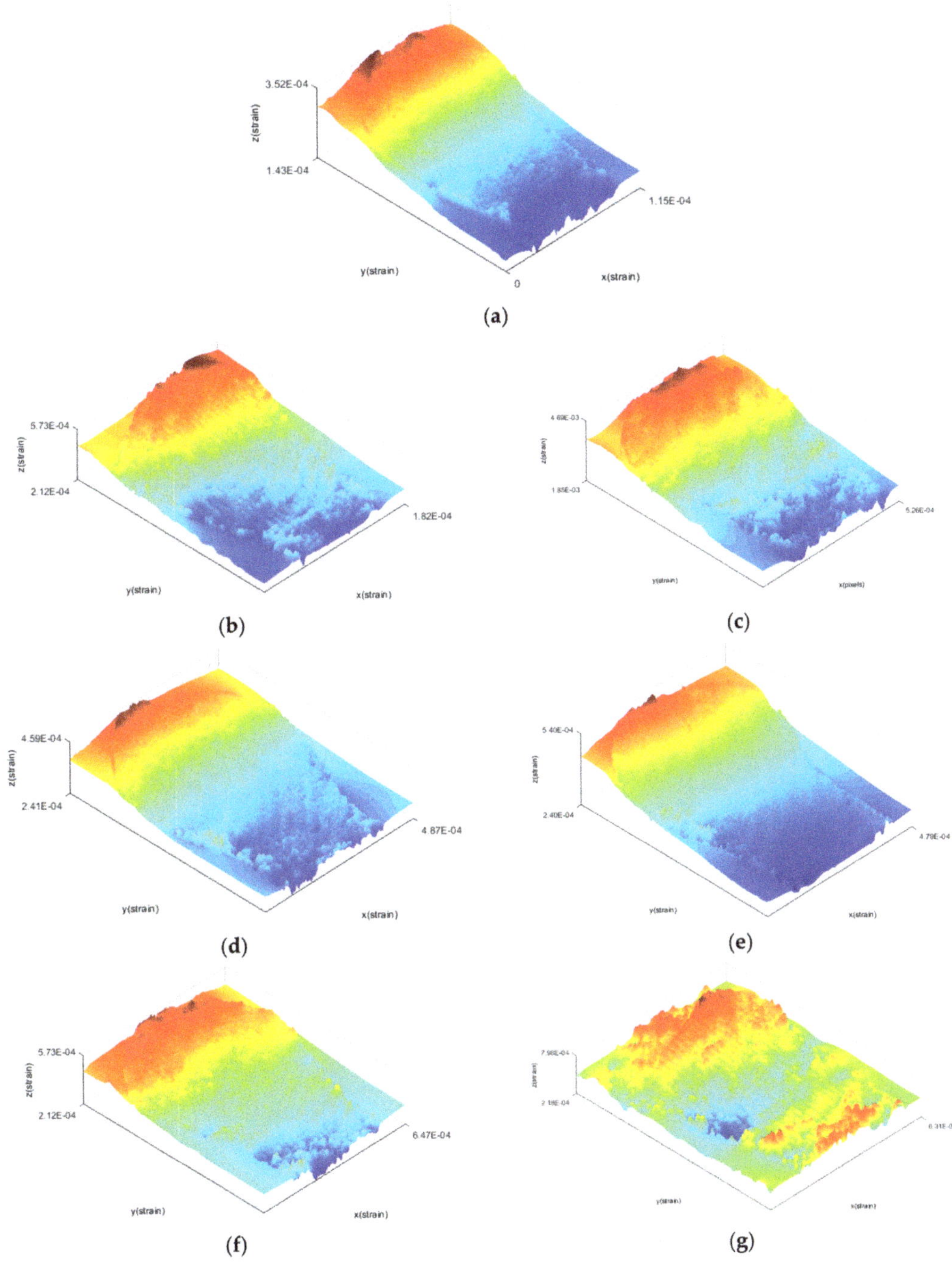

Figure 13. *Cont.*

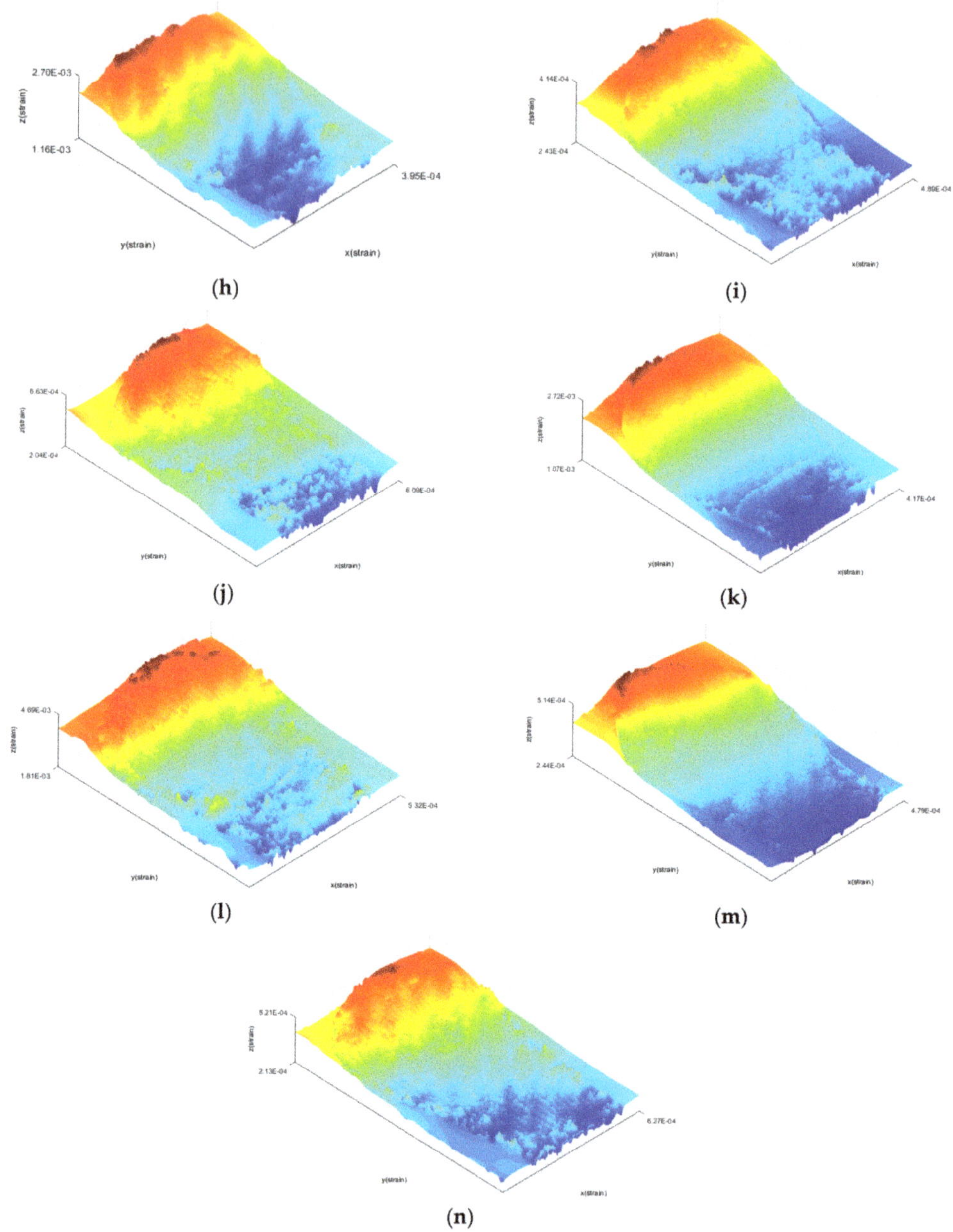

Figure 13. 3D color maps representing the strain in the Z-axis direction for 100 N loading: (**a**) tooth H healthy; (**b**) tooth H healthy oblique load 15°; (**c**) tooth AL-MOD amalgam + lesion oblique load 15°; (**d**) tooth A-MOD amalgam; (**e**) tooth ALR-MOD amalgam + restore lesion; (**f**) tooth A-MOD amalgam oblique load 15°; (**g**) tooth ALR-MOD amalgam + restore lesion oblique load. 15°; (**h**) tooth AL-MOD amalgam + lesion; (**i**) tooth R-MOD resin; (**j**) tooth R-MOD resin oblique load 15°; (**k**) tooth RL-MOD resin + lesion; (**l**) tooth RL-MOD resin + lesion oblique load 15°; (**m**) tooth RLR-MOD resin + restore lesion; (**n**) tooth RLR-MOD resin + restore lesion oblique load 15°.

The moiré experimental tests were done for the same specified cases of FEM numerical simulation. The geometric shape and strains were obtained for specified cases. Figure 13 shows, as an example, 3D color maps representing the strain in the Z-axis direction only for 100 N loading.

The experimental moiré projection technique strain average results, in the Z-axis direction and for all cases mentioned, are plotted in the graph presented in Figure 14.

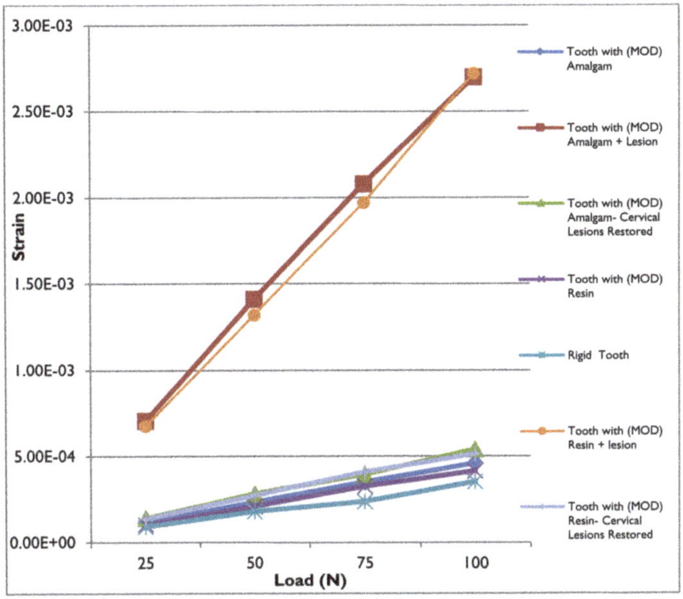

Figure 14. Moiré technique strain results in the Z-axis direction for all cases analyzed.

3. Results and Discussion

An understanding and knowledge of the behavior of healthy and restored teeth submitted to axial loads and tangencies are important in dental preparation planning and is also an indication of restorative material [22].

The analysis of the healthy dental structure was the standard for the analysis of the behavior of the tooth after the preparation in order to verify the areas to be reinforced or altered during the preparation in order to overcome the same possible deficiencies. As the experiments were performed on the same tooth in MEF and moiré, important data were obtained.

From the results obtained through the analysis by finite elements, it was possible to detect that the teeth that present a greater deformation are those where the load was applied at 15°, as can be seen in Figure 11f,l corresponding to tooth AL-MOD amalgam + lesion and tooth RL-MOD resin + lesion, whose values are 415×10^{-6} and 410×10^{-6}, respectively. While, on the other hand, the least deformations occur in the cases of tooth A-MOD-amalgam and tooth R-MOD-resin, with 98×10^{-6} and 99×10^{-6}, as presented in Figure 11c,i, respectively. For the cases of tooth AL-MOD amalgam + lesion and tooth RL-MOD resin + lesion similar behaviors are presented, with values of 211×10^{-6} and 207×10^{-6}, as observed in Figure 11e,k. In the same way, lesions with restorations with different materials are presented, tooth ALR-MOD amalgam + restore lesion and tooth RLR-MOD resin + restore lesion, whose deformation is 100×10^{-6} in both cases, according to Figure 11g,m.

From the analysis carried out and from the results obtained using the moiré method, the cases with the greatest deformation are tooth with (MOD) amalgam + lesion and tooth with (MOD) resin + lesion, both with a value of 2.7×10^{-3}. In the same way, behaviors are presented in, tooth with (MOD) resin-cervical lesions restored and tooth with (MOD) amalgam-cervical lesions restored, whose maximum values are 5.0×10^{-4} and 5.5×10^{-4}, respectively; as well as the case with tooth (MOD) amalgam and tooth with (MOD) resin, with a maximum deformation value of 4.0×10^{-4} for both. All of these results are presented in Figure 14.

The qualitative and quantitative analysis of each test specimen in each of the experimental situations can help in the understanding and evolution of dental preparations and the improvement of restorative materials. These results may indicate the need to change tooth wear during preparation by sparing healthy structures when using adhesive restorations. It is believed that these data may also inform us about some possibility of changing some form of preparation for amalgam.

The study of the reduction of dental structure is the main factor modulating the deformation and resistance to fracture of the dental element. This reduction can occur in different regions of the dental structure, such as inside the dental crown, reducing cusp height, inside the root canal, and in the cervical region [16].

The FEM and moiré techniques are totally non-intrusive, non-destructive, and allow global, non-contact measurement of displacement fields with sub-micrometric resolution. The deformation of the dental structure of each sample was analyzed in different phases and sequentially, that is, each sample was analyzed in the rigid phase without the occlusal box preparation. Afterwards it received sequential wear reducing the height of the axial wall. Each phase is measured separately. The preparations were made with a diamond tip mounted on a standard machine for cavity preparation [16].

The presence of the lesion increased shear values for both restorative materials subjected to axial loading. In oblique loading, there were no significant differences in both resin and amalgam. However, oblique loading increased compliance values with respect to axial loading, mainly for the AL group, since it presented higher strain values when compared to the other groups. The cervical lesions restored with composite and amalgam recovered biomechanical behavior similar to that of uninjured samples.

Some authors describe how under axial loading, the biomechanical behavior of the LA group has higher strain levels as compared with the RL. This can be explained by the amalgam elasticity modulus value, which is lower than the modulus of the composite.

The biomechanical analysis of these endodontically treated premolar teeth restored with amalgam, subjected to oblique loading, and with cervical lesion revealed behavior that could jeopardize the tooth structures and restorative materials. The tooth AL-MOD amalgam + lesion and tooth RL-MOD resin + lesion showed about five times the level of strain, when compared with all the other cases. On the other hand, when the cervical lesions are restored with amalgam or resin, the biomechanical behaviors are similar to those of healthy teeth.

The biomechanical results obtained by FEM and moiré show that the healthy teeth presented a lower level of normal maximum strain in any of the three directions in the *X*-, *Y*-, and *Z*-axis, for all different applied loads. For all analyzed cases, the application of oblique loading induces higher strain levels.

4. Conclusions

Even in the geometrically complex structures, such as human upper premolar teeth, the very low displacement and strain levels can be measured by using the moiré projecting technique, as detailed in this study.

The results obtained by both techniques show that healthy teeth presented the lower level of normal maximum strain for all different applied loads. In the meantime, oblique loading induced higher strain levels for all analyzed cases. In addition, it was noticed that the cervical lesions restored with composite and amalgam recovered biomechanical behavior similar to that of uninjured samples.

The results obtained by FEM simulation can be considered perfectly validated by the results presented by the experimental moiré projection analysis, demonstrating that the FEM numerical analysis can be used to evaluate the biomechanical behavior of healthy and endodontic treated teeth. Developing an alternative protocol to generate FEM three-dimensional models will lead to a ready and inexpensive tool since there is no need for costly equipment, for the prognosis of teeth extraction.

Author Contributions: Data curation, J.L.V.R., P.V.S. and H.F.L.; Formal analysis, J.O.R.P.; Funding acquisition, H.F.L.; Investigation, F.A.P.; Methodology, G.F.P.R. and H.F.L.; Project administration, G.G.d.P.; Resources, E.G.; Software, G.F.P.R. and H.F.L.; Supervision, J.O.R.P. and M.V.F.; Visualization, G.M.; Writing—original draft, J.L.V.R.; Writing—review & editing, J.O.R.P., M.V.F. and A.R.S. All authors have read and agreed to the published version of the manuscript.

Funding: The funds for the execution of this research project come from the Research Support Foundation of the State of Sao Paulo, Brazil.

Institutional Review Board Statement: It was selected and collected from the Faculty of Dentistry Bank of Teeth, University of Sao Paulo, twenty one (21) human healthy pre upper molars (1st and 2nd premolars), with indication for extraction. This selection was approved by the Research Ethics Committee: Final Analysis no. 520/11 of the Research Ethics Committee for the Protocol Registration CEP/UFU No. 171/11 and Final Analysis 372/11 of the Research Ethics Committee for the Protocol Registration CEP/UFU 065/11 the Research Ethics Committee of the Federal University of Uberlandia.

Informed Consent Statement: Not applicable.

Data Availability Statement: Not applicable.

Conflicts of Interest: The authors declare no conflict of interest.

References

1. Shilpa-Jain, D.P.; Krithikadatta, J.; Kowsky, D.; Natanasabapathy, V. Effect of cervical lesion centered access cavity restored with short glass fibre reinforced resin composites on fracture resistance in human mandibular premolars—An in vitro study. *J. Mech. Behav. Biomed. Mater.* **2021**, *122*, 104654. [CrossRef] [PubMed]
2. Iaculli, F.; Rengo, C.; Lodato, V.; Patini, R.; Spagnuolo, G.; Rengo, S. Fracture resistance of endodontically-treated maxillary premolars restored with different type of posts and direct composite reconstructions: A systematic review and meta-analysis of in vitro studies. *Dent. Mater.* **2021**, *37*, e455–e484. [CrossRef] [PubMed]
3. Alshiddia, I.F.; Aljinbaz, A. Fracture resistance of endodontically treated teeth restored with indirect composite inlay and onlay restorations–An in vitro study. *Saudi Dent. J.* **2016**, *28*, 49–55. [CrossRef]
4. Zheng, Y.; Bashandeh, K.; Shakil, A.; Jha, S.; Polycarpou, A.A. Review of dental tribology: Current status and challenges. *Tribol. Int.* **2022**, *166*, 107354. [CrossRef]
5. Bajunaid, S.O.; AlSadhan, N.O.; AlBuqmi, N.; Alghamdi, R. Influence of type of final restoration on the fracture resistance and fracture mode of endodontically treated premolars with occluso-mesial cavities. *Saudi Dent. J.* **2021**, 33, 316–321. [CrossRef]
6. Baldi, A.; Comba, A.; Tempesta, M.R.; Carossa, M.; Pereira, G.K.R.; Valandro, L.F.; Paolone, G.; Vichi, A.; Goracci, C.; Scotti, N. External Marginal Gap Variation and Residual Fracture Resistance of Composite and Lithium-Silicate CAD/CAM Overlays after Cyclic Fatigue over Endodontically-Treated Molars. *Polymers* **2021**, *13*, 3002. [CrossRef] [PubMed]
7. Shahrbaf, S.; Mirzakouchaki, B.; Oskoui, S.S.; Kahnamoui, M.A. The effect of marginal ridge thickness on the fracture resistance of endodontically-treated, composite restored maxillary premolars. *Oper. Dent.* **2007**, *32*, 285–290. [CrossRef] [PubMed]
8. Silva-Sousa, A.C.; Moris, I.C.M.; Simões Barbosa, A.F.; Corrêa Silva-Sousa, Y.T.; Sousa-Neto, M.D.; Ferreira Pires, C.R.; Alves Gomes, E. Effect of restorative treatment with endocrown and ferrule on the mechanical behavior of anterior endodontically treated teeth: An in vitro analysis. *J. Mech. Behav. Biomed. Mater.* **2020**, *112*, 104019. [CrossRef]
9. Soares de Toubes, K.M.; Soares, C.J.; Villamarim Soares, R.; de Souza Côrtes, M.I.; Quadros Tonelli, S.; Fernandes Borém Bruzinga, F.; Ferreira Silveira, F. The Correlation of Crack Lines and Definitive Restorations with the Survival and Success Rates of Cracked Teeth: A Long-term Retrospective Clinical Study. *J. Endod.* **2022**, *48*, 190–199. [CrossRef]
10. Garcia, P.P.; Wambier, L.M.; de Geus, J.L.; da Cunha, L.F.; Correr, G.; Gonzaga, C.C. Do anterior and posterior teeth treated with post-and-core restorations have similar failure rates? A systematic review and meta-analysis. *J. Prosthet. Dent.* **2019**, *121*, 887–894.e4. [CrossRef]
11. Jiang, W.; Bo, H.; YongChun, G.; LongXing, N. Stress distribution in molars restored with inlays or onlays with or without endodontic treatment: A three-dimensional finite element analysis. *J. Prosthet. Dent.* **2010**, *103*, 6–12. [CrossRef]
12. Scolavino, S.; Paolone, G.; Orsini, G.; Devoto, W.; Putignano, A. The Simultaneous Modeling Technique: Closing gaps in posteriors. *Int. J. Esthet. Dent.* **2016**, *11*, 58–81. [PubMed]

13. Siang Lin, G.S.; Kit Chan, D.Z.; Lee, H.Y.; Low, T.T.; Suvanpratum Laer, T.; Muralitharan Pillai, M.P.; Yew, Y.Q.; Syed Saadun Tarek Wafa, S.W.W. Effectiveness of Resin Infiltration in Caries Inhibition and Aesthetic Appearance Improvement of White-spot Lesions: An Umbrella Review. *J. Evid. Based Dent. Pract.* **2022**, 101723. [CrossRef]
14. Al Jeaidi, Z.A. Influence of resin cements and root canal disinfection techniques on the adhesive bond strength of fibre reinforced composite post to radicular dentin. *Photodiagn. Photodyn. Ther.* **2021**, *33*, 102108. [CrossRef] [PubMed]
15. Kadiyala, S.V.; Raj, J.D. Recent advances and modifications of dental restorative materials—A review. *Int. J. Recent Adv. Multidiscip. Res.* **2016**, *3*, 1609–1616.
16. Valin Rivera, J.V.; Gonçalves, E. *Analysis of the Biomechanical Behavior of Posterior Teeth with Different Levels of Reduction of Tooth Structure, Using Optical Interferometric Methods*; Visiting Researcher Grant—International in Polytechnic School (EP); University of São Paulo (USP): São Paulo, SP, Brazil, 2012.
17. Rivera, J.L.V.; Soares, P.V.; de Almeida Milito, G.; Fernández, F.P.; Roque, G.F.P.; Pérez, J.O.R.; Fernández, M.V. Development of the Moiré method of projecting fringes for the evaluation of deformations in upper premolars. *Ing. Mec.* **2017**, *20*, 22–30.
18. Conserva, E.; Consolo, U.; Sancho, A.G.; Foschi, F.; Paolone, G.; Giovarrusscio, M.; Sauro, S. Stress distribution in carbon-post applied with different composite core materials: A three-dimensional finite element analysis. *J. Adhes. Sci. Technol.* **2017**, *31*, 2435–2444. [CrossRef]
19. Shah, E.H.; Shetty, P.; Aggarwal, S.; Sawant, S.; Shinde, R.; Bhol, R. Effect of fibre-reinforced composite as a post-obturation restorative material on fracture resistance of endodonticall treated teeth: A systematic review. *Saudi Dent. J.* **2021**, *33*, 363–369. [CrossRef]
20. Yikilgan, I.; Bala, O. How can stress be controlled in endodontically treated teeth? A 3D finite element analysis. *Sci. World J.* **2013**, *2013*, 426134. [CrossRef]
21. Pereira, F.A.; Zeola, L.F.; Milito, G.D.A.; Reis, B.R.; Pereira, R.D.; Soares, P.V. Restorative material and loading type influence on the biomechanical behavior of wedge shaped cervical lesions. *Clin. Oral Investig.* **2016**, *20*, 433–441. [CrossRef]
22. Babaei, B.; Shouha, P.; Birman, V.; Farrar, P.; Prentice, L.; Prusty, G. The effect of dental restoration geometry and material properties on biomechanical behaviour of a treated molar tooth: A 3D finite element analysis. *J. Mech. Behav. Biomed. Mater.* **2022**, *125*, 104892. [CrossRef] [PubMed]

sensors

MDPI

Article

A Comparative Study of Parameter Identification Methods for Asymmetric Nonlinear Systems with Quadratic and Cubic Stiffness

Shibo Wang [1,2] **and Bin Tang** [1,2,*]

[1] Key Laboratory of Ocean Energy Utilization and Energy Conservation of Ministry of Education, Dalian University of Technology, Dalian 116023, China

[2] Institute of Internal Combustion Engine, Dalian University of Technology, Dalian 116023, China

[*] Correspondence: btang@dlut.edu.cn; Tel.: +86-(0)411-84706816

Abstract: Understanding the nonlinear dynamic characteristics of engineering structures is challenging, especially for the systems that exhibit asymmetric nonlinear behavior. This paper compared four parameter identification methods for asymmetric nonlinear systems incorporating quadratic and cubic stiffness nonlinearities. Hilbert transform, zero-crossing, direct quadrature, and wavelet transform were used to obtain the backbone, envelope, and restoring force curves from the free vibration time history. A nonlinear curve-fitting method was then applied to estimate the stiffness parameters of the asymmetric systems, and a linear least square fitting approach was utilized to estimate the damping parameters of the asymmetric systems. We used the Helmholtz–Duffing oscillator as a numerical example and a nonlinear vibration absorber with geometric imperfections to verify the feasibility and accuracy of these methods. The advantages and disadvantages of these methods and the deviations in estimated results are discussed.

Keywords: quadratic and cubic stiffness nonlinearity; nonlinear system identification; envelope; instantaneous frequency; nonlinear vibration absorber

Citation: Wang, S.; Tang, B. A Comparative Study of Parameter Identification Methods for Asymmetric Nonlinear Systems with Quadratic and Cubic Stiffness. *Sensors* **2022**, *22*, 5854. https://doi.org/10.3390/s22155854

Academic Editor: Stefano Lenci

Received: 17 July 2022
Accepted: 3 August 2022
Published: 5 August 2022

1. Introduction

Asymmetric nonlinear engineering structures, such as a mistuning quasi-zero-stiffness vibration isolator [1,2], cables [3], and geometric imperfect plate-like designs [4–7], have attracted widespread attention. The dynamic characteristics of these asymmetric systems are more complicated to analyze than those of the symmetric systems. For example, constant drift can occur in the response when there is little linear stiffness in the quasi-zero-stiffness vibration isolator. The nonlinear isolation system exhibits a mixed softening and hardening characteristic [1], which results in the multiple jump phenomena and hysteretic behavior [2]. Multivaluedness of the response curves occurs with different features depending on cables (or plates) and excitation force parameters [3–7]. To fully understand the nonlinear aspect of asymmetric systems, nonlinear parameter identification is one of the crucial procedures [8–10].

In recent years, backbone curves have been used to identify stiffness parameters of asymmetric systems. The calculation methods of the backbone curves can be classified into two types: analytical and numerical methods. Analytical methods include harmonic balance, multi-scale [11], and normal form methods [12,13]. The comparison of these analytical methods can be referred to [14]. Some software packages have implemented numerical algorithms based on a nonlinear normal mode framework [15–17]. Common experimental methods for extracting backbone curves include the resonance decay method [18], the control-based continuation method [19,20] and phase-locked loops [21]. For multi-degree-of-freedom systems, the nonlinear normal modes of interest are usually isolated by the force appropriation method [22–24]. The resonance decay method is then used to estimate the backbone curves. Breunung and Haller [25] recently studied the backbone curves of the forced-damped nonlinear mechanical systems. Cenedese and Haller [26] summarized approaches for constructing backbone curves of multi-degree-of-freedom systems. There are many applications to identify parameters using backbone curves, such as base-excited SDOF system [27], vibration absorber [28,29], beam-shaker system [30], aircraft wing structure [31], linear–arch composite beam piezoelectric energy harvester [32]. In addition to traditional contact measurement methods, non-contact measurement methods, such as video processing, can also be used to identify the backbone curves of an air wing prototype [33]. The restoring force curve is another useful tool that can be applied to estimate stiffness parameters of asymmetric systems [34]. Because of the asymmetric characteristics, the positive and negative parts of the asymmetric system were analyzed separately [35,36]. The bias term obtained by the signal decomposition method is the unique information in the asymmetric signal.

Hilbert transform [37], zero-crossing method [27], direct quadrature method [38], and wavelet transform [39] can be used to obtain the backbone curves and restoring force curves from the free-decay measurements. The applications of these methods to a Duffing system have been summarized [40,41]. However, these methods cannot be directly applied to asymmetric systems with quadratic and cubic stiffness nonlinearities, and need to be modified to analyze positive and negative time responses separately when obtaining the restoring force curves. The backbone curves exhibit softening-hardening nonlinear behavior and are complex to analyze. As far as the authors know, there is no relevant comparative study. Therefore, several identification methods are summarized and compared, which can better guide the understanding of the parameter identification of Helmholtz–Duffing type asymmetric engineering structures.

This paper aims to investigate the parameter identification methods applied to asymmetric systems with quadratic and cubic stiffness nonlinearities and illustrate these methods using a Helmholtz–Duffing numerical example and a vibration absorber experiment. Section 2 introduces the Helmholtz–Duffing example and its theoretical solutions for restoring force and backbone curves. In Section 3, detailed steps of the four methods to obtain different curves are introduced. Section 4 discusses the identification methods, identification results, errors of the four methods, and the corresponding advantages and disadvantages. The application in the vibration absorber experiment is described in Section 5. Finally, this paper is concluded in Section 6.

2. Asymmetric Model: A Helmholtz–Duffing Oscillator

The asymmetric engineering structures can be simplified to a Helmholtz–Duffing oscillator, and its equation of motion is given by

$$m\ddot{x} + c\dot{x} + k_1 x + k_2 x^2 + k_3 x^3 = f(t) \tag{1}$$

where m is the mass, c is the damping coefficient, $f(t)$ is the excitation force, k_1, k_2, and k_3 are the linear, quadratic, and cubic nonlinear stiffness coefficients, respectively. If the system vibrates freely with an initial displacement, $f(t) = 0$. Then, dividing both sides of Equation (1) by the mass m yields

$$\ddot{x} + 2h\dot{x} + \omega_0^2(t)x = 0 \tag{2}$$

where $h = c/2m$ is the damping factor and $\omega_0(t) = \sqrt{(k_1 + k_2 x + k_3 x^2)/m}$ is the instantaneous frequency. They have the units of $1/\mathrm{s}$. It can be seen that the analytical solution for the viscous damping force per unit mass is written as $f_c(x) = 2h\dot{x}$ when the range of velocity is given, and the analytical solution for the restoring force per unit mass is given by

$$f_k(x) = \frac{k_1}{m}x + \frac{k_2}{m}x^2 + \frac{k_3}{m}x^3 \tag{3}$$

where the regime of displacement is determined.

Backbone curves for the Helmholtz–Duffing oscillator are more complex than those for the symmetric systems [42]. Using the harmonic balance method and substituting the approximate harmonic solution $x(t) = A_0 + A_1\cos(\omega t + \theta)$ into Equation (1) yield

$$\omega^2 = \frac{k_1}{m} + \frac{2k_2}{m}A_0 + \frac{3k_3}{m}A_0^2 + \frac{3k_3}{4m}A_1^2 \tag{4}$$

where $A_0 = \left(\frac{f_0}{2k_3} + \left(\frac{f_0^2}{4k_3^2} + \left(\frac{A_1^2}{2} + \frac{\beta}{3k_3}\right)^3\right)^{1/2}\right)^{1/3} + \left(\frac{f_0}{2k_3} - \left(\frac{f_0^2}{4k_3^2} + \left(\frac{A_1^2}{2} + \frac{\beta}{3k_3}\right)^3\right)^{1/2}\right)^{1/3}$

$-\frac{k_2}{3k_3}$ is the bias term, $A_1^2 = -(k_3A_0^3 + k_2A_0^2 + k_1A_0)/(3k_3A_0/2 + k_2/2)$ is the first harmonic term, $\beta = k_1 - k_2^2/3k_3$, and $f_0 = k_1k_2/3k_3 - 2k_2^3/27k_3^2$. Adding more harmonic terms to the approximate harmonic solution can yield a more accurate solution, but the calculation of the backbone curve is complicated. Substituting the approximate harmonic solution $x(t) = A_0 + A_1\cos(\omega t + \theta) + A_2\cos(2\omega t + \varphi)$, where the second harmonic term A_2 is included, into Equation (1) for the lightly damped case, gives

$$\omega^2 \approx \frac{k_1}{m} + \frac{2k_2}{m}A_0 + \frac{3k_3}{m}A_0^2 + \frac{3k_3}{4m}A_1^2 + \frac{k_2}{m}A_2 + \frac{k_3}{m}\left(\frac{3}{2}A_2^2 + 3A_0A_2\right) \tag{5}$$

which can be compared well with the numerical results of the backbone curves [42]. The comparison between analytical solutions and the numerical results is shown in Section 3.

3. Identification Methods for the Characteristic Curves

This section introduces four methods for obtaining the characteristic curves of the Helmholtz–Duffing oscillator, which are restoring force curves, damping force curves, envelopes, and backbone curves. These curves are combined with the analytical solutions given in Section 2 to estimate the stiffness and damping parameters of asymmetric structures. The details are discussed in Sections 4 and 5. For the sake of simplicity, the parameters for the oscillator are m = 0.1 kg, c = 0.4 Ns/m, k_1 = 4000 N/m, $k_2 = -10^7$ N/m^2, and $k_3 = 10^{10}$ N/m^3. The initial displacement and velocity are $x(0) = 0.0018$ m and $\dot{x}(0) = 0$ m/s. The free-decay response is numerical integration calculated by using the fourth-order Runge-Kutta method, and the sampling frequency is f_s = 2000 Hz.

For the asymmetric system, when the free decay response is measured, the restoring force curve is constructed separately from the positive and negative signal parts [35], which is given by

$$f_k(x) = \begin{cases} \omega_c^2(t)A_c(t) & x > 0 \\ \omega_c^2(t)A_c(t) & x \le 0 \end{cases} \tag{6}$$

where A_c and ω_c included positive and negative congruent envelopes and congruent natural frequency. The viscous damping force is approximately given by

$$f_c(x) \approx \begin{cases} 2h(t)A_{\dot{x}}(t) & \dot{x} > 0 \\ -2h(t)A_{\dot{x}}(t) & \dot{x} < 0 \end{cases} \tag{7}$$

where $h(t)$ is the instantaneous damping factor and $A_{\dot{x}}(t)$ is the congruent envelope of velocity. The backbone curves can be obtained using the relationship between instantaneous frequency and amplitude of each harmonic term.

3.1. Hilbert Transform

Hilbert transform (HT) has been widely used in the parameter identification and signal decomposition of nonlinear systems. Feldman applied this method to identify free and forced vibration systems, and further applied the nonparametric identification method to the asymmetric systems [43–46]. The complex analytic form of a free-decay response is given by $X(t) = x(t) + j\tilde{x}(t) = A(t)e^{j\varphi(t)}$, where $\tilde{x}(t)$ is the Hilbert transform of the signal $x(t)$, the envelope $A(t) = \sqrt{x^2(t) + \tilde{x}^2(t)}$, and the instantaneous phase $\varphi(t) = \arctan[\tilde{x}(t)/x(t)]$. The instantaneous undamped natural frequency is given by

$$\omega_0^2(t) = \omega^2 - \frac{\ddot{A}}{A} + \frac{2\dot{A}^2}{A^2} + \frac{\dot{A}\dot{\omega}}{A\omega} \tag{8}$$

and the instantaneous damping factor is given by

$$h(t) = -\frac{\dot{A}}{A} - \frac{\dot{\omega}}{2\omega} \tag{9}$$

where $\dot{A}$, $\ddot{A}$ and $\dot{\omega}$ are the first and second derivatives of envelope and frequency, respectively.

Hilbert vibration decomposition (HVD) is a time-varying vibration decomposition method based on the Hilbert transform [45]. The main harmonic components of asymmetric systems can be obtained by using the HVD. By weighted summing the decomposed harmonic components, the congruent envelope, and the congruent frequency can be obtained. The congruent envelope A_c is given by

$$A_c(t) = \sum_{l=1}^{N} A_l(t) \cos\phi_l(t) \tag{10}$$

where A_l is the envelope of the lth order component, and ϕ_l is the phase angle between the primary component and lth order component. Instantaneous natural frequency ω_0 can be decomposed into a sum of high-order intrinsic components. The congruent natural frequency ω_c is given by

$$\omega_c(t) = \sum_{l=1}^{N} \omega_{0l}(t) \cos\phi_{\omega l}(t) \tag{11}$$

where ω_{0l} is the envelope of the lth order instantaneous natural frequency, and $\phi_{\omega l}$ is the phase angle between the primary component and lth order component.

The Hilbert transform method and the relevant Matlab programs [46] are used in this paper. Free-decay response of the Helmholtz–Duffing oscillator and its envelopes are shown in Figure 1a, where the positive and negative congruent envelopes are obtained by Equation (10). The instantaneous natural frequency and their envelopes are obtained by Equations (8) and (11), as shown in Figure 1b. Figure 1c shows the first fourth components of the free vibration obtained by using HVD. In order to remove the end effect of the Hilbert transform, only the analyzed results between 0.2 to 1.7 s are chosen. The restoring force curves are shown in Figure 1d, where the analytical solution is given by Equation (3), and the numerical results are obtained by Equation (6). The backbone curves of the first harmonic term A_1 and bias term A_0 are plotted in Figure 1e–f, respectively, where the analytical solutions are given by Equation (5), and the numerical results are obtained by using HVD. The logarithmic form of the envelope, instantaneous damping factor obtained by Equation (9), and damping force curve obtained by Equation (7) are shown in Figure 1g–i, respectively. For a weakly nonlinear system, the analytical solution of the logarithmic envelope is approximately given by $-ct/2m + \ln x(0)$, as shown in Figure 1g. The damping factor is $h = c/2m = 21/\text{s}$ and the analytical damping force is $f_c(x) = 2h\dot{x}$, which are plotted in Figure 1h,i, respectively.

3.2. Zero-Crossing

The zero-crossing (ZC) method is a simple method for estimating the instantaneous frequency and amplitude of free decay signals [27,36]. Londoño et al. [27] combined the zero-crossing and peak picking methods to obtain the backbone curves of symmetric nonlinear systems. Ondra et al. [36] extended the zero-crossing method to obtain the backbone and restoring force curves of asymmetric systems. This method is explained in detail in Figure 2. The ith zero-crossing point t_i, the pth positive peak points A_p, and the nth negative peak points A_n are first obtained using the zero and peak picking procedure. The instantaneous frequency at the zero-crossing point t_i is

$$\omega(t_i) = \frac{2\pi}{t_{i+1} - t_{i-1}} \tag{12}$$

The positive and negative envelopes at the zero-crossing points are then obtained by linear interpolating the positive and negative peak points, respectively. The corresponding instantaneous frequencies for positive and negative envelope points are $\omega_p = \pi/T_p$ and $\omega_n = \pi/T_n$, respectively.

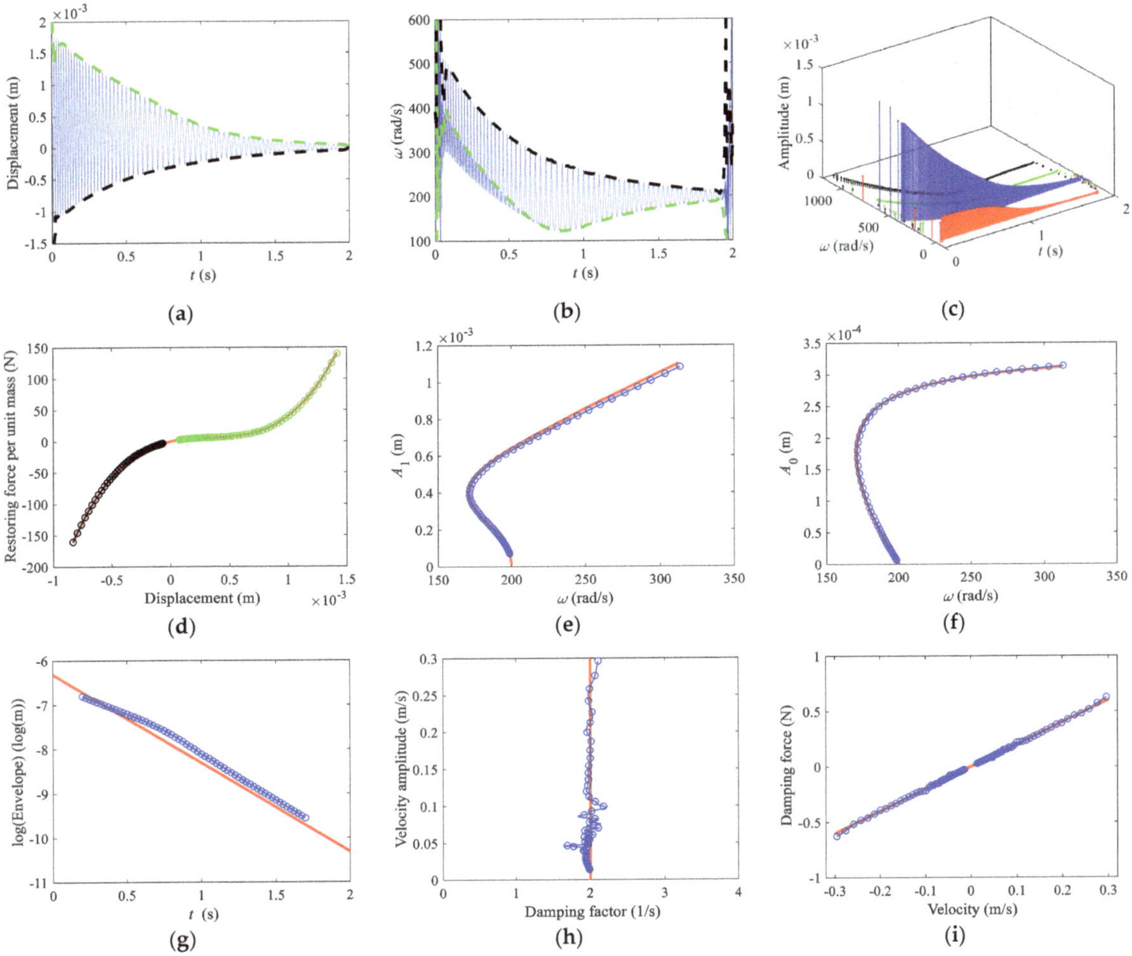

Figure 1. Hilbert transform. (**a**) Time domain response and envelopes, (**b**) frequencies, (**c**) HVD, (**d**) restoring force curve, (**e**) backbone curve of first harmonic term A_1, (**f**) backbone curve of bias term A_0, (**g**) logarithmic form of the envelope, (**h**) damping factor, (**i**) damping force curve. In (**a**,**b**), thin solid lines, time domain response, and instantaneous natural frequency; dashed lines, congruent envelope, and congruent modal frequency. In (**d**–**i**), solid lines, analytical solutions; lines with circles, numerical results.

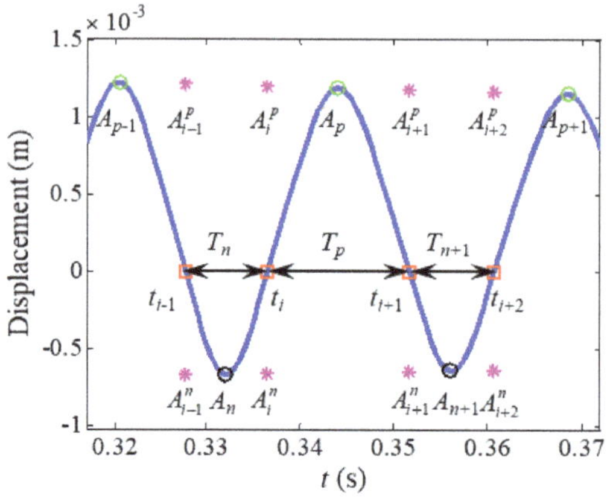

Figure 2. Zero-crossing with peak picking method for the asymmetric signal. Solid line, time-domain response; □, zero-crossing points; ○, positive peak points; O, negative peak points; *, envelopes at zero-crossing points obtained by interpolation.

The amplitudes of the first harmonic term A_1 and bias term A_0 are given by

$$A_0 = \frac{A_i^p + A_i^n}{2},\ A_1 = \frac{A_i^p - A_i^n}{2} \tag{13}$$

The free decay response is given in Figure 3a again. Then, the zero-crossing points and positive and negative peak points can be obtained and plotted in Figure 3a. The instantaneous frequencies for the positive and negative envelopes are shown in Figure 3b. Substituting the positive and negative envelopes and instantaneous frequencies in Equation (6) yields the restoring force curves, as shown in Figure 3c. It should be noticed that the time history response of asymmetric systems normally contains multiple harmonic components. Unlike the HVD, the backbone curves in the zero-crossing method are calculated using Equations (12) and (13), where the high order harmonic terms induce estimation errors, especially for the bias term. Before calculating the instantaneous frequency and amplitudes, the high-order harmonic terms should be filtered to make sure that the response can be approximately written as $x(t) \approx A_0 + A_1 \cos(\omega t + \theta)$. For the numerical example in this section, the free decay signal is passed through a low-pass filter with a cut-off frequency of 50 Hz. The backbone curves are shown in Figure 3d,e. It can be seen that the backbone curves obtained from the filtered response are closer to the analytical backbone curves. The logarithmic form of the envelope, instantaneous damping factor, and damping force curve are shown in Figure 3f–h. Numerical results for the damping factor and damping force are obtained by Equations (7) and (9), where the envelope of velocity is obtained by the peak picking method. To unify several methods, the data used in the restoring force curve, logarithmic envelope, and damping curve are also approximately taken from 0.2 to 1.7 s. The data used in backbone curves are relatively short because part of the response in the large amplitude regime is filtered out. The analytical solutions shown in Figure 3 are obtained using the similar approaches mentioned in Section 3.1.

3.3. Direct Quadrature

The direct quadrature (DQ) method was proposed by Huang et al. [38]. Firstly, because the Hilbert transform of a product of functions is limited by the Bedrosian theorem [47], a normalization scheme was proposed to separate amplitude modulation (AM) and frequency modulation (FM) of the signal. Secondly, according to the Nuttall theorem [48], it is not applicable for all signals to obtain their quadrature forms using the Hilbert transform. Therefore, the direct quadrature method is used. The direct quadrature method has been applied to the symmetric signal. However, when it is applied to the asymmetric signal, some modifications are required.

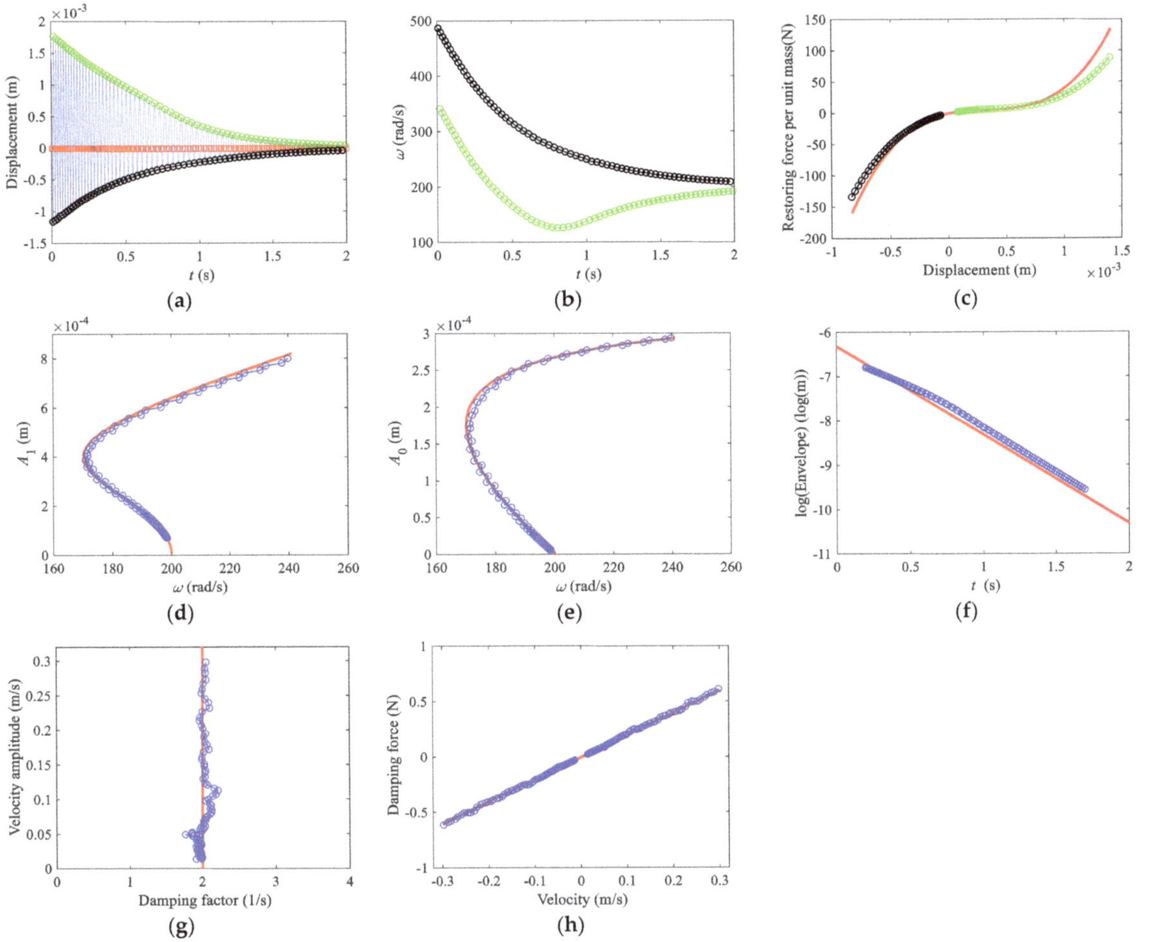

Figure 3. Zero-crossing with peak picking method. (**a**) Time domain response, (**b**) frequencies, (**c**) restoring force curve, (**d**) backbone curve of first harmonic term A_1, (**e**) backbone curve of bias term A_0, (**f**) logarithmic form of the envelope, (**g**) damping factor, (**h**) damping force curve. In (**a**,**b**), solid line, time domain response; red squares, zero-crossing points; lines with circles, positive and negative envelopes, and frequencies. In (**c**–**h**), solid lines, analytical solutions; lines with circles, numerical results.

The first step is to find the positive and negative peak points and use the cubic spline function to obtain the positive and negative envelopes A_p and A_n. Then the normalization process is carried out. The positive time-domain response is normalized by $y_p(t_p) = x_p(t_p)/A_p$ and the negative time-domain response is normalized by $y_n(t_n) = x_n(t_n)/A_n$, where $x_p(t_p)$ and $x_n(t_n)$ are the positive and negative time-domain responses, $y_p(t_p)$ and $y_n(t_n)$ are the normalized positive and negative responses. Repeat the above steps until the normalized results are all in $[-1,1]$. The FM part of the signal is $F(t) = y^l(t) = \left[y_p^l(t_p), y_n^l(t_n)\right]$, where l is the number of iterations.

After normalization, the AM part is given by

$$\widetilde{A}_p(t_p) = \frac{x_p(t_p)}{y_p^l(t_p)},\ \widetilde{A}_n(t_n) = \frac{x_n(t_n)}{y_n^l(t_n)} \tag{14}$$

Then using the cubic spline function, the positive envelope $\widetilde{A}_p(t)$ and negative envelope $\widetilde{A}_n(t)$ in the entire time domain are obtained. The amplitudes of the first harmonic and the bias terms are given by

$$A_1 = \frac{\widetilde{A}_p(t) - \widetilde{A}_n(t)}{2},\ A_0 = \frac{\widetilde{A}_p(t) + \widetilde{A}_n(t)}{2} \tag{15}$$

The time domain response and its positive and negative envelopes are shown in Figure 4a. The FM part can be regarded as a sinusoid, so the instantaneous frequency is

$$\omega(t) = \frac{d}{dt}[\arccos(F(t))] \tag{16}$$

The FM part whose absolute values are less than 0.9 is used to calculate the instantaneous frequency using Equation (16), and the rest of the instantaneous frequency points are interpolated using a cubic spline. The result is shown in Figure 4b. The restoring force curve obtained by Equation (6) is shown in Figure 4c. Similar to the zero-crossing method, the amplitudes of the first harmonic and the bias terms are also sensitive to the high-frequency components of the free decay response. Before calculating the backbone curves, the free decay response is passed through a low-pass filter with a cut-off frequency of 50 Hz. The estimated backbone curves of A_1 and A_0 are shown in Figure 4d,e, where amplitudes are obtained by Equation (15), and the frequency is the filtered instantaneous frequency. The logarithmic form of the envelope, instantaneous damping factor, and damping force curve are shown in Figure 4f–h. The analytical solutions are also shown in Figure 4.

3.4. Wavelet Transform

Wavelet transform (WT) is a time-frequency analysis tool that can automatically adjust the size of the analysis window with the change of frequency. With the development in recent years, wavelet analysis has been widely applied in nonlinear system identification [39,49–51].

For the free decay response of the Helmholtz–Duffing oscillator shown in Figure 5a, the frequency spectrum can be obtained from the Matlab *cwt* function using the Morlet wavelet and is shown in Figure 5b. From this figure, the envelope of the signal is given by

$$A(b) = \frac{2|W_x(a(b), b)|}{\sqrt{a(b)}} \tag{17}$$

where a is the scale parameter, b is the translation parameter, $|W_x(a(b), b)| = \max_a |W_x(a, b)|$ is the maximum value of wavelet coefficients at each time point. The instantaneous frequency is obtained from the frequency points corresponding to the maximum value of the wavelet coefficients at each time point. In order to obtain the smooth envelope and instantaneous frequency, the results are passed through a low-pass filter with a cut-off frequency of 20 Hz. The results are shown in Figure 5a,b. The backbone curve of A_1 constructed by the instantaneous amplitude and frequency is shown in Figure 5c. The bias term is obtained by the wavelet decomposition using the Matlab function *wavedec*, as shown in Figure 5d. The logarithmic form of the envelope, instantaneous damping factor, and damping force curve are shown in Figure 5e–g. It can be seen that the numerical solutions can be reasonably compared well with the analytical solutions.

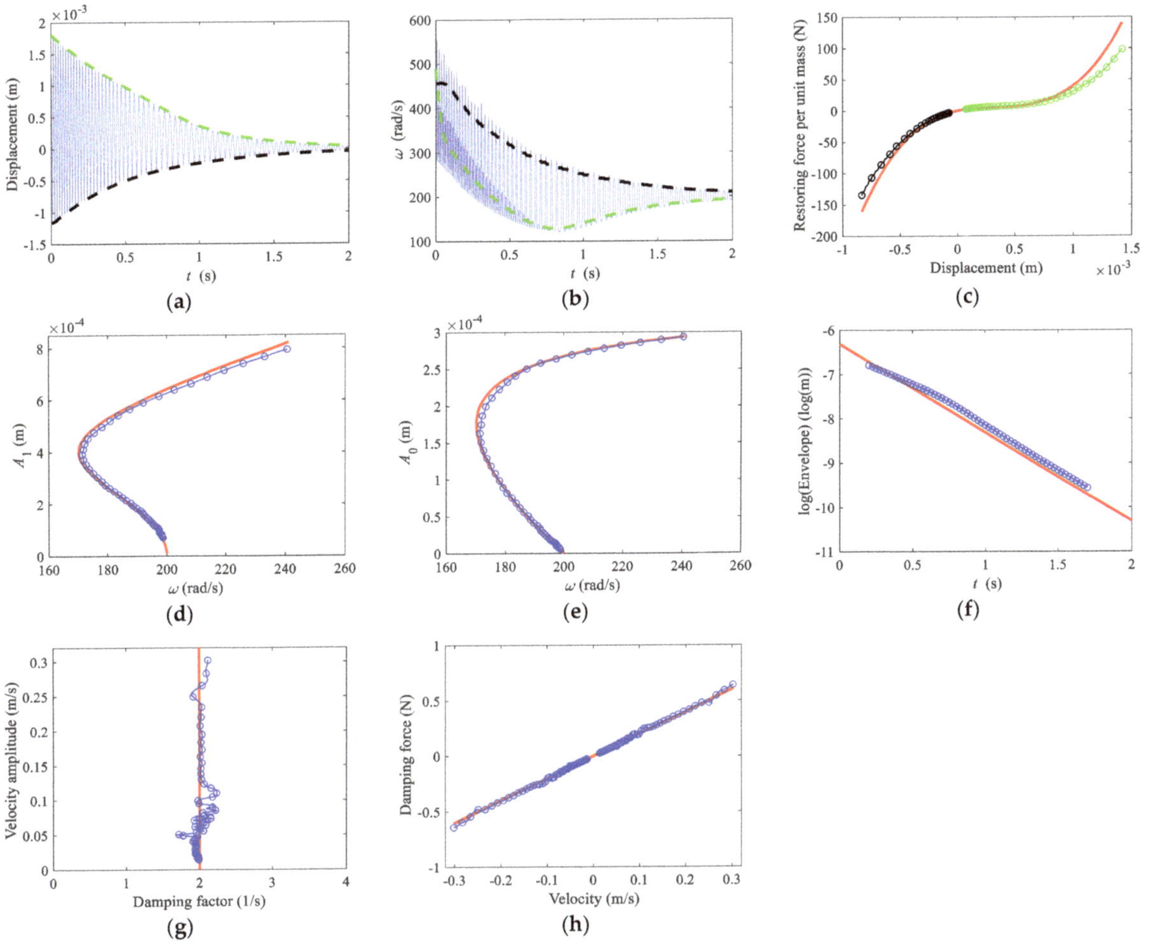

Figure 4. Direct quadrature. (**a**) Time domain response and envelopes, (**b**) frequencies, (**c**) restoring force curve, (**d**) backbone curve of first harmonic term A_1, (**e**) backbone curve of bias term A_0, (**f**) logarithmic form of the envelope, (**g**) damping factor, (**h**) damping force curve. Thin solid lines, time domain response in (**a**) and instantaneous frequency in (**b**); dashed lines, positive and negative envelopes in (**a**) and frequencies in (**b**). In (**c**–**h**), thick solid lines, analytical solutions; lines with circles, numerical results.

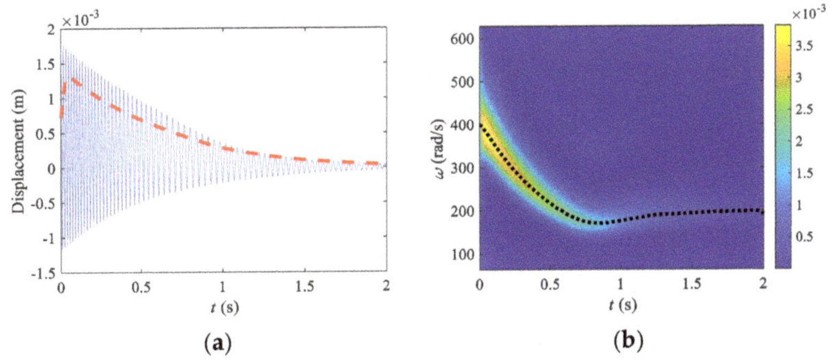

Figure 5. *Cont.*

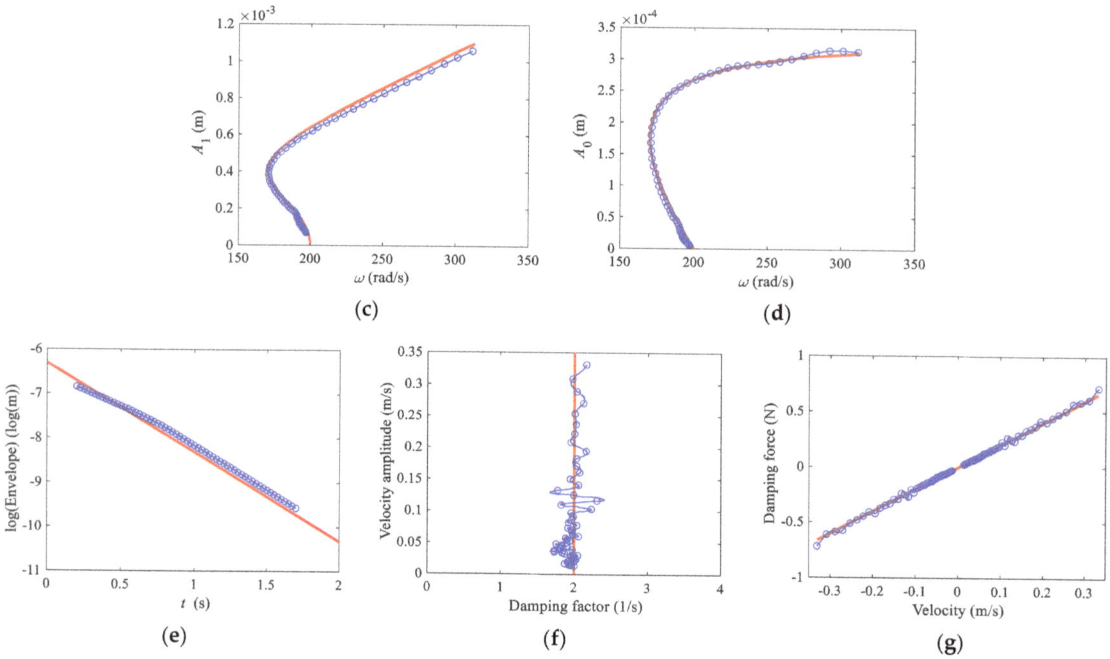

Figure 5. Wavelet transform. (**a**) Time-domain response and envelope, (**b**) spectrogram, (**c**) backbone curve of first harmonic term A_1, (**d**) backbone curve of bias term A_0, (**e**) logarithmic form of the envelope, (**f**) damping factor, (**g**) damping force curve. Thin solid line, time domain response; dashed line, envelope; dotted line, instantaneous frequency; thick solid lines, analytical solutions; lines with circles, numerical results.

4. Parameter Estimation and Discussion

In this section, the stiffness and damping parameters of the asymmetric systems are estimated from the characteristic curves of the Helmholtz–Duffing oscillator obtained in Section 3. The stiffness parameters are obtained by polynomial fitting the restoring force curve shown in Equation (3). The Matlab function *polyfit* computes the least square polynomial. This method is called the restoring force curve method (RFCM). Although the backbone curve of the first harmonic term obtained by using Equation (4) deviates from the numerical result obtained by using HVD in the bending regime [42], which is the curved part, we can also estimate the stiffness parameters from this curve. Matlab function *fminsearch* is used here to find the optimal stiffness parameters. This identification method is called the backbone curve method (BCM). The estimated stiffness parameters are shown in Table 1.

The viscous damping can be estimated by linear fitting the damping force curve, called the damping force curve method (DFCM). For a weakly nonlinear system, the envelope is approximately given by $\ln A(t) = -ct/2m + \ln A_0$. Therefore, the natural logarithm of the envelope can also be used to estimate the damping [28], called the logarithmic envelope method (LEM). Matlab function *polyfit* is utilized for computing the least square linear coefficient. The estimated damping coefficients are shown in Table 2.

Table 1. Estimated results and errors of the stiffness parameters.

	k_1 (N/m)		k_2 (N/m^2)		k_3 (N/m^3)	
	RFCM	BCM	RFCM	BCM	RFCM	BCM
HT	3940.7 (1.48%)	4082.0 (2.05%)	-1.002×10^7 (0.20%)	-1.007×10^7 (0.73%)	9.998×10^9 (0.02%)	1.023×10^{10} (2.31%)
ZC	4051.5 (1.29%)	4050.6 (1.27%)	-8.381×10^6 (16.19%)	-9.873×10^6 (1.27%)	7.213×10^{10} (27.87%)	1.003×10^{10} (0.31%)
DQ	4052.5 (1.31%)	4066.5 (1.66%)	-8.273×10^6 (17.27%)	-9.923×10^6 (0.77%)	7.262×10^9 (27.38%)	1.008×10^{10} (0.76%)
WT		3989.1 (0.27%)		-9.977×10^6 (0.23%)		1.043×10^{10} (4.28%)
True value	4000		-10^7		10^{10}	

Table 2. Estimated results and errors of the damping coefficient.

c (Ns/m)	LEM	DFCM
HT	0.381 (4.80%)	0.402 (0.37%)
ZC	0.377 (5.76%)	0.406 (1.58%)
DQ	0.381 (4.82%)	0.408 (2.02%)
WT	0.370 (7.60%)	0.405 (1.30%)
True value	0.4	

The estimated stiffness parameters are obtained by using the four methods. The Hilbert transform and Hilbert vibration decomposition combined with the low pass filter can give accurate estimation results. This method can not only decompose the signal and obtain the backbone curve of each harmonic term, but also combine harmonic terms to construct the restoring force. However, the Hilbert transform method has an end effect. The data at the beginning and end need to be removed. For zero-crossing with the peak picking method, it is straightforward to implement. Even for the asymmetric signal, this method can analyze the positive and negative time domain separately and construct the corresponding restoring force. However, when this method is applied to the signal contaminated with noise, the zero-crossing points and peak points are difficult to obtain accurately. The signal should be properly filtered to solve this issue. Also, there are only fewer points to extract for the short-time signal. In this case, the interpolation method can be used in the whole time domain to obtain more points. For the direct quadrature method, part of instantaneous frequency points are interpolated using a cubic spline, so the obtained instantaneous frequency is not accurate enough, especially for the signal with a low sampling rate or a large amount of normalized data is over 0.9. For the wavelet transform, the wavelet function needs to be selected carefully and appropriately.

The nonlinear stiffness parameters estimated by the restoring force curve using the zero-crossing and direct quadrature methods are not accurate. The estimated results of the backbone curve method are all well. The backbone curve method estimates the entire backbone curve, so the estimated stiffness parameters are comprehensively affected by the deviation of each amplitude regime.

For the damping coefficient, the four methods seem to achieve similar results. The estimated solutions of the logarithmic envelope method are all less than the actual value because it uses the analytical solution of the linear system. But this method is simple and easy to estimate for weakly damped systems. The error of the damping force curve method is small. The derivatives of the envelope, instantaneous frequency, and displacement should be obtained for the damping force curve method, so the result is disturbed by noise easily. A proper filter can be used to deal with the influence of the noise, and the Bayesian approach is a good way to measure the uncertainty of the identification results [29].

5. Experiment

5.1. Experimental Description

The test rig is shown in Figure 6, where Figure 6a is the photo of the test rig, and Figure 6b,c are the elevation and plan views of the nonlinear vibration absorber. The vibration absorber consists of a 4.86 g mass attached to a thin circle brass plate of 0.15 mm thickness. Figure 6c also shows the contour plot for the measured geometric imperfections obtained by moving the laser sensor through the translation surface of the plate. It can be seen that the plate is not flat, and has a certain initial deflection, as shown in Figure 6b.

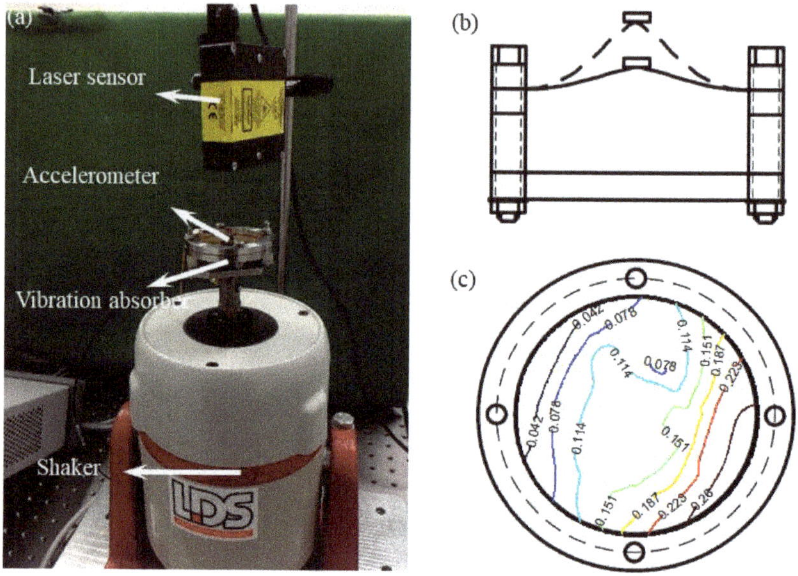

Figure 6. Test rig and schematic model of the vibration absorber. (**a**) Photo of the test rig, (**b**) elevation view, (**c**) Plan view and contour plot for the measured geometric imperfections. Relative deviations are in millimeters.

The excitation signal was generated by the LDS V406 shaker, then measured by the B&K 4517 accelerometer and Microtrak™ 3 LTS-050-10 laser sensor, respectively. The sampling frequency was 2000 Hz. Because the linear natural frequency of the vibration absorber was much higher than the natural frequency of the shaker, the absorber and the shaker can be regarded as a single-degree-of-freedom system [28]. The excitation signal supplied to the shaker was from Agilent 33512B signal generator and passed through an LDS PA500L power amplifier. The measured signals were sampled by a NI PXIe-4492 acquisition system after passing through a B&K 2693 conditioner. The equation of motion of the plate with geometric imperfections can be simplified as a Helmholtz–Duffing oscillator, as discussed in [4,28]. Therefore, the mathematical model of the free-decay response of the nonlinear vibration absorber is given by

$$m_{\text{EQ}}\ddot{x} + c\dot{x} + k_1 x + k_2 x^2 + k_3 x^3 = 0 \tag{18}$$

where $m_{\text{EQ}} = (m_s + m_v)(m_a + m)/(m_s + m_v + m_a + m) = 5.44$ g is the equivalent mass of the system. m_s, m_v, m_a, and m are the mass of armature, support structure of the absorber, accelerometer, and absorber mass, respectively. x is the vibration response of the experimental system. Damping c, linear stiffness k_1, quadratic and cubic nonlinear stiffness k_2 and k_3 are the parameters to be estimated.

5.2. Estimation and Discussion

Before the free decay experiment, several slow frequency sweep experiments from low to high frequency were carried out to determine the system's jump-down frequency, which was about 201 Hz. Therefore, the excitation signal was switched off at 200 Hz. The circular plate exhibited large stiffness nonlinearity, and the vibration modes of the system were well-separated. In order to exclude the influence of the higher harmonics and high order modes, the displacement of the mass measured by the laser sensor was passed through a low pass filter with a cut-off frequency of 270 Hz, as shown in Figure 7a. The specific experimental instruments and testing procedures were described in [42]. The data used in identification is approximately taken from 0.205 to 0.37 s.

Figure 7b shows the restoring force curve. Because the response decayed fast, cubic spline interpolation was used to construct envelope and frequency at each time point for the zero-crossing method. The first several large-amplitude points were removed due to the deviation. The restoring force was estimated by using Equation (3). The estimated stiffness parameters are shown in Table 3. The estimated results of the zero-crossing and direct quadrature methods are close to each other.

Table 3. Estimated stiffness parameters of the experimental system.

	k_1 (N/m)		k_2 (N/m^2)		k_3 (N/m^3)	
	RFCM	**BCM**	**RFCM**	**BCM**	**RFCM**	**BCM**
HT	4837.7	4272.9	-2.252×10^7	-2.155×10^7	5.705×10^{10}	6.346×10^{10}
ZC	5452.3	4265.7	-2.461×10^7	-2.109×10^7	4.822×10^{10}	6.302×10^{10}
DQ	5597.5	4949.9	-2.193×10^7	-2.776×10^7	5.123×10^{10}	8.614×10^{10}
WT		4137.9		-2.152×10^7		7.096×10^{10}

Figure 7c shows the backbone curve. The backbone curve obtained using the Hilbert transform is compared well with the backbone curve obtained using the zero-crossing method. The backbone curve obtained by using the wavelet transform shows a slight deviation from the above two methods, while the backbone curve obtained by using the direct quadrature method shows the most significant deviation. The estimated stiffness parameters are shown in Table 3. It can be seen that, except for the direct quadrature method, the estimated results of other methods are compared well. The identification results of the Hilbert transform can be used to reasonably reconstruct the experimental system's restoring force and backbone curves. The stiffness parameters are also estimated well by the zero crossing method and wavelet transform using the backbone curve.

The natural logarithm of the envelope and the damping force curve are shown in Figure 7d,e. The results of the four methods are similar, as shown in Table 4. It can be seen that the damping coefficient estimated by the logarithmic envelope method is less than that estimated by the damping force curve method. The logarithmic envelope method uses the analytical solution of the envelope of the linear system. However, the damping of the experimental system is not strictly linear. It can be noticed from the logarithmic envelope and damping force curve that the damping factor is amplitude dependant and decreases with time.

Table 4. Estimated damping coefficient of the experimental system.

c (Ns/m)	LEM	DFCM
HT	0.234	0.320
ZC	0.232	0.317
DQ	0.232	0.325
WT	0.230	0.311

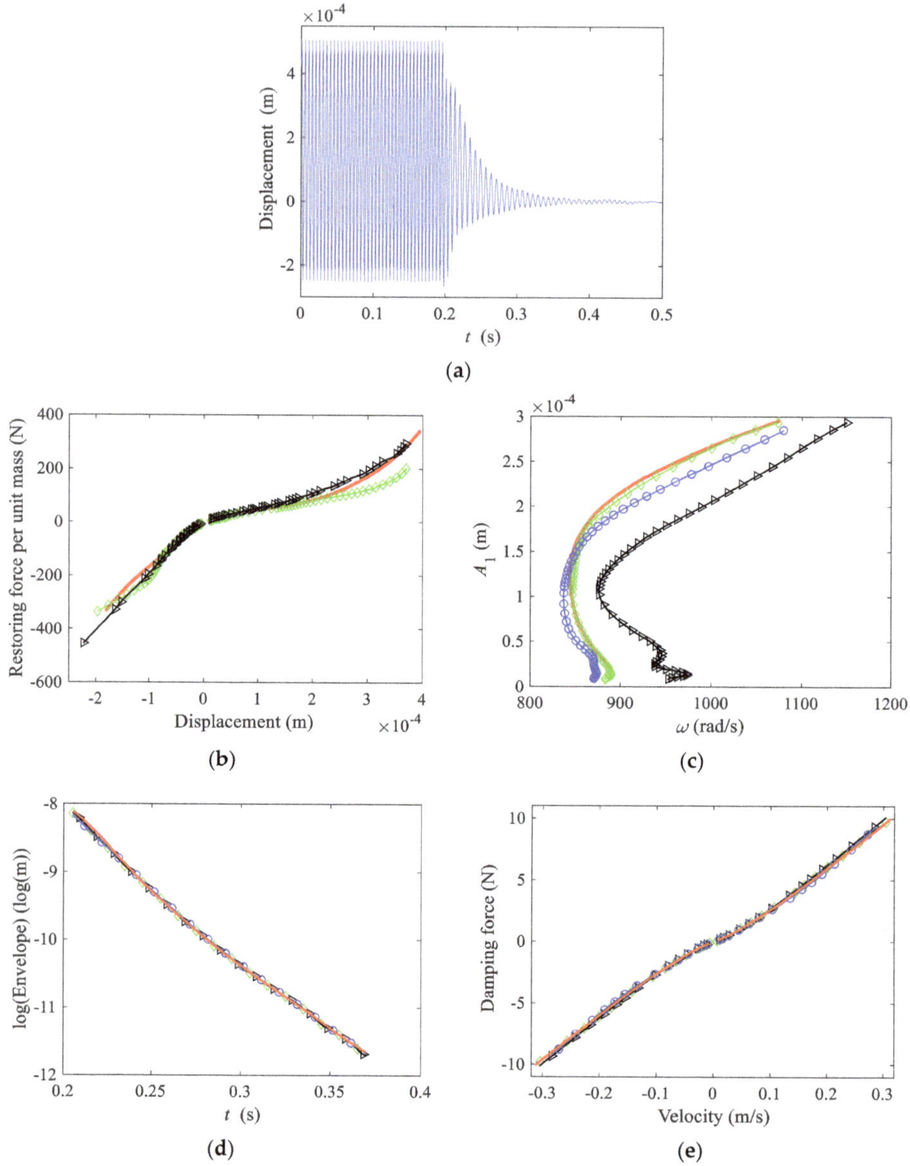

Figure 7. Measured (**a**) time domain response, (**b**) restoring force curve, (**c**) backbone curve, (**d**) logarithmic form of the envelope, (**e**) damping force curve. Thin solid line, time domain response; thick solid lines, Hilbert transform; lines with diamonds, zero-crossing; lines with triangles, direct quadrature; lines with circles, wavelet transform.

To determine whether the identification procedure is successful, a reconstructed free decay response is compared with the measured response as shown in Figure 7a. The approximate solution of Equation (18) is given by

$$x(t) = A_0(t) + A_1(t)\cos[\phi(t)] \tag{19}$$

where $A_0(t)$ is the time-dependent amplitude (or envelope) of bias term, $A_1(t)$ is the time-dependent amplitude of the first harmonic term, and $\phi(t)$ is the time-dependent phase. The time-dependent phase is obtained by integrating the time-dependent damped natural frequency

$$\phi(t) = \int_0^t \omega(t)dt \tag{20}$$

The time-dependent damped natural frequency, namely the backbone curve, is given by Equation (4). The parameters obtained by the Hilbert transform method using the backbone curve and the measured envelopes are used in reconstructing the free decay response. The results are shown in Figure 8a.

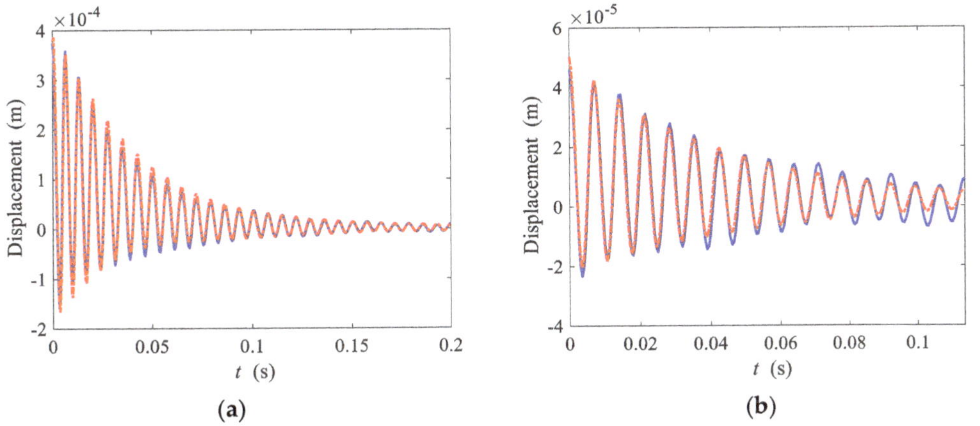

Figure 8. Experimental and reconstructed time domain response for (**a**) large initial displacement, and (**b**) small initial displacement. Solid lines, measured response; dashed-dotted lines, reconstructed response.

For the small amplitude regime, using the multiple-scales method [11], the time domain response is approximately given by

$$x(t) = A_0(t) + A_1 e^{-\zeta\omega_n t} \cos\left[\omega_n t + \frac{3}{16\zeta}\frac{k_3}{k_1}A_1^2(1 - e^{-2\zeta\omega_n t})\right] \tag{21}$$

where $A_0(t)$ is the measured envelope of bias term, A_1 is the initial amplitude of the first harmonic term. $\omega_n = \sqrt{k_1/m_{\mathrm{EQ}}}$ and $\zeta = c/2\sqrt{m_{\mathrm{EQ}}k_1}$ are the undamped natural frequency and the damping ratio of the underlying linear system respectively. Damping obtained by the logarithmic envelope method is used in reconstructing the time response. The results are shown in Figure 8b. It can be seen that the reconstructed responses obtained by the two reconstruction methods are compared well with the measured responses. The reconstruction method is described in detail in [28], which is applied to identify a Duffing type nonlinear vibration absorber.

6. Conclusions

This paper compared four identification methods for the stiffness and damping parameters of asymmetric systems with square and cubic nonlinearities. We verified the feasibility and accuracy of these methods by a Helmholtz–Duffing numerical example and a nonlinear vibration absorber with geometric imperfections. Hilbert vibration decomposition decomposes the asymmetric signal to obtain the backbone curve of each harmonic term. The asymmetric restoring force curve is constructed by the positive and negative congruent envelopes and frequencies obtained by using a weighted summing of the decomposed harmonic components. The obtained curves compare well with the analytical solution, but the disadvantage of the Hilbert transform is the end effect. Zero-crossing with the peak picking method extracts positive and negative peak points and obtains instantaneous frequency more simply. The bias term and first harmonic term are obtained by using the sum and difference of the positive and negative envelopes. The direct quadrature method also analyzes the positive and negative signal parts separately and uses the same method as the zero-crossing method to obtain the bias term and the first harmonic term. The two methods are more sensitive to noise. It is essential for the wavelet transform to select wavelet function appropriately.

The nonlinear stiffness parameters estimated are not accurate in the numerical example when the restoring force curve is obtained using the zero-crossing method and direct quadrature method. The identification errors for nonlinear stiffness parameters are about 20%. The other methods can estimate the stiffness parameters accurately, and the identification errors are less than 5%. For the experimental system, the identification results of the Hilbert transform can describe the experimental system. The stiffness parameters estimated by the restoring force curve using the zero-crossing method and the direct quadrature method are close to each other. The backbone curve obtained by using the direct quadrature method deviates from those obtained by the other methods.

The logarithmic envelope can be used to estimate damping. The estimated results are the same, and all are less than the actual values. The other tool is the damping force curve, which is constructed by the damping factor and envelope of the velocity. The estimated results are better than those of the logarithmic envelope. However, the damping force curve needs the derivative of the envelope, instantaneous frequency, and displacement, so it has poor robustness against noise. It can be seen from the logarithmic envelope and damping force curve that the damping of the experimental system is not strictly linear.

This paper is to identify the parameters of a predetermined system model. Restoring the force curve method and damping force curve method are nonparametric identification methods. They produce the best functional representation of the system without a priori assumption about the system model. However, it is necessary for the backbone curve method to know the nonlinear system model and the theoretical solution of the backbone curve in advance. The logarithmic envelope method is suitable for lightly damped systems and is an approximate estimation method.

Author Contributions: S.W.: Methodology, Software, Data curation, Writing—original draft. B.T.: Conceptualization, Methodology, Investigation, Supervision, Funding acquisition, Writing—review & editing. All authors have read and agreed to the published version of the manuscript.

Funding: This research was funded by the National Natural Science Foundation of China (grant number 11672058).

Institutional Review Board Statement: No applicable.

Informed Consent Statement: No applicable.

Conflicts of Interest: The authors declare no conflict of interest.

References

1. Abolfathi, A.; Brennan, M.J.; Waters, T.P.; Tang, B. On the effects of mistuning a force excited system containing a quasi-zero-stiffness vibration isolator. *J. Vib. Acoust.* **2015**, *13*, 44502. [CrossRef]
2. Kovacic, I.; Brennan, M.J.; Lineton, B. Effect of a static force on the dynamic behaviour of a harmonically excited quasi-zero stiffness system. *J. Sound Vib.* **2009**, *325*, 870–883. [CrossRef]
3. Benedettini, F.; Rega, G. Nonlinear dynamics of an elastic cable under planar excitation. *Int. J. Non-Linear Mech.* **1987**, *22*, 497–509. [CrossRef]
4. Amabili, M. *Nonlinear Vibrations and Stability of Shells and Plates*; Cambridge University Press: New York, NY, USA, 2008. [CrossRef]
5. Amabili, M.; Carra, S. Thermal effects on geometrically nonlinear vibrations of rectangular plates with fixed edges. *J. Sound Vib.* **2009**, *321*, 936–954. [CrossRef]
6. Chia, C.Y. *Nonlinear Analysis of Plates*; McGraw-Hill: New York, NY, USA, 1980.
7. Chia, C.Y. Geometrically nonlinear behavior of composite plates: A review. *Appl. Mech. Rev.* **1988**, *41*, 439–451. [CrossRef]

8. Kerschen, G.; Worden, K.; Vakakis, A.F.; Golinval, J.C. Past, present and future of nonlinear system identification in structural dynamics. *Mech. Syst. Signal Process.* **2006**, *20*, 505–592. [CrossRef]
9. Noël, J.P.; Kerschen, G. Nonlinear system identification in structural dynamics: 10 more years of progress. *Mech. Syst. Signal Process.* **2017**, *83*, 2–35. [CrossRef]
10. Al-hababi, T.; Cao, M.S.; Saleh, B.; Alkayem, N.F.; Xu, H. A critical review of nonlinear damping identification in structural dynamics: Methods, applications, and challenges. *Sensors* **2020**, *20*, 7303. [CrossRef]
11. Nayfeh, A.H.; Mook, D.T. *Nonlinear Oscillations*; Willey: New York, NY, USA, 1995. [CrossRef]
12. Neild, S.A.; Wagg, D.J. Applying the method of normal forms to second-order nonlinear vibration problems. *Proc. R. Soc. A Math. Phys. Eng. Sci.* **2011**, *467*, 1141–1163. [CrossRef]
13. Neild, S.A.; Champneys, A.R.; Wagg, D.J.; Hill, T.L.; Cammarano, A. The use of normal forms for analysing nonlinear mechanical vibrations. *Philos. Trans. R. Soc. A* **2015**, *373*, 20140404. [CrossRef]
14. Elliott, A.J.; Cammarano, A.; Neild, S.A. Comparing analytical approximation methods with numerical results for nonlinear systems. In Proceedings of the Conference and Exposition on Structural Dynamics, Garden Grove, CA, USA, 30 January–2 February 2017; Volume 1, pp. 37–49. [CrossRef]
15. Kerschen, G.; Peeters, M.; Golinval, J.C.; Vakakis, A.F. Nonlinear normal modes, Part I: A useful framework for the structural dynamicist. *Mech. Syst. Signal Process.* **2009**, *23*, 170–194. [CrossRef]
16. Peeters, M.; Viguié, R.; Sérandour, G.; Kerschen, G.; Golinval, J.C. Nonlinear normal modes, Part II: Toward a practical computation using numerical continuation techniques. *Mech. Syst. Signal Process.* **2009**, *23*, 195–216. [CrossRef]
17. Renson, L.; Kerschen, G.; Cochelin, B. Numerical computation of nonlinear normal modes in mechanical engineering. *J. Sound Vib.* **2016**, *364*, 177–206. [CrossRef]
18. Naylor, S.; Platten, M.F.; Wright, J.R.; Cooper, J.E. Identification of multi-degree of freedom systems with nonproportional damping using the resonant decay method. *J. Vib. Acoust.* **2004**, *126*, 298–306. [CrossRef]
19. Renson, L.; Gonzalez-Buelga, A.; Barton, D.A.W.; Neild, S.A. Robust identification of backbone curves using control-based continuation. *J. Sound Vib.* **2016**, *367*, 145–158. [CrossRef]
20. Renson, L.; Shaw, A.D.; Barton, D.A.W.; Neild, S.A. Application of control-based continuation to a nonlinear structure with harmonically coupled modes. *Mech. Syst. Signal Process.* **2019**, *120*, 449–464. [CrossRef]
21. Denis, V.; Jossic, M.; Giraud-Audin, C.; Chomette, B.; Renault, A.; Thomas, O. Identification of nonlinear modes using phase-locked-loop experimental continuation and normal form. *Mech. Syst. Signal Process.* **2018**, *106*, 430–452. [CrossRef]
22. Peeters, M.; Kerschen, G.; Golinval, J.C. Dynamic testing of nonlinear vibrating structures using nonlinear normal modes. *J. Sound Vib.* **2011**, *330*, 486–509. [CrossRef]
23. Peeters, M.; Kerschen, G.; Golinval, J.C. Modal testing of nonlinear vibrating structures based on nonlinear normal modes: Experimental demonstration. *Mech. Syst. Signal Process.* **2011**, *25*, 1227–1247. [CrossRef]
24. Ehrhardt, D.A.; Allen, M.S. Measurement of nonlinear normal modes using multi-harmonic stepped force appropriation and free decay. *Mech. Syst. Signal Process.* **2016**, *76–77*, 612–633. [CrossRef]
25. Breunung, T.; Haller, G. Explicit backbone curves from spectral submanifolds of forced-damped nonlinear mechanical systems. *Proc. R. Soc. A* **2018**, *474*, 20180083. [CrossRef]
26. Cenedese, M.; Haller, G. Constructing backbone curves from free-decay vibrations data in multi-degrees of freedom oscillatory systems. In Proceedings of the Conference and Exposition on Structural Dynamics, Houston, TX, USA, 10–13 February 2020; Volume 1, pp. 221–223. [CrossRef]
27. Londoño, J.M.; Neild, S.A.; Cooper, J.E. Identification of backbone curves of nonlinear systems from resonance decay responses. *J. Sound Vib.* **2015**, *348*, 224–238. [CrossRef]
28. Tang, B.; Brennan, M.J.; Gatti, G.; Ferguson, N.S. Experimental characterization of a nonlinear vibration absorber using free vibration. *J. Sound Vib.* **2016**, *367*, 159–169. [CrossRef]
29. Tang, B.; Wang, S.B.; Brennan, M.J.; Feng, L.Y.; Chen, W.C. Identifying the stiffness and damping of a nonlinear system using its free response perturbed with Gaussian white noise. *J. Vib. Control.* **2020**, *26*, 830–839. [CrossRef]
30. Tang, B.; Brennan, M.J.; Lopes, V., Jr.; da Silva, S.; Ramlan, R. Using nonlinear jumps to estimate cubic stiffness nonlinearity: An experimental study. *Proc. Inst. Mech. Eng. Part C* **2016**, *230*, 3575–3581. [CrossRef]
31. Londoño, J.M.; Cooper, J.E.; Neild, S.A. Identification of systems containing nonlinear stiffnesses using backbone curves. *Mech. Syst. Signal Process.* **2017**, *84*, 116–135. [CrossRef]
32. Zhang, X.H.; Zhang, C.; Wang, L.; Chen, L.Y.; Xu, D.M.; Fan, H.W.; Zhu, F.L. A method for parameter identification of composite beam piezoelectric energy harvester. *Sensors* **2021**, *21*, 7213. [CrossRef]
33. Civera, M.; Fragonara, L.Z.; Surace, C. Using Video processing for the full-field identification of backbone curves in case of large vibrations. *Sensors* **2019**, *19*, 2345. [CrossRef]
34. Masri, S.F.; Caughey, T.K. A nonparametric identification technique for nonlinear dynamic problems. *J. Appl. Mech. Trans. ASME* **1979**, *46*, 433–447. [CrossRef]
35. Feldman, M. Nonparametric identification of asymmetric nonlinear vibration systems with the Hilbert transform. *J. Sound Vib.* **2012**, *331*, 3386–3396. [CrossRef]
36. Ondra, V.; Sever, I.A.; Schwingshackl, C.W. A method for non-parametric identification of non-linear vibration systems with asymmetric restoring forces from a resonant decay response. *Mech. Syst. Signal Process.* **2019**, *114*, 239–258. [CrossRef]

37. Feldman, M. Hilbert transform in vibration analysis. *Mech. Syst. Signal Process.* **2011**, *25*, 735–802. [CrossRef]
38. Huang, N.E.; Wu, Z.H.; Long, S.R.; Arnold, K.C.; Chen, X.Y.; Blank, K. On instantaneous frequency. *Adv. Adapt. Data Anal.* **2009**, *1*, 177–229. [CrossRef]
39. Staszewski, W.J. Identification of non-linear systems using multi-scale ridges and skeletons of the wavelet transform. *J. Sound Vib.* **1998**, *214*, 639–658. [CrossRef]
40. Ondra, V.; Riethmueller, R.; Brake, M.R.W.; Schwingshackl, C.W.; Polunin, P.M.; Shaw, S.W. Comparison of nonlinear system identification methods for free decay measurements with application to MEMS devices. In Proceedings of the Conference and Exposition on Structural Dynamics, Garden Grove, CA, USA, 30 January–2 February 2017; Volume 5, pp. 29–46. [CrossRef]
41. Jin, M.S.; Brake, M.R.W.; Song, H.W. Comparison of nonlinear system identification methods for free decay measurements with application to jointed structures. *J. Sound Vib.* **2019**, *453*, 268–293. [CrossRef]
42. Wang, S.B.; Tang, B. Estimating quadratic and cubic stiffness nonlinearity of a nonlinear vibration absorber with geometric imperfections. *Measurement* **2021**, *185*, 110005. [CrossRef]
43. Feldman, M. Non-linear system vibration analysis using Hilbert transform-I. Free vibration analysis method 'Freevib'. *Mech. Syst. Signal Process.* **1994**, *8*, 119–127. [CrossRef]
44. Feldman, M. Non-linear system vibration analysis using Hilbert transform-II. Forced vibration analysis method 'Forcevib'. *Mech. Syst. Signal Process.* **1994**, *8*, 309–318. [CrossRef]
45. Feldman, M. Time-varying vibration decomposition and analysis based on the Hilbert transform. *J. Sound Vib.* **2006**, *295*, 518–530. [CrossRef]
46. Feldman, M. *Hilbert Transform. Applications in Mechanical Vibration*; Wiley: Chichester, UK, 2011. [CrossRef]
47. Bedrosian, E. A product theorem for Hilbert transforms. *Proc. IEEE* **1963**, *51*, 868–869. [CrossRef]
48. Nuttall, A.H.; Bedrosian, E. On the quadrature approximation to the Hilbert transform of modulated signals. *Proc. IEEE* **1966**, *54*, 1458–1459. [CrossRef]
49. Staszewski, W.J. Analysis of non-linear systems using wavelets. *Proc. Inst. Mech. Eng. Part C* **2000**, *214*, 1339–1353. [CrossRef]
50. Le, T.P.; Argoul, P. Continuous wavelet transform for modal identification using free decay response. *J. Sound Vib.* **2004**, *277*, 73–100. [CrossRef]
51. Ta, M.N.; Lardis, J. Identification of weak nonlinearities on damping and stiffness by the continuous wavelet transform. *J. Sound Vib.* **2006**, *293*, 16–37. [CrossRef]

applied sciences

MDPI

Article

MEMS High Aspect Ratio Trench Three-Dimensional Measurement Using Through-Focus Scanning Optical Microscopy and Deep Learning Method

Guannan Li [1], Junkai Shi [1,*], Chao Gao [1,2], Xingjian Jiang [1], Shuchun Huo [1], Chengjun Cui [1], Xiaomei Chen [1] and Weihu Zhou [1,2]

1 Institute of Microelectronics of Chinese Academy of Sciences, Beijing 100029, China
2 University of Chinese Academy of Sciences, Beijing 100049, China
* Correspondence: shijunkai@ime.ac.cn

Abstract: High-aspect-ratio structures have become increasingly important in MEMS devices. In situ, real-time critical dimension and depth measurement for high-aspect-ratio structures is critical for optimizing the deep etching process. Through-focus scanning optical microscopy (TSOM) is a high-throughput and inexpensive optical measurement method for critical dimension and depth measurement. Thus far, TSOM has only been used to measure targets with dimension of 1 μm or less, which is far from sufficient for MEMS. Deep learning is a powerful tool that improves the TSOM performance by taking advantage of additional intensity information. In this work, we propose a convolutional neural network model-based TSOM method for measuring individual high-aspect-ratio trenches on silicon with width up to 30 μm and depth up to 440 μm. Experimental demonstrations are conducted and the results show that the proposed method is suitable for measuring the width and depth of high-aspect-ratio trenches with a standard deviation and error of approximately a hundred nanometers or less. The proposed method can be applied to the semiconductor field.

Keywords: optical micro-nano metrology; three-dimensional non-destructive measurement; high-aspect-ratio structure; through-focus scanning optical microscopy; convolutional neural network

Citation: Li, G.; Shi, J.; Gao, C.; Jiang, X.; Huo, S.; Cui, C.; Chen, X.; Zhou, W. MEMS High Aspect Ratio Trench Three-Dimensional Measurement Using Through-Focus Scanning Optical Microscopy and Deep Learning Method. *Appl. Sci.* **2022**, *12*, 8396. https://doi.org/10.3390/app12178396

Academic Editor: Andrea Li Bassi

Received: 26 July 2022
Accepted: 22 August 2022
Published: 23 August 2022

1. Introduction

Micro-electro-mechanical systems (MEMS) comprise both mechanical and electronic components manufactured on a common silicon substrate, and have gained significant attention in the last three decades. Due to the miniaturized size, low cost, high manufacturing throughput, and expectance of superior performance, MEMS are used in a wide range of consumer, automotive, medical, and industrial products [1,2]. With the increase in the integration density and the reduction in the feature size, high-aspect-ratio (HAR) structures in silicon play an increasingly important role in micro-sensors and actuators [3,4]. Most MEMS HAR structures are fabricated by the dry etching process, especially deep reactive-ion etch (DRIE) with the Bosch process [5,6]. The precisely controlled profile of HAR structures is required for better device performance and higher resolution and reliability [7]. Due to aspect-ratio-dependent etching and microloading effects, there is a complex interplay during etching between the device design (topology dimensions and shape) and etch parameters [8]. Therefore, in situ, real-time critical dimension and depth measurement for HAR structures is critical for optimizing the Bosch deep etching process.

To date, various methods have been used to measure the HAR structures of MEMS [9,10]. Scanning electron microscopy (SEM) is a commonly used tool with high resolution and high magnification, but it is costly, inefficient, and not suitable for online real-time detection. Optical methods, such as scatterometry and infrared reflectometry, are reported to measure repeated structures with high throughput, but are limited in their ability for the dimensional analysis of isolated HAR structures. Moreover, a silicon oxide film usually exists in etched MEMS device, which makes it more problematic for SEM and scatterometry. Through-focus scanning optical microscopy (TSOM) is a novel fast and low-cost optical measurement method with nanometer sensitivity, and is achieved with bright-field optical microscopy [11]. The conventional TSOM method is a model-based metrology method analyzing the scattering intensity reflected from the target and the intensity change as the target performs through-focus scans along the optical axis. The dimensional information of the target is extracted by matching the intensity information with the simulation results in the database. For improving the performance, great attention has been paid to minimizing the dissimilarities between the intensity information acquisition and simulation conditions [12–19]. In the meantime, model-based TSOM utilizes only the optical intensity range and the mean-square intensity information of the TSOM image; hence, large amounts of intensity distribution information correlated to the dimensional information of the targets is ignored. To date, TSOM has only been used to measure targets with critical dimensions and depth of 1 μm or less, which is far from sufficient for MEMS.

Deep learning (DL) is a powerful artificial intelligence technology and has made important progress in pattern recognition, computer vision, automatic speech recognition, natural language processing, and other fields. Training DL algorithms directly with measurement data is a new approach to TSOM, an alternative to establishing a simulation database. It improves the TSOM performance by taking advantage of additional intensity information and avoiding the adverse effect of the dissimilarities between information acquisition and simulation conditions [20]. A convolutional neural network (CNN), as one of the most representative DL algorithms, can automatically learn the deep correlations between the scattering intensity information and dimensional information of the targets [21], and is often used in the field of image processing.

In this paper, we describe a CNN learning model-based TSOM method for measuring the individual HAR trench on a silicon substrate with width up to 30 μm and depth up to 440 μm. The TSOM image is mapped to a normalized image of appropriate size, and the width and depth measured by SEM are taken as labels for iterative training to obtain a DL regression model that is capable of accurately predicting the width and depth. The well-trained model can predict the width and depth of the HAR trench on the silicon substrate with a silicon oxide film, and the standard deviation and error of a few tens of nanometers are achieved. These results indicate that the DL-based TSOM method has great application prospects for the micro-nano three-dimensional measuring field.

The structure of the rest of this paper is as follows: the second section introduces the materials and methods, including the experimental setup, the acquisition of the dataset, and the structure of the CNN model. The third section introduces our experimental results and discussion. The fifth section introduces our conclusions.

2. Materials and Methods

2.1. TSOM Setup

A schematic of the TSOM setup is shown in Figure 1a. The TSOM setup is mainly composed of a commercial bright-field optical microscope with a Kohler illumination system (Shanghai Guangmi, Shanghai, China). A halogen lamp is used as the light source, which has an illumination peak wavelength of 589 nm with a 3 dB spectral bandwidth of 160 nm. A Piezoelectric Transducer (PZT, Coremorrow, Harbin, China) is placed under the sample stage to achieve the through-focus scanning. The scattered light intensity distribution information is acquired by a charge-coupled device (CCD) camera (Thorlabs, Newton, NJ, USA) with a resolution of 1280 px × 1024 px and pixel size of 4.65 μm × 4.65 μm. Figure 1b provides a photograph of the TSOM setup.

Figure 2 shows the MEMS trench section diagram and its geometric parameters that require metrology. The HAR trench samples are etched on the silicon substrate by DRIE with the Bosch process, and the silicon oxide film is retained. Owing to the characteristics of DRIE, there is a width difference between the silicon oxide film and the silicon layer. In this paper, width is defined as the width at the 50% thickness position of the silicon oxide film, and depth is defined as the distance from the surface of the silicon oxide film to the bottom of the HAR trench. With the prepared TSOM setup, the width and depth of the trench can be obtained easily.

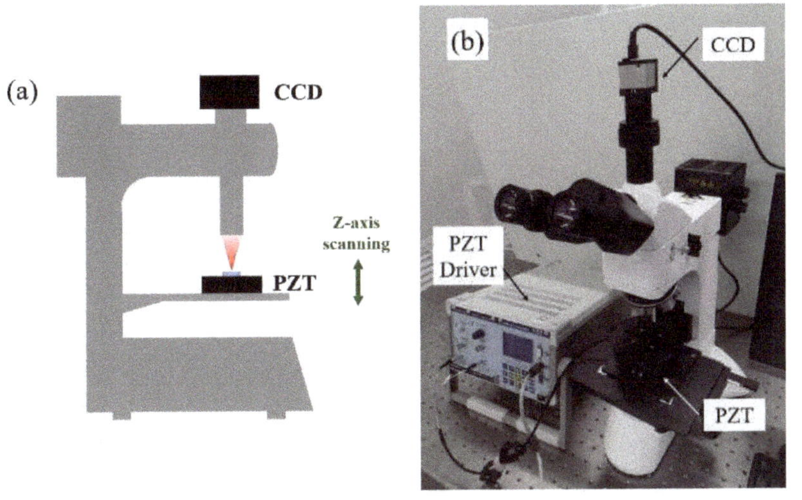

Figure 1. (**a**) Schematic and (**b**) photograph of TSOM setup. CCD, charge-coupled device; PZT, Piezoelectric Transducer.

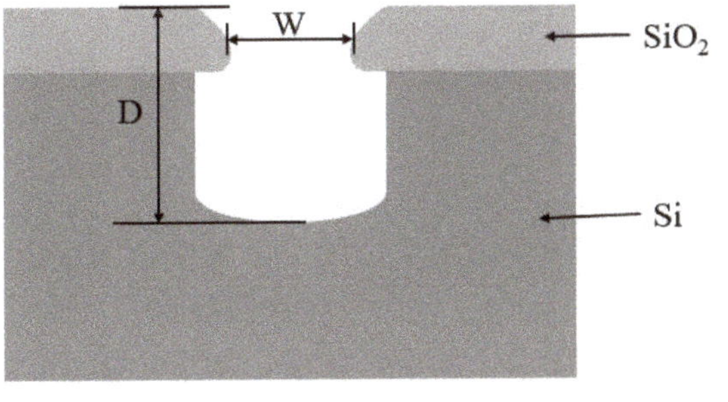

Figure 2. MEMS trench section diagram and its geometric parameters. W, width; D, depth.

2.2. The Dataset

The traditional metrology method based on optical microscopy utilizes only the focal plane image, and the nonfocal plane images are discarded as useless information. For geometric dimensions extracted from the focal plane image, the measurement accuracy is limited by the optical resolution and focal depth of the microscope. The light intensity information out of the focal plane is also related to the sample shape, and reflects the geometric characteristics on some level. Figure 3 shows the process to construct a TSOM image. The TSOM setup captures a series of images as the sample scans through the focus of the microscope along the optical axis. The image sequence consists of an in-focus image and a series of defocused images of the sample. The acquired optical images are stacked at their corresponding scan positions, and there is a slight difference between any two adjacent images. The optical intensities in a vertical plane passing through the trench on the target can be easily extracted to construct a TSOM image. The X axes, Y axes, and grayscale represent the spatial position, the focus position, and the optical intensity, respectively. Figure 4 shows the SEM images of trench sections with different widths and depths. The insets are the corresponding TSOM images, which show obvious differences in intensity distribution. The width and depth can be extracted by establishing the one-to-one correspondence between geometric characteristics and TSOM images. Therefore, TSOM can break the limits of the optical resolution and focal depth.

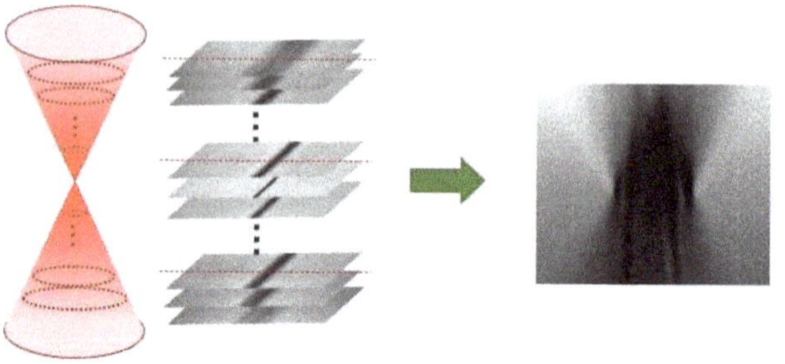

Figure 3. Process to construct a TSOM image.

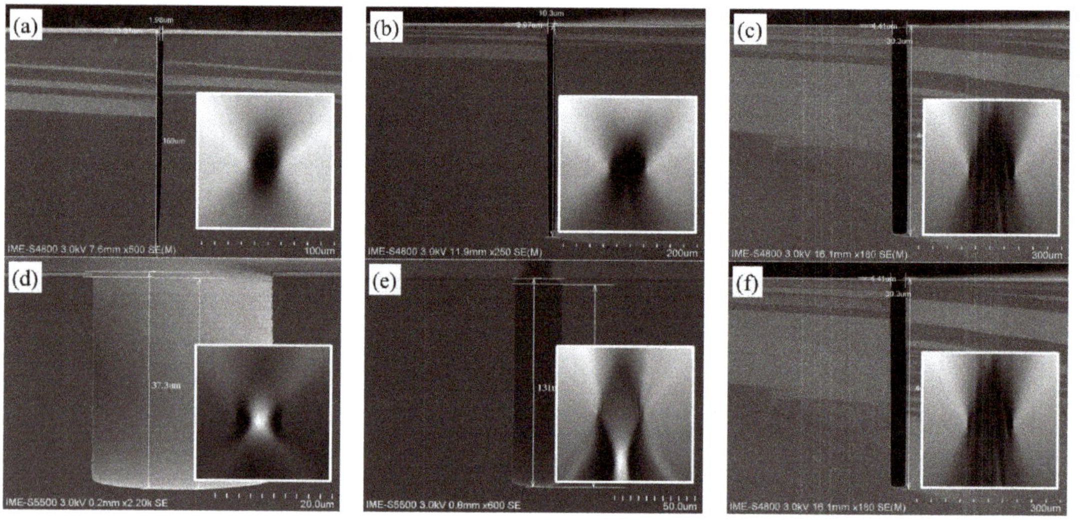

Figure 4. SEM images of MEMS HAR trench sections with width and depth of (**a**) 1.98 μm and 169 μm, (**b**) 10.3 μm and 321 μm, (**c**) 30.3 μm and 443 μm, (**d**) 30.3 μm and 37.3 μm, (**e**) 30.3μm and 131 μm, (**f**) 30.3 μm and 443 μm (inset: corresponding TSOM images).

For the samples, each trench is 5 mm long. When collecting the TSOM image, 10 positions are randomly selected over each trench, as shown in Figure 5, and the part of the trench that is contaminated or close to the ends should be avoided. Each position is scanned through-focus to obtain a series of images. The scanning range is 250 μm with a step of 2 μm. Multiple lines are selected over each image with equal space, and we construct multiple TSOM images.

For the HAR trench etch, the reactive-ion etch lag leads to differences in trench depths with different trench widths [6]. For example, the depths of 2 μm, 10 μm, and 30 μm wide trenches etched simultaneously are 169 μm, 321 μm, and 443 μm, respectively. It is difficult to obtain trenches of the same depth with different widths in a large range (e.g., 2 μm~30 μm). To alleviate the impact of the depth difference, the trench sets for width prediction consist of 4 trenches with widths around 2 μm, 10 μm, and 30 μm, respectively, and the corresponding depths are approximately 169 μm, 321 μm, and 443 μm. The depth prediction trench sets consist of 5 trenches with widths of 1.93 μm, 10.3 μm, and 30.3 μm, respectively. The corresponding depth ranges are from 24 μm to 169 μm, 34 μm to 321 μm, and 37 μm to 443 μm. The highest aspect ratios are approximately 88:1, 31:1, and 15:1, respectively.

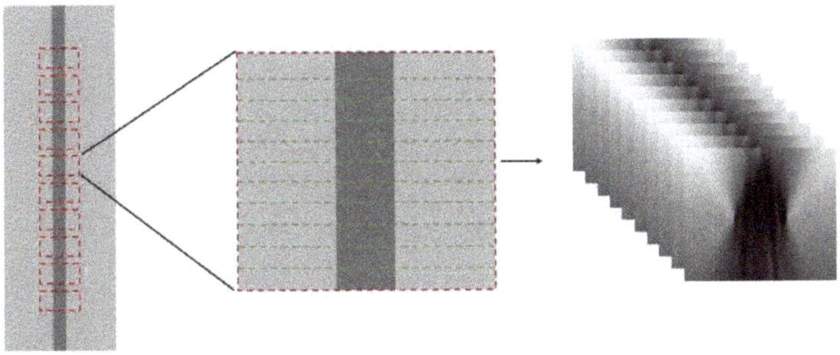

Figure 5. Process to collect TSOM image dataset.

2.3. The Structure of the CNN

To reveal the correspondence between geometric characteristics and TSOM images, we resize the TSOM image to 256 × 64 pixels, and then construct a suitable CNN model for training. CNN models usually consist of a convolution layer, pooling layer, activation layer, and full connection layer. An improved CNN model is designed based on the traditional CNN model, as shown in Figure 6. The TSOM image is a grayscale image, so the single-channel image can be convolved. There are 5 convolution layers (including two 3 × 5 kernels and three 3 × 3 kernels), 4 pooling layers (being applied to the first four convolution layers), and 4 full connection layers in the CNN model. All convolution layers contain zero padding. The ReLU function is used as the activation function, and the maximum pooling function is used as the pooling layer. The parameters of the CNN model can be found in Figure 6. The mean-square loss is used as the final loss function to calculate the mean-square error between the output values and the actual values. The loss function equation is:

$$loss(x_i, y_i) = (x_i - y_i)^2 \quad (1)$$

where x_i indicates the true value of the target and y_i indicates the predicted value of the CNN model.

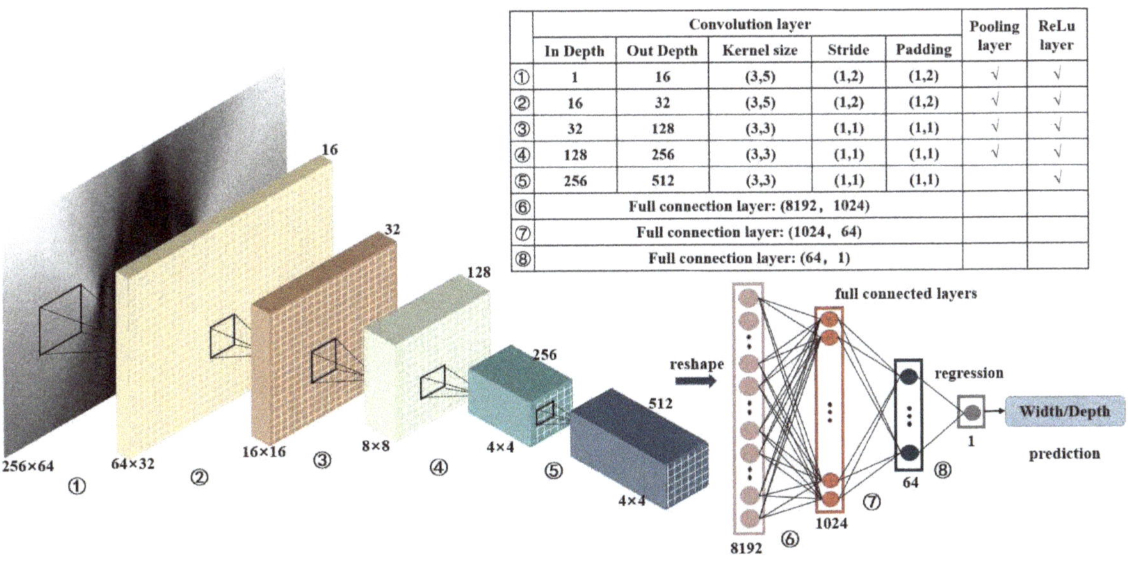

	Convolution layer					Pooling layer	ReLu layer
	In Depth	Out Depth	Kernel size	Stride	Padding		
①	1	16	(3,5)	(1,2)	(1,2)	√	√
②	16	32	(3,5)	(1,2)	(1,2)	√	√
③	32	128	(3,3)	(1,1)	(1,1)	√	√
④	128	256	(3,3)	(1,1)	(1,1)	√	√
⑤	256	512	(3,3)	(1,1)	(1,1)		√
⑥	Full connection layer: (8192, 1024)						
⑦	Full connection layer: (1024, 64)						
⑧	Full connection layer: (64, 1)						

Figure 6. Schematic diagram of improved CNN model.

The training image set and test image set contain 3000 and 200 TSOM images, respectively, which are generated from one trench set. SEM is used to measure the width and depth by the cross-section of the trench center position, and the SEM result is taken as labels. The training data set and SEM results are applied to train the CNN model for learning the correspondence of TSOM images and trench parameters. The trained model is evaluated by the test data set.

3. Results and Discussion

To verify the parameter prediction performance of the trained CNN model, the position close to the trench center was through-focus scanned continually 10 times. The center lines of a series of microscope images were selected to construct prediction TSOM image sets, and the average of 10 measurement results was considered as the predicted value.

Figure 7a–c show a comparison between the true (black triangles) and predicted (red circles) widths of HAR trenches with widths of around 2 μm, 10 μm, and 30 μm, respectively. Figure 7e shows a comparison between the actual (black triangles) and predicted (red circles) depths of HAR trenches with widths of 1.93 μm, 10.3 μm, and 30.3 μm, respectively. The comparison results show that the predicted value and true value match well. In order to further analyze the results shown in Figure 4, the standard deviation and error were used to evaluate the prediction performance. The definitions of the standard deviation and error are given in Equations (2) and (3), respectively:

$$SD = \sqrt{\frac{1}{n-1}\sum_{i=1}^{n}\left(f_i - \overline{f}\right)^2} \tag{2}$$

$$E = \left|\overline{f} - f_t\right| \tag{3}$$

where f_i is the predicted value, $\overline{f}$ is the average of the predicted values, f_t is the true value, and n is the number of measurements, which is 10 in this paper.

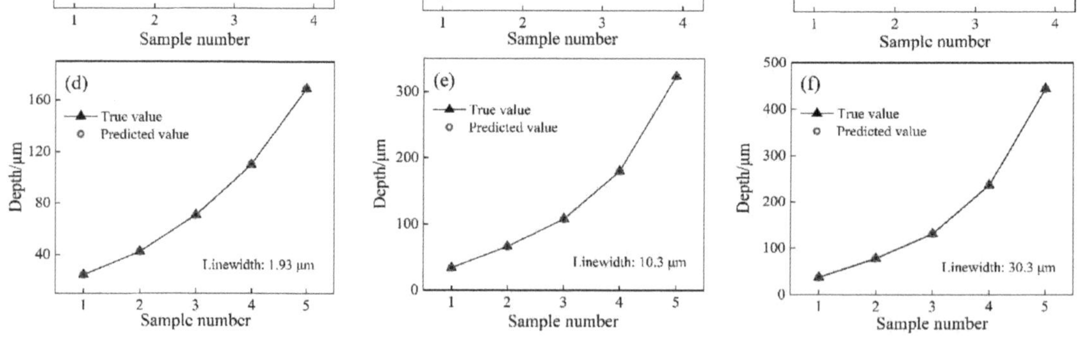

Figure 7. Width and depth prediction results. Comparison between the actual (black triangles) and predicted (red circles) widths of HAR trenches with widths of around (**a**) 2 μm, (**b**) 10 μm, and (**c**) 30 μm; comparison between the actual (black triangles) and predicted (red circles) depths of HAR trenches with widths of (**d**) 1.93 μm, (**e**) 10.3 μm, and (**f**) 30.3 μm.

The standard deviation and error of the width prediction results are shown in Figure 8. As a comparison, the results predicted by the traditional machine learning (ML) method demonstrated in Reference [22] are shown in order to verify the superiority of the CNN model. The adopted ML model is a support vector regression (SVR) model combined with a Histogram of Oriented Gradients (HOG) feature extractor [23–25]. The three trench sets for width prediction were irregularly spaced, with width intervals of a few hundred nanometers around 2 μm, 10 μm, and 30 μm. The standard deviation and error were less than 10 nm for 2 μm and 10 μm wide trenches. The corresponding ML-predicted results were 40 nm and 180 nm for 2 μm and 10 μm wide trenches, which were much higher than the DL-predicted results, especially for 10 μm wide trenches. For 30 μm wide trenches, the standard deviation and error were significantly increased, but still less than 60 nm and 80 nm, respectively. The increase in the standard deviation and error was mainly due to the etching error, which increased as the width increased. Moreover, the depth of the 30 μm wide trench was 443 μm. A large scanning range was necessary for a deep trench to obtain the complete TSOM image. Incomplete TSOM images also led to poorer performance. The corresponding ML-predicted results were 170 nm and 250 nm, which were approximately three times the DL-predicted results. The comparison results confirmed the superior prediction ability of the DL model for width prediction.

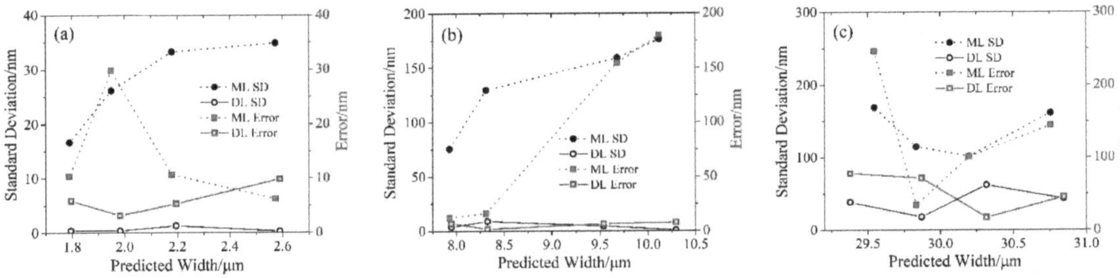

Figure 8. Standard deviation and error of width results for widths around (**a**) 2 μm, (**b**) 10 μm, and (**c**) 30 μm, predicted by machine learning model (dot line) and deep learning model (solid line). ML, machine learning; DL, deep learning; SD, standard deviation.

Figure 9 illustrates the standard deviation and error of the depth prediction results. The 1.93 μm, 10.3 μm, and 30.3 μm wide trench sets for depth prediction were also irregularly spaced. As the depth increased, the depth interval also increased from a few tens of micrometers to more than 200 μm. The standard deviation and error were 110 nm and 360 nm for 1.93 μm wide trenches. For 10.3 μm wide trenches, the results were 90 nm and 410 nm. The corresponding ML-predicted results were around a few micrometers, and were approximately an order of magnitude higher than the DL-predicted results. For 30.3 μm wide trenches, the standard deviation and error were 280 nm and 300 nm, respectively. The corresponding ML-predicted results were 2.4 μm and 850 nm, which were much higher than the DL-predicted results.

For ML depth prediction results, the predicted performance was degraded obviously as the trench depth increased or width decreased. This can be understood by the fact that less light reached the bottom and carried the depth information for narrow or deep trenches. Less depth information collected resulted in a larger standard deviation and greater error. Moreover, the scanning range was much smaller than the depth of the trench. This resulted in incomplete TSOM images, which could lead to poorer performance. However, the DL depth prediction performance did not change obviously as the trench depth or width changed. It was proven that the DL model extracted more information from less input data compared with the ML model. It was also noticed that the standard deviation and error of the depth prediction results were obviously higher than those of the width prediction results. This was mainly because the depth interval of the depth training trench set was much larger than the width interval of the width training trench set, which was only a few hundred nanometers. The prediction performance could be further improved the proper adoption of parameter intervals.

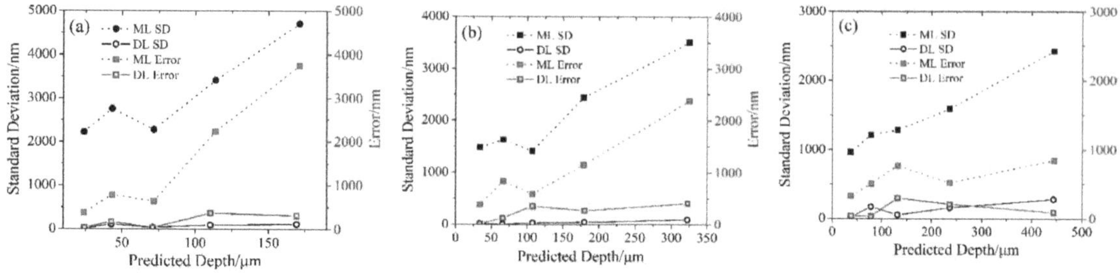

Figure 9. Standard deviation and error of depth results for widths of (**a**) 1.93 μm, (**b**) 10.3 μm, and (**c**) 30.3 μm, predicted by machine learning model (dot line) and deep learning model (solid line). ML, machine learning; DL, deep learning; SD, standard deviation.

Previously reported TSOM systems had achieved a nanometer measurement resolution, which was much higher than the result shown in this paper. This was mainly due to the accuracy of the SEM values. The SEM value was taken as the true value, and directly affected the prediction ability of the CNN model. In our study, the SEM results were rounded to the nearest 0.01 μm for dimensions less than 10 μm, 0.1 μm for dimensions from 10 μm to 100 μm, and 1 μm for dimensions greater than 100 μm, and this was not sufficient for dimensional measurement with nanometer accuracy. Moreover, the training trench set and trench set for SEM were etched in the same wafer, but with different positions, so etching errors also degraded the prediction ability of the CNN model.

4. Conclusions

In this work, HAR trenches with widths from 2 μm to 30 μm and depths from 20 μm to 440 μm were measured by a CNN model-based TSOM method. The TSOM image was mapped to a normalized image of appropriate size, and the width and depth measured by SEM were taken as labels for iterative training to obtain a DL regression model that was capable of accurately predicting the width and depth. The standard deviation and error were around a hundred nanometers or less. The superiority of the convolutional neural network model-based TSOM compared to machine learning-based TSOM was experimentally investigated and discussed. This method has the potential to achieve nanometer precision, and its application fields can be extended to integrated circuit metrology, such as through-silicon via.

Author Contributions: Conceptualization, J.S. and X.C.; methodology, G.L. and J.S.; software, G.L.; validation, G.L. and C.G.; formal analysis, S.H.; investigation, G.L.; resources, C.C.; data curation, G.L.; writing—original draft preparation, G.L. and J.S.; writing—review and editing, J.S. and X.C.; visualization, X.J.; supervision, W.Z.; project administration, X.C.; funding acquisition, W.Z. All authors have read and agreed to the published version of the manuscript.

Funding: This research was funded by the National Key Research and Development Program of China (No. 2019YFB2005600), National Key Laboratory of Precision Testing Techniques and Instrument, Tsinghua University (No. TH20-01), National Natural Science Foundation of China (No. 51905528), State Key Laboratory of Precision Measuring Technology and Instruments (Tianjin University) (No. pilab2102), and Key Laboratory of Microelectronic Devices Integrated Technology, Chinese Academy of Sciences.

Institutional Review Board Statement: Not applicable.

Informed Consent Statement: Not applicable.

Data Availability Statement: The data presented in this study are available from the corresponding author, J.S., upon reasonable request.

Conflicts of Interest: The authors declare no conflict of interest.

References

1. Wang, P.; Lu, Q. Evolutionary design optimization of MEMS a review of its history and state-of-the-art. *Clust. Comput.* **2019**, *22*, 9105. [CrossRef]
2. Shinde, P.; Shiurkar, U. MEMS for detection of environmental pollutants: A review pertains to sensors over a couple of decades in 21st century. *Mater. Today* **2021**, *44*, 615–624. [CrossRef]
3. Bosch, S.; Xu, J. Real time etch depth measurements of mems devices. *J. Microelectromech. Syst.* **2002**, *11*, 111–117. [CrossRef]
4. Xie, J. Fabrication challenges and test structures for high-aspect-ratio SOI MEMS devices with refilled electrical isolation trenches. *Microsyst. Technol.* **2015**, *21*, 1719–1727. [CrossRef]
5. Kim, H.; Kim, D. Photo-assisted electrochemical etching of a nano-gap trench with high aspect ratio for MEMS applications. *J. Micromech. Microeng.* **2006**, *16*, 906–913. [CrossRef]
6. Gerlt, M.; Laubli, N. Reduced etch lag and high aspect ratios by deep reactive ion etching (DRIE). *Micromachines* **2021**, *12*, 542. [CrossRef]
7. Laermer, F.; Urban, A. MEMS at Bosch–Si plasma etch success story, history, applications, and product. *Plasma Process. Polym.* **2019**, *16*, 1800207. [CrossRef]
8. Baklykov, D.; Andronic, M. Self-Controlled Cleaving Method for Silicon DRIE Process Cross-Section Characterization. *Micromachines* **2021**, *12*, 534. [CrossRef]
9. Attota, R.; Kavuri, P. Nanoparticle size determination using optical microscopes. *Appl. Phys. Lett.* **2014**, *105*, 163105. [CrossRef]
10. Attota, R.; Kang, H. Nondestructive shape process monitoring of three-dimensional, high-aspect-ratio targets using through-focus scanning optical microscopy. *Meas. Sci. Technol.* **2018**, *29*, 125007. [CrossRef]
11. Attota, R.; Germer, T. Through-focus scanning-optical-microscope imaging method for nanoscale dimensional analysis. *Opt. Lett.* **2008**, *33*, 1990–1992. [CrossRef] [PubMed]
12. Ryabko, M.; Shchekin, A. Through-focus scanning optical microscopy (TSOM) considering optical aberrations: Practical implementation. *Opt. Express* **2015**, *23*, 32215–32221. [CrossRef] [PubMed]
13. Attota, R. Noise analysis for through-focus scanning optical microscopy. *Opt. Lett.* **2016**, *41*, 745–748. [CrossRef]
14. Attota, R. Fidelity test for through-focus or volumetric type of optical imaging methods. *Opt. Express* **2018**, *26*, 19100–19114. [CrossRef]
15. Park, S.; Park, G. Through-focus scanning optical microscopy with the Fourier modal method. *Opt. Express* **2018**, *26*, 11649–11657. [CrossRef] [PubMed]
16. Park, S.; You, B. Metrological sensitivity improvement of through-focus scanning optical microscopy by controlling illumination coherence. *Opt. Express* **2019**, *27*, 1981–1990. [CrossRef] [PubMed]
17. Rim, M.; Agocs, E. Detecting nanoscale contamination in semiconductor fabrication using through-focus scanning optical microscopy. *J. Vac. Sci. Technol. B Nanotechnol. Microelectron. Mater. Process. Measur. Phenom.* **2020**, *38*, 050602. [CrossRef]
18. Peng, R.; Jiang, J. Lateral movement and angular illuminating non-uniformity corrected TSOM image using Fourier transform. *Opt. Express* **2020**, *28*, 6294–6305. [CrossRef]
19. Lee, J.; You, B. Motion-free TSOM using a deformable mirror. *Opt. Express* **2020**, *28*, 16352–16362. [CrossRef]
20. Qu, Y.; Hao, J. Machine-learning models for analyzing TSOM images of nanostructures. *Opt. Express* **2019**, *27*, 33978–33998. [CrossRef]
21. Kwak, H.; Ryu, S. Non-destructive thickness characterisation of 3D multilayer semiconductor devices using optical spectral measurements and machine learning. *Light: Adv. Manuf.* **2021**, *2*, 9–11. [CrossRef]
22. Li, G.; Shi, J. Through-focus scanning optical microscopy measurement based on machine learning. *Chin. Opt.* **2022**, *15*, 703–711. (In Chinese)
23. Dalal, N.; Triggs, B. Histograms of oriented gradients for human detection. In Proceedings of the IEEE Conference on Computer Vision and Pattern Recognition, San Diego, CA, USA, 20–25 June 2005.
24. Basak, D.; Srimanta, P. Support Vector Regression. *Neur. Inform. Proc.* **2007**, *11*, 203.
25. Yan, G.; Zhu, Y. Parameters Selection Method for Support Vector Machine Regression. *Comput. Eng.* **2009**, *35*, 218. (In Chinese)

sensors

MDPI

Article

Research on Measurement of Tooth Profile Parameters of Synchronous Belt Based on Point Cloud Data

Zijian Zhang, Mao Pang * and Chuanchao Teng

School of Mechanical and Energy Engineering, Zhejiang University of Science and Technology, Hangzhou 310012, China
* Correspondence: palmy@zust.edu.cn

Abstract: Accurately detecting the tooth profile parameters of the synchronous belt is crucial for the transmission's load distribution and service life. However, the existing detection methods have low efficiency, are greatly affected by the manual experience, and cannot realize automatic detection. A measurement method based on point cloud data is proposed to solve this issue. The surface space points of the synchronous belt are acquired by a line-structured light sensor, and the raw point clouds are preprocessed to remove outliers and reduce the number of points. Then, the point clouds are divided into plane and arc regions, and different methods are used for fitting. Finally, the parameters of each tooth are calculated. The experimental results show that the method has high measurement accuracy and reliable stability and can replace the original detection method to realize automatic detection.

Keywords: synchronous belt; 3D measurement; normal vector; point cloud segmentation

Citation: Zhang, Z.; Pang, M.; Teng, C. Research on Measurement of Tooth Profile Parameters of Synchronous Belt Based on Point Cloud Data. *Sensors* **2022**, *22*, 6372. https://doi.org/10.3390/s22176372

Academic Editors: Bogusław Łazarz, Grzegorz Peruń and Tangbin Xia

Received: 26 July 2022
Accepted: 22 August 2022
Published: 24 August 2022

1. Introduction

1.1. Measurement of Synchronous Belt

The synchronous belt is an important part of power transmission [1]. It has the advantages of an accurate transmission ratio, compact structure, and good wear resistance [2]. It is widely used in machine tools, light industries, chemical industries, automobiles, and other industries [3,4]. The geometric structure of the synchronous belt plays an important role in the load transmission, and the deviation of the tooth profile parameters will affect the service life and reliability. A proper tooth profile will reduce the interference between the synchronous belt and the sprocket, causing a better distribution of the load [5,6]. Conversely, the uneven load will increase the wear of the tooth surface and generate noise and vibration during operation, which leads to mechanical failures more frequently [7]. Therefore, it is very important to detect the tooth profile parameters of the synchronous belt.

The tooth profile deviation of the synchronous belt has several factors [8]: the wear of the mold, the dirt of the mold, the molding temperature, etc. During the production process, each tooth of the synchronous belt needs to be inspected. The current detection methods are mainly [9] the projection method and the manual method. Both methods require human operation, which is greatly affected by the manual experience and cannot work for a long time. The detection efficiency is low, and the cost is high.

1.2. Three-Dimensional Measurement

The definition of 3D measurement: acquiring the 3D data of the object and measuring the object in all directions. The current methods of 3D measurement can be classified as contact and non-contact [10]. Contact measurement methods include manual measurement using fixtures and three-coordinate measuring machines [11]. However, these methods have low measurement efficiency and limited measurement range, which means it is difficult to measure large-sized and special-shaped parts. Meanwhile, these methods need to contact the target to be measured, which easily leads to scratches and deforms the surface of non-rigid parts.

With the continuous development of computer technology and optical equipment, various non-contact 3D measurement methods continue to emerge, such as projected structured light [12], laser rangefinders [13], visual inspection technology [14], etc. Among these non-contact measurement methods, the structured light projection has become a prevalent non-contact 3D shape measurement technology [15]. Essentially, the 3D measurement method of structured light detects the visible surface of the object by configuring an additional light source generator to form the 3D shape of the object. The output of a 3D imaging system is usually a set of points with (x, y, z) coordinates for each measurement point in a Cartesian coordinate system. The advantages of the method are the simple structure, high measurement accuracy, and better stability. It is widely used in industrial inspection, agricultural inspection, 3D profile measurement, and other fields [16–20].

To rapidly obtain information on the 3D shape of the gear tooth flank, a 3D point cloud measurement system based on a line structure light sensor was proposed by Guo [21]. The measured 3D point cloud data were used to calculate the profile error and pitch error. The results showed that the 3D point cloud measurement system could perform rapidly and accurately in the 3D measurements of gears. Long [22] aimed to ensure the quality of aircraft assembly, and a method based on an unstructured scanned point cloud was proposed. Their method first detected and segmented the seam area based on the local point density discrepancy. A projection operation is applied based on the segmentation results to convert 3D measurement to 2D measurement to reduce the computational complexity and improve the accuracy of the measurement. Fan [23] proposed an approach to extracting curve profiles from a scanned point cloud. A slice set was constructed to extract section line points from the point cloud. Based on the normal angle of adjacent points, three methods were proposed to identify the curve-profile points on the three kinds of section lines. The results showed that extracted curve-profile points deviate from reference data within 0.1 mm with standard deviations smaller than 0.07 mm. Xu [24] proposed a method of identifying the longitudinal rip through 3D point cloud processing. The results indicated that this method could identify the longitudinal rip accurately in real-time and simultaneously characterize it.

1.3. Motivation

As can be seen, the 3D point cloud measurement methods have been widely used, and there have been better developments in size measurement. This work aims at the problems of low detection efficiency and human factors in the detection of the tooth profile parameters of synchronous belts. A measurement method based on point cloud data is proposed. Specifically, a line-structured light sensor is used to acquire point clouds on the surface of the synchronous belt. Remove outliers and simplify the point cloud through pre-preprocessing. Then, the point clouds are divided into plane and arc regions. Fitting models suitable for them are established by different methods. Finally, the detection of each tooth of the synchronous belt is completed.

Overall, our contributions are summarized as follows:

- We propose an effective method to detect the tooth profile parameters of synchronous belts from 3D point clouds and establish a point cloud acquisition system on the surface of synchronous belts;
- We design a point cloud segmentation algorithm based on the angle of the normal vector, which accurately obtains the measurement area of the tooth surface arc and establishes a fitting model;
- The 3D point cloud measurement system has better stability, and the measurement error is less than 0.03 mm.

This paper is organized as follows: Section 2 describes the proposed method in detail. The results of these tests are presented in Section 3 and they are discussed in Section 4. Finally, the paper is concluded in Section 5.

2. Methods

2.1. Methodology of Synchronous Belt Geometry Measurement

The detected object is an arc tooth of the synchronous belt, the tooth shape code is High Torque Drive (HTD). The pitch (Pb) of the synchronous belt will cause a pitch error during the meshing process of the synchronous belt and the synchronous wheel. The belt tooth cannot fully enter the gear tooth slot if the pitch error is too large, resulting in an unstable transmission system. In severe cases, the phenomenon of tooth climbing and tooth skipping will occur, which leads to the interruption of the power transmission. The tooth height (ht) and belt height (hs) of the synchronous belt also affect its strength and load-carrying capacity. It is important to detect the parameters of the synchronous belt.

Measurement plan: Pb is the distance between the center of the arc at the top of the tooth, so the Pb measurement needs to obtain the position of the centers; ht is the distance from the center to the plane B and plus the arc radius. The hs data can be obtained in the same way. The parameters of the synchronous belt are shown in Figure 1.

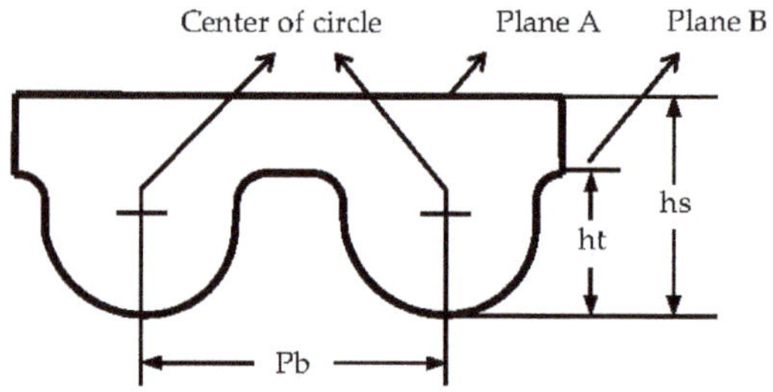

Figure 1. Synchronous belt parameters.

2.1.1. Point Cloud Model

An object contour point cloud acquisition system based on the principle of laser triangulation [25] is shown in Figure 2. The laser is vertically irradiated on the surface of the measured object and diffusely reflected [26]. The reflected light is projected on the photosensitive chip by the camera lens. The imaged data signal will be transmitted to the controller and stored as a point cloud with coordinate information. With the movement of the mobile platform, the point clouds of the measured object are obtained.

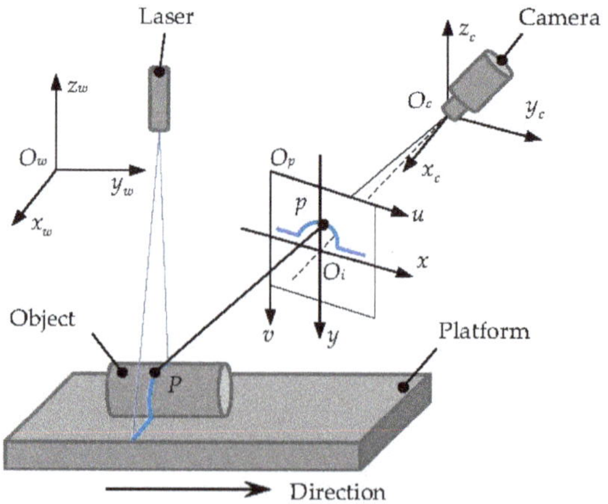

Figure 2. Point cloud acquisition system model.

As shown in Figure 2, O_w-$x_w y_w z_w$ is the measurement coordinate system, that is the world coordinate system; where the O_w-$x_w y_w$ plane is parallel to the working plane of the mobile platform, O_w-$x_w z_w$ is coplanar with the structured light plane and the $O_w y_w$ direction is consistent with the movement of the mobile platform. O_c-$x_c y_c z_c$ is the camera coordinate system. The O_c is the imaging perspective center. O_p-uv is the pixel coordinate system, which is the image plane of the camera. O_i-xy is the image coordinate system, $O_i(u_0, v_0)$ is the coordinate of the projected position of the camera's optical axis in the image plane. $P(x_w, y_w, z_w)$ is a certain point where the structured light plane intersects the surface of the object to be measured, and the intersection of the straight line PO_c and the image plane is set to $p(u, v)$. According to the perspective projection relationship of the pinhole imaging model, the mathematical model of the line-structured light 3D scanning system is established, as shown in Equation (1):

$$\lambda \begin{bmatrix} u \\ v \\ 1 \end{bmatrix} = \begin{bmatrix} f_x & 0 & u_0 \\ 0 & f_y & v_0 \\ 0 & 0 & 1 \end{bmatrix} \begin{bmatrix} \boldsymbol{R} & \boldsymbol{T} \\ 0^T & 1 \end{bmatrix} \begin{bmatrix} x_w \\ y_w \\ z_w \\ 1 \end{bmatrix} \tag{1}$$

where λ is the scale factor; (u, v) denotes the coordinates in the pixel coordinate system; f_x and f_y denote the changing relationship between the image coordinate system and the pixel coordinate system; (u_0, v_0) denotes the coordinates of the projection of the optical axis of the camera lens in the pixel coordinate system; $\boldsymbol{R}$ is the 3×3 rotation matrix; $\boldsymbol{T}$ is the 3×1 translation matrix, and both represent the transformational relationship between the camera coordinate system and the world coordinate system.

2.1.2. Point Cloud Pre-Processing

In the process of point cloud collection, due to the influence of external conditions and scanning equipment [27], the collected point clouds often contain many noise and outlier points which are not related to the surface information of the object and are usually useless. These will affect the measurement accuracy of the data, and the calculation process will take a long time [28]. The outliers are sparsely distributed in space, and statistical filtering [29] is effective for eliminating points with an obvious distribution. Therefore, according to the characteristics of outliers, it can be defined as noise when a point cloud dataset is less than a certain density. Statistical analysis is performed on the average distance from each point to its nearest k points, and the distances of all points should show a Gaussian distribution whose shape is determined by the mean μ and standard deviation σ. Let the coordinate of the n in the point cloud be $P_n(X_n, Y_n, Z_n)$, the distance S_i from this point to any point $P_m(X_m, Y_m, Z_m)$ is calculated in Equation (2):

$$S_i = \sqrt{(X_n - X_m)^2 + (Y_n - Y_m)^2 + (Z_n - Z_m)^2} \tag{2}$$

The average of the distance S_i can be expressed as Equation (3):

$$\mu = \frac{1}{n}\sum_{i=1}^{n} S_i \tag{3}$$

The standard deviation of the distance S_i can be expressed as Equation (4):

$$\sigma = \sqrt{\frac{1}{n}\sum_{i=1}^{n} (S_i - \mu)^2} \tag{4}$$

Set *std* as the standard deviation multiple. The k and *std* need to be input during the implementation of the algorithm. When the average distance of the k neighbors of a point is within the standard range $(\mu - \sigma \cdot std, \mu + \sigma \cdot std)$, the point is retained; otherwise, it is defined as an outlier.

Statistical filtering is performed on the raw point cloud. The calculations show that the outliers in the raw point cloud can be removed well, and relatively complete parts characteristics are retained when the k is 20 and the *std* is 3. As shown in Figure 3a, the raw point clouds of the synchronous belt are collected, and the number of the point clouds is 22,025. Figure 3b is the point cloud to remove outliers by statistical filtering. The number of the point clouds is 21,891, and 123 outliers are removed.

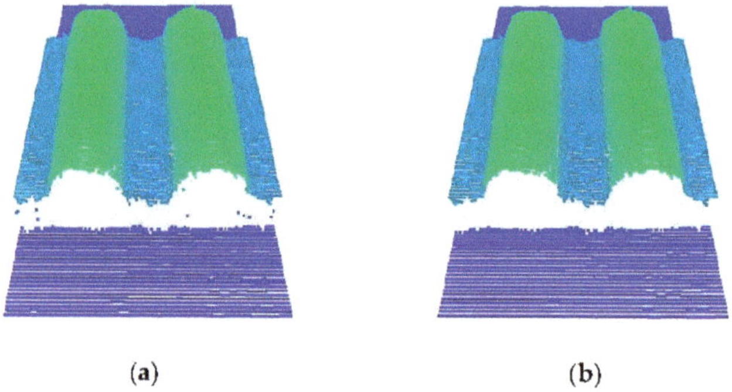

Figure 3. Point cloud before and after statistical filtering: (**a**) raw point cloud and (**b**) point cloud after statistical filtering.

2.1.3. Point Cloud Voxel Down-Sampling

Due to the huge amount of raw point clouds obtained by the 3D scanning equipment, even after statistical filtering, the number of the point cloud is still large [30]. To improve the measurement efficiency, it is necessary to simplify the point cloud. Point cloud simplification should reduce the size of the data and eliminate redundancy, maintain the basic shape of the object, and highlight the key features of the object [31], such as protrusions, depressions, etc.

We compared four commonly used point cloud down-sampling methods [32], analyzed the advantages and disadvantages of different methods, and selected an appropriate point cloud down-sampling method according to the actual measurement requirements. Taking the Bunny of Stanford University as the comparison model, four different methods were used to down-sample the Bunny model, the point cloud down-sampling effect is shown in Figure 4.

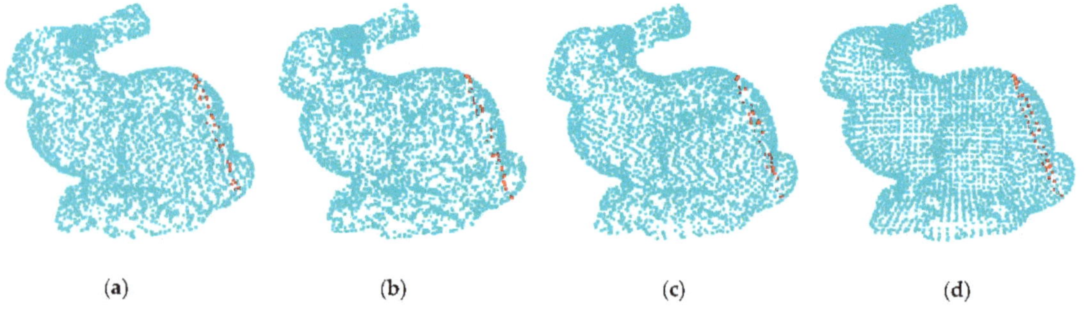

Figure 4. Point cloud down-sampling effect: (**a**) curvature down-sampling; (**b**) random down-sample; (**c**) uniform down-sample; and (**d**) voxel down-sample.

Figure 4a shows the curvature of the point cloud used as the down-sampling basis; the larger the curvature, the more down-sampling points. Figure 4b is the result of random down-sampling, randomly sampling a certain number of points in the raw point cloud. Figure 4c shows the result of uniform down-sampling, the point clouds were filtered by constructing a sphere with a specified radius, and the point closest to the center of a sphere was used as the output point of the down-sampling. Figure 4d shows the result of voxel down-sampling, by creating a three-dimensional voxel grid from the raw point cloud. the center of gravity of all points within each voxel was used to approximate the other in the voxel. These four methods have simplified point clouds, but the point cloud position of the surface contour of the object has been changed. To further study the positional relationship of the object surface contour point cloud, the same position was intercepted for the simplified point cloud. The position of the intercepted slice was the red point cloud, as shown in Figure 4, and the contour point cloud projections are shown in Figure 5.

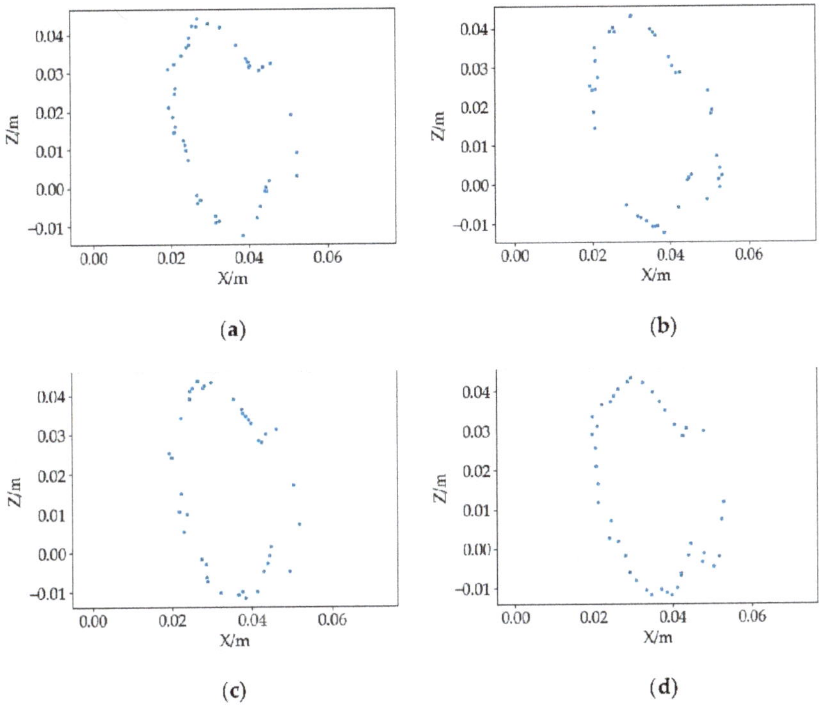

Figure 5. Contour point cloud projection: (**a**) curvature down-sampling; (**b**) random down-sample; (**c**) uniform down-sample; and (**d**) voxel down-sample.

Among the three down-sampling methods of (a–c), some point cloud contours are missing, and the distribution is scattered. While the voxel down-sampling point cloud is relatively complete and evenly distributed, the contour features of the object are preserved well. Therefore, we use voxel down-sampling to simplify the point cloud. The process of the method is as follows:

(1) In the coordinate set of point cloud data, find the value $X_{max}, Y_{max}, Z_{max}$ and $X_{min}, Y_{min}, Z_{min}$ on the three axes of X, Y and Z, obtain the side length of the minimum bounding box of the point cloud l_x, l_y, l_z by Equation (5). Set the side length of the voxel to r. After that, it can find the size of the voxel grid d_x, d_y, d_z from Equation (6).

$$\begin{cases} l_x = X_{max} - X_{min} \\ l_y = Y_{max} - Y_{min} \\ l_z = Z_{max} - Z_{min} \end{cases} \quad (5)$$

$$\begin{cases} d_x = \lfloor l_x / r \rfloor \\ d_y = \lfloor l_y / r \rfloor \\ d_z = \lfloor l_z / r \rfloor \end{cases} \quad (6)$$

(2) Calculate the barycenter of all points in each voxel to approximately represent other points in the voxel. As shown in Figure 6, all points in the voxel can be represented by a barycentric point. Assume that the set of the local points in each voxel is $P\{x_i, y_i, z_i\}(i = 1, 2 \ldots k)$, the centroid point P_g can be found by Equation (7):

$$P_g = \frac{1}{k}\sum_{i=1}^{k}(x_i, y_i, z_i) \tag{7}$$

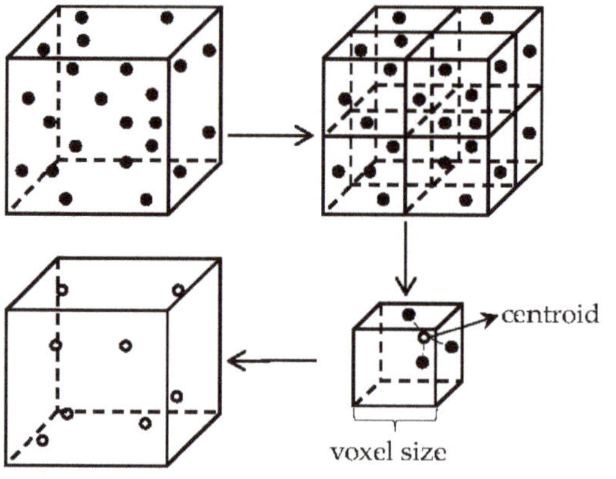

Figure 6. Process of voxel down-sampling.

The side length of the voxel grid is set according to the size of the point clouds. The longer the side length of the voxel, the more points contained in each voxel. The resulting point clouds are sparser, and the details of the object are less. As shown in Figure 7, the density of the point cloud of the synchronous belt was compared in different voxel grid side lengths. Measuring different point cloud densities found that when r was equal to 0.1 mm, the point cloud down-sampling effect was the best, which could not only meet the measurement accuracy requirements but also significantly improved the operation speed.

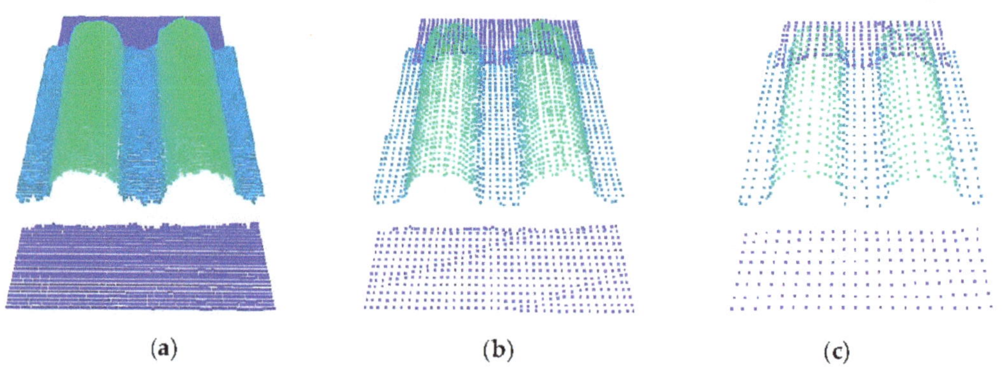

Figure 7. Voxel down-sampling of point cloud with different side lengths: (**a**) r = 0.1 mm; (**b**) r = 0.3 mm; and (**c**) r = 0.5 mm.

2.1.4. Point Cloud Plane Fitting

At present, the most common and simple method for fitting a plane is the least squares fitting, but the fitting accuracy is easily affected by noise. The random sample consensus (RANSAC) algorithm [33] can reduce the influence of noise by the iterative fitting and improve the fitting accuracy. In the RANSAC algorithm, first, a minimum number of random points are selected to define the geometric model of interest. The parameters of the model equation are calculated using the randomly selected points. The model parameters are applied to all points, whether they fit the model or not. If the appropriate number of points is found for the model to be defined, the parameters of the model and the error rate are calculated to check whether the established model is the best one. The algorithm runs iteratively until the best model is reached [34]. The process is as follows:

(1) In this work, the feature model of interest is a plane. It can be defined using at least three points which are randomly selected from the point cloud. Equation (8) is as follows:

$$a \cdot x + b \cdot y + c \cdot z + d = 0 \tag{8}$$

where a, b, c, d are the parameters of the plane, and (x, y, z) are the 3D coordinates of any point.

(2) Calculate the distance d_i from the remaining point to the plane by Equation (9):

$$d_i = \frac{|ax_i + by_i + cz_i + d|}{\sqrt{a^2 + b^2 + c^2}} \tag{9}$$

where (x_i, y_i, z_i) denotes the remaining point of the point cloud.

If the distance d_i is less than the threshold, this point is regarded as a point in the plane, and the number of interior points under the plane is counted and recorded.

$$(d_i)_{stay} \leq T \tag{10}$$

where T denotes the thresholds for segmenting the points of the extracted planes.

(3) Continue with steps 1~2. If the number of points in the plane exceeds the maximum number of interior points, recalculate the plane parameters with the saved interior points;

(4) Repeat steps 1 to 3 until there are no points that meet the conditions, and finally calculate the plane parameters again to obtain the best plane model.

According to the synchronous belt measurement plan, it is necessary to solve and fit plane A and plane B. The fitting result is shown in Figure 8. The red regions are the underside of the synchronous belt, and the blue regions are the point cloud of the tooth's bottom surface. After the algorithm fitting, the plane A and plane B models are obtained.

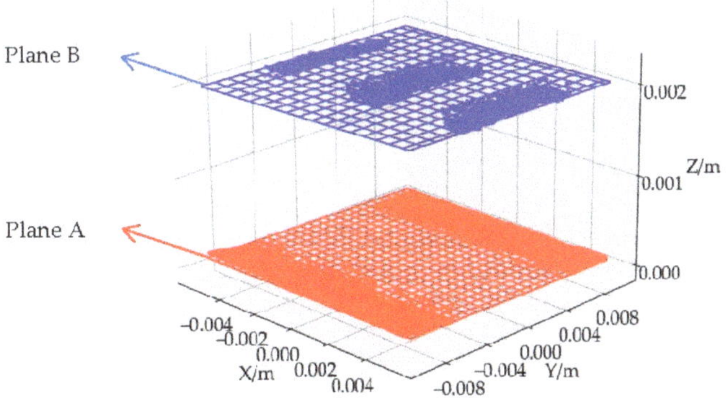

Figure 8. Fitting result of plane A and plane B.

2.1.5. Point Cloud Curve Fitting

To improve the measurement accuracy of the synchronous belt tooth surface arc, this paper proposes a surface feature extraction method based on the point cloud normal vector. Based on the normal vector of the point cloud [35], set a reference vector and calculate the angle value between the normal vector of each point in the point cloud set and the reference vector; determine its value and complete the adjustment of the tooth surface arc of the synchronous belt. The process of feature extraction is as follows:

(1) Normal vector estimation. Firstly, estimate the normal vector of the point cloud, any point p_i in the point cloud, search for its nearest neighbors k, and then use the least squares method to calculate the local plane P_l of this point. Equation (11) is as follows:

$$P_l(\vec{n}, d) = \underset{(\vec{n}, d)}{\operatorname{argmin}} \sum_{i=1}^{k} (\vec{n} \cdot p_i - d)^2 \tag{11}$$

where $\vec{n}$ is the normal vector of the plane P_l, and d is the distance from P_l to the origin of the coordinate.

Then, it can be considered that the normal vector of the plane fitted by k nearest points is approximated as the normal vector of the point p_i, and the estimated normal vector of the plane P_l is obtained by performing PCA (principal component analysis). It can decompose the eigenvalues of a covariance matrix M by Equation (12). The eigenvector corresponding to the smallest eigenvalue of the matrix is the normal vector of the plane P_l, which is also the normal vector of the point p_i.

$$M = \frac{1}{k} \sum_{i=1}^{k} (p_i - p_0)(p_i - p_0)^T \tag{12}$$

where p_0 is the centroid of the k points.

(2) Normal vector orientation. The normal vector is ambiguous [36]. The previous calculation only obtains the straight line where the normal vector is located but does not determine the final direction of the normal vector [37], as shown in Figure 9a. Before using the normal vector angle value for surface feature extraction, it is necessary to ensure normal vector consistency. Set a reference vector $\vec{a}$, and then traverse the normal vectors $\vec{b_i}$ of other points. If $\vec{a} \cdot \vec{b_i} < 0$, then the vector direction is opposite. Flip the normal vector $\vec{b_{i}}$,; otherwise, it remains unchanged. The adjustment result is shown in Figure 9b.

(a)

(b)

Figure 9. Normal vector orientation: (**a**) ambiguity of point cloud normal vector and (**b**) consistency of point cloud normal vector.

(3) The region of interest curve. The tooth surface profile of the synchronous belt is composed of two arcs, which cannot be directly fitted and solved. Therefore, it is necessary to extract the effective arc area of the tooth surface profile to improve the measurement accuracy. In the feature area, the deviation of the normal vector direction of the point cloud is large, and the angle of the sampling point can be used for identification and extraction. The extraction method is as follows: based on the consistency of the normal vector of the point cloud, Equation (13) can calculate the angle θ between the reference vector $\vec{a}$ and the normal vector $\vec{b_i}$ of the sampling point and set the discrimination threshold. If $\theta > f$, mark the point as a curve feature point dataset; otherwise, mark it as a non-curve feature point set.

$$\theta = \cos\left\langle \vec{a}, \vec{b} \right\rangle = \frac{\vec{a} \cdot \vec{b}}{\left|\vec{a}\right| \cdot \left|\vec{b}\right|} \tag{13}$$

After experimental verification, when the threshold is set to $f = 0.65$, the error of the measurement result is the smallest, and the segmentation effect is shown in Figure 10. The red area is the region of the interest curve.

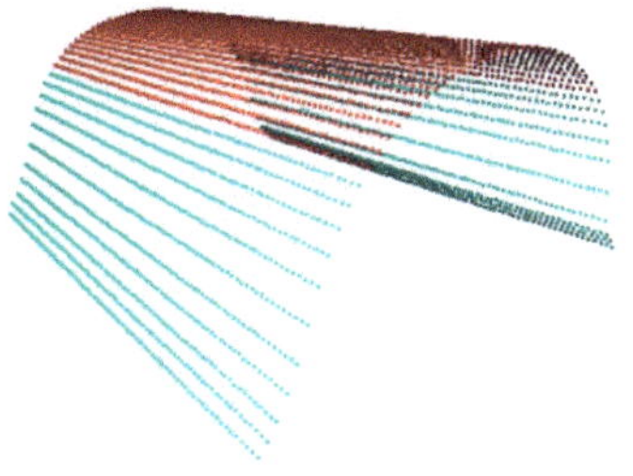

Figure 10. Region of Interest Curve.

(4) The tooth surface arc fitting. Through the above steps, the effective arc area of the tooth surface profile is extracted, but the point cloud dataset is spatial three-dimensional data. The fitting accuracy is not high, and the efficiency is low. To fit the contour of the surface accurately and efficiently, the slicing method [38] is introduced to process the point cloud datasets and realize the dimensionality reduction in the point cloud datasets. The process is as follows: the essence of point cloud slicing is the intersection of a set of planes and point cloud datasets. By setting the slice width d, the point clouds are divided into a certain number of point cloud datasets and then projected to the plane to obtain a two-dimensional point map of the object outline, as shown in Figure 11a. The least squares method [39] is used to improve the fitting accuracy of slices, and the fitting process is as follows:

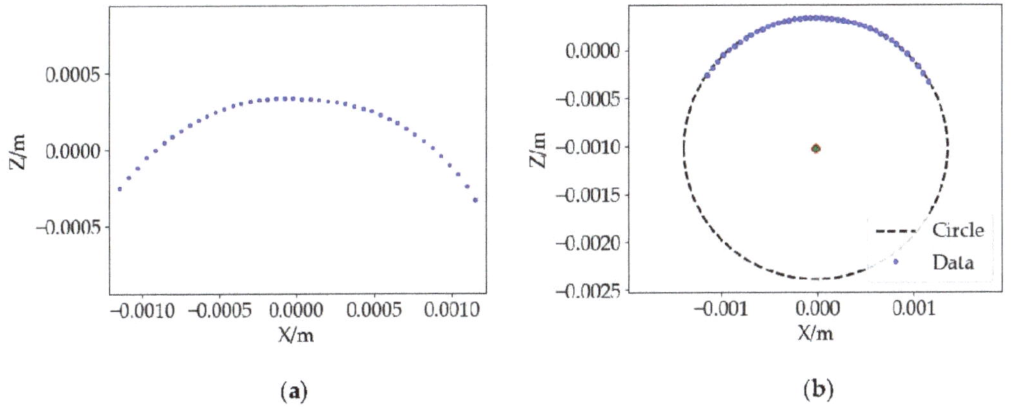

Figure 11. Tooth surface arc fitting: (**a**) contour point cloud projection, and (**b**) schematic diagram of least squares fitting circle.

(1) Build the circle Equation (14):

$$(x - A)^2 + (y - B)^2 = R^2 \tag{14}$$

(2) The difference between the distance from the point $(X_i, Y_i)i \in (1, 2 \cdots N)$ to the center of the circle and the radius is shown in Equation (15):

$$\begin{aligned} d_i &= (X_i - A)^2 + (Y_i - B)^2 - R^2 \\ &= X_i^2 + Y_i^2 - 2AX_i - 2BY_i + A^2 + B^2 - R^2 \end{aligned} \tag{15}$$

(3) Let $a = -2A, b = -2B, c = A^2 + B^2 - R^2$, substitute them into Equation (14) and simplify, then find Equation (16).

$$d_i = X_i^2 + Y_i^2 + aX_i + bY_i + c \tag{16}$$

(4) Let $Q(a, b, c) = \sum d_i^2$, because $Q(a, b, c)$ is much greater than 0. The function has a minimum value greater than or equal to 0 and a maximum value of infinity. Find the partial derivative for a, b, and c, and set the partial derivative equal to 0. The following can be obtained:

$$\begin{cases} \frac{\partial Q(a,b,c)}{\partial a} = \sum 2(X_i^2 + Y_i^2 + aX_i + bY_i + c)X_i = 0 \\ \frac{\partial Q(a,b,c)}{\partial b} = \sum 2(X_i^2 + Y_i^2 + aX_i + bY_i + c)Y_i = 0 \\ \frac{\partial Q(a,b,c)}{\partial c} = \sum 2(X_i^2 + Y_i^2 + aX_i + bY_i + c) = 0 \end{cases} \tag{17}$$

solve Equation (17) to obtain the extreme point, compare the function values of all extreme points. The parameter corresponding to the minimum value is the best fitting circle parameter, and the fitting diagram is shown in Figure 11b. The fitting circle is close to the sampling point, which indicates the fitting result is well.

3. Results

3.1. Test Environment

The composition of the synchronous belt tooth profile experimental is shown in Figure 12. The collection of point clouds is completed by the line laser profile sensor. According to the belt thickness range of the synchronous belt, the distance between the sensor and the mobile platform is 174 mm; the moving speed of the mobile platform is 350 mm/s; The line laser profile sensor is connected to the computer through Ethernet, and the point cloud datasets are transferred to the computer for the following process. The software platform and library are Python V3.8 and Open3D point cloud library V0.15.0.

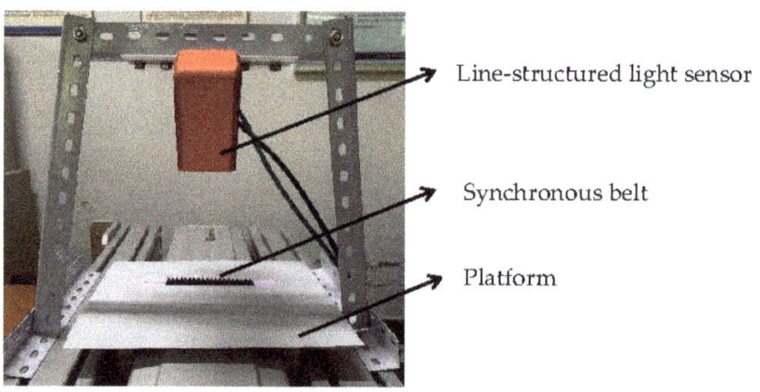

Figure 12. Synchronous belt tooth profile measurement system composition.

3.2. Measurement Data

The arc tooth synchronous belt is placed on the platform, and the point cloud on the surface of the arc tooth synchronous belt is obtained by the line structured light sensor. The raw point clouds are preprocessed, and the plane and region of the interest curve are obtained by dividing the point cloud. As in the measurement plan in Section 2, the tooth profile parameter measurement of the synchronous belt is completed.

To verify the accuracy and stability of our method, the result of the projection and the manual method were compared. Taking the manual measurement data as the standard, the error analysis of the projection method and the method in this paper is carried out. The measurement results and errors are shown in Table 1. It can be seen from the table that the error of our method is smaller than that of the projection method, the measurement data are closer to the manual measurement data, and the error range meets the measurement requirements.

Table 1. Measurement data of tooth profile parameters.

Number	Projection Method			Manual Method			Our Method		
	Pb/mm	ht/mm	hs/mm	Pb/mm	ht/mm	hs/mm	Pb/mm	ht/mm	hs/mm
1	4.960	2.000	4.140	5.061	2.110	4.109	5.009	2.020	4.149
2	5.020	2.060	4.120	5.043	2.108	4.114	5.020	1.983	4.142
3	4.980	1.980	4.080	5.025	2.099	4.100	4.979	1.990	4.159
4	4.960	2.020	4.120	5.033	2.051	4.125	5.004	1.998	4.146
5	5.000	2.060	4.100	5.053	2.109	4.115	5.022	1.987	4.145
6	4.980	2.040	4.140	5.035	2.073	4.112	5.004	2.001	4.159
7	5.020	2.000	4.160	5.037	2.108	4.108	5.003	1.995	4.133
8	5.040	2.020	4.140	5.023	2.064	4.122	4.991	1.991	4.159
9	4.980	2.040	4.180	5.054	2.073	4.111	5.020	2.015	4.143
10	4.980	2.040	4.160	5.043	2.104	4.117	4.997	2.006	4.134
Average	4.992	2.026	4.134	5.041	2.090	4.113	5.005	1.999	4.147
Error				0.049	0.064	0.021	0.013	0.027	0.013

4. Discussion

For the measurement data of synchronous belts, manual measurement is closer to the actual size, but its measurement speed is slow, the stability is poor, and it cannot work for a long time, which is difficult to process and produce automatically. It is gradually replaced by other methods.

The projection method places the light source on the back of the object. It is suitable for the detection of the outline of the opaque object. However, it is more sensitive to the intensity of the light source. When the light source is weak, the edge contour of the object is affected by burrs, and the contour of the object cannot be accurately extracted; when the light source is strong, some details of the contour are corroded, and the contour features are lost, and the actual size of the object cannot be measured accurately. As shown in Figures 13 and 14, the Pb and ht measured by the projection method are slightly larger than others. In contrast, it can be seen from Figure 15 that the measurement data of the hs are slightly smaller. This is due to the strong light source that depicts the projection profile as smaller than the actual contours.

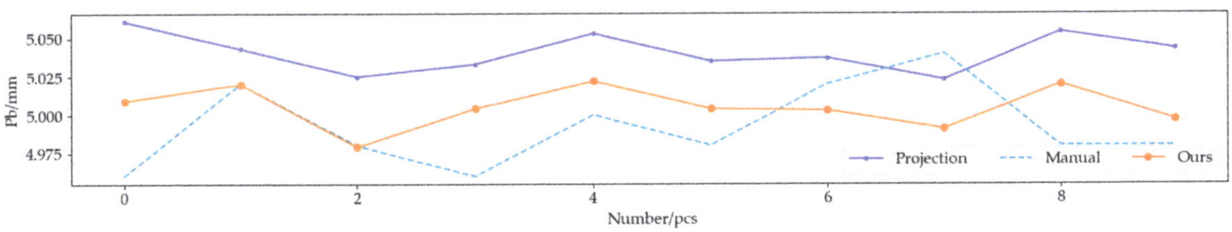

Figure 13. Graph of Pb measurement data.

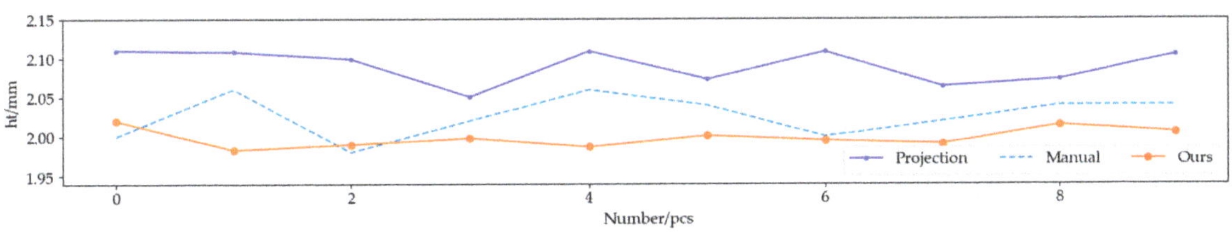

Figure 14. Graph of ht measurement data.

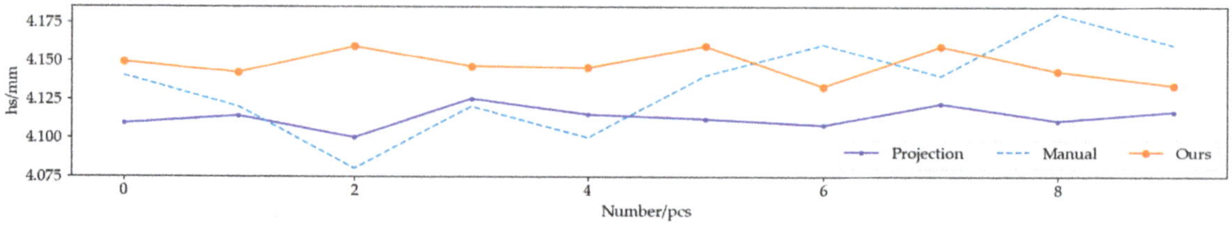

Figure 15. Graph of hs measurement data.

With our method, the surface profile information of the object is obtained by the line-structured light sensor, and the tooth profile parameters of the synchronous belt are measured based on the point cloud segmentation and fitting technology. Since the influence of the light source does not need to be considered in the process of point cloud acquisition, the measurement data are closer to the actual size, with less error than the projection method, and the stability is better than that of the manual method, which can meet the requirements of continuous and stable work and is conducive to the realization of automatic processing and production work.

5. Conclusions

In this work, a 3D point cloud measurement system for synchronous belts is built using a line-structured light sensor. Firstly, the contour point clouds of the synchronous belt are collected. To improve the accuracy and efficiency of measurement, the raw point clouds are filtered, and the number of the point cloud is reduced. Then, referring to the measurement plan of the synchronous belt, the point clouds are divided into plane and arc areas. In the plane areas, it can effectively reduce the fitting error by using the RANSAC algorithm. To obtain the effective area of the arc, a segmentation method based on a normal vector is proposed, which can effectively segment the target area. Subsequently, a curve fitting model suitable for the arc tooth profile is established by the least squares method. Finally, the parameters of each tooth are calculated. The experimental results show that the method in this paper can accurately and quickly detect the tooth profile parameters of the synchronous belt. The average errors of this method in parameters are less than 0.03 mm. This method provides a useful solution with a much smaller degree of error in the stability of the tooth profile. It is an obvious advantage over the classic, common methods. Therefore, the detection scheme has great application prospects and promotion value. The main work in this paper focuses on point cloud noise reduction, point cloud simplification, point cloud segmentation, and size measurement. However, the inspection of surface defects on the synchronous belt has not been thoroughly studied. Therefore, further research can be carried out on surface defects.

Author Contributions: Z.Z. designed the experiment and wrote the article; Z.Z. and M.P. performed the experiments; Z.Z. and C.T. analyzed the data. All authors have read and agreed to the published version of the manuscript.

Funding: This research was supported by the Natural Science Foundation of Zhejiang Province, grant number LQY19E050001.

Institutional Review Board Statement: Not applicable.

Informed Consent Statement: Not applicable.

Data Availability Statement: The data used to support the findings of this study are included within the article.

Conflicts of Interest: The authors declare no conflict of interest. The funders had no role in the design of the study; in the collection, analyses, or interpretation of data; in the writing of the manuscript, or in the decision to publish the results.

References

1. Bortnowski, P.; Kawalec, W.; Król, R.; Ozdoba, M. Types and causes of damage to the conveyor belt-review, classification and mutual relations. *Eng. Ail. Anal.* **2022**, *140*, 106520. [CrossRef]
2. Shi, Y.; Li, Z.; Zhao, T.; Yu, X.; Yin, C.; Bai, Y. Fault Diagnosis of Synchronous Belt of Machine Tool Based on Improved Back Propagation Neural Network. *J. Nanoelectron. Optoelectron.* **2021**, *16*, 1972–1979. [CrossRef]
3. Fedorko, G.; Molnar, V.; Grincova, A.; Dovica, M.; Toth, T.; Husakova, N.; Taraba, V.; Kelemen, M. Failure analysis of irreversible changes in the construction of rubber–textile conveyor belt damaged by sharp-edge material impact. *Eng. Ail. Anal.* **2014**, *39*, 135–148. [CrossRef]
4. Chamorro, J.; Vallejo, L.; Maynard, C.; Guevara, S.; Solorio, J.A.; Soto, N.; Newell, B. Health monitoring of a conveyor belt system using machine vision and real-time sensor data. *CIRP J. Manuf. Sci. Technol.* **2022**, *38*, 38–50. [CrossRef]
5. Chen, G.; Zheng, H.; Qatu, M. Decomposition of noise sources of synchronous belt drives. *J. Sound. Vib.* **2013**, *332*, 2239–2252. [CrossRef]
6. Kagotani, M.; Ueda, H. Influence of installation tension on transmission error due to resonance in a synchronous belt. *J. Mech. Design.* **2015**, *137*, 083301. [CrossRef]
7. Shi, Y.; Chen, G.; Li, Z.; Li, Q.; Tang, W. Simulation Analysis Method of Synchronous Belt Transmission Noise Prediction. *J. Nanoelectron. Optoelectron.* **2020**, *15*, 1004–1011. [CrossRef]
8. Perdan, B.; Bračun, D.; Diaci, J. Automated detection of surface defects on power transmission belts. *Int. J. Adv. Manuf. Tech.* **2011**, *53*, 267–273. [CrossRef]
9. Kim, S.H.; Hwang, Y. A survey on deep learning based methods and datasets for monocular 3D object detection. *Electronics* **2021**, *10*, 517. [CrossRef]
10. Yuan, Q.; Luo, Y.; Wang, H. 3D point cloud recognition of substation equipment based on plane detection. *Results Eng.* **2022**, *15*, 100545. [CrossRef]
11. Mitsuyuki, T.; Hiekata, K.; Kasahara, T. Development of manufacturing support system for ship curved shell plate using laser scanner. *Results Eng.* **2020**, *7*, 100157. [CrossRef]
12. Gao, H.; Aoyama, T.; Takaki, T.; Ishii, I. A Self-projected Structured Light System for Fast Three-dimensional Shape Inspection. *Int. J. Robot. Autom.* **2015**, *30*, 458–470. [CrossRef]
13. Li, Y.; Qiu, Y.; Chen, Y.; Guan, K. A novel orientation and position measuring system for large & medium scale precision assembly. *Opt. Laser. Eng.* **2014**, *62*, 31–37. [CrossRef]
14. See, J. Visual inspection reliability for precision manufactured parts. *Hum. Factors* **2015**, *57*, 1427–1442. [CrossRef] [PubMed]
15. Li, C.; Gao, F.; Han, X.; Zhang, B. A New Density-Based Clustering Method Considering Spatial Distribution of Lidar Point Cloud for Object Detection of Autonomous Driving. *Electronics* **2021**, *10*, 2005. [CrossRef]
16. Biglia, A.; Zaman, S.; Gay, P.; Aimonino, D.R.; Comba, L. 3D point cloud density-based segmentation for vine rows detection and localisation. *Comput. Electron. Agric.* **2022**, *199*, 107166. [CrossRef]
17. Sun, Y.; Luo, Y.; Chai, X.; Zhang, P.; Zhang, Q.; Xu, L.; Wei, L. Double-threshold segmentation of panicle and clustering adaptive density estimation for mature rice plants based on 3D point cloud. *Electronics* **2021**, *10*, 872. [CrossRef]
18. Chen, X.; Ravikumar, N.; Xia, Y.; Attar, R.; Diaz-Pinto, A.; Piechnik, S.K.; Frangi, A.F. Shape registration with learned deformations for 3D shape reconstruction from sparse and incomplete point clouds. *Med. Image. Anal.* **2021**, *74*, 102228. [CrossRef]
19. Yang, Y.; Li, M.; Ma, X. An Advanced Vehicle Body Part Inspection Scheme Based on Scattered Point Cloud Data. *Appl. Sci.* **2020**, *10*, 5379. [CrossRef]
20. Zhou, Y.; Wang, S.; Mei, X.; Yin, W.; Lin, C.; Hu, Q.; Mao, Q. Railway tunnel clearance inspection method based on 3D point cloud from mobile laser scanning. *Sensors* **2017**, *17*, 2055. [CrossRef]
21. Guo, X.; Shi, Z.; Yu, B.; Zhao, B.; Li, K.; Sun, Y. 3D measurement of gears based on a line structured light sensor. *Precis. Eng.* **2020**, *61*, 160–169. [CrossRef]
22. Long, K.; Xie, Q.; Lu, D.; Wu, Q.; Liu, Y.; Wang, J. Aircraft skin gap and flush measurement based on seam region extraction from 3D point cloud. *Measurement* **2021**, *176*, 109169. [CrossRef]
23. Fan, J.; Ma, L.; Sun, A.; Zou, Z. An approach for extracting curve profiles based on scanned point cloud. *Measurement* **2020**, *149*, 107023. [CrossRef]
24. Xu, S.; Cheng, G.; Pang, Y.; Jin, Z.; Kang, B. Identifying and Characterizing Conveyor Belt Longitudinal Rip by 3D Point Cloud Processing. *Sensors* **2021**, *21*, 6650. [CrossRef]
25. Liang, J.; Gu, X. Development and application of a non-destructive pavement testing system based on linear structured light three-dimensional measurement. *Constr. Build. Mater.* **2020**, *260*, 119919. [CrossRef]
26. Zhang, D.; Zou, Q.; Lin, H.; Xu, X.; He, L.; Gui, R.; Li, Q. Automatic pavement defect detection using 3D laser profiling technology. *Autom. Constr.* **2018**, *96*, 350–365. [CrossRef]
27. Ren, Y.; Li, T.; Xu, J.; Hong, W.; Zheng, Y.; Fu, B. Overall Filtering Algorithm for Multiscale Noise Removal from Point Cloud Data. *IEEE Access.* **2021**, *9*, 110723–110734. [CrossRef]
28. Rusu, R.B.; Marton, Z.C.; Blodow, N.; Dolha, M.; Beetz, M. Towards 3D point cloud based object maps for household environments. *Robot. Auton. Syst.* **2008**, *56*, 927–941. [CrossRef]
29. Shuuji, M.; Song, X.; Liao, Z.; Chen, P. Low-speed bearing fault diagnosis based on improved statistical filtering and convolutional neural network. *Meas. Sci. Technol.* **2021**, *32*, 115009. [CrossRef]

30. Wang, S.; Hu, Q.; Xiao, D.; He, L.; Liu, R.; Xiang, B.; Kong, Q. A New Point Cloud Simplification Method with Feature and Integrity Preservation by Partition Strategy. *Measurement* **2022**, *197*, 111173. [CrossRef]
31. Xu, X.; Li, K.; Ma, Y.; Geng, G.; Wang, J.; Zhou, M.; Cao, X. Feature-preserving simplification framework for 3D point cloud. *Sci. Rep.* **2022**, *12*, 9450. [CrossRef] [PubMed]
32. Lv, C.; Lin, W.; Zhao, B. Approximate intrinsic voxel structure for point cloud simplification. *IEEE. Trans. Image Process.* **2021**, *30*, 7241–7255. [CrossRef]
33. Xu, G.; Pang, Y.; Bai, Z.; Wang, Y.; Lu, Z. A fast point clouds registration algorithm for laser scanners. *Appl. Sci.* **2021**, *11*, 3426. [CrossRef]
34. Xu, M.; Lu, J. Distributed RANSAC for the robust estimation of three-dimensional reconstruction. *IET Comput. Vis.* **2012**, *6*, 324–333. [CrossRef]
35. Han, H.; Han, X.; Sun, F.; Huang, C. Point cloud simplification with preserved edge based on normal vector. *Optik* **2015**, *126*, 2157–2162. [CrossRef]
36. Zeybek, M.; Şanlıoğlu, İ. Point cloud filtering on UAV based point cloud. *Measurement* **2019**, *133*, 99–111. [CrossRef]
37. Gao, H.; Xu, G.; Zhang, Z.; Zhou, W.; Wu, Q. A novel probability iterative closest point with normal vector algorithm for robust rail profile registration. *Optik* **2021**, *243*, 166936. [CrossRef]
38. Chen, W.; Li, X.; Ge, H.; Wang, L.; Zhang, Y. Trajectory planning for spray painting robot based on point cloud slicing technique. *Electronics* **2020**, *9*, 908. [CrossRef]
39. Nurunnabi, A.; Sadahiro, Y.; Laefer, D.F. Robust statistical approaches for circle fitting in laser scanning three-dimensional point cloud data. *Pattern. Recogn.* **2018**, *81*, 417–431. [CrossRef]

applied sciences

MDPI

Article

Development of a Dual-Modality Gamma-ray/Fast Neutron Imaging System for Air Cargo Inspection

Jae Yeon Park *, Jungho Mun, Jae Hyun Lee, Yeong-Heum Yeon, Moonsik Chae, Minwoong Lee and Nam-Ho Lee

Advanced Radiation Technology Institute, Korea Atomic Energy Research Institute, 29 Geum gu-gil, Jeongup-si 56212, Jeollabuk-do, Korea
* Correspondence: jaeyeon@kaeri.re.kr; Tel.: +82-63-570-3412

Abstract: High-energy radiation sources have provided a strong security inspection capability using a non-invasive imaging system. The use of multiple radiation sources in one imaging system can also lead to a more powerful system that can classify various materials compared to using a single radiation source. The Advanced Radiation Technology Institute of Korea Atomic Energy Research Institute has developed an air cargo inspection system using multiple radiation sources such as fast neutrons and gamma-rays to classify the plastics, metals, and organics among various sample materials. The fast neutron beam with an energy of 14.1 MeV, generated using the D-T neutron generator, and the gamma-ray beam with an energy of 6 MeV, generated by an electron linear accelerator, are projected onto the vertically aligned scintillator-based radiation detectors. The neutron and gamma-ray images of a cargo container moved by a motorized linear translation stage are acquired, and the image data processing shows good material classification results. In this paper, we describe a multi-radiation imaging system for air cargo inspection and investigate its material classification capability using various sample materials.

Keywords: dual-modality imaging; gamma-ray; fast neutron; air cargo inspection

Citation: Park, J.Y.; Mun, J.; Lee, J.H.; Yeon, Y.-H.; Chae, M.; Lee, M.; Lee, N.-H. Development of a Dual-Modality Gamma-ray/Fast Neutron Imaging System for Air Cargo Inspection. *Appl. Sci.* **2022**, *12*, 9775. https://doi.org/10.3390/app12199775

Academic Editors: Bogusław Łazarz, Grzegorz Peruń and Tangbin Xia

Received: 24 August 2022
Accepted: 26 September 2022
Published: 28 September 2022

1. Introduction

Cargo inspection systems have been developed in the security area using various radiation sources, such as radioactive materials [1], electron linear accelerators [2], and neutron generators [3]. Their high penetration power is important for examining the inner container without opening the gate [4]. The first cargo inspection system was developed using a gamma-ray source for the cargo inspection of the carrier and container. The gamma-ray inspection system detects metallic weapons such as guns [5], knives [6], and bombs [7]. Non-invasive measurements that use gamma-rays are convenient and safe for cargo users as much as for criminal investigators. However, gamma-ray systems are limited in detecting objects made of low-density materials, such as drugs or plastic explosives. Neutron sources have been used to analyze low-density materials in the fields of physics [8], nuclear sciences [9], and the battery industries [10]. Neutrons with high energies of several mega-electron volts have a higher penetrating power, similar to gamma-rays. Therefore, the combined use of a gamma-ray and neutrons provides a powerful effect to extend the material discrimination.

Neutrons and gamma-rays each have their advantages and disadvantages for inspecting cargo, but combining them into one system compensates for these shortcomings. Recently, a cargo inspection system using a Deuterium-Tritium (D-T) neutron generator and Co-60 irradiators was developed by the Australia group in China [3], and it can differentiate 16 materials. The D-T generator radiates neutrons with an energy of 14.1 MeV in all directions (4π), using a fusion reaction with deuterium and tritium. The Co-60 emits characteristic gamma-rays of 1.2 and 1.3 MeV, which are lighter and smaller than electronically generated gamma-rays. However, the radioactive matter has a limited use time and low dose rates. A gamma-ray source with higher energy and a higher dose rate can be generated for a linear electron accelerator. The electron accelerator produces gamma-rays with energies of a few MeV by the bremsstrahlung effect. This energy can be higher than that of Co-60.

An air cargo inspection system requires high-energy radiation sources. The radiations with higher energy have the higher penetrating power. For instance, the penetrating lengths of iron with 5.995×10^{-2} cm^2/g at 1 MeV and 3.057×10^{-2} cm^2/g at 6 MeV of a gamma-ray are calculated to about 2.122 cm and 4.162 cm, respectively [11].

The high penetrating power of radiation adversely affects the detector owing to its low measurement efficiency. Plastic scintillators and silicon photomultiplier-based radiation readout circuit boards are well-known radiation detectors for high-energy radiation measurements [12,13]. Plastic scintillators have been widely used to convert high-energy radiation to visible light. It is very attractive because of its nontoxicity, stability at varying temperatures, and simpler geometrical control, unlike liquid scintillators and others [14–16]. A silicon photomultiplier was used to produce electrical pulses from the photon generated in the scintillator. The radiation detector modules were vertically arranged, and the moving cargo container was scanned.

The neutron and gamma-ray system outputs the neutron and gamma-ray transmittance images of the cargo container. The transmittance of the radiation has multiple pieces of information related to the material density, material thickness, and radiation beam conditions. Transmittance data should be calculated to extract the radiation absorption of materials and obtain their intrinsic characteristics for material discrimination.

In this paper, we report the development of an air cargo inspection system using fast neutron and gamma-ray sources to investigate the ability of material classification for security inspection. These radiation sources have a high penetration power, and the combined imaging provides a powerful result for material classification compared to using a single radiation source. Neutron and gamma-ray generators were installed, and their signals were measured using scintillator-based silicon photomultiplier electric circuits. Multiple detectors were arranged vertically to scan the air cargo using a motorized linear translator. The acquired neutron and gamma-ray images were used to calculate the R-value, which represented the ratio of neutron and gamma-ray attenuation coefficients. The data processing method using R-value images can categorize materials such as metals, plastics, and organic materials as a color map.

2. Materials and Methods

The air cargo inspection system developed at the Advanced Radiation Technology Institute of Korea Atomic Energy Research Institute is based on a radiation imaging system consisting of radiation sources, sample translators, and radiation detectors. Two types of radiation sources are installed: neutrons and gamma-rays. They radiate in the same direction from different positions, as shown in Figure 1. Radiation detectors were placed on the opposite side of the cargo container from the source. Radiation sources and detectors are surrounded by radiation-shielding materials, except for the emission and detection areas. The entire area of the air cargo inspection system was shielded. The machine control room was located sufficiently far from the inspection area, and there is no risk of irradiation exposure. Between the radiation sources and detectors, the cargo is moved on a motorized linear translation stage. The data acquisition system accumulates imaging data and communicates with the machine control commands from a user-controlled computer via a network.

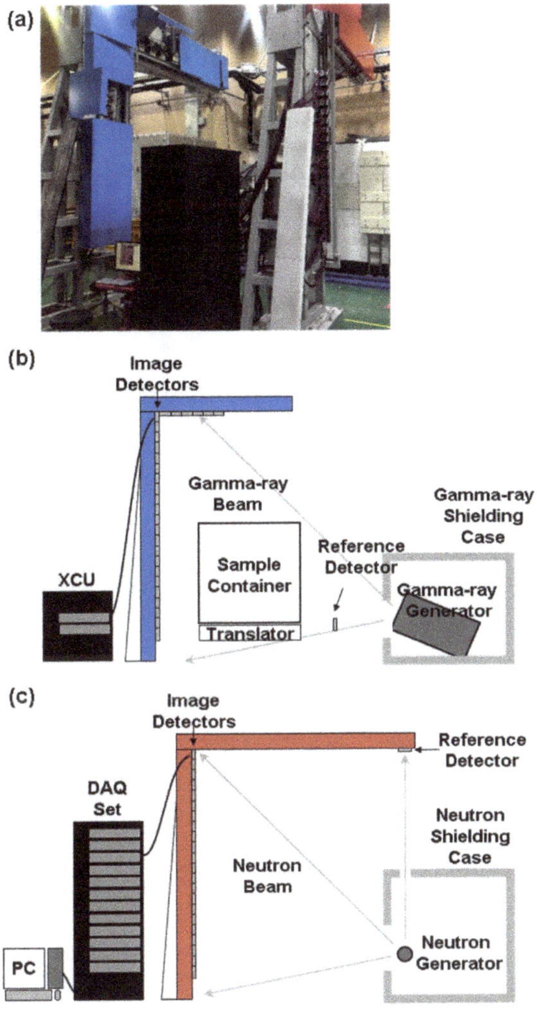

Figure 1. Air cargo inspection system: (**a**) image of the system, (**b**) diagram of the gamma-ray imaging system, and (**c**) diagram of the neutron imaging system.

2.1. Gamma-ray Generation and Detection

The gamma-ray beam is produced using an electron linear accelerator system (LINAC) to output gamma-rays. The LINAC is composed of an electron gun, radio frequency (RF) generator, RF cavity, vacuum system, cooling system, high-voltage power system, tungsten target, and collimator, as shown Figure 2. The generated electrons were accelerated in a gamma-ray generator system with a size of 965 $\times$ 1515 $\times$ 1115 mm^3 (width $\times$ length $\times$ height) and a repetition rate of 200 Hz for a tungsten target with a thickness of 2 mm. The gamma-rays produced from the tungsten target by the bremsstrahlung process have an energy of 6 MeV and a dose rate of 1.25 Gy/min. The gamma-ray beam passes through the slit collimator in front of the tungsten target and radiates towards the gamma-ray detectors.

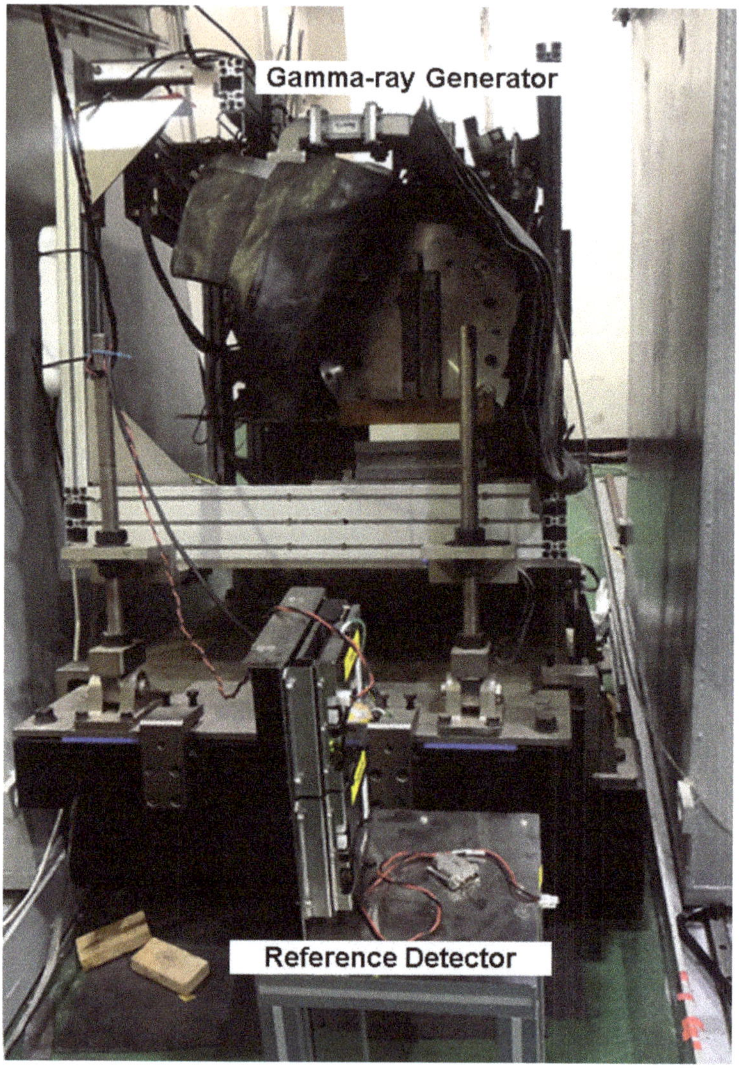

Figure 2. Gamma-ray generator. The gamma-ray beam is produced from the linear accelerator (LINAC) in the tungsten target by the bremsstrahlung interaction. In front of the LINAC, the reference detector is installed.

The gamma-ray radiation that passed through the air cargo container was detected using a gamma-ray detection tower. The structure of the gamma-ray detection tower is built from a steel frame, as shown in Figure 1b, to vertically arrange the 30 gamma-ray detection modules. They are connected to the gamma-ray control units and transmit the gamma-ray image pixel values to a user-controlled computer. One gamma-ray detection module combines CdWO4 scintillators (PSs, EPIC Crystal Co., 30,000 photons/MeV in scintillation efficiency), a Si-PIN photodiode (Si-PIN), and a control board. The PSs with a size of $4.6 \times 7 \times 30$ mm^3 (width $\times$ length $\times$ height) convert gamma-rays to visible light with a wavelength of 480 nm, and the Si-PINs detect the visible light and generate electrical signals. The control board in the gamma-ray detection module counts the number of electrical signals and restructures the pixel values of the images. The number of vertically arranged image pixel values in one gamma-ray detection module was 32. The image frame was composed of the arrangement of the image pixel values transmitted from the gamma-ray detection modules. The other gamma-ray detection module was installed in front of the gamma-ray generator to detect a reference signal with pure gamma-ray beam information with no air cargo signals. The reference signal was useful for gamma-ray image correction.

2.2. Neutron Generation and Detection

Fast neutrons were generated in nuclear fusion reactions of deuterium and tritium in a D-T generator that emits an isotropic and mono-energetic neutron beam with an energy of 14.1 MeV. The D-T generator acts as a small particle accelerator composed of an ion source, acceleration stage, target, and pressure stabilizer in a vacuum-sealed case. The ions were extracted from the plasma formed in the ion source using a single-stage acceleration optic system. The extracted beam was then focused on the target shown in Figure 3, which was constantly supplied at 150 kV. To reduce radiation risks, the D-T generator was covered with high-purity polyethylene panels and blocks, and 5 cm slits were open on the neutron shielding cover to emit the neutron beam.

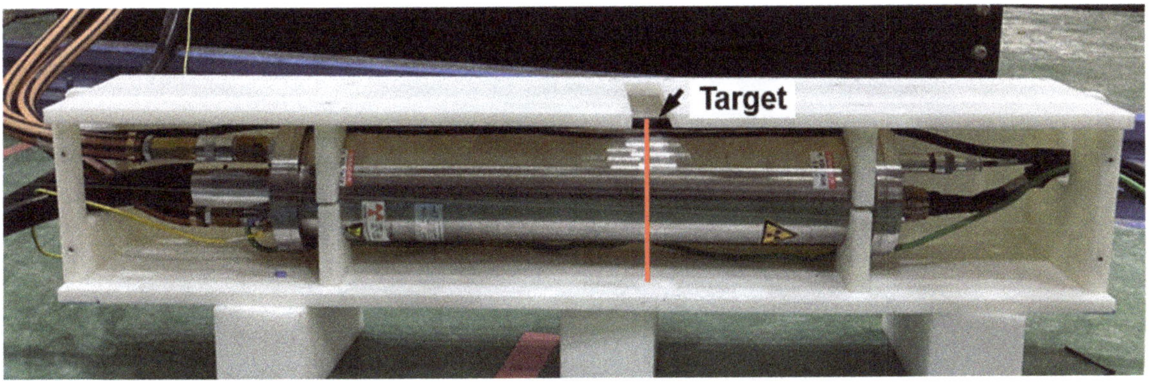

Figure 3. D-T generator. The red line in the center of the D-T generator indicates the position of the target.

During operation, the D-T neutron generator emits highly penetrating radiation, including neutrons and gamma-rays. Among the radiation, only neutron property data for air cargo must be collected. The radiation was detected using a plastic scintillator on the neutron detector. The neutron detector module in Figure 4 shows the composition of the plastic scintillator (PS), silicon photomultiplier (SiPM), analog-to-digital converter (ADC) board, and visible shielding case. A polyvinyl toluene-based plastic scintillator (PS, EJ-276G from Eljen Technology [15]) with a size of $10 \times 10 \times 30$ mm^3 was selected to develop a fast neutron detector with PSs arranged in a 2×8 array. PS converts the high-penetrating radiation to the visible region at a wavelength of approximately 490 nm. This scintillator has a scintillation efficiency of 8000 photons/MeV. A silicon photomultiplier (SiPM, S13360-6025CS, Hamamatsu) with a photosensitive area of 6×6 mm^2 senses visible light. The photon detection efficiency of this SiPM is 25%. The contact surface between the PS and SiPM is bonded to titanium oxide grease with a refractive index similar to that of PS to reduce optical losses. The SiPM counts a single photon generated from the PS and generates electrical output pulses. The ADC board measuring the electrical pulse was designed to sample it at a frequency of 125 MHz and a 14-bit resolution.

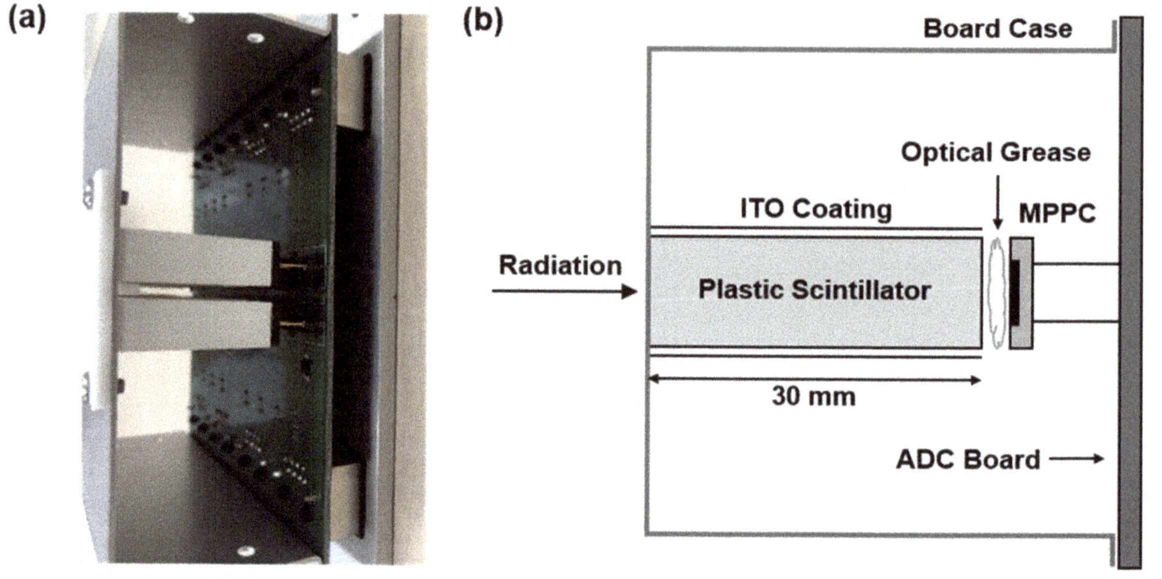

Figure 4. Neutron detector module: (**a**) side view of the neutron detector module, (**b**) block diagram of the neutron detector module. The plastic scintillator for the thickness of 30 mm renders high-energy radiations visible, the multi-pixel photon counter (MPPC) generates electrical pulses, and the analog-to-digital (ADC) board sends a signal to the DAQ system.

The electrical pulse shape is characterized by recoil protons and electrons excited by radiation. In general, electrical pulses arising from neutron signals tend to have higher peaks and longer tails than those arising from photon signals, as shown in Figure 5. The peak means the highest intensity of the pulse, and the tail is the integral of the post-peak time and the pulse end time. Using this feature, the pulse shape discrimination (PSD) technique differentiates neutron signals from various highly penetrating radiation signals [17,18]. With over 100,000 detections, the distribution of the electrical signals can be spread over a spot plot with peak and tail coordinates. Under stable measurement conditions, the PSD spot plot showed two regions: neutrons and gamma-rays. This result can also be displayed as a histogram of the ratio between the pulse peak and tail. The PSD histogram shows two Gaussian distributions, neutrons, and gamma-rays. In the PSD histogram, the valley of the two Gaussian distributions is determined by the area separating the neutron and gamma-ray signals.

The neutron detector modules were vertically arranged to create an image frame similar to the gamma-ray detector tower. Vertically arranged neutron detector modules were placed in front of the neutron generator. The neutron detector modules were connected to the data acquisition boards, and they supplied high-voltage power and transmitted the electrical pulse data.

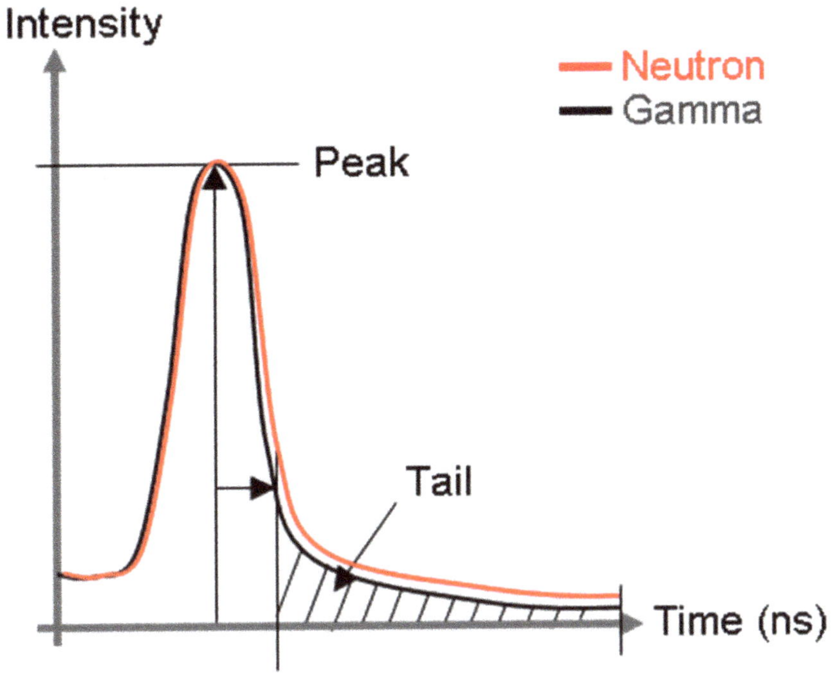

Figure 5. Example plot of normalized electrical pulses measured from the detectors. The peak is the highest intensity of the electrical pulse, and the tail is the integral value after the peak and the delay time.

2.3. Imaging Process

The gamma-ray and neutron detectors were connected to the data acquisition (DAQ) system to record the radiation image and communicate the machine control commands with a user computer. The number of electrical pulses obtained from the radiation detector during the measurement time represents the image pixel value. For instance, when the gamma-ray detector is activated for 1 s and the image pixel value is 1, and when the detector functions for longer, the image pixel value increases. The radiation detector modules and image pixel values were arranged vertically. The vertically arranged image pixels formed an image frame. By repeating the measurements at regular intervals, the image frames are horizontally accumulated, and the radiation image is expanded in two dimensions. Under this measurement condition, the container is displayed in the radiation image when the cargo container moves between the radiation generator and the radiation detector.

The radiation imaging system in the air cargo inspection system produces gamma-ray and neutron images with transmission properties. In the transmission characteristics of the radiation image, various information on the air, container, and beam, as well as the sample material, is present. However, security inspection requires only the intrinsic properties of the sample material. The reference and gain signals are used to extract the intrinsic properties of the sample material. Because the reference signal is measured before the radiation passes through the cargo container, it only has information about the pure beam intensity of radiation. Image correction using the reference signal can remove beam intensity fluctuations in the radiation image. The gain signal indicates the radiation signals passing through the air or container with no information of the sample material. This image correction using the gain signal separates the background transmission characteristics from the mixed transmission characteristics of the sample material. When used in an optical system, the gain correction removes the background information and characterizes the absorption of the sample material. The absorption coefficient is an intrinsic property of a sample material. The optical path length is required to calculate the absorption coefficient from the transmission data. However, because security inspection proceeds in an unopened state, the optical path length cannot be measured, and another method is required.

The unique information of the sample material inside the air cargo container can be extracted from the ratio of radiation absorption properties: the neutron and gamma-ray absorption values of the sample material. The ratio of the neutron and gamma-ray attenuation coefficients is:

$$R = \frac{\ln\left(I_n / I_n^0\right)}{\ln\left(I_g / I_g^0\right)} \quad (1)$$

where I_n is the measured neutron intensity through the object, $I_n{}^0$ is the measured neutron intensity with the object removed, I_g is the measured gamma-ray intensity through the object, and $I_g{}^0$ is the measured gamma-ray intensity with the object removed. The R-value does not require the calculation of the optical path length because the radiation has the same sample material optical path. Hence, calculating the R-values of the sample materials made material classification using the air cargo inspection system possible. The R-value is defined as the ratio of neutron absorption to gamma-ray absorption.

The R-values of the sample materials extracted from the radiation images were re-painted onto the radiation image as a color map. The R-value image shows the differentiation of the materials. It is important to display the R-value image so that the operators can ensure that the positions of carriages match their R-values in real airports or cargo inspection rooms. To match the R-values in the radiation image, the pixel numbers and object positions must be identical between the neutron and gamma-ray images. Because the number of pixels in the gamma-ray image is greater than that in the neutron image, the pixel values of the neutron image are interpolated. Object position matching between the two radiation images was performed using the image registration method.

2.4. Air Cargo Container Translator

During air cargo inspection, a radiation image is created by the linear motion of the cargo container. The linear motion is supported by a motorized linear stage with a translation speed of 0.1 to 1000 mm/s. The motorized stage can move between the neutron source and the detector. The transfer movement is remotely controlled to change and adjust the speed, direction, and distance. An air cargo container with a size of approximately $233 \times 162 \times 153$ cm^3 was fabricated to be of the type used at airports. The interior of the container contained three layers of shelves for loading the material samples.

2.5. Sample Material Preparation

To investigate the material classification of the air cargo inspection system, 30 materials listed in Table 1 were prepared. Their density and depth are different and in the range of 0.41 to 19.25 g/cm^3 and 17 to 430 mm, respectively. The materials had the same front size of 20×25 cm^2. The metallic samples were machined, and the powder samples were contained in an acrylic case. A maximum of six samples were placed on a shelf in the cargo container. The shelves with samples were arranged at different heights and depths to ensure that the radiation paths did not overlap. Radiation imaging experiments were conducted with various combinations of sample materials in different locations.

Table 1. Sample material list with density and depth.

No.	Item	Density (g/cm^3)	Depth (mm)
1	Bean Flour	0.41	300
2	Wheat Flour	0.70	250
3	Paper	0.80	210
4	Paraffin Wax	0.90	150
5	Polyethylene	0.96	150
6	Water	1.00	200

Table 1. *Cont.*

No.	Item	Density (g/cm^3)	Depth (mm)
7	Polyurethane	1.05	240
8	Nylon	1.15	280
9	Acetamide	1.16	150
10	Ammonium Acetate	1.17	200
11	Lucite	1.19	150
12	Rubber	1.20	170
13	Rice	1.29	200
14	Polyester	1.36	330
15	Green Tea Powder	1.37	120
16	Rayon	1.53	430
17	Melamin	1.57	365
18	Sugar	1.59	220
19	Animal Bone	1.90	220
20	Teflon	2.20	150
21	Graphite	2.25	150
22	Glass	2.50	120
23	Concrete	2.50	120
24	Aluminum	2.70	150
25	Ceramic Powder	4.00	70
26	Titanium	4.50	80
27	Iron	7.87	60
28	Bronze	8.89	45
29	Lead	11.34	40
30	Tungsten	19.25	17

3. Results

The detected radiation is of neutrons and gamma-rays of various energies. Figure 6a shows the representative single pulses of gamma-rays and neutrons. The tail regions of normalized electrical pulses of neutrons have higher values. A distribution trend of the main pulse peak intensity and tail region integration was obtained by exposing the neutron detector with the D-T generator activated for 60 s, as shown in Figure 6b. The scatter plot in Figure 6 shows two groups with different ratios of peak intensity and tail integration, and these two groups are separated by a linear line. Electrical pulses with lower intensities were considered as noise signals and were excluded from the scatter plot. The linear line in Figure 6b is the reference value for separating the neutron signals. This reference value can be calculated from the histogram plot of the ratio between the tail and peak in Figure 6c. The tail and peak ratio histograms can be fitted to two Gaussian fits, and the minimum in the valley of the Gaussian fitted plots is the reference value. In addition, the figure of merit (FOM) can be calculated from the two Gaussian fitted plots. The FOM is a value that establishes the ability to discriminate neutrons and other radiation and is defined as [14,19]

$$\text{FOM} = \text{Dpp}/(\text{GFWHM} + \text{NFWHM}),$$

where Dpp is the distance between the peaks of the Gaussian fitted plots, and GFWHM and NFWHM are the full widths at half maximum (FWHM) of the gamma and neutron peaks, respectively. The FOM of the neutron detector is approximately 1.12.

The number of electrical pulses attributed to neutrons or gamma-rays corresponds to the image pixel value. The image pixel values are transformed to the gray color map to display the radiation image. The radiation image is obtained with the horizontal translation of the cargo container.

Several experiments were conducted using different combinations of sample materials to acquire radiation images. The translation speed was adjusted from 16 mm/s to 200 mm/s. The radiation dose rates were 1.25 Gy/min for gamma-rays and 10^8 n/s for neutrons. The radiation images were captured for 30 s and 90 s for the gamma-ray and neutron images, respectively. Figure 7a,b show one of the results of the gamma-ray and neutron images. These images were corrected with reference and gain signals, and they presented the absorption properties. Figure 7c shows a visible image of the material sample inside the air cargo container.

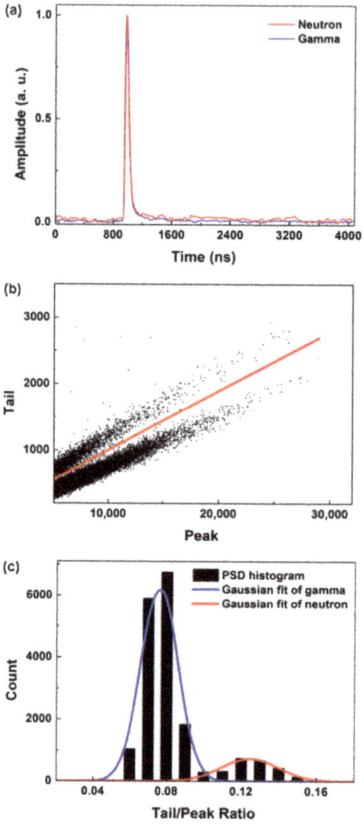

Figure 6. Signals detected with the neutron detector. (**a**) Pulse light detected from the neutron detector modules. (**b**) PSD scatter plot with a PSD threshold (red line). (**c**) PSD histogram with Gaussian fitting plots.

Using the gamma-ray and neutron images and the absorption properties of the radiation, the R-values of the 30 sample materials were calculated, as shown in Figure 8. The R-values were in the range of 0.26 and 1.26 for the prepared sample materials listed in Table 1. The R-values of the metals were mainly below 0.6, but their values were sufficiently different. Plastics have higher R-values at lower densities, and there were large differences between the R-values for materials of different atomic compositions. Organics had R-values greater than 0.6, except for animal bones, rubber, and melamine, most of which had values near 1. It is considered that the R-value is affected by the atomic compound because organics with R-values near 1 are vegetable substances, only metals have R-values under 0.6, and plastics have different R-values for different polymer families.

Figure 9 shows the R-value image of the air cargo container loaded with samples of various materials. The R-value image is obtained from the color mapping result by matching the sample positions of the gamma-ray and neutron absorption images and calculating the R-value for each pixel. The mapping colors of the R values were colors of the rainbow, as shown in the legend of Figure 9. The hollow frame and empty area in the container are filled to the air, so their R-values are almost zero, and their areas are displayed to the black color. Metals with R values of less than 0.6 have reddish colors and are clearly distinguished from other materials. Animal bones, Teflon, and rubber in the air container had higher R-values than the metals. Acetamide and polyethylene had the highest R values in Figure 9. The color map image cannot give detailed R values to distinguish objects of the same materials; however, it can differentiate between the material classes of metals, plastics, and organics.

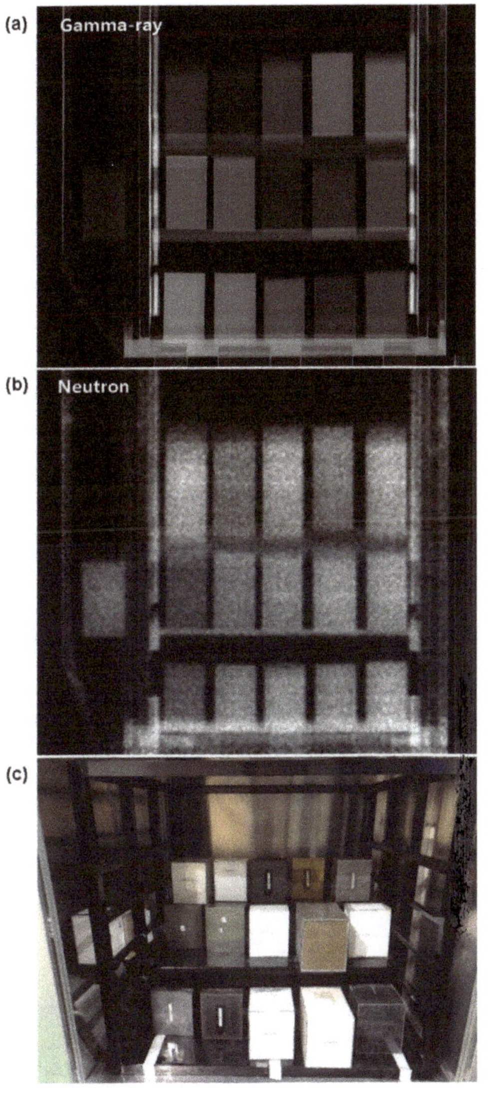

Figure 7. (**a**) Gamma-ray and (**b**) neutron image results of (**c**) material samples arranged in the air cargo container.

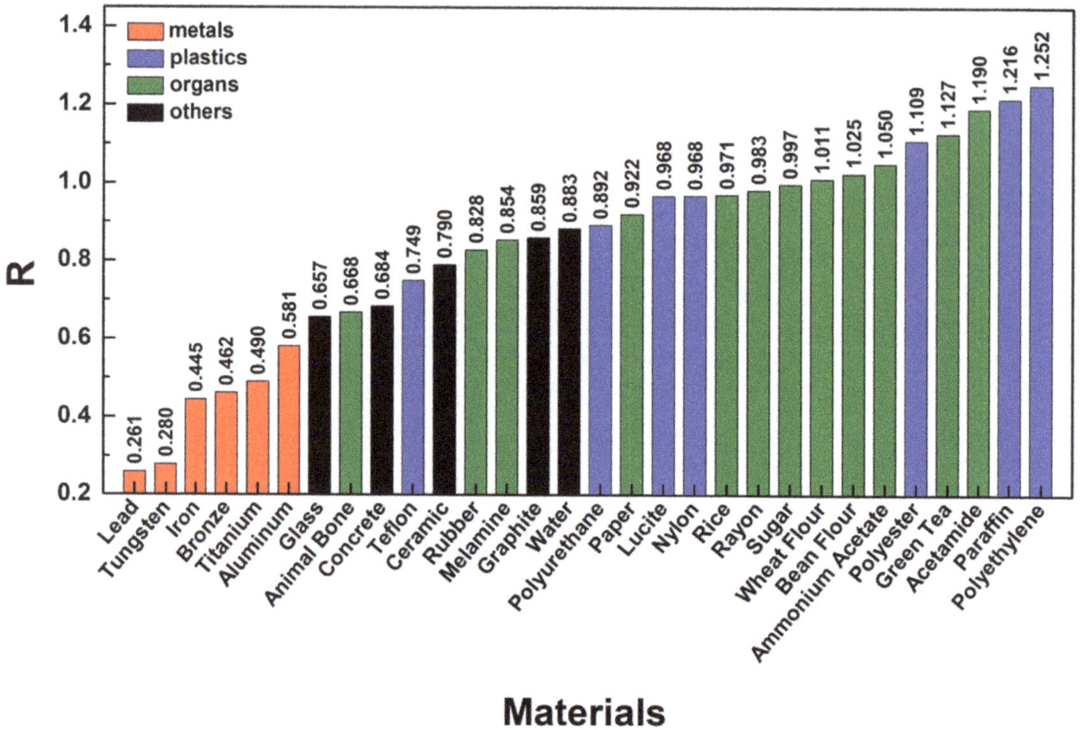

Figure 8. R-values of material samples.

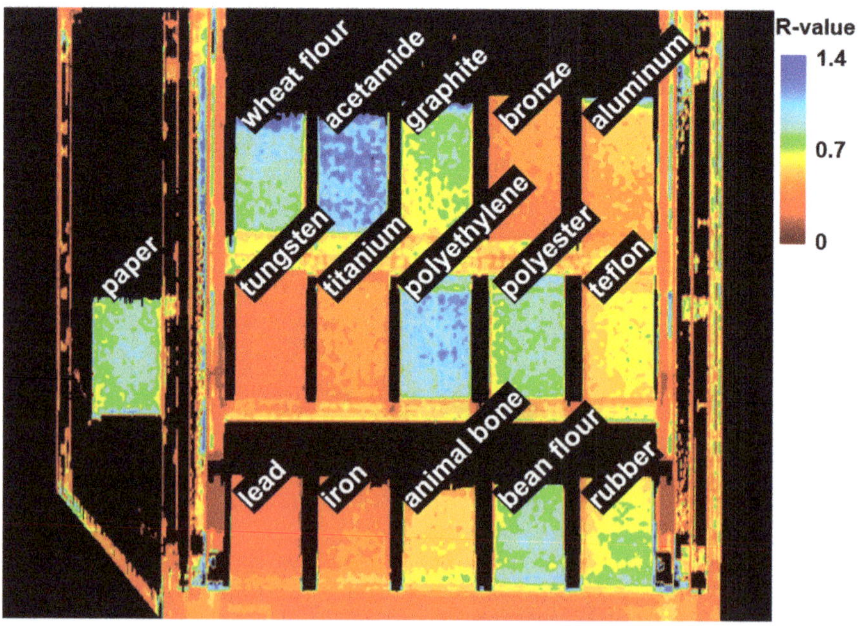

Figure 9. R-value image for an air cargo container with material samples.

4. Conclusions

We developed an air cargo inspection system using a D-T neutron generator and accelerator-based gamma-ray source. The neutron source generates neutrons with an energy of 14.1 MeV and a flux of 108 n/sec. The gamma-ray source produced a gamma-ray beam with an energy of 6 MeV and a dose of 1.25 Gy/min. A neutron and gamma-ray imaging system was installed to acquire radiation transmittance images of 30 materials loaded in an air cargo container. These radiations can penetrate materials several tens of centimeters thick within an air cargo container. A linear motorized translator moved the air cargo container between the radiation sources and detectors. The radiation detectors were based on plastic scintillators, silicon photomultipliers, and electric circuits. They were arranged vertically to scan the moving stage, including the cargo container. The radiation signals were counted as image pixel values, according to the PSD method. As a result, the radiation imaging system outputs neutron and gamma-ray images with transmittance. Data processing with image interpolation and registration was used to calculate R values from the transmittance data and displayed them on the color map image to differentiate between 30 types of materials: metals, plastics, organics, and others. This process validates the material classification capability.

The R-value image is powerful in distinguishing more diverse materials than the use of a single radiation source. However, the industrial fields require more complicated systems for use in the mixed or packed material classes in the real cargo container. In the imaging system, the three-dimensional imaging system is required because the mixed materials are superposed on the beam path. Our further step in the development of the security inspection system is a tomographic inspector that uses gamma-rays and neutrons.

Author Contributions: Conceptualization, J.Y.P. and J.M.; methodology, J.Y.P. and J.H.L.; software, J.Y.P.; validation, Y.-H.Y. and M.C.; formal analysis, J.Y.P.; investigation, J.M.; resources, M.L.; data curation, J.Y.P.; writing—original draft preparation, J.Y.P.; writing—review and editing, J.Y.P.; visualization, J.Y.P.; supervision, J.Y.P.; project administration, N.-H.L.; funding acquisition, M.L. All authors have read and agreed to the published version of the manuscript.

Funding: This work has been carried out under the Nuclear R&D Program of the Ministry of Science and ICT(MIST) of Korea (Project NO. 523420-22 and NRF NO. 2019M2A2A4A05031483).

Institutional Review Board Statement: Not applicable.

Informed Consent Statement: Not applicable.

Data Availability Statement: Not applicable.

Conflicts of Interest: The authors declare no conflict of interest.

References

1. Kansouh, W.; Ahmed, M.; Bashter, I.; Megahid, R. Effectiveness of X and Gamma Rays for Scanning Cargo Containers. *Am. Acad. Sci. Res. J. Eng. Technol. Sci.* **2016**, *21*, 99–108.
2. Kutsaev, S.; Agustsson, R.; Arodzero, A.; Berry, R.; Boucher, S.; Diego, A.; Gavryushkin, D.; Hartzell, J.; Lanza, R.; Smirnov, A.Y. Linear accelerator for security, industrial and medical applications with rapid beam parameter variation. *Radiat. Phys. Chem.* **2021**, *183*, 109398. [CrossRef]
3. Eberhardt, J.; Rainey, S.; Stevens, R.; Sowerby, B.; Tickner, J. Fast neutron radiography scanner for the detection of contraband in air cargo containers. *Appl. Radiat. Isot.* **2005**, *63*, 179–188. [CrossRef] [PubMed]
4. Xia, L.; Ouyang, X.; Wang, Q.; Kang, K.; Tan, X. Measurement of the gamma-ray sensitivity and signal-to-noise ratio of a new scattered-electron detector. *Rev. Sci. Instrum.* **2008**, *79*, 093303. [CrossRef] [PubMed]
5. Franzel, T.; Schmidt, U.; Roth, S. Object detection in multi-view X-ray images. In Proceedings of the Joint DAGM (German Association for Pattern Recognition) and OAGM Symposium, Graz, Austria, 29–31 August 2012; pp. 144–154.
6. Zhang, Y.; Kong, W.; Li, D.; Liu, X. On using XMC R-CNN model for contraband detection within X-ray baggage security images. *Math. Probl. Eng.* **2020**, *2020*, 1823034. [CrossRef]
7. Hättenschwiler, N.; Mendes, M.; Schwaninger, A. Detecting bombs in X-ray images of hold baggage: 2D versus 3D imaging. *Hum. Fact.* **2019**, *61*, 305–321. [CrossRef] [PubMed]
8. Allen, P.B. Neutron spectroscopy of superconductors. *Phys. Rev. B* **1972**, *6*, 2577. [CrossRef]

9. Bowman, C.; Bilpuch, E.; Newson, H. S-and P-wave neutron spectroscopy: Part VIII. Subshell effect on nuclear level spacing near A = 50. *Ann. Phys.* **1962**, *17*, 319–378. [CrossRef]
10. Senyshyn, A.; Mühlbauer, M.; Nikolowski, K.; Pirling, T.; Ehrenberg, H. "In-operando" neutron scattering studies on Li-ion batteries. *J. Power Sour.* **2012**, *203*, 126–129. [CrossRef]
11. Hubbell, J.H.; Seltzer, S.M. *Tables of X-ray Mass Attenuation Coefficients and Mass Energy-Absorption Coefficients 1 keV to 20 MeV for Elements Z = 1 to 92 and 48 Additional Substances of Dosimetric Interest*; NIST: Gaithersburg, MD, USA, 1995.
12. Sweany, M.; Marleau, P.; Hammon, S.; Kallenbach, G.; Polack, J.K. *Design and Evaluation of a Pixelated PSD-Capable Scintillator Detector with SiPM Readout*; Sandia National Lab. (SNL-CA): Livermore, CA, USA, 2019.
13. Bourne, M.M.; DeBruin, S.G.; Clarke, R. SiPM-based compact time-of-flight detector for neutron spectrometry of fission samples. *Nucl. Instrum. Methods Phys. Res. Sect. A Accel. Spectromet. Detect. Assoc. Equip.* **2020**, *954*, 162000. [CrossRef]
14. Pozzi, S.; Bourne, M.; Clarke, S. Pulse shape discrimination in the plastic scintillator EJ-299-33. *Nucl. Instrum. Methods Phys. Res. Sect. A Accel. Spectromet. Detect. Assoc. Equip.* **2013**, *723*, 19–23. [CrossRef]
15. Kotasidis, F.; Matthews, J.; Angelis, G.; Markiewicz, P.; Lionheart, W.; Reader, A. Impact of erroneous kinetic model formulation in Direct 4D image reconstruction. In Proceedings of the IEEE Nuclear Science Symposium and Medical Imaging Conference (NSS/MIC), Valencia, Spain, 23–29 October 2011.
16. Grodzicka-Kobylka, M.; Szczesniak, T.; Moszyński, M.; Brylew, K.; Swiderski, L.; Valiente-Dobón, J.; Schotanus, P.; Grodzicki, K.; Trzaskowska, H. Fast neutron and gamma ray pulse shape discrimination in EJ-276 and EJ-276G plastic scintillators. *J. Instrum.* **2020**, *15*, P03030. [CrossRef]
17. Roush, M.L.; Wilson, M.; Hornyak, W.F. Pulse shape discrimination. *Nucl. Instrum. Methods* **1964**, *31*, 112–124. [CrossRef]
18. Cester, D.; Lunardon, M.; Nebbia, G.; Stevanato, L.; Viesti, G.; Petrucci, S.; Tintori, C. Pulse shape discrimination with fast digitizers. *Nucl. Instrum. Methods Phys. Res. Sect. A Accel. Spectromet. Detect. Assoc. Equip.* **2014**, *748*, 33–38. [CrossRef]
19. Lempicki, A.; Wojtowicz, A.; Berman, E. Fundamental limits of scintillator performance. *Nucl. Instrum. Methods Phys. Res. Sect. A Accel. Spectromet. Detect. Assoc. Equip.* **1993**, *333*, 304–311. [CrossRef]

applied sciences

MDPI

Article

GisaxStudio—An Open Platform for Analysis and Simulation of GISAXS from 3D Nanoparticle Lattices

Igor Mekterović [1], Gabrijela Svalina [2], Senad Isaković [3] and Maja Mičetić [2,*]

1 Faculty of Electrical Engineering and Computing, University of Zagreb, 10000 Zagreb, Croatia
2 Ruđer Bošković Institute, 10000 Zagreb, Croatia
3 Faculty of Science, University of Sarajevo, 71 000 Sarajevo, Bosnia and Herzegovina
* Correspondence: maja.micetic@irb.hr; Tel.: +385-1-468-0224

Featured Application: This paper presents the open software platform GisaxStudio for analysis of GISAXS maps from 3D lattices of nanoparticles.

Abstract: Grazing-incidence small-angle X-ray scattering (GISAXS) is a powerful method for the structural analysis of ordered arrays of nanoparticles, quantum dots, or similar objects. However, for the correct interpretation of the measured GISAXS intensity distributions, a proper data analysis, including a suitable model, is required. Here, we demonstrate a software platform, GisaxStudio, aimed at the analysis and simulation of 2D GISAXS intensity distributions from ordered lattices of different nanoparticles. It contains several models that satisfactorily describe the GISAXS from 3D lattices or crystals of nanoparticles prepared by the self-assembly processes, pre-pattering, or ion-beam interaction with the material within their tracks. It also supports different shapes of nanoparticles, including core-shell structure with the center of core possibly displaced from the center of the shell. The software is very useful for fast and accurate GISAXS data analysis.

Keywords: GISAXS; GisaxStudio; nanoparticle lattices; quantum dot crystal; quantum dots; nanoparticles; structure

Citation: Mekterović, I.; Svalina, G.; Isaković, S.; Mičetić, M. GisaxStudio—An Open Platform for Analysis and Simulation of GISAXS from 3D Nanoparticle Lattices. *Appl. Sci.* **2022**, *12*, 9773. https://doi.org/10.3390/app12199773

Academic Editors: Bogusław Łazarz, Grzegorz Peruń and Tangbin Xia

Received: 31 August 2022
Accepted: 26 September 2022
Published: 28 September 2022

1. Introduction

The production of materials containing ordered lattices of nanoparticles (NPs) or other nano-objects (NOs) is of great importance today, due to their very interesting properties and many relevant applications in modern nanotechnology [1–3]. For example, semiconductor quantum dots (QDs) are the focus of modern science because of their unique optical and electrical properties, tunable by their size and internal structure due to the confinement effects [1–5]. Especially interesting are QDs with a core-shell structure, in which confinement effects are very strong in both the core and the shell [6–9]. They strongly influence the current carrier and optical properties, ensuring that such materials are highly applicative in solar cells, detectors sensors, and many other devices. Quantum wires (QWs) are another very interesting material due to their specific shape, high surface to volume ratio, and multiple exciton generation effects, which are also highly explored in many photo-sensitive devices [10,11]. As all these properties are highly size- and shape-sensitive, it is very important to develop effective tools for their accurate determination.

In addition to the nanoparticle shape and size, another important factor for the application of such materials is their ordering on the surface or within the matrix. The regular ordering is usually desirable because it allows for controlling the separation of NPs, which greatly influences the material electrical properties. The regularity in ordering can also induce some novel effects in the material, such as mini-band formation, or specific conduction in the desired spatial direction, as well as many others [12,13].

Therefore, the tools for the detailed characterization of size, shape, and arrangement of NPs in a material are of great importance. Grazing-incidence small-angle X-ray scattering

(GISAXS) is the leading technique for this purpose because it provides data in reciprocal space with excellent statistics [14–19]. The materials with ordered lattices of NPs show characteristic peaks in GISAXS, whose positions and widths are determined by the ordering type of NPs and the degree of regularity in their ordering. Due to that fact, even a visual inspection of the GISAXS map enables a rough estimation of the materials structure, including NP arrangement, separation, and shape. However, for obtaining detailed information about the systems structure, a numerical analysis of a GISAXS map should be performed.

The main problem with the analysis of GISAXS intensity distributions is the proper choice of a model that describes NP ordering, especially the deviations from the ideal NP positions given by the lattice type. The deviations from the ideal NP positions are very important because they significantly affect the position and the width of the peaks in the GISAXS map. Therefore, only the proper choice of the model will lead to the yielding of accurate data from GISAXS analysis. We have analyzed different types of NP ordering in 3D lattices and proposed theoretical models for the description of GISAXS in such systems. These models provide accurate structural data [20]. In another work [21], we have analyzed GISAXS from ion-beam treated materials in which the ion beam forms NPs along its tracks. Finally, some examples are given in [22]. Although these models are published and have proven to be efficient for the analysis of the GISAXS maps, they are usable only by very few experts in GISAXS analysis and modeling. Many users from various disciplines (physics, chemistry, biology, medicine, industry) are in need of the GISAXS technique, but most of them do not know how to analyze the data. Therefore, the software platform which incorporates these models and which is usable by a wide audience is still missing.

Several excellent platforms for description and analysis of GISAXS exist [23–29], while a full list of the available software for GISAXS analysis can be found in [30]. One of the best known is IsGISAXS [23], aimed at the analysis of different types of islands supported on a substrate. It supports different ordering types of islands, from random distribution via paracrystal, to fully ordered systems. Another frequently used platform is NANOCELL [24], which enables the simulation of 2D diffraction patterns from single-crystals for GISAXS/GISANS geometry. Further, FitGISAXS contains models for the description of GISAXS for monodisperse, polydisperse, and interacting particles with various size distributions, form factors, and structure factors [25]. HipGISAXS is suitable for the treatment of a wide range of structures, shapes, or morphologies, including multilayered polymer films and nanoparticles on top of, or embedded in, a substrate or polymer film layers [26]. DPDAK is a software for the processing and reduction of large amounts of data from synchrotron scattering experiments [27]. It also contains tools for calibration and correction of raw data, one- or two-dimensional integration, and the fitting and further analysis of the data, including the extraction of certain parameters. The most recently developed software is BornAgain, an excellent software for the simulation and fitting of X-ray and neutron reflectometry, off-specular scattering, and grazing-incidence small-angle scattering (GISAS) [28,29]. This platform includes most of the options available in IsGISAXS, along with many new features. BornAgain has a very appealing user-interface and it is, being fairly generic, usable for scientists from different disciplines.

Although currently existing platforms can be used to extract some important features of the 3D assemblies of NPs produced by self-assembly processes, they lack detailed analysis of the NP ordering properties. As stated above, and also discussed in Ref. [20], the NP ordering properties, or more precisely, the type of deviations of the NP positions from the ideal ones, can differ in various spatial directions in the same material. This fact strongly influences the GISAXS intensity distribution, and simulating the intensity distribution by the improper models usually results in incorrect structural parameters, even if the simulated GISAXS map agrees well with the measured one. All things considered, the current software options provide many options, but they are not detailed enough for suitable analysis of the materials based on 3D-ordered NP lattices, nor for specific core-shell structures with the displaced core-shell origins.

In this paper, we present the software platform GisaxStudio for the analysis of thin films consisting of 3D lattices of NPs produced by self-assembled growth or ion-beam irradiation [31]. This software is a result of GISAXS data analysis and suitable model formulation of the aforementioned thin films that has been developed over the last 20 years. It enables the analysis and fitting of 2D GISAXS intensity distributions from 3D NP crystals, 3D lattices of NPs having spheroidal or core-shell shape, and from the NP arrays formed within ion-beam tracks. By fitting of the entire 2D GISAXS map, or even some of its parts, the GisaxStudio enables the determination of the NP's shape, size, and 3D arrangement properties, including the crystal lattice parameters and disorder type along each of the basis vectors of the 3D lattice. It is also possible to determine statistical distributions of all these parameters. The software includes theoretical models published in [20–22], which take into account the NP ordering properties along each of the basis vectors of the 3D NP lattice. The models and the software are already successfully applied in the GISAXS analysis of many systems, including 3D NP lattices [5,7–9,11,13,20,21,32–36] and NPs formed within ion tracks [22,37–39], where fitting of the full 2D measured map was performed. We show how the platform is constructed, reveal its main features, and demonstrate its performance. The process of preparation for fitting and its main steps are demonstrated. The platform is free for non-commercial use, available at a link given in [31], and is very useful to a broad audience for the analysis of the measured GISAXS maps or for the prediction of measurements using its simulation option.

2. Materials and Methods

GisaxStudio is a modular, multi-platform program for GISAXS analysis of various nanostructured materials. It is written in Java programming language, featuring a graphical user interface (GUI), built-in optimization algorithms, and visualization. It stores all data in the relational database, which facilitates data exchange and reproducibility. Figure 1 shows the architecture of the GisaxStudio application. It is important to note that GisaxStudio has been developed in a modular fashion to enable extendibility and cooperation between the physicist and the computer scientist. Thus, physical models (see Section 3.1) are completely decoupled from the optimization algorithms, GUI, visualization, etc.

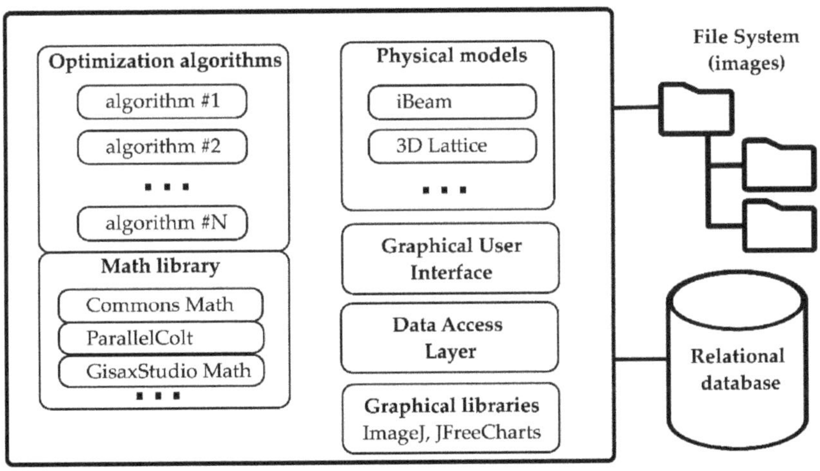

Figure 1. GisaxStudio architecture. Physical models are decoupled from the generic optimization algorithms. Images are stored in file system, and project data is stored in a JDBC-compliant relational database.

Physicists deal only with the models, and in order to produce a new model, they merely provide two Java classes:

- **Parameters class** which extends `AbstractGisaxsParam` and describes the model's parameters. For instance, instead of the existing iBeam model, suppose our model is a line $y = ax + b$. The parameter class would then have only two double members: x and y. Members must be annotated in order to expose them to the user via GUI, so that the parameters class would contain simply:

```
public class LinearModelParams extends AbstractGisaxsParam {
      @GxParam(defval = 1.00, absLimits = {-1e5, 1e30}, name = "Slope (a)", group
      = "Slope and intercept", tab = "First tab", ordinal = 1)
      public double a;
      @GxParam(defval = 0, absLimits = {-1e30, 1e5}, name = "y-intercept (b)",
      group = "Slope and intercept", tab = "First tab", ordinal = 2)
      public double b;
}
```

  Through the use of reflection, GisaxStudio will analyze its own code at runtime and dynamically build the GUI dialog for the selected model (e.g., Figure 5a for the iBeam model). Besides primitive types, the parameter class also supports complex numbers, lists of values (manifested as dropdowns for the user), etc. Of course, besides fittable parameters, the parameters class can also include constants and other non-fittable parameters that are easily assigned by the user.

- **Model class,** which implements the `IGisaxsModel` interface, defines 16 methods, the most important being `compute(double[])`, which computes the model given an array of fittable parameters. In our example, compute would receive a and b via the double array and simply return $a*x + b$, where x is a vector that has already been assigned:

```
@Override
public double[] compute(double[] params) {
      LinearModelParams pObj = (LinearModelParams)
                              fitVectorToParamObject(params);
      return _xFit
            .times(pObj.a)
            .plus(pObj.b)
            .toArray();
}
```

  `LinearModel` is actually included as a proof-of-concept in the GisaxStudio code, and the code shown above is taken from those two classes (although it is not shown in Figure 1). Helper method `fitVectorToParamObject` is used to instantiate and populate the params object so that the programmer could use meaningful variable names (a, b) in the code and autocompletion features instead of working with `params[0]` and `params[1]`.

In conclusion, the physicist deals only with the model and the math, having at their disposal math libraries such as Apache Commons Math [40], ParallelColt [41], and a custom developed matrix and vector library which enables Matlab-like expressions, such as the `.times()` and `.plus()` methods shown above. GisaxStudio could be used to fit anything which is manifested/measured as a matrix (image). Apart from these two classes and two parameters in the main application screen ($y0$ and specular shown in Figure 4), there is nothing GISAXS-specific in the remainder of the application—everything is built dynamically via reflection.

The fitting (optimization) process, whose simplified flow is shown on Figure 2, treats the model as a "black box" and varies the fittable parameters according to some algorithm, calling each time the compute method, in an attempt to find the global minimum in the least squares sense. As GISAXS models are complex and non-linear, it is not reasonable to pick "one best" optimization algorithm. Therefore, in accordance with the generic nature of the application, we have opted for the following approach: GisaxStudio provides a number of different optimization algorithms, and the user can try them all and decide to use those that perform best for their model. Thus, for the time being, the following eight optimization algorithms are included: BobyQA, Bounded Hooke and Jeeves, CMA-ES, Hooke and Jeeves, Nelder–Mead, Powell, Simulated Annealing, and Steady State GA (Figure 2). New algorithms can be easily added, much like the GISAXS model procedure described above: it is simply a matter of implementing the assigned interface and registering the algorithm; the application will then pick up the algorithm and include it in the fitting process.

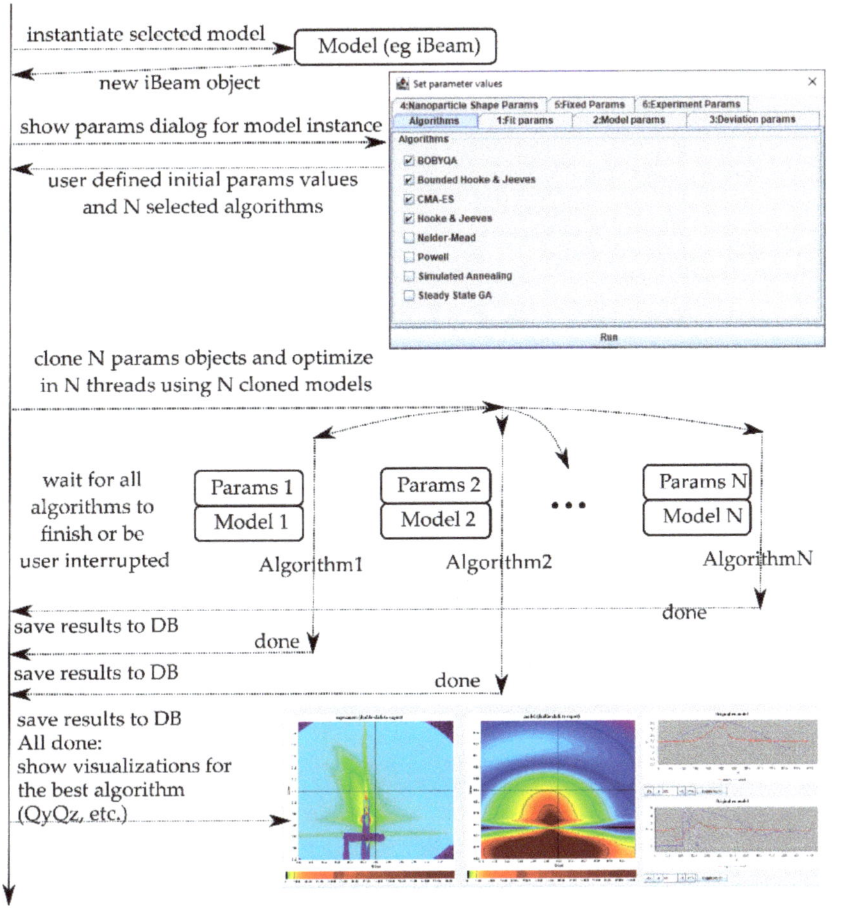

Figure 2. Fitting procedure in GisaxStudio: after the user selects the model and assigns initial values of the corresponding model, the application spawns N threads (corresponding to the number of chosen optimization algorithms) and initiates a multi-threaded optimization. When all the threads are finished (or interrupted by the user), the application visualizes the resulting winning algorithms.

GisaxStudio instantiates the selected model's class and retrieves the parameter objects from that instance, which is then inspected via reflection to dynamically build the parameters dialog GUI. This dialog presents the last used values or default values, if it is being used for the first time. After the user sets the initial parameter values and selects the optimization algorithms (e.g., N = 4 in the Figure 2), the application creates N threads, one for each algorithm, provides them with copies of the model and parameters object, and starts the multi-threaded optimization. The main thread receives notifications from the optimization threads and waits for them to finish or to be interrupted by the user. The results are saved in the database as they arrive. Finally, when all the threads are finished, the best optimization result is selected, and a new tab with the visualizations of the winning algorithm, i.e., parameters, is automatically created.

GisaxStudio features a project organization where one project, defined and named by the user, comprises of an arbitrary number of GISAXS images, which are stored in the projects/images subfolder of the application. All other data is stored in a JDBC-compliant (Java DataBase Connectivity) relational database: metadata about the images, regions of interest (ROIs), fits and parameters, results obtained with fits, etc. By default, the portable in-memory H2 Database Engine is used [42], although, any other JDBC-compliant database could be used. Note that a shared environment could also be easily configured by pointing the GisaxStudio to the shared database example, such as PostgreSQL in a local network or shared file system for images. However, for the sake of the simplicity, the default setup is single-user. Such setup allows for portability by simply copying the GisaxStudio folder, which usually contains images and database subfolders, to another computer, even if that computer features a different operating system.

3. Results

3.1. Main Features

GisaxStudio is a platform for the GISAXS analysis of two types of NP-containing materials: 3D lattices of nanoparticles (experiment type 3dLattice) and ion beam modified materials (iBeam). Figure 3 shows the main window of the software with the choice of the mentioned options. The theory describing the model and functions for the 3dLattice option is published in [20], while the theory for the iBeam option is given in [22].

Figure 3. Starting window of GisaxStudio enabling the choice of the experiment type: the 3dLattice option should be used for the analysis of GISAXS from 3D ordered lattices of NPs, while iBeam is aimed at the analysis of NPs formed within ion-beam induced tracks.

After the project type is chosen, the window (shown in Figure 4) with the main options for the simulation or analysis of the measured GISAXS maps is opened. The tab *'Project'* enables setting up new project and loading the set of image files that needs to be analyzed, or loading previous projects. The loaded files are displayed at the rightmost column of the window. GisaxStudio enables some simple image processing of the experimental GISAXS maps. For example, the *Preprocess* button removes the detector's 'dead' lines (black lines in the GISAXS maps), while the *Smooth* button automatically smooths out the GISAXS map. The *Replace to fit* button sets the processed image as the one that will be used for further fitting and data analysis. The tab *'Reports'* provides direct access to the relational database used by the program, so that data can be exported as CSV, and advanced users can even execute custom *select* and *update* SQL queries.

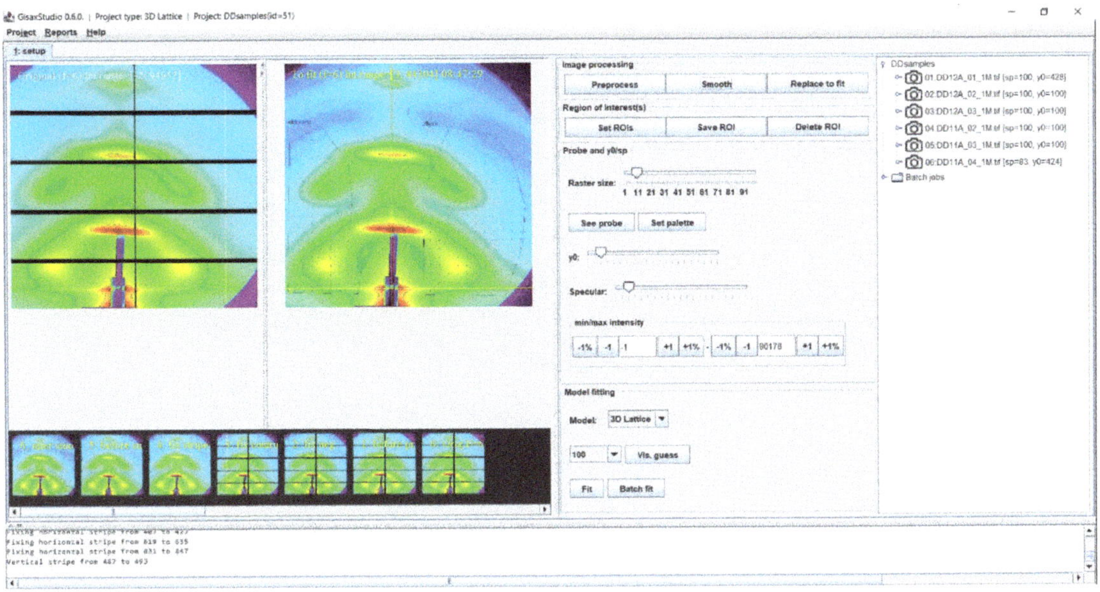

Figure 4. Window for setup the analysis options for 3dLattice experiment type. It contains the GISAXS map which should be analysed, as well as the options for image processing and simulation in the middle columns, followed by the list of loaded files in the current project on the right side.

One of the main steps in the fitting procedure of the measured data is to choose an arbitrary number of ROIs that will be fitted. ROIs can be chosen using the button *Set ROI* which opens the open-source ImageJ application incorporated into GisaxStudio. Choosing particular ROIs is required because the measured GISAXS map always contains the contribution of coherent scattering, and it usually contains contributions from the surface roughness close to the $Q_y = 0$ plain, which we do not wish to include in the fit because it cannot be accounted for by the model. In addition, the measured maps often contain some shadows of the vacuum tube, beam-stoppers, or similar items which we do not want to fit. Therefore, we choose the areas of the map (dotted parts in Figure 4) that contain NPs contributions only. These selected parts of the map will be used for fitting.

GisaxStudio enables the simple setting of the incident angle and the lateral beam position directly at the measured map. The $y0$ parameter, represented by the vertical line on the processed map, should be positioned at the center of the measured signal, while *specular* (represented by the horizontal line) should be positioned at the reflected beam.

3.2. Visual Estimate Option

In general, a program fits the given set of parameters. These parameters have reasonable defaults set by the GisaxStudio, which can be manually tweaked by the user. Therefore, before the fitting process, it is very useful to use the option *Vis guess*, which calculates the GISAXS map from the current parameters in the list. The same parameters will be remembered and used as initial parameters for the subsequent fit. This is an optional step, but it enables an experienced user to help the fitting process by moving the initial default parameters closer to the optimal values. The number on the left of the tab *'Vis Guess'* sets the resolution of the simulated map, as it significantly affects the time needed to calculate the map—lower resolutions will, of course, yield faster results. The window that appears after pressing the *Vis guess* button is shown in Figure 5. This window allows for setting the desired model and its main parameters. These are specified in six tabs: *Fit, Model, Deviation, Nanoparticle shape, Fixed Params,* and *Experiment Params.*

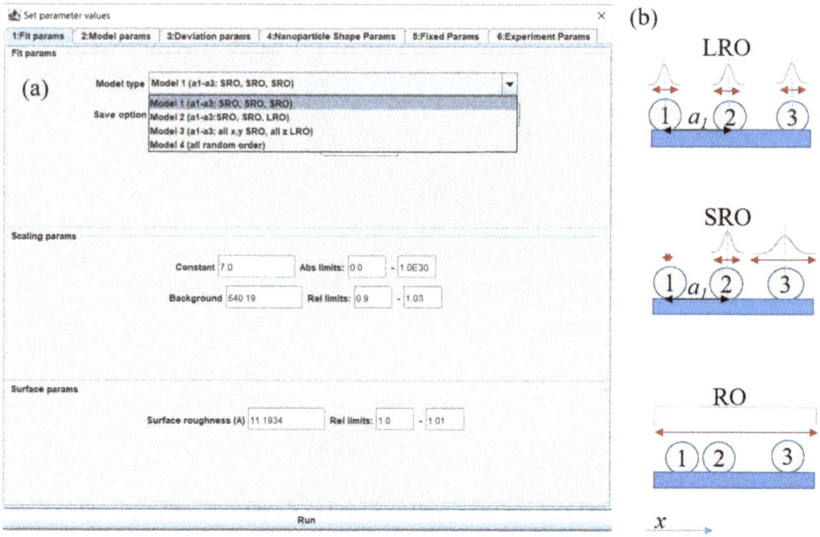

Figure 5. (**a**) Window for setting the parameter values that appears for the '*Vis Guess*' option in the main window. The opened window contains six tabs: *Fit, Model, Deviation, Nanoparticle shape, Fixed params* and *Experiment params*. The chosen tab allows for setting the model type, scaling, and surface parameters. (**b**) Scheme of possible ordering types along one of the basis vectors ($\boldsymbol{a}_1$) of the NP lattice: long range ordering (LRO), short-range ordering (SRO), and random ordering (RO).

The *Fit params* tab (Figure 5a) allows us to choose the model most suitable for our system. This tab, in addition to the model, allows us to choose the parameters, including the overall constant, which depends on the intensity collection time, background, and surface roughness of the sample. The choice of the model is crucial for the accuracy of the analysis. Therefore, three main types of the NP ordering, demonstrated in Figure 5b, are available. The program is suitable for the analysis of NP lattices with three basic types of disorder: long range ordering (LRO), short-range ordering (SRO), and random ordering (RO). The disorder describes deviations from the ideal positions of the NPs in a specific 3D lattice. In the LRO model, the positions of the NPs fluctuate around predefined ideal lattice positions, while the separation between the NPs fluctuate in the SRO model. Therefore, the fluctuation probability is increasing from the first to the last NP along the given basis vector. Random ordering shows NP correlations in the NP positions (these are random).

The disorder type can differ in various spatial directions. Because of this, in the description of the 3D lattice, the final model is a combination of three ordering types, which can be different. Each ordering type is associated with one of the three basis vectors that defines the ordering of NP in a 3D lattice, or with some of its components. The model can be set by the variable *Model type* in the tab '*Fit parameters*', as shown in Figure 5a. For example, the system that is grown by the self-assembly process, without any restrictions, usually shows SRO in all spatial directions. Therefore, in order to properly describe it, Model 1 (SRO, SRO, SRO) should be chosen. However, for the deposition of NPs in a multilayer, the vertical positions of the layers where the NP form are predefined by the deposition process. Here, the vertical (z component) of all basis vectors should be LRO, while all x, y components that define deviations in the plane within the layers are SRO. Therefore, the appropriate choice for a multilayer system is Model 3 (all x, y SRO, all z LRO). In some cases, the positions of the NPs are predefined along one basis vector (assumed $\boldsymbol{a}_3$). For these types of systems, Model 2 is the appropriate choice. Finally, in some cases, the positions of the NPs are random. In this case, we mainly see, as shown on the GISAXS map, the contribution of the NP shape, and the appropriate choice is Model 4 (all random).

The tab *'Model params'*, shown in Figure 6a, enables setting the lattice type that defines the ideal positions of the NPs in a 3D lattice. The ordering is described by three basis vectors $\boldsymbol{a}_1$–$\boldsymbol{a}_3$, whose length and the number of periods along each vector ($N_{x,y,z}$) should be specified. The main features of the basis vectors $\boldsymbol{a}_1$–$\boldsymbol{a}_3$ are illustrated in Figure 6b, where the body centered tetragonal (BCT) lattice is shown. The presented features are the same for other 3D lattices available in the software. The basis vectors $\boldsymbol{a}_1$ and $\boldsymbol{a}_2$ are in the plane parallel to the films surface/substrate (*x*-*y* plane), and they describe the ordering in that plane. The basis vector $\boldsymbol{a}_3$ describes the vertical ordering of the NPs in a 3D lattice. The multilayer period (denoted by vertical period) is given by the *z* component of basis vector $\boldsymbol{a}_3$. This is well visible in Figure 6c, where a cross-section (*x*-*z* plane) through the BCT lattice of the NPs is given.

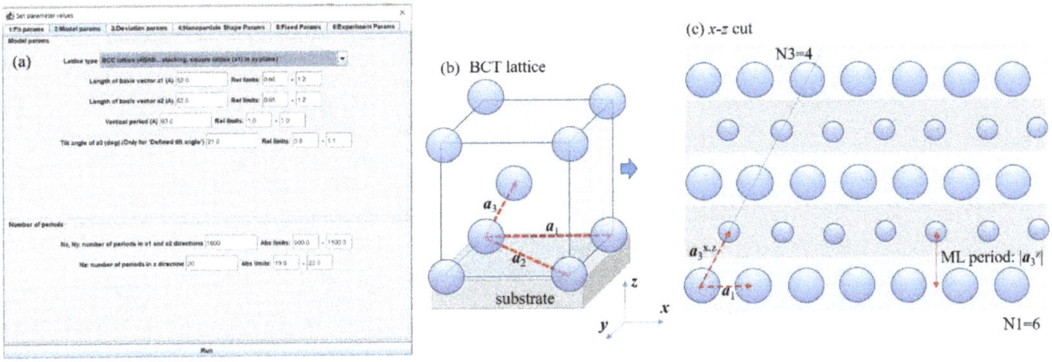

Figure 6. (**a**) Tab *Model params* with the corresponding parameters describing the lattice type and its main parameters. (**b**) Scheme of ordering in a body centered tetragonal (BCT) lattice defined by basis vectors $\boldsymbol{a}_1$–$\boldsymbol{a}_3$. (**c**) Cross-section of the 3D lattice of NPs having BCT unit cell with $N1 = 6$ and $N3 = 4$ periods along basis vectors $\boldsymbol{a}_1$ and $\boldsymbol{a}_3$, respectively.

The parameters that describe the disorder of the NP positions, with respect to the ideal positions (given by the basis vectors $\boldsymbol{a}_1$–$\boldsymbol{a}_3$), are given in the tab *'Deviation params'*, shown in Figure 7a. The deviations should be set for each spatial direction (*x*, *y*, *z*) of each basis vector. Since the basis vectors $\boldsymbol{a}_1$ and $\boldsymbol{a}_2$ are in the plane parallel to the substrate surface, they usually have the same disorder parameters ($\sigma^{1,2}{}_{x,y}$). Therefore, the software uses the same values of disorder for vectors $\boldsymbol{a}_1$ and $\boldsymbol{a}_2$ in all models, except for Model 2. For Model 2, each basis vector has its own disorder. The drawback is that if the number of fitting parameters is increased, the fitting process becomes more complex. The proper determination of the deviation parameters is often the most difficult step in the fitting process because they strongly influence the shape and width of the GISAXS map peaks caused by the regularity in the NP positions. Figure 7b shows the scheme of the corresponding NP disorder distributions. As mentioned previously, the parameters $\sigma^{1,2}{}_{x,y}$ describe disorder in the *x*-*y* plane, while the disorder parameters $\sigma^{3}{}_{x,y}$ describe the deviations of each new layer of NPs from the ideal position, as illustrated in the right section of Figure 7b. Therefore, these parameters actually define the regularity of the formed 3D lattice, and it is the measure of quality of the self-assembly process in the multilayer deposition preparation method.

The vertical (*z*) components of disorder $\sigma^{1,2,3}{}_{z}$ often have different properties than the lateral components. Here again it is assumed that the disorder related to basis vectors $\boldsymbol{a}_1$ and $\boldsymbol{a}_2$ is the same ($\sigma^{1}{}_{z} = \sigma^{2}{}_{z}$). These parameters describe the vertical fluctuation of each NP in a layer parallel to the surface. The parameter $\sigma^{3}{}_{z}$ describes the vertical deviation of the entire layer, as illustrated on the right in Figure 7c. If the material is produced by the self-assembly process during multilayer deposition, the vertical disorders $\sigma^{1,2,3}{}_{z}$ are usually much smaller than the lateral ones $\sigma^{1,2,3}{}_{x,y}$ because the NP forms within the layers of a multilayer. Some examples are given in [20,21].

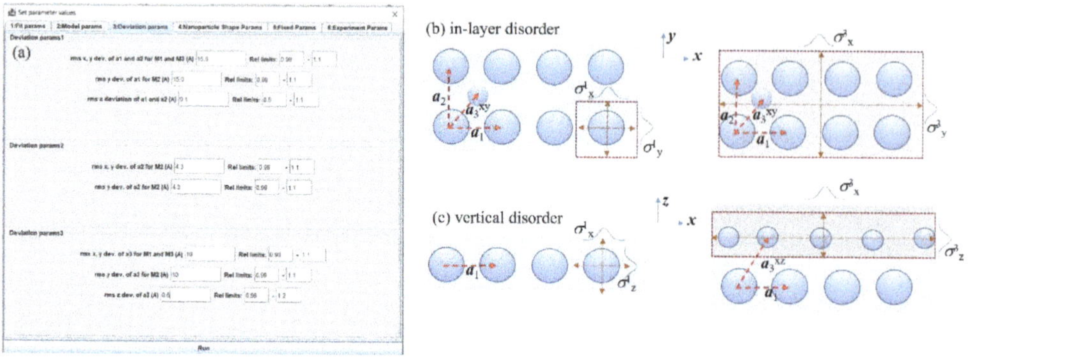

Figure 7. (**a**) Tab *Deviation params* with the corresponding parameters. (**b**) Illustration of disorder parameters $\sigma^{1,2,3}{}_{xy}$. The disorder properties within each layer parallel to the surface ($\sigma^{1,2}{}_{xy}$) are usually the same (left), and they describe the deviations of each NP within a layer from the ideal positions. The parameters $\sigma^{3}{}_{xy}$ (right) describe the disorder of each layer from the ideal position, and therefore, are a measure of the quality of the self-assembly process that leads to formation of the 3D NP lattice. (**c**) The disorder parameters in the vertical direction (perpendicular to the surface). These parameters describe the deviation of each NP within a layer ($\sigma^{1,2}{}_{z}$), and of each layer ($\sigma^{3}{}_{z}$) from the ideal positions in the vertical (z) direction.

GisaxStudio currently supports five basic types of the NP shape-structures: sphere, spheroid, ellipsoid, tilted ellipsoid, and core-shell spheroid NPs, as shown in Figure 8. Their parameters can be set in tab '*Nanoparticle Shape Params*', shown in Figure 8a. The shapes and structures are schematically illustrated in Figure 8b. The main parameters of the NP shape are their radii along the x, y, and z directions $R_{x,y,z}$ (valid for *Ellipsoid*). The option *Sphere* uses only the radius in the x direction (R_x) for all directions, while the option *Spheroid* uses R_x for radii parallel to the surface and R_z for the direction vertical to the surface. The option *Ellipsoid tilted along* a_3 assumes the same radii values as explained for *Ellipsoid*, but the entire NP is then assumed to be tilted along the basis vector $\boldsymbol{a}_3$, as shown in Figure 8b.

The option *Core-Shell* refers to a spheroid with core-shell structure. It consists of different core and shell materials, with radii R_{core} and R_{shell}, respectively. Their origins can be shifted form each other by value d, as shown in Figure 8c. Such structures are often formed by the production of core-shell QDs by thin layer deposition [7–9]. The entire system can be scaled in the z direction by factor F, which enables the transition from a spherical to a spheroidal shape. Therefore, actual core-shell radii in the vertical direction are obtained by the multiplication of R_{core} and R_{shell} by factor F.

The distribution of all NP shapes is assumed to be gamma distribution, as described by parameter γ, and the NP radii. All directions are assumed to have the same order of the distribution. The refraction indices of the NPs and their real and imaginary parts should also be set in the '*Nanoparticle Shape Params*'.

The tab *Fixed Params* includes the refraction indices of the matrix in which the NPs are embedded. It also includes the starting and final averaging angle, along with the step. The averaging is assumed to be performed for the rotation of the 3D lattice around the surface normal (z direction). This is needed because the regular 3D ordering usually appears in domains that are randomly rotated around the surface normal when the material is produced by the self-assembly process. If there is pre-patterning on the substrate that acts as nuclei for the 3D lattice growth, or some other mechanism that predefines the directions of basis vectors $\boldsymbol{a}_1$ and $\boldsymbol{a}_2$, then there is no need for averaging.

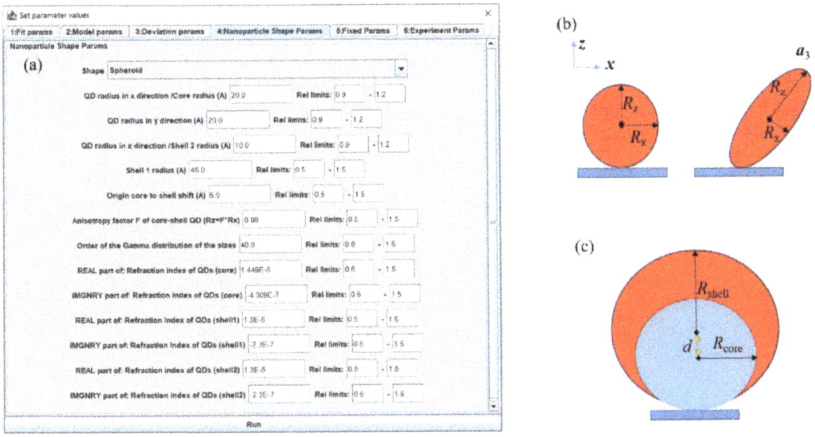

Figure 8. (**a**) Tab *Nanoparticle Shape Params* and its parameters. (**b**) Some of NP shapes currently available in the software: spheroid defined by lateral (in *x*-*y* plane) and vertical (*z* direction) radii R_x and R_z, respectively. Option *Ellipsoid* has the additional radius R_y in the *y* direction; tilted ellipsoid with the tilt along basis vector $\boldsymbol{a}_3$. (**c**) Core-shell spheroid with the core and shell radii R_{core} and R_{shell}, respectively, and the shift of the shell origin *d*. The spheroidal shape is defined by factor *F*, so that the actual vertical radii are the product of R_{core} and R_{shell} and factor *F*.

The tab *Experiment Params* contains the details about sample position (sample-detector distance), the vertical position of the primary beam on the detector that may be negative if the beam is below the detector, wavelength of the x-ray beam used for the measurement, and the pixel size.

As previously mentioned, after the parameters in '*Vis guess*' are set, the *Run* button activates the calculation of the GISAXS intensity distribution for the given set of parameters. The result appears as a new tab in the application. We recommend running the *Vis guess* option with different sets of the parameters until the simulated GISAXS map is similar to the measured one. For example, some parameters, such as multilayer period, the NP shape, or NP approximate separation, could be estimated from some other microscopy measurements. After we are satisfied with the *Vis guess* calculated map, we can start the fitting process.

3.3. Fitting Options

The fitting offers eight different fitting algorithms under the *Algorithms* tab, as visible in Figure 9a, where one or more options can be chosen for fitting. The other tabs that appear after the *Fit* button (Figure 4) is clicked contain the same parameters as the *Vis guess* option. It is important to check the limits for each parameter, which can be set in an absolute or relative fashion. They limit the fitting of the particular parameter. If the limits are too broad, it is more probable that the fitting will go in the wrong direction. On the other hand, if they are too narrow, then the fit will possibly not be able to reach the correct value. However, for most of the parameters, it is possible to estimate the limits correctly. Some of them can be estimated from the GISAXS map, such as the multilayer period, or lateral separation of the NPs (parameter $|\boldsymbol{a}_1|$). The background can also be easily estimated. The width of the Bragg spots related to the 3D lattice formation can be used for the estimation of the disorder parameters. The *Vis Guess* option is very helpful for this. Setting the limits is very important because the model contains numerous parameters, and many local minima exist in the fitting function. Therefore, if the initial parameters and the limits are not set correctly, it is possible that the fit finishes in some local minima. When the fitting process is started, the software updates the user about the progress of the fit via the status text window, as visible in Figure 9b. We can stop any of the started fitting algorithms at any time using the options *stop* or *kill*. Some optimization algorithms can be benignly stopped, and can report the best-found values before they were ended, while others can only be killed, and cannot report any data. Finally, when the fitting procedure is finished, a new tab (Figure 9c) pops up. This tab shows the simulated GISAXS map and the parameters of the fit, along with the line profiles showing the intensity distribution at the chosen cross section.

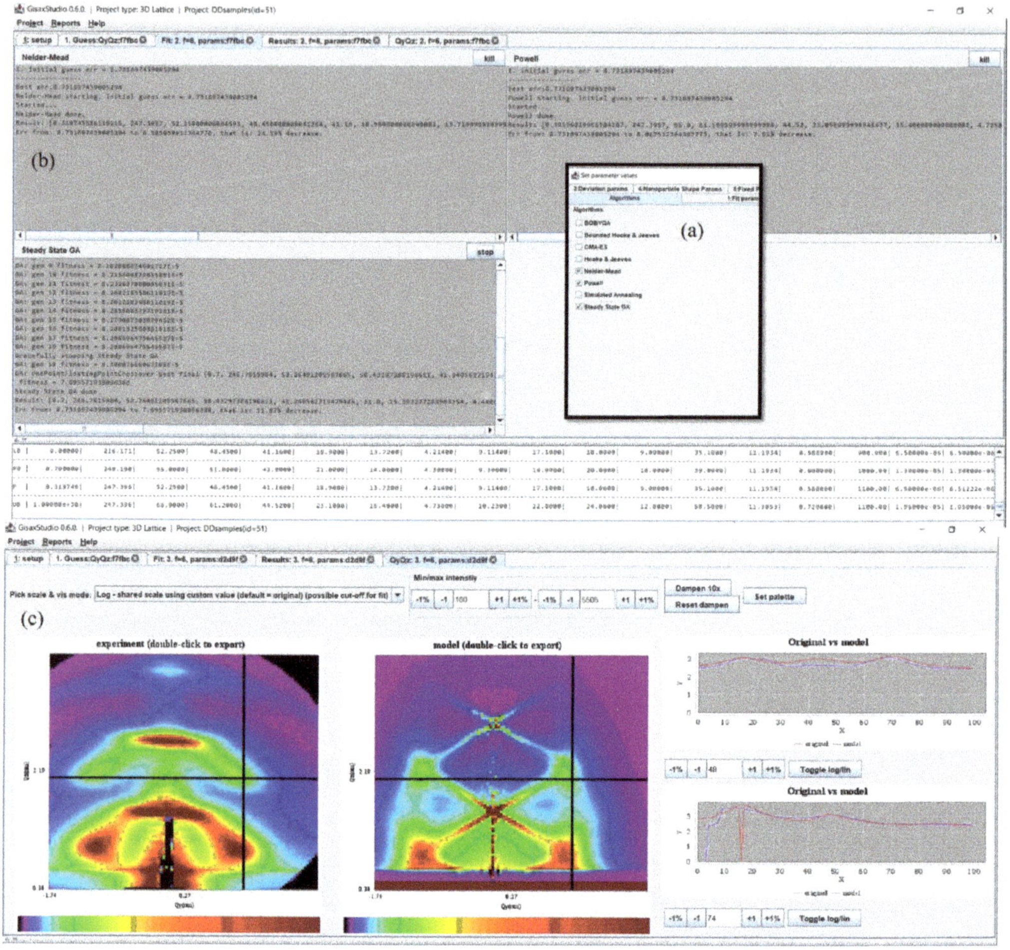

Figure 9. Tabs related to the fitting of the 2D GISAXS map. (**a**) The window with fitting parameters, including the same parameters as in *Vis guess* option and an additional *Algorithms* tab that enables the choice of the fitting algorithms. (**b**) Window that appears after starting of the fit, showing its progress. After the fitting is finished, the parameters are shown in the window at the bottom. (**c**) Results of the fitting process, including the original and simulated GISAXS map using the best-fit parameters, along with the line profiles at the right.

Finally, the software offers the *Batch fit* option. Sometimes, fitting can be a time-consuming operation, especially if the number of parameters is large or when user wants to check many variations. In these cases, it is possible to generate an arbitrary number of initial parameter sets ("batch items"), store them in the database, and then launch the batch fitting process, for instance, overnight. Figure 10a shows the dialog for setting the parameters for batch fit. Multiple initial values of each parameter, along with lower and upper limits, can be set by typing in comma delimited values (e.g., "1, 3, 8"), or by using a rudimentary "for-loop" scripting command (e.g., "for((1, 1); (3, 3); (1, 1))" that will produce complex numbers "(1, 1), (2, 2), (3, 3)". Since parallel execution on a desktop or laptop could require too many resources, GisaxStudio will process batch items in a sequential fashion, as follows (Figure 10b):

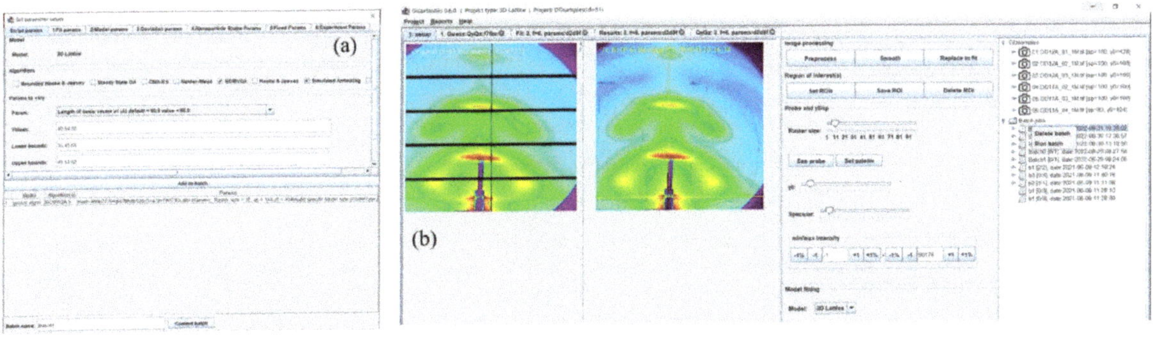

Figure 10. (**a**) Tab related to *Batch fit* option. (**b**) Start or delete options of *Batch fit* in the main window.

Repeat until there are unfinished batch items:

- ○ load the first unprocessed batch item (parameter set) from list
- ○ execute the batch item (fit)
- ○ store the results to the database

A beneficial feature of the program is that batch execution can be interrupted at any time and resumed afterwards. For instance, assume that we run a 40-batch item. The program progresses to the, e.g., 20th item, and then the computer powers off (or a user simply kills the program). A total of 19 items have been executed, and their results have been stored in the database. The 20th item was interrupted, and those results are lost. However, when the same batch is run afterwards, it will simply resume from the 20th item and finish the whole batch. When the batch fit is completed, the user can inspect and visualize the results.

In the above text, we have used the option 3dLattice of GisaxStudio. The other option is iBeam, and it has all the same features as 3dLattice, except for the models describing the NP ordering properties within the ion tracks. We have used all the models given in Ref. [22], so we avoid repeating them here.

4. Conclusions

The software platform GisaxStudio, which is aimed at the simulation, analysis, and fitting of GISAXS maps from different types of 2D or 3D lattices of NPs or QDs, is demonstrated. It is specifically developed for GISAXS measured on thin films consisting of 3D lattices of NPs grown by the self-assembly processes, and for the NPs formed within ion-tracks during ion beam irradiation. Therefore, it contains two main groups of models: 3dLattice for application in analysis of GISAXS from NP lattices formed by self-assembled growth, and iBeam aimed at the NPs formed during the ion-beam irradiation of materials. Each group contains several models of the NP arrangement that depend on the material production conditions, i.e., the type of disorder along each of the basis vectors of the formed 3D NP lattice. The main NP shapes supported by GisaxStudio are ellipsoid, spheroid, and sphere with full or core-shell structure. The core-shell structure includes the option of single or double shells, as well as the possibly displaced center of the core with respect to the center of the shell. The models of the NP ordering and their main features available in the software, along with the NP shapes, are discussed. The options for the fitting are provided and supported by examples. The platform is suitable for use by broad audience and non-experts in the GISAXS field.

Author Contributions: Conceptualization M.M.; methodology, I.M. and M.M.; software, I.M.; formal analysis, G.S., S.I. and M.M.; data curation, I.M. and M.M.; writing—review and editing, all authors.; visualization, I.M. and M.M.; project administration, M.M.; funding acquisition, I.M. All authors have read and agreed to the published version of the manuscript.

Funding: This research was funded by the Croatian Science Foundation (no. IP-2018-01-3633) and the European Regional Development Fund under the grant KK.01.1.1.01.0009 (DATACROSS). The authors acknowledge the CERIC-ERIC Consortium for the access to experimental facilities and financial support.

Institutional Review Board Statement: Not applicable.

Informed Consent Statement: Not applicable.

Data Availability Statement: Not applicable.

Acknowledgments: The authors are thankful to J. Erceg for assistance in the sample preparation and D. Mičetić for GISAXS measurements. The authors acknowledge the CERIC-ERIC Consortium for the access to the experimental SAXS facilities and financial support.

Conflicts of Interest: The authors declare no conflict of interest. The funders had no role in the design of the study, in the collection, analyses, or interpretation of data, in the writing of the manuscript, or in the decision to publish the results.

References

1. Baig, N.; Kammakakam, I.; Falath, W. Nanomaterials: A review of synthesis methods, properties, recent progress, and challenges. *Mater. Adv.* **2021**, *2*, 1821–1871. [CrossRef]
2. Chandrakala, V.; Aruna, V.; Angajala, G. Review on metal nanoparticles as nanocarriers: Current challenges and perspectives in drug delivery systems. *Emergent Mater.* **2022**. [CrossRef] [PubMed]
3. de Arquer, F.P.G.; Talapin, D.V.; Klimov, V.I.; Arakawa, Y.; Bayer, M.; Sargent, E.H. Semiconductor quantum dots: Technological progress and future challenges. *Science* **2021**, *373*, eaaz8541. [CrossRef]
4. Ossicini, S.; Amato, M.; Guerra, R.; Palummo, M.; Pulci, O. Silicon and Germanium Nanostructures for Photovoltaic Applications: Ab-Initio Results. *Nanoscale Res. Lett.* **2010**, *5*, 1637–1649. [CrossRef]
5. Nekić, N.; Šarić, I.; Salamon, K.; Basioli, L.; Sancho-Parramon, J.; Grenzer, J.; Hübner, R.; Bernstorff, S.; Petravić, M.; Mičetić, M. Preparation of non-oxidized Ge quantum dot lattices in amorphous Al_2O_3, Si_3N_4 and SiC matrices. *Nanotechnology* **2019**, *30*, 335601. [CrossRef] [PubMed]
6. de Oliveira, E.L.; Albuquerque, E.L.; de Sousa, J.S.; Farias, G.A.; Peeters, F.M. Configuration-Interaction Excitonic Absorption in Small Si/Ge and Ge/Si Core/Shell Nanocrystals. *J. Phys. Chem. C* **2012**, *116*, 4399–4407. [CrossRef]
7. Nekić, N.; Sancho-Parramon, J.; Radovic, I.B.; Grenzer, J.; Huebner, R.; Bernstorff, S.; Ivanda, M.; Buljan, M. Ge/Si core/shell quantum dots in alumina: Tuning the optical absorption by the core and shell size. *Nanophotonics* **2017**, *6*, 1055–1062. [CrossRef]
8. Buljan, M.; Radić, N.; Sancho-Paramon, J.; Janicki, V.; Grenzer, J.; Radovic, I.B.; Siketić, Z.; Ivanda, M.; Utrobičić, A.; Hübner, R.; et al. Production of three-dimensional quantum dot lattice of Ge/Si core–shell quantum dots and Si/Ge layers in an alumina glass matrix. *Nanotechnology* **2015**, *26*, 065602. [CrossRef]
9. Basioli, L.; Sancho-Parramon, J.; Despoja, V.; Fazinić, S.; Radović, I.B.; Mihalić, I.B.; Salamon, K.; Nekić, N.; Ivanda, M.; Dražić, G.; et al. Ge Quantum Dots Coated with Metal Shells (Al, Ta, and Ti) Embedded in Alumina Thin Films for Solar Energy Conversion. *ACS Appl. Nano Mater.* **2020**, *3*, 8640–8650. [CrossRef]
10. Kes, H.; Okan, S.; Aktas, S. The excitons in infinite potential centered multilayered coaxial quantum wire and the magnetic field effects on their properties. *Superlattices Microstruct.* **2020**, *139*, 106421. [CrossRef]
11. Basioli, L.; Tkalčević, M.; Bogdanović-Radović, I.; Dražić, G.; Nadazdy, P.; Siffalovic, P.; Salamon, K.; Mičetić, M. 3D Networks of Ge Quantum Wires in Amorphous Alumina Matrix. *Nanomaterials* **2020**, *10*, 1363. [CrossRef] [PubMed]
12. Lazarenkova, O.L.; Balandin, A.A. Miniband formation in a quantum dot crystal. *J. Appl. Phys.* **2001**, *89*, 5509–5515. [CrossRef]
13. Buljan, M.; Roshchupkina, O.; Šantić, A.; Holý, V.; Baehtz, C.; Mücklich, A.; Horák, L.; Valeš, V.; Radić, N.; Bernstorff, S.; et al. Growth of a three-dimensional anisotropic lattice of Ge quantum dots in an amorphous alumina matrix. *J. Appl. Crystallogr.* **2013**, *46*, 709–715. [CrossRef]
14. Müller-Buschbaum, P. A Basic Introduction to Grazing Incidence Small-Angle X-Ray Scattering. In *Applications of Synchrotron Light to Scattering and Diffraction in Materials and Life Sciences*; Lecture Notes in Physics; Chapter 3; Springer: Berlin/Heidelberg, Germany, 2009; pp.61–89. [CrossRef]
15. Renaud, G.; Lazzari, R.; Leroy, F. Probing surface and interface morphology with Grazing Incidence Small Angle X-Ray Scattering. *Surf. Sci. Rep.* **2009**, *64*, 255–380. [CrossRef]
16. Cristofolini, L. Synchrotron X-ray techniques for the investigation of structures and dynamics in interfacial systems. *Curr. Opin. Colloid Interface Sci.* **2014**, *19*, 228–241. [CrossRef]

17. Hexemer, A.; Müller-Buschbaum, P. Advanced grazing-incidence techniques for modern soft-matter materials analysis. *IUCrJ* **2015**, *2*, 106–125. [CrossRef]
18. Müller-Buschbaum, P. GISAXS and GISANS as metrology technique for understanding the 3D morphology of block copolymer thin films. *Eur. Polym. J.* **2016**, *81*, 470–493. [CrossRef]
19. Smilgies, D. GISAXS: A versatile tool to assess structure and self-assembly kinetics in block copolymer thin films. *J. Appl. Polym. Sci.* **2021**, *60*, 1023–1041. [CrossRef]
20. Buljan, M.; Radić, N.; Bernstorff, S.; Dražić, G.; Bogdanović-Radović, I.; Holý, V. Grazing-incidence small-angle X-ray scattering: Application to the study of quantum dot lattices. *Acta Crystallogr. Sect. A Found. Crystallogr.* **2012**, *68*, 124–138. [CrossRef]
21. Basioli, L.; Salamon, K.; Tkalčević, M.; Mekterović, I.; Bernstorff, S.; Mičetić, M. Application of GISAXS in the Investigation of Three-Dimensional Lattices of Nanostructures. *Crystals* **2019**, *9*, 479. [CrossRef]
22. Buljan, M.; Karlušić, M.; Nekić, N.; Jerčinović, M.; Bogdanović-Radović, I.; Bernstorff, S.; Radić, N.; Mekterović, I. GISAXS analysis of ion beam modified films and surfaces. *Comput. Phys. Commun.* **2017**, *212*, 69–81. [CrossRef]
23. Lazzari, R. *IsGISAXS*: A program for grazing-incidence small-angle X-ray scattering analysis of supported islands. *J. Appl. Crystallogr.* **2002**, *35*, 406–421. [CrossRef]
24. Tate, M.P.; Urade, V.N.; Kowalski, J.D.; Wei, T.-C.; Hamilton, B.D.; Eggiman, B.W.; Hillhouse, H.W. Simulation and Interpretation of 2D Diffraction Patterns from Self-Assembled Nanostructured Films at Arbitrary Angles of Incidence: From Grazing Incidence (Above the Critical Angle) to Transmission Perpendicular to the Substrate. *J. Phys. Chem. B* **2006**, *110*, 9882–9892. [CrossRef]
25. Babonneau, D. *FitGISAXS*: Software package for modelling and analysis of GISAXS data using IGOR Pro. *J. Appl. Crystallogr.* **2010**, *43*, 929–936. [CrossRef]
26. Chourou, S.T.; Sarje, A.; Li, X.S.; Chan, E.R.; Hexemer, A. *HipGISAXS*: A high-performance computing code for simulating grazing-incidence X-ray scattering data. *J. Appl. Crystallogr.* **2013**, *46*, 1781–1795. [CrossRef]
27. Benecke, G.; Wagermaier, W.; Li, C.; Schwartzkopf, M.; Flucke, G.; Hoerth, R.; Zizak, I.; Burghammer, M.; Metwalli, E.; Müller-Buschbaum, P.; et al. A customizable software for fast reduction and analysis of large X-ray scattering data sets: Applications of the new *DPDAK* package to small-angle X-ray scattering and grazing-incidence small-angle X-ray scattering. *J. Appl. Crystallogr.* **2014**, *47*, 1797–1803. [CrossRef]
28. Durniak, C.; Ganeva, M.; Pospelov, G.; Van Herck, W.; Wuttke, J. BornAgain—Software for Simulating and Fitting X-ray and Neutron Small-Angle Scattering at Grazing Incidence. In Proceedings of the 26th International Conference on Amorphous and Nanocrystalline Semiconductors, Aachen, Germany, 13–18 September 2015. Available online: http://juser.fz-juelich.de/record/255761 (accessed on 28 August 2022).
29. Pospelov, G.; Van Herck, W.; Burle, J.; Loaiza, J.M.C.; Durniak, C.; Fisher, J.M.; Ganeva, M.; Yurov, D.; Wuttke, J. *BornAgain*: Software for simulating and fitting grazing-incidence small-angle scattering. *J. Appl. Crystallogr.* **2020**, *53*, 262–276. [CrossRef]
30. Available online: http://gisaxs.com/index.php/Software/ (accessed on 21 September 2022).
31. GisaxStudio. Available online: http://homer.zpr.fer.hr/gisaxstudio/ (accessed on 30 August 2022).
32. Tkalčević, M.; Boršćak, D.; Periša, I.; Bogdanović-Radović, I.; Janković, I.; Petravić, M.; Bernstorff, S.; Mičetić, M. Multiple Exciton Generation in 3D-Ordered Networks of Ge Quantum Wires in Alumina Matrix. *Materials* **2022**, *15*, 5353. [CrossRef]
33. Shaji, A.; Micetic, M.; Halahovets, Y.; Nadazdy, P.; Matko, I.; Jergel, M.; Majkova, E.; Siffalovic, P. Real-time tracking of the self-assembled growth of a 3D Ge quantum dot lattice in an alumina matrix. *J. Appl. Crystallogr.* **2020**, *53*, 1029–1038. [CrossRef]
34. Tkalčević, M.; Basioli, L.; Salamon, K.; Šarić, I.; Parramon, J.S.; Bubaš, M.; Bogdanović-Radović, I.; Bernstorff, S.; Fogarassy, Z.; Balázsi, K.; et al. Ge quantum dot lattices in alumina prepared by nitrogen assisted deposition: Structure and photoelectric conversion efficiency. *Sol. Energy Mater. Sol. Cells* **2020**, *218*, 110722. [CrossRef]
35. Krause, M.; Buljan, M.; Mücklich, A.; Möller, W.; Fritzsche, M.; Facsko, S.; Heller, R.; Zschornak, M.; Wintz, S.; Endrino, J.L.; et al. Compositionally modulated ripples during composite film growth: Three-dimensional pattern formation at the nanoscale. *Phys. Rev. B* **2014**, *89*, 085418. [CrossRef]
36. Karlušić, M.; Mičetić, M.; Kresić, M.; Jakšić, M.; Šantić, B.; Bogdanović-Radović, I.; Bernstorff, S.; Lebius, H.; Ban-D'Etat, B.; Rožman, K.; et al. Nanopatterning surfaces by grazing incidence swift heavy ion irradiation. *Appl. Surf. Sci.* **2021**, *541*, 148467. [CrossRef]
37. Karlušić, M.; Bernstorff, S.; Siketić, Z.; Šantić, B.; Bogdanović-Radović, I.; Jakšić, M.; Schleberger, M.; Buljan, M. Formation of swift heavy ion tracks on a rutile TiO_2 (001) surface. *J Appl Crystallogr.* **2016**, *49*, 1704–1712. [CrossRef] [PubMed]
38. Radović, I.B.; Buljan, M.; Karlušić, M.; Jerčinović, M.; Dražič, G.; Bernstorff, S.; Boettger, R. Modification of semiconductor or metal nanoparticle lattices in amorphous alumina by MeV heavy ions. *New J. Phys.* **2016**, *18*, 093032. [CrossRef]
39. Karlušić, M.; Kozubek, R.; Lebius, H.; Ban-D'Etat, B.; Wilhelm, R.A.; Buljan, M.; Siketić, Z.; Scholz, F.; Meisch, T.; Jakšić, M.; et al. Response of GaN to energetic ion irradiation: Conditions for ion track formation. *J. Phys. D: Appl. Phys.* **2015**, *48*, 325304. [CrossRef]
40. Commons Math: The Apache Commons Mathematics Library. Available online: https://commons.apache.org/proper/commons-math/ (accessed on 30 August 2022).
41. Parallel Colt. A Multithreaded Version of Colt—A library for High Performance Scientific Computing in JAVA. Available online: https://github.com/rwl/ParallelColt (accessed on 30 August 2022).
42. H2 Database Engine. Available online: https://www.h2database.com/ (accessed on 30 August 2022).

sensors

MDPI

Article

SF_6 High-Voltage Circuit Breaker Contact Status Detection at Different Currents

Ze Guo [1,2,*], Linjing Li [1,2], Weimeng Han [1,2] and Zixuan Guo [1,2]

1 State Key Laboratory of Reliability and Intelligence of Electrical Equipment, Hebei University of Technology, Tianjin 300401, China
2 Key Laboratory of Electromagnetic Field and Electrical Apparatus Reliability of Hebei Province, Hebei University of Technology, Tianjin 300401, China
* Correspondence: guoze92@hebut.edu.cn

Abstract: Currently, the online non-destructive testing (NDT) methods to measure the contact states of high-voltage circuit breakers (HVCBs) with SF_6 gas as a quenching medium are lacking. This paper aims to put forward a novel method to detect the contact state of an HVCB based on the vibrational signal. First, for a 40.5-kV SF_6 HVCB prototype, a mechanical vibration detection system along with a high-current generator to provide the test current is designed. Given this, vibration test experiments are carried out, and the vibration signal data under various currents and corresponding contact states are obtained. Afterward, a feature extraction method based on the frequency is designed. The state of the HVCB contacts is then determined using optimized deep neural networks (DNNs) along with the method of adaptive moment estimation (Adam) on the obtained experimental data. Finally, the hyperparameters for the DNNs are tuned using the Bayesian optimization (BO) technique, and a global HVCB contact state recognition model at various currents is proposed. The obtained results clearly depict that the proposed recognition model can accurately identify five various contact states of HVCBs for the currents between 1000 A and 3500 A, and the recognition accuracy rate is above 96%. The designed experimental and theoretical analysis in our study will provide the references for future monitoring and diagnosis of faults in HVCBs.

Keywords: SF_6 HVCB; non-destructive testing; contact status detection; vibration signal; DNN; Bayesian optimization

Citation: Guo, Z.; Li, L.; Han, W.; Guo, Z. SF_6 High-Voltage Circuit Breaker Contact Status Detection at Different Currents. *Sensors* **2022**, *22*, 8490. https://doi.org/10.3390/s22218490

Academic Editors: Bogusław Łazarz, Grzegorz Peruń and Tangbin Xia

Received: 14 October 2022
Accepted: 1 November 2022
Published: 4 November 2022

1. Introduction

Mechanical vibrations that occur in HVCBs during the current interruption operation can be collected by installing acceleration sensors on the HVCB housing. The obtained vibration signals are then used to identify the mechanical state of the HVCB [1–3]. In this process, due to the influence of the HVCB's flow size, field noise and the complex structure of the HVCB itself, the acquired vibration signal may be distorted, which increases the difficulty of finding an accurate and efficient contact state recognition model [4]. An HVCB's contacts will wear out under the impact of the mechanical closing force and contact stress during current interruption, resulting in contact deterioration [5,6]. Establishing a state-of-the-art HVCB contact status identification model can reduce its maintenance costs and avoid potentially devastating consequences before a severe accident occurs [7,8].

At present, the contact status of an HVCB can only be identified by measuring its contact resistance before the equipment is put into operation, and there are only a few research studies that are available on the online contact state identification of in-service HVCBs. The increasing studies on vibration detection methods in NDT for identifying the conditions of industrial equipment (e.g., applications in bearings) are attracting significant attention, owing to their optimized and controlled strategies. For example, scholars have studied the fault diagnosis methods of bearings based on vibration signals, including wavelet packet analysis [9], Complete Ensemble Empirical Mode Decomposition with

Adaptive Noise (CEEMDAN) [10], the Teager–Kaiser Energy Operator [11], support vector machines (SVMs) [12,13], neural networks (NNs) [14,15] and enhanced differential product weighted morphological filtering [16]. Given this, it was observed that when the HVCB fails, there is also an abnormal vibration signal which can be used to analyze the HVCB's running state. In this regard, scholars have put their efforts toward analyzing the HVCB vibration signal for HVCB fault diagnosis [17–21]. Previous studies have indicated that the contact mode of an HVCB changes due to deteriorated contacts [22–24]. The flowing current through these poor contacts will cause the abnormal vibration of HVCBs. In actual practice, it is necessary to achieve real-time and efficient monitoring and judgment of the contact state of an HVCB at different currents, and it is necessary to further explore the relationship between the vibration signal and the contact state of an HVCB. In terms of HVCB fault recognition, this mainly includes three types of methods that are based on qualitative empirical knowledge, statistical analysis and artificial intelligence (AI) techniques. The deployment of the first two methods is difficult to implement, owing to the complex structure of HVCBs, the strong uncertainty and the large amount of required data. Currently, experts and scholars generally use AI algorithms to achieve fault diagnosis of HVCBs to overcome the above problems, and the algorithms used in fault diagnosis of HVCBs mainly include support vector machines (SVMs) [25–27], neural networks (NNs) [28,29], clustering [30], random forest [31], autoencoder networks [32] and so on. The previously presented fault recognition model for HVCBs mostly uses manual parameter adjustment methods for hyperparameter optimization. The learning rate of NNs, the number of hidden layers and the number of neurons in each layer may be problem-specific and related to the complexity of the data. Manual parameter adjustment methods are used to set a fixed learning rate, the number of hidden layers, the number of hidden layer nodes and the dropout rate for the model, which are difficult to adapt due to the complex and varying working conditions of the actual HVCB [33–35]. Therefore, based on this problem, it is an essential need to establish an NN recognition model with hyperparameter optimization for detecting the contact state of the HVCB while considering various currents.

Due to its remarkable arc-quenching capability and dielectric insulating properties, SF_6 is widely used as a switching and insulating gas in high-voltage apparatuses [36,37]. In addition, it is chemically inert, non-toxic and non-flammable [38]. Therefore, using SF_6 as the quenching medium, an HVCB with compact structure, strong environmental adaptability and high working reliability can be manufactured. Nevertheless, SF_6 has the major drawback of presenting a global warming potential (GWP) of 23,500 [39]. The search for an alternative gas for SF_6 continues unceasingly. Research was conducted in the past few decades on substitutes to SF_6, covering different candidates including common gases (nitrogen, air and CO_2) and vacuum and gas mixtures [40]. Traditional gases such as dry air, nitrogen, CO_2 or their mixtures have the advantage of a low global warming potential but very limited dielectric strength compared with SF_6. A vacuum is widely used in the medium voltage domain as a current interruption medium, but the application of a vacuum at high voltages does not appear to be economically competitive. Moreover, mixtures of SF_6 and other gases such as SF_6-CF_4, SF_6-C_2F_6 and SF_6-N_2 are also considered potential SF_6 substitutes [41–45]. They can be used to reduce costs and the environmental impact of SF_6 gas to a certain extent. At present, HVCBs with SF_6 gas as a quenching medium will still occupy a large share of the market for a long time. In this paper, we chose SF_6 HCVB as our research object.

In this paper, a feature extraction method based on frequency is proposed, a recognition model composed of a DNN and BO is constructed, and an NDT for the contact state of the SF_6 HVCB is proposed.

The main contributions of this paper can be summarized as follows:

1. The vibration test platform of the SF_6 HVCB is set up, and the vibration signal of the HVCB in different contact states is obtained while considering a current between 1000 A and 3000 A.

2. Using the method of a DNN, the amplitudes of the signal in frequency multiples of 50 Hz are used as the eigenvalue matrix to complete the recognition of the contact state of the HVCB.
3. The BO is used for hyperparameter optimization of the NNs so that the proposed model can be applied globally to SF_6 HVCB while considering different currents while having a wider range of applications.

The remainder of the paper is organized as follows. Section 2 explains the experimental platform of the HVCB vibration test, experimental procedures, vibration mechanism and vibration signal analysis. The HVCB contact status recognition model is presented in Section 3. The model recognition results and discussions are presented in Section 4. Section 5 concludes the paper.

2. Experiments and Vibration Data Analysis

2.1. HVCB Vibration Detection Experimental System

Vibration detection in an HVCB was carried out by considering a prototype design of a 40.5-kV SF_6 HVCB as shown in the experimental platform in Figure 1. First, a large current was injected from both ends of the HVCB interface using a high-current generator consisting of a voltage regulator with a maximum output current of 5000 A and a single-phase output. The maximum applied current in this study was 3500 A. The frequency of the applied current in the experiment was 50 Hz. The current through the HVCB was then measured using a Rogowski coil connected with an oscilloscope. In the experiment, a Rogowski coil was sheathed on an energized conductor to measure the amount of current flowing into the HVCB. The lead wires at both ends of the HVCB were connected in parallel with multiple strands of copper conductors, and the total through-current cross-sectional area was greater than 1000 mm^2. The contact state of the HVCB was changed by manually adjusting its tie rod, and the specific value of the contact resistance was confirmed by the loop resistance tester. SF_6 gas was injected into the HVCB using an air pump with a pressure of 0.35 MPa.

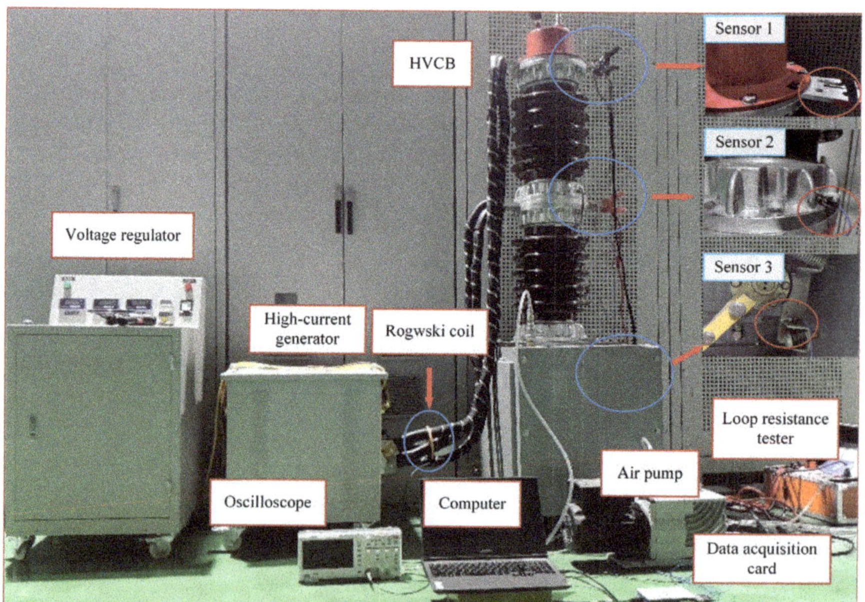

Figure 1. Experimental platform structure diagram.

A high-quality and high-precision data acquisition system was required for the subsequent signal analysis and feature extraction. The vibration signal acquisition platform consisted of signal detection sensors, a data acquisition card and a computer to process the data. The various parameters of the acquisition platform are summarized in Table 1. Considering the frequency and amplitude of the mechanical vibration, as well as the strong magnetic field for in-service HVCBs, acceleration sensors with reliable performance from a PCB company were selected. For the more precise and comprehensive detection of the vibration signal in the HVCB, the acceleration sensors were connected at three different positions and numbered as 1, 2 and 3. The different positions of the sensors were as follows: (1) in front of the HVCB contact, (2) in front of the tie rod and (3) near the operating assembly. These sensors at different positions expanded the fault monitoring range and could effectively collect the vibration signals near the contact of the HVCB, the operation link, the insulator and other different parts. The obtained data from the sensors were transferred to the data acquisition card (NI9234) and recorded on a computer for further processing.

Table 1. Parameters of the acquisition system.

Parameters	Value
Measuring range	0.5–3000 Hz
Sensitivity	10.2 mV/(m/s^2)
Temperature range	From −18 to +66 °C
Weight	4.0 g
Sampling rate	51.2 kHz
Sampling time length	45 s

2.2. HVCB Vibration Detection Experimental Procedures

First, the HVCB was filled with SF_6 gas at a pressure of 0.35 MPa. Secondly, the contact state of the HVCB was changed by manually adjusting the HVCB tie rod, and the specific value of the contact resistance was confirmed with a loop resistance tester. At this time, the HVCB contact was adjusted to a good contact state, and the contact resistance of the HVCB was measured with a loop resistance tester to be 48 μΩ. Thirdly, the HVCB vibration signal at 6 various currents between 1000 A and 3500 A with an incremental step of 500 A when the HVCB was in good contact was measured using sensors placed on the HVCB in advance. In order to ensure the safety of the experiment and ensure that the data were not affected, after each group of experiments was completed, we stopped the current flow, paused for 15 min and then started the next group of experiments. The above is the acquisition of vibration signals when the HVCB contacts were in good contact under six currents.

Next, we adjusted the HVCB tie rod appropriately to make its contact state a slightly poor contact state. The contact resistance was measured to be 78 μΩ. At this time, the vibration signal of the HVCB at 6 various currents between 1000 A and 3500 A with an incremental step of 500 A was measured and recorded. By analogy, the subsequent experiments for moderate contact defects (contact resistance of 93 μΩ), severe contact defects (contact resistance of 117 μΩ) and extremely severe contact defects (contact resistance of 150 μΩ) were continued. A total of 5 experiments under different contact states were completed, and 6 experiments at different currents were completed in each contact state, resulting in 30 sets of experimental data. The detection time of each set of data was 45 s, and the sampling frequency was set to 51.2 kHz.

2.3. HVCB Vibration Characteristics

The excitation forces of the housing during the operation of the HVCB are the mechanical forces, the electromagnetic forces and the discharge forces. This study discusses the vibration signal of the in-service HVCB and does not consider the phenomenon of the HVCB closing and discharging. Therefore, only the vibration signal generated by the electromagnetic force was considered. The electrical power between the contacts and the electrical forces received by the metal housing of the HVCB were calculated separately, and their frequencies were analyzed.

First, we calculated the electrical power received at both ends of the contact. When there is a current flowing through the equipment circuit, the current line shrinks near the contact surface of the HVCB to generate electrical power, which causes the equipment to vibrate. Suppose that the current passing through the contacts is $i = I_0 \cos(\omega t)$, where ω is the angular frequency corresponding to the power frequency. The electrical force between the contacts can be calculated using the following equation:

$$F_1 = \frac{\mu_0}{4\pi} i^2 \ln \frac{D}{2a} = \frac{\mu_0 I_0^2 \{\cos(2\omega t) + 1\}}{8\pi} \ln \frac{D}{2a} \tag{1}$$

The parameter a in the above equation can be calculated using the following equation:

$$a = \sqrt{\frac{F_j}{\pi \zeta H_b}} \tag{2}$$

where D is the diameter of the contact surface, a is the radius of the contact spot, F_j is the initial pressure acting on the contact, ζ is the material deformation coefficient between 0.3 and 1 and H_b is the Brinell strength of the transparent material.

Therefore, under normal circumstances, the force on the HVCB contact is a simple harmonic force with twice the frequency of the power supply so that the vibration of the HVCB contact caused by the electromagnetic force is a vibration of twice the frequency of the power supply.

Subsequently, the electrical power of the metal housing of the circuit breaker is calculated. Suppose that R is the diameter of the cylindrical shell of the HVCB, R_{eq} is the equivalent resistance of the current loop, μ_0 is the vacuum permeability and μ_r is the relative permeability. B_R is the magnetic induction strength of the metal cell when the contact is introduced with the current. Assuming that the base area of the metal unit divided into HVCB shells is S (where S approaches zero), the length is L, the metal element is treated as a thin shell, and the electromagnetic force on the edge of the metal element divided by the HVCB shell is approximately expressed as follows:

$$F_2 = B_R i L = \frac{-\mu_0^2 \mu_r^2 I_0^2 S \omega L}{8\pi R^2 R_{eq}} \sin(2\omega t) \tag{3}$$

Therefore, under normal circumstances, the combined force of each metal unit is a simple harmonic force with twice the frequency of the power supply so that the vibration of the HVCB shell caused by the electromagnetic force is a vibration of twice the frequency of the power frequency.

From the above theoretical calculations, when the current into the HVCB is 50 Hz, and the HVCB is in normal operation, the frequency of the vibration signal collected from the HVCB should be 100 Hz.

2.4. Vibration Data Analysis

The contact state recognition process for the SF_6 HVCB in this study included three steps: vibration signal preprocessing, feature extraction and the establishment of a contact state recognition model. Due to the influence of onsite high electromagnetic interference and the complex structure of the HVCB itself, the collected vibration signal may have contained a certain amount of noise that needed to be filtered for precise and comprehensive measurements. Figure 2 depicts the comparison of the vibration signal acquired by sensor 1 at 2000 A before and after preprocessing when the contact state of the HVCB contact was in good condition. It can be observed that the vibration signal was a stationary signal whose frequency did not change with time, which is the main reason for the subsequent selection of Fourier analysis. However, the vibration signal was not stable around the zero point of vibration, because during the operation of the HVCB, there was heat generated, making the sensors appear to have the "zero drift" phenomenon. Therefore, the DC component in the signal needed to be removed in subsequent signal processing to eliminate the impact of "zero drift" on the data analysis.

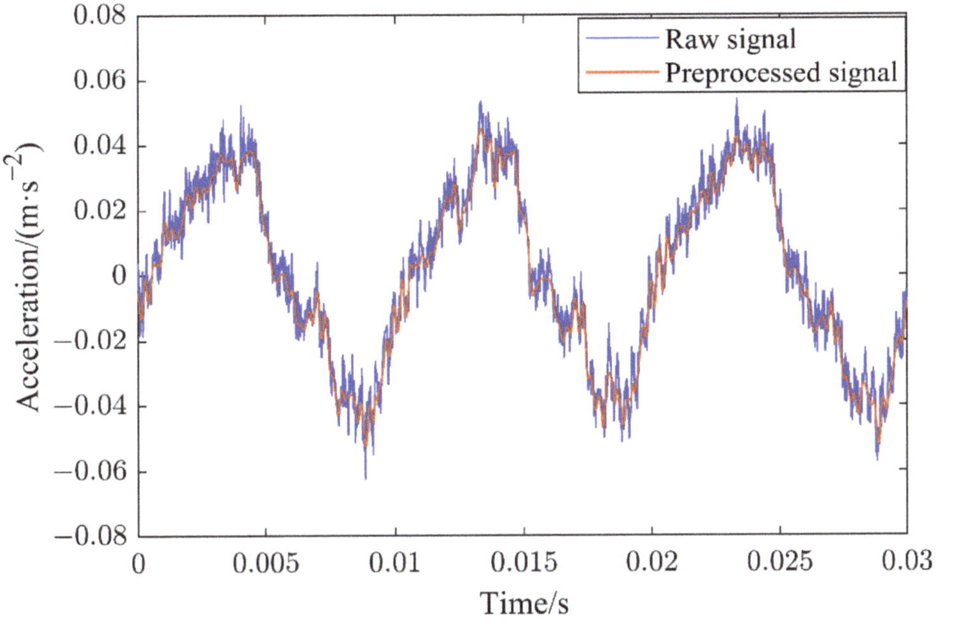

Figure 2. Vibration signal comparison before and after preprocessing.

Figure 3 is a spectrum of the vibration signal collected by sensor 1 at 2000 A when the contact state of the HVCB contact was in good condition and was decomposed by Fourier analysis. The vibration signal frequency of the collection point was concentrated at integer multiples of 50 Hz, the range was 50–1500 Hz, and the amplitude was the largest at 100 Hz, which verifies that the vibration was mainly caused by electromagnetic force, and the frequency of the electromagnetic force was 100 Hz. At the same time, it can be seen that the frequency of the vibration signal was mainly 99.83 Hz instead of the accurate 100 Hz, and it was caused by the on-site high electromagnetic interference as well as the frequency of the power supply voltage. For subsequent feature extraction, the amplitudes of the signals with frequencies near integer multiples of 50 Hz were selected as the eigenvalue. In order to avoid missing some feature points due to frequency shifting, in the actual calculation, the point with the largest amplitude was selected in the range of $(50 \times k \pm 5)$ Hz (where k is a positive integer between 1 and 30), and their horizontal and vertical coordinate values were recorded. This was used as the eigenvalue to form a matrix of eigenvalues.

The preprocessed time domain and frequency domain patterns of the vibration signals obtained by the three sensors are illustrated in Figures 4 and 5, respectively. The vibration signals were obtained at a current of 2000 A when the HVCB contacts were in good condition with a contact resistance of 48 μΩ. In Figure 5, the 3 sets of histograms represent sensors 1–3, and the columns in a set of histograms represent the amplitude of the signal collected by the sensor when the signal's frequency was 50–1500 Hz with an integer multiple of 50 Hz. It can be observed that the vibration amplitudes obtained by sensors 1 and 3 were much higher compared with the vibration amplitude obtained by sensor 2. In the actual measurement, it was found that the vibration signal obtained by sensor 3 could easily be affected by the operating mechanism. In addition, it contained a multi-frequency signal and richer information on noise that may bring severe challenges for the analysis and processing of the signal. Based on comprehensive consideration, sensor 1, which was closer to the contact, making it easier to analyze its resulting vibration signal, was selected as the subsequent analysis object.

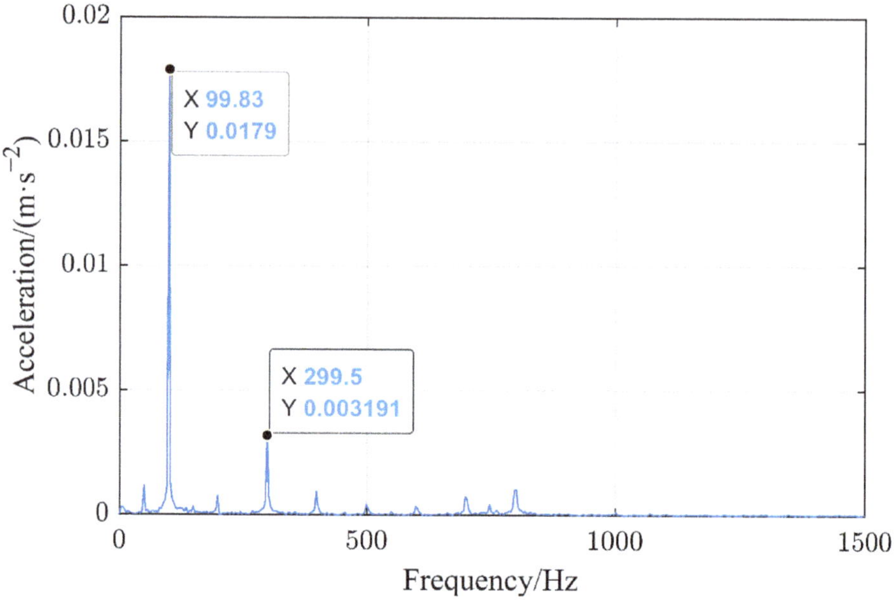

Figure 3. Spectrum of the vibration signal.

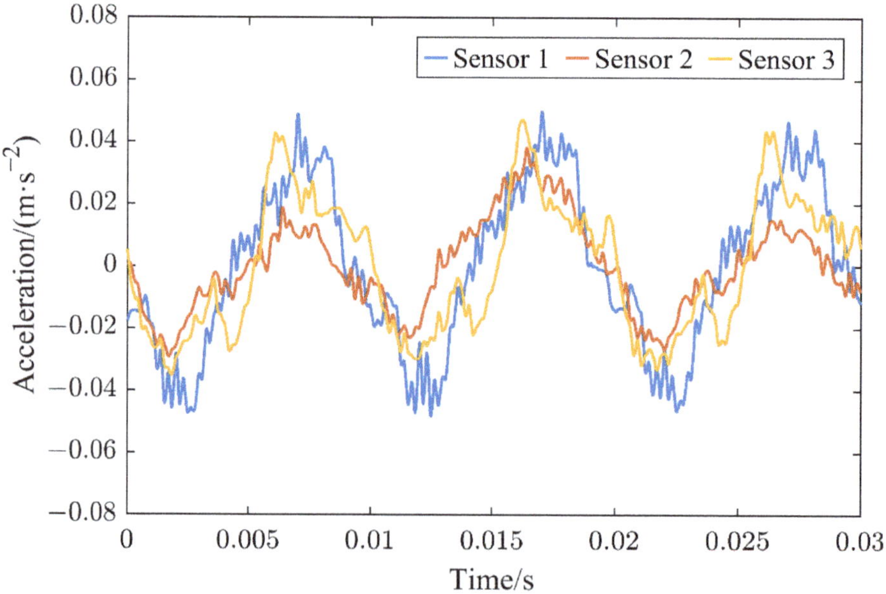

Figure 4. Time domain diagram of the vibration signals acquired by different sensors when the contacts were in good contact.

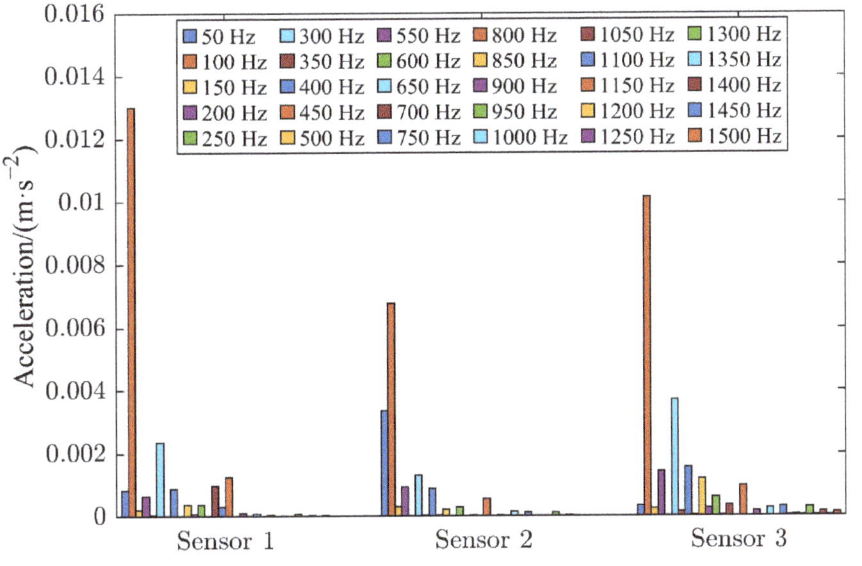

Figure 5. Frequency domain diagram of vibration signals acquired by different sensors when contacts were in good contact.

Figure 6 shows a comparison of the time domain of the vibration signal collected when the contact of the HVCB was in good condition at a current between 1000 A and 3500 A with an incremental step of 500 A. The data used herein were measured by sensor 1, as already mentioned above. After that, the obtained data were filtered and smoothed in order to analyze the results more clearly. It can be observed from Figure 6 that the vibration signal of the HVCB increased nonlinearly with the increase in the current.

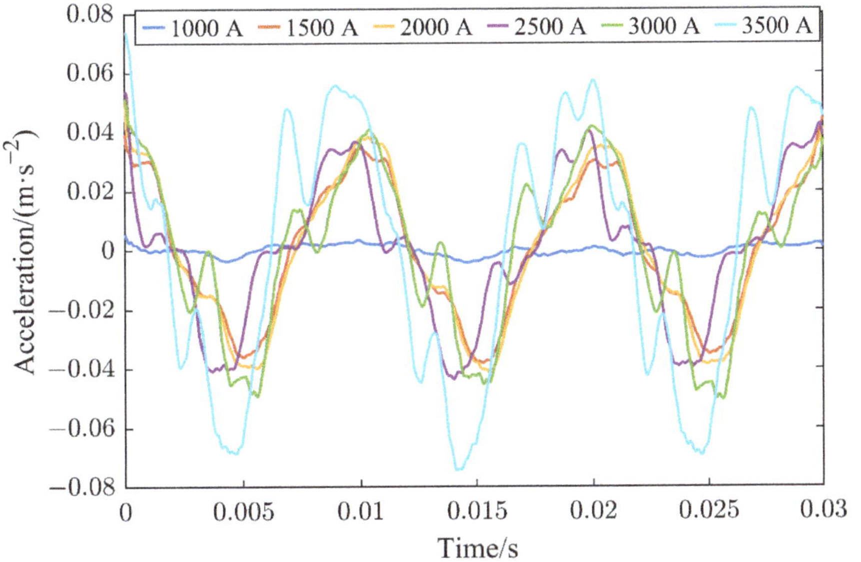

Figure 6. Time domain diagram of the vibration signals in good contact states acquired at different currents.

Similarly, Figure 7 shows a comparison of the frequency domains of the vibration signals collected when the contact of the HVCB was in good condition at a current between 1000 A and 3500 A with an incremental step of 500 A. The amplitude of the vibration signal was augmented dramatically with the increase in the current. However, the ratio of the signal amplitudes at each frequency was not static, and the comparison was most obvious from 2000 A to 3000 A. In this current range, the amplitude of the signal at a frequency of 100 Hz increased slightly, while the amplitude of the signal at 300 Hz rose rapidly. Given this, it is not feasible to say directly that for different current situations, the amplitude of the signal at each frequency is enough to determine the contact of the current HVCB. Therefore, it is necessary to introduce deep learning methods to find potential internal laws.

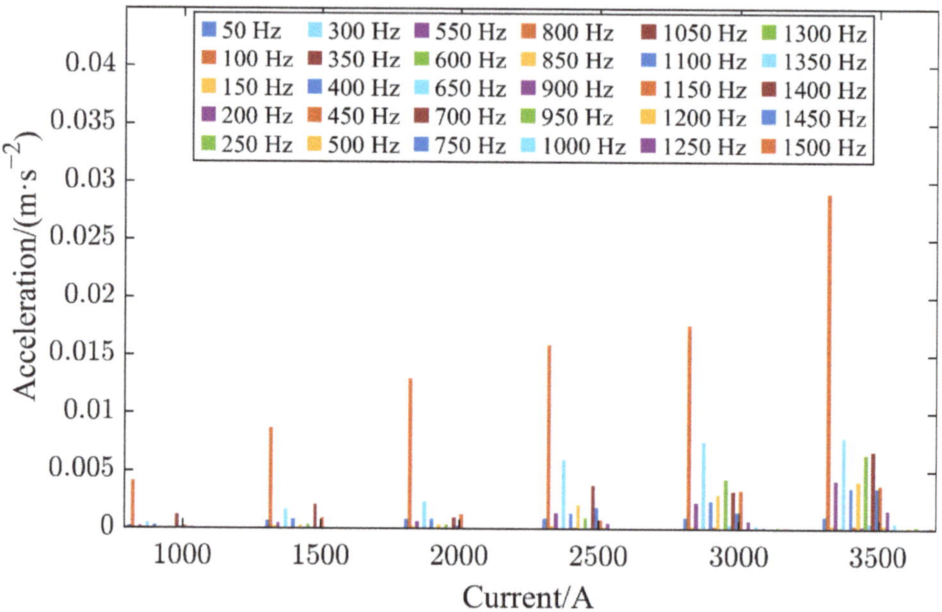

Figure 7. Frequency domain diagram of vibration signals in good contact state acquired at different currents.

As shown in Figure 8, the contact resistance was changed by adjusting the angle of the HVCB tie rod that was confirmed by the circuit resistance tester. Figure 8 shows the vibration signal collected by sensor 1 at 2000 A when the contacts of the HVCB were in poor condition with contact resistances of 78 μΩ, 93 μΩ, 117 μΩ and 150 μΩ. It is worth mentioning that in the different contact states of the HVCB, the frequency obtained by Fourier decomposition was basically located between 0 and 1500 Hz, and the amplitude at 100 Hz held the highest value. Additionally, there were some variations in the signal amplitude near 50–1500 Hz with integer multiples of 50 Hz. As the contact state deteriorated, the 100-Hz signal amplitude showed a fluctuating change, while the 300-Hz signal amplitude increased slightly. In short, the amplitude changes of different frequencies showed a certain law, but it was difficult to explore its specific law based on a simple calculation. Moreover, under actual working conditions, the current magnitude and contact resistance of the HVCB have complex and variable relations, and the amount of data is also larger, making them difficult to process using simple algorithms. Therefore, it is an essential need to further excavate the potential deep information of the vibration signal in order to conduct a more accurate and comprehensive analysis.

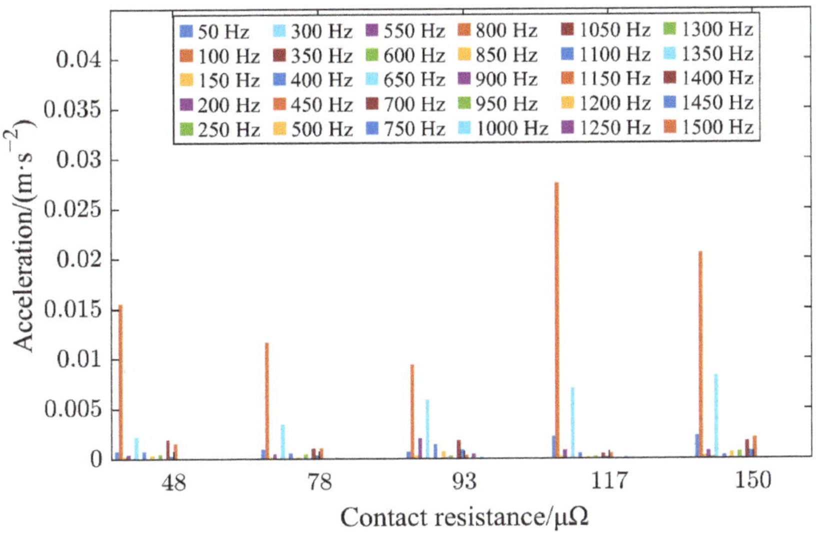

Figure 8. Frequency domain plot of vibration signals in poor contact conditions.

3. HVCB Contact Status Recognition Model Based on DNNs

3.1. Deep Neural Networks (DNNs)

A DNN is a feedforward artificial NN that differs from general NNs by having multiple layers of hidden units between the inputs and outputs. DNNs utilize many properties of the natural signal that are combinatorial hierarchies, where higher-level features are obtained by combining lower-level features [46]. DNN are divided according to the position of different layers, and can be divided into the following three categories; (i) the input layer, (ii) the hidden layer, and (iii) the output layer as shown in DNN structure diagram in Figure 9. In this study, the input parameters for the input layer are the amplitude of the extracted frequency signal, the number of hidden layers, (Hidden layer nodes will be adjusted according to the model recognition.) and the output layer, representing five classification results. The dotted line represents an idea of preventing the NN from overfitting by randomly removing elements (along with their connections) from the NN during training, which can be quantified by the dropout rate. Similarly, the dropout rate is subsequently adjusted according to the identification situation.

Forward propagation calculates the predicted value and then calculates the loss based on the difference between the predicted value and the true value, and backpropagation updates the parameters of each layer sequentially from the last layer forward according to the loss function. In the forward propagation phase, the hidden layer takes the output of the previous layer as the input of the latter layer:

$$b_k^L = \sum_{i=0}^{n} w_{ik}^L \times a_i^{L-1} + d_k^L \tag{4}$$

$$a_k^L = f(b_k^L) \tag{5}$$

In this equation, b_k^L and a_k^L represent the outputs of the k neuron in the Lth layer of the DNN before and after activation, respectively, w_{ik}^L represents the linear transfer coefficient from a_i^{L-1} to b_k^L, d_k^L represents the bias constant of the b_k^L forward propagation function and $f(x)$ is the activation function. As a classification model, the Sigmoid function was chosen as the activation function. The mathematical expression for the Sigmoid function is as follows:

$$f(x) = \frac{1}{1+e^{-x}} \tag{6}$$

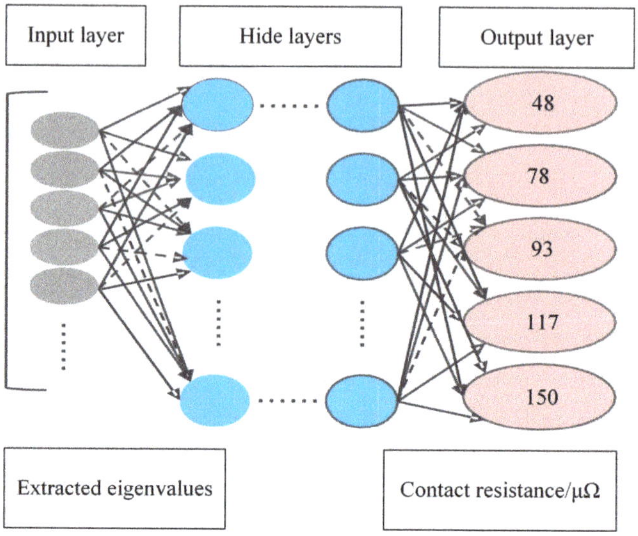

Figure 9. DNN architecture diagram.

Binary cross-entropy is often used as a loss function for classification problems, where cross-entropy is used as a loss function to assess the losses of the classification models in the prediction process. When using small batch gradient descents, this updated selection of weights leads to the risk of the loss function stagnating at the local minimum. The adaptive moment estimation (Adam) algorithm can be used to solve the optimization problem of large data volumes and high feature latitudes, and it requires only a small amount of memory in machine learning [47].

3.2. Bayesian Hyperparameter Optimization (BO)

Common automated machine learning hyperparameter methods are mesh tuning, random search and BO. The former is slower. The latter, while faster, is more likely to miss important points in the search space during processing. The BO algorithm establishes a substitution function based on the past evaluation results of the objective function to find the value of the minimized objective function. Therefore, BO is comparatively much faster, smaller in terms of iterations and more efficient [48–50]. In this paper, the BO based on a tree-structured Parzen estimator (TPE) with a good effect and speed in a high-dimensional space was used as a method of automatic machine learning hyperparameter optimization for DNNs.

BO involves an iterative process along with the proxy function and the acquisition function as its two main components. In each iteration, a probabilistic proxy model is built using TPE:

$$p(x|y) = \begin{cases} l(x), y < y^* \\ g(x), y \geq y^* \end{cases} \quad (7)$$

In the equation, x is the observation point obtained in the search spaces, $y^* = \min\{(x_1, f(x_1)), \ldots, (x_i, f(x_i))\}$ represents the optimal value on the observation domain, where $l(x)$ is the density formed using the observation $\{x_i\}$ that results in the corresponding loss $f(x_i)$ being less than y^*, and $g(x)$ is the density formed using the observed value $\{x_i\}$ to make the corresponding loss $f(x_i)$ greater than y^*.

In the TPE, the expected improvement (EI) is used as the acquisition function, and the next evaluation point for the objective function value is optimized until the maximum number of iterations is reached.

In the equation, $p(x)$ is the probability of reaching an observation point x, and $p(y)$ represents the probability that y is the optimal value. The equation for the calculation of the EI is as follows:

$$EI_{y^*}(x) = \int_{-\infty}^{y^*} (y^* - y)p(y|x)\,dy = \int_{-\infty}^{y^*} (y^* - y)\frac{p(x|y)p(y)}{p(x)}dy \quad (8)$$

3.3. Recognition Process

The hyperparameter combination after the iteration was used as the optimal input parameter combination to enter the HVCB contact state recognition model and to complete the HVCB contact state recognition. The flowchart of the HVCB contact state recognition model is shown in Figure 10.

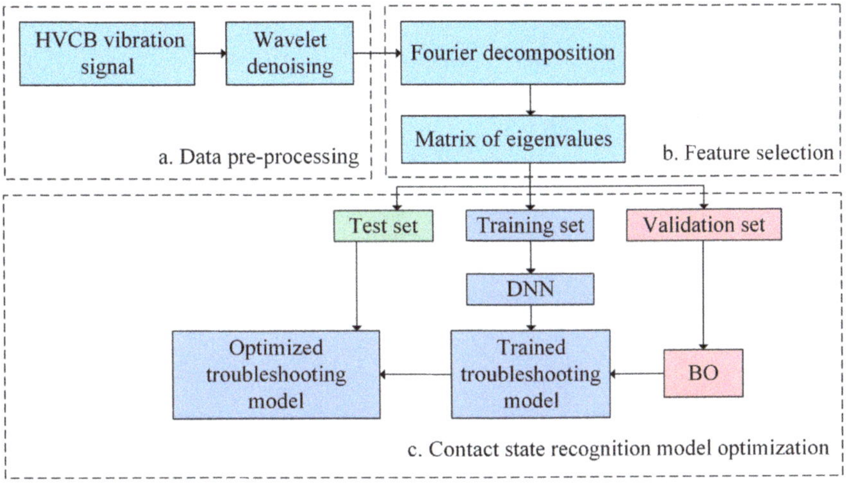

Figure 10. Contact state recognition model flowchart.

The 45-s vibration signals obtained by each set of experiments, or 2,304,000 sample points, were evenly divided into 768 sets of data. They were analyzed separately by Fourier decomposition, and the amplitude of the signal with a frequency of 50 times between 0 and 1500 Hz was saved, forming a matrix of 768 $\times$ 30, which would be fed into the DNN as eigenvalues for the next step of classification.

Based on this, the number of input layer nodes for the DNN obtained was 30, and the number of output layer nodes was 5. The experimental data were divided into three parts: a training set, a validation set and a test set, where 70% of the data in the samples in each state was randomly selected as the training set, while 20% was the verification set and 10% was the test set. The training set was used to train a DNN model. The validation set was used for BO to find suitable hyperparameters. Finally, the test set was input into the final DNN model to test the accuracy of the recognition model and verify the effectiveness of the recognition model.

4. Recognition Results and Discussions

Table 1 shows the accuracy of the state identification at different currents when using fixed parameters without BO. As can be seen in Table 2, when the DNN adopted fixed parameters, it was impossible to guarantee a high recognition rate for each current condition. At the same time, manual parameter adjustment took a long time and was inefficient, which could limit its usage for practical applications. Therefore, it is necessary to find more efficient hyperparameter optimization methods.

Table 2. Accuracy of status recognition at different currents when fixed parameters are used. (a) When the learning rate is 0.01, the number of hidden layers is 4, and the dropout rate is 0.15. (b) When the learning rate is 0.02, the number of hidden layers is 5, and the dropout rate is 0.25.

Current (A)	Accuracy (%)
(a)	
1000	20.18
1500	22.00
2000	92.65
2500	93.75
3000	97.66
3500	99.21
(b)	
1000	95.05
1500	96.88
2000	96.35
2500	82.68
3000	79.56
3500	84.38

Using BO, one can set the search space and search time in advance, which can greatly improve the accuracy of the recognition model. In the actual experiment, the manual parameter adjustment method was first used to determine the optimization space according to the recognition accuracy under a combination of different hyperparameters. The optimization space set in this study is illustrated in Table 3.

Table 3. The search space set in Bayesian optimization.

Hyperparameters	Search Space
Learning rate	$[1 \times 10^{-5}, 5 \times 10^{-3}]$
The number of hidden layer nodes	[128, 256, 512, 1024]
The number of hidden layers	[1, 10]
Dropout rate	[0.1, 0.5]

As shown in Table 4, at the current between 1000 A and 3500 A, the recognition model could obtain a satisfactory accuracy rate and accurately identify the five contact states of the HVCB.

Table 4. Recognition accuracy at different currents.

Current (A)	Accuracy (%)
1000	96.88
1500	99.74
2000	99.87
2500	99.21
3000	100
3500	100

In order to more efficiently and precisely display the results of the HVCB contact state recognition model, 384 sample points in the test set of the HVCB contact state recognition model at a current of 2000 A (At this time, only a hidden layer could be classified.) were selected, and the dimensionality reduction process was carried out by principal component analysis (PCA). Afterward, the classifications of the sample points of the input layer, the hidden layer and the output layer were mapped to a three-dimensional plane. The results are shown in Figure 11. In Figure 11, the x, y and z axes represent the three dimensions after data dimensionality reduction, and there is no specific physical meaning.

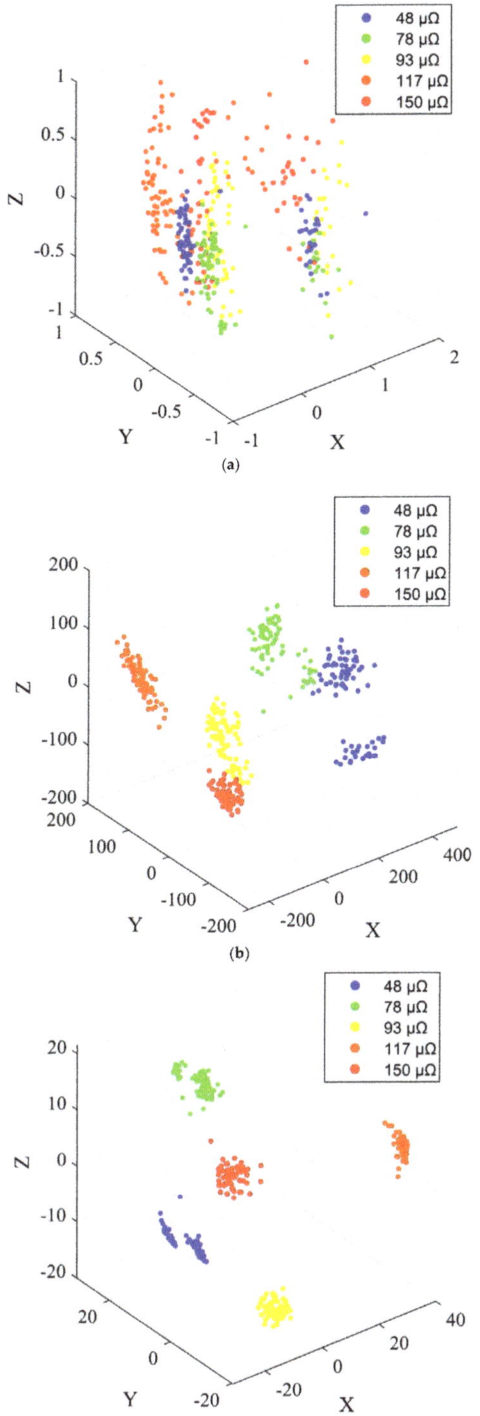

Figure 11. 3D display of recognition results: (**a**) input layer, (**b**) hidden layer and (**c**) output layer.

It can be observed form Figure 11 that at the input layer, the locations of the sample points were disorganized and scattered throughout the space. After going through a hidden layer, the positions of the sample points began to show a certain pattern, and the data points in the five cases were separated. At the output layer, the data points under different contact resistances were further separated, presenting an ordered distribution in space. This shows that the HVCB contact status recognition model can complete the classification of sample points under different contact resistances, which verifies the effectiveness of the model.

5. Conclusions

This paper studied the different states of SF_6 HVCB contacts while considering various currents for power system applications. The main conclusions of this study are summarized as follows.

With the change in the contact state of the HVCB, the spectrogram of the vibration signal collected by the sensors in the field showed a certain law. The vibration signal frequency of the acquisition point was concentrated at integer multiples of 50 Hz, the range existed in between 50 Hz and 1500 Hz, and the amplitude was the largest at 100 Hz. The vibration was mainly caused by electromagnetic force, and the frequency of the electromagnetic force was 100 Hz. The amplitude of the vibration signal increased with the increase in the interruption current. When the HVCB was in poor contact, the 300-Hz signal amplitude increased.

Considering the amplitude of the signal at a frequency of 50 Hz as the eigenvalue, the signal was used as the eigenvalue to form an eigenvalue matrix, which could fully depict the mechanical vibration characteristics of the SF_6 HVCB when different contacts were in contact. The contact state recognition model of the HVCB was established by replacing the stochastic gradient descent in the DNN with an Adam algorithm supplemented by BO, which is hyperparameter optimization. At an HVCB current of 1000–3500 A, the model could accurately identify five different contact states. In the experiment, the accuracy rate of the measurements could reach more than 96%. The contact state recognition model proposed in this paper can play a certain role in the condition detection of HVCB and promote the development of intelligent detection technology for future monitoring and diagnosis of faults in HVCBs.

Author Contributions: Conceptualization, Z.G. (Ze Guo); methodology, Z.G. (Ze Guo) and L.L.; software, L.L.; validation, L.L., W.H. and Z.G. (Zixuan Guo); formal analysis, L.L. and W.H.; investigation, L.L. and W.H.; resources, Z.G. (Ze Guo); data curation, L.L.; writing—original draft preparation, L.L.; writing—review and editing, Z.G. (Ze Guo); visualization, L.L.; supervision, W.H.; project administration, Z.G. (Zixuan Guo); funding acquisition, Z.G. (Ze Guo). All authors have read and agreed to the published version of the manuscript.

Funding: This research was supported by the National Natural Science Foundation of China (51825702), Natural Science Foundation of Hebei Province of China (E2020202216), and Key Laboratory of Special Machine and High Voltage Apparatus (Shenyang University of Technology), Ministry of Education (KFKT202110).

Institutional Review Board Statement: Not applicable.

Informed Consent Statement: Not applicable.

Data Availability Statement: Not applicable.

Conflicts of Interest: The authors declare no conflict of interest. The founding sponsors had no role in the design of the study; in the collection, analyses, or interpretation of the data; in the writing of the manuscript; or in the decision to publish the results.

References

1. Ma, H.; Liu, B.; Xu, H.; Chen, B.; Ju, P.; Zhang, L.; Qu, B. GIS mechanical state identification and defect diagnosis technology based on self-excited vibration of assembled circuit breaker. *IET Sci. Meas. Technol.* **2020**, *14*, 56–63. [CrossRef]
2. Liu, Y.; Chen, L. Application of accelerometers in vibration test of high voltage circuit breaker. In Proceedings of the 2015 5th International Conference on Electric Utility Deregulation and Restructuring and Power Technologies (DRPT), Changsha, China, 26–29 November 2015. [CrossRef]
3. Polycarpou, A.A.; Soom, A.; Swarnakar, V.; Valtin, R.A.; Acharya, R.S.; Demjanenko, V.; Soumekh, M.; Benenson, D.M.; Porter, J.W. Event timing and shape analysis of vibration bursts from power circuit breakers. *IEEE Trans. Power Deliv.* **1996**, *11*, 848–857. [CrossRef]
4. Yang, Q.; Ruan, J.; Zhuang, Z.; Peng, Z. Parameter adaptive noise reduction method for mechanical vibration signals of high-voltage circuit breakers. *High Volt. Eng.* **2021**, *47*, 4274–4287. [CrossRef]
5. Žarković, M.; Stojković, Z. Artificial intelligence SF_6 circuit breaker health assessment. *Electr. Power Syst. Res.* **2019**, *175*, 105912. [CrossRef]
6. Feizifar, B.; Usta, O. A Novel Arcing Power-Based Algorithm for Condition Monitoring of Electrical Wear of Circuit Breaker Contacts. *IEEE Trans. Power Deliv.* **2019**, *34*, 1060–1068. [CrossRef]
7. Henriquez, P.; Alonso, J.B.; Ferrer, M.A.; Travieso, C.M. Review of Automatic Fault Diagnosis Systems Using Audio and Vibration Signals. *IEEE Trans. Syst. Man Cybern. Syst.* **2014**, *44*, 642–652. [CrossRef]
8. Wang, Z.; Gordon, R.J.; Joseph, W.S.; Wang, X.; Rong, M. Spectroscopic On-Line Monitoring of Cu/W Contacts Erosion in HVCBs Using Optical-Fibre Based Sensor and Chromatic Methodology. *Sensors* **2017**, *17*, 519. [CrossRef]
9. Gao, L.; Yang, Z.; Cai, L.; Wang, H.; Chen, P. Roller Bearing Fault Diagnosis Based on Nonlinear Redundant Lifting Wavelet Packet Analysis. *Sensors* **2011**, *11*, 260–277. [CrossRef] [PubMed]
10. Hu, L.; Wang, L.; Chen, Y.; Hu, N.; Jiang, Y. Bearing Fault Diagnosis Using Piecewise Aggregate Approximation and Complete Ensemble Empirical Mode Decomposition with Adaptive Noise. *Sensors* **2022**, *22*, 6599. [CrossRef]
11. Shi, X.; Zhang, Z.; Xia, Z.; Li, B.; Gu, X.; Shi, T. Application of Teager–Kaiser Energy Operator in the Early Fault Diagnosis of Rolling Bearings. *Sensors* **2022**, *22*, 6673. [CrossRef]
12. Zhang, N.; Wu, L.; Yang, J.; Guan, Y. Naive Bayes Bearing Fault Diagnosis Based on Enhanced Independence of Data. *Sensors* **2018**, *18*, 463. [CrossRef]
13. Zhou, J.; Xiao, M.; Niu, Y.; Ji, G. Rolling Bearing Fault Diagnosis Based on WGWOA-VMD-SVM. *Sensors* **2022**, *22*, 6281. [CrossRef] [PubMed]
14. Goyal, D.; Dhami, S.S.; Pabla, B.S. Non-Contact Fault Diagnosis of Bearings in Machine Learning Environment. *IEEE Sens. J.* **2020**, *20*, 4816–4823. [CrossRef]
15. Hoang, D.T.; Tran, X.T.; Van, M.; Kang, H.J. A Deep Neural Network-Based Feature Fusion for Bearing Fault Diagnosis. *Sensors* **2021**, *21*, 244. [CrossRef] [PubMed]
16. Yan, X.; Liu, T.; Fu, M.; Ye, M.; Jia, M. Bearing Fault Feature Extraction Method Based on Enhanced Differential Product Weighted Morphological Filtering. *Sensors* **2022**, *22*, 6184. [CrossRef]
17. Runde, M.; Ottesen, G.E.; Skyberg, B.; Ohlen, M. Vibration analysis for diagnostic testing of circuit-breakers. *IEEE Trans. Power Deliv.* **1996**, *11*, 1816–1823. [CrossRef]
18. Charbkeaw, N.; Suwanasri, T.; Bunyagul, T. Mechanical Defect Detection of SF_6 High Voltage Circuit Breaker Using Wavelet Based Vibration Signal Analysis. In Proceedings of the 2008 5th International Conference on Electrical Engineering/Electronics, Computer, Telecommunications and Information Technology, Krabi, Thailand, 14–17 May 2008. [CrossRef]
19. Zhao, L.; Hong, G.; Wang, Z.; Chen, W.; Long, W.; Ren, J.; Wang, Z.; Huang, X. Research on Fault Vibration Signal Features of GIS Disconnector Based on EEMD and Kurtosis Criterion. *IEE J. Trans. Elec. Electron. Eng.* **2021**, *16*, 677–686. [CrossRef]
20. Sun, Q.; Cao, T.; Hou, Y.; Zhao, T. Detection and Analysis Based on the Abnormal Mechanical Vibration Signal of GIS. In Proceedings of the 2015 Fifth International Conference on Instrumentation and Measurement, Computer, Communication and Control (IMCCC), Qinhuangdao, China, 18–20 September 2015. [CrossRef]
21. Landry, M.; Leonard, F.; Landry, C.; Beauchemin, R.; Turcotte, O.; Brikci, F. An Improved Vibration Analysis Algorithm as a Diagnostic Tool for Detecting Mechanical Anomalies on Power Circuit Breakers. *IEEE Trans. Power Deliv.* **2008**, *23*, 1986–1994. [CrossRef]
22. Liu, Y.; Yang, J.; Jia, Y.; Song, S.; Wu, B.; Li, J. Connection state diagnosis method of GIS disconnector based on mechanical vibration. *High Volt. Eng.* **2019**, *45*, 1591–1599. [CrossRef]
23. Yang, Q.; Ruan, J.; Dao, C.; Zhuang, Z. Over-range status recognition of high-voltage circuit breaker contacts based on improved Hilbert-Huang transforms and support vector machines. *Electr. Power Autom. Equip.* **2019**, *39*, 198–204. [CrossRef]
24. Feng, J.; Sun, L.; Chen, W.; Su, Y.; Shu, Y.; Zhao, L. Vibration Characteristics of GIS Isolating Switch Under Different Operating Conditions. *High Volt. Eng.* **2021**, *47*, 4314–4322. [CrossRef]
25. Zhong, Y.; Hao, J.; Liao, R.; Wang, X.; Jiang, X.; Wang, F. Mechanical defect identification for gas-insulated switchgear equipment based on time-frequency vibration signal analysis. *High Volt.* **2021**, *6*, 531–542. [CrossRef]
26. Yuan, Y.; Ma, S.; Wu, J.; Jia, B.; Li, W.; Luo, X. Frequency Feature Learning from Vibration Information of GIS for Mechanical Fault Detection. *Sensors* **2019**, *19*, 1949. [CrossRef] [PubMed]
27. Wei, Y.; Chen, S.; Ma, Q.; Li, X.; Su, H. Mechanical Fault Diagnosis in High Voltage Vacuum Circuit Breaker Based on Improved S Transform and Support Vector Machine. In Proceedings of the 2020 IEEE International Conference on High Voltage Engineering and Application (ICHVE), Beijing, China, 6–10 September 2020. [CrossRef]

28. Huang, N.; Chen, H.; Cai, G.; Fang, L.; Wang, Y. Mechanical Fault Diagnosis of High Voltage Circuit Breakers Based on Variational Mode Decomposition and Multi-Layer Classifier. *Sensors* **2016**, *16*, 1887. [CrossRef] [PubMed]
29. Xing, Y.; Liu, M. A method of classified HV circuit breaker fault signal based on EEMD and SOM neural network. In Proceedings of the 2018 Chinese Control and Decision Conference (CCDC), Shenyang, China, 9–11 June 2018. [CrossRef]
30. Wang, X.; Lin, X.; Zhou, K.; Lu, Y. CNN based Mechanical Fault Diagnosis of High Voltage Circuit Breaker using Sound and Current Signal. In Proceedings of the 2020 IEEE International Conference on High Voltage Engineering and Application (ICHVE), Beijing, China, 6–10 September 2020. [CrossRef]
31. Lin, L.; Wang, B.; Qi, J.; Chen, L.; Huang, N. A Novel Mechanical Fault Feature Selection and Diagnosis Approach for High-Voltage Circuit Breakers Using Features Extracted without Signal Processing. *Sensors* **2019**, *19*, 288. [CrossRef]
32. Ma, S.; Chen, M.; Wu, J.; Wang, Y.; Jia, B.; Jiang, Y. High-Voltage Circuit Breaker Fault Diagnosis Using a Hybrid Feature Transformation Approach Based on Random Forest and Stacked Autoencoder. *IEEE Trans. Ind. Electron.* **2019**, *66*, 9777–9788. [CrossRef]
33. Bergstra, J.; Bardenet, R.; Bengio, Y.; Kégl, B. Algorithms for hyper-parameter optimization. In Proceedings of the 24th International Conference on Neural Information Processing Systems, Granada, Spain, 12–14 December 2011.
34. Allamanis, M.; Barr, E.T.; Devanbu, P.; Sutton, C. A Survey of Machine Learning for Big Code and Naturalness. *ACM Comput.* **2018**, *51*, 1–37. [CrossRef]
35. LeCun, Y.; Bengio, Y.; Hinton, G. Deep learning. *Nature* **2015**, *521*, 436–444. [CrossRef]
36. Christophorou, L.G.; Olthoff, J.K.; van Brunt, R.J. Sulfur hexafluoride and the electric power industry. *IEEE Electr. Insul. Mag.* **1997**, *13*, 20–24. [CrossRef]
37. Kieffel, Y.; Irwin, T.; Ponchon, P.; Owens, J. Green Gas to Replace SF_6 in Electrical Grids. *IEEE Power Energy Mag.* **2016**, *14*, 32–39. [CrossRef]
38. Li, X.; Zhao, H.; Murphy, A.B. SF_6-alternative gases for application in gas-insulated switchgear. *J. Phys. D Appl. Phys.* **2018**, *51*, 15300. [CrossRef]
39. IPCC Fifth Assessment Report: Climate Change 2013 (AR5). Available online: https://www.ipcc.ch/report/ar5/wg1/ (accessed on 28 October 2022).
40. Kieffel, Y.; Biquez, F. SF_6 alternative development for high voltage switchgears. In Proceedings of the 2015 IEEE Electrical Insulation Conference (EIC), Denver, CO, USA, 26–30 July 2015. [CrossRef]
41. Chervy, B.; Gonzalez, J.J.; Gleizes, A. Calculation of the interruption capability of SF_6-CF_4 and SF_6-C_2F_6 mixtures. II. Arc decay modeling. *IEEE Trans. Plasma Sci.* **1996**, *24*, 210–217. [CrossRef]
42. Gleizes, A.; Sakalis, I.; Razafinimanana, M.; Vacquie, S. Decay of wall stabilized arcs in SF_6-N_2 mixtures. *J. Appl. Phys.* **1987**, *61*, 510–518. [CrossRef]
43. Woo, S.Y.; Jeong, D.H.; Seo, K.B.; Kim, J.H. A study on dielectric strength and insulation property of SF_6/N_2 mixtures for GIS. *J. Int. Counc. Electr. Eng.* **2012**, *2*, 104–109. [CrossRef]
44. Safar, Y.A.; Malik, N.H.; Qureshi, A.H. Impulse breakdown behavior of negative rod-plane gaps in SF_6-N_2, SF_6-Air and SF_6-CO_2 mixtures. *IEEE Trans. Electr. Insul.* **1982**, *EI-17*, 441–450. [CrossRef]
45. Li, X.; Zhu, K.; Guo, Z.; Zhang, Y.; Jia, S.; Jiang, X. Experimental study on arc interruption characteristics of SF_6 and its mixtures with CF_4. *Proc. CSEE* **2017**, *37*, 3315–3322. [CrossRef]
46. Hinton, G.; Deng, L. Deep Neural Networks for Acoustic Modeling in Speech Recognition: The Shared Views of Four Research Groups. *IEEE Signal Process. Mag.* **2012**, *29*, 82–97. [CrossRef]
47. Kingma, D.; Ba, J. Adam: A Method for Stochastic Optimization. In Proceedings of the International Conference for Learning Representations, San Diego, CA, USA, 7–9 May 2015.
48. Hinton, G.; Osindero, S.; Teh, Y. A fast learning algorithm for deep belief nets. *Neural Comput.* **2006**, *18*, 1527–1554. [CrossRef]
49. Srivastava, N.; Hinton, G.; Krizhevsky, A.; Sutskever, I.; Salakhutdinov, R. Dropout: A simple way to prevent neural networks from overfitting. *Mach. Learn.* **2014**, *15*, 1929–1958.
50. Shahriari, B.; Swersky, K.; Wang, Z.; Adams, R.P.; de Freitas, N. Taking the Human Out of the Loop: A Review of Bayesian Optimization. *Proc. IEEE* **2016**, *104*, 148–175. [CrossRef]

applied sciences

MDPI

Article

Identification and Classification of Defects in PE Gas Pipelines Based on VGG16

Yang Wang [1], Qiankun Fu [1], Nan Lin [2,*], Huiqing Lan [3], Hao Zhang [4] and Toktonur Ergesh [5]

1 School of Mechanical Engineering, Xinjiang University, Urumqi 830046, China
2 Pressure Pipe Department, China Special Equipment Inspection and Research Institute, Beijing 100013, China
3 Laboratory of Vehicle Advanced Manufacturing, Measuring and Control Technology (Ministry of Education), Beijing 100044, China
4 China Construction Seventh Engineering Division Corp., Ltd., Zhengzhou 450004, China
5 JiHua Laboratory, Foshan 528200, China
* Correspondence: sy_linnan@163.com

Abstract: For the problem of classification and identification of defects in polyethylene (PE) gas pipelines, this paper firstly performs preliminary screening of the acquired images and acquisition efficiency of defective image acquisition was improved. Images of defective PE gas pipelines were pre-processed. Then, edge detection of the defective images was performed using the improved Sobel algorithm and an adaptive threshold segmentation method was applied to segment the defects in the pipeline images. Finally, the defect images were morphologically processed to obtain binary images. The obtained binary images were applied with VGG16 to complete the training of the defect classifier. The experimental findings show that in the TensorFlow API environment, the test set's highest accuracy reached 97%, which can achieve the identification of defect types of underground PE gas transmission pipelines.

Keywords: image pre-processing; classification; identification; VGG16; threshold segmentation; pipeline defects

Citation: Wang, Y.; Fu, Q.; Lin, N.; Lan, H.; Zhang, H.; Ergesh, T. Identification and Classification of Defects in PE Gas Pipelines Based on VGG16. *Appl. Sci.* **2022**, *12*, 11697. https://doi.org/10.3390/app122211697

Academic Editors: Grzegorz Peruń, Tangbin Xia and Bogusław Łazarz

Received: 27 October 2022
Accepted: 15 November 2022
Published: 17 November 2022

1. Introduction

In the past decades, polyethylene (PE) pipelines have been widely used in natural gas networks around the world because of their flexibility and corrosion resistance [1]. Therefore, the long-term performance of PE pipes and their materials is of great concern to date [2]. According to international natural gas pipeline accident statistics, natural gas pipeline defects are frequently caused by localized corrosion [3], operator mistakes, defective materials, and construction flaws [4]. As the use of pipelines for transporting hazardous substances becomes more popular worldwide, the possibility of major accidents caused by pipeline failures is gradually increasing [5]. For example, the explosion caused by a gas pipeline leak in a residential building in Slovakia in 2019, which killed at least seven people, reminds us that gas pipelines must be checked regularly [6]. There are many nondestructive inspection methods for gas pipelines in practice (e.g., ultrasonic-based sensors, laser-based systems, etc.), and compared to other inspection techniques commonly used for PE pipelines, due to their distinctive benefits of intuitiveness, accuracy and convenience, visual inspection techniques have been used extensively in a variety of fields [7]. Pipeline defect detection robots equipped with intra-visual inspection of image processing technology can directly collect, transmit and process images, reducing labor costs [8].

Traditionally, automatic classification of images is carried out using extracted image features, which are used to represent unclear information in the original pixel values. Convolutional neural networks (CNN) have taken the place of that approach in recent years [9]. Image pre-processing, which includes image de-noising [10], image enhancement [11], image segmentation [12], morphological operations [13], etc., is generally performed before image classification. Based on image pre-processing, Zhou et al. [14] investigated an improved spline local mean decomposition (ISLMD), proposed to be CNN-based and enabling noise reduction of images to locate pipe leakage locations. Ma et al. [15] proposed a sewer multi-defect detection system based on CNN-style GAN-SDM image pre-processing, and the proposed model's average accuracy and macro F1 score were 95.64% and 0.955, respectively. Hosseinzadeh et al. [16] presented a small and simple probe design that was used to check small-bore pipes for defects. Hua et al. [17] developed a visual recognition-based pipeline fault detection algorithm. It is capable of both autonomous localization and pipeline fault detection.

Several frameworks based on the original CNN structure have been proposed to enhance target detection performance as a result of the advancement of CNN technology and classification, such as R-CNN [18], Fast R-CNN [19], SSD [20], YOLO series [21], etc. In this paper, a CNN-based classification framework for PE pipe defect detection is proposed, which can automatically extract the abstract features of defects for accurate classification of three defects, including cracks, fractures, and holes. In this paper, three different algorithms are applied to the existing framework and the confusion matrix is used to determine which model framework has the highest accuracy. The experimental results indicate that the highest accuracy of the test set in this paper reached 97% in the environment of TensorFlow API.

2. Image Pre-Processing

The pre-processing of pipe images does more than remove noise; it also enhances the contrast between the pipeline's background and any pipeline defects, making it simpler to locate and categorize pipeline defects [22]. Figure 1 depicts the image pre-processing process used in this paper.

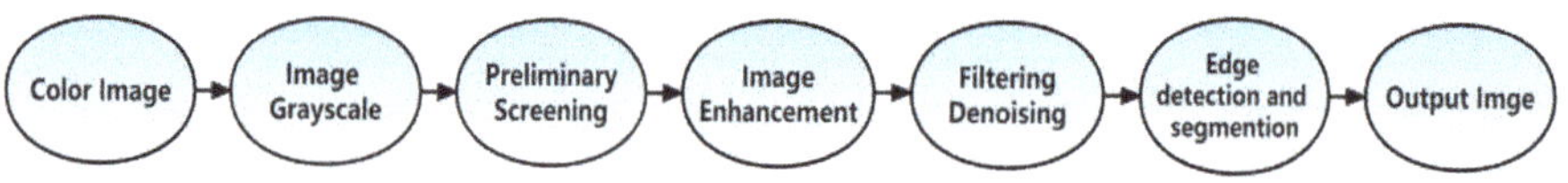

Figure 1. Image pre-processing process.

2.1. Greyscale Processing for Pipeline Images

Grayscale is an important feature to characterize the brightness and darkness of an image. In recent years, on the basis of image grayscale differences and discontinuous changes, it has been used in target recognition, image segmentation, and machine vision technique [23]. Grayscale occupies less memory and enables faster computer operations compared to color images. The mean value method, maximum value method, and weighted average method are the three commonly used techniques for transforming color photos into grayscale.

Where the maximum value method is to directly take the value of the component with the largest value among the three components of a *R,G,B* Equation (1). Red (*R*), green (*G*) and blue (*B*) are the three color channels of color images.

$$R = G = B = \max(R, G, B) \tag{1}$$

The mean method is to take the mean of the values in the three components of *R,G,B* Equation (2).

$$R = G = B = (R + G + B)/3 \tag{2}$$

The weighted average method is based on the sensitivity of the human eye for the *R,G,B*'s three colors, according to a certain weighted average, obtained in Equation (3).

$$I(u, v) = 0.3 \times I_R(u, v) + 0.59 \times I_G(u, v) + 0.11 \times I_B(u, v) \tag{3}$$

where: $I(u,v)$ denotes the gray value at coordinate, $I_R(u,v)$, $I_B(u,v)$ and $I_G(u,v)$ denote the luminance value of the pixel's three color components, respectively.

The maximum value method (Figure 2B1–B3), the average value method (Figure 2C1–C3), and the weighted average method (Figure 2D1–D3) were used to grayscale process the three original defect images (Figure 2A1–A3) of cracks, fractures and holes, respectively. From Figure 2, it can be seen that the weighted average method produces the best results for the grayscale image, and the grayscale image has moderate brightness and does not cover the characteristics of the pipe defects. As a result, the image grayscale uses the weighted averaging method.

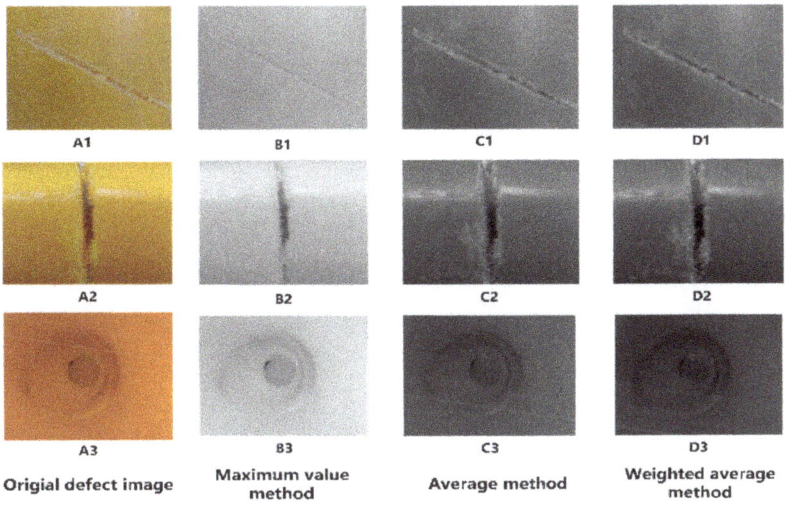

Figure 2. Comparison of grayscale processing methods for pipe defect images.

2.2. Defect Image Acquisition

We compared and examined the continuously captured PE pipeline images and discovered that: 1. the percentage of pipeline defects in the entire PE gas pipeline network system is small; 2. there is a significant grayscale discrepancy between the defective and normal parts of the grayscale processed defective images; 3. after grayscale processing of any two adjacent PE pipeline images, there is a significant grayscale discrepancy between the normal PE pipeline images and the defective PE pipeline images in the same position, while the grayscale discrepancy between the two normal PE pipeline images is very small. Thus, in order to determine whether there are defects in the pipeline images, we propose a screening method for pipeline defects to improve the detection effectiveness [24].

First, let x and y be consecutive images of any adjacent PE pipes in our pipe database, and Δ be the greyscale discrepancy between x_k and y_k, which is the correspondent pixel locations of the two images x and y, as shown in Equation (4). However, there will be some information loss during the compression and transmission of the image data [25], which will make the corresponding pixel grayscale values of the two adjacent images differ greatly, even if they are both normal, resulting in incorrect judgments of the system. To decrease this error, we improve the grayscale discrepancy Δ of two corresponding pixels to the grayscale discrepancy of the corresponding region $\overline{\Delta}$, which is the mean of the grayscale discrepancy of all pixels in the designated size region, as shown in Equation (5).

$$\Delta = |x_k - y_k| \tag{4}$$

$$\overline{\Delta} = \frac{1}{S_n} \sum_{k=1}^{n} |x_k - y_k| \tag{5}$$

where S_n denotes number of pixels in designated size region.

Then, for the minimum mean $\overline{\Delta}_1$ of the pixel grayscale discrepancy between the defective pipeline images and the normal pipeline images computed in the designated size region, a statistical method can be used; the number of pixels with the smallest defect area is calculated statistically as M, let the length of the defect area image be M_l pixels, the width be M_w pixels, as shown in Equation (6).

$$\mathrm{M} = M_w \times M_l \tag{6}$$

Finally, to further improve the accuracy of defect detection, we set the discrepancy $\overline{\Delta}_{min}$ to be somewhat less than $\overline{\Delta}_1$. When $\overline{\Delta} \geq \Delta_{min}$, it is judged as abnormal and the number of pixels in the abnormal area is counted as B, when $\overline{\Delta} < \overline{\Delta}_{min}$, it is judged as normal and the number of pixels in the normal area is counted as $B1$; calculate $B1$ and B according to Equation (6). If $B \geq M$, then there are defects in x and y, otherwise x and y are considered as normal images.

More than 9000 consecutive images of pipes were screened using the above defect screening method (Figure 3). A total of 160 out of 163 defective images were picked out, and the screening accuracy rate was up to 98.15%. Figure 3 can better help us understand the PE pipeline defect detection algorithm.

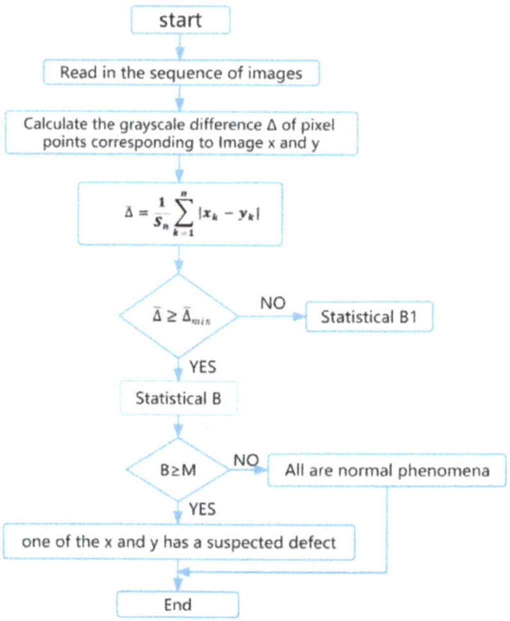

Figure 3. PE gas pipeline defect screening algorithm.

2.3. *Images Enhancement for Pipe Defects*

Image enhancement is the process of enhancing an image's display by highlighting its edges and significant texture details and suppressing the display of unimportant areas. This somewhat enhances the image's visual impact [26] or highlights some "useful" and compresses other "useless" information in the image. In this paper, global histogram equalization (Figure 4B1–B3), adaptive histogram equalization (Figure 4C1–C3), and gamma transform (Figure 4D1–D3) was applied to enhance the image of the grayscale (Figure 4A1–A3).

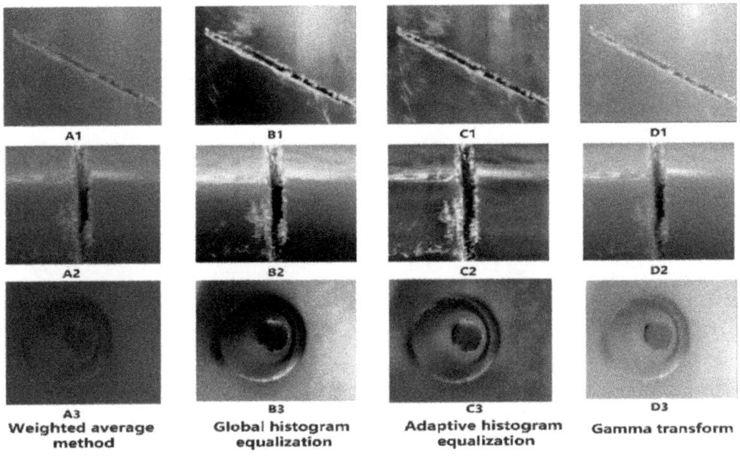

Figure 4. Comparison of pipeline defect image enhancement effect.

Figure 4 illustrates that after the gamma transform, the defect image is not distorted; additionally, the defect's edges become more noticeable and stand out against the background with a greater difference. In order to increase the contrast between the background of the pipe and the pipe defects, gamma transform was applied.

2.4. *Pipe Defect Images Filtering and Denoising*

During image transmission, the final image is often received with a lot of noise due to the interference of the environment or the limitations of the equipment. Image denoising is a classical image recovery task aiming to predict clean images from noisy observations [27]. Bilateral filtering (Figure 5B1–B3), Gaussian filtering (Figure 5C1–C3), mean filtering (Figure 5D1–D3), and adaptive median filtering (Figure 5E1–E3) were applied in this paper to denoise the images obtained above (Figure 5A1–A3).

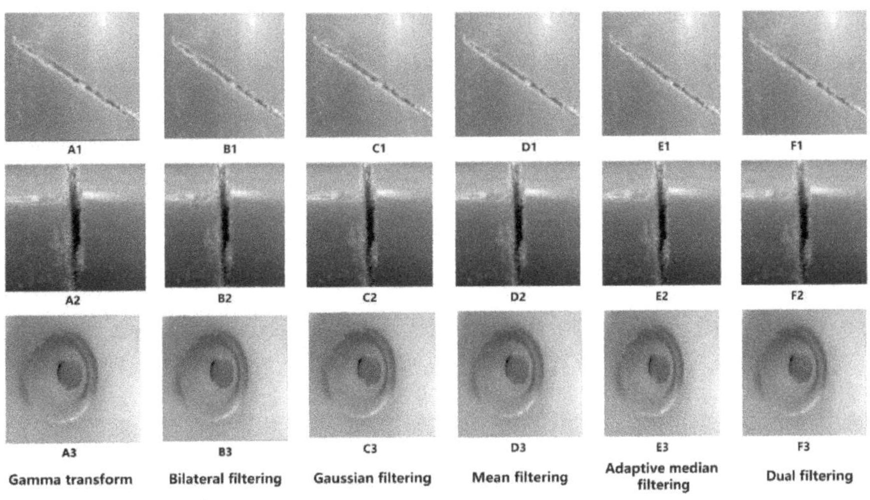

Figure 5. Comparison of pipeline defect image filtering methods.

Step-by-step processing is a very common tactic for resolving complicated noisy images [28]. One of the weighted averages used for bilateral filtering is based on Gaussian distribution, which removes Gaussian noise from the image. However, the removal of

Gaussian noise will ignore salt-and-pepper noise, and adaptive median filtering is the best algorithm to remove salt-and-pepper noise; however, it has poor results when removing Gaussian noise [24], Therefore, it was proposed to use dual filtering (Figure 5F1–F3) to remove the Gaussian noise by bilateral filtering after removing the salt-and-pepper noise by median filtering. Compared to other filtering algorithms, the dual filtering preserves the details of the edges and eliminates the noise points to achieve the effect of keeping the edges (as shown in Figure 5). Therefore, dual filtering was used for noise reduction in this paper.

3. Image Edge Detection and Segmentation for Pipe Defects

Image edges are the most basic feature of an image and using this feature the image can be segmented. In many imaging applications, it is sufficient to detect the periphery of an unknown object [29]. Image segmentation is usually performed before image feature quantization [30]. Threshold segmentation is a pixel-division technique used in region-based image segmentation, according to gray levels, into regions that have consistent properties, while neighboring regions do not have such consistent properties.

3.1. Improved Edge Detection with Sobel Operator

The conventional Sobel operator first performs a weighted average process for each pixel using a convolutional template (shown in Figure 6) and then acquires the gradient values in the X and Y directions by performing a difference process. It is challenging for the algorithm to achieve the desired detection results and the localization accuracy is not satisfactory, because the prevalent Sobel algorithm is just sensitive to the X directions and Y directions and can only assess the edges in both X directions and Y directions [24]. Various shapes and depths of the PE gas pipeline defects lead to negligible regional variations in the grayscale of the deficiency images; the collected images of PE gas pipeline defects contain a lot of interference data because of compression and real-time transmission. It is ineffective and very likely to result in missing edges, relying solely on two directional templates to identify the edges of pipe defects. In order to detect edge pixels in images more accurately, the Sobel algorithm was improved to eight directions [31] (as in Figure 7), which not only detects image edges more effectively but also increases edge detection accuracy and lowers the likelihood of incorrect edges.

−1	0	+1
−2	0	+2
−1	0	+1

Finds verticals

−1	−2	−1
0	0	0
+1	+2	+1

Finds horizontals

Figure 6. Sobel operator template.

For the pipe defect filtered image obtained above (Figure 8A1–A3), the improved Sobel algorithm (Figure 8B1–B3), the Sobel edge detection algorithm (Figure 8C1–C3) and the Prewitt algorithm (Figure 8D1–D3) are used for edge detection of the image, respectively. Figure 8 displays the outcomes of the three defects' edge detection. According to the comparison study, the improved Sobel algorithm extracts defect edges with more continuity and integrity and can completely display the defect shape characteristics. Therefore, the improved Sobel algorithm was used for edge detection in this paper.

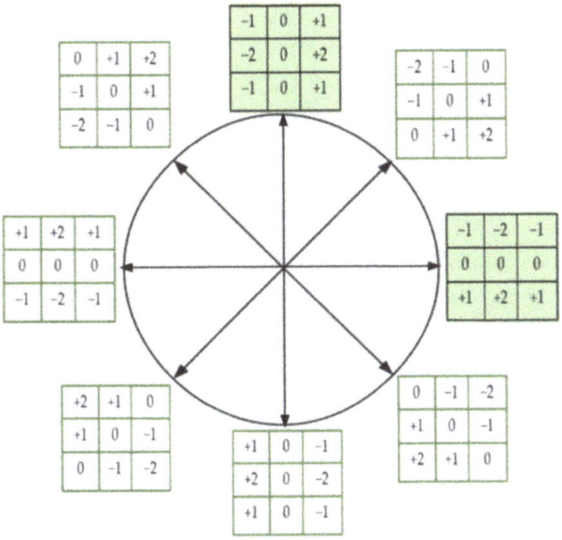

Figure 7. Improved Sobel operator.

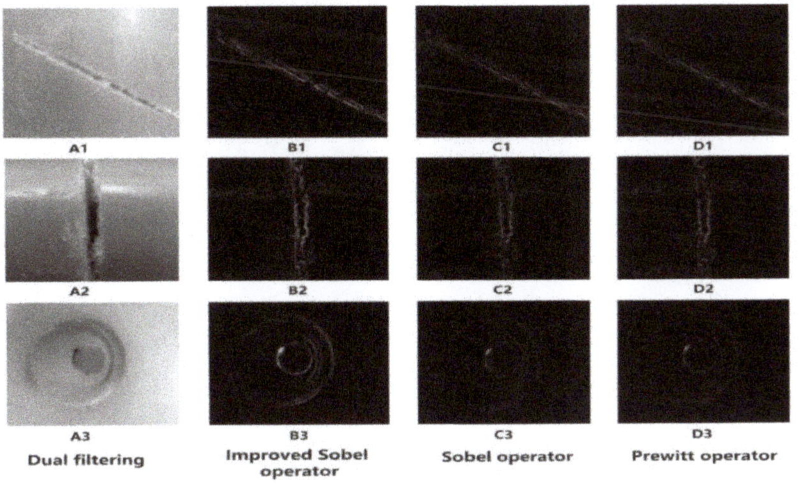

Figure 8. Comparison of pipeline defect image edge detection.

3.2. Adaptive Threshold Segmentation

In the case of inhomogeneous illumination or uneven distribution of gray values, the segmentation results obtained if global threshold is used are often unsatisfactory, and adaptive threshold (also called local segmentation) can produce good results [32]. Adaptive threshold segmentation does not use one threshold for the whole matrix as a global threshold, but has a corresponding threshold for each value at each position of the input matrix.

The images (Figure 9A1–A3) present a comparison after the above edge detection was processed by adaptive threshold (Figure 9B1–B3), global threshold (Figure 9C1–C3) and Otsu threshold segmentation (Figure 9D1–D3), respectively. According to visual observation, an adaptive threshold can be used to distinguish pipeline defects from the pipe background, with the best segmentation of PE gas pipeline defects, with complete edge segmentation and less disturbing information. Therefore, the adaptive threshold algorithm was used to segment the image in this paper.

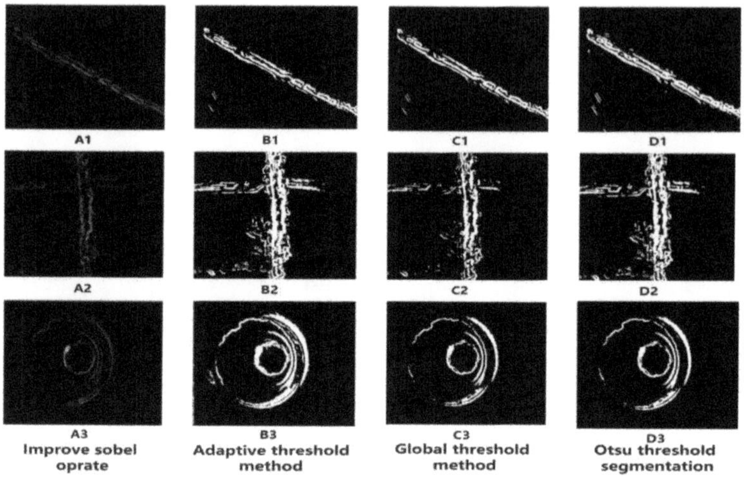

Figure 9. Comparison of pipeline defect image threshold methods.

3.3. Morphological Operation

The shape of features in an image is typically the focus of morphological operation. It has been applied to eliminate defects in a variety of shapes with the purpose of smoothing the contours and preserving the object's size and shape [33]. Erosion and dilation are the two main operations.

Let $f(m,n)$ be the input image and $g(m,n)$ be a structure element, If the set of real integers is denoted by Z, while assuming that (m,n) is an integer $f(m,n)$ from Z*Z, $g(m,n)$ is a function of a pixel's gray value with the given coordinates (m,n) and the gray value is also an integer. Namely, $g(m,n)$ to $f(m,n)$ for grayscale dilation can be defined as $f \oplus g$, which is shown as Equation (7).

$$(f \oplus g)(s,t) = \max\left\{f(s-m,t-n) + g(m,n) \mid (s-m),(t-n) \in D_f; (m,n) \in D_g\right\} \quad (7)$$

Equation (7) D_f, D_g is the definition domain of $f(m,n)$ and $g(m,n)$, respectively, $g(m,n)$ is the structural element of the morphological treatment is also a function, the displacement parameters $(s-m)$, $(t-n)$ must be in the definition domain of the function $f(m,n)$.

The dilation operation is to find the local maximum value, and the anchor point is assigned the maximum of pixels in the nucleus coverage area. The erosion operation is to find the local minimum value. The minimum value of the pixel in the kernel coverage area is assigned to the anchor point. Erosion of grayscale image is defined as Equation (8).

$$(f \ominus g)(s,t) = \min\left\{f(s+m,t+n) - g(m,n) \mid (s+m),(t+n) \in D_f; (m,n) \in D_g\right\} \quad (8)$$

Equation (8) D_f, D_g is the definition domain of $f(m,n)$ and $g(m,n)$, respectively, $g(m,n)$ is the structural element of the morphological treatment is also a function, the displacement parameters $(s+m)$, $(t+n)$ must be in the definition domain of the function $f(m,n)$.

The expressions for the opening operation and closing operation of the grayscale image have the same form as the erosion and dilation, and the structural element $g(m,n)$ performing the opening operation on the image $f(m,n)$ can be defined $f \circ g$, in Equation (9).

$$f \circ g = (f \ominus g) \oplus g \quad (9)$$

The opening operation is an erosion operation of $g(m,n)$ on $f(m,n)$ followed by an dilation operation on the result of the erosion. A similar closing operation of $g(m,n)$ on $f(m,n)$ can be defined $f \cdot g$ in Equation (10).

$$f \cdot g = (f \oplus g) \circ g \tag{10}$$

In addition to the above operations, there is the morphological gradient operation; let the gradient be denoted by h in Equation (11).

$$h = (f \oplus g) - (f \ominus g) \tag{11}$$

The basic morphological operations: closed operation (Figure 10B1–B3), dilation (Figure 10C1–C3), erosion (Figure 10D1–D3), morphological gradient (Figure 10E1–E3), and opening operation (as in Figure 10F1–F3) are applied to the threshold segmentation image obtained above (Figure 10A1–A3) to compare the results. As shown in Figure 10, The opening operation's result is to remove the image area that is slightly relative to the structure element and keeping the image area that is larger than the structure element [34]. The open operation more fully preserves the overall impact, and in this paper, we use the opening operation to fill the contours of defects in the images.

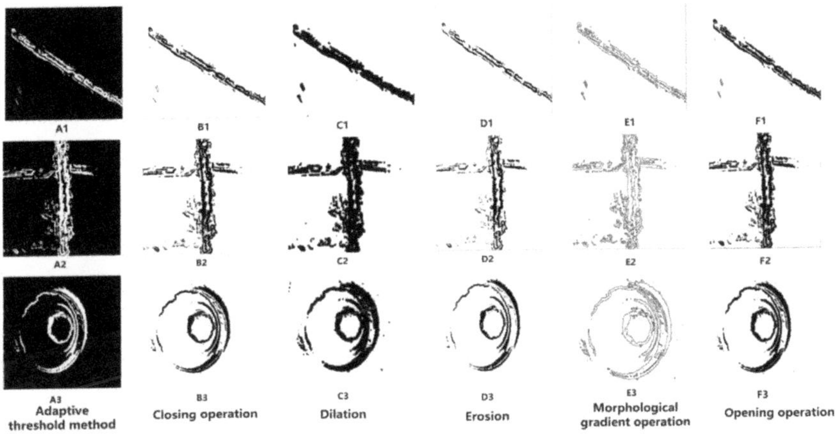

Figure 10. Comparison chart of different morphological operations.

4. CNN-Based Image Defect Classification

The CNN's primary characteristic is that its front-end input obtains image information by using multiple layers of locally interconnected neurons. The CNN can extract view invariant features [35] and takes full account of the translation, rotation, and scaling in space. Deep learning and convolutional networks have greatly improved the capability of target detection and classification using images. In this paper, three classification models, VGG16, Resnet50 and the original CNN model, are used for comparison, and the corresponding network structures are shown in Figure 11.

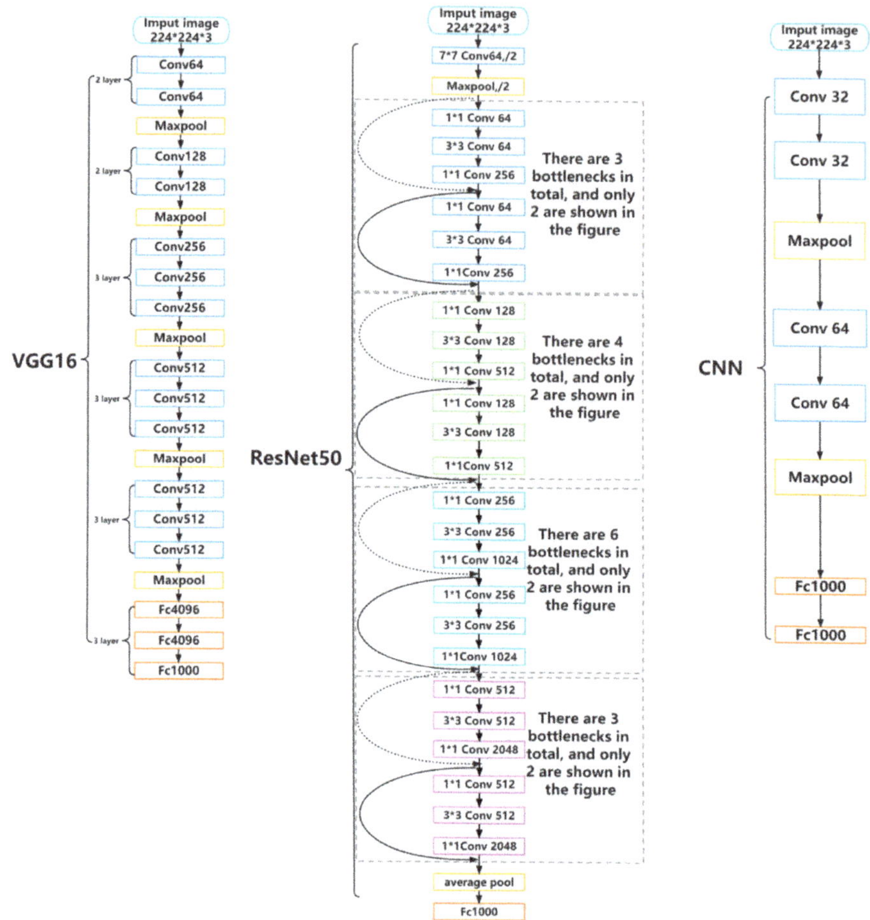

Figure 11. Comparison of three network structures.

4.1. Classification Model

VGG16 has 16 layers, including 3 fully-connected layers, 5 pooling layers, and 13 convolutional layers. Resnet50 has 50 layers, including 1 fully-connected layer, 2 pooling layers, and 49 convolutional layers. The original CNN has 6 layers, including 2 fully-connected layers, 2 pooling layers, and 4 convolutional layers. The pooling layer is not counted as a pooling layer when determining the number of layers in the network because it has no parameters.

4.2. Convolutional Layer

The convolutional layer's function is to extract the data from the source image, also referred to as image features. In the CNN architecture, the convolutional layer is typically the first layer, and the weight (w_{ij}) controls it, where i and j are the number of input and output feature mappings, respectively [36]. Thus, the entire convolution process can be defined, as in Equation (12).

$$a_j := f\left(\sum_{i \in M_j} a_j * w_{ij} + b_j\right) \quad (12)$$

In Equation (12), $*$ denotes the 2D convolution operation, a_j is the set of input feature maps, b_j denotes the jth underlying feature map, and f denotes the activation function. In the CNN, each individual convolution layer performs a linear transformation from its input to the output representation through a multichannel multidimensional convolution operation. The convolution can be represented as a matrix–vector product, where the linear transform matrix is derived from the convolution filter and the vector represents the reshaped input of the layer [37].

4.3. Pooling Layer

The pooling layer's primary function is to use specific factors to downsample the input feature mapping size, the average pool averages the features in the neighborhood to produce a blurring effect [38]. Spatial invariance can be attained by the pooling layer by lowering the feature map's resolution. Each feature map that has been combined corresponds to a feature map from the layer before. Their cells combine inputs from a small block of n $\times$ n cells. The size of the pooling window is flexible, and the windows may overlap [39]. In this paper, maximum pooling is used.

4.4. Fully Connected Layer

Each feature matrix from the pooling layer is converted by the fully connected layer into a single column feature macrovector of dimension 1 $\times$ m [40]. The most common layer in a neural network is the fully connected layer, and each of its nodes is connected to each node in the previous and next layer. The dimensionality may increase, decrease, or remain constant during the transformation process [41]. In a convolutional neural network structure for classifying images, the fully connected layer is typically placed at the very end, and for better fitting the nonlinear problem, three fully connected layers are used in this paper. The VGG16 framework is shown in Figure 12.

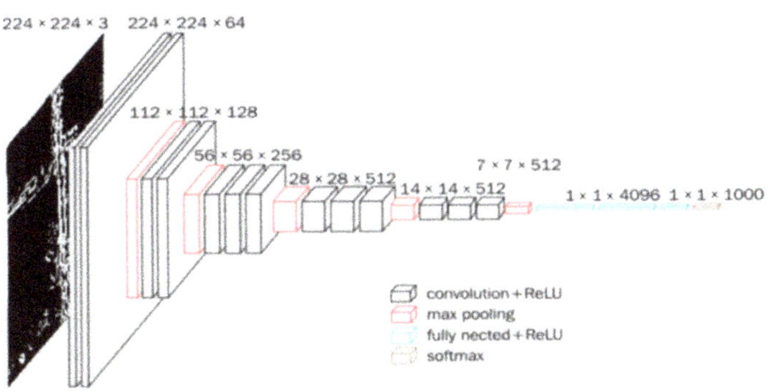

Figure 12. VGG16 framework structures.

4.5. Classification Results

This project was developed using Python and the TensorFlow API, and the graphics card used for training, verification, and testing was an NVIDIA 2080TI, 32GB, DDR4. Each training run is 100 epochs (as shown Figure 13). In this paper, to evaluate the classification effectiveness of the model, the precision and recall of our confusion matrix were applied. Equations (13) and (14) give their definitions, where, respectively, *TP*, *FP*, and *FN* stand for true positives, false positives, and false negatives.

$$Precision = \frac{TP}{TP + FP} \tag{13}$$

$$Recall = \frac{TP}{TP + FN} \tag{14}$$

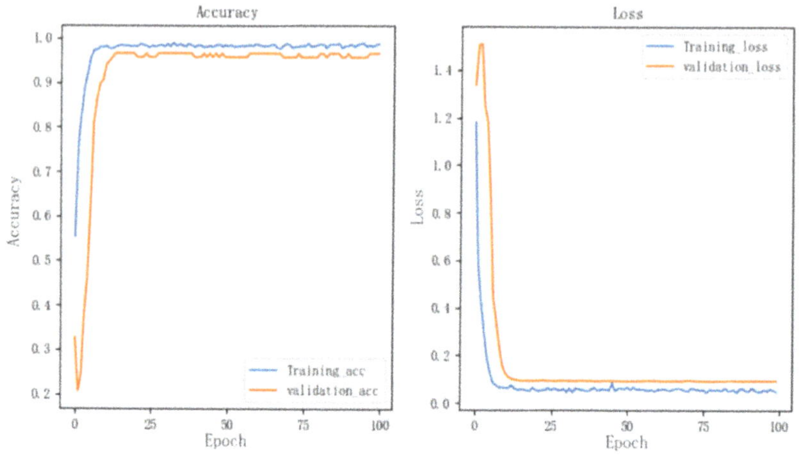

Figure 13. Accuracy and loss function of VGG16 model.

Three types of pipe defect images, 100 of each were randomly selected for each defect as a test set from Tables 1–3, made evident that the VGG16 model's recognition accuracy has improved significantly, particularly for cracks and fractures in the pipe body, where the recognition accuracy can reach an average of 98.5%, while the accuracy rate of holes is relatively low, but still can reach about 94%. From the confusion matrix (as shown in Figure 14), as can be seen, the classification algorithm in this paper has successfully identified three different types of defects with an average recognition rate of 97%. Compared with 94.96% percent of Xie et al. [36], the accuracy rate of this article has been improved by 2.04%. Compared with the 96.3% percent of Cong et al. [24], the accuracy rate of this article has been improved by 0.7%, which can be applied to the engineering practice of pipe defect detection and classification.

Table 1. VGG16 defect classification results.

Defect Type	Picture	True Positive	False Positive	False Negative	Precision	Recall
Cracks	100	99	1	5	99%	95.19%
Holes	100	94	6	2	94%	97.91%
Fractures	100	98	2	2	98%	98%

Table 2. Resnet 50 defect classification results.

Defect Type	Picture	True Positive	False Positive	False Negative	Precision	Recall
Crack	100	96	4	7	96%	93.2%
Hole	100	92	8	6	92%	93.87%
Fractures	100	95	5	4	95%	95.96%

Table 3. CNN defect classification results.

Defect Type	Picture	True Positive	False Positive	False Negative	Precision	Recall
Crack	100	95	5	13	95%	87.96%
Hole	100	87	13	10	87%	89.69%
Fractures	100	89	11	6	89%	93.68%

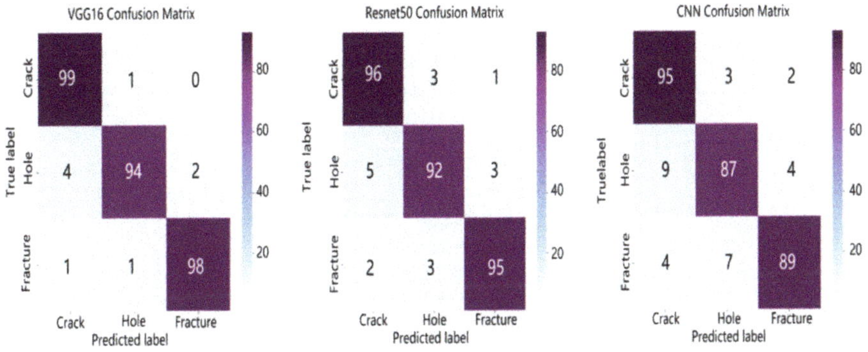

Figure 14. Comparison of three algorithms confusion matrix.

5. Conclusions

This paper proposes a framework for classifying and identifying defects of underground PE gas transmission pipes. Fast and accurate detection of defects in PE pipelines is made possible through experimental verification. The following are the main results of this experiment.

1. Based on the similarity of images of continuous pipelines, the images were first grayed out using the weighted average method and then preliminary screening of the acquired images was conducted to identify images with suspected defects, which improved the acquisition efficiency of defect image. For the defective PE gas pipeline images, contrast was enhanced using gamma transform, and finally noise was removed using the dual filtering method. The results show that the pre-processing technique can speed up image processing while retaining much of the detail of the originals.
2. In order to detect the edges of image defects, an improved Sobel algorithm was presented. An adaptive thresholding segmentation method was used to segment the defects in the image. Then, after morphologically processing the image to obtain a binary image, the feature parameters of the defects were extracted.
3. The obtained binary images were trained using VGG16, and the classifier was completed by selecting appropriate parameters for classification of various defects in images of the PE gas pipes. The experimental test results reveal that the highest accuracy of the classification approach adopted in this paper is 97% and it can be used not only for the identification of defective types of underground PE gas transmission pipes, but also for steel pipes, cast iron pipes, etc.

Author Contributions: Conceptualization, Y.W.; data curation, Q.F.; formal analysis, Y.W. and N.L.; funding acquisition, Y.W.; investigation, Y.W. and T.E.; methodology, N.L.; project administration, Y.W.; resources, Y.W. and N.L.; software, N.L.; supervision, Y.W. and H.L.; validation, H.Z.; visualization, Q.F.; writing—original draft, Q.F. and H.Z.; writing—review and editing, T.E. and H.L. All authors have read and agreed to the published version of the manuscript.

Funding: This research was funded by the National Natural Science Foundation of China (no. 11903072), the Xinjiang University Doctoral Start-up Foundation (no. 620321029) and the Science and technology planning project of State Administration for Market Regulation (no. 2022MK201).

Institutional Review Board Statement: Not applicable.

Informed Consent Statement: Not applicable.

Data Availability Statement: The data presented in this study are available on request from the corresponding author. The data are not publicly available.

Acknowledgments: The authors wish to thank all the reviewers who participated in the review during the preparation of this manuscript.

Conflicts of Interest: The authors declare no conflict of interest.

Nomenclature

x,y	consecutive images of any adjacent PE pipes in our pipe database
x_k,y_k	the grayscale discrepancy between the correspondent pixel locations of x, y
S_n	number of pixels in designated size region
u, v	coordinates of grayscale values
s,t	displacement parameters
m,n	the functions f and g are the corresponding coordinates
a_j	represents the convolution operation
b_j	represents the jth base feature map
f	activation function
w_{ij}	weights
$I(u,v)$	grayscale values
$f(m,n)$	input image
$g(m,n)$	structural elements
$f \oplus g$	dilation operation
$f \ominus g$	erosion operation
$f \circ g$	opening operation

References

1. Bachir-Bey, T.; Belhaneche-Bensemra, N. Investigation of Polyethylene Pipeline Behavior after 30 Years of Use in Gas Distribution Network. *J. Mater. Eng. Perform.* **2020**, *29*, 6652–6660. [CrossRef]
2. Zha, S.; Lan, H. Fracture behavior of pre-cracked polyethylene gas pipe under foundation settlement by extended finite element method. *Int. J. Press. Vessel. Pip.* **2021**, *189*, 104270. [CrossRef]
3. Velázquez, J.C.; Hernández-Sánchez, E.; Terán, G.; Capula-Colindres, S.; Diaz-Cruz, M.; Cervantes-Tobón, A. Probabilistic and Statistical Techniques to Study the Impact of Localized Corrosion Defects in Oil and Gas Pipelines: A Review. *Metals* **2022**, *12*, 576. [CrossRef]
4. Hou, Q.; Yang, D.; Li, X.; Xiao, G.; Ho, S.C.M. Modified leakage rate calculation models of natural gas pipelines. *Math. Probl. Eng.* **2020**, *2020*, 6673107. [CrossRef]
5. Zheng, T.; Liang, Z.; Zhang, L.; Tang, S.; Cui, Z. Safety assessment of buried natural gas pipelines with corrosion defects under the ground settlement. *Eng. Fail. Anal.* **2021**, *129*, 105663. [CrossRef]
6. Wu, J.; Bai, Y.; Fang, W.; Zhou, R.; Reniers, G.; Khakzad, N. An Integrated Qu-antitative Risk Assessment Method for Urban Underground Utility Tunnels. *Reliab. Eng. Syst. Saf.* **2021**, *213*, 107792. [CrossRef]
7. Sun, X.; Gu, J.; Tang, S.; Li, J. Research progress of visual inspection technology of steel products—A review. *Appl. Sci.* **2018**, *8*, 2195. [CrossRef]
8. Yin, X.; Chen, Y.; Bouferguene, A.; Zaman, H.; Al-Hussein, M.; Kurach, L. A deep learning-based framework for an automated defect det-ection system for sewer pipes. *Autom. Constr.* **2020**, *109*, 102967. [CrossRef]
9. Meijer, D.; Scholten, L.; Clemens, F.; Knobbe, A. A Defect Classication Methodology for Sewer Image Sets with Convolutional Neural Networks. *Autom. Constr.* **2019**, *104*, 281–298. [CrossRef]
10. Liu, H.; Hou, L.; Luo, Z.; Zhou, Y.; Jing, X.; Truong, T.-K. Image Recovery with Data Missing in the Presence of Salt-and-Pepper Noise. *Appl. Sci.* **2019**, *9*, 1426. [CrossRef]
11. Zhuang, P.; Li, C.; Wu, J. Bayesian retinex underwater image enhancement. *Eng. Appl. Artif. Intell.* **2021**, *101*, 104171. [CrossRef]
12. Houssein, E.H.; Helmy, E.D.; Elngar, A.A.; Abdelminaam, D.S.; Shaban, H. An improved tunicate swarm algorithm for global optimization and image segmentation. *IEEE Access* **2021**, *9*, 56066–56092. [CrossRef]
13. Zhang, B. Reconfigurable Morphological Processor for Grayscale Image Processing. *Electronics* **2021**, *10*, 2429. [CrossRef]
14. Zhou, M.; Pan, Z.; Liu, Y.; Zhang, Q.; Cai, Y.; Pan, H. Leak Detection and Location Based on ISLMD and CNN in a Pipeline. *IEEE Access* **2019**, *7*, 30457–30464. [CrossRef]
15. Ma, D.; Liu, J.; Fang, H.; Wang, N.; Zhang, C.; Li, Z.; Dong, J. A Multi-defect detection system for sewer pipelines based on StyleGAN-SDM and fusion CNN. *Constr. Build. Mater.* **2021**, *312*, 125385. [CrossRef]
16. Hosseinzadeh, S.; Jackson, W.; Zhang, D.; McDonald, L.; Dobie, G.; West, G.; MacLeod, C. A Novel Centralization Method for Pipe Image Stitching. *IEEE Sens. J.* **2020**, *21*, 11889–11898. [CrossRef]
17. Song, H.; Ge, K.; Qu, D.; Wu, H.; Yang, J. Design of in-pipe robot based on inertial positioning and visual detection. *Adv. Mech. Eng.* **2016**, *8*, 1687814016667679. [CrossRef]
18. Chen, K.; Hu, H.; Chen, C.; Chen, L.; He, C. An intelligent sewer defect detection method based on convolutional neural network. In Proceedings of the IEEE International Conference on Information and Automation (ICIA), Wuyishan, China, 11–13 August 2018.
19. Girshick, R. Fast R-CNN. In Proceedings of the International Conference on Computer Vision, Santiago, Chile, 7–13 December 2015; IEEE Computer Society: Washinton, DC, USA, 2015.

20. Liu, W.; Anguelov, D.; Erhan, D.; Szegedy, C.; Reed, S.; Fu, C.; Berg, A.C. Ssd: Single shot multibox detector. In Proceedings of the European Conference on Computer Vision, Amsterdam, The Netherlands, 11–14 October 2016; Springer: Cham, Switzerland, 2016; pp. 21–37.
21. Redmon, J.; Divvala, S.; Girshick, R.; Farhadi, A. You only look once: Unified, real-time object detection. In Proceedings of the IEEE Conference on Computer Vision and Pattern Recognition, Las Vegas, NV, USA, 27–30 June 2016. [CrossRef]
22. Zhang, Y. The design of glass crack detection system based on image preprocessing technology. In Proceedings of the IEEE 7th Joint International Information Technology and Artificial Intelligence Conference, Chongqing, China, 20–21 December 2014.
23. Jiang, Y.; Liu, Z.; Li, Y.; Li, J.; Lian, Y.; Liao, N.; Li, Z.; Zhao, Z. A Digital Grayscale Generation Equipment for Image Display Standardi-zation. *Appl. Sci.* **2020**, *10*, 2297. [CrossRef]
24. Li, C.; Lan, H.-Q.; Sun, Y.-N.; Wang, J.-Q. Detection algorithm of defects on polyethylene gas pipe using image recognition-Science Direct. *Int. J. Press. Vessel. Pip.* **2021**, *191*, 104381. [CrossRef]
25. Restivo, M.C.; Campbell-Washburn, A.E.; Kellman, P.; Xue, H.; Ramasawmy, R.; Hansen, M.S. A framework for constraining image SNR loss due to MR raw data compression. *Magn. Reson. Mater. Phys. Biol. Med-Icine* **2019**, *32*, 213–225. [CrossRef] [PubMed]
26. Wu, H.; Zhou, Z. Partial Color Photo Processing Method for Components Based on Image Enhancement Technology. *Wirel. Commun. Mob. Comput.* **2021**, *2021*, 4132016. [CrossRef]
27. Zhang, L.; Li, Y.; Wang, P.; Wei, W.; Xu, S.; Zhang, Y. A separation–aggregation network for image denoising. *Appl. Soft Comput.* **2019**, *83*, 105603. [CrossRef]
28. Tian, C.; Fei, L.; Zheng, W.; Xu, Y.; Zuo, W.; Lin, C.-W. Deep learning on image denoising: An overview. *Neural Netw.* **2020**, *131*, 251–275. [CrossRef] [PubMed]
29. Ren, H.; Zhao, S.; Gruska, J. Edge detection based on single-pixel imaging. *Opt. Express* **2018**, *26*, 5501–5511. [CrossRef]
30. Cui, L.; Feng, J.; Zhang, Z.; Yang, L. High throughput automatic muscle image segmentation using parallel framework. *BMC Bioinform.* **2019**, *20*, 158. [CrossRef]
31. Liu, W.; Wang, L. Quantum image edge detection based on eight-direction Sobel operator for NEQR. *Quantum Inf. Process.* **2022**, *21*, 190. [CrossRef]
32. Liao, J.; Wang, Y.; Zhu, D.; Zou, Y.; Zhang, S.; Zhou, H. Automatic segmentation of crop/background based on luminance partition correction and adaptive threshold. *IEEE Access* **2020**, *8*, 202611–202622. [CrossRef]
33. Abid Hasan, S.M.; Ko, K. Depth edge detection by image-based smoothing and morphological operations. *J. Comput. Des. Eng.* **2016**, *3*, 191–197. [CrossRef]
34. Yang, M.D.; Su, T.C. Segmenting ideal morphologies of sewer pipe defects on CCTV images for automated diagnosis. *Expert Syst. Appl.* **2009**, *36*, 3562–3573. [CrossRef]
35. Ben, X.; Zhang, P.; Lai, Z.; Yan, R.; Zhai, X.; Meng, W. A general tensor representation framework for cross-view gait recognition. *Pattern Recognit.* **2019**, *90*, 87–98. [CrossRef]
36. Xie, Q.; Li, D.; Xu, J.; Yu, Z.; Wang, J. Automatic detection and classification of sewer defects via hierarchical deep learning. *IEEE Trans. Autom. Sci. Eng.* **2019**, *16*, 1836–1847. [CrossRef]
37. Yi, X. Asymptotic Spectral Representation of Linear Convolutional Layers. *IEEE Trans. Signal Process.* **2022**, *70*, 566–581. [CrossRef]
38. Singh, P.; Raj, P.; Namboodiri, V.P. EDS pooling layer. *Image Vis. Comput.* **2020**, *98*, 103923. [CrossRef]
39. Scherer, D.; Müller, A.; Behnke, S. Evaluation of pooling operations in convolutional architectures for object recognition. In Proceedings of the International Conference on Artificial Neural Networks, Thessaloniki, Greece, 15–18 September 2010; Springer: Berlin/Heidelberg, Germany, 2010; pp. 92–101.
40. Basha, S.H.S.; Dubey, S.R.; Pulabaigari, V.; Mukherjee, S. Impact of fully connected layers on performance of convolutional neural networks for image classification. *Neurocomputing* **2020**, *378*, 112–119. [CrossRef]
41. Kalaycı, T.A.; Asan, U. Improving Classification Performance of Fully Connected Layers by Fuzzy Clustering in Transformed Feature Space. *Symmetry* **2022**, *14*, 658. [CrossRef]

sensors

MDPI

Article

Ship Trajectory Generator under the Interference of Wind, Current and Waves

Xian Ding, Hongwei Bian, Heng Ma * and Rongying Wang

The Department of Navigation Engineering, Naval University of Engineering, Wuhan 430033, China
* Correspondence: mkggrr2004@aliyun.com

Abstract: In view of the low accuracy of the motion parameters generated by the typical ship trajectory generator, and the fact that the problem of wind, current and wave interference is not considered, this paper establishes a new ship trajectory generator by analyzing the changes in the ship's attitude and speed under different motion states. Through simulation, the accuracy of the main motion parameters is significantly improved compared with the typical trajectory generator; the time-varying non-uniform wind, current and wave fields are constructed, and the interference effect of wind, current and waves on ship motion is analyzed by combining the empirical formulas of force and moment; an adaptive neuro fuzzy inference system (ANFIS) based on wind, current and wave interference is designed, and the fuzzy rules of the fuzzy system are determined by training and testing the measured data; the motion parameters of superimposed wind, current and wave interference are compared with the measured data, and the accuracy is further improved after superimposing wind, current and wave interference.

Keywords: trajectory generator; ship motion; wind, current and wave; ANFIS; inertial navigation system

Citation: Ding, X.; Bian, H.; Ma, H.; Wang, R. Ship Trajectory Generator under the Interference of Wind, Current and Waves. *Sensors* **2022**, *22*, 9395. https://doi.org/10.3390/s22239395

Academic Editors: Grzegorz Peruń, Tangbin Xia and Bogusław Łazarz

Received: 3 November 2022
Accepted: 24 November 2022
Published: 1 December 2022

1. Introduction

A trajectory generator is a tool used to generate sensor group simulation data and corresponding navigation parameters required for the simulation of an inertial navigation system (INS) and its integrated navigation system. In the algorithm research and simulation verification of INS and its integrated navigation, the research and application of trajectory generators is indispensable. The principle is to generate the motion parameters required for the simulation test of the inertial measurement unit (IMU) by simulating the maneuvering state of the carrier [1–4]. The motion of the ship is not only affected by the maneuvering, but it also subject to interference by the sea conditions, such as wind, current and waves. In order to better study the algorithm of the ship strapdown inertial navigation system (SINS) and verify the accuracy of the algorithm by simulation, the ship trajectory generator needs to be able to generate motion parameters that conform to the ship's motion characteristics and consider the interference of wind, current and wave.

At present, the research objects of the trajectory generator are mainly aircraft [5,6]. The main idea of a typical trajectory generator is to calculate the acceleration and angular velocity of the carrier by presetting ideal motion states, such as uniform linear motion, uniform rotation motion, uniform acceleration motion, etc., and then calculate the attitude, velocity and position of the carrier. The main idea is shown in the green part of Figure 1. The classical example is PROFGEN, designed by the American Air Force Avionics Laboratory [7,8]. This trajectory generator can generate attitude, velocity and position information by inputting the initial state, acceleration, angular velocity and other information. Due to the preset acceleration and angular velocity information, the generated trajectory is too ideal and does not conform to the motion characteristics of the ship, so it is difficult to meet the requirements of high-precision ship SINS simulation tests. The ship trajectory generator designed in reference [9,10] takes into account the characteristics of the ship, but it is still based on the ideal motion states, such as uniform circumference and uniform acceleration, and cannot generate complex ship trajectories. In reference [11,12], the ship motion parameter generator designed on the basis of ship measured data can generate a relatively complex ship motion trajectory, but it does not consider the interference of wind, current and waves. In the study of wind, current and waves, a variety of calculation formulas for wind, current and wave interference forces and moments have been proposed [13,14]. For example, Wu et al. [15], Zhao Qiaosheng [16] and Min Guk Seo [17] have, respectively, studied the empirical formulas for wind, current and wave forces and moments. However, the described formulas involve complicated hydrodynamic parameters and detailed ship type data, which are usually difficult to obtain and cannot be directly applied to the ship trajectory generator.

Aiming at the above problems, this paper designs a new ship trajectory generator under wind, current and wave interference. Compared with the typical trajectory generator, the main changes are as shown in the red area in Figure 1. The main ideas are as follows: (1) based on the motion state, the ship motion characteristics are analyzed, and the ship trajectory generator without wind, current and wave interference is established; (2) the time-varying wind, current and wave field is established to analyze the interference of wind, current and waves in the ship motion parameters; (3) an adaptive neuro fuzzy inference system (ANFIS) is designed; with the wind and wave information of the measured data as the input and the speed change as the output, the ANFIS rules are determined through neural network training; (4) the ship trajectory generator after superposing wind, current and wave interference is compared with the measured data to verify the rationality and superiority of ANFIS.

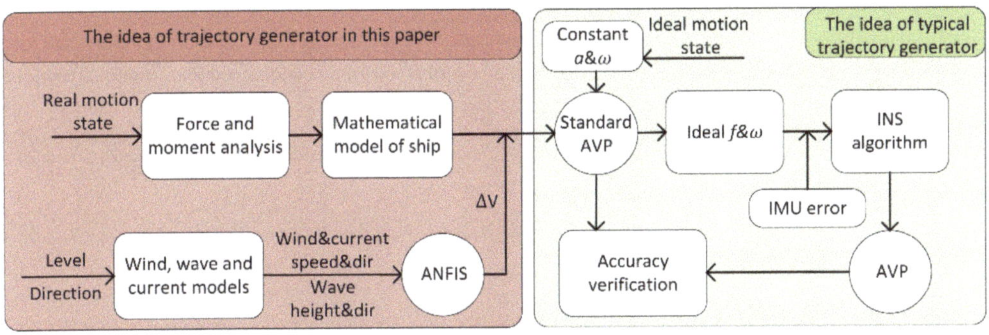

Figure 1. Flow chart of design and simulation of ship trajectory generator.

2. Mathematical Model of Ship Motion

The ship motion states can be divided into uniform linear motion, variable speed motion and steering motion. In order to simulate the ship characteristics, it is necessary to focus on the acceleration and angular velocity change analysis of variable speed motion and steering motion. The description of the ship motion variables in the carrier coordinate system is shown in Figure 2.

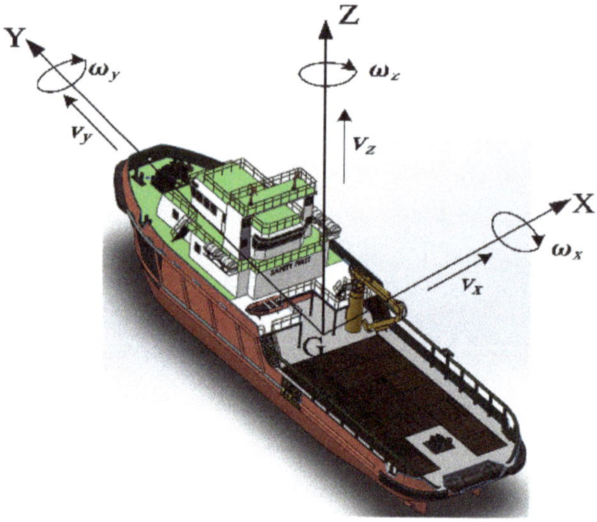

Figure 2. Schematic diagram of ship motion variables.

(1) Uniform linear motion

Without wind, current and wave interference, the acceleration and angular velocity of the ship are approximately zero when the ship sails at a constant speed.

(2) Variable speed linear motion

When there is no interference of wind, current and waves, the attitude of the straight sailing ship with variable speed will not alter, so the angular velocity can be regarded as zero. It can be expressed as $(\omega_x, \omega_y, \omega_z)^T = 0$. The variable speed motion of the ship is mainly achieved by the maneuvering propeller, and the differential equation of its motion can be expressed as [18]

$$m\dot{v}_y = F_P - R_u v_y + m v_x \omega_z \quad (1)$$

where F_P is the propulsion force and R_u is the drag coefficient along the Y-axis. F_P can be expressed as [19]

$$\begin{cases} F_P = (1 - t_P)\rho n^2 {D_P}^4 K_T(J_P) \\ J_P = (1 - v_z\omega_y) \cdot v_y / (n \cdot D_P) \\ K_T(J_P) = a_0 + a_1 J_P \end{cases} \quad (2)$$

Inserting F_P from Equation (2) into Equation (1), Equation (1) can be expressed as

$$m\dot{v}_y + \left[R_u - a_1(1 - v_z\omega_y)(1 - t_P)\rho n^2 {D_P}^4/(nD_P)\right] v_y = (1 - t_P)\rho n^2 {D_P}^4 a_0 \quad (3)$$

where t_p is the thrust reduction coefficient, n is the rotational speed of the propeller and D_p is the propeller diameter. When the ship data are fixed, Equation (3) can be simplified as follows:

$$m\dot{v}_y + Q(n)v_y = G(n) \quad (4)$$

According to the actual situation, when the ship is in a stable state at different gears, the rotational speed of the propeller is approximately a constant value. When n is a constant value, the speed variations of the ship reflect a first-order system. Considering the change in the rotational speed of the propeller when the ship is accelerating, combined with the measured data, this paper adopts the second-order overdamping function as the longitudinal speed model of the ship when the ship is accelerating. This is because the first-order system can be regarded as a special case of the second-order system, and the second-order system can also reflect the characteristics of the higher-order system.

Assume that the speed of the ship at the initial moment is v_0, and the speed increases by Δv after the propeller gear is switched; thus, the speed change can be expressed as

$$v_y(t) = v_0 + \Delta V(1 + \frac{e^{-t/T_1}}{T_2/T_1 - 1} + \frac{e^{-t/T_2}}{T_1/T_2 - 1}) \quad (5)$$

In (5), $T_1 = \frac{1}{\omega_1(\xi_1 - \sqrt{\xi_1{}^2 - 1})}$, $T_2 = \frac{1}{\omega_1(\xi_1 + \sqrt{\xi_1^2 - 1})}$; ξ_1 and ω_1 are the damping ratio and oscillating frequency, respectively. By identifying the measured data, different ξ_1 and ω_1 can be obtained, which can reflect the acceleration performance of ships in different gears [20]. Deceleration is regarded as the reverse process of acceleration and will not be analyzed in detail.

(3) Steering motion

The amplitude of pitch and heave caused by the steering motion of a ship in still water is very small, and the pitch angular velocity ω_x and heave acceleration a_z are approximately zero.

According to the transfer function of the rudder angle δ and heading angular velocity ω_z in the Nomoto response model [21],

$$\frac{\omega_z(s)}{\delta(s)} = \frac{K(1 + T_3 s)}{(1 + T_1 s)(1 + T_2 s)} \quad (6)$$

Since T_3 is very small and the zero point $s_3 = -\frac{1}{T_3}$ is far away from the negative half axis, the removal of this point has little impact, so the heading angular velocity can be expressed as

$$\omega_z(t) = K\delta(t)(1 + \frac{e^{-t/T_3}}{T_4/T_3 - 1} + \frac{e^{-t/T_4}}{T_3/T_4 - 1}) \quad (7)$$

In (7), the definition of T_3 and T_4 refers to Equation (5).

The steering movement can be divided into three stages: initial steering, constant steering and steering recovery. Assuming that the roll angle in the first stage will increase to the angle γ, the change in roll angle can be expressed as

$$\theta(t) = \gamma(1 - \frac{1}{\sqrt{1 - \xi_2{}^2}} e^{-\xi_2 \omega_2 t} \sin(\omega_2 \sqrt{1 - \xi_2{}^2 t} + \beta_1)) \quad (8)$$

In (8), ξ_2 and ω_2 are the damping ratio and oscillating frequency, respectively; β_1 is the initial phase. In the phase of steady steering, the roll angle remains unchanged, so the angular velocity ω_y is zero. In the phase of steering recovery, the change in roll angle can be regarded as the reverse process of the initial phase.

The displacement and attitude angle are denoted as $\mu = (x, y, z, \varphi, \theta, \psi)^T$, and the velocity and angular velocity are denoted as $\tau = (v_x, v_y, v_z, \omega_x, \omega_y, \omega_z)^T$. The trajectory parameters of the ship can be obtained from the following formula:

$$\dot{\mu} = \begin{bmatrix} c_b^n & 0_{3*3} \\ 0_{3*3} & J \end{bmatrix} \tau \quad (9)$$

In (9),

$$c_b^n = \begin{bmatrix} \cos\psi\cos\theta & -\sin\psi\cos\varphi + \cos\psi\sin\theta\sin\varphi & \sin\psi\sin\varphi + \cos\psi\cos\varphi\sin\theta \\ \sin\psi\cos\theta & \sin\psi\cos\varphi + \sin\psi\sin\theta\sin\varphi & -\cos\psi\sin\varphi + \sin\psi\cos\varphi\sin\theta \\ -\sin\theta & \cos\theta\sin\varphi & \cos\theta\cos\varphi \end{bmatrix},$$

$$J = \begin{bmatrix} 1 & \sin\varphi\tan\theta & \cos\varphi\tan\theta \\ 0 & \cos\varphi & -\sin\varphi \\ 0 & \sin\varphi\sec\theta & \cos\varphi\sec\theta \end{bmatrix}.$$

In order to verify the feasibility and superiority of the above model, the simulation verification is carried out based on the ship's measured data in a period of relatively calm sea conditions. The ship's initial information and motion state are shown in Tables 1 and 2.

Table 1. Initial information table.

Lng (°)	Lat (°)	v (kn)	Heading (°)
123.05516	31.06247	8.7	239.5

Table 2. Maneuver state table.

Time (s)	500	120	325	35	50	120
State	Linear	Linear	Linear	Steering	Steering	Steering
ΔV (kn)	0	8.2	0	−0.8	0.8	2
δ (°)	0	0	0	20	15	−20

Based on the position, speed and heading of the measured data, the comparison results of typical trajectory generators and the trajectory generator described in this paper are shown in Figures 3 and 4. The black line in the figure denotes the data generated by the trajectory generator, and the red line denotes the measured data.

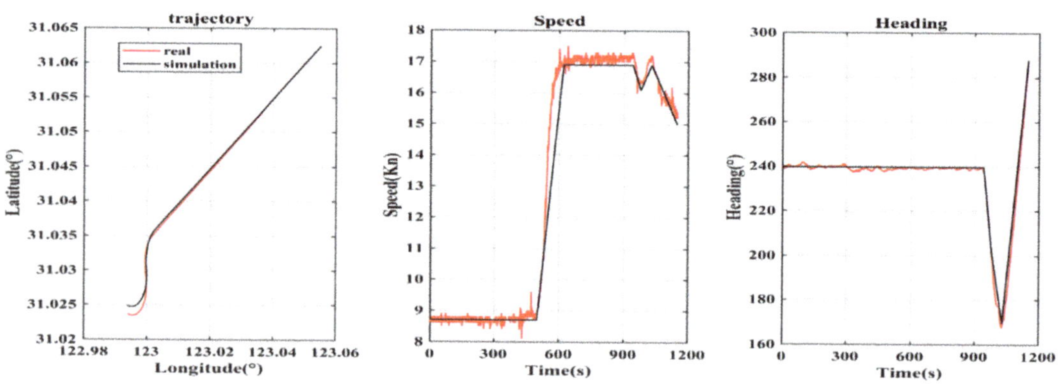

Figure 3. Comparison between measured data and typical trajectory generator.

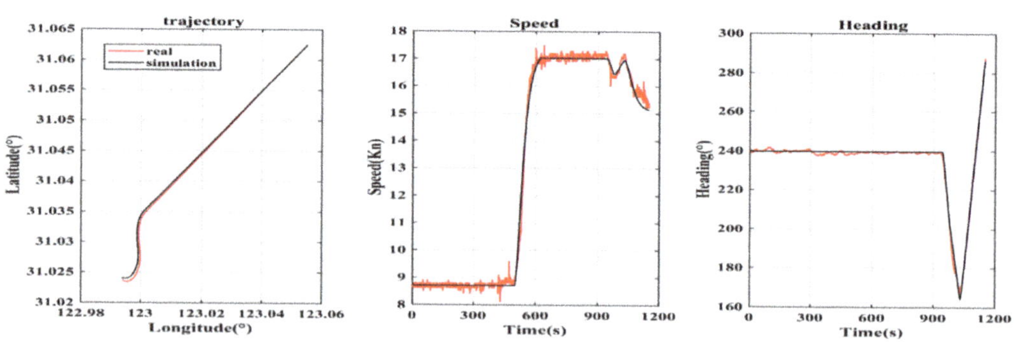

Figure 4. Comparison between measured data and the new trajectory generator.

In order to judge the accuracy of the simulation results more intuitively and verify the superiority of this model, the accuracy of position, speed and heading and root-mean-square error (RMSE) are calculated. The results are shown in Table 3.

Table 3. Accuracy and RMSE comparison results.

Parameter		Pos (m)	v (m/s)	Heading (°)
Threshold		70	0.1	1
Accuracy	Typical	52.09%	64.96%	94.00%
	New	85.48%	68.70%	94.96%
RMSE	Typical	72.4141	0.2743	2.6821
	New	50.8821	0.2441	2.1743

Accuracy refers to the accuracy within the acceptable error range (threshold), and RMSE reflects the statistical rule of error. It can be seen from Table 3 that the accuracy of the position, speed and heading generated by the trajectory generator described in this paper within the threshold is higher than that of the typical trajectory generator, and the RMSE value is also smaller. This is because the preset ideal track of the typical trajectory generator does not conform to the motion characteristics of the ship, which also shows that the ship trajectory generator described in this paper is superior to the typical trajectory generator.

Although the simulation effect of the model proposed in this paper is better than that of the typical trajectory generator, because the model is not disturbed by wind, current and waves, it is still quite different from real ship motion. In order to obtain more realistic ship motion parameters, it is necessary to analyze the wind, current and wave interference.

3. Analysis of Wind, Current and Wave Interference

In the actual navigation of a ship, the level and direction of the wind, waves and currents are constantly changing at random. It is difficult to directly address the variation in the motion parameters due to the combined interference of wind, current and waves at all times. In order to simulate the actual environment of ship navigation, it is necessary to develop time-varying inhomogeneous wind, current and wave field models that conform to the characteristics of the actual sea conditions.

3.1. Wind Field Model

Taking the combined wind speed model as the research object, the actual wind speed at sea is divided into three components: average wind, gradual wind and random wind [22].

3.1.1. Average Wind

This component reflects the average wind speed of the wind field, which generally does not change with time and can be expressed as

$$V_m = \frac{1}{2}(V_{i\max} + V_{i\min}) \tag{10}$$

where i represents the wind level.

3.1.2. Gradual Wind

Gradual wind is generally simulated by the frequency density spectrum. In this paper, the frequency spectrum of the NORSOK wind model is used, and its expression is as follows [23]:

$$\begin{cases} S_w(\omega) = 320 \cdot (V_m/10)^2 \cdot (z_{cw}/10)^{0.45} / (1 + x^{0.468})^{3.561} \\ x = 172 \cdot \omega (z_{cw}/10)^{2/3} (V_m/10)^{-0.75} \end{cases} \tag{11}$$

where V_m is the wind speed; Z_{cw} is the vertical height from the wind action point to the sea level; ω is the frequency. The gradual wind is regarded as the result of the superposition of multiple simple harmonic waves, so the gradual wind can be expressed as

$$V_{di}(t) = \sqrt{2S_w(\omega_i) \triangle \omega_i} \cos(2\pi\omega_i t + \varphi_i) \tag{12}$$

where ω_i is the frequency of the simple harmonic component; $\triangle\omega_i$ is the frequency width of the simple harmonic component; φ_i is the random phase.

3.1.3. Random Wind

Random wind mainly reflects the randomness of wind in the wind field, which can be expressed by the superposition of the average wind with the first-order Markov process.

$$\begin{cases} \dot{V}_m + a_1 V_m = w_1 \\ \dot{\psi}_w + a_2 \psi_w = w_2 \end{cases} \tag{13}$$

where a_1 and a_2 are constants greater than zero; w_1 and w_2 are white noise.

To sum up, the wind model can be expressed as

$$V_w = V_m + \sum_{i=1}^{N} V_{di} \tag{14}$$

In 10 min, the simulation of the wind field is as shown in Figure 5 when the wind speed is level 5 (8–10.7 m/s) and the average wind direction is 45°.

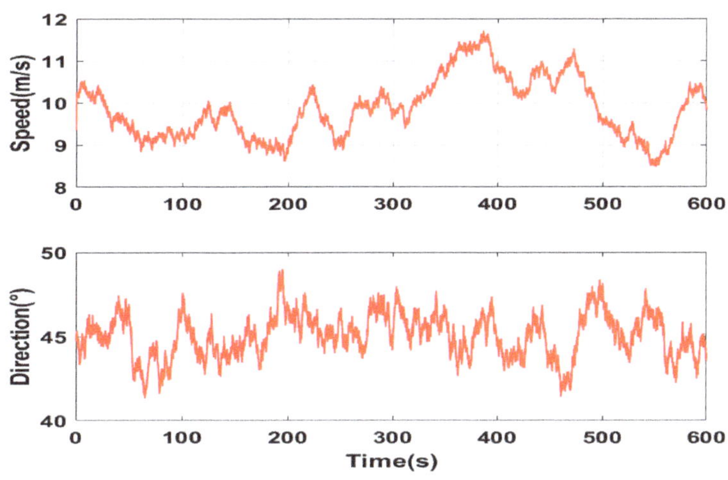

Figure 5. Simulation diagram of wind speed and direction.

3.2. *Current Field Model*

Similar to the wind field, the current speed and direction also change with time and space. In order to reflect this change, the first-order Markov process can be used to express the change in current. Its expression is similar to that in Equation (11) and will not be detailed here.

In 10 min, the simulation of the current velocity is 2 m/s and the current direction is 45°, as shown in Figure 6.

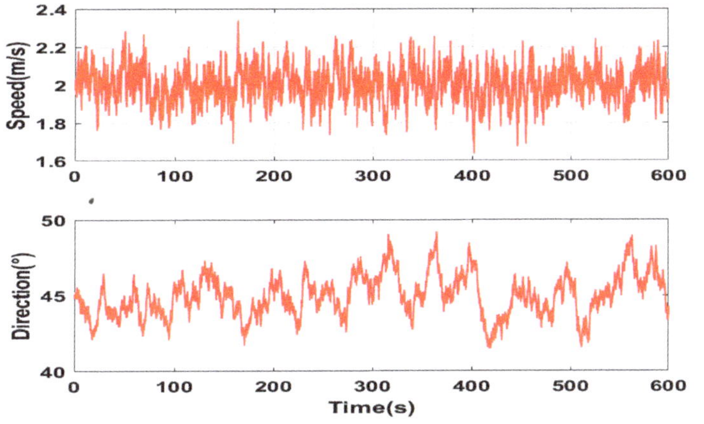

Figure 6. Simulation diagram of current speed and direction.

3.3. Wave Field Model

Generally, in practical applications, the wave energy spectrum is used to describe irregular waves, such as the PM spectrum, ITTC/ISSC spectrum, JONSWAP spectrum, etc. In order to reduce the unknown parameters, the ITTC single-parameter wave energy spectrum is adopted in this paper, which can be expressed as [24]

$$S(\omega) = \frac{A}{\omega^5} \exp(-B/\omega^4) \tag{15}$$

where $A = 8.1 * 10^{-3}\ g^2$; $B = 3.11/h^2$; ω is the wave frequency; h is the significant wave height, which can be obtained from the wave level. The wave can be expressed as

$$\xi_{wa} = \sqrt{2S_{wa}(\omega_p) \triangle \omega_p} \cos(\omega_p t + \varphi_p) \tag{16}$$

Waves are generally caused by wind, and the average wave direction is consistent with the average wind direction. In 10 min, the wave height and wave direction simulation is as shown in Figure 7 when the wave is level 6 and the average wave direction is 45°.

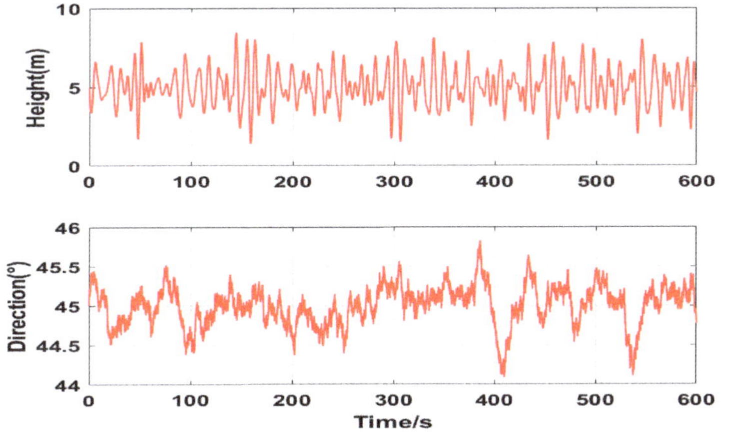

Figure 7. Simulation diagram of wave speed and direction.

3.4. Wind, Current and Wave Interference Analysis

Generally, compared with the waves, the wind has less influence on the vertical linear acceleration and pitch angular velocity of the ship, so the vertical velocity a_z and pitch angle velocity ω_x caused by it can be regarded as zero. Combined with the empirical formula of the wind force and moment, the change in the ship's linear acceleration and angular velocity caused by wind can be simplified as [25]

$$\begin{cases} a_y = K_1 {v_{rw}}^2 \cos \chi_{rw} \\ a_x = K_2 {v_{rw}}^2 \sin \chi_{rw} \\ \omega_z = K_3 {v_{rw}}^2 \sin \chi_{rw} \end{cases} \tag{17}$$

where K_1, K_2 and K_3 are wind-related proportional coefficients; $v_{\rm rw}$ is the relative wind speed; χ_{rw} is the relative wind direction angle.

It can be seen from the characteristics of the current that it has little influence on the ship's angular velocity. The interference of the current mainly exerts an influence on the longitudinal and transverse linear velocity of the ship. Assuming that the velocity of the current is $V_{\rm c}$, the direction is $\gamma_{\rm c}$, the ship's heading is ψ and the ship's longitudinal and transverse speeds are $V_{\rm y}$ and $V_{\rm x}$, respectively, and the method of speed vector synthesis is adopted, the ship's speed after the superimposed current interference is

$$\begin{cases} V_{yc} = V_y + V_c \cos(\pi - \gamma_c - \psi) \\ V_{xc} = V_x + V_c \sin(\pi - \gamma_c - \psi) \end{cases} \tag{18}$$

When the length of the ship is far less than the wave length, combined with the wave force and moment formula and the wave surface attitude, the wave interference can be simplified as follows [26]:

$$\begin{cases} a_y = K_4 h^2 \cos \chi_{wa} \\ a_x = K_5 h^2 \sin \chi_{wa} \\ a_z = \dot{h}(t) \\ \omega_y = 0 \\ \omega_x = \frac{\dot{h}(t)}{\sqrt{L^2 - (h(t))^2}} \\ \omega_z = K_6 h^2 \sin \chi_{wa} \end{cases} \tag{19}$$

where K_4, K_5 and K_6 are the wave-related proportional coefficients; L is the length of the ship; h is the wave amplitude of the wave, and its value is obtained from the wave field; $\chi_{\rm wa}$ is the relative wave direction angle.

3.5. Design of ANFIS

Due to the complexity, nonlinearity and strong coupling of the interaction between wind, waves and ship, it is impossible to establish a clear mathematical model to reflect the real-time movement state changes of the ship [27]. In order to obtain more accurate speed and attitude changes, this paper designs an ANFIS based on the double-input single-output T-S model [28,29]. The structure of the fuzzy neural network of the T-S model is shown in Figure 8. The network consists of a premise network and consequent network. The premise network is used to match the premise of fuzzy rules, and the consequent network is used to generate the consequents of the fuzzy rules. The learning algorithm of ANFIS will not be detailed here.

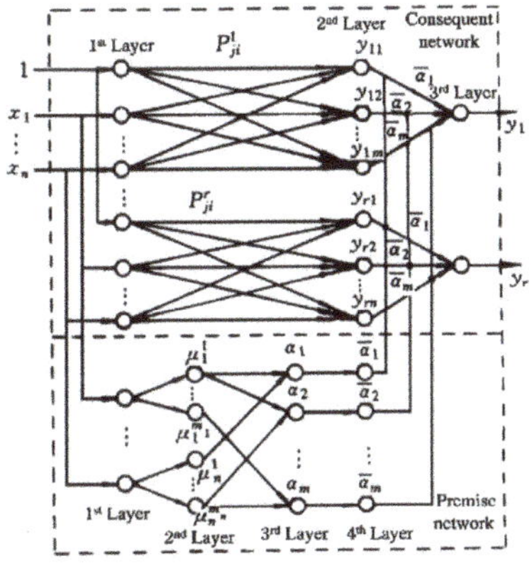

Figure 8. Fuzzy neural network structure diagram of T-S model.

In this paper, the first layer of the premise network comprises two input variables, which are the accelerations in Equation (17) and Equation (19), respectively. The membership function type of the input variables is Gauss, the number of membership functions is 5 and 7, respectively, and the output variable type is linear. The second layer is used to calculate the membership function of the input variables. Each node in the third layer represents a fuzzy rule, which is used to match the premise of the fuzzy rule and calculate the fitness of each rule. The fourth layer is used to achieve normalization. In the consequent network, the first layer is the same as the antecedent network. The second layer is used to calculate the consequent of each rule. The third layer is used to calculate the output of the ANFIS system, and the output is the speed change of the ship.

Taking the measured data of the ship as the training set, the input of the training set is the measured wind speed, wind direction, wave height and wave direction. The output is the speed change, which is the change in current speed subtracted from the ship's speed change. In order to reduce the influence of propeller and rudder forces, the measured data of the ship when sailing straight at a constant speed are selected as the training data. Because the ship is in dynamic equilibrium under this motion state, the changes in its speed and attitude can be seen to be caused by time-varying wind, current and waves. The training result of ANFIS is shown in Figure 9. After training, the average error between the system output and the training data is 0.2005, meeting the error tolerance.

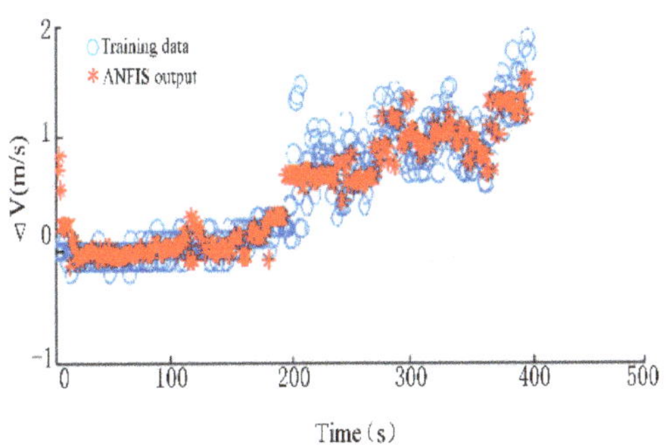

Figure 9. Comparison chart of training data and ANFIS output.

4. Simulation Analysis and Comparative Verification

Due to the time-varying nature and randomness of wind, current and waves, the time-varying non-uniform wind, current and wave field constructed in this paper cannot be consistent with the wind, current and waves encountered by the ship in actual navigation, and its impact on the ship's motion state cannot be consistent with the measured data. In order to verify the rationality and superiority of this method, the following simulation strategies are formulated:

(1) Based on the ship trajectory generator designed in this paper, the ship motion parameters without wind, current and wave interference are generated, and then the wind, current and wave interferences are superimposed to compare and analyze whether the change in motion parameters is reasonable;

(2) We select a period with a relatively large disturbance of wind and waves from the measured data; take the actual measured wind speed, wind direction, wave height and wave direction as the input of ANFIS; generate the corresponding speed change; and then add it to the ship trajectory generator in this paper to observe whether the accuracy of the main motion parameters has been improved.

4.1. Simulation Analysis

The trajectory generator described in this paper generates ship motion parameters, sets the level of wind, current and waves and generates corresponding interferences in combination with the wind, current and wave fields. The initial information, maneuvering state and detailed information of wind, current and waves are shown in Tables 4–6.

Table 4. Ship's initial condition.

Lng (°)	Lat (°)	v (kn)	Heading (°)
18.10	119.50	0	0

Table 5. Maneuver state.

Time (s)	100	50	50	50	100	50	50	50
State	Linear	Linear	Steering	Steering	Linear	Steering	Steering	Linear
ΔV (kn)	10	0	−5	5	0	−5	5	0

Table 6. Wind, current and wave conditions.

Wind		Current		Wave	
Level	Dir	Speed	Dir	Level	Dir
4	45°	1 m/s	90°	2	45°

The simulation results are shown in Figure 10. The black lines and red lines in the figure represent the simulation results without interference and with interference, respectively.

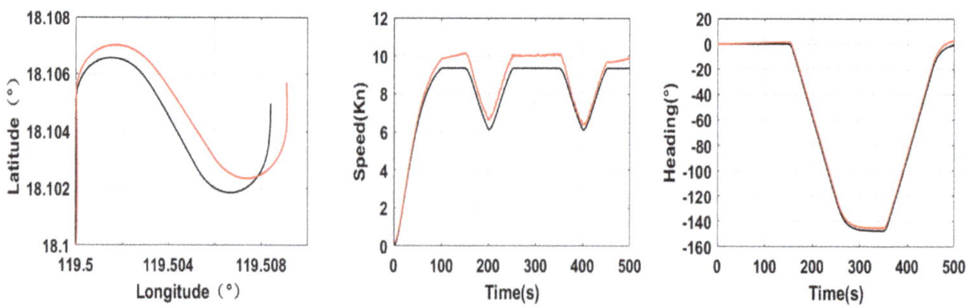

Figure 10. Comparison of simulation data without and with interference.

It can be seen from Figure 10 that after the interference is added, the movement of the ship changes as follows:

(1) The track obviously deviates to the east and north;

(2) The speed generally increases, slightly increases before steering, and then increases as a whole;

(3) The overall trend of the heading angle changes little—it decreases slightly after the first steering and increases slightly after the second steering.

According to the information in Tables 4–6, the above changes are analyzed as follows:

(1) The wind direction is 45° north by east, and the flow direction is 90° north by east. Therefore, under the combined action of wind and current, the ship's track will naturally drift to the east and north.

(2) In the wind, current and wave interference considered in this paper, the current has the greatest impact on the ship's speed. Table 6 shows that when the ship does not start to steer, the current direction is almost perpendicular to the ship's moving direction, so it has little impact on the speed. However, the speed caused by wind has an influence on the ship's speed, so the ship's speed rises slightly before steering. After the first steering, the current begins to exert an influence on the direction of the ship's motion, which causes the ship's speed to rise.

(3) As the wave level is small, it has little influence on the heading angle. It can be seen from the wind direction that the wind will cause the ship's heading angle to change to the north by east direction. The ship will move southward after the first steering, so the wind will reduce the ship's heading angle. After the second steering, the ship will move northward, so its heading angle will increase.

Therefore, the above changes are reasonable after the interference, which shows the rationality and feasibility of the wind, current and wave interference model proposed in this paper.

4.2. Simulation Comparison Verification

In order to verify the accuracy of ANFIS, based on the ship trajectory generator presented in this paper, the ship motion parameters are generated using the same initial state and maneuvering state. We compare the accuracy of the ship motion parameters with and without wind, current and wave interference. The initial information and the maneuvering status of the ship are shown in Tables 7 and 8.

Table 7. Ship's initial condition.

Lng	**Lat**	**Speed**	**Heading**
122°52′45″	30°54′05″	9.6 kn	225.9°

Table 8. Maneuver state.

Time (s)	**830**	**106**	**64**	**77**
State	Linear	Steering	Steering	Linear
ΔV (kn)	0	−5.4	2.8	1.8

The comparison results of the simulated speed and track are shown in Figure 11. The red line in the figure is the measured data; the blue line is the data generated by the ship trajectory generator with wind, current and wave interference; and the black line is the data generated by the ship trajectory generator without wind, current and wave interference.

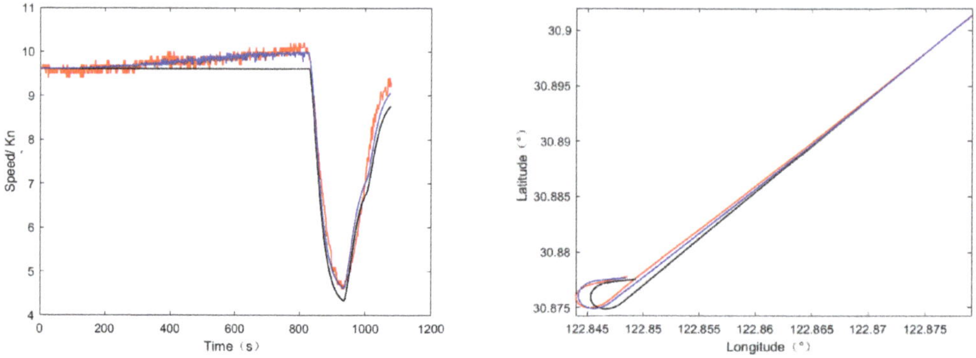

(a) (b)

Figure 11. Comparison chart with and without interference and measured data. (**a**) Speed comparison chart (**b**) Track comparison chart.

It can be seen from the above figures that, compared with the ship trajectory generator without wind, current and wave interference, the accuracy of the speed and track after adding ANFIS is significantly improved, especially in the constant speed stage, where the change in speed is more consistent with the measured data; the specific accuracy and RMSE values are shown in Table 9. Since the accuracy of the position is obviously improved, if the same threshold value as in Table 3 is adopted, the position accuracy will be 100%, so the threshold value of the position listed in Table 9 is adjusted to 35 m.

Table 9. Accuracy and RMSE comparison results.

Parameter		Pos (m)	v (m/s)
Threshold		35	0.1
Accuracy	Without ANFIS	42.99%	55.62%
	ANFIS	88.12%	87.00%
RMSE	Without ANFIS	68.9173	0.1641
	ANFIS	23.4941	0.0948

The above simulation results show that the method proposed in this paper can better simulate the impact of the interference of wind, current and waves on ship motion, and they also verify the advantages of ANFIS.

5. Discussion

The wind, current and wave field models constructed in this paper can reflect the characteristics of wind, current and waves. However, due to the randomness and time variability of wind, current and waves, there is a lack of effective evaluation and verification methods, and it is difficult to compare and verify the model with the measured data. Therefore, it is necessary to obtain appropriate evaluation methods for wind, current and wave models.

The force of ship motion at sea is complex and the coupling degree of wind, current and wave interference is high. The ANFIS designed in this paper takes the data under the condition of straight sailing at a constant speed as the training data, and the interference of wind, current and waves under the condition of variable speed or steering needs further research.

6. Conclusions

In this paper, a new mathematical model of a ship trajectory generator is established; the interference of wind, current and waves with ship motion is analyzed, and an adaptive neuro fuzzy inference system for wind, current and wave interference is designed. After simulation analysis and verification, the following conclusions are obtained:

(1) The motion parameters generated by the ship trajectory generator proposed in this paper are more accurate than those generated by the typical trajectory generator. After the wind, current and wave disturbances are superimposed, the accuracy is further improved. The ship trajectory generator has certain significance for the simulation testing and algorithm research of a ship's strapdown inertial navigation system.
(2) The wind, current and wave fields constructed in this paper can reflect the randomness and time variability of wind, current and waves on the sea. Combined with the designed ANFIS, we can simulate the corresponding interference, which has certain significance for the research of ship motion under wind, current and wave conditions.
(3) Compared with the traditional mathematical model of ship motion, the trajectory generator described in this paper does not require complicated hydrodynamic parameters, has low modeling requirements and has the advantage of high simulation accuracy, and it can be used in the field of navigation simulators.

Author Contributions: Conceptualization, H.B. and X.D.; Funding acquisition, H.B.; Investigation, R.W.; Project administration, H.B.; Resources, R.W. and H.M.; Supervision, H.B. and H.M.; Validation, X.D.; Writing—original draft, X.D.; Writing—review and editing, R.W. and H.M. All authors have read and agreed to the published version of the manuscript.

Funding: This research received no external funding.

Conflicts of Interest: The authors declare no conflict of interest.

References

1. Costley, A.; Christensen, R.; Leishman, R. Analytical Aircraft State and IMU Signal Generator from Smoothed Reference Trajectory. *IEEE Trans. Aerosp. Electron. Syst.* **2021**, *58*, 2517–2530. [CrossRef]
2. Fukuda, G.; Kubo, N. Application of Initial Bias Estimation Method for Inertial Navigation System (INS)/Doppler Velocity Log (DVL) and INS/DVL/Gyrocompass Using Micro-Electro-Mechanical System Sensors. *Sensors* **2022**, *22*, 5334. [CrossRef] [PubMed]
3. Chae, M.-S.; Cho, S.-Y.; Shin, K.-H. Development of a real trajectory-based simulator to verify the reliability of the integrated navigation system for trains. *J. Korea Inst. Electron. Commun. Sci.* **2021**, *16*, 135–144.
4. Fukuda, G.; Hatta, D.; Kubo, N. A Study on Estimation of Acceleration and Angular Velocity Data from Actual Measurements by Trajectory Generator. *J. Jpn. Inst. Navig.* **2021**, *144*, 14–20. [CrossRef]
5. Luo, Y.; Liu, Y. Design and Simulation of the Trajectory Generator Based on the Strapdown Inertial Navigation Vehicle. *J. Henan Polytech. Univ. (Nat. Sci.)* **2015**, *34*, 867–871.
6. Yang, G.; Wang, J.; Zhou, X. High-precision simulator for strapdown inertial navigation systems based on real dynamics. *J. Navig. Position.* **2015**, *3*, 27–31.
7. Chen, K.; Shen, F.; Zhou, J. Simulation Platform for SINS/GPS Integrated Navigation System of Hypersonic Vehicles Based on Flight Mechanics. *Sensors* **2020**, *20*, 5418. [CrossRef]
8. Chen, K.; Wei, F.; Zhang, Q. Trajectory generator of SINS on flight dynamics with application in hardware-in-the-loop simulation. *J. Chin. Inert. Technol.* **2014**, *22*, 486–491.
9. Zhang, B. Design of Ship Trajectory Simulation Generator Based on Strapdown Inertial Navigation. *Digit. Technol. Appl.* **2009**, *11*, 11–13.
10. Zhang, M.; Hu, B.; Chen, Q. Design of trajectory generator in marine digital north platform inertial navigation simulator. *Ship Sci. Technol.* **2021**, *43*, 146–150.
11. Bian, H.; Ding, X.; Ma, H. A 6-DOF a&ω motion parameter generator based on ship hydrodynamics. *Syst. Eng. Electron.* **2022**, *1*, 9.
12. Bian, H.; Zhu, Z.; Ma, H. Ship motion parameter generator based on linear acceleration and angular velocity six degree of freedom motion model. *Syst. Eng. Electron.* **2022**, *44*, 2628–2634.
13. Lee, J.-H.; Kim, Y. Prediction of ship operation performance in waves by seakeeping-maneuvering coupled analysis. In Proceedings of the 33rd International Workshop on Water Waves and Floating Bodies (IWWWFB), Guidel-Plages, France, 4–7 April 2018; pp. 105–108.

14. Paroka, D.; Muhammad, A.; Asri, S. Prediction of ship turning maneuvers in constant wind and regular waves. *Int. J. Technol.* **2017**, *8*, 387–397. [CrossRef]
15. Wu, G.; Zhao, X.; Sun, Y. Cooperative maneuvering mathematical modeling for multi-tugs towing a ship in the port environment. *J. Mar. Sci. Eng.* **2021**, *9*, 384. [CrossRef]
16. Zhao, Q.; He, C.; Li, D. Research on Dynamic Modeling and Simulation of Manned Submersible Operated under Ocean Current. *Shipbuild. China* **2021**, *62*, 145–152.
17. Seo, M.; Ha, Y.; Nam, B. Experimental and Numerical Analysis of Wave Drift Force on KVLCC2 Moving in Oblique Waves. *J. Mar. Sci. Eng.* **2021**, *9*, 136. [CrossRef]
18. Sun, J.; Chen, Y.; Zhou, G. The comparison and applicability analysis of two kinds of responding ship model. *Ship Sci. Technol.* **2016**, *38*, 14–19.
19. Shi, C. Study on Mathematical Model of Ship Manoeuvring Motion Simulation in Wind and Waves. Ph.D. Thesis, Harbin Engineering University, Harbin, China, March 2011.
20. Zheng, J.; Yan, M.; Li, Y. An online identification approach for a nonlinear ship motion model based on a receding horizon. *Trans. Inst. Meas. Control.* **2021**, *43*, 3000–3012. [CrossRef]
21. Nomoto, K.; Taguchi, T.; Honda, K.; Hirano, S. On the steering qualities of ships. *Int. Shipbuild. Prog.* **1957**, *4*, 354–370. [CrossRef]
22. Xie, Y. Wind Speed Prediction System based on Kalman Filter for Offshore Blade Hoisting. *World Sci. Res. J.* **2021**, *7*, 151–162.
23. Standard, N. Actions and action effects. *Edition* **2007**, *2*, 36–57.
24. Wan, B.; Peng, X.; Li, X. Real-time ocean wave simulation based on ITTC spectrum using OpenGL. In Proceedings of the 2015 10th International Conference on Computer Science & Education (ICCSE), Cambridge, UK, 22–24 July 2015; pp. 745–748.
25. Novaselic, M.; Mohovic, R.; Baric, M. Wind Influence on Ship Manoeuvrability—A Turning Circle Analysis. *The International J. Mar. Navig. Saf. Sea Transp.* **2021**, *1*, 47–51. [CrossRef]
26. Zhang, W.; Zou, Z. A numerical study on prediction of ship maneuvering in waves. *Ocean. Eng.* **2017**, *137*, 367–381. [CrossRef]
27. Mahdi, A.; Azouz, A.; Abdalla, A. IMU-Error Estimation and Cancellation Using ANFIS for Improved UAV Navigation. In Proceedings of the 2022 13th International Conference on Electrical Engineering (ICEENG), Cairo, Egypt, 29–31 March 2022; pp. 120–124.
28. Mustafa, R.; Samui, P.; Kumari, S. Reliability Analysis of Gravity Retaining Wall Using Hybrid ANFIS. *Infrastructures* **2022**, *7*, 121. [CrossRef]
29. Arora, S.; Keshari, A. Simulation of Re-Aeration Coefficient Using Anfis and Arima Models. In *Groundwater and Water Quality: Hydraulics, Water Resources and Coastal Engineering*; Jha, R., Singh, V.P., Singh, V., Roy, L.B., Thendiyath, R., Eds.; Springer International Publishing: Cham, Switzerland, 2022; pp. 53–69.

sensors

MDPI

Article

Three-Dimensional Localization of Buried Polyethylene Pipes Using Acoustic Method

William Xerri [1,2,*], Gineth Saracco [1], Alessandra Ribodetti [3], Laurent Zomero [2] and Philippe Picon [2]

1 CNRS-UMR7330 CEREGE, AMU, CdF, IRD, INRAE, OSU-Pytheas, Europole de l'Arbois, BP 80, 13545 Aix-en-Provence, CEDEX 4, France
2 MADE-SA, 167 Impasse de la Garrigue, 83210 La Farlède, France
3 IRD-UR082 CNRS-UMR7329 GEOAZUR, Campus Azur Université et Observatoire de la Côte d'Azur, 250 rue Albert Einstein CS 10269, 06905 Sophia Antipolis, France
* Correspondence: xerri@cerege.fr

Abstract: Localization of buried polyethylene pipes is an important issue for network managers. This study focuses on an acoustic method, which consists of vibrating the pipe and observing the signal with a receiver placed on the ground surface. This method provides an estimate of the path of the pipe but gives no information on the depth. We developed a multi-sensor method based on the principle of vibrating the pipe, which allows estimating the depth while being non-invasive and non-destructive and without a priori information on the propagation medium. These sensors are positioned perpendicular to the pipe. We developed a new estimator to estimate the depth and the propagation velocity in the medium, which is an important variable in our problem. This estimator is based on the MUSIC algorithm and is adapted to our choice of modeling. In this paper, two models of travel times in typical situations are presented. The first one represents the case where all sensors can be placed inside the trench (on the ground surface) in which the pipe is buried. The second one represents the case where sensors are placed inside and outside the trench. These travel time models aim to provide a fast result to allow the method to be used by field agents. They are compared with a full wavefield modeling by finite differences.

Keywords: acoustic method; buried polyethylene pipe; MUSIC algorithm; propagation time modeling; signal processing; full wavefield

Citation: Xerri, W.; Saracco, G.; Ribodetti, A.; Zomero, L.; Picon, P. Three-Dimensional Localization of Buried Polyethylene Pipes Using Acoustic Method. *Sensors* **2022**, *22*, 9433. https://doi.org/10.3390/s22239433

Academic Editors: Grzegorz Peruń, Bogusław Łazarz, Tangbin Xia and Luca De Marchi

Received: 10 October 2022
Accepted: 29 November 2022
Published: 2 December 2022

1. Introduction

Different methods exist for locating buried pipes [1] and, more generally, for investigating near-surface structures [2–7]. The choice of method to locate buried pipes depends on the context. The use of tracer wires is becoming increasingly widespread. This method consists of burying an electric wire with the pipe, which allows, by passing a current through the wire, the creation of an electromagnetic field and, therefore, the use of electromagnetic methods. Electromagnetic methods provide very good results for locating pipes when they are applicable. If there is no tracer wire, then two possible cases can be discerned. Either the pipe is metallic, or it is not. In the case of a metal pipe, an electric current can be injected into the pipe; electromagnetic methods, which work very well, can, therefore, be used. In the case of non-metallic pipes, two types of processes can be distinguished in a non-exhaustive way: (i) Ground-Penetrating Radar (GPR) [8,9] is a versatile tool for locating buried infrastructures; however, this type of tool often needs to be calibrated according to the type of soil to which it is applied. Moreover, it is difficult to differentiate a water pipe from a gas pipe or a buried electrical cable. (ii) Acoustic methods are applicable in the case of non-conductive pipes [1,10–12]. According to [1], we can refer to several types of acoustic methods: seismic wave methods [13], point vibration measurements [14] and pipe excitation methods [15]. The first two categories can use a specific source placed on the ground surface for probing. These methods do not differentiate between the pipe of interest and other pipes present. What distinguishes the last category is its ability to vibrate the pipe, and this vibration will then be diffused into the ground.

This study focuses on the pipe excitation methods that have the advantage of discriminating the pipe of interest in a dense urban environment with noise.

The principle of this is to inject an acoustic signal (called acoustic signature) into the pipe. A method that has been used for many years in the field uses a receiver (geophone) placed on the ground surface, which measures the vertical vibrations of the ground. Through successive measurements of the energy of the received signal, this method can estimate the passage of the pipe and follow its path. However, this method does not provide any information on the pipe depth. The GasTracker tool, based on this method, was developed by the company MADE-SA. In this study, a GasTracker is used to estimate, experimentally on the ground surface, the X and Y position (at ± 10 cm) of the pipe in order to position the sensor network in the area where the pipe is located.

The objective of this study is to implement a multi-sensor method, based on the differences in travel times between sensors, to estimate the depth in addition to the passage of the pipe [10,11].

The aim of this work is to estimate the depth with an accuracy below 10 cm with a non-destructive and non-invasive method and without any a priori information on the propagation medium (e.g., characteristics of the soil, which can change completely from one application area to another).

Nevertheless, some information on how the pipe is buried is known. The pipe is buried between 0.4 and 1.5 m. It is buried in a trench whose width can vary from approximately 30 cm to over 1 m. Therefore, two different media can be distinguished, the inside of the trench and the outside of the trench (vertical stratifications). Moreover, in this problem, the depth is at the meter scale.

An important aim of this study is to define a model of the travel time of the signal that provides the fastest result in order to directly estimate the depth in the field. In the travel time models presented in this paper, we consider the first arrival time of the signals. Only P-waves and converted S-waves are considered. Geophones used are single component sensors. They measure the vertical component of the received vibrations.

The travel time models that we propose are confronted using a simulation with much more complete propagation modeling than that of the complete wavefield with finite differences. This comparison aims to verify the coherence of the estimated quantities, in particular, the velocity.

A cylindrical scattering of the vibration induced by the pipe is considered [16–19]. To ensure that all sensors observe the same section of the vibrating pipe, they are placed perpendicular to the pipe. This is why the problem is represented in a plane orthogonal to the passage of the pipe. This section can be approximated by a point source.

The variables to be estimated in this problem are the position of this source, its lateral position S_X and especially its depth Sz, but also the average velocity of propagation in the ground. An average velocity is assigned to each medium of propagation.

In the first part, the sensors are assumed to be all positioned inside the trench. Therefore, a modeling of the travel times of the signal between the pipe and the receivers by considering a weakly heterogeneous medium and by assigning a mean propagation velocity is presented. In order to ensure an acceptable estimation accuracy, an estimator based on the MUSIC (Multiple Signal Classification) algorithm adapted to this modeling of travel times is developed. This estimator allows us to evaluate variables of interest (position of the source and the velocity). Measurements are carried out on a test area in order to work on real data. In the second part, the sensors are positioned inside and outside the trench. As an extension of the method developed in the first part, the modeling is adapted to this case to fit a larger number of situations. The different compaction of the soil on either side of the trench is taken into account. This requires considering a velocity specific to each of these environments. This second modeling of travel times is validated within the framework of the problem, initially through numerical simulations, then through comparison with finite differences, and finally by comparison with real data.

2. Modeling with a Single Propagation Medium (M1)

2.1. Propagation Time Modeling M1

In this section, all sensors are considered to be located inside the trench. Since no information on the characteristics of the propagation medium is known, the first approach is to consider the propagation medium as weakly heterogeneous and to assign an average propagation velocity. The problem is represented in a plane orthogonal to the passage of the pipe (Figure 1). The pipe section has been represented using a point source S, the receivers R_i were placed on the ground surface perpendicular to the pipe. An average velocity V_0 has been assigned to the propagation medium. The position of the source S (S_X; S_Z) and the velocity of the wave V_0 were unknown. The only known parameters were the coordinates of the receivers R_i (R_{iX}; $R_{iZ} = 0$).

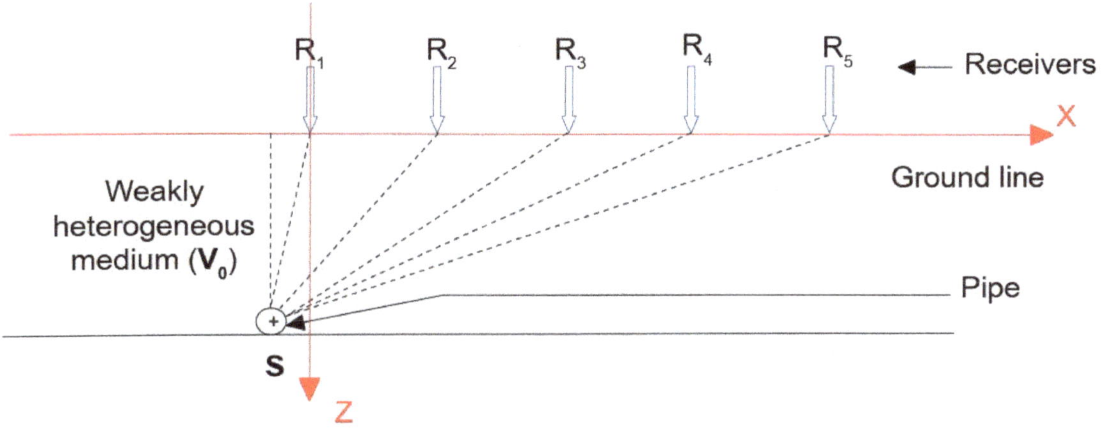

Figure 1. Scheme of Model M1: case of a single propagation medium.

We defined θ_{M1} as the vector of variables to be estimated in the case of Model M1.

$$\theta_{M1} = [S_X \quad S_Z \quad V_0]^T \tag{1}$$

Here, the symbol $[.]^T$ is the transposed operator, S_x the plumb of the pipe, S_z the pipe depth and V_0 is the average propagation velocity in the medium.

We note $\tau_i(\theta_{M1})$ as the wave travel time between the source S and the receiver R_i for Model M1.

$$\tau_i(\theta_{M1}) = \frac{|SR_i|}{V_0} \tag{2}$$

We note $\tau_{1i}(\theta_{M1})$ as the relative delay time between sensors R_1 and R_i for Model M1.

$$\tau_{1i}(\theta_{M1}) = \frac{|SR_i| - |SR_1|}{V_0} \tag{3}$$

Relative delay times are considered because, in this problem, the emission time of the source S is unknown, but the receivers are triggered synchronously.

2.2. Validation of Travel Time Estimate of M1 through Comparison with a Finite Difference Modeling of the Full Wavefield

In this section, the P-wave first arrival travel time estimation proposed in this paper is validated through comparison with full wavefield modeling using finite differences.

The acoustic and isotropic version of the code developed by Operto et al. is used [19] from the original formulation proposed by Jean Virieux [20,21].

A comparison of the travel time estimate of M1 through the finite difference modeling of the full P-wavefield is presented here, which is computationally fast and accurate in the presence of flat topography, as in the test site analyzed in this study.

The following case study has been performed: 30 sensors were aligned and spaced 0.01 m apart; sensor 1 (R_1) was placed just above the source ($S_X = 0$); the rest of the sensors were placed increasingly far away from the source; the depth of the source (S_Z) was 1.5 m, and the average velocity of propagation (V_0) was 500 m/s.

To ensure simplicity in the simulation, the origin of the reference frame was translated, and the source was placed at the top and the sensors at the bottom, but they were symmetrical. In the simulation presented here, the source was placed at S (0.35; 0), and the sensors at Ri (0.35 + (i−1) 0.01; 1.5), but this corresponds well to the situation described above. In Figure 2, we observe the physical model used for the simulation. The source used was a Ricker function, also called the "mexican hat"; this function is a second derivative of a Gaussian function.

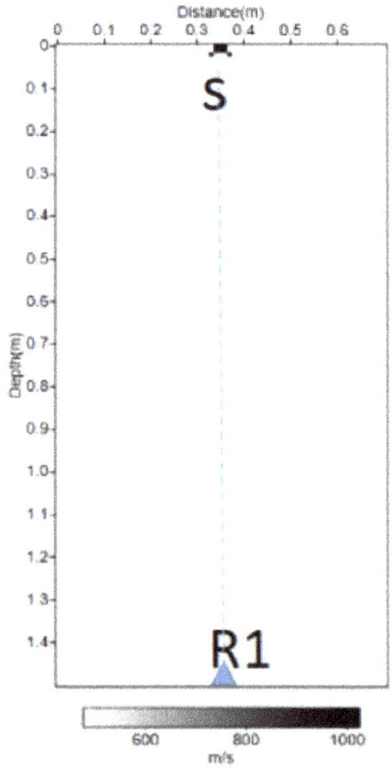

Figure 2. Physical velocity model in the case of the propagation of the full wavefield by finite differences with one propagation medium. S indicates the source at the surface. R_1 is the first receiver.

The results of this simulation are presented in Figure 3. The vertical axis represents the spatial position of the sensors, and the horizontal axis shows the recorded times. The simulation of the received signals using finite differences is shown in black, and the travel time curve calculated using Model M1 is shown in blue. At the scale of the problem, Model M1 is in agreement with the more complex full wavefield model using finite differences.

2.3. Cramer–Rao Bound from Model M1

The Cramer–Rao Bound (CRB) was calculated to obtain information on the estimation accuracy [22,23]. The CRB represents the smallest possible standard deviations of all unbiased estimates of the model variables. First, the CRB was computed, and second, simulations using the CRB were performed.

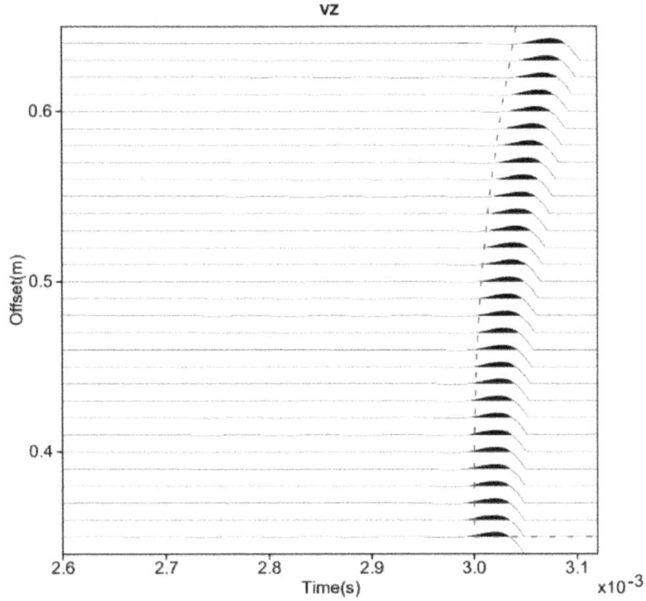

Figure 3. Comparison of travel time estimate of Model M1 (dashed blue line) with the full P-wavefield propagation model using finite differences.

2.3.1. Calculation of the Cramer–Rao Bound from Model M1

The CRB was calculated according to the Fisher information matrix, which is denoted by F.

$$\mathrm{CRB}(\theta_{M1}) = F^{-1}(\theta_{M1}/\tau_{1i}(\theta_{M1})) \tag{4}$$

where ./. represents the known operator.

The Fisher information matrix is expressed

$$F(\theta_{M1}/\tau_{1i}(\theta_{M1})) = \sum_{i=2}^{N} \frac{1}{\mathrm{Var}(\tau_{1i}(\theta_{M1}))} \nabla_{\theta_{M1}}(\tau_{1i}(\theta_{M1})) \nabla^{T}_{\theta_{M1}}(\tau_{1i}(\theta_{M1})) \tag{5}$$

where N is the number of sensors, $\mathrm{Var}(\tau_{1i}(\theta_{M1}))$ is the variance of the relative delay time between sensor 1 and i, and $\nabla_{\theta_{M1}}[.]$ is the gradient operator as a function of θ_{M1}.

Let us focus on the calculation of the gradient of $\tau_{1i}(\theta_{M1})$. The expression of this gradient can be decomposed from Equation (3),

$$\nabla_{\theta_{M1}}(\tau_{1i}(\theta_{M1})) = \nabla_{\theta_{M1}}\left(\frac{|SR_i|}{V_0}\right) - \nabla_{\theta_{M1}}\left(\frac{|SR_1|}{V_0}\right) \tag{6}$$

By developing the calculation, the expression becomes

$$\nabla_{\theta_{M1}}(\tau_{1i}(\theta_{M1})) = \begin{bmatrix} \frac{S_X - R_{iX}}{V_0|SR_i|} - \frac{S_X}{V_0|SR_1|} \\ \frac{S_Z - R_{iZ}}{V_0|SR_i|} - \frac{S_Z}{V_0|SR_1|} \\ \frac{|SR_1| - |SR_i|}{V_0^2} \end{bmatrix} \tag{7}$$

Using this result in Equation (5), the Fisher information matrix can be calculated, and by calculating its inverse, the CRB is obtained, as shown in Equation (4).

2.3.2. Numerical Simulation Using the Cramer–Rao Bound

The figures presented in this section are the results of the numerical simulations. These results are not exhaustive and are used to give an idea of the impact of the variables of the problem on the accuracy of the depth estimation. To obtain these curves, one variable was varied and the others were fixed. The impact on the possible accuracy of depth estimation was observed.

Figure 4a shows the evolution of the CRB with changes in depth as a function of the error in the relative delay times. The situation considered for this simulation is as follows: five sensors were used, the distance between sensors was 0.2 m, the plumb of the pipe (S_X) was placed at 0 m, and the average propagation velocity (V_0) was 500 m/s. Curves for two different depths (S_Z), 0.4 and 1 m, were observed. An accuracy of the relative delay times in a microsecond range was needed to obtain an accuracy of 0.1 m in the depth. In the worst case, an accuracy of the order of 0.1 μs was needed.

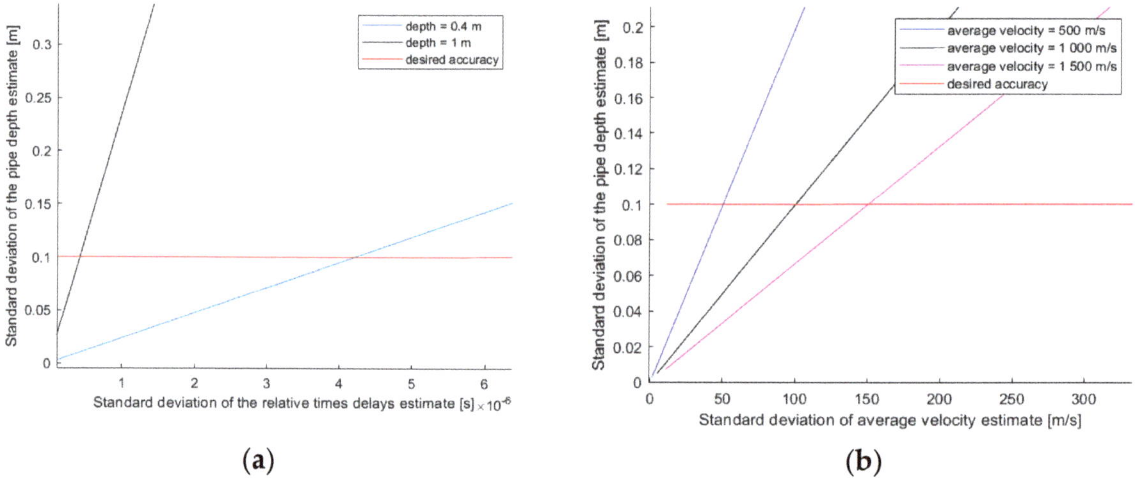

Figure 4. Depth error boundary as a function of the error of the other variables: (**a**) depth error as a function of the error in the relative delay times; (**b**) depth error as a function of the error in the propagation velocity.

Figure 4b represents the evolution of the CRB with changes in depth as a function of the error in the propagation velocity. The situation considered for this simulation is as follows: the distance between the sensors was 0.2 m, the plumb of the pipe (S_X) was 0 m, and the pipe depth (S_Z) was 0.4 m. The curves for different average propagation velocities (V_0) were observed. A propagation velocity accuracy of the 10% range was required to obtain an error of less than 0.1 m in the depth.

2.4. Adaptation of the MUSIC Algorithm for Model M1

In the previous section, a relative delay time accuracy of the order of 10^{-6} s was defined as a necessary condition. Following the choice to focus on the MUSIC algorithm (MUltiple SIgnal Classification) [24–30] for antenna processing, the so-called 'high resolution algorithm' became interesting.

2.4.1. Presentation of the MUSIC Algorithm Adapted to Our Problem

The MUSIC algorithm is usually used to discern different sources and their direction of arrival through an antenna array [24]. It can be adapted in the near field to estimate the distance of sources in addition to their direction of arrival [25–30]. In this paper, the aim was to estimate the location of a single source S in the near field and also to estimate the propagation velocity V_0 according to Model M1.

The algorithm can be divided into several steps as follow:

1. Estimate the variance–covariance matrix of the system from the signals received by sensors.
2. Decompose the variance–covariance matrix of the system into eigenvalues and eigenvectors.
3. Definition of the noise subspace, denoted by U_b, with the eigenvectors corresponding to the smallest eigenvalues.
4. Construction of a family of vectors, denoted by 'a', parametrized by the variables we want to estimate, S and V0 (Equation (1)). This family of vectors is constructed from the modeling of relative delay times (Equation (3)).

$$a(S, V_0) = \begin{bmatrix} 1 & e^{-j2\pi f_0 \frac{|SR_2| - |SR_1|}{V_0}} & \ldots & e^{-j2\pi f_0 \frac{|SR_N| - |SR_1|}{V_0}} \end{bmatrix}^T \tag{8}$$

where f_0 is the signal frequency, and N is the number of sensors.

5. Knowing that the signal subspace and the noise subspace are orthogonal, the project of a (S,V_0) on the noise subspace must be at the minimum for the values of S and V_0 corresponding best to the received signal. It is traditional to take the inverse of this projection and to look for the values of S and V_0 that maximize this criterion, which is denoted by C_{music}.

$$C_{MUSIC}(S, V_0) = \frac{1}{a(S, V_0)^H U_b U_b{}^H a(S, V_0)} \tag{9}$$

where $[.]^H$ is the conjugate transposed operator.

2.4.2. Test of Estimator Using Numerical Simulation

In this part, the algorithm used in the simulation of an ideal case is verified. The statistics of the estimator (mean and variance) using the Monte Carlo method are presented. All estimates were made with 1000 runs of the noise, and noise is applied to the propagation times. For these simulations, a situation has been fixed in which all the variables are known. The signals received are simulated as a function of these variables, and then these ones were used to run the algorithm.

The following situation is considered: a depth (S_Z) of 0.7 m, a propagation velocity (V_0) of 500 m/s, and the plumb of the pipe (S_X) at 0 m. Five sensors were placed at 0.2 m, the first one vertically above the pipe.

In Table 1, the statistics of the results of the estimator, using the Monte Carlo method, are presented. A total of 1000 runs of noise were performed. The same situation as before was considered, and the theoretical propagation times between the source S and sensors were noised with a white Gaussian noise of standard deviation σ_{noise}. The propagation times are in the millisecond range, and the propagation time differences with sensor 1 vary between 10^{-5} and 10^{-4} seconds.

For this set of variables, which is representative, the observed accuracy is lower than the desired accuracy of 0.1 m. It is interesting to note that increasing the number of sensors reduces the accuracy required on the relative delay times.

2.5. Experimental Measurements

Experimental measurements were performed on a semi-controlled test area in order to work on real data.

2.5.1. Experimental Set-Up

The measurement chain presented in this section is not innovative; this measurement chain was designed to be adapted to experimental needs. In particular, it was necessary to be able to control the emission and to change the type of signal emitted at any time.

Moreover, in the reception chain, the filter should not be so fine that it could not adapt to different types of signals that could be emitted.

Table 1. Statistics of the MUSIC estimator adapted to Model M1 using the Monte Carlo method (for 1000 runs, 5 sensors spaced 0.2 apart). Noise is applied to the propagation times, which are in the order of a millisecond.

		Mean	Standard Deviation	True Value
Numerical Simulation 1 with 5 sensors and $\sigma_{noise} = 1 \times 10^{-7}$	Depth S_Z (m)	0.7056	0.0486	0.7
	Average velocity V_0 (m/s)	494	17	500
	Plumb of the pipe S_X (m)	0.0047	0.0065	0
Numerical Simulation 2 with 6 sensors and $\sigma_{noise} = 5 \times 10^{-7}$	Depth S_Z (m)	0.6658	0.0317	0.7
	Average velocity V_0 (m/s)	499	7	500
	Plumb of the pipe S_X (m)	0.0172	0.0109	0
Numerical Simulation 3 with 7 sensors and $\sigma_{noise} = 1 \times 10^{-6}$	Depth S_Z (m)	0.7386	0.0466	0.7
	Average velocity V_0 (m/s)	501	9	500
	Plumb of the pipe S_X (m)	0.0325	0.0315	0

The emission chain (Figure 5) was composed of a computer to control the emitted signal, an amplifier, which received the signal from the computer sound card, and finally, a loudspeaker. The loudspeaker was fixed at one end of the pipe and emitted the signal inside the pipe.

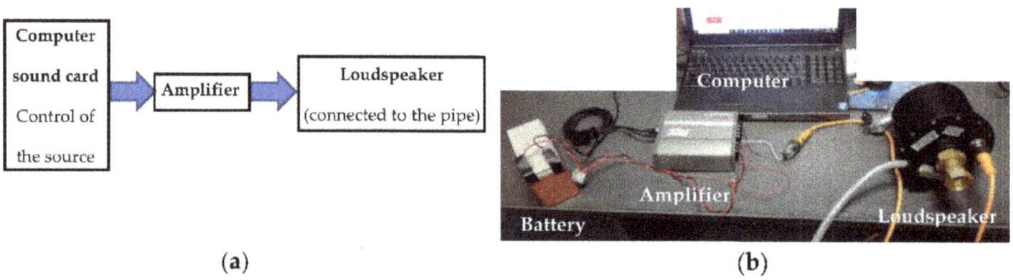

Figure 5. Experimental set-up of the acoustic emission chain: (**a**) diagram of the main components; (**b**) photograph of the material with the battery to power the amplifier.

Experiments were carried out in an anechoic room to characterize the loudspeaker (CNRS-LMA-Marseille, C. Pinhède). A white noise is sent in command to the loudspeaker, and the emitted signal is measured with a microphone placed just in front of the loudspeaker output. This experimentation allows to obtain the transfer function of the loudspeaker. The modulus and phase of its transfer function are shown in Figure 6.

The responses of the loudspeaker to the two types of signals used in this study are presented.

The first signal of interest is a Ricker function. The time response of the loudspeaker to a Ricker function control signal is presented in Figure 7 and the frequency response in Figure 8. The properties of this type of signal are used in the calculation of the propagator in the simulation of the propagation of the complete wave field by finite differences. The real measurements with a Ricker function emission are to compare the trend of the delay times with the simulations.

The second signal of interest is a monochromatic signal at 500 Hz, a burst signal. It is used to estimate the pipe depth. The time response of the loudspeaker to a burst control signal is presented in Figure 9 and the frequency response in Figure 10.

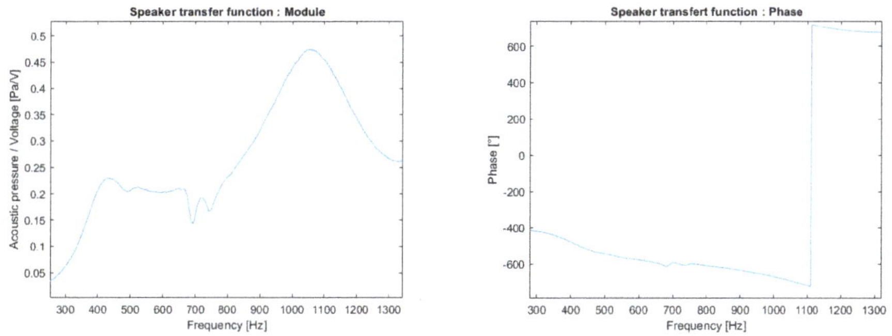

(**a**) (**b**)

Figure 6. Transfer function of the loudspeaker estimated by emitting a white noise: (**a**) modulus; (**b**) phase.

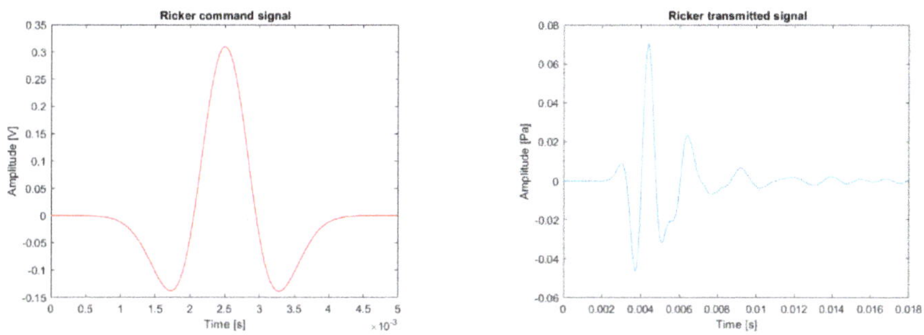

(**a**) (**b**)

Figure 7. Example of the loudspeaker response (Ricker function): (**a**) theoretical source signal (red line); (**b**) experimental source signal sent by the loudspeaker through the pipe (blue line) recorded during the controlled experiment.

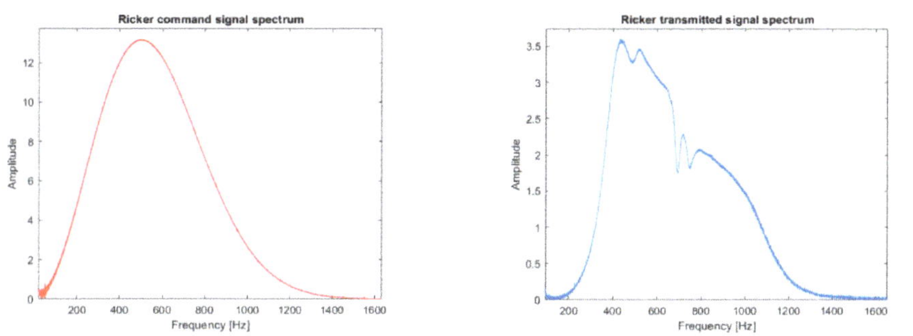

(**a**) (**b**)

Figure 8. Fourier transform of the Ricker function: (**a**) theoretical Ricker spectrum (red line) showing the central frequency of the theoretical source around 500 Hz; (**b**) experimental Ricker spectrum (loudspeaker output; blue line) showing a central frequency around 500 Hz but distorted in relation to the command.

The loudspeaker distorts the burst less than the Ricker function, because the Ricker function is richer in frequency.

The acoustic reception chain is presented in Figure 11. The sensors are geophones that measure vibrations in their vertical axis. They are cylindrical with a diameter of 2.54 cm. During the measurements, a fatty substance was added between the sensors and the ground in order to obtain a better coupling. Geophones receive the signals that are filtered and amplified by electronic cards. At the input of electronic cards, signals are filtered using a high-pass filter with a cut-off frequency of 100 Hz, and the output, using a low-pass filter with a cut-off frequency of 100 kHz. This wide analog filtering allows us to digitally refine the filtering. The amplification is adjustable between 0 and 112 dB, which allows the amplification to be adjusted in the field. At the output of the electronic cards, the signals are digitized by the acquisition card; it is an analog-to-digital converter. The digitized signals are recovered on a computer.

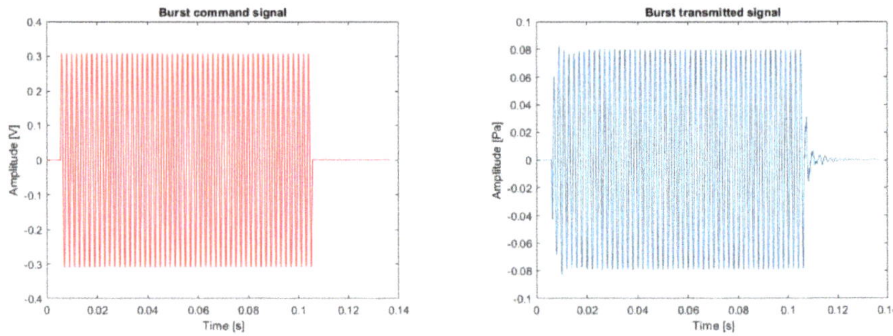

(a) (b)

Figure 9. Example of the loudspeaker response (burst of 100 ms at 500 Hz): (**a**) theoretical source signal; (**b**) experimental source signal sent by the loudspeaker through the pipe.

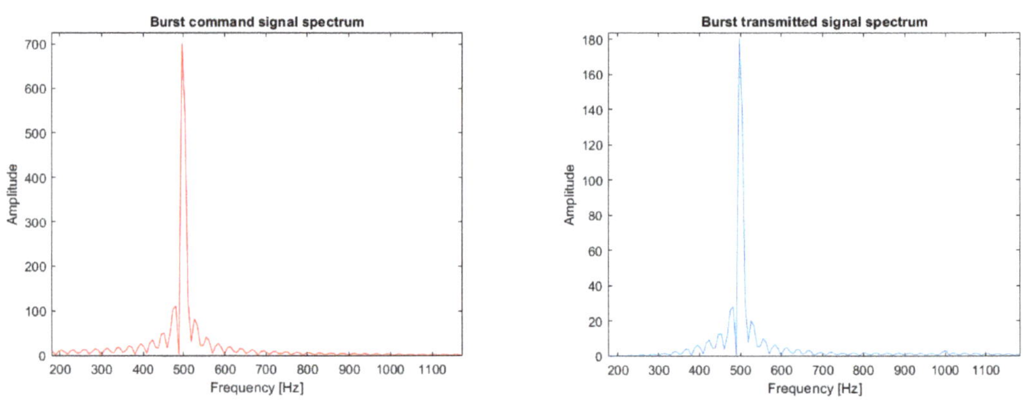

(a) (b)

Figure 10. Fourier transform of the burst of 100 ms at 500 Hz: (**a**) theoretical burst spectrum; (**b**) experimental burst spectrum (loudspeaker output).

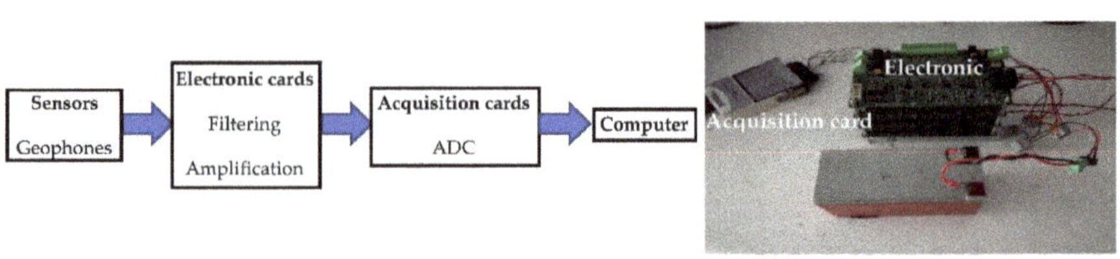

(a) (b)

Figure 11. Experimental set-up of the reception chain: (**a**) scheme of the main components; (**b**) photograph of the material.

Figure 12 shows an example of measurements taken on the semi-controlled test area. The trench and the sensors positioned perpendicular to the pipe passage can be observed. Sensors are spaced 5 cm apart.

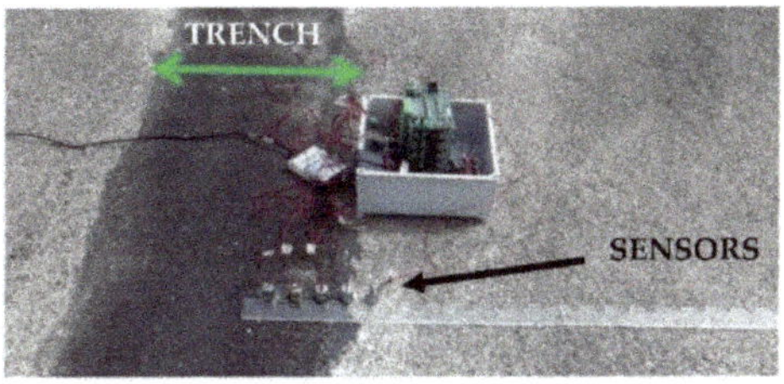

Figure 12. Example of positioning of the experimental device for signal acquisition (reception chain). The black arrow indicates the acquisition design composed of a line of five sensors and the support (graduated ruler), where the sensors are moved during the experiment. The green arrow indicates the trench where a pipe is buried. The box contains the electronic cards presented in Figure 11 and, on the left of the box, the acquisition card connected to the computer to digitalize the signals received from the buried pipe is shown.

2.5.2. Experimental Results

Experimental measurements were performed on a semi-controlled test area. In this test area, the position of the pipe was known, and it was a polyethylene pipe, but the characteristics of the propagation medium were unknown. The propagation velocity is unknown.

To take the measurements, a line of five sensors was positioned perpendicular to the pipe route. To focus on the depth, it was assumed that the plumb of the pipe (S_X) was known, and sensor 1 was placed just above the pipe. Sensors were spaced 0.2 m apart. The signal emitted in the pipe was a monochromatic signal at 500 Hz.

In Figure 13, an example of signals measured by sensors is presented. The signal emitted by the loudspeaker is a burst of 100 ms at 500 Hz repeated every second. Figure 13a shows the signals received by the sensors before digital filtering, and Figure 13b shows the same signals digitally filtered between 480 and 520 Hz. Sensor 1 is the closest to the pipe and sensor 5 the farthest. The sensors farthest from the pipe receive the most attenuated signals.

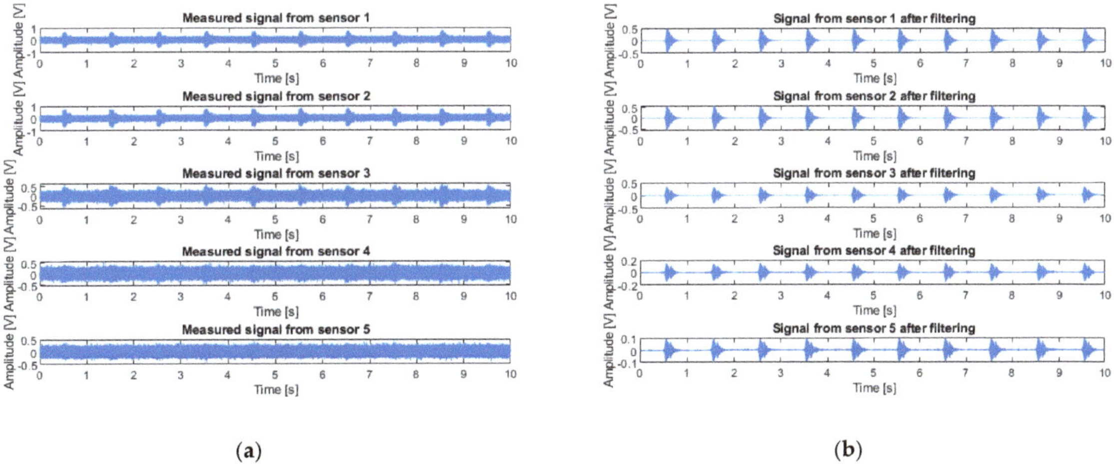

(a) (b)

Figure 13. Example of signals measured by sensors (burst emission of 100 ms at 500 Hz): (**a**) signals received by sensors before digital filtering; (**b**) signals digitally filtered between 482 and 520 Hz.

Results on real signals are presented, for which depth estimation with the desired accuracy of 0.1 m (Tables 2 and 3) is reached. For a depth of 0.42 m, an accuracy of 0.03 m (Table 2) is obtained. For a depth of 0.7 m, an accuracy of 0.05 (Table 3) is obtained.

Table 2. First example of estimation obtained with the MUSIC algorithm adapted to the problem and to Model M1 (5 sensors spaced 0.2 m apart).

	Estimate Value	Reference Value
Depth (S_Z) (m)	0.39	0.42
Average velocity (V_0) (m/s)	360	unknown

Table 3. Second example of estimation obtained with the MUSIC algorithm adapted to the problem and to Model M1 (5 sensors spaced 0.2 m apart).

	Estimate Value	Reference Value
Depth (S_Z) (m)	0.75	0.70
Average velocity (V_0) (m/s)	540	unknown

Figure 14a shows measured signals (filtered between 480 and 520 Hz) on the test area that match with Model M1 (a zoom of Figure 13b). With Model M1, the more receivers are far from the source S, the more they should receive in a delayed way the information of the sound wave propagating in the pipe. It is on real signals of this type that the results, presented in Tables 2 and 3, are obtained.

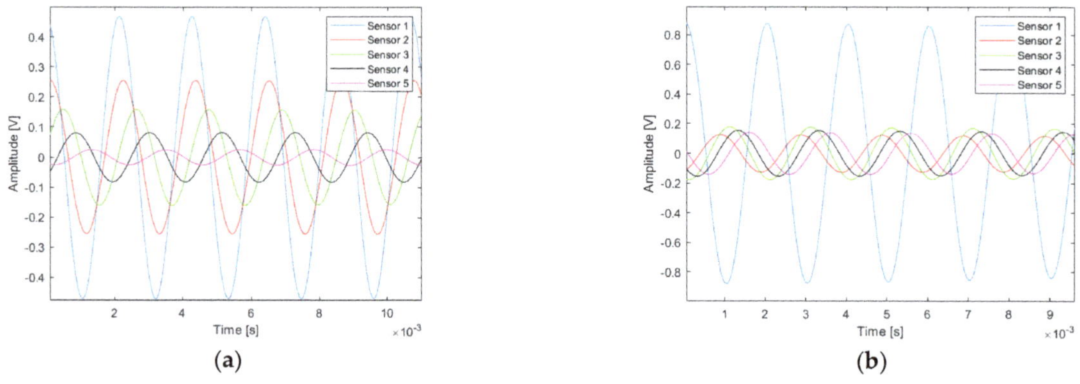

Figure 14. Example of real received signals digitally filtered between 480 and 520 Hz: (**a**) corresponding to Model M1; (**b**) not corresponding to Model M1.

Figure 14b shows measured signals on the test area that do not match with Model M1. A phenomenon can be discerned several times, in that, sensors farthest from the source S receive the signal before the closest sensors. This phenomenon shows the role played by the vertical discontinuities (the trench) not included in Model M1.

It is needed to evolve the model to cover a greater number of situations. This is why an evolution of this model is presented in Section 3. Model M1 is sufficient when the transition is smooth or the medium weakly heterogeneous.

3. Modeling with Two Propagation Media (M2)

In this part, an evolution of the model is presented. Now sensors are placed inside and outside the trench. A vertically stratified change in medium is considered, which represents the trench in which the pipe is buried. First, the modeling of the propagation times is presented, and then the calculation of the intermediate variables is presented in more detail.

3.1. Propagation Time Modeling M2

3.1.1. Presentation of the Propagation Time Model M2

In the problem modeling, a change in medium was added to take into account the trench (Figure 15). The pipe section was still represented using a point source S and the sensors R_i were arranged on the ground surface in the same way as before. An average velocity V_0 was assigned to the propagation medium inside the trench and an average velocity V_1 outside the trench. For the sensors outside the trench, new intermediate variables P_i appeared, which represented the interface points between the two media. For each R_i outside the trench, there was a corresponding P_i.

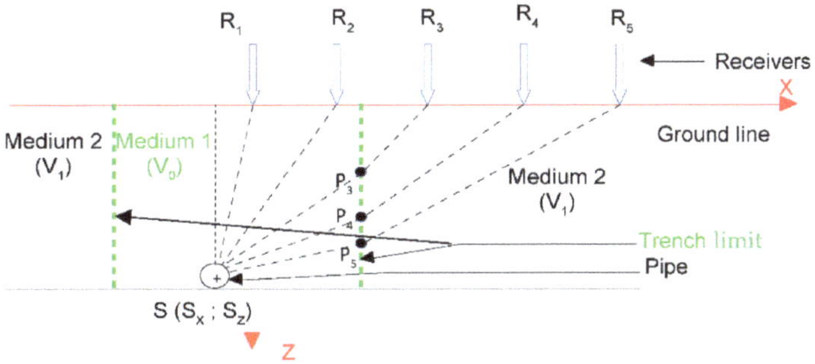

Figure 15. Scheme of the problem modeling in the case of two propagation media: Model M2.

We defined θ_{M2} as the vector of variables to be estimated in the case of Model M2.

$$\theta_{M2} = \begin{bmatrix} S_X & S_Z & V_0 & V_1 \end{bmatrix}^T \tag{10}$$

We noted $\tau_i(\theta_{M2})$ as the wave travel time between the source S and the receiver R_i for Model M2. Two cases can be discerned:

- If R_i is inside the trench, it is returned to the single propagation medium case of Model M1 (Equation (2))
- If R_i is outside the trench, then

$$\tau_i(\theta_{M2}) = \frac{|S\,P_i|}{V_0} + \frac{|P_i\,R_i|}{V_1} \tag{11}$$

In the following, only the case of R_i outside the trench is considered because the case R_i inside the trench was already covered in Section 2.

Relative delay times are also considered because the emission time of the source S is unknown. We note $\tau_{1i}(\theta_{M2})$ as the relative delay time between the sensors R_1 and R_i for Model M2.

$$\tau_{1i}(\theta_{M2}) = \frac{|S\,P_i| - |S\,R_1|}{V_0} + \frac{|P_i\,R_i|}{V_1} \tag{12}$$

The modeling proposed in this paper depended on the interface point P_i, which was unknown. We had to express P_i as a function of θ_{M2}.

3.1.2. Analytical Expression of the Interface Point P_i

The variable $\tau_{1i}(\theta_{M2})$ (Equation (12)) depended on the interface point P_i (P_{iX}; P_{iZ}). It was assumed that the X coordinate P_{iX} was known a priori (i.e., we knew the position of the medium change). However, we had no a priori information on the Z coordinate P_{iZ}. We focused on the analytical expression of P_{iZ}.

The Snell–Descartes law was considered at the interface between the two media. We can then write

$$\frac{S_Z - P_{iZ}}{|S\,P_i|} = \frac{P_{iZ}}{|P_i\,R_i|}\frac{V_0}{V_1} \tag{13}$$

After squaring and developing, the following fourth-order equation was obtained:

$$\begin{gathered} P_{iZ}^4\left[V_1^2 - V_0^2\right] \\ + P_{iZ}^3\left[-2R_{iZ}V_1^2 - 2S_ZV_1^2 + 2S_ZV_0^2\right] \\ + P_{iZ}^2\left[S_Z^2V_1^2 + R_{iZ}^2V_1^2 + V_1^2(R_{iX} - P_{iX})^2 + 4R_{iZ}S_ZV_1^2 - S_Z^2V_0^2 - V_0^2(P_{iX} - S_X)^2\right] \\ + P_{iZ}\left[-2R_{iZ}S_Z^2V_1^2 - 2S_ZR_{iZ}^2V_1^2 - 2S_ZV_1^2(R_{iX} - P_{iX})^2\right] \\ + \left[R_{iZ}^2S_Z^2V_1^2 + S_Z^2V_1^2(R_{iX} - P_{iX})^2\right] = 0 \end{gathered} \tag{14}$$

We propose

$$P_{iZ}^4\,m_1 + P_{iZ}^3\,m_2 + P_{iZ}^2\,m_3 + P_{iZ}\,m_4 + m_5 = 0 \tag{15}$$

After solving Equation (15), we obtained

$$P_{iZ} = \frac{\sqrt{A} - \sqrt{A - 2\left(g + A + \frac{u}{\sqrt{A}}\right)}}{2} - \frac{m_2}{4m_1} \tag{16}$$

with

$$A = \left(\frac{-K + \sqrt{K^2 + \frac{4J^3}{27}}}{2}\right)^{1/3} - \frac{J}{3\left(\frac{-K + \sqrt{K^2 + \frac{4J^3}{27}}}{2}\right)^{1/3}} - \frac{2g}{3}, \tag{17}$$

$$J = -\frac{g^2}{3} - 4w, \tag{18}$$

$$K = \frac{8wg}{3} - \frac{2g^3}{27} - u^2, \tag{19}$$

$$g = \frac{m_3}{m_1} - \frac{3m_2^2}{8m_1^2}, \tag{20}$$

$$u = \frac{m_4}{m_1} - \frac{m_2m_3}{2m_1^2} + \frac{m_2^3}{8m_1^3}, \tag{21}$$

$$w = \frac{m_5}{m_1} - \frac{m_2m_4}{4m_1^2} + \frac{m_2^2m_3}{16m_1^3} - \frac{3m_2^4}{256m_1^4} \tag{22}$$

m_1 to m_5 depended on θ_{M2} (Equation (10)), the position of sensors R_i (known) and the position of the medium change P_{iX} (known). The expression of the interface point depth P_{iZ} depended on m_i, so we expressed P_{iZ} as a function of variables of interest θ_{M2}.

The theoretical delay times with Model M2 from θ_{M2} can be estimated following this approach.

3.2. *Validation of Travel Time Estimate of M2 through Comparison with a Finite Difference Modeling of the Full Wavefield*

Here, the travel times of Model M2 are compared with the full P-wavefield propagation modeling using finite differences in the same way as in Section 2.2. The simulations were performed for the same case study in order to compare the results.

The following set-up was considered: sensor 1 is the origin of the reference frame; the point source S is at a depth of $S_Z = 0.7$ m; sensor 1 is in line with the source $S_X = 0$ m; sensors are spatially distributed every 0. 05 m and move away from the source; the position of the change in medium is at $P_X = 0.1$ m; the average propagation velocity inside the trench is

V_0 = 300 m/s and outside the trench is V_1 = 600 m/s. These values were proposed for the propagation velocities because they were close to those that would be estimated for the real data in the test site.

To ensure simplicity in the simulation, the origin of the reference frame was translated, and the source was placed at the top and the sensors at the bottom, but they were symmetrical. In the simulation presented here, the source was placed at S (0.75; 0) and the sensors at Ri (0.75 + (i − 1)0.05; 0.7), but this corresponds well to the situation described above. In Figure 16, the physical model used for simulation and the wavefront propagation in the case of the full wavefield propagation modeling using finite differences are observed. Each panel is taken at a different time, and the evolution of the wavefront over time is observed. In this simulation, the source used was a Ricker function (Figures 7a and 8a).

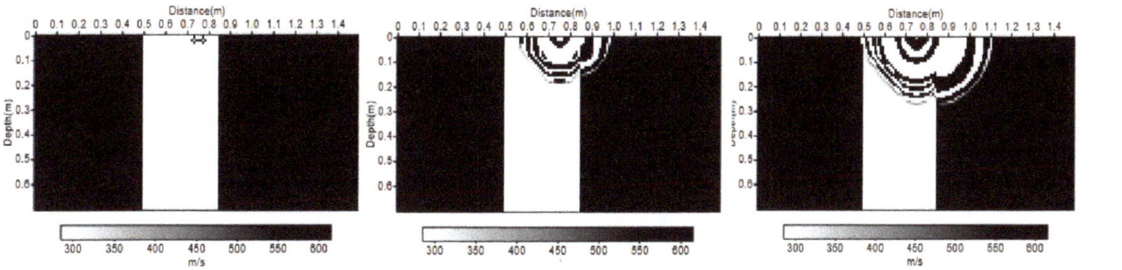

Figure 16. Physical velocity model and wavefront in the case of the full P-wavefield propagation modeling using the finite differences method.

The results of this simulation are presented in Figure 17. The abscissa represents the spatial position of the sensors, and the ordinate indicates the recorded time. The seismograms represent the simulation of the received signals using finite differences, and the travel time curve calculated from Model M2 is shown in blue. At the scale of the problem, Model M2 was in agreement with the more complex full wavefield model using finite differences. This allowed us to obtain a first validation of the travel time estimate of Model M2 through numerical simulation.

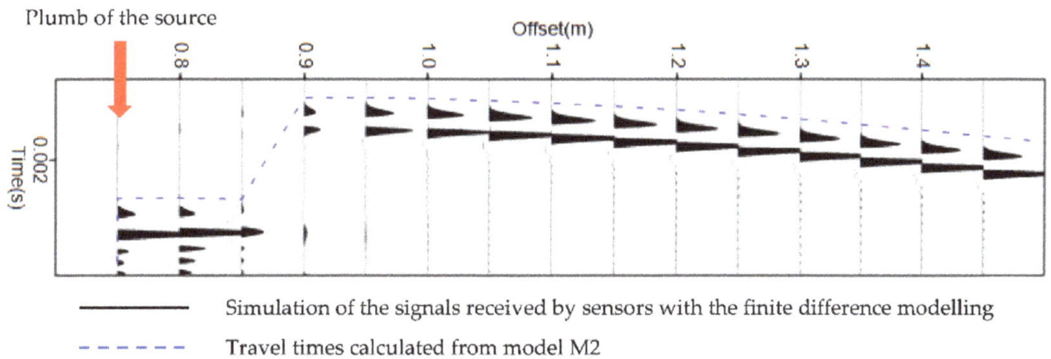

Figure 17. Seismograms obtained by full wavefield propagation using the finite differences method in the physical velocity model presented in Figure 16. The horizontal axis indicates the position of the sensors. The red arrow indicates the sensors located at the plumb of the source. The theoretical travel times calculated in the same model (M2) by the method proposed in Section 3.1 are plotted with a blue dashed line to assess the good agreement. The transmitted signal is considered as the same emitted Ricker function (Figure 7a).

The same simulation is also carried out but using as source signal the signal actually transmitted by the loudspeaker (Figures 7b and 8b). In Figure 18, a small shift between the travel times calculated with Model M2 and the simulations of the signals received using finite differences can be observed, but the trend of the curve remains correct.

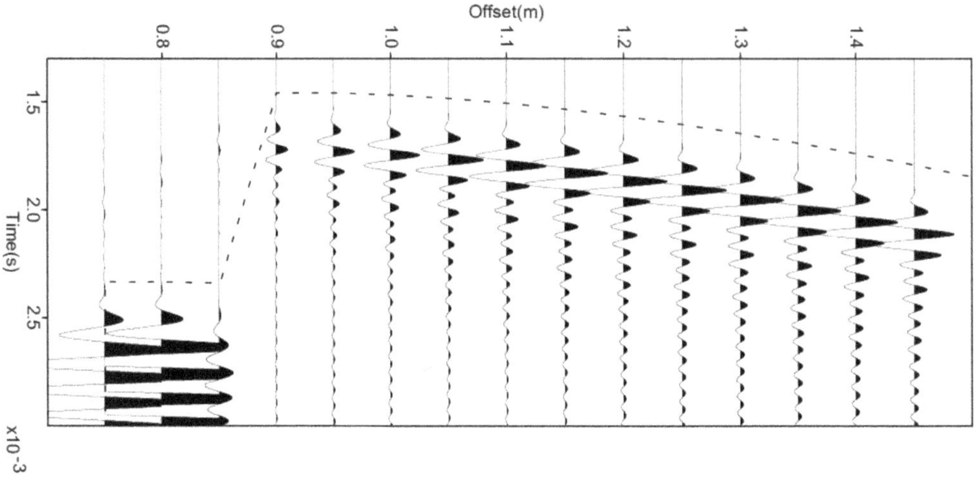

Figure 18. Seismograms obtained by full wavefield propagation using the finite differences method in the physical velocity model presented in Figure 16. The horizontal axis indicates the position of the sensors. The red arrow indicates the sensors located at the plumb of the source. The theoretical travel times calculated in the same model (M2) by the method proposed in Section 3.1 are plotted with a blue dashed line to assess the good agreement. We consider that the emitted signal was transformed by the loudspeaker (Figure 7b).

3.3. *Cramer–Rao Bound from Model M2*

In this section, as in Section 2.3, the Cramer–Rao Bound (CRB) is calculated to obtain information on the estimation accuracy.

3.3.1. Calculation of the Cramer–Rao Bound from Model M2

The CRB is calculated according to the Fisher information matrix, which is denoted by F.

$$\mathrm{CRB}(\theta_{M2}) = F^{-1}(\theta_{M2}/\tau_{1i}(\theta_{M2})) \tag{23}$$

where ./. is the known operator.

The Fisher information matrix is expressed

$$F(\theta_{M2}/\tau_{1i}(\theta_{M2})) = \sum_{i=2}^{N} \frac{1}{\mathrm{Var}(\tau_{1i}(\theta_{M2}))} \nabla_{\theta_{M2}}(\tau_{1i}(\theta_{M2})) \nabla^{T}_{\theta_{M2}}(\tau_{1i}(\theta_{M2})) \tag{24}$$

where N is the number of sensors, $\mathrm{Var}(\tau_{1i}(\theta_{M2}))$ is the variance of the relative delay time between sensor 1 and i, and $\nabla_{\theta_{M2}}[.]$ is the gradient operator as a function of θ_{M2}.

The gradient of $\tau_{1i}(\theta_{M2})$ is calculated from Equation (12),

$$\nabla_{\theta_{M2}}(\tau_{1i}(\theta_{M2})) = \nabla_{\theta_{M2}}\left(\frac{|S\,P_i|}{V_0}\right) - \nabla_{\theta_{M2}}\left(\frac{|S\,R_1|}{V_0}\right) + \nabla_{\theta_{M2}}\left(\frac{|P_i\,R_i|}{V_1}\right) \tag{25}$$

The calculation of the gradient can be decomposed as follows

$$\nabla_{\theta_{M2}}\left(\frac{|SP_i|}{V_0}\right) = \left[\frac{(S_X - P_{iX}) - (S_Z - P_{iZ})\frac{\partial}{\partial S_X}P_{iZ}}{V_0|SP_i|} \quad \frac{(S_Z - P_{iZ})\left(1 - \frac{\partial}{\partial S_Z}P_{iZ}\right)}{V_0|SP_i|} \quad \frac{-V_0(S_Z - P_{iZ})\frac{\partial}{\partial V_0}P_{iZ} - |SP_i|^2}{V_0^2|SP_i|} \quad \frac{-(S_Z - P_{iZ})\frac{\partial}{\partial V_1}P_{iZ}}{V_0|SP_i|}\right]^T \tag{26}$$

$$\nabla_{\theta_{M2}}\left(\frac{|SR_1|}{V_0}\right) = \left[\frac{S_X - R_{1X}}{V_0|S\,R_1|}\;\frac{S_Z - R_{1Z}}{V_0|S\,R_1|}\;\frac{-|S\,R_1|}{V_0^2}\;0\right]^T \tag{27}$$

$$\nabla_{\theta_{M2}}\left(\frac{|P_i\,R_i|}{V_1}\right) = \left[\frac{(P_{iZ} - R_{iZ})\,\frac{\partial}{\partial S_X}P_{iZ}}{V_1|P_i\,R_i|}\;\frac{(P_{iZ} - R_{iZ})\,\frac{\partial}{\partial S_Z}P_{iZ}}{V_1|P_i\,R_i|}\;\frac{(P_{iZ} - R_{iZ})\,\frac{\partial}{\partial V_0}P_{iZ}}{V_1|P_i\,R_i|}\;\frac{V_1(P_{iZ} - R_{iZ})\,\frac{\partial}{\partial V_1}P_{iZ} - |P_i\,R_i|^2}{V_1^2|P_i\,R_i|}\right]^T \tag{28}$$

All these expressions depended on the gradient of P_{iZ}. The gradient of P_{iZ}, $\nabla_{\theta_{M2}}P_{iZ}$, needed to be calculated to be able to calculate the CRB. This gradient calculation is presented in Appendix A to avoid overloading the text.

3.3.2. Numerical Simulation Using the Cramer–Rao Bound from Model M2

The figures presented in this section are the result of numerical simulations. These results are not exhaustive and are used to give an idea of the impact of the variables of the problem on the accuracy of the depth estimation.

Figure 19 shows the evolution of the CRB on the depth as a function of the error in the relative delay times. For these simulations, the following situation is considered: the distance between sensors is 0.2 m; the plumb of the pipe (S_X) is at 0 m; the average propagation velocity inside the trench (V_0) is 300 m/s; the average propagation velocity outside the trench (V_1) is 600 m/s; and the position of the change in medium (P_X) is at 0.15 m.

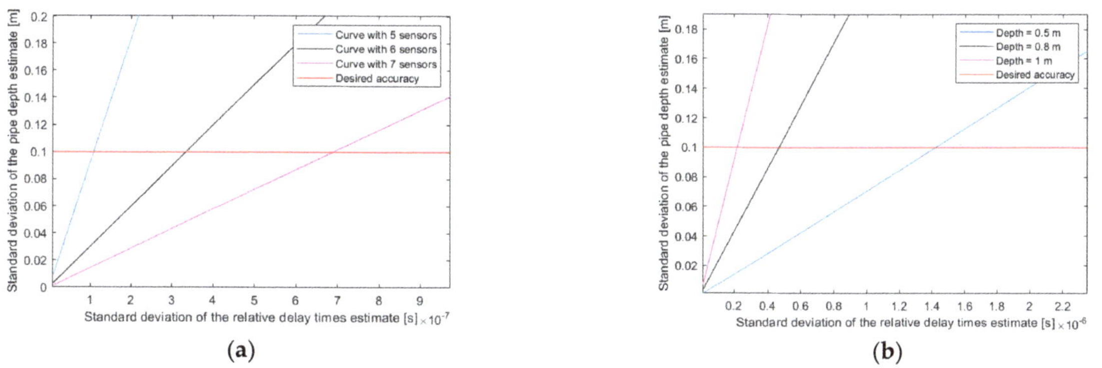

Figure 19. Depth error as a function of the error in the relative delay times: (**a**) for several numbers of sensors; (**b**) for several depths.

For Figure 19a, a depth of 0.7 m is considered, and the curves for different numbers of sensors are observed. The accuracy constraint on the relative delay times is relaxed by increasing the number of sensors.

For Figure 19b, seven sensors are considered, and the curves for different depths are observed. The estimation of relative delay times should be more accurate as the depth increases. For example, here, with seven sensors, to obtain a precision of 0.1 m for the depth, the precision of the relative delay times is needed to be in a 0.1 µs range.

3.4. Validation of the Travel Time Model M2 on Real Data

To compare travel time modeling with real data, an experiment was performed on the semi-controlled test area in a case similar to the simulation in Section 3.2. The same signal as in the simulation, a Ricker, was emitted. Sensors were placed every 0.05 m perpendicular to the pipe passage (as in Figure 15). Since we did not have a large enough number of sensors, several successive measurements were performed by leaving one sensor fixed and moving the others to mesh the space and act as if there were a large number of sensors. The time bases were recalculated with respect to the fixed sensor of each measurement to act as if only one measurement had been realized with many sensors. This added an error in the travel times, but here, we were only interested in the trend of the evolution of the relative delay times in order to validate Model M2.

Sensor 1 was taken as a reference, which was the closest to the source S, to calculate the relative delay times. Figure 20 shows the evolution of the relative delay times between sensors 1 and i in black. These relative delay times were estimated using cross-correlation between the signal received by sensor 1 and that of sensor i.

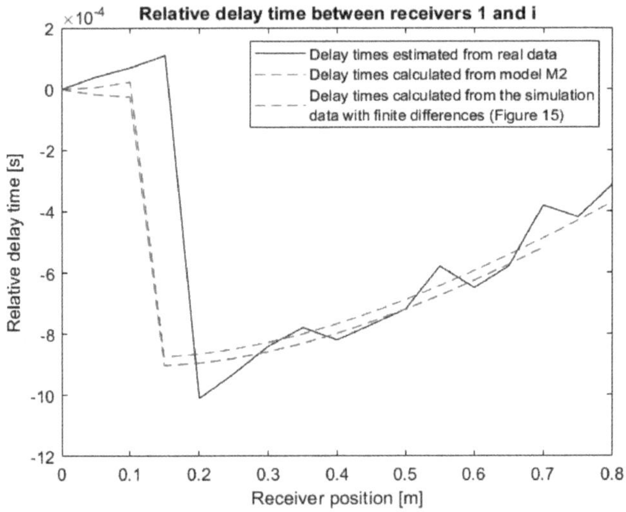

Figure 20. Comparison of the relative delay times between the real data and Model M2.

The propagation velocities were unknown, a priori, for the semi-controlled test area. The velocities V_0 and V_1 of Model M2 can be estimated a posteriori by knowing the position of the source S in the test area. After determining all the variables in the test zone a posteriori, the relative delay times of Model M2 (blue curve of Figure 20) were calculated.

In the beginning, the relative delay times increased until the break that marked the change in environment. Then, they started to increase again. In the simulation (blue curve), the position of the change in medium was fixed at 0.1 m. On the real data (black curve), the change in medium, highlighted by the break of the curve, was instead between 0.15 and 0.2 m. The Model M2 is representative of reality. It is logical and is in agreement with the observation of real data.

3.5. *Depth Estimation from Model M2*

3.5.1. Numerical Simulation with MUSIC Algorithm Adapted to Model M2

In this section, numerical simulations are performed to qualify the MUSIC estimator adapted to Model M2 (as in Section 2.4.2). The statistics of the estimator using the Monte Carlo method (1000 runs of noise) are presented. Consider the following situation 1: a depth (S_Z) of 0.7 m, a propagation velocity (V_0) of 300 m/s, a propagation velocity (V_1) of 600 m/s, the plumb of the pipe (S_X) at 0 m, and the position of the change in medium (P_X) at 0.15 m. Seven sensors were spaced 0.2 m apart, the first one directly at the plumb of the pipe.

In Table 4, the statistics of the results of the estimator using the Monte Carlo method are presented. A total of 1000 runs of noise were performed. The theoretical propagation times between the source S and sensors were noised with a white Gaussian noise of standard deviation σ_{noise}. The propagation times are in the millisecond range, and the propagation time differences with sensor 1 vary between 10^{-5} and 10^{-3} seconds. In this situation, an accuracy in a microsecond range is required on the delay times to obtain an accuracy of less than 0.1 m on the depth. These results confirm those presented in Section 3.3.2.

Table 4. Statistics of the MUSIC estimator adapted to Model M2 using the Monte Carlo method in situation 1 (7 sensors spaced 0.2 m apart). Noise is applied to the propagation times, which are in the order of a millisecond.

		Mean	Standard Deviation	True Value
Numerical Simulation 1 with $\sigma_{noise} = 5 \times 10^{-6}$	Depth S_Z (m)	0.7294	0.0849	0.7
	Average velocity V_0 (m/s)	293	19	300
	Average velocity V_1 (m/s)	592	41	600
Numerical Simulation 2 with $\sigma_{noise} = 5 \times 10^{-5}$	Depth S_Z (m)	0.6666	0.2246	0.7
	Average velocity V_0 (m/s)	261	58	300
	Average velocity V_1 (m/s)	638	128	600

Now, the case with seventeen sensors spaced 0.05 m apart is presented. Consider the following situation 2: a depth (S_Z) of 0.7 m, a propagation velocity (V_0) of 300 m/s, a propagation velocity (V_1) of 600 m/s, the plumb of the pipe (S_X) at 0 m, and the position of the change in medium (P_X) at 0.15 m. Seventeen sensors were spaced 0.05 m apart, the first one directly at the plumb of the pipe.

The results in Table 5 show that an error of the order of microseconds in delay times allows an accuracy of 0.1 m on the depth.

Table 5. Statistics of the MUSIC estimator adapted to Model M2 using the Monte Carlo method in situation 2 (17 sensors spaced 0.05 m apart). Noise is applied to the propagation times, which are in the order of a millisecond.

		Mean	Standard Deviation	True Value
Numerical Simulation 1 with $\sigma_{noise} = 5 \times 10^{-6}$	Depth S_Z (m)	0.6604	0.0704	0.7
	Average velocity V_0 (m/s)	293	18	300
	Average velocity V_1 (m/s)	611	79	600

3.5.2. Estimation on Real Data

At least seven sensors are required (Section 3.3.2) to use the MUSIC algorithm adapted to Model M2. As mentioned in Section 3.4, we only had five sensors. The sensors were moved to perform successive measurements. The different measurements are then not synchronized. Therefore, our estimator, based on MUSIC, cannot be used on these measurements.

However, results evaluated from the delay times between the sensors are presented in Figure 20 (black curve). These delay times were obtained after retiming, introducing additional errors.

The criterion presented here is the inverse of the squared error between the estimated delay times and the delay times calculated by Model M2.

The values of the criterion are obtained by varying the variables to be estimated: the depth (S_Z), the velocity inside the trench (V_0) and the velocity outside the trench (V_1). The lateral position of the pipe (S_X) and the position of the medium change (P_X) are considered known. Indeed, S_X estimated with the GasTracker tool and P_X from the temporal break are observed in Figure 20.

Figure 21 is a display of the obtained criterion. The estimation results are presented in Table 6. The error in the depth is less than 0.1 m. In addition, the velocity estimates are really small, which could be due to the large error in the delay times. The parameters S_X and P_X are fixed by an a priori estimate. It would be interesting to also vary these parameters around their estimated position.

These results obtained with least squares are encouraging, but we would like to have a more discriminating criterion. The next step of our work will be to test the MUSIC algorithm on real synchronous data in order to have a criterion with a better resolution power.

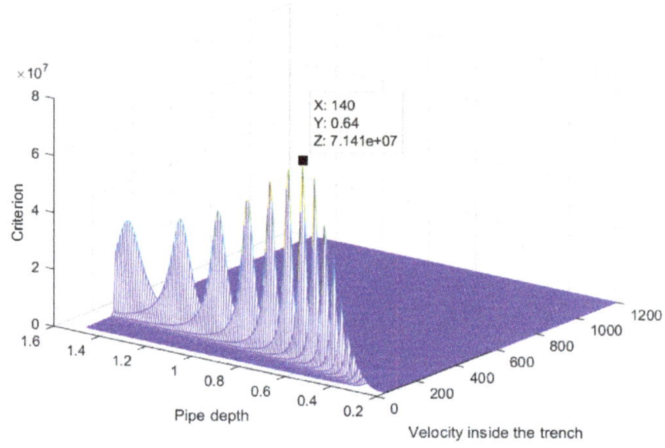

Figure 21. The inverse of the squared error between the estimated delay times and the delay times calculated by Model M2. For this display, the velocity outside the trench V1 is fixed at the estimated value.

Table 6. Estimation obtained with least squares from Model M2.

	Estimate Value	**Reference Value**
Depth (S_Z) (m)	0.64	0.70
Average velocity inside the trench (V_0) (m/s)	140	unknown
Average velocity outside the trench (V_1) (m/s)	250	unknown

4. Conclusions and Future Perspectives

We developed a method using several sensors positioned perpendicular to the passage of the pipe. This allows us to model the problem in a plane orthogonal to the passage of the pipe. The vibrating section of the pipe is represented by a point source.

In the first section, an initial model of the problem was defined by considering a single propagation medium (M1), i.e., all sensors are located at the ground surface inside the trench. This Model M1 was validated, on the scale of the problem, by comparing it to a more complete modeling (full wavefield modeling by finite differences). The Cramer–Rao bound was calculated to obtain theoretical information on the accuracy reached by the variables, in particular on the relative delay times. The error in the positioning of the sensors (e.g., line of sensors not perpendicular to the pipe) induces an error in the travel times. This is why, in these simulations and theoretical studies, the error in the depth estimation was quantified as a function of the error in the delay times. The MUSIC algorithm was adapted to Model M1, and then the algorithm was tested on real data. This allowed us to question the model and thus, to advance it. The results obtained from the M1 model show that the depth estimates reach the desired accuracy of 0.1m.

In the second section, a second model of the problem was defined by considering two vertically stratified propagation media (M2), i.e., sensors are located at the ground surface inside and outside the trench, according to the information obtained from experiments onto a test area. This model was validated in two steps, first through numerical simulation by comparing it to a more complete model, then by comparing it to real data. The Cramer–Rao bound was also calculated to obtain theoretical information on the accuracy reached by the variables. The MUSIC algorithm was adapted to this Model M2 and was tested in simulation.

Model M2 aims to cover the case of a marked change in medium (inside/outside the trench). The results obtained with the Cramér–Rao bound show that seven or eight sensors would be needed to obtain a satisfactory depth estimate. Not having real synchronous measurements realized with eight sensors, a least squares criterion between the delay times estimated from the real signals and those given by Model M2 was established. The results obtained on real data reach the desired accuracy.

In our future work, with more sensors, we will test the estimator based on the MUSIC algorithm adapted to Model M2 on real data. Depending on the feedback from these experiments, we will be able to continue to evolve the model by potentially considering other propagation media, such as the layer of sand surrounding the pipe or the layer of tar covering the ground. This aims to produce a model closer to real situations that provide the fastest possible result to be able to apply the method in the field in real-time.

Author Contributions: Conceptualization, W.X. and G.S.; methodology, W.X.; software, W.X. and A.R.; validation, W.X., G.S., A.R., P.P. and L.Z.; formal analysis, W.X.; investigation, W.X.; resources, P.P. and L.Z.; data curation, W.X.; writing—original draft preparation, W.X.; writing—review and editing, W.X., G.S. and A.R.; visualization, W.X. and A.R.; supervision, G.S.; project administration, not applicable; funding acquisition, not applicable. All authors have read and agreed to the published version of the manuscript.

Funding: This project is supported by the CIFRE contract between MADE-SA and CNRS-UMR7330, CEREGE, from the ANRT French program of CNRS (Agence National de la Recherche Technique). This project is partially supported by RICE (Research and Innovation Center of Energy), GRTgaz.

Institutional Review Board Statement: Not applicable.

Informed Consent Statement: Not applicable.

Data Availability Statement: Not applicable.

Acknowledgments: We would like to thank the Laboratory of Mechanics and Acoustics (LMA—CNRS), Marseille, France. We also express our sincere gratitude to Cédric Pinhède, research engineer at LMA, who allowed us to access the anechoic chamber of LMA and to qualify our loudspeaker. We gratefully acknowledge S. Operto for providing the wave propagation finite difference modeling code used in this study.

Conflicts of Interest: The authors declare no conflict of interest.

Appendix A

Calculation of the Gradient of P_{iZ}

In this section, the gradient of P_{iZ} is expressed with respect to θ_{M2}. The notations from Section 3.1.2 are used in the calculation of P_{iZ}.

From Equation (16),

$$\nabla_{\theta_{M2}} P_{iZ} = \frac{1}{2}\left(\frac{1}{2\sqrt{A}}\nabla_{\theta_{M2}}(A) - \frac{\nabla_{\theta_{M2}}(A) - 2\left[\nabla_{\theta_{M2}}(g) + \nabla_{\theta_{M2}}(A) + \nabla_{\theta_{M2}}\left(\frac{u}{\sqrt{A}}\right)\right]}{2\sqrt{A - 2\left(g + A + \frac{u}{\sqrt{A}}\right)}}\right) - \frac{m_1 \nabla_{\theta_{M2}}(m_2) - m_2 \nabla_{\theta_{M2}}(m_1)}{4\, m_1^2} \quad \text{(A1)}$$

The gradients of all intermediate variables are calculated (A, J, K, g, u, w, m_1, m_2, m_3, m_4, m_5) to express $\nabla_{\theta_{M2}} P_{iZ}$. The variable A (Equation (17)) is decomposed into three terms to ensure that the calculation is still readable.

$$A = A_1 + A_2 + A_3 \quad \text{(A2)}$$

with

$$A_1 = \left(\frac{-K + \sqrt{K^2 + \frac{4J^3}{27}}}{2}\right)^{1/3}, A_2 = -\frac{J}{3\left(\frac{-K + \sqrt{K^2 + \frac{4J^3}{27}}}{2}\right)^{1/3}}, A_3 = -\frac{2g}{3} \quad \text{(A3)}$$

The gradient of each term is calculated.

$$\nabla_{\theta_{M2}}(A_1) = \frac{-\nabla_{\theta_{M2}}(K) + \frac{2K\nabla_{\theta_{M2}}(K) + \frac{12J^2}{27}\nabla_{\theta_{M2}}(J)}{2\sqrt{K^2+\frac{4J^3}{27}}}}{6\left(\frac{-K+\sqrt{K^2+\frac{4J^3}{27}}}{2}\right)^{2/3}} \tag{A4}$$

$$\nabla_{\theta_{M2}}(A_2) = -\frac{1}{3}\frac{\nabla_{\theta_{M2}}(J)\left(\frac{-K+\sqrt{K^2+\frac{4J^3}{27}}}{2}\right)^{1/3} - J\nabla_{\theta_{M2}}(A_1)}{\left(\frac{-K+\sqrt{K^2+\frac{4J^3}{27}}}{2}\right)^{2/3}} \tag{A5}$$

$$\nabla_{\theta_{M2}}(A_3) = -\frac{2}{3}\nabla_{\theta_{M2}}(g) \tag{A6}$$

Once the gradient of A was calculated, the gradients of J and K can be calculated (Equations (18) and (19)).

$$\nabla_{\theta_{M2}}(J) = -\frac{2g}{3}\nabla_{\theta_{M2}}(g) - 4\nabla_{\theta_{M2}}(w) \tag{A7}$$

$$\nabla_{\theta_{M2}}(K) = \frac{8}{3}[g\,\nabla_{\theta_{M2}}(w) + w\,\nabla_{\theta_{M2}}(g)] - \frac{2g^2}{9}\nabla_{\theta_{M2}}(g) - 2u\nabla_{\theta_{M2}}(u) \tag{A8}$$

The calculation of the gradients of g, u and w can be realized (Equations (20)–(22)).

$$\nabla_{\theta_{M2}}(g) = \frac{m_1\nabla_{\theta_{M2}}(m_3) - m_3\nabla_{\theta_{M2}}(m_1)}{m_1^2} - \frac{3}{4}\frac{m_2m_1\nabla_{\theta_{M2}}(m_2) - m_2^2\nabla_{\theta_{M2}}(m_1)}{m_1^3} \tag{A9}$$

$$\nabla_{\theta_{M2}}(u) = \frac{m_1\nabla_{\theta_{M2}}(m_4) - m_4\nabla_{\theta_{M2}}(m_1)}{m_1^2} - \frac{1}{2}\frac{[m_2\nabla_{\theta_{M2}}(m_3) + m_3\nabla_{\theta_{M2}}(m_2)]m_1 - 2m_3m_2\nabla_{\theta_{M2}}(m_1)}{m_1^3} + \frac{3}{8}\frac{m_1m_2^2\nabla_{\theta_{M2}}(m_2) - m_2^3\nabla_{\theta_{M2}}(m_1)}{m_1^4} \tag{A10}$$

$$\nabla_{\theta_{M2}}(w) = \begin{cases} \frac{m_1\nabla_{\theta_{M2}}(m_5) - m_5\nabla_{\theta_{M2}}(m_1)}{m_1^2} - \frac{1}{4}\frac{[m_2\nabla_{\theta_{M2}}(m_4) + m_4\nabla_{\theta_{M2}}(m_2)]m_1 - 2m_2m_4\nabla_{\theta_{M2}}(m_1)}{m_1^3} \\ +\frac{1}{16}\frac{[m_2^2\nabla_{\theta_{M2}}(m_3) + 2m_3m_2\nabla_{\theta_{M2}}(m_2)]m_1 - 3m_3m_2^2\nabla_{\theta_{M2}}(m_1)}{m_1^4} - \frac{3}{64}\frac{m_1m_2^3\nabla_{\theta_{M2}}(m_2) - m_2^4\nabla_{\theta_{M2}}(m_1)}{m_1^5} \end{cases} \tag{A11}$$

Then, the gradients from m_1 to m_5, on which all other gradients depended, were calculated.

$$\nabla_{\theta_{M2}}(m_1) = [0 \quad 0 \quad -2V_0 \quad 2V_1]^T \tag{A12}$$

$$\nabla_{\theta_{M2}}m_2 = [0 \quad 2[V_0^2 - V_1^2] \quad 4S_ZV_0 \quad -4V_1[R_{iZ} + S_Z]]^T \tag{A13}$$

$$\nabla_{\theta_{M2}}(m_3) = [2V_0^2(P_{iX} - S_X) \quad 2S_ZV_1^2 + 4R_{iZ}V_1^2 - 2S_ZV_0^2 \quad -2V_0[S_Z^2 + (P_{iX} - S_X)^2] \quad 2V_1[S_Z^2 + R_{iZ}^2 + (R_{iX} - P_{iX})^2 + 4R_{iZ}S_Z]]^T \tag{A14}$$

$$\nabla_{\theta_{M2}}(m_4) = [0 \quad -2V_1^2[2R_{iZ}S_Z + R_{iZ}^2 + (R_{iX} - P_{iX})^2] \quad 0 \quad -4V_1S_Z[R_{iZ}S_Z + R_{iZ}^2 + (R_{iX} - P_{iX})^2]]^T \tag{A15}$$

$$\nabla_{\theta_{M2}}(m_5) = \left[0 \quad 2S_ZV_1^2[R_{iZ}^2 + (R_{iX} - P_{iX})^2] \quad 0 \quad 2S_Z^2V_1[R_{iZ}^2 + (R_{iX} - P_{iX})^2]\right]^T \tag{A16}$$

The gradients of m_1 to m_5 were calculated and expressed according to the gradients of P_{iZ}.

References

1. Liu, Y.; Habibi, D.; Chai, D.; Wang, X.; Chen, H.; Gao, Y.; Li, S. A Comprehensive Review of Acoustic Methods for Locating Underground Pipelines. *Appl. Sci.* **2020**, *10*, 1031. [CrossRef]
2. Jeong, H.S.; Arboleda, C.A.; Abraham, D.M.; Halpin, D.W.; Bernold, L.E. *Imaging and Locating Buried Utilities*; Joint Transportation Research Program, Indiana Department of Transportation and Purdue University: West Lafayette, IN, USA, 2003. [CrossRef]

3. Harmankaya, U.; Kaslilar, A.; Thorbecke, J.; Wapenaar, K.; Draganov, D. Locating near-surface scatterers using non-physical scattered waves resulting from seismic interferometry. *J. Appl. Geophys.* **2013**, *91*, 66–81. [CrossRef]
4. Filippi, C.; Leparoux, D.; Grandjean, G.; Bitri, A.; Côte, P. New robust observables on Rayleigh waves affected by an underground cavity: From numerical to experimental modelling. *Geophys. J. Int.* **2019**, *218*, 1903–1918. [CrossRef]
5. Muggleton, M.J.; Rustighi, E. Mapping the Underworld: Recent developments in vibro-acoustic techniques to locate buried infrastructure. *Géotech. Lett.* **2013**, *3*, 137–141. [CrossRef]
6. Hao, T.; Rogers, C.D.F.; Metje, N.; Chapman, D.N.; Muggleton, J.M.; Foo, K.Y.; Wang, P.; Pennock, S.R.; Atkins, P.R.; Swingler, S.G.; et al. Condition assessment of the buried utility service infrastructure. *Tunn. Undergr. Space Technol.* **2012**, *28*, 331–344. [CrossRef]
7. Dou, Q.; Wei, L.; Magee, D.R.; Atkins, P.R.; Chapman, D.N.; Curioni, G.; Goddard, K.F.; Hayati, F.; Jenks, H.; Metje, N.; et al. 3D Buried Utility Location Using a Marching-Cross-Section Algorithm for Multi-Sensor Data Fusion. *Sensors* **2016**, *16*, 1827. [CrossRef]
8. Li, J.; Guo, T.; Leung, H.; Xu, H.; Liu, L.; Wang, B.; Liu, Y. Locating Underground Pipe Using Wideband Chaotic Ground Penetrating Radar. *Sensors* **2019**, *19*, 2913. [CrossRef]
9. Pettinelli, E.; di Matteo, A.; Mattei, E.; Crocco, L.; Soldovieri, F.; Redman, J.D.; Annan, A.P. GPR Response from Buried Pipes: Measurement on Field Site and Tomographic Reconstructions. *IEEE Trans. Geosci. Remote Sens.* **2009**, *47*, 2639–2645. [CrossRef]
10. Xerri, W.; Saracco, G.; Gassier, G.; Zomero, L.; Picon, P. *Preliminary Acoustic Study of 3D Localization of Buried Polyethylene Pipe. Quantitative Nondestructive Evaluation*; American Society of Mechanical Engineers: New York, NY, USA, 2022. [CrossRef]
11. Xerri, W.; Saracco, G.; Ribodetti, A.; Zomero, L.; Picon, P. Estimating the depth of buried polyethylene pipe by comparing propagation models. In Proceedings of the French of Congress Acoustics (CFA), Marseille, France, 11–15 April 2022.
12. Liu, Y.; Habibi, D.; Chai, D.; Wang, X.; Chen, H. A Numerical Study of Axisymmetric Wave Propagation in Buried Fluid-Filled Pipes for Optimizing the Vibro-Acoustic Technique When Locating Gas Pipelines. *Energies* **2019**, *12*, 3707. [CrossRef]
13. Papandreou, B.; Rustighi, E.; Brennan, M.J. *Study into the Feasibility of Using Acoustic Techniques to Locate Buried Objects*; ISVR Technical Memorandum; Institute of Sound and Vibration Research: Southampton, UK, 2008.
14. Muggleton, J.M.; Brennan, M.J.; Rogers, C.D.F. Point vibration measurements for the detection of shallow-buried objects. *Tunn. Undergr. Space Technol.* **2014**, *39*, 27–33. [CrossRef]
15. Katz, A.; Karaa, F.; Niver, E. *Innovative and Effective Techniques for Locating Underground Conduits*; New Jersey Institute of Technology: Newark, NJ, USA, 2011.
16. McFadden, J.A. Radial Vibrations of Thick-Walled Hollow Cylinders. *J. Acoust. Soc. Am.* **1954**, *26*, 714. [CrossRef]
17. Gazis, D.C. Exact Analysis of the Plane-Strain Vibrations of Thick-Walled Hollow Cylinders. *J. Acoust. Soc. Am.* **1958**, *30*, 786–794. [CrossRef]
18. Sinha, B.K.; Plona, T.J.; Kostek, S.; Chang, S.K. Axisymmetric wave propagation in fluid-loaded cylindrical shells. I: Theory. *J. Acoust. Soc. Am.* **1992**, *92*, 1132–1143. [CrossRef]
19. Operto, S.; Virieux, J.; Ribodetti, A.; Anderson, J.E. Finite-difference frequency-domain modeling of viscoacoustic wave propagation in 2D tilted transversely isotropic media. *Geophysics* **2009**, *74*, T75–T95. [CrossRef]
20. Virieux, J. P-SV wave propagation in heterogeneous media: Velocity-stress finite-difference method. *Geophysics* **1986**, *51*, 889–901. [CrossRef]
21. Virieux, J. SH wave propagation in heterogeneous media: Velocity-stress finite-difference method. *Geophysics* **1984**, *49*, 1933–1957. [CrossRef]
22. Stoica, P.; Nehorai, A. MUSIC, maximum likelihood, and Cramer-Rao bound. *IEEE Trans. Acoust. Speech Signal Process* **1989**, *37*, 720–741. [CrossRef]
23. Scharf, L.L.; McWhorter, L.T. Geometry of the Cramer-Rao bound. *Signal Process* **1993**, *31*, 301–311. [CrossRef]
24. Schmidt, R.O. Multiple Emitter Location and Signal Parameter Estimation. *IEEE Trans. Antennas Propag.* **1986**, *34*, 276–280. [CrossRef]
25. Wang, B.; Zhao, Y.; Liu, J. Mixed-Order MUSIC Algorithm for Localization of Far-Field and Near-Field Sources. *IEEE Signal Process. Lett.* **2013**, *20*, 311–314. [CrossRef]
26. Fossati, C.; Bourennane, S.; Marot, J. Chapter 3: Localization of Buried Objects in Sediment Using High Resolution Array Processing Methods. In *Underwater Acoustics*; Bourennane, S., Ed.; InTech: Rijeka, Croatia, 2012; pp. 41–58. [CrossRef]
27. Han, D.; Fossati, C.; Bourennane, S.; Saidi, Z. *Bearing and Range Estimation Algorithm for Buried Object in Underwater Acoustics. Research Letters in Signal Processing*; Hindawi: London, UK, 2009. [CrossRef]
28. Saidi, Z.; Bourennane, S. Cumulant-Based Coherent Signal Subspace Method for Bearing and Range Estimation. *EURASIP J. Adv. Signal Process* **2007**, *2007*, 084576. [CrossRef]
29. Sahin, A.; Miller, E.L. Object-based localization of buried objects using high-resolution array processing techniques. In Proceedings of the SPIE, Orlando, FL, USA, 31 May 1996. [CrossRef]
30. Choi, J.W.; Kim, Y.H. Spherical beam-forming and MUSIC methods for the estimation of location and stength of spherical sound sources. *Mech. Syst. Signal Process* **1995**, *9*, 569–588. [CrossRef]

materials

MDPI

Article

Effect of Temperature on Ultrasonic Nonlinear Parameters of Carbonated Concrete

Jinzhong Zhao [1], Jin Wu [1,*], Xuejun Chen [2] and Ruifu Zeng [2]

[1] College of Civil Aviation, Nanjing University of Aeronautics and Astronautics, Nanjing 210016, China
[2] Eastern Airports, Nanjing 210016, China
* Correspondence: wujin@nuaa.edu.cn

Abstract: In order to explore the monitoring technique of concrete carbonation in various temperatures, longitudinal ultrasonic nonlinear parameters of carbonated concrete are measured by using an embedded composite piezoelectric transducer (ECPT) and a surface-mounted transducer. The effect of temperature from −20 °C to 40 °C with a temperature interval of 5 °C and water–cement ratio on the measurements of ultrasonic parameters for carbonated concrete is investigated. The ultrasonic transmission detection method and the second harmonic generation (SHG) technique for longitudinal waves are used in the study. Results of the experiment demonstrate that ECPT is effective in the monitoring of the changes in ultrasonic parameters of carbonated concrete. At the temperature ranging from 15 °C to 40 °C, the increasing temperature slightly increases the relative nonlinear parameters of carbonated concrete. It decreases significantly that the relative nonlinear parameters of carbonated concrete measured at 0 °C compared with that at 10 °C. The configuration in this measurement is also appropriate for the assessment of carbonated concrete during carbonation time in low-temperature environments (below 0 °C). In the same carbonation time, the relative nonlinear parameters also increase slightly when the temperature is at −20 °C to 0 °C, but it does not change too much. Furthermore, there is a more significant variation of the nonlinear parameters in the same carbonation time for the specimens with a high water–cement ratio than that with a low one.

Keywords: concrete carbonation; nonlinear parameter; temperature effects; embedded composite piezoelectric transducer; nondestructive testing

Citation: Zhao, J.; Wu, J.; Chen, X.; Zeng, R. Effect of Temperature on Ultrasonic Nonlinear Parameters of Carbonated Concrete. *Materials* **2022**, *15*, 8797. https://doi.org/10.3390/ma15248797

Academic Editors: Grzegorz Peruń, Tangbin Xia and Bogusław Łazarz

Received: 10 October 2022
Accepted: 6 December 2022
Published: 9 December 2022

1. Introduction

Ultrasonic measurement as non-destructive testing technology has been widely used to determine the characterization of concrete structures for a long time. Nonlinear resonance ultrasonic spectroscopy (NRUS) technology has been used in the testing of progressive damage occurring in concrete, such as alkali–silica reaction [1], thermal damage [2], and delayed ettringite formation in concrete [3]. These progressive changes in concrete increase the material nonlinearity, resulting in a shift of the resonance frequency and amplitude in concrete. During concrete carbonation, the material nonlinearity is also changed by the effects of the carbonation product on the porosity and microcracks in concrete. Bouchaala et al. [4], reported that resonant ultrasonic nonlinear parameters have the feasibility of measuring concrete carbonation. Eiras et al. [5] used resonance ultrasonic spectroscopy technology to study the effect of concrete carbonation on the linear and nonlinear dynamic properties of cement-based materials. The linear and nonlinear dynamic properties can be used to deem material against carbonation in laboratory tests, whereas concrete has inherent nonlinearity due to its material composition. The nonlinear response of the concrete to the incident wave distorts the waveform when a single-frequency ultrasonic wave is transmitted into the concrete. Then, the higher harmonic waves will be generated in the transmitted wave. Kim et al. [6,7] investigated the feasibility of the second harmonic generation (SHG) technology of Rayleigh surface wave to characterize the carbonation of concrete. The amplitude of the active acoustic source of SHG technology is constant, while the percussive force of NURS needs to be control.

At present, health inspection of concrete structures is usually performed by manually carrying external ultrasonic probes. However, the use of large external probes requires extensive work preparation, which not only hinders the use of this technique for monitoring but also greatly limits the measurement area of concrete structures [8]. In contrast, embedding low-cost piezoelectric transducers in concrete structures allows for a more flexible configuration of the measurement network on the one hand, and long-term stable health measurements of the structure on the other. Such kinds of transducers have been successfully used for the monitoring of concrete hydration by deducing the evolution of Young's modulus based on the flight time of the acoustic wave [9–12], the evolution of the concrete compressive strength at an early age [10,13], concrete cracking [8,14–18], water seepage [19], and mechanical properties of concrete as well as the acoustoelastic effect in compression [20–23], applying piezoelectric lead zirconate titanate (PZT) patches to the rebar by monitoring the conductance changes of the piezoelectric patches to study the effect of concrete carbonation on rebar corrosion [24,25]. In the current research of embedded transducers, it is found that cement-based and marble-based piezoelectric composite sensors have the feature of diminishing the distortion of signal and benefiting the energy transmission efficiency [26,27]. However, these large size transducers are not suitable for detecting P waves at close distances due to the carbonation of concrete developing from the surface. Qin et al. [12] and Chen et al. [28] provided two ideas for small size transducers in specimens with specific sizes. The signal transmitted by these transducers at an operating state with a high frequency and high energy carries less sensor nonlinearity than cement-based or marble-based transducers, whereas the influence of boundary reflection [6,29,30] on the signal should be considered in both time domain analysis and frequency domain analysis when the wave is propagating in small-size concrete. Otherwise, the parameters obtained from a small-size concrete are not suitable for the characterization of the large one due to the distance of boundary reflection.

In this paper, the ultrasonic transmission detection method was taken with an embedded transducer and a surface-mounted transducer. The embedded composite piezoelectric transducer (ECPT) which was composited with four materials was developed and used to detect the evolution of ultrasonic parameters in carbonated concrete by the SHG [31] technique of ultrasonic longitudinal waves to verify the effect of the concrete water–cement ratio on the relative nonlinear parameters of carbonation. This kind of ECPT had the advantage combined with the transducers used by Qin et al. [12] and Chen et al. [28], such as small size and electromagnetic shielding. Subsequently, the effect of the ambient temperature on the ultrasonic longitudinal detection parameters of carbonated concrete was also studied.

2. Materials and Methods

2.1. Specimens

In this experiment, concrete specimens (CI, CII, CIII) with different water–cement ratios were designed. The coarse aggregate used in the concrete was a continuous grade of natural gravel from 5 mm to 25 mm and the fine aggregate was natural river sand with a fineness modulus of 2.56. Table 1 shows the mixture design of the concrete specimens.

Table 1. Mixture ratio of the concrete specimen.

Category	CI	CII	CIII
OPC (kg/m^3)	342	375	415
Fine aggregate (kg/m^3)	708	695	680
Coarse aggregate (kg/m^3)	1155	1135	1110
Water (kg/m^3)	195	195	195

The structure of the embedded composite piezoelectric transducer (ECPT) is shown in Figure 1a. It consisted of a metal shielding layer, an encapsulation layer, a piezoelectric lead zirconate titanate (PZT) patch, and a composite backing layer. The model of the PZT was PSN-33 with a size of ϕ 8 mm × 0.48 mm, which was provided by the Haiying company. Table 2 shows the parameters of PZT, where d_{33} is piezoelectric strain constant, C is capacitance, $\tan\delta$ is dielectric dissipation factor, ft is resonance frequency, and Zr is impedance.

Table 2. The parameters of PZT.

Model	d_{33} (pC · N^{-1})	C (pF)	tan δ (%)	ft (MHz)	Zr (Ω)
PSN-33	377	1167	1.77	4.03	5.45

This kind of composite increased the structural strength of the transducer and made it have good waterproofing and working performance. In addition, the ECPT could resist certain electromagnetic wave interference and suppress the interference of some noise to the signal during operation. The ECPT vibrated along the poling direction to generate mechanical waves when connected with the electric field. As shown in Figure 1b, the PZT patch is fastened to the printed circuit film in this transducer. This connection method ensured the working stability of the PZT patch and improved the convenience of the manufacturing process for the embedded composite transducer. The sensitivity of ECPT was verified with the face-to-face secondary calibration method (GB/T 19801-2005/ISO 12714:1999) [32–34]. The sensors were placed in the symmetry location to transmit and receive elastic waves from the solid medium surface—whereafter, the sensitivity of the sensor was calculated by the output signals. Figure 2 shows the sensitivity of ECPT.

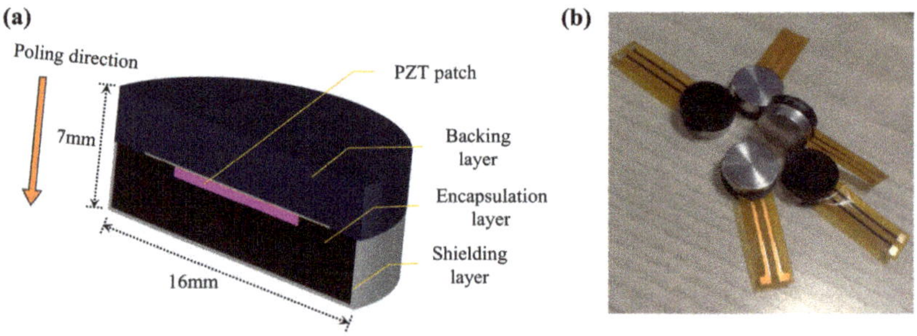

Figure 1. ECPT: (**a**) structure of ECPT; (**b**) the finished product of ECPT.

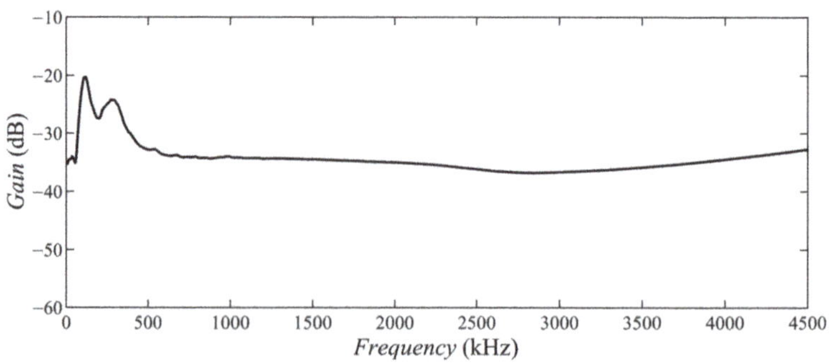

Figure 2. Response curves for ECPT.

There were three specimens for each kind of concrete. Each size of the concrete specimen was 100 mm × 100 mm × 100 mm. As shown in Figure 3a, only one ECPT is embedded in the concrete specimen as a transmitter. On the surface in which CO_2 was ingressed, a surface mounted transducer was fixed to obtain the nonlinear parameters of the ultrasonic radial propagation in carbonated concrete. The reason for this arrangement was that the received signal transmitted from inner concrete was experimentally found to be more stable than from the surface. In addition, this configuration could also be used in a large concrete structure. It was assumed that two transducers were embedded in one concrete and their poling direction pointed to the surface in which CO_2 was ingressed, and the reflected P wave would be affected by the S wave [27].

As shown in Figure 3a, Section 1-1 is the plane perpendicular to the bottom surface of the concrete and along the central axis of the ECPT. Section 2-2 is perpendicular to the plane of the ECPT shielding layer on the bottom of the concrete in Figure 3a and the projection of ECPT on this plane is a circle. Figure 3b is Section 1-1 of the concrete in which the transducer is embedded. The central axis of the ECPT is 45 mm from the bottom surface of the specimen in Figure 3a. In addition, the concrete is up to 70 mm away from the surface of the shielding layer of the ECPT. Figure 3c is a Section 2-2 view of the concrete in which the transducer embedded. It can be found from the figure that the closest distance between the central axis of the ECPT and the surface of the concrete is 42 mm.

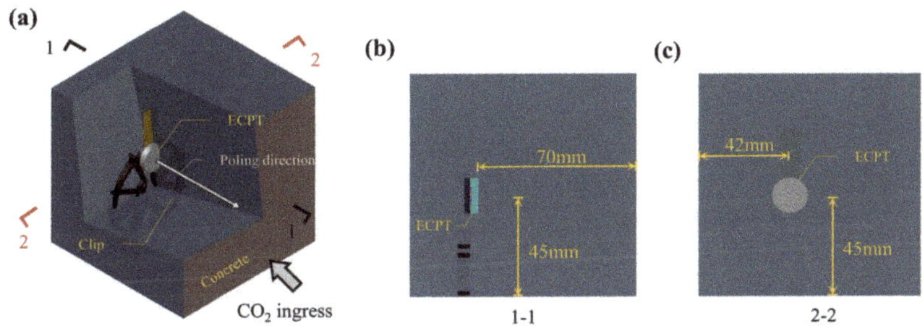

Figure 3. Arrangement of the embedded composite piezoelectric ultrasonic transducer: (**a**) partial cutaway perspective view of the specimen; (**b**) Section 1-1; (**c**) Section 2-2.

2.2. Accelerated Carbonation Experiment

Concrete specimens were cured in an environmental chamber with a temperature of 20 ± 3 °C and a relative humidity of 95%. After 28 days of curing, all surfaces of the specimens were sealed with paraffin wax except for the surface which CO_2 ingressed. The concrete specimens were placed in the carbonation chamber for concrete carbonation experiments. In this case, the ambient temperature was 20 °C, the CO_2 concentration was 20%, and the relative humidity was 70%. The carbonation depth was measured with phenolphthalein in spare samples after 0, 7, and 21 days of carbonation.

2.3. Testing Methods

The specimens were removed from the carbonation chamber after 3, 7, 14, and 21 days of concrete carbonation, respectively. In order to test the ultrasonic parameters of carbonated concrete under different temperature environments, the concrete was placed in an environment chamber at the temperature from −20 °C to 40 °C for 24 h with a temperature interval of 5 °C, and the relative humidity was controlled at 70%. Figure 4 shows the experimental setup. A function waveform generator (DG 4202) generated 15-cycle, 1 MHz frequency, 10 peak-to-peak voltage sinusoidal waveform signal pulses at a burst interval of 3 ms. Then, the signal was amplified to a peak-to-peak voltage of 120 V by a broadband power amplifier (KROHN-HITE 7602M, Brockton, MA, USA), and the ultrasonic signal was transmitted by the ECPT in the concrete. It should be noted that the receiving transducer fixed to the concrete surface had the same central axis as the ECPT. The oscilloscope (Keysight InfiniiVision DSOX 3014T, Santa Rosa, CA, USA) was connected to the receiving transducer to acquire a time domain signal with a sampling length of 16,000 at a sampling frequency of 625 MSa/s, and the acquisition mode was the average acquisition of 2048 times. At the same time, a sync output line connected the function waveform generator to the oscilloscope to receive the reference waveform. The measurements of per specimen were repeated 10 times. Finally, received signal was subjected to post-processing such as denoising (wavelet transform), interception, windowing, and fast Fourier transform (FFT) by the computer.

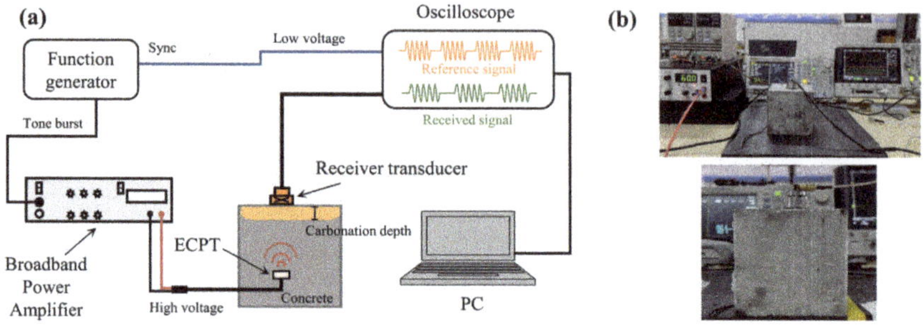

Figure 4. Experimental device: (**a**) connection of experimental instruments; (**b**) the photos of the experimental device.

The received time domain signal (the blue squares indicate the width of the Hann window) of carbonated concrete acquired by the oscilloscope is shown in Figure 5a. It should be noted that the time for the concrete boundary reflection wave to reach the receiving transducer measured in the experiment was around 28 μs. However, the pulse time length of the excitation wave was less than the time length required for the arrival of the reflected wave. In order to ensure maximum signal accuracy and avoid the influence of boundary reflection waves on the received time-domain signal, the first ten cycles of the time-domain signal were intercepted for windowing (Hanning) and subjected to FFT processing. The spectrum of the FFT processed signal is shown in Figure 5b, and the fundamental amplitude (A_1) and the second harmonic amplitude (A_2) can be observed from the figure. The products of the carbonation reaction filled the pores in concrete, resulting in an effect on the nonlinear response of concrete. Subsequently, the fundamental amplitude (A_1) and the harmonic amplitude (A_2) of the transmitted wave were changed by this effect. As such, the relative nonlinear parameter (A_2/A_1^2) could be assessed for the change in carbonated concrete.

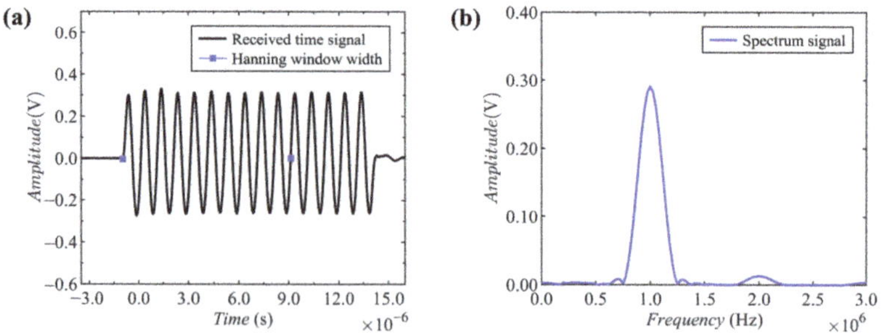

Figure 5. Time signal and spectrum for measurement: (**a**) received time domain signal; (**b**) spectrum signal after FFT (A_1: 1MHz, A_2: 2MHz).

3. Results and Discussion

The carbonation depth was measured with phenolphthalein in spare samples after 0, 7, and 21 days of carbonation. Table 3 shows the average carbonation depth (cd). The carbonation depth of concrete with different water cement ratio was different at the same carbonation time.

Table 3. The average carbonation depth.

	0 Day–cd (mm)	7 Day–cd (mm)	21 Day–cd (mm)
CI $w/c = 0.57$	0.51	6.41	11.83
CII $w/c = 0.52$	0.32	5.17	9.51
CIII $w/c = 0.47$	0.11	3.65	7.12

During the concrete carbonation progress, the existing pores and microcracks were deposited by the carbonation product $CaCO_3$ [5,35]. The molar volume of carbonated products was higher than that of hydrates, resulting in a decrease in the porosity of concrete and an increase in the density of concrete. These microstructural changes altered the ultrasonic parameters of the concrete. After 3, 7, 14, and 21 days of concrete carbonation, the concrete specimens were removed for ultrasonic measurement at 20 °C. It can be observed from Figure 6 that the ultrasonic parameters change with the carbonation time of concrete. As shown in Figure 6b, the fundamental amplitude (A_1) of concrete increases with the concrete carbonation time. Meanwhile, the amplitude of the second harmonic (A_2) shows a decreasing trend in general. Hence, the relative nonlinear parameters (A_2/A_1^2) of concrete are reduced by the change of the fundamental amplitude (A_1) and the second harmonic amplitude (A_2). The variation trend of relative nonlinear parameters with the carbonation progress measured by the configuration at 20 °C in this study is similar to that obtained by the second harmonic generation (SHG) technology of Rayleigh surface wave taken by Kim et al. [6,7]. The relative nonlinear parameters decrease with the carbonation progress. Although the water–cement ratios of CI, CII, and CIII are different, the variation of their ultrasonic parameters of carbonated concrete follows this trend.

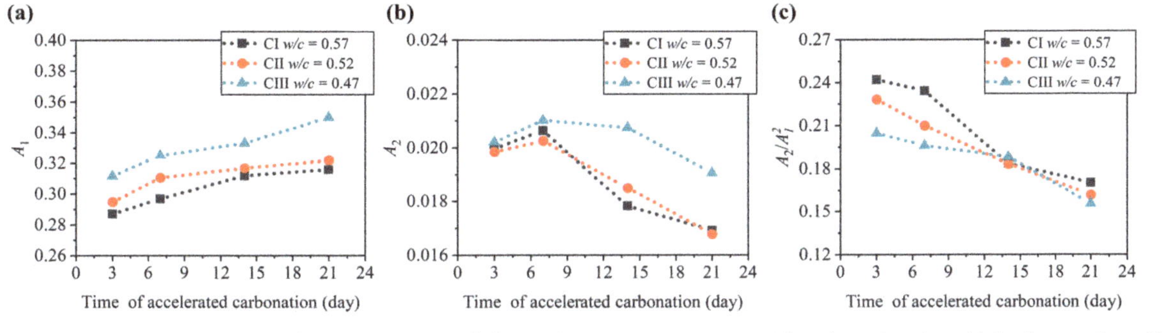

Figure 6. Variation of ultrasonic average parameters with carbonation time: (**a**) fundamental amplitude A_1; (**b**) second harmonic amplitude A_2; (**c**) nonlinear parameter A_2/A_1^2.

After 3, 7, 14, and 21 days of concrete carbonation, specimens were placed in an environment chamber at the temperature from −20 °C to 40 °C for 24 h with a temperature interval of 5 °C, all with a relative humidity of 70%. Figures 7 and 8 show the average fundamental amplitudes and second harmonic amplitudes of the ultrasonic longitudinal wave measured at different ambient temperatures after 14 days of concrete carbonation. The fundamental amplitudes and second harmonic amplitudes of the three kinds of concrete are different at various ambient temperatures. As shown in Figure 7, the fundamental amplitude is also affected by the increasing temperature from −20 °C to 40 °C. It can be seen that the fundamental amplitude measured at 15 °C is the lowest. The fundamental amplitude (A_1) measured at 40 °C increased significantly compared with that at 15 °C. Then, the fundamental amplitude also rises in the temperature decreasing from 15 °C to 0 °C. This may be due to the trend of water freezing in concrete pores increasing the acoustic impedance of concrete. The fundamental amplitude (A_1) at −20 °C does not change much from that at 0 °C. It may be the reason that the water in the concrete pores at 0 °C is in the state of a mixture of ice and water. Whereafter, the water in the concrete pores froze into ice at −20 °C. As shown in Figure 8, the temperature reduction can not increase the second harmonic amplitudes (A_2). The second harmonic amplitudes (A_2) also rise after temperature increases ranging from 0 °C to 40 °C. Interestingly, the rising trend of the second harmonic amplitudes (A_2) is greater when the temperature ranges from 0 °C to 10 °C than when the temperature ranges from 10 °C to 40 °C. It can be seen clearly that the second harmonic amplitudes (A_2) of the three kinds of specimens decrease quickly with the temperature reduction. The amplitudes of the second harmonic decrease slightly at a temperature between −20 °C and 0 °C. Significantly, the amplitude of the second harmonic at low-temperature decreases by nearly 80% compared with that at 10 °C.

It can be seen from Figures 9–12 that the relative nonlinear parameters (A_2/A_1^2) of carbonated concrete are changed by the influence of temperature during the carbonation time. The relative nonlinear parameters (A_2/A_1^2) measured at any temperature decrease with carbonation time. The relative nonlinear parameters of carbonated concrete measured at 0 °C compared with that at 10 °C decrease significantly (around 80 ∼ 90%), whereas the relative nonlinear parameters of carbonated concrete measured at 15 °C to 40 °C increase slightly. The relative nonlinear parameters also increase slightly when the temperature is at −20 °C to 0 °C, but it does not change too much. Table 4 shows the measured relative nonlinear parameters in different temperatures at the carbonation day 14.

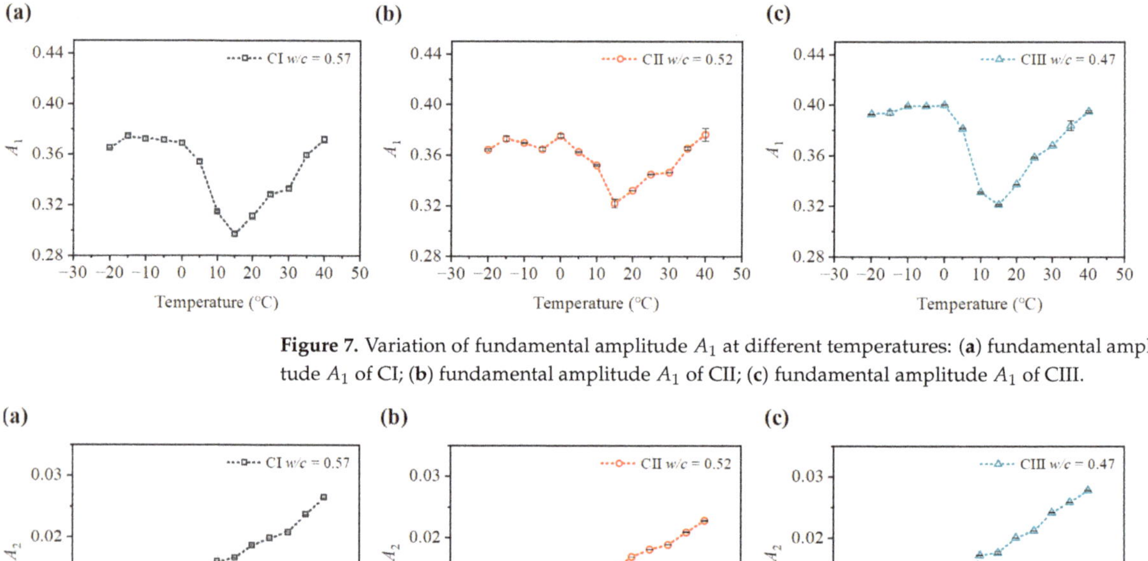

Figure 7. Variation of fundamental amplitude A_1 at different temperatures: (**a**) fundamental amplitude A_1 of CI; (**b**) fundamental amplitude A_1 of CII; (**c**) fundamental amplitude A_1 of CIII.

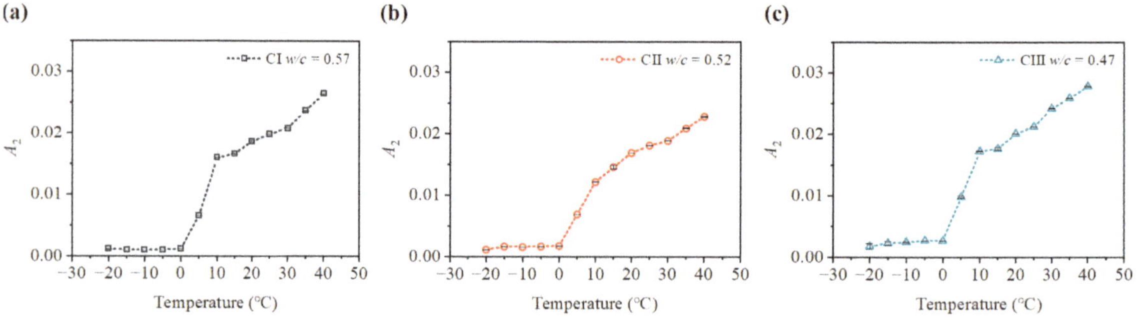

Figure 8. Variation of second harmonic amplitude A_2 at different temperatures. (**a**) second harmonic amplitude A_2 of CI; (**b**) second harmonic amplitude A_2 of CII; (**c**) second harmonic amplitude A_2 of CIII.

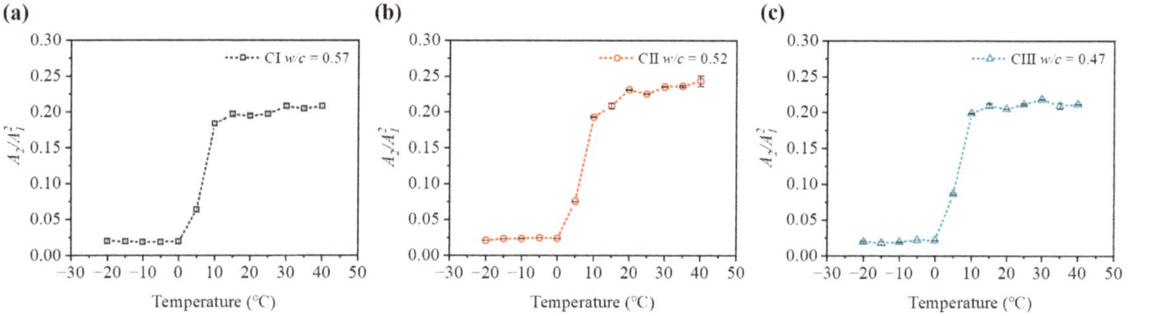

Figure 9. Variation of nonlinear parameter A_2/A_1^2 at different temperatures (day 3): (**a**) nonlinear parameter A_2/A_1^2 of CI; (**b**) nonlinear parameter A_2/A_1^2 of CII; (**c**) nonlinear parameter A_2/A_1^2 of CIII.

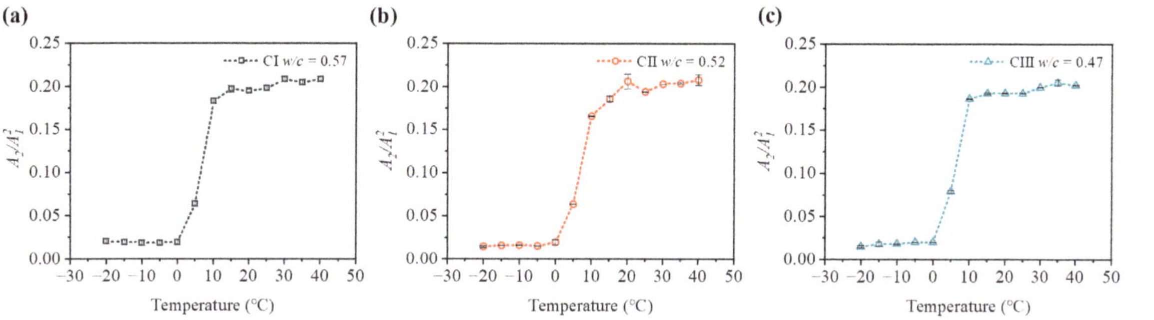

Figure 10. Variation of nonlinear parameter A_2/A_1^2 at different temperatures (day 7): (**a**) nonlinear parameter A_2/A_1^2 of CI; (**b**) nonlinear parameter A_2/A_1^2 of CII; (**c**) nonlinear parameter A_2/A_1^2 of CIII.

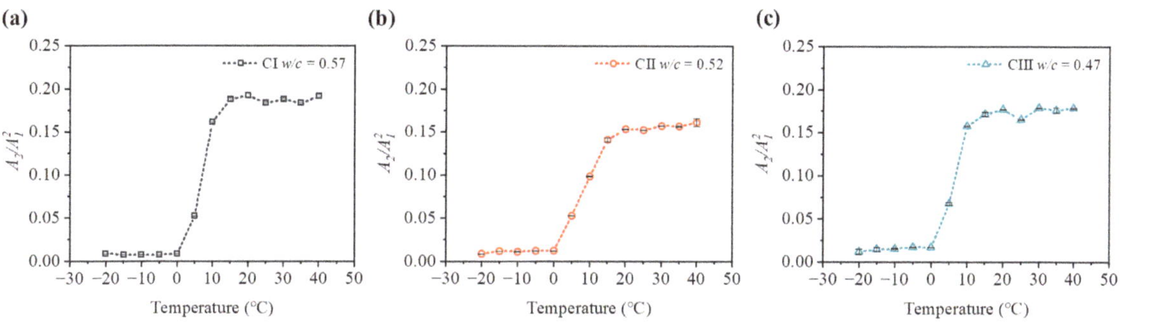

Figure 11. Variation of nonlinear parameter A_2/A_1^2 at different temperatures (day 14): (**a**) nonlinear parameter A_2/A_1^2 of CI; (**b**) nonlinear parameter A_2/A_1^2 of CII; (**c**) nonlinear parameter A_2/A_1^2 of CIII.

Table 4. The measured average relative nonlinear parameters.

Temperature (°C)	−20	−15	−10	−5	0	5	10	15	20	25	30	35	40
CI w/c = 0.57	0.00867	0.00842	0.00752	0.00843	0.00951	0.05216	0.16068	0.18743	0.19172	0.18288	0.18733	0.18284	0.19091
CII w/c = 0.52	0.01052	0.01169	0.01128	0.01135	0.01236	0.05288	0.0984	0.14073	0.15304	0.15215	0.15699	0.15659	0.16099
CIII w/c = 0.47	0.01362	0.01453	0.01512	0.01587	0.01564	0.06761	0.15755	0.17139	0.17661	0.16495	0.17859	0.17563	0.17809

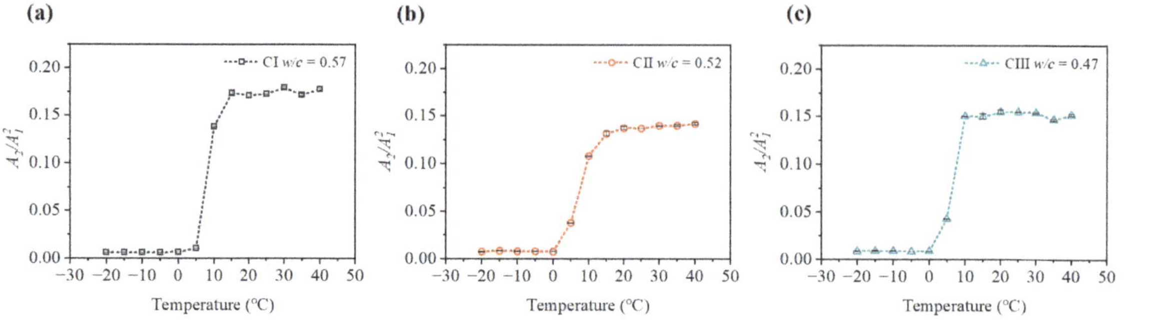

Figure 12. Variation of nonlinear parameter A_2/A_1^2 at different temperatures (day 21): (**a**) nonlinear parameter A_2/A_1^2 of CI; (**b**) nonlinear parameter A_2/A_1^2 of CII; (**c**) nonlinear parameter A_2/A_1^2 of CIII.

As shown in Figures 9–12, the values of the relative nonlinear parameters are more minor at 0 °C than at 10 °C, whereas the measured nonlinear parameters for the three concrete specimens change obviously with carbonation time in low-temperature environments as shown in Figures 13 and 14. In the same carbonation environment, the carbonation rate of concrete with high water–cement ratios is faster than that with low water–cement ratios [36]. The data in Table 3 follow this trend and can also be reflected from the changes of the relative nonlinear parameters in Figures 13 and 14. In addition, the variational degree of nonlinear parameter is greater for CI (w/c = 0.57) than for CII (w/c = 0.52) and CIII (w/c = 0.47). It means that the detection method in this study is suitable for the assessment of carbonated concrete in low-temperature environments as well.

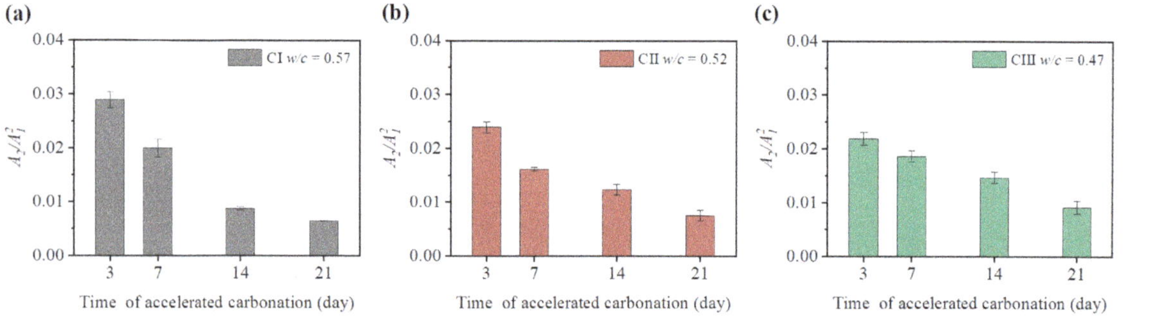

Figure 13. Variation of nonlinear parameter A_2/A_1^2 at 0 °C: (**a**) nonlinear parameter A_2/A_1^2 of CI; (**b**) nonlinear parameter A_2/A_1^2 of CII; (**c**) nonlinear parameter A_2/A_1^2 of CIII.

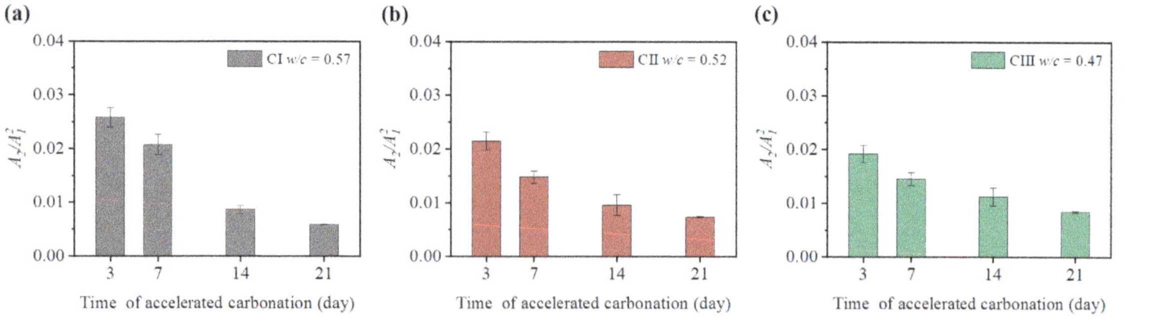

Figure 14. Variation of nonlinear parameter A_2/A_1^2 at −20 °C: (**a**) nonlinear parameter A_2/A_1^2 of CI; (**b**) nonlinear parameter A_2/A_1^2 of CII; (**c**) nonlinear parameter A_2/A_1^2 of CIII.

4. Conclusions

In this paper, the longitudinal nonlinear ultrasonic parameters of carbonated concrete were detected using an embedded composite piezoelectric transducer (ECPT) as the transmitter, and the effect of temperature on the nonlinear parameters of carbonated concrete is investigated. Conclusions are drawn from the experimental results as follows:

- Though only one ECPT was embedded in the concrete specimen as a transmitter, the nonlinear parameters of the ultrasonic radial propagation in carbonated concrete could also be obtained by the surface-mounted transducer fixed on the surface in which CO_2 was ingressed. This configuration can detect concrete carbonation with a monitoring network for a long time without damage when the surface-mounted transducer is permanently fixed on the structural surface.
- The relative nonlinear parameter of carbonated concrete can be affected by the water–cement ratio of the concrete. There is a more significant variation of the nonlinear parameters in the same carbonation time for the specimens with a high water–cement ratio than that with a low one. It indicates that the change of relative nonlinear parameters is closely related to the concrete carbonation depth.
- At the same ambient relative humidity (RH = 0.7), the fundamental amplitudes (A_1) and the second harmonic amplitudes (A_2) are increased by the increasing temperature from 15 °C to 40 °C, and the relative nonlinear parameters (A_2/A_1^2) between the two temperatures are also increased slightly. The fundamental amplitude of carbonated concrete at temperatures below 15 °C increases dramatically, while the second harmonic amplitude decreases sharply below 0 °C compared with that at 10 °C (fell around 80%). The relative nonlinear parameters of carbonated concrete measured at 0 °C compared with that at 10 °C decreases significantly (around 80~90%). The relative nonlinear parameters also increase slightly when the temperature is at −20 °C to 0 °C, but it does not change too much. As such, the relative nonlinear parameters in low temperatures below 0 °C are much smaller than at temperatures above 10 °C. It may be related to the water condensation in the concrete. In addition, the detection method in this study is also appropriate for the measurement of carbonated concrete in low-temperature environments, since the measured nonlinear parameters for the three concrete specimens change obviously with carbonation time.
- Considering the effect of temperature variations on the relative parameters of carbonated concrete, temperature compensation should be performed for the relative nonlinear parameters when continuously monitoring the concrete carbonation process.

It can be seen from the results that the relative nonlinear parameters change significantly at the temperature from 0 °C to 10 °C. The variation of the relative nonlinear parameter in this temperature interval needs to be studied further. Moreover, future studies could investigate the quantitative association between the relative nonlinear parameters and concrete carbonation depth for different kinds of concrete. In addition, the other factors in the measurements for carbonated concrete nonlinear parameters will be investigated in more detail.

Author Contributions: Conceptualization, J.Z., J.W., X.C. and R.Z.; methodology, J.Z. and J.W.; software, J.Z., X.C. and R.Z.; writing—original draft preparation, J.Z. and J.W.; writing—review and editing, J.Z., J.W., X.C. and R.Z. All authors have read and agreed to the published version of the manuscript.

Funding: This research received no external funding.

Institutional Review Board Statement: Not applicable.

Informed Consent Statement: Not applicable

Data Availability Statement: Not applicable.

Conflicts of Interest: The authors declare no conflict of interest.

References

1. Chen, J.; Jayapalan, A.R.; Kim, J. Rapid evaluation of alkali–silica reactivity of aggregates using a nonlinear resonance spectroscopy technique. *Cem. Concr. Res.* **2010**, *40*, 914–923. [CrossRef]
2. Payan, C.; Garnier, V.; Moysan, J.; Johnson, P.A. Applying nonlinear resonant ultrasound spectroscopy to improving thermal damage assessment in concrete. *J. Acoust. Soc. Am.* **2007**, *121*, EL125–EL130. [CrossRef] [PubMed]
3. Florian, O.B.; Jesús, N.E.; Vincent, G.; Cédric, P.; Narintsoa, R.; Benoît, D.; Christophe, M. Linear and nonlinear resonant ultrasonic techniques applied to assess delayed ettringite formation on concrete samples. *Constr. Build. Mater.* **2021**, *275*, 121545. [CrossRef]
4. Bouchaala, F.; Payan, C.; Garnier, V. Carbonation assessment in concrete by nonlinear ultrasound. *Cem. Concr. Res.* **2011**, *41*, 557–559. [CrossRef]
5. Eiras, J.; Kundu, T.; Popovics, J. Effect of carbonation on the linear and nonlinear dynamic properties of cement-based materials. *Opt. Eng.* **2015**, *55*, 011004. [CrossRef]
6. Kim, G.; Kim, J.; Kurtis, E. Quantitative evaluation of carbonation in concrete using nonlinear ultrasound. *Mater. Struct.* **2016**, *49*, 399–409. [CrossRef]
7. Kim, G.; In, C.; Kurtis, E. Nondestructive detection and characterization of carbonation in concrete. *AIP Conf. Proc.* **2014**, *1581*, 805–813. [CrossRef]
8. Dumoulin, C.; Karaiskos, G.; Deraemaeker, A. 8—Monitoring of crack propagation in reinforced concrete beams using embedded piezoelectric transducers. In *Acoustic Emission and Related Non-Destructive Evaluation Techniques in the Fracture Mechanics of Concrete*; Ohtsu, M., Ed.; Woodhead Publishing: Oxford, UK, 2015; pp. 161–175. [CrossRef]
9. Kong, Q.; Hou, S.; Ji, Q. Very early age concrete hydration characterization monitoring using piezoceramic based smart aggregates. *Smart Mater. Struct.* **2013**, *22*, 085025. [CrossRef]
10. Annamdas, V.; Yang, Y.; Soh, C. Impedance based concrete monitoring using embedded PZT sensors. *Int. J. Civ. Struct. Eng.* **2010**, *1*, 414–424. [CrossRef]
11. Deraemaeker, A.; Dumoulin, C. Embedding ultrasonic transducers in concrete: A lifelong monitoring technology. *Constr. Build. Mater.* **2019**, *194*, 42–50. [CrossRef]
12. Qin, L.; Liao, Z. Monitoring of cement hydration using embedded piezoelectric transducers. *Smart Mater. Struct.* **2008**, *17*, 055005. [CrossRef]
13. Gu, H.; Song, G.; Dhonde, H. Concrete early-age strength monitoring using embedded piezoelectric transducers. *Smart Mater. Struct.* **2007**, *15*, 1837–1845. [CrossRef]
14. Wang, D.S.; Zhu, H.P. Monitoring of the strength gain of concrete using embedded PZT impedance transducer. *Constr. Build. Mater.* **2011**, *25*, 3703–3708. [CrossRef]
15. Song, G.; Gu, H.; Mo, Y. Concrete structural health monitoring using embedded piezoceramic transducers. *Smart Mater. Struct.* **2007**, *16*, 959–968. [CrossRef]
16. Kocherla, A.; Duddi, M.; Subramaniam, K.V.L. Embedded PZT sensors for monitoring formation and crack opening in concrete structures. *Measurement* **2021**, *182*, 109698. [CrossRef]
17. Zhao, X.; Li, H.; Du, D. Concrete structure monitoring based on built-in piezoelectric ceramic transducers. *Proc. SPIE* **2008**, *6932*, 693208. [CrossRef]
18. Zhao, G.Q.; Zhang, D.; Zhang, L. Detection of Defects in Reinforced Concrete Structures Using Ultrasonic Nondestructive Evaluation with Piezoceramic Transducers and the Time Reversal Method. *Sensors* **2018**, *18*, 4176. [CrossRef]
19. Zou, D.; Liu, T.; Huang, Y. Feasibility of water seepage monitoring in concrete with embedded smart aggregates by P-wave travel time measurement. *Smart Mater. Struct.* **2014**, *23*, 067003. [CrossRef]
20. Li, Z.; Qin, L.; Huang, S. Embedded piezo-transducer in concrete for property diagnosis. *J. Mater. Civ. Eng.* **2009**, *21*, 085025. [CrossRef]
21. Liu, T.; Zou, D.; Du, C. Influence of axial loads on the health monitoring of concrete structures using embedded piezoelectric transducers. *Smart Mater. Struct.* **2017**, *16*, 202–214. [CrossRef]
22. Liao, W.; Wang, J.; Song, G. Structural health monitoring of concrete columns subjected to seismic excitations using piezoceramic-based sensors. *Smart Mater. Struct.* **2011**, *20*, 125015. [CrossRef]
23. Kaur, N.; Bhalla, S. Combined Energy Harvesting and Structural Health Monitoring Potential of Embedded Piezo-Concrete Vibration Sensors. *J. Energy Eng.* **2014**, *141*, D4014001. [CrossRef]
24. Talakokula, V.; Bhalla, S.; Ball, R.J. Diagnosis of carbonation induced corrosion initiation and progression in reinforced concrete structures using piezo-impedance transducers. *Sens. Actuators A* **2016**, *242*, 79–91. [CrossRef]
25. Chen, Y.Z.; Xue, X.W. Advances in the Structural Health Monitoring of Bridges Using Piezoelectric Transducers. *Sensors* **2018**, *18*, 4312. [CrossRef] [PubMed]
26. Zhang, H.; Li, J.; Kang, F. Real-time monitoring of humidity inside concrete structures utilizing embedded smart aggregates. *Constr. Build. Mater.* **2022**, *331*, 127317. [CrossRef]
27. Lim, Y.Y.; Smith, S.T.; Soh, C.K. Wave propagation based monitoring of concrete curing using piezoelectric materials: Review and path forward. *NDT E Int.* **2018**, *99*, 50–63. [CrossRef]
28. Chen, J.; Yang, C.; Wang, Q.; Xu, N.; Zhang, T. Nonlinear Ultrasonic Second Harmonic Assessment of Concrete Defects Based on Embedded Piezoelectric Sensors. *Res. Nondestruct. Eval.* **2020**, *4*, 254–270. [CrossRef]
29. Zhou, S.; Shul, Y. Nonlinear reflection of bulk acoustic waves at an interface. *J. Appl. Phys.* **1992**, *72*, 5070. [CrossRef]

30. Bender, F.A.; Kim, J.Y.; Jacobes, L.J.; Qu, J. The generation of second harmonic waves in an isotropic solid with quadratic nonlinearity under the presence of a stress-free boundary. *Wave Motion* **2013**, *50*, 146–161. [CrossRef]
31. Jhang, K.Y. Nonlinear Ultrasonic Techniques for Non-destructive Assessment of Micro Damage in Material: A Review. *Int. J. Precis. Eng. Manuf.* **2009**, *10*, 123–135. [CrossRef]
32. *GB/T 19801-2005/ISO 12714:1999*; Non-Destructive Testing—Secondary Calibration of Acoustic Emission Testing Transducer. Chinese Standard: Beijing, China, 2005.
33. Ono, K. Calibration Methods of Acoustic Emission Sensors. *Materials* **2016**, *9*, 508. [CrossRef]
34. Xiao, D.; Zhang, Z.; Ren, H.; Yang, B.; Pan, Q.; Xu, C. Calibration Principle for Acoustic Emission Sensor Sensitivity. In *Advances in Acoustic Emission Technology*; Springer: Berlin/Heidelberg, Germany, 2015; pp. 33–45.
35. Chang, J.J.; Yeih, W.; Huang, R. Mechanical properties of carbonated concrete. *Opt. Eng.* **2003**, *26*, 513–522. [CrossRef]
36. Venkat Rao, N.; Meena, T. A review on carbonation study in concrete. *IOP Conf. Ser. Mater. Sci. Eng.* **2017**, *263*, 032011. [CrossRef]

applied sciences

MDPI

Article

Non-Destructive Evaluation of the Quality of Adhesive Joints Using Ultrasound, X-ray, and Feature-Based Data Fusion

Elena Jasiūnienė [1,2,*], Bengisu Yilmaz [1,3], Damira Smagulova [1,2], Gawher Ahmad Bhat [1], Vaidotas Cicėnas [1], Egidijus Žukauskas [1] and Liudas Mažeika [1,2]

1 Ultrasound Research Institute, Kaunas University of Technology, K. Barsausko Str. 59, LT-51423 Kaunas, Lithuania
2 Department of Electronics Engineering, Kaunas University of Technology, Studentu St. 50, LT-51368 Kaunas, Lithuania
3 Acoustic and Electromagnetic Methods Division, Bundesanstalt für Materialforschung und -prüfung (BAM), 12205 Berlin, Germany
* Correspondence: elena.jasiuniene@ktu.lt

Abstract: The aim of this work is to achieve reliable nondestructive evaluation (NDE) of adhesively bonded aerospace components by developing novel multidimensional data fusion techniques, which would combine the information obtained by ultrasonic and X-ray NDE methods. Separately, both NDE techniques have their advantages and limitations. The integration of data obtained from pulse echo immersion ultrasound testing and radiography holds immense potential to help improve the reliability of non-destructive evaluation. In this study, distinctive features obtained from single techniques, traditional ultrasonic pulse echo testing, and radiography, as well as fused images, were investigated and the suitability of these distinctive features and fusion techniques for improving the probability of defect detection was evaluated. For this purpose, aluminum single lap joints with brass inclusions were analyzed using ultrasound pulse echo and radiography techniques. The distinctive features were extracted from the data obtained, and images of features obtained by both techniques were fused together. Different combinations of features and fusion algorithms were investigated, considering the desire to automate data evaluation in the future.

Keywords: data fusion; ultrasonics; radiography; adhesive joints; adhesive bond; non-destructive evaluation; interface defects; multiple reflections; signal modeling

Citation: Jasiūnienė, E.; Yilmaz, B.; Smagulova, D.; Bhat, G.A.; Cicėnas, V.; Žukauskas, E.; Mažeika, L. Non-Destructive Evaluation of the Quality of Adhesive Joints Using Ultrasound, X-ray, and Feature-Based Data Fusion. *Appl. Sci.* **2022**, *12*, 12930. https://doi.org/10.3390/app122412930

Academic Editors: Bogusław Łazarz, Grzegorz Peruń and Tangbin Xia

Received: 14 November 2022
Accepted: 14 December 2022
Published: 16 December 2022

1. Introduction

Recently, adhesive bonding technology has gained considerable attention in various industrial sectors, especially in the aviation, automobile, civil, and marine industries due to the weight reduction of the structure, uniform distribution of mechanical stresses, fairly high strength-to-weight ratio, and ability to join dissimilar materials [1–3]. Adhesive joints also provide improved performance of the structure and reduce fuel consumption, which benefits the economic sector. All the advantages and abilities of adhesive joints make this technology the most effective and practical for joining material components, in comparison to traditional bonding techniques, such as riveting, welding, etc. Adhesives are used to bond metal and composite materials to produce different structural components and advanced materials. In the aerospace industry, a substantial number of parts are fabricated using the adhesive bonding of metal and composite materials; however, the manufacturing processes are still unfavorably manual, resulting in various types of defects in the components [4,5]. Factors that can affect the durability of adhesive joints and cause the occurrence of different defects are moisture, poor curing, contamination, a constant structural load, impact force, and others [2,3]. Distinct types of defects, such as delamination, debonding, and weak or kissing bonds, can appear in adhesive joints. These internal defects are extremely critical for structural integrity due to their invisibility and

the strong influence on bonding quality. As a result, it becomes difficult to predict the mechanical behavior of the joints in the case of the presence of internal defects [6]. The presence of flaws or discontinuities in the bonded joints should be avoided at all costs for implementing them in critical aerospace components to provide environmental safety, safety of human life and health, and economic benefits.

Recently, many different investigations were carried out using different non-destructive testing (NDT) methods, such as ultrasonic [7], thermography [8,9], eddy current stimulated thermography [10], and others for the inspection of adhesive joints [11–14]. However, each non-destructive method has its own advantages and limitations upon which the selection of the appropriate method depends. Moreover, the choice of method depends on the structure to be inspected, since most of the methods become impractical after the structure is assembled due to the geometrical complexity of the structure, one-sided access, or even coverage with other layers of functional material [15].

The ultrasonic testing technique is widely and successfully used for the inspection of different structures. Ultrasonic non-destructive testing is advantageous and widely used in identifying and locating defects and can be utilized to evaluate the coherence of the adhesive bonded joints. This method was used by many researchers for the characterization of adhesion quality [16,17]. Adhesive joints can have complex geometry, roughness of the surface, curved surface, or interface layers of the component, which in addition can be not parallel, with varying thicknesses of the materials and their types. All these factors influence the ability of the technique used to detect defects. Zhang et al. [18] investigated the detection of disbonds in multi-layered structures using the laser ultrasonic technique. The technique used was pulse-echo mode for defect detection, and quantitative evaluation was used for defect sizing. Numerical simulations and ultrasonic inspection were also used for the detection of fatigue debonding in composite lap joints in the work of Liu et al. [19]. Liu et al. [20] used numerical and experimental investigations of ultrasonic guided waves. Disbonds were detected because of the variation in the arrival time of the ultrasonic waves. Many other researchers suggest the use of different ultrasonic methods for the detection of interface defects in multi-layered structures, including guided waves, laser ultrasonics, and non-linear ultrasonic evaluation [13,21–24]. Samaitis et al. proposed a novel approach based on the classical pulse echo technique and machine learning for the classification of weak bonds [25] As a result, ultrasonic techniques are extremely popular and more suitable for the inspection of multi-layered structures for detecting debonding-type defects.

X-ray radiography can offer significant spatial resolution and can investigate the overall thickness of the adhesive. For this reason, X-rays can image and quantify voids over the entire adhesively bonded joint, being an attractive alternative for different inspection techniques [26]. However, in terms of feature detection, contrast is required [27]. Good contrast is achieved when the attenuation coefficient of the X-rays differs significantly in the material inspected in contrast to the defect and/or there is a significant difference in the propagation path length of X-rays in the various materials (defect/non-defect) [28]. Therefore, X-ray radiography is especially good for the detection of volumetric defects, such as large voids [29] or inclusions. In the case of the inspections of thin structures, such as lap joints, where the thickness of the adhesive is small, this poses difficulties in detecting the defects in the joints where thickness along the propagation path of the X-rays is negligibly small (such as in the case of disbonds) and where the attenuation of X-rays in good/bad joints are almost the same.

In addition, it is difficult to detect interface defects with high reliability, since there are many influencing factors depending on the techniques used, samples under investigation, and interface defect types. Therefore, it is significantly important to select and develop a high-performance NDT technique that will satisfy the requirements of testing adhesive joints for high-reliability internal defect detection. The limitation of identifying appropriate NDT techniques has had a considerable influence on the general usage of adhesives, as it is widely assumed that if NDT techniques were reliable enough to ensure bond line integrity, adhesive bonding would be utilized significantly more than it is now.

Given that each non-destructive evaluation technique has varying sensitivity to different defects depending on the component inspected, multiple non-destructive evaluation methods are frequently applied to achieve improved, more reliable, and precise results, demonstrating the need for data fusion of multiple techniques [30]. X. E. Gros published the first book on the applications of NDT data fusion [31]. It provides a variety of research studies reflecting on how multi-sensor NDT is competent to integrate redundant data, distinguishing flaws more precisely and achieving a better signal-to-noise ratio (SNR) [32]. Later, a state-of-the-art survey of NDE data fusion techniques was provided by Liu et al. [33]. Each individual non-destructive method has its own advantages and limitations. For example, ultrasonic waves are best for the detection of delamination or disbond type of defects, while radiography is best for the detection of changes in density. Therefore, the most suitable single NDT method has to be selected according to the requirements to achieve the evaluation goal. Furthermore, it is observed that none of the single NDE techniques can fully characterize the object inspected, and it has its own uncertainty of the obtained result. Moreover, accuracy and reliability requirements are constantly increasing, especially in the field of aerospace and nuclear power industries. Therefore, to meet such requirements, multiple inspection techniques are needed. The fusion of multiple inspections benefits from the advantages of all individual non-destructive inspection methods and diminishes the uncertainty of the measurement result, as well as enhancing the signal. In addition, combining fusion techniques with advanced post-processing algorithms can improve the evaluation results. Fusion can help reduce the uncertainty of the results [33]. Currently, data fusion extends to a wide range of diverse applications [34]. B. Yilmaz et al. illustrated the implementation of the data fusion algorithm in advanced NDT techniques for the identification of defects in single lap composite adhesive joints [35]. Wang et al. have performed terahertz and X-ray image fusion to improve the evaluation of damage of composite materials [36].

In addition, when NDE is applied in industry, there is always a desire to automate the evaluation of data to avoid human factor and increase the probability of detection [37]. Therefore, one of the objectives of this work is to develop the process of determination and extraction of features for different inspection results for their further fusion, as well as to create a fusion algorithm to develop a reliable NDT technique for the investigation of multi-layered structures with bonding defects.

The aim of this work is to achieve reliable non-destructive evaluation (NDE) of adhesively bonded aerospace components by developing novel multidimensional data fusion techniques, which would combine the information obtained by ultrasonic and X-ray NDE methods. Separately, both NDE techniques have their advantages and limitations. For instance, ultrasonic NDE methods are sensitive to elastic properties and density and are good at detecting planar defects, such as lack of bonding/delamination; however, they may suffer from attenuation or diffraction. X-rays are very sensitive to changes in density and volumetric defects, but their performance on planar defects is limited. The integration of data obtained from pulse-echo immersion ultrasound testing and radiography holds immense potential for helping to improve the reliability of non-destructive evaluation. The objective of this investigation was to study and compare distinctive features obtained from single techniques, traditional ultrasonic pulse echo testing and radiography, and fused images, as well as to evaluate the suitability of different features and fusion techniques to improve the probability of defect detection.

On the basis of the data fusion algorithms, this article presents feature-based image fusion of ultrasonic and radiography testing. For this purpose, single lap aluminum joints with brass inclusions were analyzed using ultrasound pulse echo and radiography techniques. Then, different features were extracted from the obtained data. Then, images of the features of both techniques were fused together. Different combinations of features and fusion algorithms were investigated considering the desire to automate data evaluation in the future. The discussion of the qualitative and quantitative analysis obtained is presented in the discussion section, followed by the conclusions.

2. Materials and Methods

The workflow of the NDE of the adhesive joints using ultrasonic, radiography, and data fusion of both is presented in Figure 1. It consists of the following steps:

- Data acquisition (using ultrasonic and radiography techniques);
- Modelling of ultrasonic signal to determine the zone of interest in ultrasonic A-scans;
- Feature extraction;
- Feature-based data fusion;
- Evaluation of results.

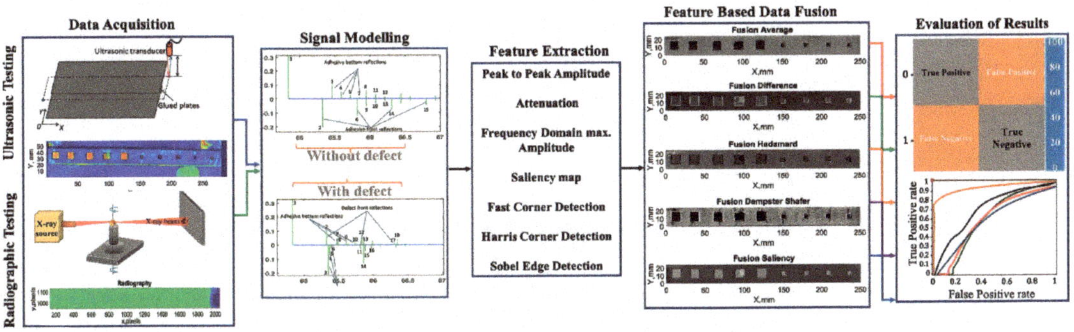

Figure 1. Schematic algorithm of the workflow.

2.1. Sample Description

The aluminum–epoxy adhesive single lap joints were manufactured by COTESA GmBH, Germany. For aluminum–epoxy joints, adherend was selected as 1.6 mm sheet 2024 aerospace aluminum alloy. The aluminum plate shown in Figure 2 had dimensions of 280 × 215 × 1.6 mm. A 3M Scotch-Weld AF 163-K red structural adhesive epoxy film with a thickness of 0.16 mm was used as adhesive. The lap joint was 25 mm wide. Following the necessary surface preparation, the epoxy film was placed on the adherend. There were 10 brass inclusions placed on the adhesive of different dimensions—5 large inclusions with a size of 12.7 mm, while the space between the consecutive defects is approximately 12.75 mm. On the other side, five smaller brass inclusions of size 6.35 mm were equally distributed with the space between the consecutive defects, approx. 18.05 mm. In further investigations, four out of five smaller defects were investigated because the holder was shadowing one of the defects. The schematics of the sample with the size and position of the defects is shown below in Figure 2. The thickness of the double-sided brass inclusion was 0.05 mm.

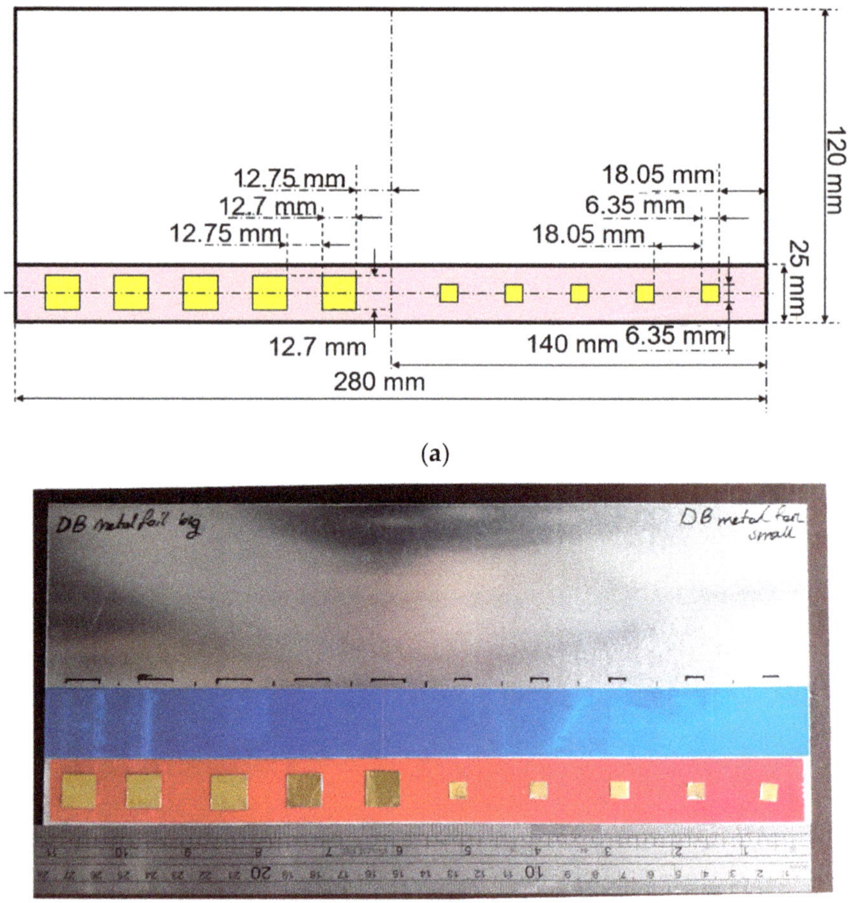

Figure 2. Single lap adhesive joints with brass inclusions prior to bonding: (**a**) schematic; (**b**) aluminum–epoxy joint.

2.2. *Investigation Techniques*

The aluminum single lap joint was examined using pulse-echo immersion ultrasonic and radiographic testing, and then the fusion of the results obtained by these two techniques was performed.

2.2.1. Immersion Ultrasonic Testing

The investigation of the aluminum adhesive lap joint with brass inclusion was carried out in an immersion tank using an Olympus V328-SU focused transducer with a central frequency of 15 MHz and a diameter of 10 mm. The far-field distance between the transducer and the sample top surface was kept fixed at 48 mm. The transducer was positioned in such a way that the emitted beam was perpendicular to the sample surface. Investigation of the single-lap joint was carried out on the entire bonded region with a 0.2 mm scan step and 1422 × 289 scanning points. The schematics of the ultrasonic pulse-echo immersion testing is shown in Figure 3.

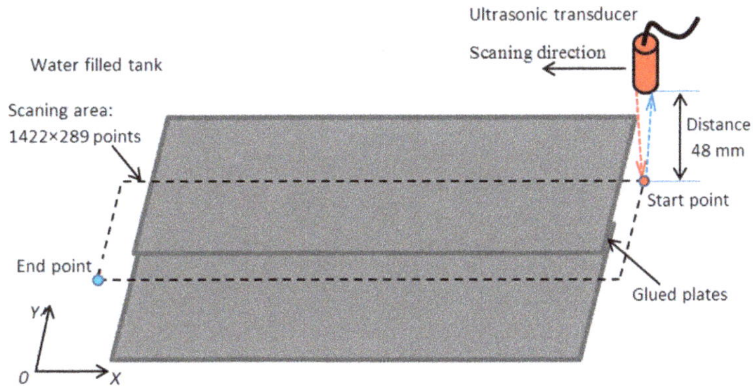

Figure 3. Experimental set-up of ultrasonic pulse-echo immersion.

2.2.2. Radiographic Testing

The adhesive bonded single lap joint was scanned using radiography as well. For this, the Rayscan 3D CT system was used. In this investigation, a 225 kV microfocus tube was used. The detector of the computed tomography system was a 2048 × 2048 flat matrix detector equipped with a scintillator. The schematic of the experimental set-up of the radiographic testing is shown in Figure 4. During the measurements, an X-ray source irradiated the sample at a voltage of 150 kV and a current of 300 µA with the cone beam and a 2D image was recorded at the detector with the integration time of 1500 ms. The focal spot was 47 µm, while the voxel size was 145 µm.

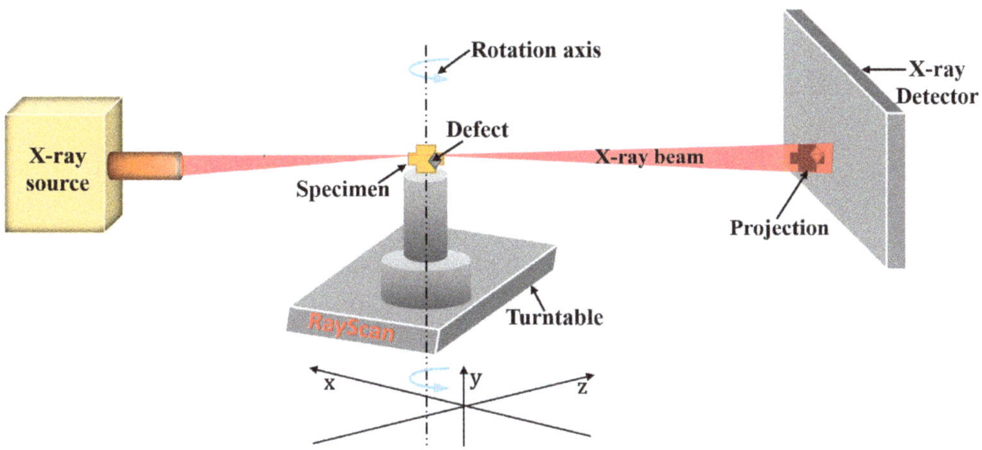

Figure 4. Schematic of experimental set-up for radiographic inspection.

2.3. *Ultrasonic Signal Modelling*

Modelling was performed to study the behaviour of ultrasonic waves propagating through the layers of adhesively bonded aluminum to aluminum with brass inclusion defects located between metal and adhesive, as well as to study their propagation and reflection paths from the sample interfaces to determine time instances of each for determination of gates to be used when selecting the part of the signal of interest for determination of features. According to our previous work [3], it was discovered that the analysis of multiple interface reflections improves the probability of defect detection. Therefore, time instances of ultrasonic impulses that correspond to multiple interface reflections were studied in this part.

The model that complies with the sample under investigation was modelled in MatLab software. Sample and medium characteristics are presented in Table 1. The distance between the focused transducer and the sample surface was 48 mm.

Table 1. Characteristics of the sample materials and medium.

Material	Thickness, mm	Density, kg/m^3	Ultrasound Velocity, m/s
Water	48.5	998	1496
Aluminum	1.6	2780	6532
Adhesive	0.16	1214	2433
Brass inclusion	0.05	8730	4430

As a result of modelling, pulse responses of each reflection from the structure boundary were obtained. Images with impulse plots characterizing the arrival time of each signal reflected from different sample boundaries with and without defects are presented below in Figure 5. Additionally, in Figure 6 the propagation paths of the ultrasonic wave for each signal impulse reflected from the sample interfaces, as well as their time moments for the non-defective and defective options, are presented.

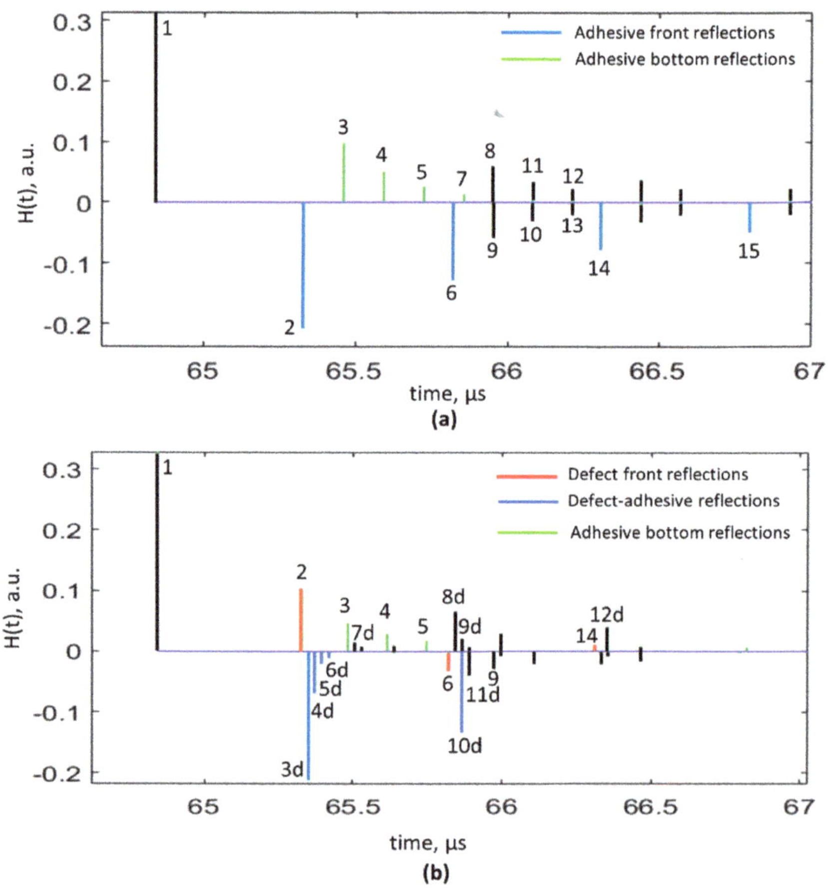

Figure 5. Impulses characterizing each signal reflection from the boundaries of the sample: (**a**) without defect; (**b**) with defect.

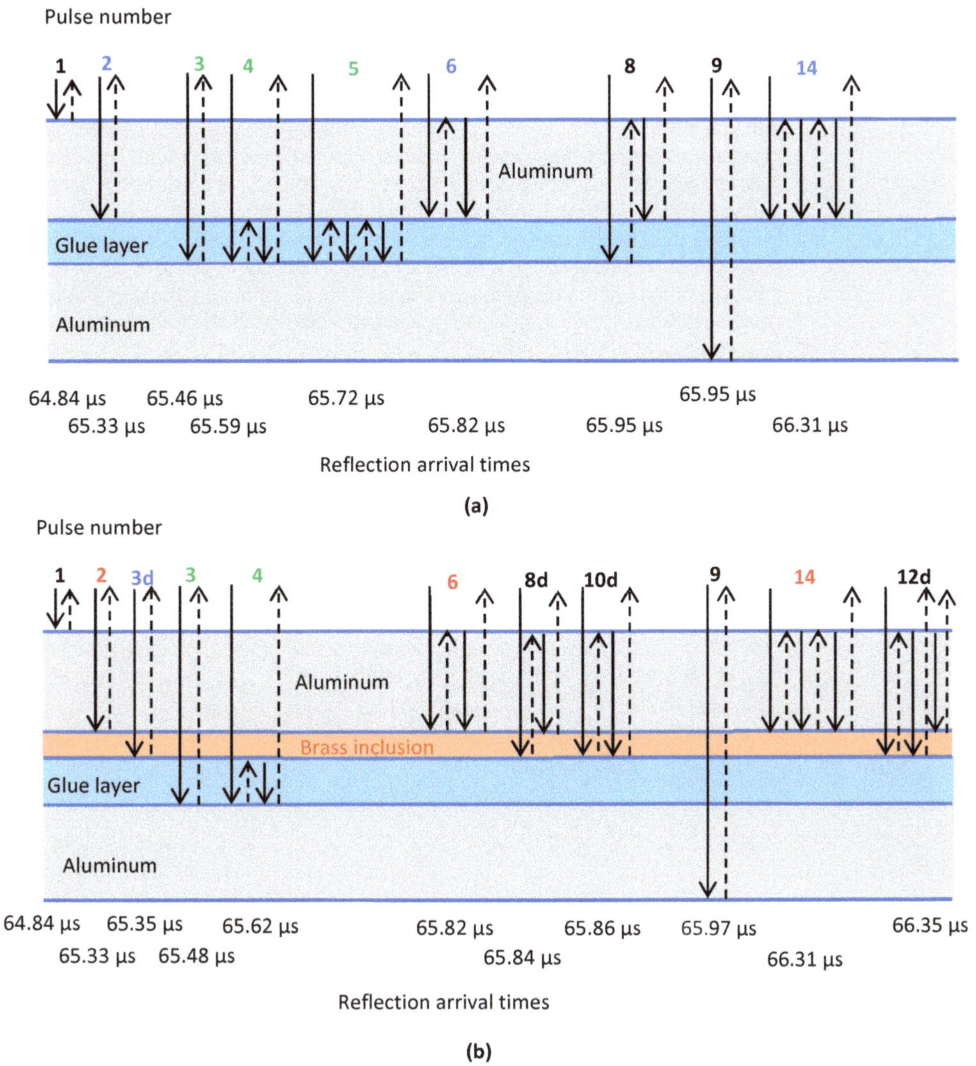

Figure 6. Ultrasonic wave propagation paths through the boundaries and time instances for: (**a**) not defected and (**b**) defected options.

According to obtained plots of impulses, the time moments of multiple reflected signals from the sample interfaces coincided: aluminum–adhesive interface (impulses 2, 5, 7) in the case of the sample without defect and aluminum–brass inclusion (impulses 2, 5, 7) for the sample with defect. In addition, for the sample with defect, there was a change in phase, for multiple interface reflections: aluminum–brass inclusion, which differed from the option of sample without defect. In addition, additional wave reflections from the brass inclusion bottom appear and overlap with other reflections from the boundaries.

To obtain time diagrams of the signals propagating in the investigated structure, the convolution of the pulse response and the excitation signal was performed. The excitation signal was a sine burst of 15 MHz frequency 2 periods with a Gaussian envelope. The time diagrams obtained in defect-free and defective objects are presented in Figure 7a,b, respectively. In the case of defect-free objects, individual pulses reflected from different material boundaries were easily distinguished. Pulse no.1 was surface reflection, pulse no.2 was reflection obtained from aluminum–adhesive boundary, pulses no.3–5 were multiple reflections in adhesive layer, and pulse no.6 was the second reflection from aluminum–adhesive boundary. However, in the case of a defective sample, the analysis of the reflected signals (Figure 7b) was much more complicated: pulse no.1 was surface reflection, while pulse no.2 was the result of interference of pulses reflected from aluminum–brass, brass–adhesive, and adhesive–aluminum boundaries (see Figure 5b pulses 2–7). It can be observed that it was impossible to separate signals reflected from separate boundaries. However, there is a particularly important feature that allows us to determine the presence of the defect—the phases of second reflections were opposites. In case of defect-free sample wave, it reflected from aluminum–adhesive boundary, i.e., from material with lower acoustic impedance. In this case, the wave was reflected with opposite phase. In case of defective sample ultrasonic wave, it reflected from aluminum–brass boundary, i.e., from material with higher acoustic impedance. In this case, the wave is reflected with the same phase.

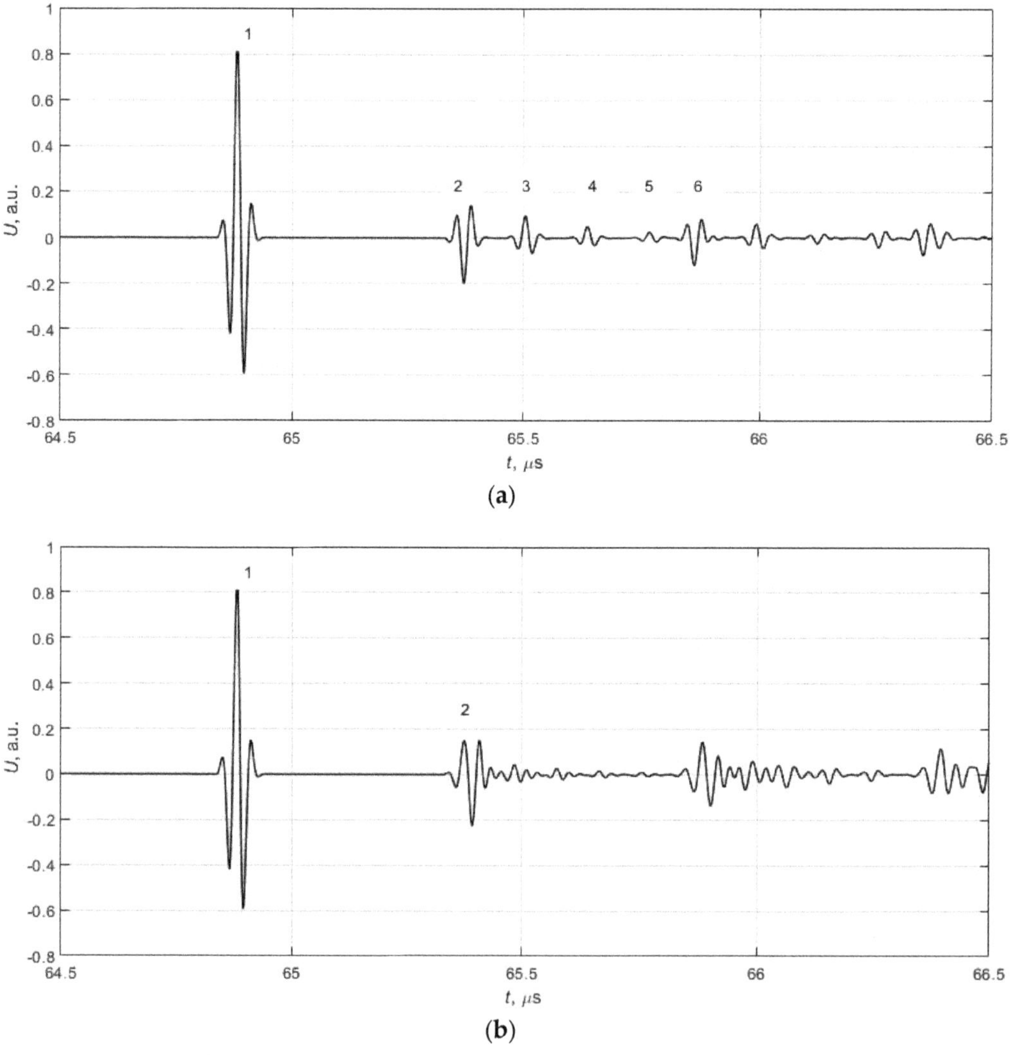

Figure 7. Impulses characterizing each signal reflection from the boundaries of the sample: (**a**) without defect, (**b**) with defect.

In terms of the impulse response, Figure 7 presents the main signals reflected from the different layers of the sample, whereas in Figure 8 the experimental data are presented with an overlayed theoretically calculated impulse response. Red spikes denote pulses with positive phase, blue with negative, and green is the combination of two pulses with the same flight time but opposite phases, as can be seen in Figure 8. The measured data were normalized according to the first plate backwall and adhesive interface reflection.

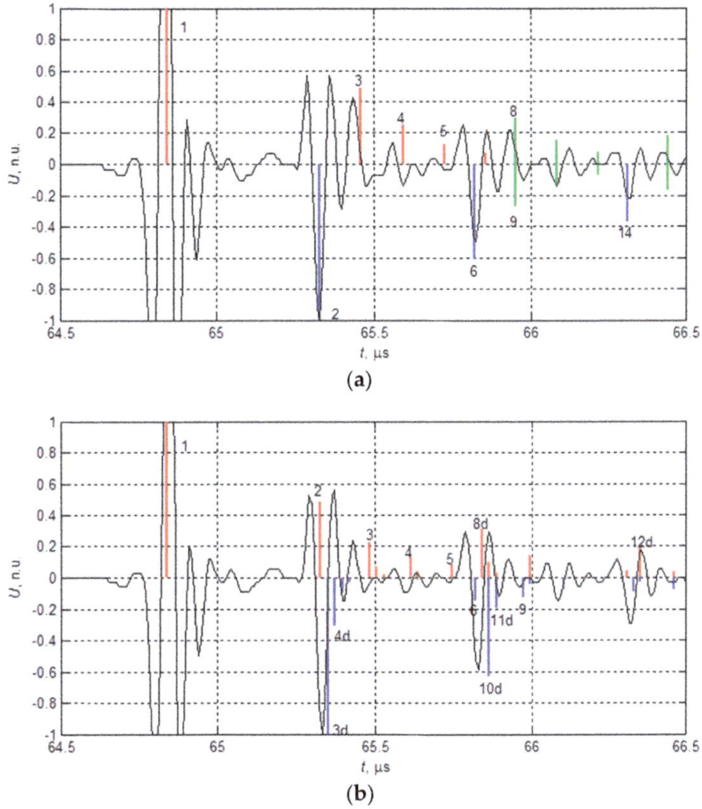

Figure 8. Measured signal and impulse response function overlay: sample part without defect (**a**) and defected (**b**). The black line denotes the measured ultrasonic signal, and the color bars denote the reflected impulse time moments presented in Figure 7.

According to the analysis of impulses and waves propagation paths, most the information of the defect was from time moment of wave reflection from the aluminum–defect interface. From the previous study [3], analysis of multiple reflections increases the probability of defect detection. Therefore, time intervals starting from the moment of aluminum–defect reflection to the next multiple reflections were used as a feature for setting the gates in the step of data fusion.

2.4. Data Fusion

It is not always possible to extract all relevant data from a single sensor and detect defects in a component. Often, there is a need to have information from two or more sensors/techniques for the better evaluation of the results. Data fusion can be used at various levels: pixel, signal, feature level [38]. In this work, feature-based fusion was chosen for the fusion of ultrasonic and radiographic testing results. In Figure 9, the workflow of the procedures applied is shown. Using two different measurement techniques, the different features were extracted. Then, different combinations of ultrasonic and radiographic features were supplied to a variety of fusion algorithms to test the suitability of features and fusion algorithms for automation of the defect detection in the future.

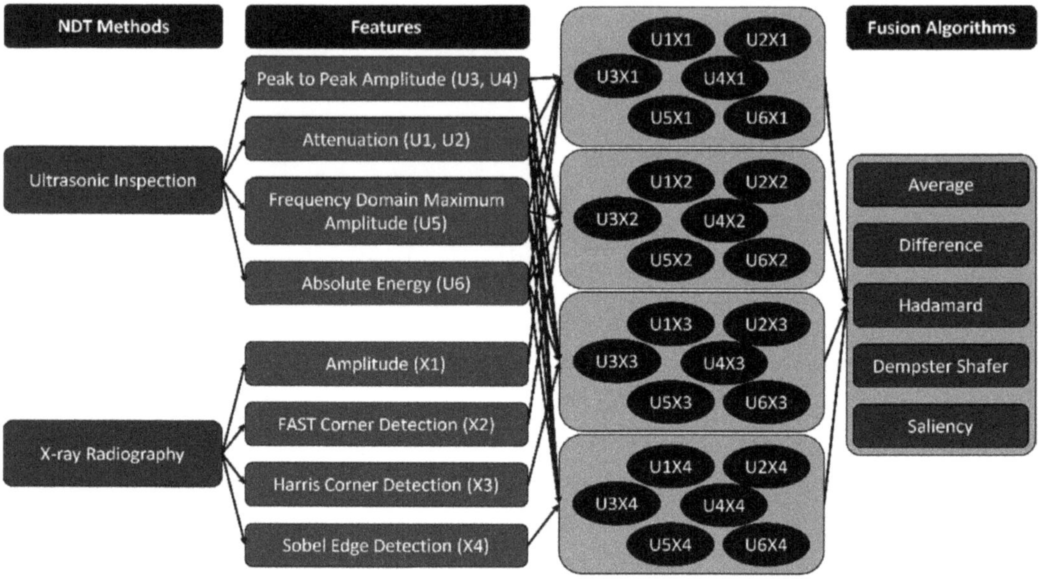

Figure 9. Workflow of data fusion.

2.4.1. Preparation of Data for Data Fusion

Before the fusion, the data needed to be prepared, undergoing several steps of preparation: noise reduction, coordinate matching, registration/interpolation, and normalization (see Figure 10). It was important to align the data obtained from different NDE techniques before the implementation of the data fusion. The raw data obtained from the experimental investigation of the single lap adhesive joint using ultrasonic and radiographic testing were aligned based on different parameters, such as noise removal, coordinate matching, interpolation, and normalization. The multi-step algorithm describing the pre-processing on the experimental data of two different techniques of non-destructive testing is shown in Figure 10.

The evaluation of bonding quality in the aluminum single-lap joint was conducted using separate experimental set-ups, and as such the data obtained had a local coordinate system. To perform data fusion on such varied systems, the coordinates of the experimental data from the investigation had to be matched. The data were aligned with the position of the bonding edges of the adhesive joint to match the coordinates. The data from immersion ultrasonic inspection had to be flipped to match the radiographic testing data and further interpolated to match the coordinates of the newly registered data shown. Finally, the values of the radiographic and ultrasonic data were normalized using linear scaling (0 to 1).

2.4.2. Selection of Features

Several distinctive features for both the ultrasonic and radiographic experimental data were used as input for data fusion. Previous work on the reliability of detecting debonding defects with ultrasonic techniques in adhesive joints showed that in addition to the amplitude of the ultrasonic echo, the attenuation and the amplitude in frequency domain play important roles in the defect contrast [2], allowing higher probability of defect detection. Therefore, in this study, in addition to the peak-to-peak amplitude, we also selected signal attenuation, frequency domain maximum amplitude, and absolute energy as features of the ultrasonic inspection results, as shown in Table 2. The features are extracted by applying the specified formula to 3D ultrasonic data (x, y, t), hence the dimensions are reduced to 2D (x, y). On the other hand, we applied 2D image processing techniques to obtain 2D (x, y) X-ray radiography results to create additional features, namely, FAST (features from accelerated segment test), Harris corner detection, and Sobel edge detection.

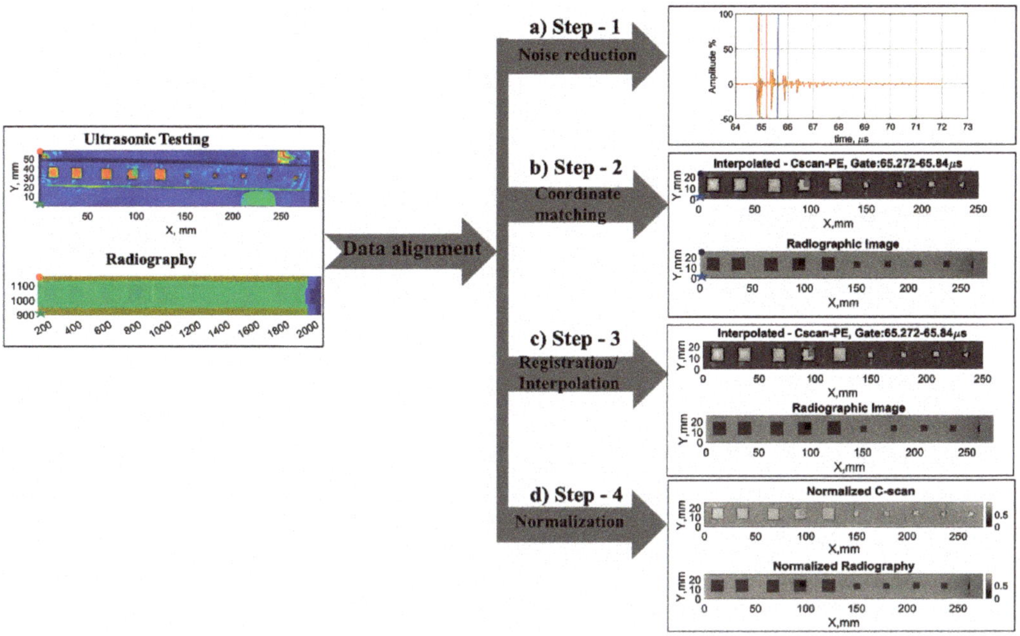

Figure 10. Measured data preparation for data fusion.

Table 2. Different features extracted from the measured data.

NDT Technique	Feature	Formula
Ultrasonic	Peak-to-peak amplitude	$\max(A(t)) - \min(A(t))$
Ultrasonic	Attenuation	$\frac{(\max(A(t_0))-\min(A(t_0)))-(\max(A(t_k))-\min(A(t_k)))}{\max(A(t_0))-\min(A(t_0))}$
Ultrasonic	Frequency domain maximum amplitude	$\max(FFT(A(t)))$
Ultrasonic	Absolute energy	$\sum_t A(t)^2$
X-ray	Amplitude	none
X-ray	FAST	$S_{p\to x} = \begin{cases} d, & I_{p\to x} \leq I_p - T \\ s, & I_p - T < I_{p\to x} < I_p + T \\ b, & I_p + T \leq I_{p\to x} \end{cases}$
X-ray	Harris Corner Detection	$E(u,v) = \sum_{x,y} w(x,y)\,[I(x+u, y+v) - I(x,y)]^2$
X-ray	Sobel Edge Detection	$G_x = \begin{bmatrix} 1 \\ 2 \\ 1 \end{bmatrix} * ([+1\,0-1] * A),$ $G_y = \begin{bmatrix} +1 \\ 0 \\ -1 \end{bmatrix} * ([1\,2\,1] * A),$ $\Theta = atan2(G_x,\ G_y)$

2.4.3. Data Fusion Algorithms

The main purposes of employing data fusion in different testing environmental conditions are to reduce the probability of error and improve the reliability of defect detection. A wide range of fusion algorithms can be used for the fusion of two-dimensional data, including wavelet-based combinations, Bayesian theory, Kalman filter, Dempster Shafer [39], and artificial neural networks [40]. This piece of work focused on the application of multiple feature-based fusion algorithms, namely averaging, difference, Hadamard, Dempster Shafer rule of combination, and pyramid fusion with saliency detection.

Simplest of all fusion algorithms, average, is performed at a pixel level—the average of feature matrices obtained from two sources is calculated. On the other hand, difference fusion is estimated based on the feature value difference on each pixel value from different NDT results. Hadamard fusion is calculated with the pixel-wise multiplication of same-size matrices.

The Dempster–Shafer theory was initially proposed by Arthur Dempster in the 1960s for reasoning uncertainties and later developed by Glenn Shafer in the 1970s [41,42]. The Dempster–Shafer (DS) theory is applicable to instances wherein non-probability uncertainty is evident. The DS evidence theory is based on mass, belief, and plausibility functions. The DS rule of combination is used to calculate the unique evidence mass (m) for the hypothesis by incorporating the evidence masses (m_1, m_2). In this study, the hypothesis was categorized as positive in case of defects and negative in case of non-defects.

Saliency detection is a hot topic that has recently piqued the interest of researchers from a wide range of disciplines. In our investigation, source images of ultrasonic and radiographic testing were decomposed into approximation base layer coefficients and detail layer coefficients using the multi-scale decomposition algorithm [43]. The salient features of the X-ray and ultrasonic features are extracted using a basic saliency map detection algorithm [44].

The data fusion algorithms used, their short descriptions, and mathematical formulas are provided in Table 3.

Table 3. Data fusion algorithms.

Fusion Algorithm	Description	Mathematical Formula
Average	at the pixel level, the average of feature matrices from two sources: UT [1] and RT [2]	$\frac{F_{UT}+F_{RT}}{2}$
Difference	at the pixel level, the difference of feature matrices from two sources: UT [1] and RT [2]	$F_{UT} - F_{RT}$
Hadamard t	pixel-wise multiplication of same-size feature matrices of two sources UT [1] and RT [2]	$(F_{UT} \circ F_{RT})_{ij} = (F_{UT})_{ij}(F_{RT})_{ij}$
Dempster –Shafer	evidence theory based on mass, belief, and plausibility functions [3]	$(m_1 \oplus m_2)(A) = \frac{1}{K-1} \sum\limits_{F_{UT} \cap F_{RT} = A \neq \varnothing} m_1(F_{UT})m_2(F_{RT})$
Saliency	the salient features of the X-ray and ultrasonic features are extracted using a basic saliency map detection algorithm	$\chi_m = \lvert \kappa_\eta(p,q) - \kappa_\mu(p,q) \rvert$

[1] UT stands for feature matrix of ultrasonic testing. [2] RT stands for feature matrix of radiographic testing. [3] F—feature matrix; κ_η is the mean filter output, and κ_μ is the median filter output of the source images p, q; $K = \sum\limits_{B \cap C = \varnothing} m_1(B)m_2(C)$.

3. Results

The aluminum adhesive lap joint with brass inclusions of varied sizes has been investigated using the techniques described above: ultrasonic immersion testing and radiography.

3.1. Features

From the results obtained, different feature images were extracted to be used later for data fusion. The selection of features was based on our previous investigations of adhesive joints [35]. The features described in the previous sections were used. In Figure 11, images for the following ultrasonic features are presented: peak-to-peak amplitude, attenuation, and frequency. In the case of radiographic testing, the following features were analyzed (see Figure 12): amplitude, FAST corners, Harris corners, Sobel edges.

3.2. Data Fusion

Data fusion was applied to the combination of ultrasonic and X-ray-based features, as shown in Table 2, using five fusion algorithms. The selected combinations of the results are presented in Figure 13.

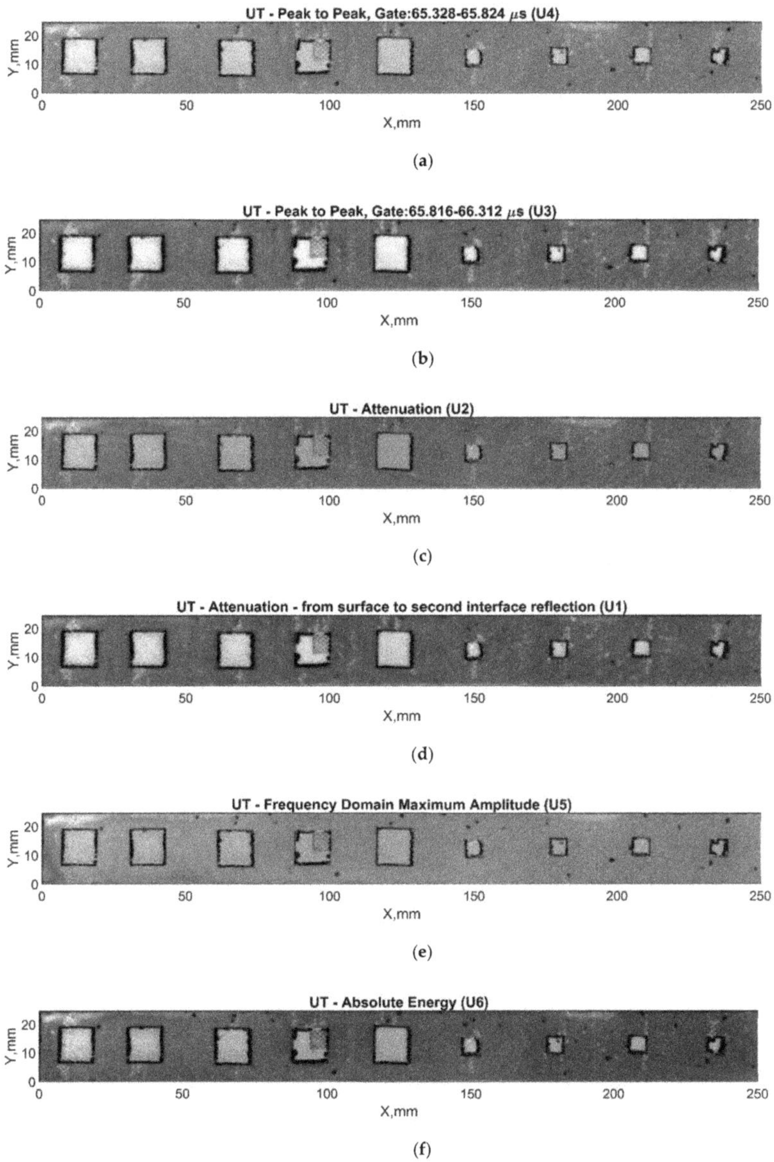

Figure 11. C-scans of different ultrasonic features: (**a**) peak-to-peak amplitude in the region from 2 to 11; (**b**) peak-to-peak amplitude in the region from 11 to 17; (**c**) attenuation in between gates 2 and 11; (**d**) attenuation in between gates 11 and 17; (**e**) frequency domain max amplitude; (**f**) absolute energy.

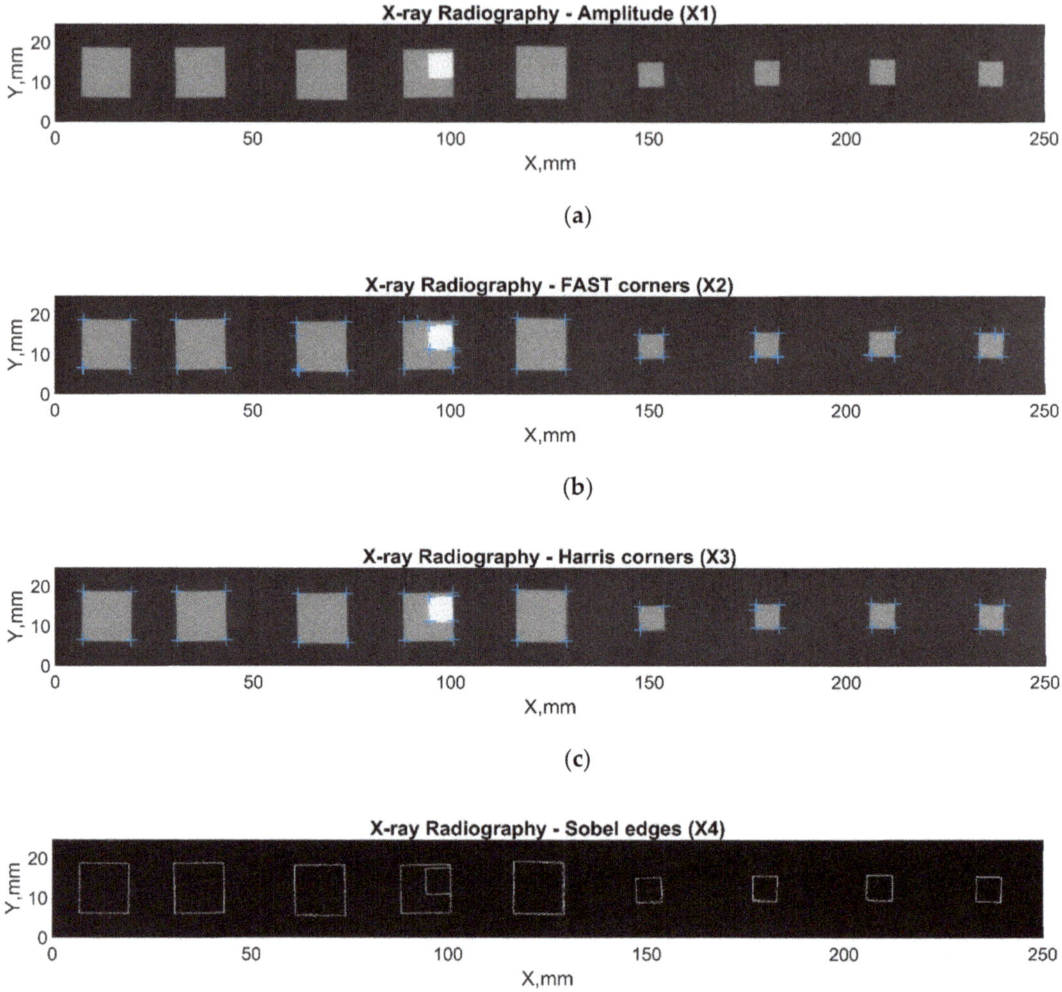

(**a**)

(**b**)

(**c**)

(**d**)

Figure 12. Features extracted from radiography images: (**a**) radiography; (**b**) FAST corners; (**c**) Harris corners; (**d**) Sobel edges.

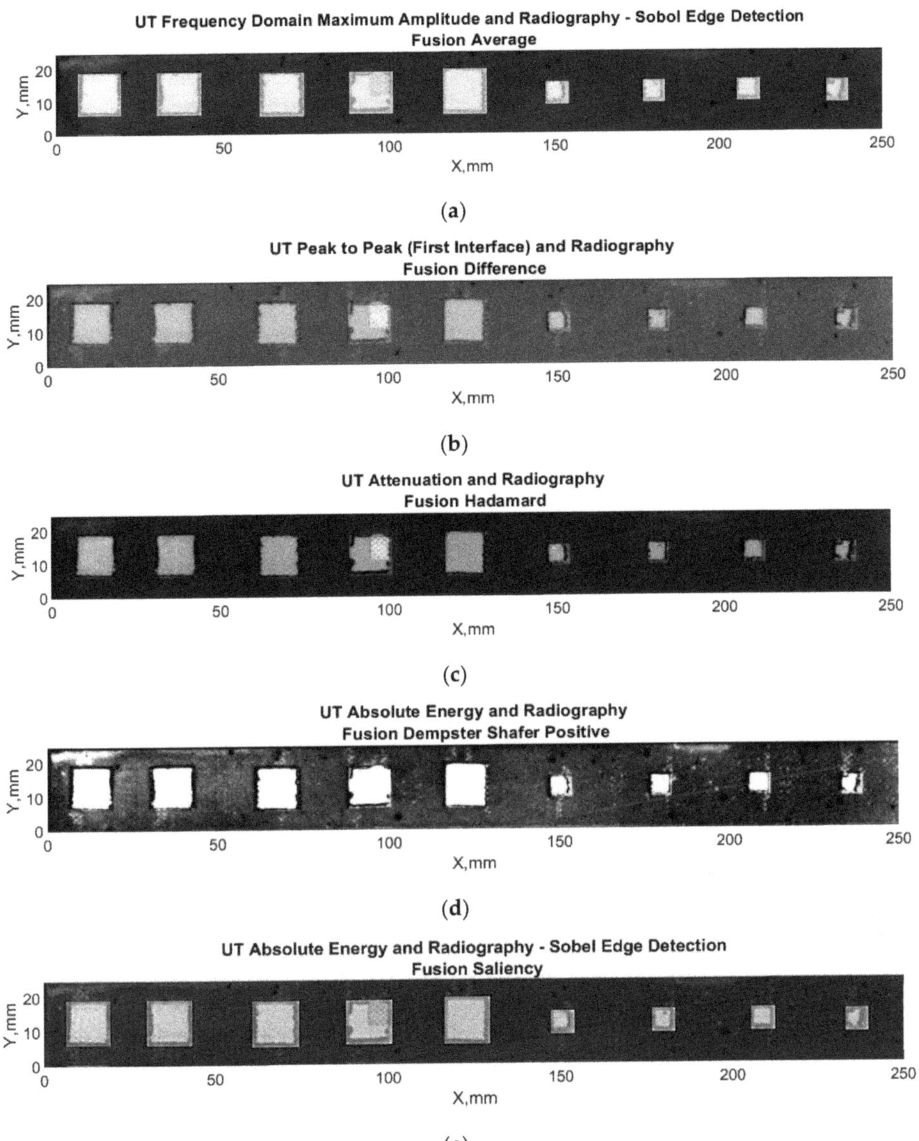

Figure 13. Examples of different combinations of fused features and used fusion algorithms: (**a**) U5X4 average; (**b**) U4X1 difference; (**c**) U2X1 Hadamard; (**d**) U6X1 Dempster Shafer Positive; (**e**) U6X4 saliency.

4. Comparison of Different Techniques and Discussion

The features and fused images of the X-ray and ultrasonic testing were analyzed using the response operating characteristics (ROC) curves and the area under curve (AUC) values. The photo of the sample with defects before bonding, as shown in Figure 2, was used as the reference image to evaluate the quality of the fusion results: the form and position of the defects was extracted from this photo to evaluate the accuracy of the features and fused images, in terms of inclusion-type defect recognition. First, registration (including interpolation and normalization) and coordinate matching have been applied to the image, considering the fusion data size. Then, the defect positions have been determined via postprocessing of the edge detection algorithm. The fusion results of each defect have been evaluated separately, considering this image. The results are evaluated using the area under curve estimations of response operating characteristic curves.

An arbitrary reference matrix was constructed based on the known location of the defects, with defected regions given the value 1 and perfectly bonded regions—0. By knowing the position of the defects, it was possible to create a mathematical formulation of a defect and a non-defect matrix that was used for comparing techniques. Every reference matrix was converted to binary so that the ROC curve could be used for the evaluation. Histogram-based segmentation was carried out. The segmented pixels in the matrix were then categorized into one of four categories: true positive (TP) indicates the presence of a defect in the defect position, false positive (FP) indicates the defect in a perfect bond region, true negative (TN) represents the defect-free good bond region, and false negative (FN) represents the defect-free position. The true positive and false positive rates for each segmented matrix were calculated in accordance with specific and sensitivity information and could be mathematically expressed as follows:

$$R_{TP} = \frac{N_{TP}}{N_{TP} + N_{FN}} \quad (X)$$

$$R_{FP} = \frac{N_{FP}}{N_{FP} + N_{TN}} \quad (X)$$

where R_{TP} is the true positive rate, R_{FP} is the false positive rate, N_{TP} is the number of true positives, N_{FN} is the number of false negatives, N_{FP} is the number of false positives, and N_{TN} is the number of true negatives.

Figure 14 shows the average performance of the distinctive features, according to the ROC curves and AUC values. It could be observed that in the case of the inclusion type of the defect, all features extracted from radiography data perform better according to the ROC curves and the AUC values.

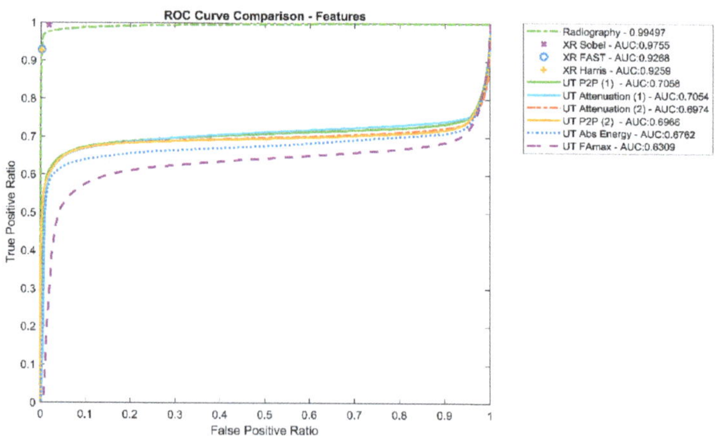

Figure 14. Overall defect detection performance of different features, according to ROC curves and AUC values.

The fused results were evaluated with distinctive features that were extracted from the results obtained by different techniques and are presented in Figure 15. The analysis of the results shows that from the fusion algorithms used, simple average and saliency gave the best results in the case of a brass inclusion type of defect. It is surprising that the performance of the Hadamard and Dempster–Shafer algorithms was much worse than expected. For ultrasonic features in terms of suitability for fusion, the best performance was achieved when using following features: ultrasonic signal attenuation (U2), peak-to-peak (U4), and absolute energy (U6). Although all radiographic features performed similarly, the final result was more influenced by the fusion algorithm.

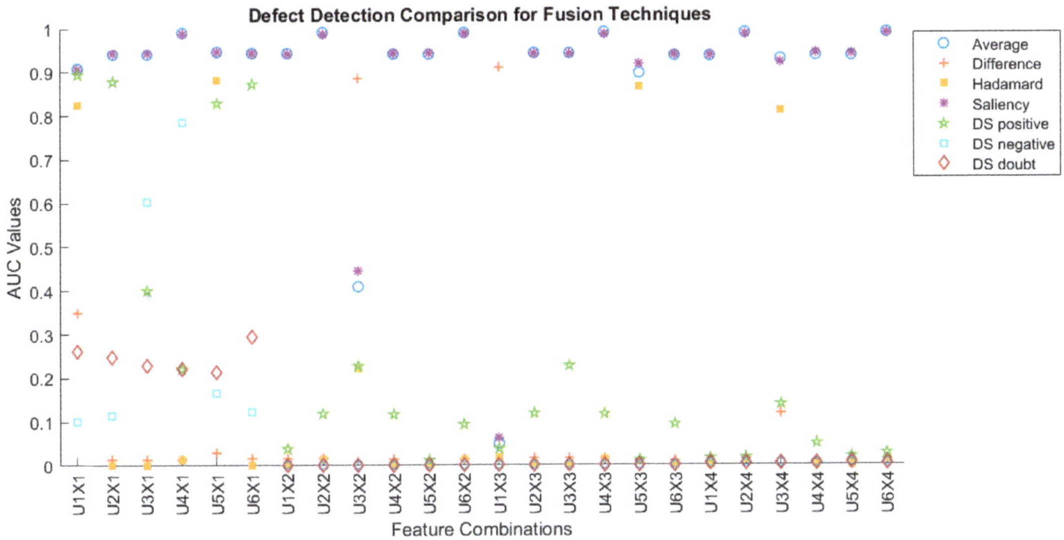

Figure 15. Evaluation of results using receiver operating characteristics (ROC) curve.

The AUC values for each feature are given in Figure 14, and the AUC values presented for each feature and fusion combination are presented in Figure 15. The combination of features refers to the values given in Figure 9. It could be observed that fusion with averaging and saliency performs best in almost all fusion combinations. Additionally, the defect detection performance of radiographic features (except for the radiography amplitude feature) can be improved by the fusion application of the best performing ultrasonic signals (U2, U4, U6).

In addition, the visual examination of the fusion results presented in Figure 13 reveals that fused results give additional information about the structure. It is obvious that radiography reveals the inclusion type defects very well; however, ultrasonic results provide additional information about the structure—additional disbond regions around the edges of the inclusions are visible.

5. Conclusions

The purpose of this investigation was to develop a feature-based data fusion technique to improve the probability of defect detection in the adhesive lap joints. Data were experimentally obtained using two common NDE techniques: ultrasonic pulse-echo and radiography. For this investigation, single lap adhesive-bonded aluminum joint with different sizes of brass inclusions were used. For the selection of the zone of interest in ultrasonic A-scans, theoretical analysis and modelling of impulse responses from interfaces in the sample were performed and multiple reflections inside the sample were analyzed, as well as different propagation paths. The modelling results were confirmed by experimental investigations of the propagation of ultrasonic waves. For the analysis, several distinctive features were extracted from the experimental ultrasonic and radiographic data and used as inputs for the data fusion. The comparison of extracted features shows that, at least in the case of the inclusion type of defect, the most promising features giving the highest contrast in terms of the defect/non-defect region from ultrasonic images are signal attenuation, peak-to-peak amplitude, and absolute energy; however, other features, especially frequency domain related, should not be disregarded. Feature-based image fusion of both techniques was also applied and evaluated using different data fusion algorithms. This helps to overcome the limitation of multimodal imaging and improves the prediction possibility and reconstruction of the missing information from the images of the single techniques. The fused image i provides intricate details of the defects in the sample on a finer scale. Although radiographic image contributes to the detection of the inclusion type of defect the most, the ultrasonic image provides extra information about additional types of defects, such as disbonds around the edges of the inclusions, which are not visible in the radiographic images. Of the fusion algorithms used, the best performance was achieved using average and saliency algorithms—AUC values of more than 0.99 are achieved.

Author Contributions: Conceptualization, E.J.; methodology, E.J.; software, B.Y., L.M., V.C. and G.A.B.; validation, E.J.; formal analysis, B.Y, D.S. and G.A.B.; investigation, E.J., D.S., E.Ž. and G.A.B.; resources, L.M. and E.J.; data curation, B.Y.; writing—original draft preparation, G.A.B. and D.S.; writing—review and editing, E.J., B.Y., V.C. and E.Ž.; visualization, B.Y., D.S., V.C., E.Ž. and G.A.B.; supervision, E.J.; project administration, E.J.; funding acquisition, E.J., L.M., V.C., E.Ž., D.S. and B.Y. All authors have read and agreed to the published version of the manuscript.

Funding: This research has received funding from the Research Council of Lithuania (LMTLT), agreement No S-MIP-22-5 and agreement No S-PD-22-25.

Institutional Review Board Statement: Not applicable.

Informed Consent Statement: Not applicable.

Data Availability Statement: The data presented in this study are available on request from the corresponding author.

Conflicts of Interest: The authors declare no conflict of interest.

References

1. Tserpes, K.; Sioutis, I.; Floros, G.; Moutsompegka, E. Numerical simulation of debonding of a composite-to-metal adhesive joint subjected to centrifugal load. *Eng. Fail. Anal.* **2022**, *136*, 106131. [CrossRef]
2. Yilmaz, B.; Smagulova, D.; Jasiuniene, E. NDT and E International Model-assisted reliability assessment for adhesive bonding quality evaluation with ultrasonic NDT. *NDT E Int.* **2022**, *126*, 102596. [CrossRef]
3. Smagulova, D.; Mazeika, L.; Jasiuniene, E. Novel Processing Algorithm to Improve Detectability of Disbonds in Adhesive Dissimilar Material Joints. *Sensors* **2021**, *21*, 3048. [CrossRef] [PubMed]
4. Bannister, M. Challenges for composites into the next millennium—A reinforcement perspective. *Compos. Part A Appl. Sci. Manuf.* **2001**, *32*, 901–910. [CrossRef]
5. Heinecke, F.; Willberg, C. Manufacturing-Induced Imperfections in Composite Parts Manufactured via Automated Fiber Placement. *J. Compos. Sci.* **2019**, *3*, 56. [CrossRef]
6. Fame, C.M.; Wu, C.; Feng, P.; Tam, L.-H. Numerical investigations on the damage tolerance of adhesively bonded pultruded GFRP joints with adhesion defects. *Compos. Struct.* **2022**, *301*, 116223. [CrossRef]
7. Yilmaz, B.; Asokkumar, A.; Jasiūnienė, E.; Kažys, R.J. Air-coupled, contact, and immersion ultrasonic non-destructive testing: Comparison for bonding quality evaluation. *Applied. Sci.* **2020**, *10*, 1–22. [CrossRef]
8. Tighe, R.C.; Dulieu-barton, J.M.; Quinn, S. International Journal of Adhesion & Adhesives Identi fi cation of kissing defects in adhesive bonds using infrared thermography. *Int. J. Adhes. Adhes.* **2016**, *64*, 168–178.
9. Palumbo, D.; Tamborrino, R.; Galietti, U.; Aversa, P.; Tatì, A.; Luprano, V.A.M. NDT & E International Ultrasonic analysis and lock-in thermography for debonding evaluation of composite adhesive joints. *NDT* **2016**, *78*, 1–9.
10. Yi, Q.; Tian, G.Y.; Yilmaz, B.; Malekmohammadi, H.; Laureti, S.; Ricci, M.; Jasiuniene, E. Evaluation of debonding in CFRP-epoxy adhesive single-lap joints using eddy current pulse-compression ther-mography. *Compos. Part B Eng.* **2019**, *178*, 107461. [CrossRef]
11. Sam-Daliri, O.; Faller, L.-M.; Farahani, M.; Zangl, H. Structural health monitoring of adhesive joints under pure mode I loading using the electrical impedance measurement. *Eng. Fract. Mech.* **2021**, *245*, 107585. [CrossRef]
12. Chen, H.; Nie, X.; Gan, S.; Zhao, Y.; Qiu, H. Interfacial imperfection detection for steel-concrete composite structures using NDT techniques: A state-of-the-art review. *Eng. Struct.* **2021**, *245*, 112778. [CrossRef]
13. Wojtczak, E.; Rucka, M. Damage imaging algorithm for non-destructive inspection of CFRP/steel adhesive joints based on ultrasonic guided wave propagation. *Compos. Struct.* **2022**, *297*, 115930. [CrossRef]
14. Zhang, K.; Zhou, Z. Quantitative characterization of disbonds in multilayered bonded composites using laser ultrasonic guided waves. *NDT E Int.* **2018**, *97*, 42–50. [CrossRef]

15. Nsengiyumva, W.; Zhong, S.; Lin, J.; Zhang, Q.; Zhong, J.; Huang, Y. Advances, limitations and prospects of nondestructive testing and evaluation of thick composites and sandwich structures: A state-of-the-art review. *Compos. Struct.* **2020**, *256*, 112951. [CrossRef]
16. Maev, R.G.; Titov, S. *Pulse-echo Ultrasonic NDE of Adhesive Bonds in Automotive Assembly*; ECNDT: Paris, France, 2006; pp. 1–8.
17. Vine, K.; Cawley, P.; Kinloch, A.J. The Correlation of Non-Destructive Measurements and Toughness Changes in Adhesive Joints during Environmental Attack. *J. Adhes.* **2014**, *77*, 37–41. [CrossRef]
18. Zhang, K.; Li, S.; Zhou, Z. Detection of disbonds in multi-layer bonded structures using the laser ultrasonic pulse-echo mode. *Ultrasonics* **2019**, *94*, 411–418. [CrossRef]
19. Liu, Y.; Zhang, X.; Lemanski, S.; Nezhad, H.Y.; Ayre, D. Experimental and numerical study of process-induced defects and their e ff ect on fatigue debonding in composite joints. *Int. J. Fatigue* **2019**, *125*, 47–57. [CrossRef]
20. Liu, M.; Chen, S.; Zheng, Z.; Yao, K.; Cui, F. In situ disbond detection in adhesive bonded multi-layer metallic joint using time-of-fl ight variation of guided wave. *Ultrasonics* **2020**, *102*, 106062. [CrossRef]
21. Titov, S.A.; Maev, R.G.; Bogachenkov, A.N. Pulse-echo NDT of adhesively–bonded joints in automotive assemblies. *Ultrasonics* **2008**, *48*, 537–546. [CrossRef]
22. Kumar, R.V.; Bhat, M.; Murthy, C. Some studies on evaluation of degradation in composite adhesive joints using ultrasonic techniques. *Ultrasonics* **2013**, *53*, 1150–1162. [CrossRef] [PubMed]
23. Shui, G.; Wang, Y.-S.; Huang, P.; Qu, J. Nonlinear ultrasonic evaluation of the fatigue damage of adhesive joints. *NDT E Int.* **2015**, *70*, 9–15. [CrossRef]
24. Kumar, S.A.; Sudheer, G. Influence of the oxide layer on the quality of bonding in adhesively bonded metallic structures by ultrasonic guided waves. *Int. J. Adhes. Adhes.* **2021**, *111*, 102981. [CrossRef]
25. Samaitis, V.; Yilmaz, B.; Jasiuniene, E. Adhesive bond quality classification using machine learning algorithms based on ul-trasonic pulse-echo immersion data. *J. Sound Vib.* **2022**, *546*, 117457. [CrossRef]
26. Ehrhart, B.; Valeske, B.; Bockenheimer, C. Non-destructive evaluation (NDE) of aerospace composites: Methods for testing adhesively bonded composites. In *Non-Destructive Evaluation (NDE) of Polymer Matrix Composites*; Woodhead Publishing: Sawston, UK, 2013; pp. 220–237. [CrossRef]
27. Garcea, S.; Wang, Y.; Withers, P. X-ray computed tomography of polymer composites. *Compos. Sci. Technol.* **2018**, *156*, 305–319. [CrossRef]
28. Rimasauskas, M.; Jasiuniene, E.; Kuncius, T.; Rimašauskienė, R.; Cicenas, V. Investigation of influence of printing parameters on the quality of 3D printed composite structures. *Compos. Struct.* **2022**, *281*, 115061. [CrossRef]
29. Zukauskas, E.; Jasiuniene, E.; Dragatogiannis, D.A.; Koumoulos, E.P.; Charitidis, C.A. Investigation of dissimilar metal joints with nanoparticle. *NDT E Int.* **2017**, *92*, 122–129.
30. Eleftheroglou, N.; Zarouchas, D.; Loutas, T. Prognostics of composite structures utilizing structural health monitoring data fusion. In Proceedings of the Conference Paper, European Workshop on Structural Health Monitoring, Manchester, UK, 10–13 July 2018.
31. Gros, X.E. *Applications of NDT Data Fusion*; Springer: New York, NY, USA, 2001; 277p, ISBN 978-1-4615-1411-4. [CrossRef]
32. Poularikas, A. (Ed.) ; *Handbook of Multisensor Data Fusion Theory and Practice*; The Electrical Engineering and Applied Signal Processing Series; CRC Press: Boca Raton, FL, USA, 2017; 813p.
33. Liu, Z.; Forsyth, D.S.; Komorowski, J.P.; Hanasaki, K.; Kirubarajan, T. Survey: State of the Art in NDE Data Fusion Techniques. *IEEE Trans. Instrum. Meas.* **2007**, *56*, 2435–2451. [CrossRef]
34. Harris, C.J.; Bailey, A.; Dodd, T.J. Multi-sensor data fusion in defence and aerospace. *Aeronaut. J.* **1998**, *102*, 229–244.
35. Yilmaz, B.; Ba, A.; Jasiuniene, E.; Bui, H.-K.; Berthiau, G. Evaluation of Bonding Quality with Advanced Nondestructive Testing (NDT) and Data Fusion. *Sensors* **2020**, *20*, 5127. [CrossRef]
36. Wang, J.; Xu, T.; Zhang, L.; Chang, T.; Zhang, J.; Yan, S.; Cui, H.-L. Nondestructive damage evaluation of composites based on terahertz and X-ray image fusion. *NDT E Int.* **2022**, *127*, 102616. [CrossRef]
37. Brierley, N.; Tippetts, T.; Cawley, P. Data fusion for automated non-destructive inspection. *Proc. R. Soc. A Math. Phys. Eng. Sci.* **2014**, *470*, 20140167. [CrossRef] [PubMed]
38. Goshtasby, A.A.; Nikolov, S. Image fusion: Advances in the state of the art. *Inf. Fusion* **2007**, *8*, 114–118. [CrossRef]
39. Khan, M.N.; Anwar, S. Paradox Elimination in Dempster—Shafer Combination Rule with Novel Entropy Function: Application in Decision-Level Multi-Sensor Fusion. *Sensors* **2019**, *19*, 4810. [CrossRef]
40. Wu, R.T.; Jahanshahi, M.R. Data fusion approaches for structural health monitoring and system identification: Past, present, and future. *Struct. Health Monit.* **2018**, *19*, 552–586. [CrossRef]
41. Shafer, G. *A Mathematical Theory of Evidence*; Princeton University Press: Princeton, NJ, USA, 2020; 314p, ISBN 9780691100425.
42. Dempster, A.P. Upper and Lower Probabilities Induced by a Multivalued Mapping. *Ann. Math. Statist.* **1967**, *38*, 325–339. [CrossRef]
43. Baohua, Z.; Xiaoqi, L.; Haiquan, P.; Ying, Z. Infrared Physics & Technology A fusion algorithm for infrared and visible images based on saliency analysis and non-subsampled Shearlet transform. *Infrared Phys. Technol.* **2015**, *73*, 286–297.
44. Bavirisetti, D.P.; Dhuli, R. Infrared Physics & Technology Two-scale image fusion of visible and infrared images using saliency detection. *Infrared Phys. Technol.* **2016**, *76*, 52–64.

applied sciences

MDPI

Article

Late Shelf Life Saturation of Golden Delicious Apple Parameters: TSS, Weight, and Colorimetry

Salma Kassebi [1], Csaba Farkas [2,3,*], László Székely [4], Attila Géczy [3] and Péter Korzenszky [5,*]

1 Doctoral School of Mechanical Engineering, Hungarian University of Agriculture and Life Sciences, 2100 Gödöllő, Hungary
2 Department of Vehicle Technology, Institute of Technology, Hungarian University of Agriculture and Life Sciences, 2100 Gödöllő, Hungary
3 Department of Electronics Technology, Faculty of Electrical Engineering and Informatics (VIK), Budapest University of Technology and Economics, 1111 Budapest, Hungary
4 Department of Mathematics and Modelling, Institute of Mathematics and Basic Science, Hungarian University of Agriculture and Life Sciences, 2100 Gödöllő, Hungary
5 Department of Farm and Food Machinery, Institute of Technology, Hungarian University of Agriculture and Life Sciences, 2100 Gödöllő, Hungary
* Correspondence: farkas.csaba2@vik.bme.hu (C.F.); korzenszky.peter.emod@uni-mate.hu (P.K.)

Abstract: This work aims to estimate the shelf life of Golden Delicious apple fruit stored at room temperature by determining the changing trend in color every week using a non-destructive measurement method. Moreover, the study will measure the changes in weight loss (Δm) and the total soluble solids (TSS) contained in the apple. The research focuses on the last stage of ripening and the effect of shelf life affecting consumer behavior; therefore, the examined fruits were picked at the end of the season and were stored at an ambient temperature and in controlled laboratory conditions for six weeks, at 24 °C under 60% RH relative. Color measurements were performed with a portable color sensor, which provided a simple and effective examination method in the case of an appropriate number of fruit samples. The findings showed a significant increase in TSS and weight loss over time. Color varying (ΔE) and chroma (C^*) parameters increased with prolonged storage duration, meaning that the color of the apples became darker and more color-saturated at the end of storage. While weight loss and TSS follow a linear tendency in the given storage period, the color changes in deteriorating fruits were characterized by exponential asymptotic regression. It was found that although the moisture content reaches its limit value during the drying process of apples, the saturation of color coordinates allows for determination of the limit parameters of consumption in the linear stage of weight loss. The saturation limits (70.97; 12.77; 56.34 CIELAB L^*; a^*; b^*), the dominant color part (b^*), and the growth rate of the saturation curves allow an accurate characterization of ripening from the visual aspect, thus determining the limit parameters of shelf life and improving the critical analysis factors affecting the life of fruits after harvest. By assessing color characteristics using a non-destructive technology, customers may quickly evaluate the quality of apples and make better decisions during their purchase.

Keywords: Golden Delicious apple; storage; weight loss; color; total soluble solids; shelf life; food engineering

Citation: Kassebi, S.; Farkas, C.; Székely, L.; Géczy, A.; Korzenszky, P. Late Shelf Life Saturation of Golden Delicious Apple Parameters: TSS, Weight, and Colorimetry. *Appl. Sci.* **2023**, *13*, 159. https://doi.org/10.3390/app13010159

Academic Editor: Grzegorz Peruń

Received: 17 November 2022
Revised: 14 December 2022
Accepted: 16 December 2022
Published: 23 December 2022

1. Introduction

Apple (*Malus × domestica* Borkh) belongs to the family of Rosaceae and sub-family Pomoideae and is one of the most widely produced and economically significant fruits worldwide. In 2020, the production of apples around the world reached 86,442,716 tons [1], and they are typically most appealing and health-promoting when gathered at their optimum ripeness. The apple is one of those fruits whose quality changes quickly over time while being stored, which causes variations in customer acceptance. First, it is assessed based on its appearance, including color and size, and then its total soluble solids content.

Over the years, research on post-harvest storage of fruits and vegetables has focused on studying fruit's appearance, quality, and mechanical properties under different conditions such as temperature or shelf life to meet consumer expectations [2–4].

Fruits include numerous critical nutrients and phytochemicals that may help prevent and reduce the risk of, for example, chronic illnesses such as heart disease, obesity, diabetes, some forms of cancer, inflammation, stroke, and septic shock [5,6]. It is popular among producers and consumers because of its excellent nutritional content and remarkable ecological resilience [7–9]. Fruits are a good source of water, soluble solids, sugars, organic acids, and dietary fiber [10,11]. For instance, an apple is a "healthy fast food" that replenishes the body's stores of vitamins, minerals, and trace elements. Over 600 essential and intermediate chemicals beneficial to human nutrition and health are found in apple fruits. [12].

Several widespread methods are applied to evaluate fruit quality. Visual examination and destructive procedures are generally used to assess product quality. [13] In recent years, specific spectroscopic and imaging techniques have been effectively used in food quality inspection to deal with destructive methods' challenges [14].

These non-contact imaging techniques are made possible by technological advancements in cameras and robust computer hardware, software, and analysis tools. They are now a crucial instrument in applications for quality evaluation [15,16].

Total soluble solids (TSS) are generally a crucial quality indicator linked to the texture and composition of apples and other fruits [17,18].

Water loss causes the product's weight to drop, causing the quality of the fruit to deteriorate. On the exterior of the apple fruit, there is a natural wax coating that acts like a protective barrier or coating, preserving the fruit as it grows. Fruit weight loss is determined by the skin's structure and the waxes' composition on the fruit's surface [19]. The moisture loss decreases the visual quality and contributes to the loss of turgor pressure and subsequent softening; in industrial terms, moisture loss is also an economic [20,21] Therefore, fruit weight measurement is critical in fruit quality evaluation too.

Although fruit may be kept fresh for a long time in cold storage, fruit storage after harvesting is one of the significant difficulties in the world today. The availability of cold storage facilities is entirely restricted in contrast to fruit production, which raises the price of the produce. Additionally, customers can buy the fruits at their small local market, where they are often kept at room temperature. As a result, it is crucial to evaluate the apple's quality while it is being stored at room temperature. The non-destructive method to measure the color can help consumers to assess the quality of fruit during purchasing [22]. Consumers' first delightful experience with fruits is a healthy, fresh, and colorful appearance, influencing the remainder of the observed sensations. The color is the initial point of contact [23] between the product and the consumer, which is crucial in purchasing [24]. Using a sophisticated picture processing system and RGB and hyperspectral photography, researchers [24] carefully graded and tracked the color progression of the subjects. Compared to RGB imaging, the hyperspectral approach showed more promise for objectively differentiating apple ripeness during shelf life.

The fruit's surface chlorophyll degradation occurs throughout the ripening phase; while-colored pigments, such as carotenoids (yellow) and anthocyanins (red), are created. Two colors may appear: one is the background color, yellow, and the other is the surface color, red. Golden Delicious apples have a faint pinkish tint ranging from green to yellow-green tones [24].

Some cultivars, such as Golden Delicious, have a long shelf life of roughly four weeks at room temperature [25], although the fruit softens and the peel color changes from green to yellow [26]. Analysis of the green-to-yellow skin color shift in Golden Delicious apples revealed that carotenoid production and chlorophyll degradation are responsible for the alterations [27,28]. Rutkowski et al. [29] proposed a non-destructive determination of the Golden Delicious ripening stage. However, they confirmed in the end that their development is still a destructive estimation due to their chlorophyll content assay of the apple peel.

Even though numerous research has been conducted during storage, there is little information about fruit color changes and the classification of ripening stages during shelf life. Cárdenas-Pérez et al. [26] showed that color changes during storage could be used as a classification approach. Apple quality deteriorates significantly during storage, impacting consumer acceptance [30,31]. According to the findings in the literature above, our approach focuses on highlighting the post-harvest late shelf life saturation of Golden Delicious color parameters, based on weight, TSS, and colorimetry data, which is raised as a novel approach in the field.

Due to its resemblance to the human perception of color, CIELAB color space is an appropriate choice to describe the color ingredients of fruits and vegetables [5,32]. The CIELAB color scale comprises L*, a*, and b* coordinates, with L* denoting brightness, a* denoting red or green, and b* denoting yellow or blue color perception.

Complex computer vision systems require sensor-based measurements, but flexible and frequent in-situ investigations are a continuous requirement during ambient storage. A calibrated portable device (in a laboratory environment at this research stage) can also be implemented to examine the color changes of different ripening stages, taking multiple, simple, and fast measurements on the surface of given fruits.

Besides examining fresh products, it is also essential to evaluate the results mentioned in the last stage of ripening, where the limiting parameters of shelf life can be observed with the saturation limits of the investigated changes in the crucial color parameters. In the case of room temperature storage, these circumstances can be achieved if the examined fruits are selected from the end of the season.

This study aims to characterize the ripening process with an applied sensor-based colorimetric approach. It aims to identify the changes in the weight, color, and TSS properties of Golden Delicious apple fruit during storage at room temperature, focusing on the changes in TSS and color parameters that objectively indicate the shelf life limitations of the given cultivar based on asymptotic regression. For the description of the observed limits of the parameters, the approach of exponential asymptotic regression was considered instead of power function [25] to characterize the weekly ripening process, and a color sensor-based measurement method was applied to improve the effectivity of such investigations further.

2. Materials and Methods

A fresh "Golden Delicious" variety of apple fruit was collected directly from a farm near Kecskemét, Hungary, in 2021. The fruits were all picked at a maturity stage corresponding to the commercial harvest date.

Golden Delicious apples were placed in six boxes, and each one separately contained 24 apple samples. All apples have an identity code (Example: B1A1, B for box A for apple). The apples were sorted manually for approximately similar sizes and a lack of any external damage. The average weight of the total fruit sample was 150 ± 12 g, and the average diameter size of the apple samples were d = 70 ± 2 mm; three measurements were applied for each sample for geometry and weight measurement. The apples were stored at ambient conditions (24 ± 1 °C) under relative humidity (RH 60%) for six weeks. The samples in the first three groups were examined with non-destructive methods: weight measurement (g) and color parameters (L*, a*, b* in CIELab color space [5,26,32]) acquisition. The fourth, fifth, and sixth batches were prepared to measure the total soluble solids (TSS), which required a destructive approach.

2.1. Measurement of Weight Loss

All of the apples were weighed before, during, and after the storage period, and three measurements were applied for each sample for the weight loss measurement. The same samples were assessed for weight loss at weekly intervals until the end of the experiment.

Weight loss Δm (%) was calculated using the following equation:

$$\Delta m = \frac{m_0 - m}{m_0} \cdot 100, \ [\%] \tag{1}$$

where m is the mass of apples (g) during storage and m_0 is apples' beginning mass (g). The samples of apple fruits in the first three groups (72 apples) were taken and weighed on a precision scale type KERN (KERN & SHON GmbH, Balingen, Germany, KERN PCB 3500-2, max.: 3500 g ± 0.01 g) to calculate the average fruit weight.

2.2. Measurement of Color

Color parameters were measured using a Nix Pro wireless color sensor (Nix Sensor Ltd., Hamilton, Ont, Canada, NixPro Mini). The device blocks ambient light, provides its own calibrated LED-based light source during measurement, and works with a proprietary sensor module. The Nix Pro Device has a circular shape with a 15 mm diameter and 45/0° measuring geometry. Two high-CRI LEDs designed explicitly for color reproduction provided the light source for the measurements. The instrument has <0.1 DE2000 repeatability.

The color data are used to match physical colors. The results were investigated using the CIELAB color system from the Nix pro color sensor application (App version v1.33, Nix Sensor Ltd., Hamilton, ON, Canada), which displays the RGB, CMYK, HEX, CIELAB, and XYZ values for each color. Three repeated measurement data of 72 apples were obtained for each storage time (from week one to week six) using the HunterLab colorimeter (illuminant D50 and 2° standard observer). The color was recorded using a CIELAB uniform color space (Lab), where L* indicates lightness, a* indicates chromaticity on a green (−) to red (+) axis and b* chromaticity on a blue (−) to yellow (+) axis. We calculated color parameters from the L*, a*, and b* data. ΔE is the difference between two colors represented by two points in the CIELAB color space and chroma (C*), which expresses the degree of color for an area viewed with its brightness.

$$\Delta E = \sqrt{(L_0^* - L^*)^2 + (a_0^* - a^*)^2 + (b_0^* - b^*)^2}, \quad (2)$$

$$C^* = \sqrt{a^{*2} + b^{*2}}, \quad (3)$$

where L_0^*, a_0^*, and b_0^* represent the color data based on the samples, and L*, a*, and b* indicate the color of the measured instantaneous data.

At the end of the storage process, the color of the apples will always turn into some shade of brown, which can be considered a final state or limiting color. Since the a* and b* coordinates of green and yellow shades are less than the coordinates of brown shades, the color change process can thus be considered as a saturation or limited growth process, pointing to asymptotic regression analysis.

2.3. Measurement of Total Soluble Solids (TSS)

The apple juice was extracted from the fruit without peeling using a hand mixer (BRAUN, Rubi, Barcelona, Spain, MODEL: MR-5550 BC-HC, 600 W + Turbo, 13,500 RPM) with stainless steel blades to measure the total soluble solids content. The extraction juice takes one minute for all fruit samples. The TSS of juice was measured using a refractometer with automatic temperature compensation. The digital hand refractometer (Ebro Electronic GmbH, Ingolsadt, Germany, Model: DR-10) has a measurement range between 0 and 54 °Brix within the +5 ... +60° sample temperature; the instrument works between the +5 ... +40° temperature range with ±0.2 °Brix/ ±1 °C accuracy and 0.1 °Brix/0.1 °C resolution. The results were expressed in °Brix (±0.2).

2.4. Statistical Analysis

All the quality attributes in each experiment on every apple were measured three times. The results were compared by one-way analysis of variance (ANOVA). Deviations between groups were tested with Levene's test. For post hoc tests, Dunnett's T3 test or Tukey's test was applied, respectively, depending on whether the variances between groups were significant or not. The whole process is presented in detail only in the case of the examination of weight loss, and then the process is consequential for the rest of the cases.

The exponential asymptotic regression can be described with the following differential equation

$$y' = c(a - y), \tag{4}$$

where a is the limit coordinate and c is proportional to the growth rate. The differential equation has the solution:

$$y(w) = a - be^{-cw}, \tag{5}$$

which was fitted for a* and b* (and C*). It is similar to the von Bertalanffy growth function and the differential equation describing the process [33].

Regression analysis was carried out for all investigated variables as a function of time (measured in weeks). The fitted functions were linear (of the form $Y = aw + b$) or based on exponential functions (of the form $Y = a - (a - b)e^{-cw}$ for saturation-like processes or of the form $Y = a + be^{-cw}$ for a process with exponential decay). The method of least squares was used for model fitting. An important parameter of this analysis was the limit where the saturation is nearing the asymptote.

All analyses were performed using Excel and IBM SPSS Statistics v27 software, and differences at the 5% level ($p < 0.05$) were considered statistically significant.

3. Results

3.1. Weight Loss (%)

The percentage cumulative weight loss (Δm) of apple fruit during storage under ambient temperature (24 ± 1 °C, φ = 60RH%) over six weeks is presented in Figure 1.

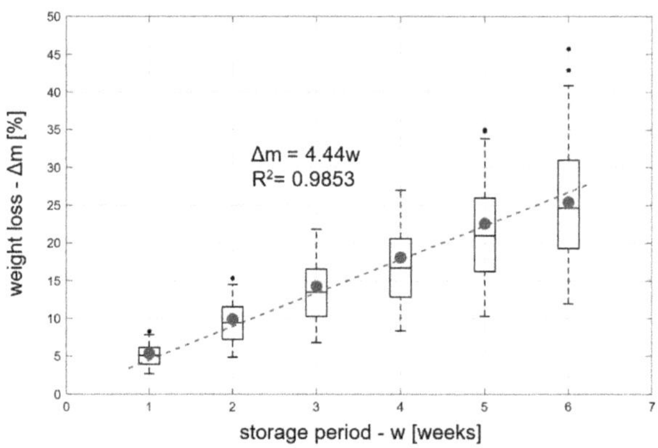

Figure 1. Effect of ambient temperature on the weight loss of Golden Delicious apples during storage.

Apple fruit weight loss increased strongly throughout storage at an ambient temperature. A linear function with a coefficient of determination, $R^2 = 0.9853$, fitted reasonably well to the data regarding the weight loss percentage of the apples during storage under ambient conditions. The moisture and subsequent weight loss in fruits changed linearly with storage duration due to water loss and respiration [34]. The average weight reduction was 5.29% in the first week of measurement, reaching 25.32% at the end of the sixth week (Figure 1). The weight loss percentage grew linearly with each additional week of storage, reaching 22.51 and 25.32 per cent in the fifth and sixth weeks, respectively.

Using ANOVA, this study showed significant ($F = 121.371$, $p < 0.001$) differences in fruit weight loss among the storage durations. Based on Levene's test, there was an important ($F = 8.615$, $p < 0.001$) difference between the variances of groups (that is, the data belonging to each week) for the post-hoc test. Dunnett's T3 test was applied. In Table 1, the p values are presented. Since the table is symmetric and the main diagonal is out of interest, only the values above the main diagonal are included.

Table 1. p values results from one-way analysis of variance ANOVA.

Week	2	3	4	5	6
1	$p < 0.001$	$p < 0.001$	$p < 0.001$	$p < 0.001$	$p < 0.001$
2		$p < 0.001$	$p < 0.001$	$p < 0.001$	$p < 0.001$
3			$p < 0.001$	$p < 0.001$	$p < 0.001$
4				$p = 0.018$	$p < 0.001$
5					$p = 0.025$

It shows that the weight loss in weeks one, two, and three is statistically different from all other weeks, but it is statistically the same between weeks four and five and five and six. A change describes this phenomenon within the ripening process. The next stage in weight loss tendency is developing to move to an asymptotic model, similar to the moisture content characteristics during the given fruits' drying process [35]. The changes in the fruit during ripening are related to an increase in ethylene production and respiration. Ethylene is considered the primary signal for regulating fruit ripening, which begins just before the onset of this process and determines its correct evolution [36]. Exogenous ethylene speeds up the ripening process in climacteric fruit such as apples, whether it comes from other climacteric fruit or is created artificially by substances such as 2-chloroethyl-phosphonic acid [37].

Interestingly, in line with this, the deterioration of the apples was observed in the samples after the third week and during handling as well. The rapid increase in deviance (from Figure 1) also emphasizes this observation.

In short, the detailed presentation of the post-hoc tests is ignored for the latter examined quantities, and only a summary and discussion of the results are included. In addition, also in short, we state that for all other quantities, Levene's test was significant except in the case of TSS; furthermore, ANOVA showed a significant difference between at least two of the groups (weeks) for all quantities, as noted above.

3.2. Color Analysis

Across the entire storage process, a vast difference in color from week one to week six could be observed, changing from qualitative smooth green to yellow and brown spotted apples (Figure 2).

a) b)

Figure 2. Color of a Golden Delicious apple at week one (**a**) and week six (**b**).

For quantitatively evaluating color changes in apple skin during storage, L*, a*, and b* are plotted in Figures 3–5.

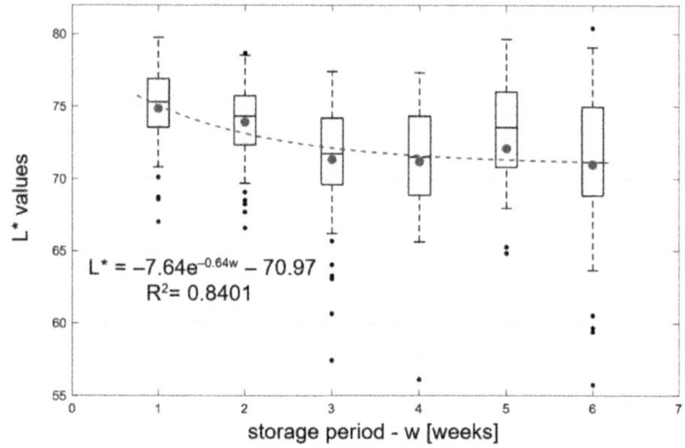

Figure 3. Effect of ambient temperature on L* values of apple during storage: lightness factor decreases to a specific limit.

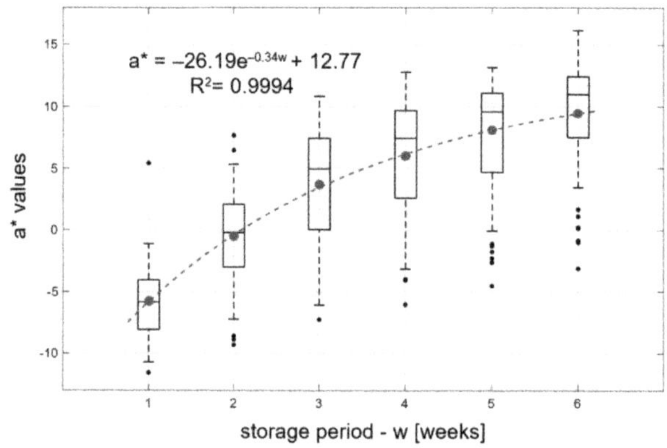

Figure 4. Effect of ambient temperature on a* values of apples during storage: the red color parameter increases to a specific limit.

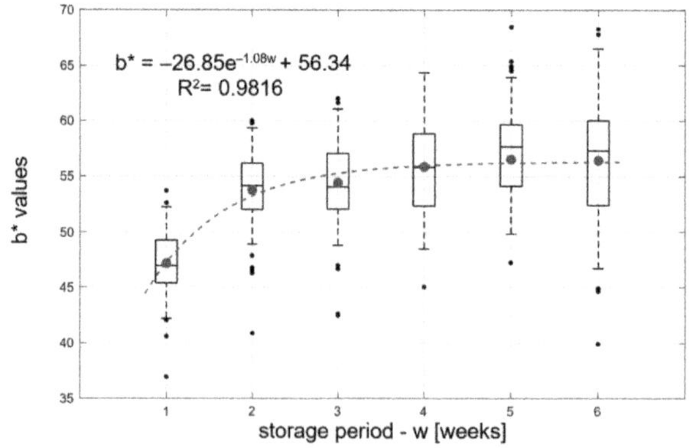

Figure 5. Effect of ambient temperature on b* color values of apple during storage: the yellow color parameter increases to a specific limit.

The L* value decreases from week one to week six. This result could be associated with the change from green to bright green and then yellow with brown spots. L* values change

4.81 in week one to 71.27 in week six. L* variation is related to PPO (polyphenol activity, which alters the tissue surface and reduces apple pigments [38]. study statistically confirmed a significant difference between this parameter's and end state ($p < 0.05$).

The a* and b* values increase during storage at room temperature. This increase is related to the chlorophyll breakdown processes that lead to its decrease and fruit ripening [38]. The a* values change from −5.77 in week one to 9.47 in week six, and the b* values change from 47.15 in week 1 to a maximum in week 6 of 56.45. For parameters a* and b*, it can also be stated that there is a significant difference between the data measured in the first week and the data measured in the last week ($p < 0.05$).

According to Itle and Kabelka, higher b* values are associated with high carotenoids and xanthophylls and the loss of chlorophylls in the pericarp during climacteric fruit ripening of pumpkins and squash [39]. These changes result from the decrease in the greenness of the apple and an increase in its yellowness. In addition, it must be noted that the yellow (b*) color parameter reaches the limiting asymptote much earlier than the red (a*). This could be because Golden Delicious can develop a red color, even at the storage stage, where both light (blue-violet and UV) and temperature affect this color change [40]. However, we showed that this change is slower than for the other color parameter.

It is also interesting to see how outliers are grouped from Figures 3–5, meaning that these values can be attributed to one outlier sample, apples (giving meaningful outlier results on three different sampling points per apple). It must be noted that the outliers and the deviances generally increase with time, meaning that the investigated parameters vary more with the imminent degradation of the samples, which can also affect consumer preference.one 1 to 25.82 on week six (Figure 6). The higher the Delta E value, the lower the color accuracy. The increment in delta E values confirmed the samples' visual color changes.

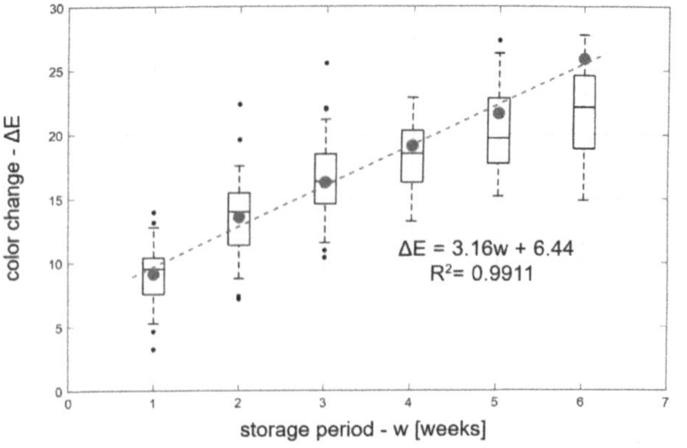

Figure 6. Apple color changes (ΔE) during storage at an ambient temperature: the yellow color parameter increases to a specific limit.

Chroma C* is the distance from the axis of lightness (L) to the origin of color and begins at zero in the center. It describes the degree of color for an area seen with its brightness; it becomes increasingly relevant as color coordinates go away from the origin. Values for C* and are plotted in Figure 7. With regards to the chroma parameter, it is seen to increase from 47.61 to 57.45 across the whole ripening period; during storage, the color of the apple became less bright and more color-saturated than at the time of harvest [26].

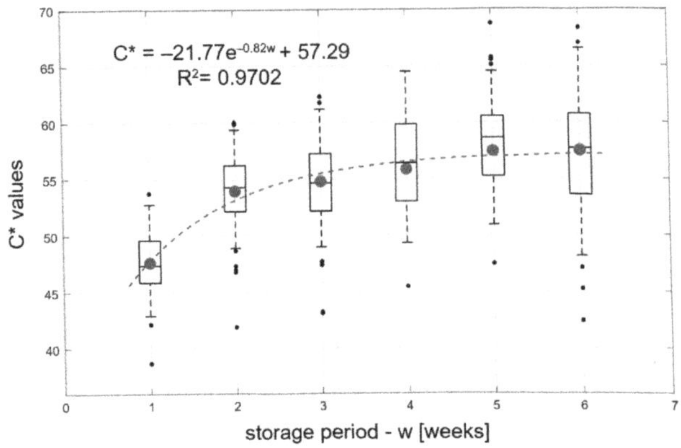

Figure 7. Chroma (C*) values during storage at the ambient temperature of apple.

Apples lose their green skin and flesh color during maturity as chlorophyll is lost. Before maturation, chlorophyll is continuously regenerated; once growth begins, the rate of chlorophyll production slows, initially causing a loss of green coloration. As the chlorophyll disappears further, other pigmentation, often yellow, dominates [41].

Chlorophyll depletion, the synthesis of new pigments such as carotenoids and anthocyanins and the revealing of other pigments previously created throughout the growth of the fruit are all factors in the color changes that occur in many fruits. Such mechanisms are intimately associated with increased respiration, oxygen generation, and starch degradation in fruits with a climacteric ripening pattern [42].

3.3. Total Soluble Solids

As shown in Figure 8, the TSS of apples increased week by week. At the initial week of storage, the TSS was 13.61 °Brix. The maximum TSS attained during storage was about 14.78 °Brix.

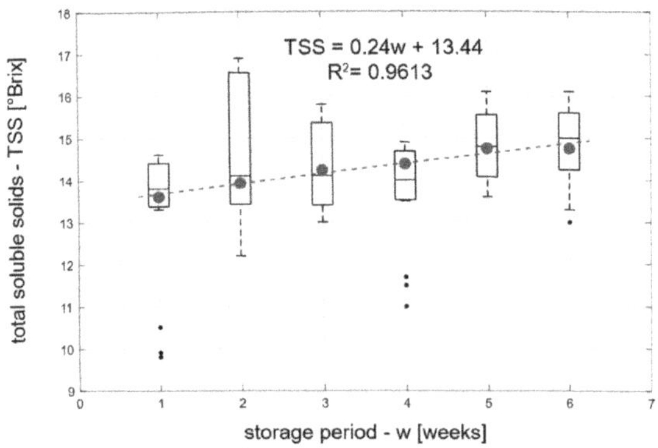

Figure 8. The effect of ambient temperature on the total soluble solids content (TSS) of Golden Delicious apples during storage.

The analysis of the variance homogeneity test showed that the variances for the parameters TSS are the same ($p = 0.053$). Due to the identical standard deviations in the TSS case, Tukey's b-test can be used to distinguish the two groups in terms of means. Tukey's b-test shows that week one is significantly different from weeks five and six ($\alpha = 0.05$).

For TSS, a different (non-linear) trend was reported by Cárdenas-Pérez et al. [26]. They also observed an increase in TSS; our results align with our initial results with weight loss, which is in a linear stage. The rise in TSS could be attributed to moisture loss, leading to the concentration of sugars inside the fruit [43]. The other possible reason for the increased TSS contents is the breakdown of starch into sugars or the hydrolysis of cell wall polysaccharides [44]. The existence of a substrate is recognized to be essential to the respiration process. This substrate serves as a store for carbohydrates such as starch in numerous tissues. These complicated compounds are converted into simple sugars and enter the respiratory route to supply energy for plant vital nutrients. Different quantities of oxygen are used relative to the amount of CO_2 generated when specific substrates are used and fully oxidized. The lower discovered TSS levels of apples in early periods rather than the last weeks may be attributed to the slower respiration process, which reveals the lower total sugar content in apples' tissue. Starch hydrolysis begins during fruit ripening and storage. This pattern may also be explained by considering that the TSS increases with increasing weight loss values. The metabolic processes involved in fruit ripening, including the rapid conversion of starch into sugar and the rise in fruit sweetness during storage, may be the cause of the direct association between TSS and weight loss [38].

In addition, these results back up the studies of Crouch and Ali et al. [45,46]. When several cultivars of apples were maintained at room temperature, they saw an increase in total soluble solids, starch, and sugar content.

According to the current research, total soluble solids in apple fruit increased during storage at an ambient temperature.

4. Discussion

The regression equations and their corresponding coefficients of determination that perfectly represented the weight loss, color values L*, a*, b*, ΔE, C*, and TSS are given in Table 2.

Table 2. Regression equations regarding the storage period with their corresponding coefficient of determination (R^2) for measured quality parameters.

Color Change			
Regression Equations	**Coefficient of Determination (R^2)**	**A-Limit Parameter**	**Consumption Limit (90% Saturation Level)**
$L^* = -7.64\ e^{-0.64w} - 70.97$	0.840	70.97 (decay)	71.35
$a^* = -26.19\ e^{-0.34w} + 12.77$	0.999	12.77 (association)	10.92
$b^* = -26.85\ e^{-1.08w} + 56.34$	0.981	56.34 (association)	55.42
$C^* = -21.77\ e^{-0.82w} + 57.29$	0.970	57.29 (association)	56.32
$\Delta E = 3.16\ w + 6.44$	0.991	-	-
Weight Loss			
Regression Equations	**Coefficient of Determination (R^2)**	**A-Limit Parameter**	**Consumption Limit (90% Saturation Level)**
$\Delta m = 4.44\ w$	0.985	-	-
Total Soluble Solids (TSS)			
Regression Equations	**Coefficient of Determination (R^2)**	**A-Limit Parameter**	**Consumption Limit (90% Saturation Level)**
$TSS = 0.24\ w + 13.44$	0.961	-	-

During the storage period, apples become less acceptable with time from the aspect of consumer evaluation because of color change. The increasing tendency of a* and b* coordinates results in a color shift toward brown and yellow tones on the surface of the examined Golden Delicious specimens. Jha et al., Vieira et al., and Mizrach et al. have also observed similar variations in the values of L*, a*, and b* in the case of Golden Delicious apples and different varieties; however, they used image processing methodologies and other storage conditions [30,47,48]. As a novel and important finding, we can connect the significant changes in TSS in the fifth and sixth weeks, where L* and b* (yellowness) and C* values were nearing saturation.

During the six weeks, the 90% saturation level is defined as the last consumption stage before the total degradation of stored fruits. Table 3 shows the percentage of examined fruits exceeding this quality level each week.

Table 3. The proportion of fruit specimens beyond the consumption limit in the case of color coordinates and TSS.

Storage Time (Weeks)	Parameters beyond Consumption Limit [%]				
	L*	a*	b*	C*	TSS
1	15.3	0.0	0.0	0.0	5.6
2	11.1	0.0	31.9	25.0	16.7
3	44.4	0.0	40.3	36.1	30.6
4	50.0	11.1	55.6	51.4	25.0
5	36.1	34.7	66.7	67.1	27.8
6	58.3	55.6	63.9	60.0	63.9

By investigating the time constant (τ) of the a* and b* values (4.216 and 2.012 weeks, respectively), it must be noted that the b* constant reaches this value significantly earlier, rendering it the dominant part of the visual degradation based on CIELAB color space. It must be pointed out that all asymptotic curves present their τ in the investigated time region (Figure 9).

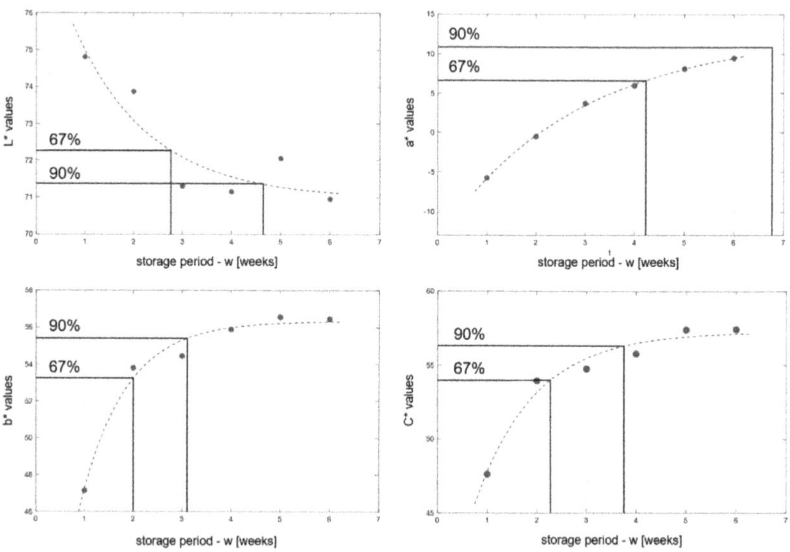

Figure 9. Time constants of color coordinates at 67% and 90% saturation levels during ripening (top left: L*, top right: a*, bottom left: b*, bottom right: C*).

Apples have a short shelf life because they are perishable foods with a rapid metabolism and respiration rate. Because losses during post-harvest handling might reach 25–28% [49], improvements in the post-harvest handling of fresh fruit are crucial for boosting food supply [50].

Apples contain a variety of nutritional benefits; however, they are destroyed during processing after harvest. Both internal and external circumstances influence losses that happen during storage. The most significant effects on apple storage effectiveness, fruit quality traits, and shelf life are caused by temperature and relative humidity during post-harvest handling activities [51]. For instance, they impact the fruit's firmness, weight loss, color, and total soluble solids. Fruit's shelf life is affected by the pectin degradation process that occurs during ripening and plays a significant part in breaking the fruit's cell walls [52].

Proper post-harvest treatment requires changing the product's natural environment to extend the product's post-harvest life. A new approach must be developed to increase the shelf life of fruit stored at room temperature and to prevent softening and senescence [53].

Physical means can satisfy this demand (high-voltage electric field, UV–C light, IR radiation, and microwave radiation).

5. Conclusions

Apple fruit undergoes a post-harvest ripening process that leads to generalized softening and a decline in quality, noticeable in physical and biochemical parameters such as a decrease in weight, color change, and increased sugar content in the juice. All these changes primarily affect the shelf life. The changes in the fruit during ripening are related to an increase in ethylene synthesis; the respiration process and the process of pectin degradation play an essential role in cell wall degradation, affecting the shelf life.

In this study, we combined the changes in weight loss, TSS, and color properties (L*, a*, b*, ΔE and C*) of Golden Delicious apple fruit during storage under room temperature in a novel way to improve their assessment during ambient shelf life with a novel multidisciplinary approach. Analysis using IBM SPSS Statistics v27 software showed a significant difference between the values measured at the beginning and the end state for all parameters during the six weeks of the apple storage experiment. We found all measured apple quality parameters to vary during storage at 24 °C during the study.

The weight loss of samples was increased, indicating that the internal transpiration of the fruit mainly influenced the weight loss. Although the moisture content reaches its limit value during the drying process of apples, the saturation of color coordinates allows for determination of the limit parameters of consumption in the linear weight loss stage. In the laboratory, we applied a flexible and effective measurement method, which can easily be used in a commercial environment. These results improve the provision of helpful information regarding apples' quality changes during storage, which influences customer behavior.

Author Contributions: Conceptualization, S.K. and P.K.; methodology, L.S. and A.G.; investigation, S.K., P.K., L.S. and C.F.; data curation, L.S.; writing—original draft preparation, S.K.; writing—review and editing, C.F. and A.G.; visualization, C.F. and A.G.; supervision, P.K. All authors have read and agreed to the published version of the manuscript.

Funding: This research received no external funding.

Institutional Review Board Statement: Not applicable.

Informed Consent Statement: Not applicable.

Data Availability Statement: Data are available on request from the authors.

Acknowledgments: This work was supported by the Stipendium Hungaricum Program and by the Mechanical Engineering Doctoral School, Hungarian University of Agriculture and Life Sciences, Gödöllő, Hungary, and was supported by the ÚNKP-21-4 New National Excellence Program of the Ministry for Innovation and Technology from the source of the National Research, Development and Innovation Fund.

Innovációs és Technológiai Minisztérium

Conflicts of Interest: The authors declare no conflict of interest.

References

1. FAOSTAT. Available online: https://www.fao.org/faostat/en/#home (accessed on 28 February 2022).
2. Ziv, C.; Fallik, E. Postharvest Storage Techniques and Quality Evaluation of Fruits and Vegetables for Reducing Food Loss. *Agronomy* **2021**, *11*, 1133. [CrossRef]
3. Kassebi, S.; Korzenszky, P. The Effect of Post-Harvest Storage on the Weight of Golden Delicious Apples. *Sci. Technol. Innov.* **2021**, *13*, 7–11. [CrossRef]
4. Farkas, C.; Fenyvesi, L.; Petróczki, K. Multiple Linear Regression Model of Golden Apple's Failure Characteristics under Repeated Compressive Load. *Potravin. Slovak J. Food Sci.* **2019**, *13*, 793–799. [CrossRef] [PubMed]

5. Cömert, E.D.; Mogol, B.A.; Gökmen, V. Relationship between Color and Antioxidant Capacity of Fruits and Vegetables. *Curr. Res. Food Sci.* **2020**, *2*, 1–10. [CrossRef] [PubMed]
6. Majdan, M.; Bobrowska-Korczak, B. Active Compounds in Fruits and Inflammation in the Body. *Nutrients* **2022**, *14*, 2496. [CrossRef]
7. Butkeviciute, A.; Abukauskas, V.; Janulis, V.; Kviklys, D. Phenolic Content and Antioxidant Activity in Apples of the 'Galaval' Cultivar Grown on 17 Different Rootstocks. *Antioxidants* **2022**, *11*, 266. [CrossRef]
8. Peace, C.P.; Bianco, L.; Troggio, M.; van de Weg, E.; Howard, N.P.; Cornille, A.; Durel, C.-E.; Myles, S.; Migicovsky, Z.; Schaffer, R.J.; et al. Apple Whole Genome Sequences: Recent Advances and New Prospects. *Hortic. Res.* **2019**, *6*, 59. [CrossRef]
9. Chen, Z.; Yu, L.; Liu, W.; Zhang, J.; Wang, N.; Chen, X. Research Progress of Fruit Color Development in Apple (Malus Domestica Borkh.). *Plant Physiol. Biochem.* **2021**, *162*, 267–279. [CrossRef]
10. Musacchi, S.; Serra, S. Apple Fruit Quality: Overview on Pre-Harvest Factors. *Sci. Hortic.* **2018**, *234*, 409–430. [CrossRef]
11. Milošević, T.; Milošević, N.; Mladenović, J. Role of Apple Clonal Rootstocks on Yield, Fruit Size, Nutritional Value and Antioxidant Activity of 'Red Chief® Camspur' Cultivar. *Sci. Hortic.* **2018**, *236*, 214–221. [CrossRef]
12. Milošević, T.; Milošević, N.; Mladenović, J. The Influence of Organic, Organo-Mineral and Mineral Fertilizers on Tree Growth, Yielding, Fruit Quality and Leaf Nutrient Composition of Apple Cv. *'Golden Delicious Reinders.' Sci. Hortic.* **2022**, *297*, 110978. [CrossRef]
13. Caballero, D.; Antequera, T.; Caro, A.; del Mar Ávila, M.; G Rodríguez, P.; Perez-Palacios, T. Non-Destructive Analysis of Sensory Traits of Dry-Cured Loins by MRI–Computer Vision Techniques and Data Mining. *J. Sci. Food Agric.* **2017**, *97*, 2942–2952. [CrossRef] [PubMed]
14. Caballero, D.; Pérez-Palacios, T.; Caro, A.; Amigo, J.M.; Dahl, A.B.; ErsbØll, B.K.; Antequera, T. Prediction of Pork Quality Parameters by Applying Fractals and Data Mining on MRI. *Food Res. Int.* **2017**, *99*, 739–747. [CrossRef] [PubMed]
15. Ma, J.; Sun, D.-W.; Qu, J.-H.; Liu, D.; Pu, H.; Gao, W.-H.; Zeng, X.-A. Applications of Computer Vision for Assessing Quality of Agri-Food Products: A Review of Recent Research Advances. *Crit. Rev. Food Sci. Nutr.* **2015**, *56*, 113–127. [CrossRef]
16. Ekramirad, N.; Adedeji, A.; Alimardani, R. A Review of Non-Destructive Methods for Detection of Insect Infestation in Fruits and Vegetables. *Innov. Food Res.* **2016**, *2*, 6–12.
17. Yang, S.; Meng, Z.; Li, Y.; Chen, R.; Yang, Y.; Zhao, Z. Evaluation of Physiological Characteristics, Soluble Sugars, Organic Acids and Volatile Compounds in 'Orin' Apples (Malus Domestica) at Different Ripening Stages. *Molecules* **2021**, *26*, 807. [CrossRef]
18. Kamiloğlu, Ö. Influence of Some Cultural Practices on Yield, Fruit Quality and Individual Anthocyanins of Table Grape CV. *"Horoz Karasi." J. Anim. Plant Sci.* **2011**, *21*, 240–245.
19. Yang, Y.; Zhou, B.; Zhang, J.; Wang, C.; Liu, C.; Liu, Y.; Zhu, X.; Ren, X. Relationships between Cuticular Waxes and Skin Greasiness of Apples during Storage. *Postharvest Biol. Technol.* **2017**, *131*, 55–67. [CrossRef]
20. Khodaei, D.; Hamidi-Esfahani, Z.; Rahmati, E. Effect of Edible Coatings on the Shelf-Life of Fresh Strawberries: A Comparative Study Using TOPSIS-Shannon Entropy Method. *NFS J.* **2021**, *23*, 17–23. [CrossRef]
21. Sinha, S.R.; Singha, A.; Faruquee, M.; Jiku, A.S.; Rahaman, A.; Alam, A.; Kader, M.A. Post-Harvest Assessment of Fruit Quality and Shelf Life of Two Elite Tomato Varieties Cultivated in Bangladesh. *Bull. Natl. Res. Cent.* **2019**, *43*, 185. [CrossRef]
22. Mardziah, O.; Mat Jafri, M.Z.; Abdul Aziz, A.; Omar, A. Non-Destructive Quality Evaluation of Fruit by Color Based on RGB LEDs System. In Proceedings of the 2014 2nd International Conference on Electronic Design, ICED 2014, Penang, Malaysia, 19–21 August 2015; pp. 230–233.
23. Mesa, A.R.; Chiang, J.Y. Multi-Input Deep Learning Model with RGB and Hyperspectral Imaging for Banana Grading. *Agriculture* **2021**, *11*, 687. [CrossRef]
24. Garrido-Novell, C.; Pérez-Marin, D.; Amigo, J.M.; Fernández-Novales, J.; Guerrero, J.E.; Garrido-Varo, A. Grading and Color Evolution of Apples Using RGB and Hyperspectral Imaging Vision Cameras. *J. Food Eng.* **2012**, *113*, 281–288. [CrossRef]
25. Cen, H.; Lu, R.; Mendoza, F.; Beaudry, R.M. Relationship of the Optical Absorption and Scattering Properties with Mechanical and Structural Properties of Apple Tissue. *Postharvest Biol. Technol.* **2013**, *85*, 30–38. [CrossRef]
26. Cárdenas-Pérez, S.; Chanona-Pérez, J.; Méndez-Méndez, J.V.; Calderón-Domínguez, G.; López-Santiago, R.; Perea-Flores, M.J.; Arzate-Vázquez, I. Evaluation of the Ripening Stages of Apple (Golden Delicious) by Means of Computer Vision System. *Biosyst. Eng.* **2017**, *159*, 46–58. [CrossRef]
27. Khorshidi, J. Storage Temperature Effects on the Postharvest Quality of Apple (Malus Domestica Borkh. Cv. Red Delicious). *N. Y. Sci. J.* **2019**, *3*, 67–70.
28. de AlmeidaTeixeira, G.H.; Meakem, V.; Morais, C. de L.M. de; Lima, K.M.G. de; Whitehead, S.R. Conventional and Alternative Pre-Harvest Treatments Affect the Quality of 'Golden Delicious' and 'York' Apple Fruit. *Environ. Exp. Bot.* **2020**, *173*, 104005. [CrossRef]
29. Rutkowski, K.; Michalczuk, B.; Konopacki, P. Nondestructive Determination of "Golden Delicious" Apple Quality and Harvest Maturity. *J. Fruit Ornam. Plant Res.* **2008**, *16*, 39–52.
30. Jha, S.N.; Rai, D.R.; Shrama, R. Physico-Chemical Quality Parameters and Overall Quality Index of Apple during Storage. *J. Food Sci. Technol.* **2012**, *49*, 594–600. [CrossRef]
31. Bavisetty, S.C.B.; Venkatachalam, K. Physicochemical Qualities and Antioxidant Properties of Juice Extracted from Ripe and Overripe Wax Apple as Affected by Pasteurization and Sonication. *J. Food Process. Preserv.* **2021**, *45*, e15524. [CrossRef]

32. Dobrzański, B.; Rybczyński, R. Colour Change of Apple as a Result of Storage, Shelf-Life, and Bruising. *Int. Agrophysics* **2002**, *16*, 261–268.
33. Calrson, J.K.; Goldman, K.J. (Eds.) *Special Issue: Age and Growth of Chondrichthyan Fishes: New Methods, Techniques and Analysis*, 1st ed.; Springer: Dordrecht, The Nederlands, 2006.
34. Li, M.; Feng, F.; Cheng, L. Expression Patterns of Genes Involved in Sugar Metabolism and Accumulation during Apple Fruit Development. *PLoS ONE* **2012**, *7*, e33055. [CrossRef] [PubMed]
35. Alibas, I.; Zia, M.P.; Yilmaz, A.; Asik, B.B. Drying Kinetics and Quality Characteristics of Green Apple Peel (Mallus Communis L. Var. "Granny Smith") Used in Herbal Tea Production. *J. Food Process. Preserv.* **2020**, *44*, e14332. [CrossRef]
36. Verde, A.; Míguez, J.M.; Gallardo, M. Melatonin Stimulates Postharvest Ripening of Apples by Up-Regulating Gene Expression of Ethylene Synthesis Enzymes. *Postharvest Biol. Technol.* **2023**, *195*, 112133. [CrossRef]
37. Chen, Y.; Wang, T.; Zhou, M.; Hou, H.; Xue, Y.; Wang, H. Rice Husk and Sewage Sludge Co-Combustion Ash: Leaching Behavior Analysis and Cementitious Property. *Constr. Build. Mater.* **2018**, *163*, 63–72. [CrossRef]
38. Sadat Razavi, M.; Golmohammadi, A.; Nematollahzadeh, A.; Ghanbari, A.; Davari, M.; Rovera, C.; Carullo, D.; Farris, S. Impact of Bacterial Cellulose Nanocrystals-Gelatin/Cinnamon Essential Oil Emulsion Coatings on the Quality Attributes of 'Red Delicious' Apples. *Coatings* **2022**, *12*, 741. [CrossRef]
39. Itle, R.A.; Kabelka, E.A. Correlation Between L*a*b* Color Space Values and Carotenoid Content in Pumpkins and Squash (Cucurbita Spp.). *HortScience* **2009**, *44*, 633–637. [CrossRef]
40. Fruit Color—Promoting Red Color Development in Apple. Available online: https://extension.psu.edu/fruit-color-promoting-red-color-development-in-apple (accessed on 13 December 2022).
41. Kingston, C.M. Maturity Indices for Apple and Pear. In *Hortical Reviews*; Janick, J., Ed.; John Wiley & Sons, Ltd.: Hoboken, NJ, USA, 2010; Volume 13, pp. 407–432.
42. Łysiak, G.; Kurlus, R.; Zydlik, Z.; Walkowiak-Tomczak, D. Apple Skin Colour Changes during Harvest as an Indicator of Maturity. *Acta Sci. Pol. Hortorum Cultus* **2014**, *13*, 71–83.
43. Arendse, E.; Fawole, O.A.; Opara, U.L. Influence of Storage Temperature and Duration on Postharvest Physico-Chemical and Mechanical Properties of Pomegranate Fruit and Arils. *CyTA J. Food* **2014**, *12*, 389–398. [CrossRef]
44. Ahmad, F.; Zaidi, S.; Arshad, M. Postharvest Quality Assessment of Apple during Storage at Ambient Temperature. *Heliyon* **2021**, *7*, e07714. [CrossRef]
45. Ali, M.A.; Raza, H.; Khan, M.A.; Hussain, M. Effect of Different Periods of Ambient Storage on Chemical Composition of Apple Fruit. *Int. J. Agric. Biol.* **2004**, *6*, 568–571.
46. Crouch, I. 1-Methylcyclopropene (Smartfresh Tm) as an Alternative to Modified Atmosphere and Controlled Atmosphere Storage of Apples and Pears. In Proceedings of the 8th International Controlled Atmosphere Research Conference, Rotterdam, The Netherlands, 8–13 July 2001; pp. 433–436.
47. Vieira, F.G.K.; Borges, G.D.S.C.; Copetti, C.; Amboni, R.D.D.M.C.; Denardi, F.; Fett, R. Physico-Chemical and Antioxidant Properties of Six Apple Cultivars (Malus Domestica Borkh) Grown in Southern Brazil. *Sci. Hortic.* **2009**, *122*, 421–425. [CrossRef]
48. Mizrach, A.; Lu, R.; Rubino, M. Gloss Evaluation of Curved-Surface Fruits and Vegetables. *Food Bioprocess. Technol.* **2009**, *2*, 300–307. [CrossRef]
49. Saletnik, B.; Zaguła, G.; Saletnik, A.; Bajcar, M.; Słysz, E.; Puchalski, C. Method for Prolonging the Shelf Life of Apples after Storage. *Appl. Sci.* **2022**, *12*, 3975. [CrossRef]
50. Ghazanfar, M.; Khan, M.; Khan, C.; Bhatti, R. Post Harvest Losses in Apple and Banana During Transport and Storage. *Pak. J. Agric. Sci.* **2007**, *44*, 534–539.
51. Tano, K.; Oulé, M.K.; Doyon, G.; Lencki, R.W.; Arul, J. Comparative Evaluation of the Effect of Storage Temperature Fluctuation on Modified Atmosphere Packages of Selected Fruit and Vegetables. *Postharvest Biol. Technol.* **2007**, *46*, 212–221. [CrossRef]
52. Gwanpua, S.G.; Mellidou, I.; Boeckx, J.; Kyomugasho, C.; Bessemans, N.; Verlinden, B.E.; Hertog, M.L.A.T.M.; Hendrickx, M.; Nicolai, B.M.; Geeraerd, A.H. Expression Analysis of Candidate Cell Wall-Related Genes Associated with Changes in Pectin Biochemistry during Postharvest Apple Softening. *Postharvest Biol. Technol.* **2016**, *112*, 176–185. [CrossRef]
53. González-Aguilar, G.A.; Ayala-Zavala, J.F.; Olivas, G.I.; de la Rosa, L.A.; Álvarez-Parrilla, E. Preserving quality of fresh-cut products using safe technologies. *J. Verbr. Lebensm.* **2010**, *5*, 65–72. [CrossRef]

applied sciences

MDPI

Article

Comparison of Measurement Possibilities by Non-Invasive Reflectometric Sensors and Invasive Probes

Magdalena Paśnikowska-Łukaszuk [1], Magda Wlazło-Ćwiklińska [1], Jarosław Zubrzycki [2,*] and Zbigniew Suchorab [3]

[1] Fundamentals of Technology Faculty, Lublin University of Technology, ul. Nadbystrzycka 38, 20-618 Lublin, Poland
[2] Mechanical Engineering Faculty, Lublin University of Technology, ul. Nadbystrzycka 36, 20-618 Lublin, Poland
[3] Faculty of Environmental Engineering, Lublin University of Technology, ul. Nadbystrzycka 40B, 20-618 Lublin, Poland
* Correspondence: j.zubrzycki@pollub.pl

Abstract: The measurement of the moisture content of building materials is of key importance both in the process of building structures and in their subsequent operation. In engineering practice, indirect techniques of moisture measurement, mainly, resistance and capacitive, are the most popular. The main objective of this research work was to compare the classic TDR measurement technique to the non-invasive, surface TDR sensors. Moisture measurements were carried out on samples made of cellular concrete with density class of 400 and 600. These samples were moist to various degrees, from 0 to 69% (400 c.c.) and from 0 to 55 (600 c.c.). For each sample, five measurements were carried out. Both the RMSE and the expanded uncertainty values were more favorable for the TDR FP/mts probe and were consistent with the literature data. Compared to them, the measurement result for the 400 c.c. samples with the S1 probe was 154.6%, and that with the S2 probe was 87.03% of the values obtained with the invasive probe. When measuring the 600 c.c. samples, we found values of 122.16% for S1 and of 120.1% for S2 of those obtained with the invasive probe. The use of surface TDR sensors provided an easy and quick measurement without damaging the surface and structure of the tested material, as there was no need to introduce the probe actuators inside the tested material.

Keywords: TDR sensor; moisture; porous materials; calibration model

Citation: Paśnikowska-Łukaszuk, M.; Wlazło-Ćwiklińska, M.; Zubrzycki, J.; Suchorab, Z. Comparison of Measurement Possibilities by Non-Invasive Reflectometric Sensors and Invasive Probes. *Appl. Sci.* **2023**, *13*, 665. https://doi.org/10.3390/app13010665

Academic Editor: Giuseppe Lacidogna

Received: 26 October 2022
Revised: 14 December 2022
Accepted: 21 December 2022
Published: 3 January 2023

1. Introduction

Measurement research has been refined over the years by developing various data acquisition methods. The study of environmental factors involves many measurement aspects. One of such factors is moisture, which affects the building materials, but also the quality of life of the building users [1]. The moisture of building materials depends primarily on their absorption properties, as well as on the operating conditions and on independent external factors, which are often difficult to detect also due to the improper construction of objects, faults or the lack of anti-moisture insulation [2,3]. Moisture tests are also carried out in order to determine all the factors that lead to the multiplication of fungi, especially molds, that threaten human health [2]. Moisture can cause physical, chemical and biological damage [4]. A moistened material is subject to faster corrosion. With increased fungal infection, it can also be a substrate for the development of other microorganisms, additionally losing its thermal insulation properties [5]. If the moisture condensed in the porous space freezes, the structure of the material is destroyed. The processes of freezing and thawing lower the strength parameters of building materials [6]. An important element when examining building structures is the assessment of moisture parameters and the identification of harmful salts that may be present in the building material in various proportions [7]. The sources of moisture are often unnoticeable and may be hidden under the insulation or located deep in the building material, which may make their detection difficult [8,9]. Capillary water located deep in a wall is especially harmful. This water destroys the wall from the inside and is invisible on the wall surface, unlike condensation water that appears on the surface and is the result of improper ventilation of a room [10].

For the quick detection of moisture in a wall, indirect techniques are the most useful, enabling to estimate the moisture of a partition on the basis of other physical parameters, the values of which indirectly depend on moisture. Most often, these are electrical techniques based on the measurement of electrical conductivity or dielectric permittivity [11]. An advantageous feature of the electrical methods is that most often these methods are non-invasive and allow the testing of building materials without disturbing their structure and shape [12–14]. The basic electrical method is the resistance technique, which consists in measuring the conductivity or electrical resistance of the tested materials in porous materials depending on the degree of their moisture [15]. The measures based on this method are most often calibrated to the mass moisture of the material [16] and are most often used to determine the moisture of construction wood. However, the environmental parameters of walls or partitions are commonly determined using other methods [17,18]. These include the capacitive Frequency Domain (FD) method, which is considered to be better than the resistance technique. The FD method consists in measuring the capacitance of a properly constructed capacitor with alternating voltage. In FD methods, the conductive electrodes in the area of the test medium are treated as capacitor plates. The dielectric of this capacitor is the material to be measured. The value of the apparent permittivity of the material affects the capacity of the capacitor thus formed. The measurement of this capacity allows the assessment of the material's moisture [19,20]. Another non-invasive method, i.e., microwaves, allows measuring the phase shift and the degree of electromagnetic wave attenuation in the tested material, which allows the further determination of the moisture content. With the help of microwave moisture measurements, it is possible to determine the degree of moisture in building partitions inside their structure and on their surface. Microwave radiation is absorbed by matter through ionic conductivity or through the phenomenon of dielectric losses resulting from dipole polarization [21–24]. Water is a dipole which, while appearing in the structure of another material, still maintains the asymmetric nature of its molecules [25].

A separate group of measurement methods are the direct methods that enable to measure the presence of water. These are mostly invasive methods that interfere with the structure or shape of the tested element. The invasive measurement method which is most frequently used to determine the moisture in walls is the gravimetric method [26]. The basic instrument used in this method is a moisture analyzer, which can be used to test the material taken from a wall. The test material is weighed and then dried and reweighed in order to determine the weight differences [27]. It is a method that requires the destruction of the material structure.

Destructive methods also include the carbide method (CM) [28]. This is an indirect method that also involves sampling and testing with a CM hygrometer using the chemical process of calcium carbide decomposition by water. During decomposition, acetylene gas is released, which causes an increase in pressure in the device [29]. Other invasive methods supporting the moisture measurement process also include measurements with a Peltier probe, which determines a sample's suction potential in a given moisture state. Instead of the material moisture, the psychrometric Peltier probe measures the water potential, which is an indicator of the ability of a porous material to bind water [30]. Thus, the relative moisture of a porous material can be calculated [31]. In order to conduct invasive psychrometric measurements, the probe must be placed in the tested material by drilling a hole, which, if the material is loose, can be done by pressing the probe [31]. It is a standard form of measurement, additionally used to determine parameters that can support the process of measuring the moisture content of building materials.

A measuring technique that can combine the advantages of destructive and non-invasive techniques is Time Domain Reflectometry (TDR), which uses the measurement of the propagation time of a short electromagnetic pulse in a material sample to determine the apparent permittivity of porous materials, which is strongly dependent on the water presence, i.e., material moisture [21,32–41]. The apparent permittivity of the medium determines the velocity of signal propagation along the waveguide. The TDR sensor can also be used for the non-invasive monitoring of changes in moisture content in rigid porous materials [42]. The design of a TDR sensor determines the shape of the reflected signal, and the accuracy of the propagation time measurement depends on its design details [43]. Common applications of the TDR technique are based on the use of typical invasive probes introduced into the tested porous medium; this technique is most often used in soil science. With reference to building materials, its use involves the destruction of samples or walls [44]. Therefore, in the case of hard and rigid media, which include building materials, surface sensors are an alternative to traditional probes [42,45,46]. In the measurement of the basic environmental factors of building partitions, which include moisture, the TDR technique should be treated as a method at the stage of implementation for the purpose of testing walls. In order to improve this measurement technology, the obtained test results should be properly analyzed using typical invasive and non-invasive sensors and compared to the reference values obtained by direct testing [11]. The aim of this article was to determine the possibility of measuring moisture in building materials with non-invasive TDR sensors and to compare the measuring potential of this method to that of the TDR method using traditional invasive sensors.

2. Materials and Methods

A set of cellular concrete samples was used for the measurements. The devices that were utilized for sample preparation were the following: a 06-DZ-3BC laboratory oven (Chemland, Stargard, Poland), an SBS-LW-3000N laboratory scale (Steinberg Systems, Zielona Góra, Poland), TDR equipment including a LOM laboratory multimeter (ETest, Lublin, Poland), traditional TDR FP/mts probes described in detail in Section 2.1.1 (ETest, Lublin, Poland), TDR surface sensors (manufactured in Lublin University of Technology) described in detail in Section 2.1.2 and a Personal Computer serving for TDR multimeter control and data management.

2.1. Materials

2.1.1. TDR Meter Description

The measurements were conducted with a TDR multimeter emitting a needle peak signal with rise time equal to 300 ps, produced by ETest manufacturer, Lublin, Poland. The emitted signal propagated along the coaxial cable to the sensor where reflections occurred on the characteristic points of the propagation line. Those reflections derived from both the beginning of the sensor and its termination and served as measurement markers. The time differences between those reflections were read by the TDR meter and could be automatically or manually recalculated into the apparent permittivity value that depends on a material's moisture.

2.1.2. FP/Mts Sensor

The FP/mts sensor presented in Figure 1 was intentionally designed for the in situ evaluation of soil moisture but for several years has been successfully applied for the moisture evaluation of soft building materials [42].

Its main functional elements are the two 10 cm long sharpened acid-resistant steel rods (2 mm in diameter, separated by 14 mm), a sensor support made from a section of a PVC tube of 2 cm outer diameter and length, and a coaxial cable of length from 1.5 to 6 m, from the sensor to the terminating SMA connector.

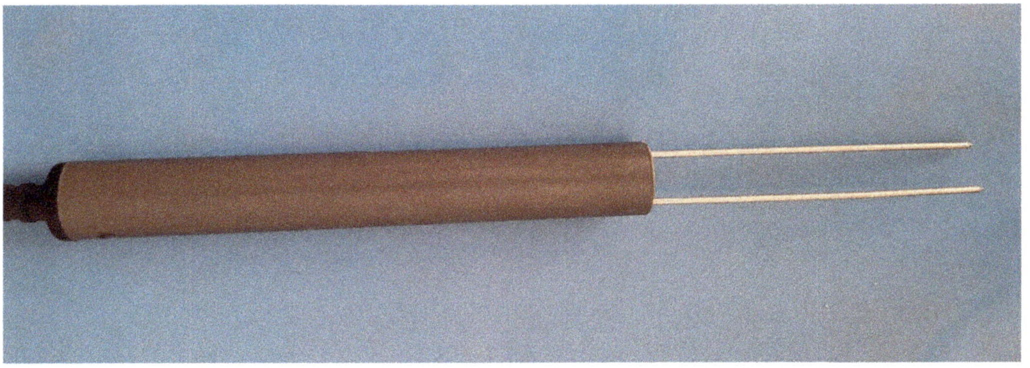

Figure 1. FP/mts TDR sensor applied in this research (ETest, Lublin, Poland).

2.1.3. Non-Invasive TDR Sensors

For the experiment, two non-invasive TDR sensors manufactured in Lublin University of Technology were applied. The sensors were previously described in articles by Suchorab et al. [42,43].

2.1.4. S1 Non-Invasive Sensor

The S1 non-invasive sensor (Figure 2a,b) is made of black polyoxymethylene, characterized by an apparent permittivity value of 3.8 [-] [47]. The length of the measuring probe is 200 mm, and its width is 50 mm. The measuring rods are made of a brass flat bar with a cross section of 2 × 10 mm. The sensor communicated with the TDR meter via an angled BNC connector that was soldered to the printed circuit board that linked the measuring rods to the connector. In the design of this sensor, a flat bar was placed in a dielectric.

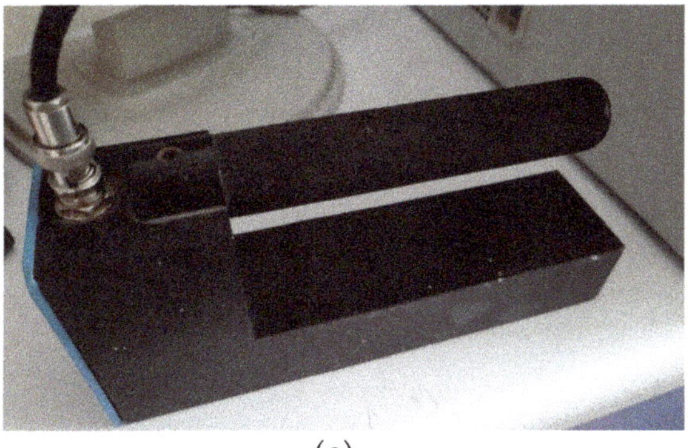

(a)

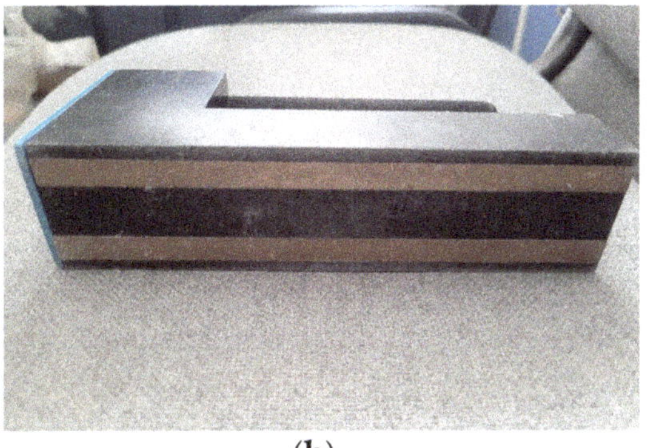

(b)

Figure 2. S1 non-invasive TDR sensor [42]. (**a**) Isometric view, (**b**) view from the bottom.

2.1.5. S2 Non-Invasive Sensor

The S2 sensor shown in Figure 3a,b is similar in construction and made of the same material as the S1 sensor. Its length is 200 mm, and its width is 100 mm. As in the case of the S1 sensor, the waveguides of the probes were made of a brass flat bar with a cross section of 2 mm × 10 mm.

(a) (b)

Figure 3. S2 non-invasive TDR sensor [42]. (**a**) Isometric view, (**b**) view from the bottom.

2.2. Description of the Tested Material

Cellular concrete was used for the tests as the building material. In the presented research, samples of 400 kg/m^3 and 600 kg/m^3 of cellular concrete were used.

2.3. Samples and Preparations

2.3.1. Samples for Invasive Measurements

Due to the FP/mts invasive probe geometry and its sensitivity range (ETest, Lublin, Poland), a set of samples of cellular concrete was prepared. The samples with dimensions equal to 5 × 5 × 12 cm were prepared in the amount of 40 pieces in the case of the 400 kg/m^3 cellular concrete (400 c.c.) and of 35 pieces in the case of the 600 kg/m^3 cellular concrete (600 c.c.). They were dried to a constant mass at a temperature of 105 °C. Then, the samples were saturated to the desired moisture by dosing the appropriate amount of water until the expected moisture was obtained, in steps of 10% by mass, until the state of full saturation (69% mass in the case of cellular concrete 400, and 55% mass in the case of cellular concrete 600).

2.3.2. Samples for Non-Invasive Measurements

The dimensions of the samples used in non-invasive research were 220 mm × 120 mm × 40 mm. Similarly to the samples for invasive measurements, they were dried to a constant mass and gradually moistened to achieve 69% and 55% of moisture. The samples were weighed to check the moisture status. Then, the samples were examined with the non-invasive S1 sensor and subsequently with the S2 sensor in order to obtain the measurement results.

2.4. Methods

Description of the Measurement Procedure

The research consisted in measuring the apparent permittivity of the material with different moisture values. The measurements were made on dry samples (5 readouts for statistical purposes) and then on samples of increasing moisture until saturation. The tests were carried out under constant conditions of temperature (20 °C) and relative air moisture (50%).

In the case of the invasive FP/mts probes, they were introduced into the structure of the tested material.

In the case of the non-invasive sensors, the measurements were carried out using the touch method on the samples with dimensions of 220 mm × 120 mm × 40 mm.

2.5. Data Analysis Method

As a result of this research, we obtained measurements based on the relationship between the apparent permittivity readings obtained with the TDR sensors and the mass moisture content of the material of the tested samples. The uncertainty of the measurements was assessed with the use of appropriate regression models. A polynomial relationship was used for the calibration based on the formula:

$$\hat{w} = \beta_1 \cdot \varepsilon^2 + \beta_2 \cdot \varepsilon + \beta_3 + \in \quad (1)$$

where $\hat{w}$ is the mass moisture value estimated using a polynomial model [%mass], ε is the apparent permittivity measured using TDR, ϵ is the random error of normal distribution, β_{1-3} are the model estimators.

The parameters defining the measurement uncertainty of the methods are the mean-square error (RMSE) and the coefficient of determination R^2, which allows for the assessment of the regression model adaptation and the quality of the model fit in relation to the measured data.

Additionally, an extended assessment of the measurement uncertainty (standard and extended uncertainty) of the invasive FP/mts TDR sensors and the non-invasive S1 and S2 sensors was performed based on the GUM guidelines [48].

The uncertainty assessment included the complex uncertainty assessment, which is a combination of two types of uncertainty, i.e., type A, that is statistical uncertainty depending on the quality of the adopted model fit, and type B, which depends on the uncertainty and resolution of the individual device. With a complex standard uncertainty, the expanded measurement uncertainty was estimated [43].

For the applied measurement methods, type B uncertainties were much lower than type A uncertainties and were ignored in the calculations; therefore, the following elements were taken into account for the estimation of uncertainty: the estimators β_0, β_1, β_2 and the relative permeability (ε):

$$w = f(\beta_1, \beta_2, \beta_3, \varepsilon) \quad (2)$$

The composite standard uncertainty of the measurement mentioned above, which included both type A and type B uncertainties, can be described by the formula:

$$u_C(w) = \sqrt{\left(\frac{\partial w}{\partial \varepsilon} u(\varepsilon)\right)^2 + \sum_{i=0}^{2} \left(\frac{\partial w}{\partial \beta_i} u(\beta_i)\right)^2 + 2\sum_{i=0}^{2} \sum_{j=i+1}^{2} \frac{\partial w}{\partial \beta_i} \frac{\partial w}{\partial \beta_j} u(\beta_i, \beta_j)} \quad (3)$$

The expanded uncertainty can be described with the following formula [47].

$$U(w) = k_p \cdot u_c(w) \quad (4)$$

where k_p is the coverage factor that depends on the number of degrees of freedom, whose value is approximately 2.

3. Results

Tables 1–6 show the permittivity readings by the different sensor types (i.e., invasive FP/mts and non-invasive S1 and S2) for the two classes of cellular concrete with different bulk moisture (400 and 600). The first column of each table shows the mass moisture w, and the following columns show the individual moisture readings for each tested sample, ε_1–ε_5.

Table 1. Results of the apparent permittivity ε measurements with the FP/mts invasive probe for different values of material moisture w; 400 c.c. sample.

Moisture w [%]	Apparent Permittivity ε [-]				
	ε_1	ε_2	ε_3	ε_4	ε_5
0	1.18	1.16	1.14	1.2	1.18
10	2.14	2.01	2.22	1.84	2.41
20	5.31	5.82	5.09	5.42	5.47
30	7.17	6.87	7.55	7.25	7.02

Table 1. *Cont.*

Moisture w [%]	Apparent Permittivity ε [-]				
	ε_1	ε_2	ε_3	ε_4	ε_5
40	9.32	9.47	9.37	9.27	9.07
50	12.54	11.91	12.35	12.01	12.78
60	15.55	15.38	14.84	15.83	15.26
69	20.35	20.27	20.33	20.55	20.11

Table 2. Results of the apparent permittivity ε measurements with the S1 surface sensor for different values of moisture w; 400 c.c. sample.

Moisture w [%]	Apparent Permittivity ε [-]				
	ε_1	ε_2	ε_3	ε_4	ε_5
0	3.63	3.73	3.54	3.44	3.65
10	4.06	3.97	4.05	3.921	4.17
20	5.00	5.09	5.05	5.09	5.09
30	6.05	6.00	6.05	5.67	5.43
40	7.54	7.53	7.42	7.77	7.48
50	7.98	7.54	7.81	7.77	7.95
60	8.94	9.13	9.07	8.88	9.21
69	12.92	13.16	12.92	12.64	12.71

Table 3. Results of the apparent permittivity ε measurements with the S2 surface sensor for different values of moisture w; 400 c.c. sample.

Moisture w [%]	Apparent Permittivity ε [-]				
	ε_1	ε_2	ε_3	ε_4	ε_5
0	3.56	3.59	3.67	3.79	3.71
10	4.18	3.88	4.01	4.25	4.28
20	5.38	5.16	5.34	5.02	5.20
30	6.04	6.24	5.90	6.38	6.17
40	7.26	7.30	6.84	6.69	7.10
50	8.31	7.86	7.69	7.69	7.96
60	9.62	9.56	10.22	10.22	10.09
69	12.25	11.86	11.53	11.67	11.88

Table 4. Results of the apparent permittivity ε measurements with the FP/mts invasive probe for different values of moisture w; 600 c.c. sample.

Moisture w [%]	Apparent Permittivity ε [-]				
	ε_1	ε_2	ε_3	ε_4	ε_5
0	1.06	1.07	1.11	1.22	1.04
10	2.75	2.38	2.91	2.83	2.64
20	5.48	5.21	5.84	5.49	5.65
30	6.74	7.20	6.91	6.59	7.07
40	11.45	11.31	10.94	11.24	11.67
50	16.57	16.24	16.97	16.85	16.78
55	19.20	20.41	19.48	20.30	20.21

Table 5. Results of the apparent permittivity ε measurements with the S1 surface sensor for different values of moisture *w*; 600 c.c. sample.

Moisture *w* [%]	Apparent Permittivity ε [-]				
	ε_1	ε_2	ε_3	ε_4	ε_5
0	3.33	3.42	3.25	3.32	3.50
10	3.82	4.05	3.94	3.88	3.77
20	5.27	5.44	5.28	5.29	5.20
30	6.27	6.58	6.50	6.39	6.58
40	9.05	8.79	8.58	8.83	8.85
50	9.78	10.27	9.82	10.05	9.84
55	12.05	12.25	12.31	12.62	12.35

Table 6. Results of the apparent permittivity ε measurements with the S2 surface sensor for different values of moisture *w*; 600 c.c. sample.

Moisture *w* [%]	Apparent Permittivity ε [-]				
	ε_1	ε_2	ε_3	ε_4	ε_5
0	3.11	3.32	3.27	3.31	3.43
10	3.72	3.76	3.87	3.94	3.84
20	5.11	5.24	5.42	5.30	5.36
30	6.75	6.81	6.99	6.85	6.86
40	8.82	8.96	9.10	9.12	8.96
50	10.53	10.28	10.05	9.96	10.18
55	12.65	12.49	12.88	12.72	12.57

4. Discussion

4.1. Regression Model

The graphs in Figure 4 present the relationships obtained on the basis of the results presented in Tables 1–6. The graphs show the average permittivity readings from the five measurements for all examined moisture contents and the confidence intervals.

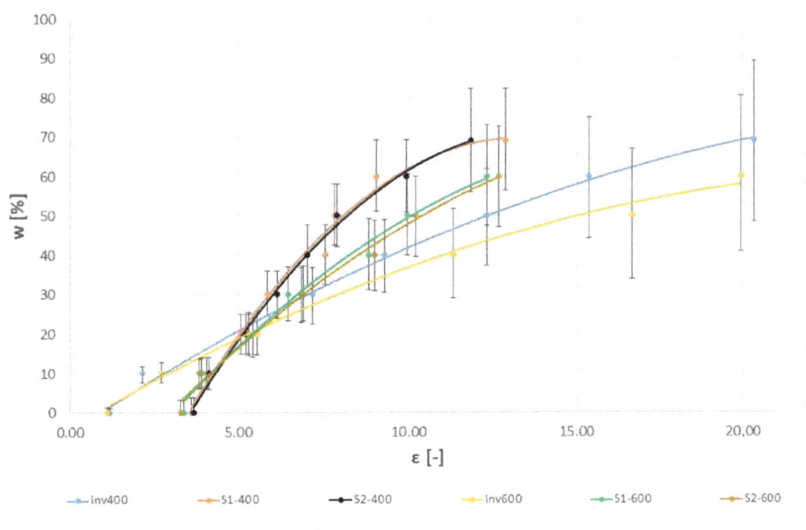

Figure 4. Relationship between the apparent permittivity reads determined with the use of the FP/mts invasive probe and the non-invasive probe and the moisture content of class 400 c.c. and 600 c.c. samples.

The graphs in Figure 4 confirmed the relationship between the apparent permittivity and the moisture content of the material. The effect of the apparent bulk density of the material was not high, and the different regression courses resulted from the fact that the two materials had different water absorption capacity, i.e., a maximum of 55% for the aerated concrete 600 and a maximum of 69% for the aerated concrete 400, which was a consequence of the different structure of the porous medium.

There were differences in the permittivity readings between the invasive probes and the surface sensors, which resulted from their construction and the type of contact between the measuring element and the material. In the case of the invasive probes, the measuring rods were inserted into the material, whereas in the case of the surface sensors, they were in contact with the surface of the material. The invasive probes read the permittivity of the tested material, and the apparent permittivity read by the sensors S1 and S2 was the average value of the apparent permittivity of the tested material as well as of the housing from which the sensor was made. As a consequence, in the case of the invasive sensors for dry material, the average apparent permittivity value was 1.17 [-] for the cellular concrete 400 and 1.1 [-] for the cellular concrete 600. However, in the case of the S1 and S2 probes, they were 3.60 [-] and 3.66 [-] for the concrete 400 and 3.36 [-] and 3.29 [-] for the concrete 600. For the maximum saturation of the 400 cellular concrete sample, the apparent permittivity result when measured with an invasive probe exceeded 20 [-].

On the basis of the measurements, the obtained data and the adopted polynomial regression model described by formula (1), the values of the β_1 β_2 and β_3 estimators were calculated for the individual sensor models and the classes of cellular concrete. These values are summarized in Tables 7 and 8.

Table 7. Estimator values of the adopted calibration models for the measuring probes used (400 c.c. sample).

Sensor	β_1	β_2	β_3	R^2	F Statistic	RSE [%mass]	RMSE [%mass]
Invasive	−0.098 **	5.660 ***	−5.005	0.994	419.593 *** (df = 2; 5)	2.212	1.748806
Non-invasive (S1)	−0.733 **	19.419 ***	−58.949 ***	0.986	174.069 *** (df = 2; 5)	3.42	2.703889
Non-invasive (S2)	−0.684 **	18.86 ***	−58.64 ***	0.996	555.122 *** (df = 2; 5)	1.925	1.521514

df—degree of freedom, p—critical level of significance (* $p < 0.05$; ** $p < 0.01$; *** $p < 0.001$).

Table 8. Estimator values of the adopted calibration models of the measuring probes used (600 c.c. sample).

Sensor	β_1	β_2	β_3	R^2	F Statistic	RSE [%mass]	RMSE [%mass]
Invasive	−0.127 **	5.482 ***	−4.827	0.993	286.774 *** (df = 2; 4)	2.089	1.579467
Non-invasive (S1)	−0.458 *	13.056 **	−36.376 **	0.990	191.525 *** (df = 2; 4)	2.552	1.929395
Non-invasive (S2)	−0.379 *	11.670 **	−31.802 **	0.990	555.122 *** (df = 2; 5)	2.509	1.896724

df—degree of freedom, p—critical level of significance (* $p < 0.05$; ** $p < 0.01$; *** $p < 0.001$).

In Table 7, it can be seen that the measurement error expressed as RMSE in the case of 400 c.c. was the highest when using the S1 non-invasive sensor. On the other hand, the values of the determination coefficient R^2 were similar for all analyzed sensors and exceeded the value of 0.98, which proved a very good fit of the adopted model to the dependence tested. In turn, in the case of class 600 c.c., the RMSE error values presented in

Table 8 were the highest for the invasive measurement by the FP/mts probe. As in the case of the 400 c.c. sample, the values of the R^2 coefficient were similar in all measurements for the class 600 c.c. samples and, in each case, exceeded 0.99.

In the case of the cellular concrete 400, the critical significance levels of the β_2 and β_3 estimators assumed a value lower than 0.001 for all sensors, which means that they were statistically significant. In turn, the critical significance levels of β_1 for all sensors assumed a value of less than 0.01. In the case of the aerated concrete 600, the β_2 and β_3 estimators were less than 0.1, and the β_1 estimator was less than 0.05. This means that their significance levels were lower for all sensors. On the other hand, the analysis of the F statistic ($p < 0.001$) in all cases confirmed the statistical significance of the adopted regression models.

4.2. Uncertainty Analysis

The results obtained from the measurements made it possible to describe the relationship between the apparent permittivity and the moisture content of the material with the second-grade polynomial regression model according to formula (2) and to determine the quality of the model fit to the obtained data using the determination coefficients R^2, as well as the standard errors of measurement expressed as Residual Standard Error (RSE) and Root-Mean-Square Error (RMSE). Additionally, based on the GUM [49], the analysis of the measurement uncertainty and expanded uncertainty in the material moisture function was performed. The calculations were made on the basis of formulas (3) and (4). The obtained results are presented in the form of graphs in Figure 5.

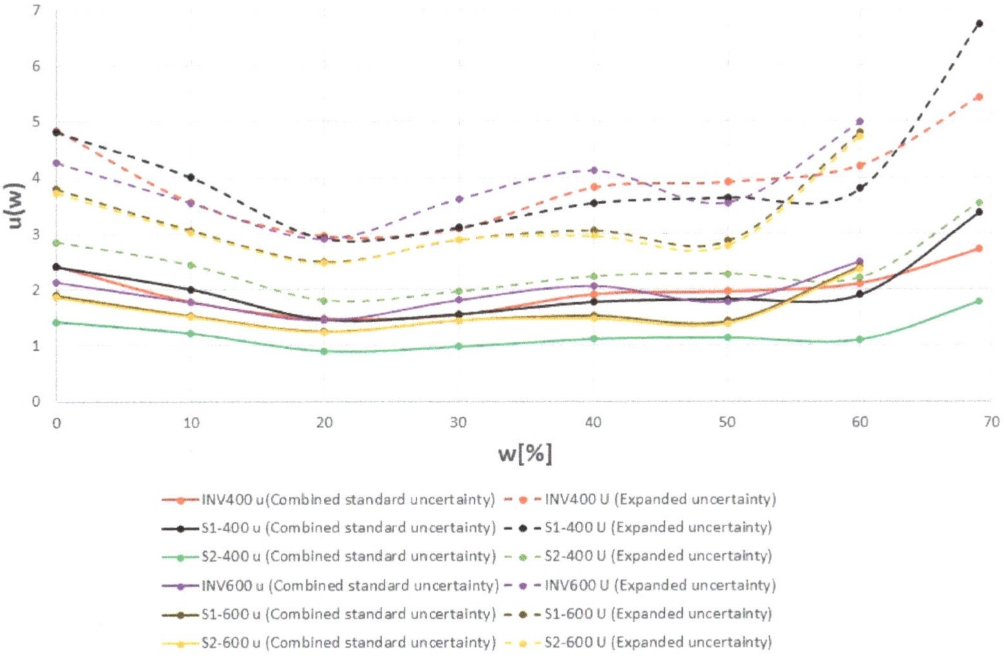

Figure 5. Uncertainty dependence on material moisture: invasive, S1 and S2 sensor, 400 and 600 c.c. samples.

4.3. Discussion of the Calibration Results and Uncertainty Calculations

In the case of dry samples and a moisture content below 0.05 cm^3/cm^3, the apparent permittivity determined by means of the surface sensors ranged from 3 to 4. This was a consequence of the value of the apparent permittivity of the solid phase of the tested material and the apparent permittivity of polyoxymethylene of 3.8 [-] [43]. In the higher moisture ranges, the TDR surface sensor readings showed higher moisture values than the conventional invasive probe.

This was also confirmed by the statistical characteristics of the model used, mainly, the coefficient of determination (R^2) equal to 0.986 and 0.996 for the S1 and S2 TDR surface sensors and to 0.988 for the invasive probe. The RSE values were equal to 1.925 and 3.42 for the non-invasive TDR sensors and to 3.183 for the invasive probe. Similar observations were made in the case of the RMSE. It was equal to 2.70- and 1.52 for the TDR surface sensors and to 2.52 for the invasive probe. The RSE and RSME values were lower for the non-invasive S2 probe, while R^2 was the highest for this probe among the tested sensors [42,49].

Due to the fact that in the measurements of building materials and partitions the most frequently used quantity is the mass moisture (in %), calibration models were developed for these units. Unfortunately, in the case of the TDR technique, most of the available literature expresses a material's moisture content as volumetric moisture; therefore, for the purposes of literature discussion, the obtained values were converted to cm^3/cm^3 according to the formula presented in [11,42].

Tables 7 and 8 show that the RMSE values obtained for all probe models ranged from 1.521% mass to 2.7% mass (0.0091 to 0.0162 cm^3/cm^3) for the cellular concrete 400 and from 1.579% mass to 1.929% mass (0.0063 to 0.0077 cm^3/cm^3) for the aerated concrete 600. The lowest values of this coefficient, 1.521% mass (0.0091 cm^3/cm^3), for the concrete 400 were recorded when measured with the non-invasive S2 sensor. In the case of the aerated concrete 600, the lowest value of this coefficient was achieved in the case of measurement with an invasive probe and was 1.579% mass (0.0063 cm^3/cm^3).

In turn, the maximum values of the RMSE coefficient for the cellular concrete 400 were obtained with the non-invasive S1 sensor (1.58% mass, 0.0162 cm^3/cm^3) and for the cellular concrete 600, again with the same sensor (1.93% mass, 0.0077 cm^3/cm^3).

According to the data presented by [49], the application of Topp et al. [49] in relation to selected soil centers resulted in uncertainties expressed as RMSE in the range of 0.01–0.066 cm^3/cm^3. In the case of the model proposed by Roth et al. [50], the RMSE ranged from 0.008 to 0.037 cm^3/cm^3. The RMSE for the popular calibration formula proposed by Malicki et al. (1996) was set at 0.03 cm^3/cm^3. It should be borne in mind that in most of the cited literature sources, the proposed models were universal. For this reason, the quality of the fit to the measured data was lower. The obtained RMSE values were comparable and in many cases lower than the values determined by the team of Udawatta et al. for regression models developed individually for each material (0.008–0.034 cm^3/cm^3) [51].

When analyzing the characteristics of the influence of moisture on the measurement uncertainty (Figure 5), it can also be noticed that the measurement uncertainties for the sensors S1 and S2 were smaller compared to those obtained for classic invasive probes. It was also found that the uncertainties read at low and high material moisture levels were greater than those obtained at middle moisture levels, which is a typical observation for many measuring devices. It also resulted from the adopted regression model [11,27].

In turn, in the case of the measurement uncertainty, it can be referred to the expanded uncertainty U. As its value is mainly influenced by the measurement uncertainty of type A, related to the quality of fit of the adopted regression model, the correlation between the U and the RMSE values was clearly visible. Only the U values from the middle measuring range were analyzed. The lowest value of expanded uncertainty was observed for the S2 sensor (concrete 400 class) corresponding to 1.799% mass. In the case of the measurement in concrete 600, the same values were also obtained with the S2 sensor, i.e., 2.468%mass.

Various literature sources indicate the following values of measurement uncertainties of models developed for soil media and invasive probes: according to [52] and [53], this uncertainty was in the range of 0.022–0.023 cm^3/cm^3; according to [54], it was 0.0269 cm^3/cm^3, according to [55], it was 0.004–0.018 cm^3/cm^3, and according to [50], it was 0.011–0.013 cm^3/cm^3. Most of these values are higher or comparable to those obtained in the course of the experimental research performed in this work.

5. Conclusions

The use of TDR surface sensors (noninvasive S1 and S2 type) showed that the Time Domain Reflectometry technique can be successfully used for non-invasive research to determine the moisture content of rigid porous materials used in construction. The TDR sensors provide very good responses similar to the measurements made with traditional invasive sensors.

Since, in contrast to the invasive probes, the electrodes of the TDR sensors do not require to be immersed into the tested material, the measurements are significantly simplified, and potential damage to both the sensor and the tested object is avoided.

When switching to the use of TDR sensors, however, one should be aware that they require individual calibration, which may be technically difficult to perform, especially in the case of partitions of objects, the characteristics of which are not known before starting the test.

It was also noticed that the RMSE of the tested sensor was higher for the almost dry and almost saturated states of the measured material. The obtained test results allow concluding that more accurate measurement results are achieved with the use of invasive sensors. The extended measurement uncertainty U(w) for measurements with an invasive probe (middle range of material moisture w) was equal to 2.955 [-] for the cellular concrete 400, while, for the cellular concrete 600, it was 2.897 [-]. In the case of the non-invasive sensors S1 it was for the 400 c.c. sample, 3.114 [-], and for the 600 c.c. sample, 2.889 [-]. In the case of S2 and the cellular concrete 400, it was 1.956 [-], and for the cellular concrete 600, it was 2.893 [-]. Finally, it should be stated that the expanded uncertainty for the FP/mts probe was 94.1% of that for S1 and 151% of that for S2 (cellular concrete 400); 99.72% of that for S1 and 99.86% of that for S2 (cellular concrete 600).

This does not change the fact that the surface sensors, despite the slightly different results compared to those obtained with the invasive probe, have a number of advantages. The following can be mentioned among them: (1) a simple and fast implementation allowing for more measurements per unit of time; (2) the time needed to prepare the device and the test object itself is much shorter than in the case of measurements with invasive probes; (3) the surface TDR sensors allow measuring the moisture of the tested objects without damaging them. The lack of need to introduce elements deep into the tested object allows maintaining the continuity of the structure of the object. Therefore, it is possible to measure fragile objects, e.g., objects with historic value or objects whose structure must not be destroyed.

Author Contributions: Z.S.: Conceptualization, methodology, data curation. M.P.-Ł.: software, investigation, writing—original draft preparation. M.W.-Ć.: visualization, funding acquisition. J.Z.: formal analysis, writing—review and editing. All authors have read and agreed to the published version of the manuscript.

Funding: This research was funded by Lublin University of Technology grant's number's M/KIRP/FD-20/IM-5/142/2022 and FD-20/IS-6/025.

Institutional Review Board Statement: Not applicable.

Informed Consent Statement: Not applicable.

Data Availability Statement: Not applicable.

Conflicts of Interest: The authors declare no conflict of interest.

References

1. Bobociński, A. Wpływ wilgotności sorpcyjnej na przewodność cieplną betonów komórkowych. *Pr. Inst. Tech. Bud.* **2003**, *32*, 35–47.
2. Dutkiewicz, J.; Górny, R.L. Biologiczne Czynniki Szkodliwe Dla Zdrowia–Klasyfikacja i Kryteria Oceny Narażenia. *Med. Pr.* **2002**, *53*, 29–39. [PubMed]
3. Ważny, J.; Karyś, J. *Ochrona Budynków Przed Korozją Biologiczną: Praca Zbiorowa.*; Arkady: Warsaw, Poland, 2001; ISBN 83-213-4205-1.

4. Wiszniewska, M.; Walusiak, J.; Gutarowska, B.; Żakowska, Z.; Pałczyński, C. Grzyby Pleśniowe w Środowisku Komunalnym i w Miejscu Pracy—Istotne Zagrożenia Zdrowotne. *Med. Pr.* **2004**, *55*, 257–266. [PubMed]
5. Orlik-Kożdoń, B.; Steidl, T.; Rubim, J.A. Wybrane Problemy Związane z Termoizolacją Ścian Budynków Zabytkowych. *Izolacje* **2018**, *23*, 36–40.
6. Guz, Ł.; Sobczuk, H. Water Potential Measurement in Porous Materials with Peltier Psychrometer. 2008. Available online: https://www.researchgate.net/publication/262674645_Water_potential_measurement_in_porous_materials_with_Peltier_psychrometer (accessed on 12 December 2022).
7. Monczyński, B. Diagnostyka Zawilgoconych Konstrukcji Murowych. *Izolacje* **2019**, *24*, 89–93.
8. Buczkowski, W.; Wiśniewska, J.; Żelazny, R. Analiza porównawcza wyników pomiarów wilgotności uzyskiwanych przy zastosowaniu różnych metod i przyrządów. *Apar. Badaw. I Dydakt.* **2003**, *8*, 16–22.
9. Wiebe, H.H.; Campbell, G.S.; Gardner, W.H.; Rawlins, S.L.; Cary, J.W.; Brown, R.W. Measurement of Plant and Soil Water Status. Available online: https://eprints.nwisrl.ars.usda.gov/id/eprint/1250/ (accessed on 30 September 2022).
10. Marten, J. Methods of Measuring the Moisture Diffusivity at High Moisture Levels. Bachelor's Thesis, Lund Unversity, Lund, Sweden, 1997.
11. Suchorab, Z.; Majerek, D.; Koci, V.; Cerny, R. Time Domain Reflectometry Flat Sensor for Non-Invasive Monitoring of Moisture Changes in Building Materials. *Measurement* **2020**, *165*, 108091. [CrossRef]
12. Maser, K. Integration of Ground Penetrating Radar and Infrared Thermography for Bridge Deck Condition Evaluation. In Proceedings of the NDTCE'09, Non-Destructive Testing in Civil Engineering Conference, Nantes, France, 30 June–3 July 2009; p. 8.
13. Maldague, X.P.V. *Nondestructive Evaluation of Materials by Infrared Thermography*; Springer: Berlin/Heidelberg, Germany, 1993; ISBN 978-1-4471-1995-1.
14. Vollmer, M.; Mollmann, K.-P. *Infrared Thermal Imaging: Fundamentals, Research and Applications*, 2nd ed.; Wiley: Hoboken, NJ, USA, 2018; ISBN 978-3-527-41351-5.
15. Trochonowicz, M.; Witek, B.; Chwiej, M. Impact analysis of humidity and temperature on the value of thermal conductivity λ coefficient of insulating materials used inside buildings in polish. *Bud. I Archit.* **2013**, *12*, 165–176. [CrossRef]
16. Pasierb, B. Techniki Pomiarowe Metody Elektrooporowej. *Środowisko* **2012**, *109*, 191–199.
17. Farahiyah, D.; Purnama, B.W. Design Prototype of Temperature and Humidity Control and Monitoring on Weaver Ant Cage Based on Internet of Things. *J. Ilm. Tek. Elektro Komput. Dan Inform.* **2021**, *7*, 326–337. [CrossRef]
18. Bogdan, M. How to Use the DHT22 Sensor for Measuring Temperature and Humidity with the Arduino Board. *Acta Univ. Cibiniensis. Tech. Ser.* **2016**, *68*, 22–25. [CrossRef]
19. Łobzowski, A.; Opęchowski, A.; Szkolnikowski, W. Pomiar Wilgotności Gleby, Materiałów Stałych i Sypkich. *POMIAR* **2008**, *4*. Available online: https://www.label.pl/po/wilgotn_mater_stalych.html (accessed on 12 December 2022).
20. Lin, C.-P. Frequency Domain versus Travel Time Analyses of TDR Waveforms for Soil Moisture Measurements. *Soil Sci. Soc. Am. J.* **2003**, *67*, 720–729. [CrossRef]
21. Suchorab, Z.; Tabiś, K.; Brzyski, P.; Szczepaniak, Z.; Rogala, T.; Susek, W.; Łagód, G. Comparison of the Moist Material Relative Permittivity Readouts Using the Non-Invasive Reflectometric Sensors and Microwave Antenna. *Sensors* **2022**, *22*, 3622. [CrossRef]
22. Liu, Q.; Zhou, J. Research on Microwave Humidity Testing of Ceramic Products. *J. Wuhan Univ. Technol.-Mater. Sci. Ed.* **2000**, *15*, 45–48.
23. Oliveira, J.G.D.; Junior, J.G.D.; Pinto, E.N.M.G.; Neto, V.P.S.; D'Assunção, A.G. A New Planar Microwave Sensor for Building Materials Complex Permittivity Characterization. *Sensors* **2020**, *20*, 6328. [CrossRef]
24. Adhiwibowo, W.; Daru, A.F.; Hirzan, A.M. Temperature and Humidity Monitoring Using DHT22 Sensor and Cayenne API. *J. Transform.* **2020**, *17*, 209–214. [CrossRef]
25. Gąsiorski, A.; Posyłek, Z.; Dróżdż, T. Mikrofalowa metoda określania wilgotności lignocelulozowych paliw stałych. *Electrotech. Rev.* **2015**, *1*, 101–103. [CrossRef]
26. Brown, A.; Yardley, R.; Quincey, P.; Butterfiels, D. Studies of the Effect of Humidity and Other Factors on Some Different Filter Materials Used for Gravimetric Measurements of Ambient Particulate Matter. *Atmos. Environ.* **2006**, *40*, 4670–4678. [CrossRef]
27. Trochonowicz, M.; Szostak, B.; Lisiecki, D. Analiza porównawcza badań wilgotnościowych metodą chemiczną w stosunku do badań grawimetrycznych wybranych materiałów budowlanych. *Bud. I Archit.* **2016**, *15*, 163–171. [CrossRef]
28. Hola, A. Measuring of the Moisture Content in Brick Walls of Historical Buildings—The Overview of Methods. *IOP Conf. Ser. Mater. Sci. Eng.* **2017**, *251*, 012067. [CrossRef]
29. Trochonowicz, M.; Szostak, B.; Lackorzyński, M. Accuracy of the CM Carbide Method When Testing Counter-Floors Used in the Construction Industry. *Civ. Environ. Eng. Rep.* **2020**, *30*, 64–73. [CrossRef]
30. Chow, T.L.; Vries, J. de Dynamic Measurement of Soil and Leaf Water Potential with a Double Loop Peltier Type Thermocouple Psychrometer. *Soil Sci. Soc. Am. J.* **1973**, *37*, 181–188. [CrossRef]
31. Guz, Ł.; Sobczuk, H. Badanie Potencjału Wody w Materiałach Budowlanych za Pomocą Sondy Psychrometrycznej. *Proc. ECOpole* **2008**, *2*, 185–190.
32. Calamita, G.; Brocca, L.; Perrone, A.; Piscitelli, S.; Lapenna, V.; Melone, F.; Moramarco, T. Electrical Resistivity and TDR Methods for Soil Moisture Estimation in Central Italy Test-Sites. *J. Hydrol.* **2012**, *454–455*, 101–112. [CrossRef]

33. Hailong, H.; Kailin, A.; Min, L.; Jinghui, X.; Wenyi, S.; Jones, S.; Gonzalez-Teruel, J.; Robinson, D.; Horton, R.; Bristow, K.; et al. A Review of Time Domain Reflectometry (TDR) Applications in Porous Media. *Adv. Agron.* **2021**, *168*, 83–155. [CrossRef]
34. Krishnapillai, M.; Ranjan, R.S. Non-destructive Monitoring of Nitrate Concentration in a Laboratory Flow Experiment Using Time Domain Reflectometry (TDR). *Environ. Technol.* **2008**, *30*, 101–109. [CrossRef]
35. Topp, G.C.; Davis, J.L. Measurement of Soil Water Content Using Time-Domain Reflectrometry (TDR): A Field Evaluation. *Soil Sci. Soc. Am. J.* **1985**, *49*, 19–24. [CrossRef]
36. Goetzke-Pala, A.; Hoła, J. Testing of the moisture content of saline brick walls using the non-destructive microwave method in polish. *Mater. Bud.* **2016**, *9*, 104–106. [CrossRef]
37. Gorzelańczyk, T.; Schabowicz, K. Nieniszczące badania wilgotności płyt włóknisto-cementowych metodą dielektryczną; Non-destructive testing of fiber cement board moisture using dielectric method. *Weld. Technol. Rev.* **2014**, *86*, 1. [CrossRef]
38. Bi, H.; Yin, K.; Xie, X.; Ji, J.; Wan, S.; Sun, L.; Terrones, M.; Dresselhaus, M.S. Ultrahigh Humidity Sensitivity of Graphene Oxide. *Sci. Rep.* **2013**, *3*, 2714. [CrossRef]
39. Leciejewski, P. Wykorzystanie Metody TDR Do Ciaglego Pomiaru Zmian Warunkow Wilgotnosciowych i Termicznych w Profilu Glebowym. *Stud. I Mater. Cent. Edukac. Przyr.-Leśnej* **2009**, *11*, 235–246.
40. Mojid, M.A.; Cho, H. Evaluation of the Time-Domain Reflectometry (TDR)-Measured Composite Dielectric Constant of Root-Mixed Soils for Estimating Soil-Water Content and Root Density. *J. Hydrol.* **2004**, *295*, 263–275. [CrossRef]
41. Watanabe, K.; Wake, T. Measurement of Unfrozen Water Content and Relative Permittivity of Frozen Unsaturated Soil Using NMR and TDR. *Cold Reg. Sci. Technol.* **2009**, *59*, 34–41. [CrossRef]
42. Suchorab, Z.; Widomski, M.; Łagód, G.; Barnat-Hunek, D.; Majerek, D. A Noninvasive TDR Sensor to Measure the Moisture Content of Rigid Porous Materials. *Sensors* **2018**, *18*, 3935. [CrossRef] [PubMed]
43. Suchorab, Z.; Malec, A.; Sobczuk, H.; Łagód, G.; Gorgol, I.; Łazuka, E.; Brzyski, P.; Trník, A. Determination of Time Domain Reflectometry Surface Sensors Sensitivity Depending on Geometry and Material Moisture. *Sensors* **2022**, *22*, 735. [CrossRef] [PubMed]
44. Barnat-Hunek, D.; Smarzewski, P.; Suchorab, Z. Effect of Hydrophobisation on Durability Related Properties of Ceramic Brick. *Constr. Build. Mater.* **2016**, *111*, 275–285. [CrossRef]
45. Majcher, J.; Kafarski, M.; Wilczek, A.; Szypłowska, A.; Lewandowski, A.; Woszczyk, A.; Skierucha, W. Application of a Dagger Probe for Soil Dielectric Permittivity Measurement by TDR. *Measurement* **2021**, *178*, 109368. [CrossRef]
46. Brzyski, P.; Suchorab, Z. Capillary Uptake Monitoring in Lime-Hemp-Perlite Composite Using the Time Domain Reflectometry Sensing Technique for Moisture Detection in Building Composites. *Materials* **2020**, *13*, 1677. [CrossRef]
47. Skierucha, W.; Wilczek, A.; Alokhina, O. Calibration of a TDR Probe for Low Soil Water Content Measurements. *Sens. Actuators A Phys.* **2008**, *147*, 544–552. [CrossRef]
48. *Guide to the Expression of Uncertainty in Measurement—Part 6: Developing and Using Measurement Models*; Joint Committee for Guides in Metrology: Sevres, France, 2020; p. 103.
49. Topp, G.C.; Davis, J.L.; Annan, A.P. Electromagnetic Determination of Soil Water Content: Measurements in Coaxial Transmission Lines. *Water Resour. Res.* **1980**, *16*, 574–582. [CrossRef]
50. Roth, K.; Schulin, R.; Flühler, H.; Attinger, W. Calibration of Time Domain Reflectometry for Water Content Measurement Using a Composite Dielectric Approach. *Water Resour. Res.* **1990**, *26*, 2267–2273. [CrossRef]
51. Udawatta, R.P.; Anderson, S.H.; Motavalli, P.P.; Garrett, H.E. Calibration of a Water Content Reflectometer and Soil Water Dynamics for an Agroforestry Practice. *Agrofor. Syst.* **2011**, *82*, 61–75. [CrossRef]
52. TOPP, G.C.; ZEBCHUK, W.D.; DAVIS, J.L.; BAILEY, W.G. The Measurement of Soil Water Content Using a Portable Tdr Hand Probe. *Can. J. Soil. Sci.* **1984**, *64*, 313–321. [CrossRef]
53. Amato, M.; Ritchie, J.T. Small Spatial Scale Soil Water Content Measurement with Time-Domain Reflectometry. *Soil Sci. Soc. Am. J.* **1995**, *59*, 325–329. [CrossRef]
54. Černý, R. Time-Domain Reflectometry Method and Its Application for Measuring Moisture Content in Porous Materials: A Review. *Measurement* **2009**, *42*, 329–336. [CrossRef]
55. Malicki, M.A.; Plagge, R.; Roth, C.H. Improving the Calibration of Dielectric TDR Soil Moisture Determination Taking into Account the Solid Soil. *Eur. J. Soil Sci.* **1996**, *47*, 357–366. [CrossRef]

applied sciences

MDPI

Article

An Inversion Algorithm for the Dynamic Modulus of Concrete Pavement Structures Based on a Convolutional Neural Network

Gongfa Chen [1], Xuedi Chen [1], Linqing Yang [2,3], Zejun Han [2,*] and David Bassir [4,5,*]

1 School of Civil and Transportation Engineering, Guangdong University of Technology, Guangzhou 510006, China
2 School of Civil Engineering, Sun Yat-Sen University, Guangzhou 510275, China
3 School of Architectural Engineering, Guangzhou Institute of Science and Technology, Guangzhou 510540, China
4 Centre Borelli, ENS-University of Paris-Saclay, 91190 Gif-sur-Yvette, France
5 UTBM, IRAMAT UMR 7065-CNRS, Rue de Leupe, CEDEX, 90010 Belfort, France
* Correspondence: zjhangdgy@gdut.edu.cn (Z.H.); david.bassir@utbm.fr (D.B.); Tel.: +86-132-2998-2506 (Z.H.)

Abstract: Based on the spectral element method (SEM) and a convolutional neural network (CNN), an inversion algorithm for the dynamic modulus of concrete pavement structures is proposed in this paper. In order to evaluate the service performance of pavement structures more systematically and accurately via the existing testing techniques using a falling weight deflectometer (FWD), it is necessary to obtain accurate dynamic modulus parameters of the structures. In this work, an inversion algorithm for predicting the dynamic modulus is established by using a CNN which is trained with the dynamic response samples of a multi-layered concrete pavement structure obtained through SEM. The gradient descent method is used to adjust the weight parameters in the network layer by layer in reverse. As a result, the accuracy of the CNN can be improved via iterative training. With the proposed algorithm, more accurate results of the dynamic modulus of pavement structures are obtained. The accuracy and numerical stability of the proposed algorithm are verified by several numerical examples. The dynamic modulus and thickness of concrete pavement structure layers can be accurately predicted by the CNN trained with a certain number of training samples based on the displacement curve of the deflection basin from the falling weight deflectometer. The proposed method can provide a reliable testing tool for the FWD technique of pavement structures.

Keywords: pavement structure; dynamic modulus; convolution neural network; inversion algorithm; falling weight deflectometer

Citation: Chen, G.; Chen, X.; Yang, L.; Han, Z.; Bassir, D. An Inversion Algorithm for the Dynamic Modulus of Concrete Pavement Structures Based on a Convolutional Neural Network. *Appl. Sci.* **2023**, *13*, 1192. https://doi.org/10.3390/app13021192

Academic Editors: Bogusław Łazarz, Grzegorz Peruń and Tangbin Xia

Received: 10 November 2022
Revised: 8 January 2023
Accepted: 10 January 2023
Published: 16 January 2023

1. Introduction

In recent years, with the vigorous development of nondestructive testing techniques for pavements, falling weight deflectometer (FWD) has been widely applied. Simultaneously, the inversion analysis of the dynamic modulus of pavement structures has also become one of the most critical techniques used in their service performance evaluation [1]. The method for solving the dynamic responses of a multi-layered pavement under impact loads is limited by the boundary conditions and computational efficiency; thus, in the early development stage of FWD instruments, most inversion software was based on the quasi-static analysis method [2]. However, this method cannot simulate the dynamic effect of vehicle loads. Moreover, the static inversion method is prone to causing the phenomenon of "modulus transferring", which leads to the failure of the inversion results [3]. Although the dynamic effect of loads can be considered in the dynamic response analysis of the pavement structure system by the finite element method (FEM), its solution process requires meshing the pavement structures, which greatly reduces its solution efficiency and seriously affects its application in the inversion of the dynamic modulus of the pavement structures [4,5]. In recent years, with the deepening of research on the dynamic responses of layered semi-infinite media, increasingly more dynamic response algorithms have emerged, most notably the integral equation transformation method [6–9]. These algorithms can obtain the dynamic displacement and stress responses of layered soil under dynamic load in the frequency domain and can reveal the dynamic characteristics of the layered soil. However, when solving the displacement responses of road surfaces under FWD loads, it is necessary to solve the infinite integral problem; thus, the solution efficiency is greatly reduced. As an algorithm for solving the dynamic responses of layered soil under axisymmetric loads, the spectral element method (SEM) not only can consider the dynamic effect of FWD loads to obtain accurate calculation results, but also has high solution efficiency [10,11]. Thus, it has been successfully applied to the back-calculation of the dynamic modulus of pavement structures [3,12–17].

With the development of forward algorithms for the dynamic responses of layered pavement structures, a variety of dynamic modulus inversion software has been developed based on nondestructive testing techniques using FWD equipment. However, most existing inversion software is based on the peak deflection of the road surface, which only reflects the overall bearing capacity of the pavement structure. The deflection basin displacement time–history curve obtained by FWD testing equipment contains comprehensive pavement structure information, which cannot be completely reflected by its peak deflection. In the inversion process, the insufficient input conditions usually lead to the non-unique problem of parameter inversion. It is thus difficult for the inversion results to objectively characterize the actual performance of each structural layer of the pavement [13–15]. Zhang et al. [13] used SEM as the forward model, and used the backpropagation neural network (BPNN) and real-coded multi-population genetic algorithm to predict the structural parameters of pavements. While the input conditions of the pavement mechanical model were increased (the interlayer contact conditions and transverse isotropic properties of materials), the non-unique problem could not be solved by relying only on the maximum deflection of the road surface. Cao et al. [3] also used SEM as a forward analysis method, constructed an optimization objective function using the measured deflection time–history curve and theoretical calculation results, and carried out the dynamic inversion of the mechanical parameters of a pavement via numerical optimization methods. Compared with traditional static methods, better inversion results were achieved, and the phenomenon of "modulus transferring" was effectively avoided, but the inversion is not efficient and the results are not accurate. Li et al. [18] overcame the low search efficiency and heavy calculation workload by applying the conventional particle swarm optimization (PSO) [19] algorithm to the inversion of geotechnical engineering parameters, and proposed an improved PSO algorithm. However, the inversion based on peak deflection was still characterized by the problem of non-unique solutions, while the inversion based on the time–history curve involved a large calculation workload. Moreover, a single objective function cannot reflect the complex nonlinear function optimization problem.

The existing inversion algorithms are restricted by various factors. To solve the current challenges, an accurate and effective inversion method is needed. The convolutional neural network (CNN) has a strong feature extraction ability [20,21]. It is a potential scheme to realize pavement detection based on vibration signals, and can effectively solve the challenges faced by the above research. Due to its unique mechanism of partial connection and weight sharing, the CNN has exceptional advantages in processing massive data. In the field of damage detection based on vibration signals, Teng et al. [22] used a CNN to determine the damage location of a steel structure. Furthermore, Teng et al. [23] presented a novel and efficient approach for detection of structural damage from real-time vibration signals via a CNN. In terms of the prediction of the mechanical properties of materials by CNNs, Hu et al. [24] carried out research on the mechanical properties of hot-rolled steel. To further improve the computational efficiency of the CNN, Zheng et al. [25] established a fault diagnosis method for cylindrical roller bearing cages based on vibration signals and a 1-D CNN. These applications all confirm the unique advantages of CNNs in the field of diagnosis and prediction, as well as their potential application prospects for the task of processing pavement vibration signals and the inversion of pavement structural parameters.

For this reason, this study develops an accurate and effective inversion algorithm for the parameter identification of multi-layered pavement structures based on a CNN with SEM as the forward calculation model. As an algorithm for solving the dynamic responses of pavement structures, SEM can accurately simulate the dynamic effect of FWD loads and avoid the phenomenon of "modulus transferring". In addition, most of the inversion calculations use the maximum value of the deflection as the training samples, which do not reflect the overall dynamic characteristics of the pavement structures. As a result, the inversion results suffer from the non-unique problem, and the calculation accuracy cannot be guaranteed. In this paper, the deflection curves are used to replace the peak values of the deflection curves as the training samples, which can reflect the dynamic characteristics of the pavement structure more comprehensively. The CNN model is used to predict the dynamic modulus and thickness of the pavement structure layers. The proposed inversion algorithm is verified by the assumed numerical model of pavement structures, and relevant parameters in the algorithm are analyzed.

2. Forward Modeling of Pavement Structures

2.1. Pavement Structure System

It is assumed that a layered pavement structure is placed on a layered semi-infinite soil (Figure 1), both of which are isotropic and homogeneous, and the bottom soil can be an elastic or rigid bedrock. As displayed in Figure 1, a rectangular coordinate system is established. The coordinate origin O is set on the surface of the layered pavement structure, with the z-axis vertically downward and the x-axis in the horizontal direction. An FWD load is applied to a disc with a radius of Δ, and the center of the disk is at the origin of the coordinate system. The load is axisymmetric and expressed as $P(r,\ t) = S(r)F(t)$, where $S(r)$ represents the spatial distribution of the load and $F(t)$ represents its change with time. As the wave propagation under the load is axisymmetric, a cylindrical coordinate system is then established in Figure 2. In this study, SEM is used to solve the displacement responses of the deflection basin under FWD loads and provide an efficient and accurate forward model for the inversion calculation of the dynamic modulus and thickness of the layered pavement structure.

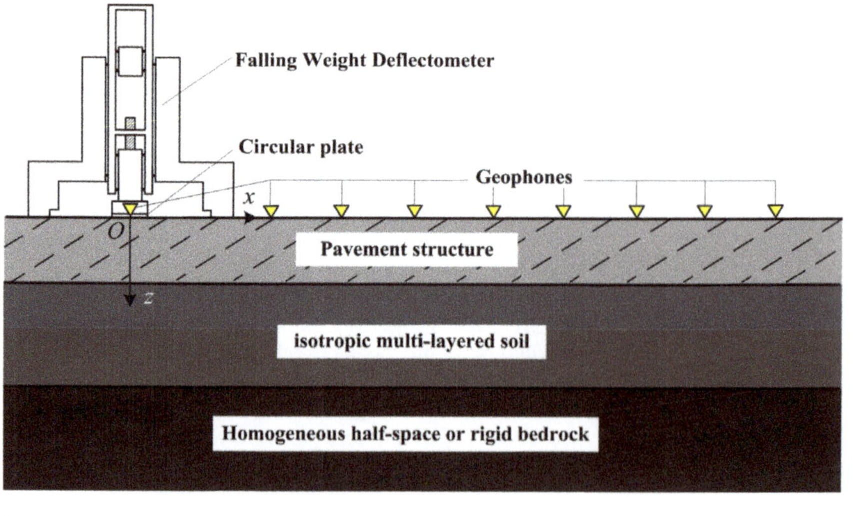

Figure 1. The schematic diagram of the FWD system.

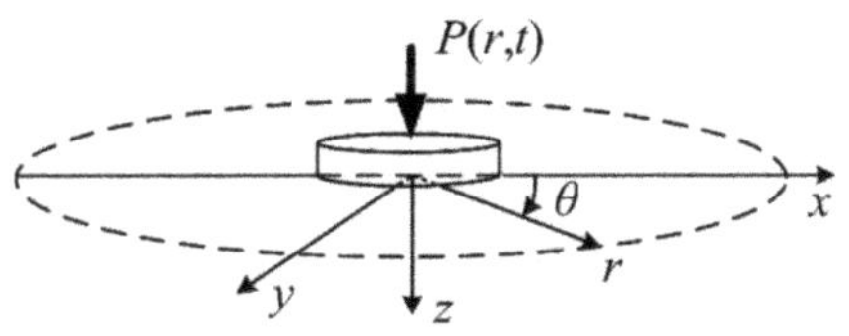

Figure 2. The wave propagation in the cylindrical coordinate system.

2.2. Spectral Element Method

For the axisymmetric wave problem, only the radial and vertical displacements are considered. The wave propagation formula of the continuous medium in the time–spatial domain under the cylindrical coordinate system is

$$\begin{aligned} \frac{\partial \sigma_r}{\partial r} + \frac{\partial \tau_{zr}}{\partial z} + \frac{\sigma_r - \sigma_\theta}{r} = \rho \frac{\partial^2 u_r}{\partial t^2}, \\ \frac{\partial \tau_{zr}}{\partial r} + \frac{\partial \sigma_z}{\partial z} + \frac{\tau_{zr}}{r} = \rho \frac{\partial^2 u_z}{\partial t^2}, \end{aligned} \tag{1}$$

where σ_i $(i = r, \theta, z)$ is the normal stress in the subscript direction, τ_{zr} is the shear stress in the z-r plane, ρ represents the mass density of the medium, and u_r and u_z represent the radial and vertical displacements of the medium, respectively.

In Helmholtz decomposition, the displacement field of the medium can be expressed as the gradient of scalar potential α and the curl of vector potential $\boldsymbol{\beta}$, i.e.,

$$u = \nabla \alpha + \nabla \times \boldsymbol{\beta}, \tag{2}$$

where ∇ represents a partial differential operator. In the axisymmetric wave problem, vector potential $\boldsymbol{\beta}$ has only one component β_θ, so the problem to be solved can be simplified as a scalar potential problem. For simplicity, β is used to represent β_θ. As it is an axisymmetric problem, the displacement of the medium in the circumferential direction θ is zero, and the radial and vertical displacement can be expressed as follows [10]:

$$u_r = \frac{\partial \alpha}{\partial r} - \frac{\partial \beta}{\partial z}, \quad u_z = \frac{\partial \alpha}{\partial z} + \frac{1}{r}\frac{\partial (r\beta)}{\partial r}. \tag{3}$$

The relationship between stress and displacement in the cylindrical coordinate system is

$$\begin{aligned} \sigma_r &= (\lambda + 2\mu)\frac{\partial u_r}{\partial r} + \mu\left(\frac{u_r}{r} + \frac{\partial u_z}{\partial z}\right), \\ \sigma_\theta &= (\lambda + 2\mu)\frac{u_r}{r} + \mu\left(\frac{\partial u_r}{\partial r} + \frac{\partial u_z}{\partial z}\right), \\ \sigma_z &= (\lambda + 2\mu)\frac{\partial u_z}{\partial z} + \mu\left(\frac{\partial u_r}{\partial r} + \frac{u_r}{r}\right), \\ \tau_{zr} &= \mu\left(\frac{\partial u_z}{\partial r} + \frac{\partial u_r}{\partial z}\right), \end{aligned} \tag{4}$$

where λ and μ are the Lamé constants, and

$$\lambda = \frac{\nu E}{(1+\nu)(1-2\nu)}, \ \mu = \frac{E}{2(1+\nu)}, \tag{5}$$

where E is the Young's modulus and ν is the Poisson's ratio.

Equation (4) is substituted into Equation (1) and the displacement is expressed by the scalar potential in Equation (3). After algebraic simplification, the following equations are satisfied:

$$(\lambda + 2\mu)\left(\frac{\partial^2 \alpha}{\partial r^2} + \frac{\partial^2 \alpha}{\partial z^2} + \frac{\partial \alpha}{r\partial r}\right) = \rho \frac{\partial^2 \alpha}{\partial t^2} \tag{6}$$

$$\mu\left(\frac{\partial^2 \beta}{\partial z^2} + \frac{\partial^2 \beta}{\partial r^2} + \frac{\partial \beta}{r\partial r} - \frac{\beta}{r^2}\right) = \rho \frac{\partial^2 \beta}{\partial t^2} \tag{7}$$

It is evident that Equations (6) and (7) represent two forms of waves, namely P and SH waves, respectively, the wave velocities of which are $c_P = \sqrt{\frac{(\lambda+2\mu)}{\rho}}$ and $c_{SH} = \sqrt{\frac{\mu}{\rho}}$, respectively. For isotropic layered media, the two waves can be decoupled and solved separately. The potential function α in Equation (6) in the frequency–wavenumber domain can be obtained by using the variable separation method:

$$\hat{\alpha}(r,z) = A_n e^{-ik_{pz}z} J_0(kr), \tag{8}$$

where superscript "^" represents the function of the frequency–wavenumber domain, A_n represents an unknown constant which is determined by the boundary conditions, J_0 is the first kind of Bessel function, and k is the wavenumber. The expression of k_{pz} is

$$k_{pz} = \left(\frac{\omega_n^2}{c_P^2} - k^2\right)^{\frac{1}{2}}, \tag{9}$$

where ω_n is the frequency.

To introduce the boundary conditions in the radial direction, the amplitude of the oscillation at the radial boundary $r = R$ (far away from the load source) is considered to vanish. That requires the item $J_0(kR)$ to be equal to 0. This condition can be satisfied at the infinite positive roots $k_m R$ of the J_0 function. Hence, the m functions,

$$\hat{\alpha}_{mn}(r,z) = A_{mn} e^{-ik_{pz}z} J_0(k_m r), \tag{10}$$

In the same way, the solution of the potential function β in the frequency–wavenumber domain in Equation (7) is

$$\hat{\beta}_{mn}(r,z) = B_{mn} e^{-ik_{sz}z} J_1(k_m r), \tag{11}$$

where B_{mn} is a constant determined by the boundary conditions and J_1 is the first kind of Bessel function. Moreover, k_{sz} represents the vertical shear wavenumber, and is expressed as

$$k_{sz} = \left(\frac{\omega_n^2}{c_{SH}^2} - {k_m}^2\right)^{\frac{1}{2}} \tag{12}$$

Because Equations (6) and (7) are linear homogeneous formulas, the superposition principle is applicable, i.e., the solutions of the potential functions α and β can be obtained by the superposition of the infinitely many solutions of $\hat{\alpha}_{mn}$ and $\hat{\beta}_{mn}$. The inversion of the integral transformation is usually solved using infinite numerical integration [6,26]. A high number of wavenumbers must be calculated to approximate the infinite integration. In order to obtain an accurate result, the step of the integration needs to be divided very small, which can reduce the computational efficiency significantly. In SEM, the double-sum of the discrete angular frequency ω_n $(n = 1, 2, \ldots, N)$ and wavenumber k_m $(m = 1, 2, \ldots, M)$ is used to obtain all arrays of the whole system dissipated at $r = R$ (R is a large finite value). The summation over the wave numbers in SEM is discontinuous, and the wave number is finite. Therefore, the general solution of the wave formula of an axisymmetric system can be obtained by superimposing the special solution [10], as follows.

$$\alpha(r,z,t) = \sum_N \sum_M A_{mn} e^{-ik_{pz}z} J_0(k_m r) e^{-i\omega_n t} \tag{13}$$

$$\beta(r,z,t) = \sum_N \sum_M B_{mn} e^{-ik_{sz}z} J_1(k_m r) e^{-i\omega_n t} \tag{14}$$

The finite values N and M in the general solution Equations (13) and (14) can be determined by the time and spatial amplitude spectra of the excitation load. The sum of N frequencies can be obtained by using the fast Fourier transform (FFT), and the sum of M

wavenumbers can be obtained by using the Fourier–Bessel series. The sum of wavenumbers replaces the integral from 0 to infinity in the conventional numerical integration method, which is crucial to improving the efficiency of the algorithm. For the infinite integral, when the damping ratio of the medium is zero, the integrand has singular points, which makes its integral calculation more expensive. The sum of the wavenumbers in double-sum form can effectively avoid this problem.

The general solutions of the potential functions α and β (Equations (13) and (14)) are substituted into Equation (3) to obtain the displacements of the axisymmetric problem, as follows:

$$\begin{Bmatrix} u_r(r,z,t) \\ u_z(r,z,t) \end{Bmatrix} = \sum_n \sum_m \begin{Bmatrix} \hat{G}_{rz}(k_m,z)J_1(k_m r) \\ \hat{G}_{zz}(k_m,z)J_0(k_m r) \end{Bmatrix} \hat{F}_m \hat{F}_n e^{i\omega_n t}, \tag{15}$$

where $\hat{G}(k_m,z)$ represents the flexibility coefficient of the medium in the conversion domain, and its solution is evident [10], and $\hat{F}_m$ is the Fourier–Bessel transformation coefficient of the excitation load $S(r)$, which can be obtained by the Fourier–Bessel transformation of the excitation load $S(r)$. For the contact boundary condition at the interface between the adjacent layers, the displacements and stresses meet the continuous conditions. If the layered soil is underlain by an elastic semi-infinite space, the radiation condition should be considered. Two spectral elements were developed in the literature [10] to illustrate the solution of $\hat{G}(k_m,z)$ as well as the contact boundary condition and radiation condition.

Assuming that the FWD load is applied to the disc with radius Δ, $S(r)$ can be expressed as follows:

$$S(r) = \begin{cases} q & \text{for } 0 \leqslant r \leqslant \Delta \\ 0 & \text{for } r > \Delta \end{cases}, \tag{16}$$

The Fourier–Bessel transformation of $S(r)$ is

$$S(r) = \sum_M \hat{F}_m J_0(k_m r), \tag{17}$$

where $\hat{F}_m$ can be parsed as

$$\hat{F}_m = \frac{2}{R^2 J_1^2(\alpha_m)} \int_0^a r J_0\left(\frac{\alpha_m}{R} r\right) dr = \frac{2\Delta}{\alpha_m R J_1^2(\alpha_m)} J_1\left(\frac{\alpha_m}{R}\Delta\right), \tag{18}$$

where $\alpha_m = \frac{k_m}{R}$, and $\hat{F}_n$ is the Fourier transform coefficient of excitation load $F(t)$, which can be obtained by FFT. Under the discrete Fourier transform state, the relationship between $F(t)$ and $\hat{F}_n$ is

$$\begin{aligned} F(j\Delta t) &= \tfrac{1}{N} \sum_n \hat{F}_n\left(\tfrac{n}{N\cdot\Delta t}\right) e^{\frac{i2\pi nj}{N}} \; (n = 1,2,\ldots,N-1), \\ \hat{F}_n\left(\tfrac{n}{N\cdot\Delta t}\right) &= \sum_j F(j\Delta t) e^{-\frac{i2\pi nj}{N}} \; (j = 1,2,\ldots,N-1), \end{aligned} \tag{19}$$

where Δt is the load time step.

2.3. Forward Model Validation

First, the accuracy of SEM was verified by selecting a three-layer pavement structure model placed on a homogeneous semi-infinite soil. Grenier et al. [27] studied a general pavement in Quebec, Canada, including a 150 mm thick asphalt layer, a 300 mm thick granular material base course, and a 600 mm thick subbase sand layer set on the semi-infinite soil, as shown in Figure 3. The generic pavement was isotropic, and its material parameters, including its Young's modulus, Poisson's ratio, mass density, and damping ratio, are reported in Table 1.

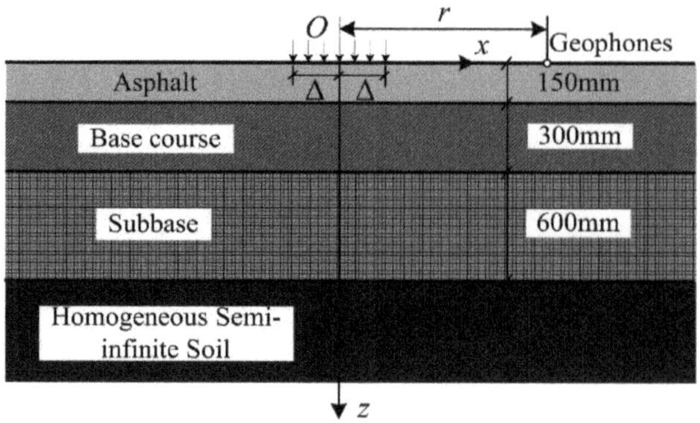

Figure 3. A generic multi-layered pavement structure resting on a homogeneous half-space.

Table 1. The properties of the generic pavement.

Layers	h (m)	E (MPa)	μ	ρ (kg/m^3)	ζ (%)
Asphalt	0.15	3000	0.35	2400	5
Base course	0.30	300	0.35	2300	2
Subbase	0.60	150	0.35	1900	2
Subgrade	-	75	0.35	1800	2

The FWD load applied on the pavement is shown in Figure 4; the load amplitude was 40 kN, the duration was 30 ms, the number of Fourier transform samples $N = 2048$, and the number of wavenumber samples $M = 200$. The load acted within a circle of radius 0.15 m, i.e., the radius of the drop hammer was 0.15 m. The displacement of the deflection point was calculated at the radial distances from the load center point $r = 0, 0.25, 0.375, 0.50, 0.625, 0.75, 1.00, 1.25, 1.50$, and 2.00 m. The oscillation dissipation distance was assumed to be $R = 150$ m. The calculated displacement time–history curve is shown in Figure 5a, and the calculated maximum displacement curve at the deflection point is shown in Figure 5b. The calculation results were compared with the published results in the reference [27]. The figure reveals that the two results were basically completely matched, thus verifying the accuracy of the calculation results of the proposed algorithm in the time domain.

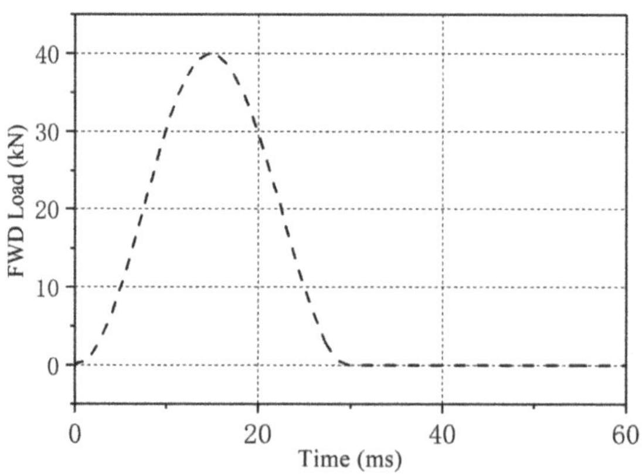

Figure 4. The FWD load pulse on the circular disk.

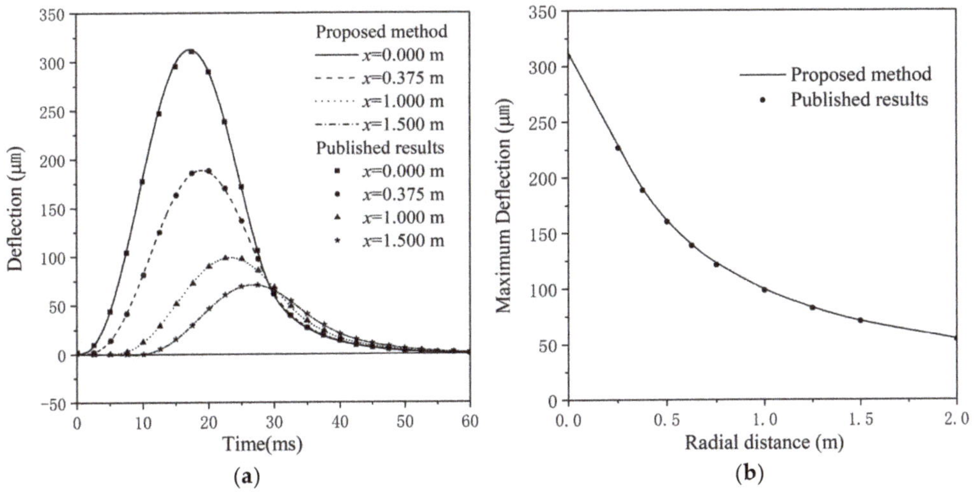

Figure 5. The response of the generic multi-layered pavement system [27]. (**a**) Displacement-time curve at deflection points; (**b**) Maximum displacement curve at deflection points.

3. Inversion Method for the Dynamic Modulus of Pavement Structures

3.1. Convolutional Neural Network

A CNN can effectively reduce the number of network parameters, alleviate the model overfitting problem, and reduce the memory occupied by deep networks. A standard fitting CNN is usually composed of convolution layers, pooling layers, activation layers, fully connected layers, and regression layers. In practical applications, the convolution and activation layers are often referred to together as convolution layers. The data input (network input) transmits through a series of layers, and is finally mapped to the fitting value corresponding to the original data through regression layers. In particular, the input of a 1-D CNN is a $1 \times N$ or $N \times 1$ matrix. Figure 6 presents the basic structure of the CNN. The feature extraction is carried out through the convolution layer and the pooling layer, and the fitting value (parameter value of soil layers 1–3) is finally obtained in the regression layer.

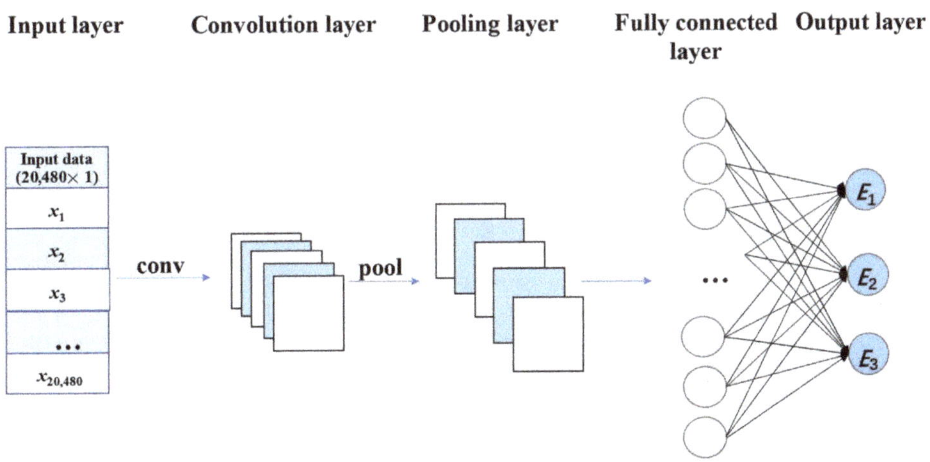

Figure 6. The basic structure of the CNN.

A 1-D CNN is often used in sequence models. Figure 7a presents a typical 1-D CNN operation, in which the original data of the convolution layer are an input matrix whose size is 1 × 6, and the convolution kernel is a matrix whose size is 1 × 3. The convolution operation process is to multiply each element in the convolution kernel with the corresponding element in a sub-region (e.g., orange box or purple dotted box) of the original data of the convolution layers, and then add the results to obtain an element in the feature vector. Each time the convolution operation is performed, the convolution kernel moves down with a certain step size (the step size is 1 in Figure 7a), and the operation process is repeated until all elements of the original data are traversed. Finally, after the convolution operation, a new array is formed (for example, the feature vector [8 23 6 29]).

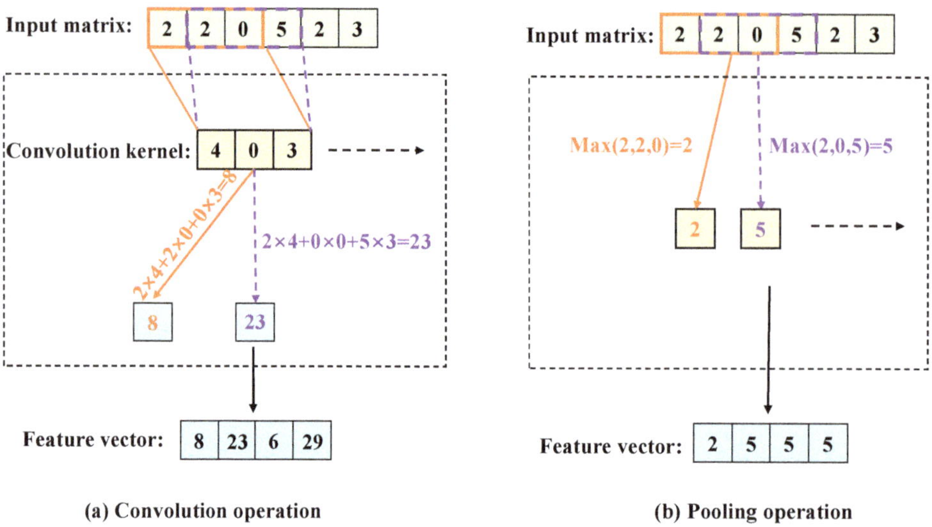

Figure 7. The basic principle of the convolution and pooling operations.

The pooling operation is a down sampling technique. By reducing the characteristic dimensions of the output after convolution, the network parameters can be effectively reduced, thus accelerating the calculation process. This can greatly improve the computational speed of the CNN and effectively prevent overfitting. During pooling, the input matrix is divided into several regions, and the maximum value is output for each subregion. There are usually two different pooling methods, namely maximum pooling and average pooling. For the purposes of this study, maximum pooling is superior to average pooling. Figure 7b shows the maximum pooling operation.

The activation function introduces nonlinear factors into the neural network and improves its fitting ability to learn complex features. Thus, the neural network can approximate any nonlinear function, and the convergence of the network can also be accelerated. Common activation functions include the Sigmoid, Tanh, ReLU, and Leaky ReLU. The activation function used in this study is Leaky ReLU, which is similar to ReLU and helps to expand the applicability of the ReLU function. This function retains the characteristic information that the negative values may contain. It has greater application significance than the ReLU function. Its form is

$$\text{Leaky ReLU}(x) = \begin{cases} x, \ x \geq 0 \\ \alpha x, x < 0 \end{cases} \tag{20}$$

where x is the input value of the network, and α is 0.01 by default.

3.2. Inversion Process

The training process of the CNN is divided into two stages, namely forward propagation and backward propagation. In this study, the time–history curve of pavement deflection obtained by using SEM is used as the CNN input, and the parameter value of the dynamic modulus of the three-layer pavement structure is used as the CNN output. In the forward propagation process, the pavement deflection time–history curve is calculated via a series of convolution–activation–pooling to obtain the dynamic modulus of the three-layer pavement structures, i.e., the parameters of the soil layers are predicted through the response. By comparing the difference between the predicted and exact values, the training error of the CNN is obtained. When the error is greater than the expected value, the backward propagation process is carried out, the error is sent back to the network to readjust the weight, and the training is continued to reduce the error. When the error is equal to or less than the expected value (close to 0, reaching convergence), the network ends the training.

The data set of the CNN prediction model used in this study consisted of a training set, a validation set, and a testing set. The training set was used to train the parameters in the neural network, and the validation set was used to verify the performance of the training model. Whether the network training reached convergence was judged according to the root mean square error (RMSE) and the loss function of the validation set. The testing set was used to evaluate the performance of the trained neural network.

To ensure the accuracy of the inversion results, two groups of comparative training were conducted. According to the elastic modulus E of the three-layer pavement structure (as shown in Figure 3), different values were selected to establish the sample set. The value range of E_1 of the first layer was selected as 1500–4500 MPa, the value range of E_2 of the second layer was 150–450 MPa, and the value range of E_3 of the third layer was 75–225 MPa. The elastic modulus of each layer was considered to be 11 different values at equal intervals. Therefore, 1331 samples were established as the data set of the neural network. A certain number of samples were randomly selected from the 1331 samples for training and testing.

To avoid random and accidental errors in the experiment process, both the testing and validation sets are 100 randomly selected discrete samples from the 1331 samples.

The first group of comparative training aimed to find the optimal sample input dimension. The specific division of the training data set is shown in Table 2. The first sampling method was to take the peak values of the 10 FWD deflection basin curves, and the second sampling method was to take the entire curves. The sample input and output dimensions are reported in Table 3.

Table 2. The division of the first group of comparative training data sets.

Project	Training Set	Validation Set	Testing Set	Total
Quantity	950	100	100	1150

Table 3. The input and output dimensions of the first group of comparative training.

Sampling Method	Input Dimension	Output Dimension
Peak value of 10 deflection curves	10	3
All values of 10 deflection time history curves	20,480	3

The purpose of the second group of comparative training was to determine the optimal number of training samples and carry out a comparative test in which the number of training samples in eight groups varied from 600 to 950. The specific division of the second group of comparative training data sets is exhibited in Table 4.

Table 4. The division of the second group of comparative training data sets.

The Best Training Sample Quantity Comparison Training	Training Set	Validation Set	Testing Set	Total
①	600	100	100	800
②	650	100	100	850
③	700	100	100	900
④	750	100	100	950
⑤	800	100	100	1000
⑥	850	100	100	1050
⑦	900	100	100	1100
⑧	950	100	100	1150

Based on the divided data set, the specific parameter settings of the convolution layer after CNN model optimization are presented in Table 5. The pooled window size of a pooled layer was 1 × 2, and the pooled window moved in two steps. The activation function used by the activation layer was the Leaky ReLU function.

Table 5. The parameter settings of the convolution layer.

Convolution Layer	Size	Step	Number of Convolution Kernels	Padding
1	1 × 2	1	60	0

The performance evaluation indicators of common regression models mainly include the RMSE, mean absolute error (MAE), mean square error (MSE), and R^2 value. Among them, the RMSE is a typical evaluation indicator of regression models, and is used to represent the size of the error generated in the model prediction process. In this study, the RMSE and relative error were used as the evaluation indicators of the CNN model, and their expressions are

$$\text{RMSE} = \sqrt{\frac{1}{N}\sum_{t=1}^{N}(observed_t - predicted_t)^2}, \tag{21}$$

$$\text{Relative error} = \frac{observed_t - predicted_t}{observed_t} * 100\% \tag{22}$$

where N is the number of samples.

4. Analysis of the Inversion Results

4.1. *Influence of Different Input Dimensions on the Inversion Results*

The trained CNN was used to predict the validation sample, and the RMSE values are reported in Table 6. The RMSE obtained by the first input condition (peak value of 10 deflection curves in Table 3) was found to be far greater than that obtained by the second input condition (all values of 10 deflection time history curves in Table 3). Therefore, the inversion of the entire deflection curves was more accurate than their peak values. The first input condition does not reflect the overall dynamic characteristics of the pavement structures. As a result, the inversion results have a non-unique problem. In this example, the deflection curves are used to replace the maximum values of the deflection curves as the training samples, which can reflect the dynamic characteristics of the pavement structure more comprehensively. The calculation results also show the high accuracy of the algorithm. The main reason why existing algorithms seldom use the entire deflection time–history curve for inversion is that they cannot cope with huge data sets. The CNN inversion model used in this study has better regression performance and can overcome this problem more effectively.

Table 6. The comparison of the training effects of CNN models with two kinds of input data.

Sampling Method	Peak Values of 10 Deflection Curves	All Values of 10 Deflection Curves
RMSE of validation samples	0.1564	0.0417

4.2. *Influence of Different Sample Numbers on the Inversion Results*

The final calculation results of the second comparison group are exhibited in Figure 8. With the increase in the number of training samples, the RMSE of the validation samples decreased gradually. Figure 8 shows that when the number of training samples was 950, the RMSE of the validation samples reached the minimum value of 0.0417, and the validation accuracy of the CNN model reached about 96%. To further illustrate the accuracy of this algorithm for the inversion of the dynamic modulus of the pavement structure, the evaluation metrics (RMSE, R^2 value, MAPE (mean absolute percentage error), and MAE) of the algorithm in this paper are shown in Table 7 when the training set samples are taken as 950.

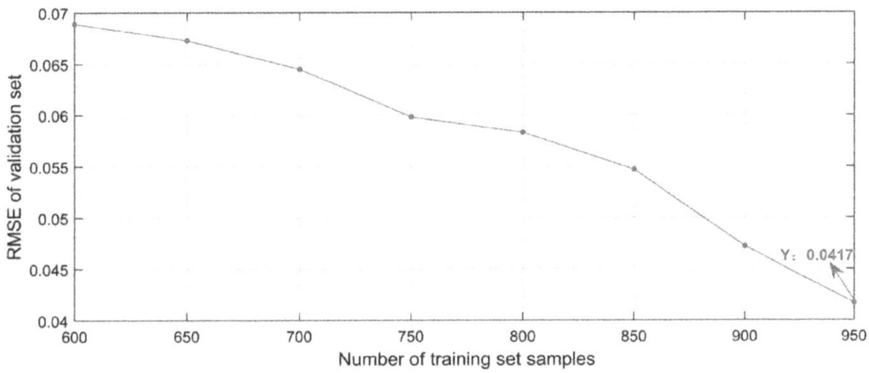

Figure 8. The RMSE change of the validation samples.

Table 7. Training result.

Training Set	RMSE	R^2	MAPE	MAE
950	0.0417	0.9809	[4.11% 3.18% 3.34%]	[101.0193 8.6192 4.5244]

4.3. *The Inversion Calculation Results of the Dynamic Modulus for Pavement Structures*

To further illustrate the accuracy of the proposed algorithm for the inversion of the dynamic modulus of the pavement structure, the 950 training samples used in Section 3.2 were used for exploration. The resulting changes in the RMSE and training loss curve of the training samples predicted by the CNN model are shown in Figure 9. During the first 300 iterations of the CNN training, the loss curve showed a downward trend with the improvement of prediction accuracy. After 3000 iterations at a certain stage of training, the loss value reached the minimum, the loss curve tended to be stable, and the CNN model converged. The changing trend of the RMSE curve of the training samples decreased greatly at first and then tended to be stable.

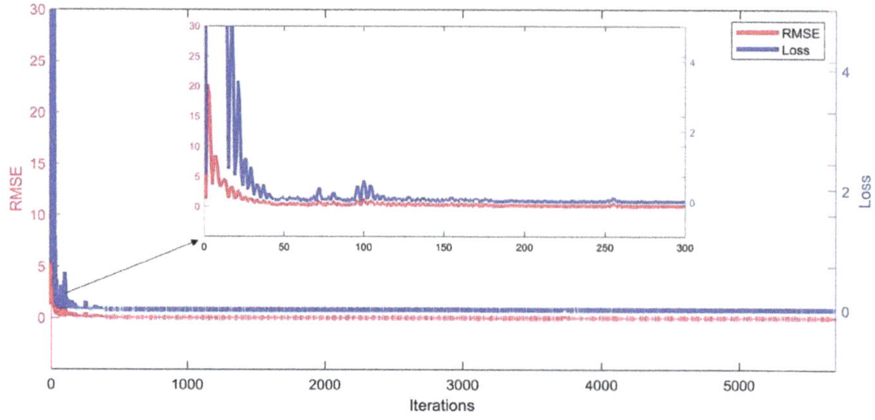

Figure 9. The RMSE and loss change curve of the training samples.

To reflect the prediction effect of the proposed inversion model more intuitively for the dynamic moduli of the pavement structure layers, the inversion results of the validation samples were compared with the exact values, as exhibited in Figure 10. Figure 10a–c represent the comparison results of the dynamic elastic moduli E_1, E_2, and E_3 of the pavement structure layers, respectively. The predicted and exact values of the dynamic elastic modulus of the first layer of the pavement were found to be basically completely matched. Furthermore, the dynamic elastic moduli of the second and third layers were also well-matched, with only a few samples having certain errors. As the FWD load is applied on the road surface, the deflection curves obtained contain more information on the dynamic characteristics of the surface layer than that of the deep pavement layers. That is, with the increase in the depth, the correlation of the pavement structure properties to the displacement of the road surface decreases; thus, the relative error of its dynamic modulus gradually increases. The relative errors of the inversion results of the dynamic elastic moduli E_1, E_2, and E_3 of the three-layer pavement structure are shown in Figure 11. The MAPE of the first layer was 4.11%, and the maximum relative error was 9.53%. The MAPE of the second and third layers were basically within 3.34%, and their maximum relative errors were 11.33% and 11.91%, respectively. It is worth noting that only 950 samples were selected for training in this example. More accurate inversion results can be achieved if the number of training samples is increased.

4.4. The Inversion Calculation of the Dynamic Modulus and Depth for Pavement Structures

As with the dynamic modulus, the thickness of the pavement structures also has a significant effect on the pavement deflection. In this example, the same pavement profile as in Section 4.3 was used with the assumption that, in addition to the dynamic modulus, the thicknesses of the three pavement structure layers were also unknown. Hence, the number of the unknown inversion variables is twice as large as the example in Section 4.3. In order to obtain an accurate result, more samples are needed for the training than the previous example. A total of 3000 samples are calculated, of which 2400 samples were used for training, 300 samples for validation, and 300 samples for testing. The dynamic modulus and thickness are randomly generated in these 3000 samples. The value ranges of these inversion variables are shown in Table 8.

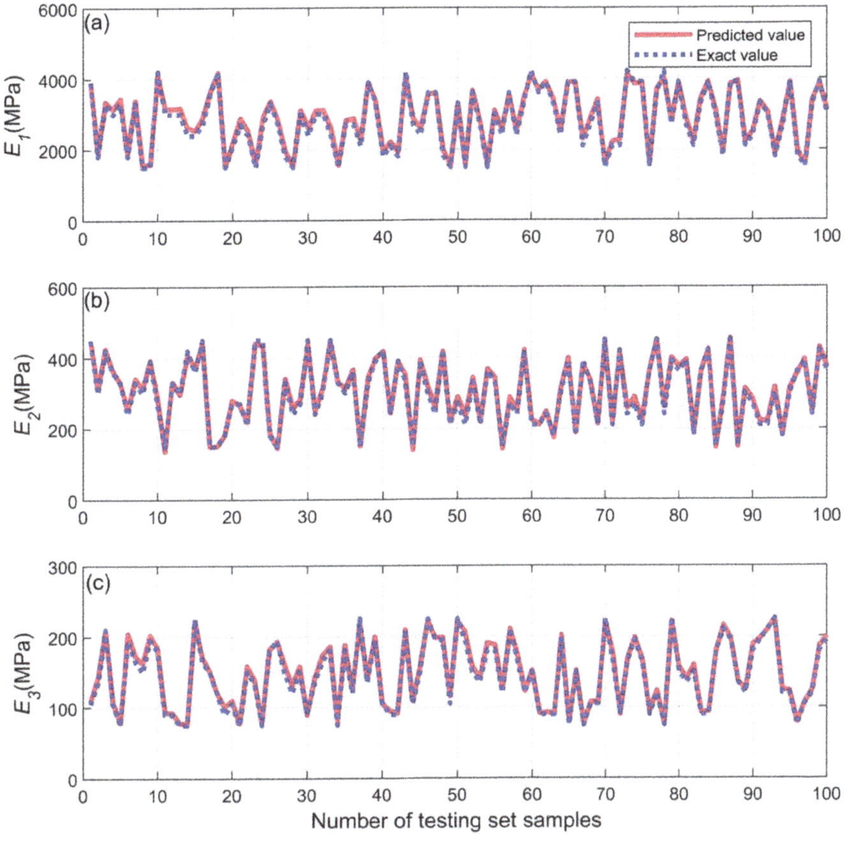

Figure 10. The curves of the predicted and measured values of the testing samples. (**a**) Dynamic elastic modulus of the first layer of the pavement structure; (**b**) Dynamic elastic modulus of the second layer of the pavement structure; (**c**) Dynamic elastic modulus of the third layer of the pavement structure.

Table 8. The value range of the inversion variables.

Inversion Variables	E_1 **(MPa)**	E_2 **(MPa)**	E_3 **(MPa)**	h_1**(m)**	h_2**(m)**	h_3**(m)**
Value range	1500–4500	150–450	75–225	0.075–0.225	0.15–0.45	0.30–0.90
average relative error	7.96%	3.52%	6.90%	10.58%	8.51%	11.97%

The inversion results of the validation samples were compared with the exact values, as exhibited in Figures 12 and 13. Figure 12a–c represent the comparison results of the dynamic elastic moduli E_1, E_2, and E_3 of the pavement structure layers, respectively. Figure 13a–c represent the comparison results of the dynamic elastic moduli h_1, h_2, and h_3 of the pavement structure layers, respectively. The average relative error of the six inversion variables is listed in Table 8. As the figures show, the predicted modulus and depth obtained by the presented inversion algorithm have a good agreement with exact modulus and depth. The non-unique problem in the inversion calculation of the pavement structure parameters is solved by the proposed algorithm in engineering accuracy requirements.

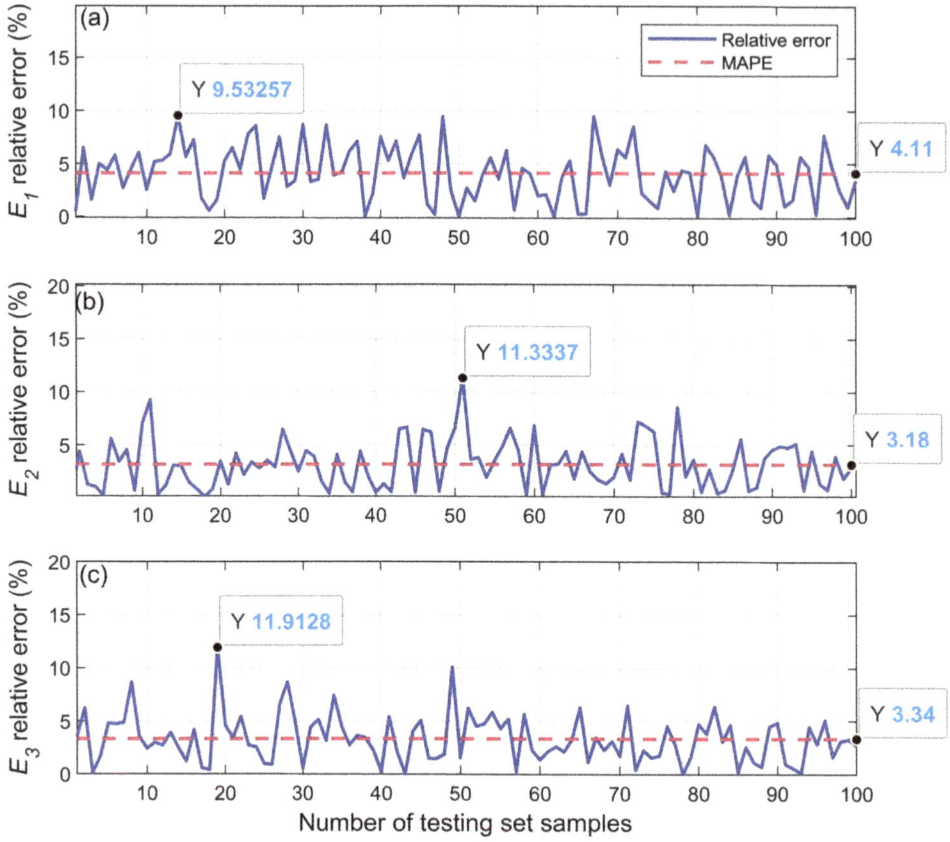

Figure 11. The relative error curve of the testing samples. (**a**) Relative error of the first layer of the pavement structure; (**b**) Relative error of the second layer of the pavement structure; (**c**) Relative error of the second layer of the pavement structure.

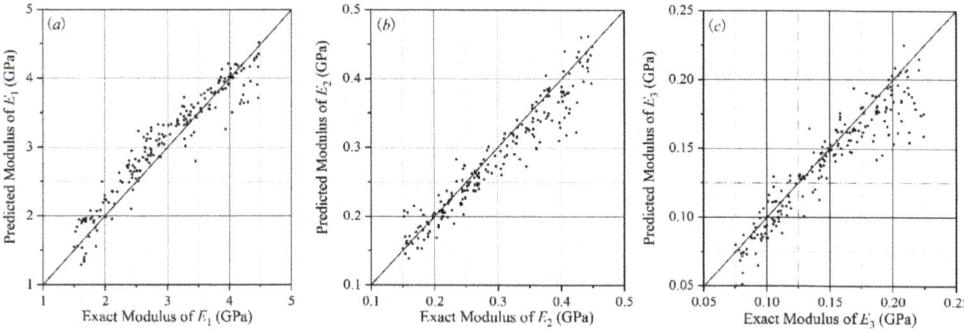

Figure 12. Comparison between the predicted and exact dynamic modulus of the testing samples. (**a**) Dynamic elastic modulus of the first layer of the pavement structure; (**b**) Dynamic elastic modulus of the second layer of the pavement structure; (**c**) Dynamic elastic modulus of the third layer of the pavement structure.

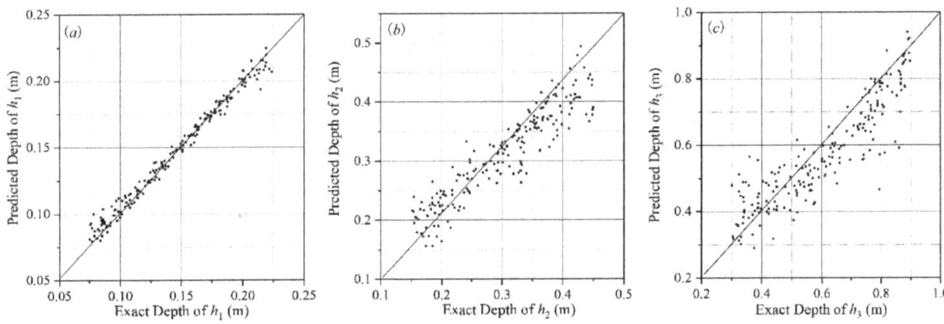

Figure 13. Comparison between the predicted and exact depth of the testing samples. (**a**) Thickness of the first layer of the pavement structure; (**b**) Thickness of the second layer of the pavement structure; (**c**) Thickness of the third layer of the pavement structure.

5. Conclusions

In this study, an algorithm for parameter identification of concrete pavement structures has been presented. SEM was used as the forward algorithm for the dynamic responses of layered pavement structures, and the time–history curve of the pavement deflection basin displacement responses under FWD loads was obtained. The accuracy of the forward algorithm was verified via a comparison with the existing literature. Based on this, the CNN was adopted, and the deflection basin displacement time–history curves were taken as the input condition to establish a numerical model for the parameter identification of layered pavement structures. The main conclusions of this research are as follows.

1. Compared with inversion using the peak values of the pavement deflection basin displacement as the input condition, more accurate inversion results can be obtained by using the entire time–history curves of the pavement deflection basin displacement as the input condition. The phenomenon of "modulus transferring" is avoided, as well as the non-unique problem.
2. With the increase in the number of samples, the accuracy of the inversion results increases gradually. For the inversion model of the dynamic modulus of the three-layer pavement structure, when the number of training samples reached 950, the RMSE was only 0.0417, which proves the high accuracy of the proposed inversion algorithm. For the inversion of the dynamic modulus of the pavement structure, the MAPE of the inversion results was found to be basically controlled within 4.11%, and the MAPE of the inversion results of the base and subbase layers were also basically controlled within 3.34%. Thus, inversion results were found to have adequate numerical stability.
3. The algorithm can also be used for the inversion of the thickness and dynamic modulus of the pavement structure layer. As the number of inversion variables increases, the inversion accuracy will decline to a certain extent. The algorithm in this paper is of a high solution accuracy for the inversion of the dynamic elastic modulus of the pavement structure surface. The average relative error of the inversion results is basically controlled within 8%, while the inversion accuracy for the thickness is relatively low. The average relative error of the inversion results is basically controlled within 12%, and the inversion results have adequate numerical stability.
4. The limitation of this study is that the contact boundary conditions between the pavement structure and the subgrade are not considered, and all the results obtained are purely simulation results without considering the impact of noise. Previous research shows that the interface between the pavement structure and the subgrade will have a certain degree of relative slip, which has a certain impact on the dynamic response of the pavement surface. Therefore, the inversion accuracy of pavement structure dynamic modulus will also be affected. The contact boundary conditions

will be considered in future research, and the proposed algorithm will be applied to the actual tests and engineering practice.

Author Contributions: Methodology, G.C. and Z.H.; Software, X.C. and L.Y.; Validation, X.C. and L.Y.; Investigation, D.B.; Data curation, X.C. and L.Y.; Supervision, D.B.; Funding acquisition, D.B. All authors have read and agreed to the published version of the manuscript.

Funding: This research was funded by Guangdong Province Universities and Colleges Characteristic Innovation Project of China (2022KTSCX171).

Institutional Review Board Statement: Not applicable.

Informed Consent Statement: Not applicable.

Data Availability Statement: Some or all data, models, or codes generated or used during the study are available from the corresponding author by request.

Conflicts of Interest: The authors declare no conflict of interest.

References

1. Aashto. *Guide for Design of Pavement Structures*; AASHTO: Washington, DC, USA, 1993.
2. Kutay, M.E.; Chatti, K.; Lei, L. Backcalculation of Dynamic Modulus Mastercurve from Falling Weight Deflectometer Surface Deflections. *Transp. Res. Rec.* **2011**, *2227*, 87–96. [CrossRef]
3. Cao, D.D.; Zhao, Y.Q.; Kong, F.S.; Wang, G.Z. Dynamic Backcalculation for Parameters of Asphalt Pavement with Rigid Base. *China J. Highw. Transp.* **2018**, *31*, 66–73.
4. Dubois, F.; Moutou-Pitti, R.; Picoux, B.; Petit, C. Finite element model for crack growth process in concrete bituminous. *Adv. Eng. Softw.* **2012**, *44*, 35–43. [CrossRef]
5. Wang, H.; Li, M.Y. Comparative Study of Asphalt Pavement Responses under FWD and Moving Vehicular Loading. *J. Transp. Eng.* **2016**, *142*, 4016069. [CrossRef]
6. Lin, G.; Han, Z.; Li, J. An efficient approach for dynamic impedance of surface footing on layered half-space. *Soil Dyn. Earthq. Eng.* **2013**, *49*, 39–51. [CrossRef]
7. Rajapakse, R.; Wang, Y. Green's functions for transversely isotropic elastic half space. *J. Eng. Mech.* **1993**, *119*, 1724–1746. [CrossRef]
8. Ai, Z.Y.; Li, Z.X.; Cang, N.R. Analytical layer-element solution to axisymmetric dynamic response of transversely isotropic multilayered half-space. *Soil Dyn. Earthq. Eng.* **2014**, *60*, 22–30. [CrossRef]
9. Khojasteh, A.; Rahimian, M.; Pak, R.Y.S.; Eskandari, M. Asymmetric Dynamic Green's Functions in a Two-Layered Transversely Isotropic Half-Space. *J. Eng. Mech.* **2008**, *134*, 777–787. [CrossRef]
10. Al-Khoury, R.; Scarpas, A.; Kasbergen, C.; Blaauwendraad, J. Spectral element technique for efficient parameter identification of layered media. I. Forward calculation. *Int. J. Solids Struct.* **2001**, *38*, 1605–1623. [CrossRef]
11. Han, Z.; Zhang, J.; Xue, L.; Fang, H.; Xiao, Z. Dynamic simulation of FWD tests on flexible transversely isotropic pavements with imperfect interfaces. *Comput. Geotech.* **2021**, *130*, 103914. [CrossRef]
12. Zhao, Y.; Cao, D.; Chen, P. Dynamic backcalculation of asphalt pavement layer properties using spectral element method. *Road Mater. Pavement Des.* **2015**, *16*, 870–888. [CrossRef]
13. Junhui, Z.; Haishan, F.; Shiping, Z.; Jie, L. Analytical Solution for the Dynamic Responses and Parameter Inversion of Pavement Structures Considering the Condition of Interlayer Contact. *China J. Highw. Transp.* **2021**, *34*, 11–23.
14. Chun-Ying, W.; Xiu-Run, G.; Xue-Yan, L.; Scarpas, A. Inverse dynamic system for pavement structure based on 3D spectral elements. *Chin. J. Geotech. Eng.* **2007**, *14*, 1060–1064.
15. Al-Khoury, R.; Kasbergen, C.; Scarpas, A.; Blaauwendraad, J. Spectral element technique for efficient parameter identification of layered media: Part II: Inverse calculation. *Int. J. Solids Struct.* **2001**, *38*, 8753–8772. [CrossRef]
16. Lai, J.; Liu, J.; Huang, C. The Application of Frequency-Temperature Superposition Principle for Back-Calculation of Falling Weight Deflectometer. *Appl. Sci.* **2020**, *10*, 132. [CrossRef]
17. Svilar, M.; Peško, I.; šešlija, M. Model for Estimating the Modulus of Elasticity of Asphalt Layers Using Machine Learning. *Appl. Sci.* **2022**, *12*, 10536. [CrossRef]
18. Li, X.L.; Wang, F.M.; Li, X.N. Improved Particle Swarm Optimization for Elastoplastic Back Analysis in Geotechnical Engineering. *J. Min. Saf. Eng.* **2009**, *26*, 50–54.
19. Wei, G. Back analysis algorithm in geotechnical engineering based on particle swarm optimization. *Rock Soil Mech.* **2006**, *20*, 795–798.
20. Cha, Y.; Choi, W.; Büyüköztürk, O. Deep Learning-Based Crack Damage Detection Using Convolutional Neural Networks. *Comput.-Aided Civ. Inf.* **2017**, *32*, 361–378. [CrossRef]
21. Zhong, K.; Teng, S.; Liu, G.; Chen, G.; Cui, F. Structural Damage Features Extracted by Convolutional Neural Networks from Mode Shapes. *Appl. Sci.* **2020**, *10*, 4247. [CrossRef]

22. Teng, S.; Chen, G.; Wang, S.; Zhang, J.; Sun, X. Digital image correlation-based structural state detection through deep learning. *Front. Struct. Civ. Eng.* **2022**, *16*, 45–56. [CrossRef]
23. Teng, Z.; Teng, S.; Zhang, J.; Chen, G.; Cui, F. Structural Damage Detection Based on Real-Time Vibration Signal and Convolutional Neural Network. *Appl. Sci.* **2020**, *10*, 4720. [CrossRef]
24. Hu, S.X.; Li, W.G.; Yang, W. Mechanical property prediction of hot-rolled strip based on convolutional neural network. *J. Wuhan Univ. Sci. Technol.* **2018**, *41*, 338–344.
25. Zheng, Y.Z.; Niu, L.K.; Xiong, X.Y.; Qi, H.W.; Ma, X.X. Fault diagnosis of cylindrical roller bearing cage based on 1D convolution neural network. *J. Vib. Shock* **2021**, *40*, 230–238.
26. Banerjee, S.; Pol, C.B. Theoretical modeling of guided wave propagation in a sandwich plate subjected to transient surface excitations. *Int. J. Solids Struct.* **2012**, *49*, 3233–3241. [CrossRef]
27. Grenier, S.; Konrad, J.; Lebœuf, D. Dynamic simulation of falling weight deflectometer tests on flexible pavements using the spectral element method: Forward calculations. *Can. J. Civ. Eng.* **2009**, *36*, 944–956. [CrossRef]

sensors

MDPI

Article

The Signal Characteristics of Oil and Gas Pipeline Leakage Detection Based on Magneto-Mechanical Effects

Bin Liu, Qian Ge *, Zihan Wu, Zheng Lian, Lijian Yang and Hao Geng

College of Information Science and Engineering, Shenyang University of Technology, Shenyang 110870, China
* Correspondence: q13352444326@163.com; Tel.: +86-18-150-3017-2474

Abstract: In order to solve the problem of the quantification of detection signals in the magnetic flux leakage (MFL) of defective in-service oil and gas pipelines, a non-uniform magnetic charge model was established based on magnetic effects. The distribution patterns of magnetic charges under different stresses were analyzed. The influences of the elastic load and plastic deformation on the characteristic values of MFL signals were quantitatively assessed. The experimental results showed that the magnetic charge density was large at the edges of the defect and small at the center, and approximately decreased linearly with increasing stress. The eigenvalues of the axial and radial components of the MFL signals were compared, and it was found that the eigenvalues of the radial component exhibited a larger decline rate and were more sensitive to stress. With the increase in the plastic deformation, the characteristic values of the MFL signals initially decreased and then increased, and there was an inflection point. The location of the inflection point was associated with the magnetostriction coefficient. Compared with the uniform magnetic charge model, the accuracy of the axial and radial components of the MFL signals in the elastic stage of the improved magnetic charge model rose by 17% and 16%, respectively. The accuracy of the axial and radial components of the MFL signals were elevated by 9.15% and 9%, respectively, in the plastic stage.

Keywords: magnetic flux leakage detection; magnetic charge model; force-magnetic coupling; magnetic flux leakage signal

Citation: Liu, B.; Ge, Q.; Wu, Z.; Lian, Z.; Yang, L.; Geng, H. The Signal Characteristics of Oil and Gas Pipeline Leakage Detection Based on Magneto-Mechanical Effects. *Sensors* **2023**, *23*, 1857. https://doi.org/10.3390/s23041857

Academic Editors: Bogusław Łazarz, Grzegorz Peruń and Tangbin Xia

Received: 22 December 2022
Revised: 29 January 2023
Accepted: 2 February 2023
Published: 7 February 2023

1. Introduction

With the rapid development of the global economy and the improvement of people's living standards, the demand for different energy resources, such as oil and natural gas, is gradually increasing. Due to the advantages of the cheap and convenient transportation of long-distance oil and gas pipelines, they are widely used in the oil and gas industries [1,2]. Long-distance oil and gas pipelines in long-term service are prone to accidents due to aging, corrosion, and other risk factors, which pose a serious threat to the global economy and people's safety. Therefore, the safety of pipelines cannot be overlooked [3]. At present, of the majority of in-service pipelines in China are at the aging accident stage. In order to ensure the safe operation of oil and gas pipelines, they must be tested and monitored regularly [4,5]. The in-pipeline testing technology is the main nondestructive testing method, which has been developed for several decades, mainly including piezoelectric ultrasonic testing, electromagnetic ultrasonic testing, magnetic flux leakage (MFL) testing, etc. [6,7].

The MFL internal detection technology has shown an outstanding diagnostic performance [8,9], which can detect crack defects, weld defects, corrosion pits, metal losses, and other abnormal operating conditions of pipelines [10]. Compared with other detection technologies, the MFL internal detection technology possesses the advantages of non-pollution, non-coupling, high speed, and high reliability [11]. It is more appropriate for the rapid online detection of long-haul oil and gas pipelines, and it is globally recognized as the most effective detection method for the safety detection of pipelines [12,13]. The quantization of the pipeline MFL signal is the ultimate goal of MFL detection [14]. Most of the quantization methods are to carry out a large number of calibrations of experimental data [15,16], but the failure to take into account the complex situation in the pipeline will lead to poor accuracy and impracticality of the quantization method; therefore, the quantization of the theoretical model has become a hot research topic [17,18]. Under the influence of the internal pressure of the medium and the surrounding environment, there is a large stress concentration area at the defect [19,20]. The existing quantization model does not consider the influence of the stress concentration area on the magnetic leakage signal [21,22], resulting in poor quantization accuracy. Therefore, it is of great significance for the establishment of a quantitative model of magnetic leakage internal detection and quantitative calculation of the magnetic leakage signal [23,24].

In the present study, the force-magnetic coupling relationship between the non-uniform magnetic charge model and the analytical model of compound magnetic charge was developed. The distribution of the magnetic charges on the side wall under different stresses was analyzed. The influences of the elastic load and plastic deformation on the MFL signals and their characteristic values were studied. Compared with the experimental data, the reliability and accuracy of the model were verified. The research could provide a scientific basis for the quantitative detection of defects in long-distance oil and gas pipelines.

2. Model Development

2.1. Classical Model

The pipeline is magnetized by the excitation device of the MFL internal detection. If there is no defect in the pipeline, the magnetic lines of force pass through the inside of the pipe wall, and the probe cannot detect the magnetic leakage field in the air. When there are defects in the pipeline, the magnetic field lines in the wall will leak into the air because the permeability of the defect is small and the reluctance is large, thereby forming a magnetic leakage field, as shown in Figure 1a.

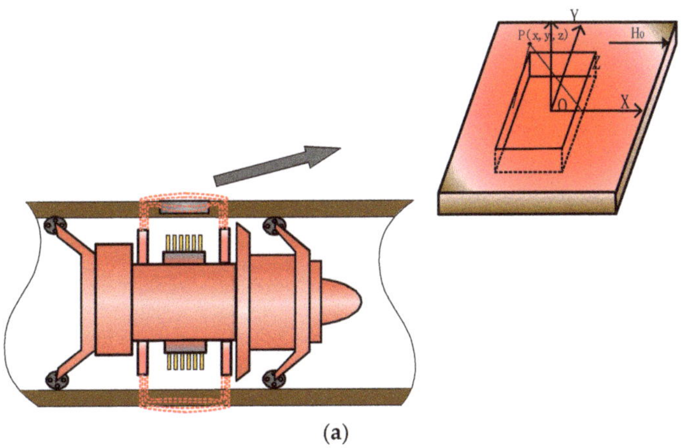

(a)

Figure 1. *Cont.*

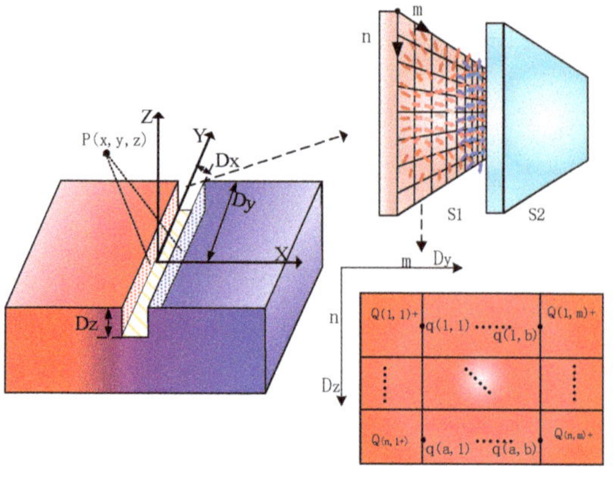

(b)

Figure 1. Schematic diagram of defects in pipeline and distribution of magnetic charges. (**a**) Schematic diagram of MFL internal detection of defects in pipeline. (**b**) Schematic diagram of non-uniform distribution of magnetic charges on the side wall of the defect.

At present, magnetic charge models are widely used in the analysis of magnetic leakage fields [25,26]. As shown in Figure 1b, magnetic charges are generated at the defects and distributed on the side walls. The magnetic leakage field, when the micro-element surface dydz on the side wall is distant from point p, is formulated as follows:

$$\mathrm{d}H = \frac{\rho dydz}{4\pi\mu_0 r^3} \tag{1}$$

where the uniform magnetic charge density is $\rho = 5.3\left(\frac{D_z/D_x+1}{D_z/(D_x\mu)+1}\right)H_0$ [27]. The magnetic charge at the defect is affected by the shape of the defect, and the magnetic charge in the stable state presents a non-uniform distribution under the action of Coulomb force [28,29], which is difficult to overcome. As shown in Figure 1b, the rectangular defect with the defect side length dy × dz is divided into m × n sub-regions, and the side length of each sub-region is $\frac{D_y}{n} \times \frac{D_z}{m}$. Therefore, only the magnetic charge carried by each sub-region is required, and the distribution of the magnetic charge density in each sub-region can be calculated as [30]:

$$\rho_{ij} = \frac{Q_{ij}}{s} \tag{2}$$

where i = 1, 2, 3... n, j = 1, 2, 3... m, p is the magnetic charge density in the subregion, $Q_{(i,j)}$ is the magnetic charge in the subregion, and s is the area of the side wall of the defect. As illustrated in Figure 1b, the mathematical model is established in the S1 plane. The magnetic charge is repulsive in the S1 plane and attracted by the magnetic charge in the S2 plane. When the magnetic charge of the side wall of the defect is in a stable state, the vector sum of the magnetic force received by the magnetic charge is 0; thus, the Coulomb force received by the magnetic charge of any subregion is formulated as follows [31,32]:

$$F = \frac{Kq_{(a,b)}Q_{(i,j)}}{r_3{}^2} + \frac{Kq_{(a,b)}Q_{(i,j)-}}{r_4{}^2} = 0 \tag{3}$$

where $Q_{(i,j)+}$ is the magnetic charge of any subregion in the S1 plane, $Q_{(i,j)-}$ is the magnetic charge of any subregion in the S2 plane, and q (a, b) is the unit point magnetic charge in the plane region (a = 1, 2, 3... n − 1, b = 1, 2, 3... m − 1, a ≠ b), $r_3 = \sqrt{\left(\frac{D_y}{m}\right)^2\left(\frac{2b+1}{2} - j\right)^2 + \left(\frac{D_z}{n}\right)^2\left(\frac{2a+1}{2} - i\right)^2}$ is the distance between the magnetic charges on the same plane, $r_4 = \sqrt{\left(\frac{D_y}{m}\right)^2\left(\frac{2b+1}{2} - j\right)^2 + \left(\frac{D_z}{n}\right)^2\left(\frac{2a+1}{2} - i\right)^2 + {D_x}^2}$ is the distance between the magnetic charges on different planes, and K is the Coulomb constant. Therefore, the horizontal and vertical components of any $q_{(a,b)}$ unit magnetic charge subjected to the Coulomb force in the subregion $Q_{(i,j)}$ are expressed as:

$$F_m = \frac{\left(\frac{2b+1}{2} - j\right)}{r_3} \times \frac{Kq_{(a,b)}Q_{(i,j)+}}{{r_3}^2} + \frac{\left(\frac{2b+1}{2} - j\right)}{r_3} \times \frac{Kq_{(a,b)}Q_{(i,j)-}}{r_4} = 0 \quad (4)$$

$$F_n = \frac{\left(\frac{2a+1}{2} - i\right)}{r_3} \times \frac{Kq_{(a,b)}Q_{(i,j)+}}{{r_3}^2} + \frac{\left(\frac{2a+1}{2} - i\right)}{r_3} \times \frac{Kq_{(a,b)}Q_{(i,j)-}}{r_4} = 0 \quad (5)$$

As there are several magnetic charge regions in the plane, the force analysis of several magnetic charge regions is combined into a system of equations, as follows:

$$\begin{cases} \sum\limits_{i=1}^{n}\sum\limits_{j=1}^{m}\sum\limits_{a=1}^{n-1}\sum\limits_{b=1}^{m-1} F_m = 0 \\ \sum\limits_{i=1}^{n}\sum\limits_{j=1}^{m}\sum\limits_{a=1}^{n-1}\sum\limits_{b=1}^{m-1} F_n = 0 \end{cases} = \begin{cases} \left(\frac{(\frac{3}{2}-1)Kq_{(1,1)}Q_{(1,1)+}}{{r_3}^{3/2}} + \frac{(\frac{3}{2}-1)Kq_{(1,1)}Q_{(1,1)-}}{r_3 \times r_4} + \cdots\cdots + \frac{(\frac{3}{2}-j)Kq_{(1,1)}Q_{(i,j)+}}{{r_3}^{3/2}} + \frac{(\frac{3}{2}-j)Kq_{(1,1)}Q_{(i,j)-}}{r_3 \times r_4} = 0\right) \\ \vdots \\ \left(\frac{(\frac{3}{2}-1)Kq_{(1,1)}Q_{(1,1)+}}{{r_3}^{3/2}} + \frac{(\frac{3}{2}-1)Kq_{(1,1)}Q_{(1,1)-}}{r_3 \times r_4} + \cdots\cdots + \frac{(\frac{3}{2}-i)Kq_{(1,1)}Q_{(i,j)+}}{{r_3}^{3/2}} + \frac{(\frac{3}{2}-i)Kq_{(1,1)}Q_{(i,j)-}}{r_3 \times r_4} = 0\right) \\ \vdots \end{cases} \quad (6)$$

Through the above-mentioned equations, the magnetic charge of $Q_{(n,m)+}$ in any subregion can be solved. By substituting the magnetic charge in the magnetic region obtained from Equation (6) into Equation (2), the magnetic charge density expression in the S1 plane can be formulated as follows:

$$\begin{cases} \rho_{(1,1)} = 0.136\frac{{D_y}^2{D_z}^2}{{D_x}^2} \\ \rho_{(1,2)} = 0.131\frac{{D_y}^2{D_z}^2}{{D_x}^2} \\ \vdots \end{cases} \quad (7)$$

The non-uniform distribution of the magnetic charge on the side wall of the defect can be solved by Equation (7), as shown in Figure 2, which is the schematic diagram of the non-uniform distribution of the magnetic charge on the side wall of the S1 plane.

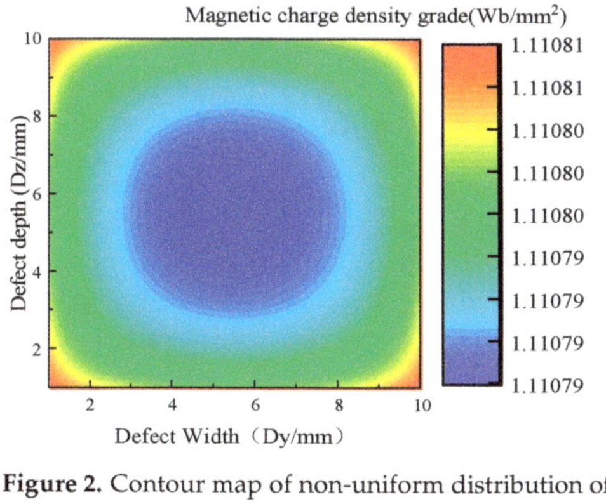

Figure 2. Contour map of non-uniform distribution of magnetic charges.

The magnetic charge density on the side wall of the defect presents a non-uniform distribution, and the magnetic charge density in the middle is small, whilst it gradually increases near the edge, and the magnetic charge density at the edge is the largest (Figure 2).

2.2. An Improved Magnetic Dipole Model

An in-service pipeline is affected by the internal pressure of the pipeline, and a stress concentration area is formed at the defect. Therefore, the establishment of the computational model of the magnetic charge density at the defect under the action of stress is more compatible with the actual condition. The magnetic field intensity at the defect can be expressed as [33]:

$$H_{\text{eff}} = H + H_\sigma^e + H_\sigma^p \tag{8}$$

where H is the strength of the excitation magnetic field, H_{eff} represents the effective magnetic field, H_σ^e denotes the effective magnetoelastic field, and H_σ^p is the effective plastic field. In the leakage magnetism, H_σ^e can be expressed as [34]:

$$H_\sigma^{\text{e}} = \frac{3\sigma\lambda_{\text{s}}}{\mu_0 M_{\text{s}}}\left(\cos^2\theta - v\sin^2\theta\right) \tag{9}$$

where θ is the angle between the direction of magnetization and the direction of stress, and v is Poisson's ratio. In the leakage magnetism, H_σ^p is the effective plastic field, which can be expressed as [35]:

$$H_\sigma^{\text{p}} = \frac{6\text{k}E\varepsilon_p\lambda_s}{\mu_0 M_s} - k'\varepsilon_p^2 M_{\text{s}} \tag{10}$$

Equations (9) and (10) are substituted into Equation (8) to achieve:

$$H_{\text{eff}} = H + \frac{3\lambda_s}{\mu_0 M_{_s}}\left(2\text{k}E\varepsilon_p + \sigma\left(\cos^2\theta - v\sin^2\theta\right)\right) - k'\varepsilon_p^2 M_{\text{s}} \tag{11}$$

According to the relationship between the magnetization of ferromagnetic materials and the magnetic charge density, it can be expressed as [36,37]:

$$M = M_{\text{s}}[\coth\left(\frac{H_{eff}}{a}\right) - \frac{a}{H_{eff}}] \tag{12}$$

$$\rho'_{(\text{i,j})} = \mu_0 M_{(\text{i,j})} = \rho_{(\text{i,j})}[\coth\left(\frac{H_{eff}}{a}\right) - \frac{a}{H_{eff}}] \tag{13}$$

By substituting Equation (13) into Equation (1), the magnetic charge density under stress can be obtained. Then, the magnetic charge model under the enhanced stress can be formulated as follows:

$$\text{d}H = \sum_{i=1}^{\text{n}}\sum_{\text{j}=1}^{m}\frac{\rho'_{(\text{i,j})}dydz}{4\pi\mu_0 r^3} \tag{14}$$

According to the non-uniform magnetic charge model under stress, this study presented a method to quantify the MFL signals of defects [38,39].

3. Calculation and Analysis

In this study, the control variable method was used to assess the variations of the magnetic signal in the elastoplastic stage.

The test specimen used was a X70 pipe. The material parameter k is the ratio of elastic energy to magnetic energy, k' is the average density of the pinpoint in the unit volume, the Young's modulus (E) is 207 GPa, the Poisson's ratio (υ) is 0.3, the saturation magnetization (Ms) is $1.585 \times 106\text{A/m}{-1}$, the saturation magnetostriction (λs) is 1.2 ppm, σ is the stress, the angle between the stress and magnetic field (θ) is 0°, and ε_{p} is the amount of plastic deformation.

3.1. Relationship between Internal Pressure and Stress

In the process of pipeline transportation, the internal pressure of the pipeline is within the range of 5~10 MPa, and the stress concentration area at the defect is remarkably larger than that in the other areas [40]. Therefore, the corresponding relationship between the stress at the defect and the internal pressure is formulated as follows [41,42]:

$$\sigma = \frac{F\sqrt{\pi D_y / Q} pR}{t} \tag{15}$$

where Q is expansion coefficient, p is the internal pressure, R is the radius of pipeline, t is the wall thickness of the pipeline, and D_y is the defect depth. In addition, F is the stress intensity factor at the crack tip and only depends on the crack size and internal pressure. The radius R of the pipelines was 259 mm, the wall thickness t was 12 mm, the defect depth was 10 mm, the axial length was 1 mm, and the internal pressure of 0.5~10 MPa was applied to the pipeline. The calculated stress is shown in Figure 3.

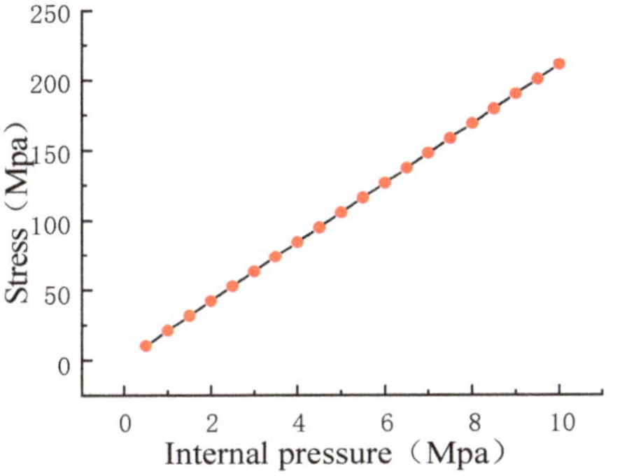

Figure 3. Relationship between pressure and stress in pipeline.

3.2. Effects of Stress on Magnetic Charge Density

The stress concentration area at the defect leads to the change in the magnetization of the side wall of the defect. In the present study, the defect sizes were designed with length $D_x = 1$ mm, width $D_y = 10$ mm, and depth $D_z = 10$ mm, and tensile forces of 0, 10, 20, 30, 40, and 50 MPa were applied to analyze the variation pattern of the magnetic charge density under different stresses. Figure 4 shows the distribution of the magnetic charge density on the side wall of the defects under stress.

With the increase in the stress, the magnetic charge density on the side wall of the defect gradually decreased, and the overall magnetic charge presented a non-uniform distribution (Figure 4). The corresponding relationship between the stress and magnetic charge density is shown in Figure 5.

The magnetic charge density approximately decreased linearly with the increase in the stress. The rates of the change in the magnetic charge density were 6.25%, 6.67%, 4.28%, 3.73%, and 3.87%, respectively.

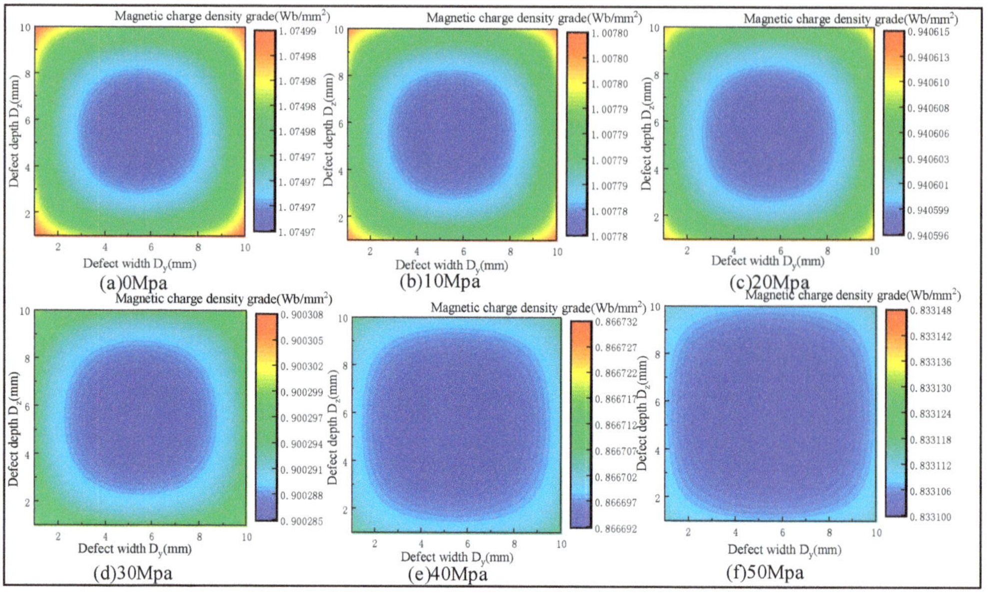

Figure 4. Distribution of magnetic charge density on the side wall of defects under stress.

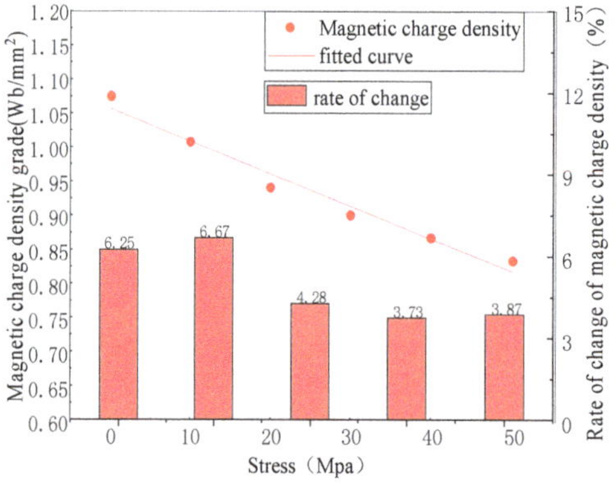

Figure 5. Relationship between stress and magnetic charge.

3.3. *The Effects of Elastic Strain on the MFL Signals*

When the same defect size is associated with different stresses, the MFL signals are significantly different. Therefore, the MFL signals of the defect were analyzed under different stresses, the defect size was designed with length $D_x = 10$ mm, width $D_y = 1$ mm, and depth $D_z = 1$ mm, and the stresses of 0, 10, 20, 30, and 40 kN were applied (Figure 6).

Figure 6 shows the MFL signals with defects under different stresses. It was revealed that the axial component had the maximum point and the radial component had peaks and troughs. The great eigenvalues of the axial component and the peak and valley eigenvalues of the radial component were extracted, as shown in Figure 7, for the correspondence between the stress and eigenvalue.

The results showed that the eigenvalues of the MFL signals decreased nonlinearly with the increase in the stress, where the rates of the change in the axial component eigenvalues were 20.36%, 6.5%, 16.7%, and 9.2%, and the rates of the change in the radial component were 23.24%, 7.82%, 18.12%, and 5.51%, yielding a larger rate of change in the radial component eigenvalues, which exhibited more sensitivity to stress changes (Figure 7).

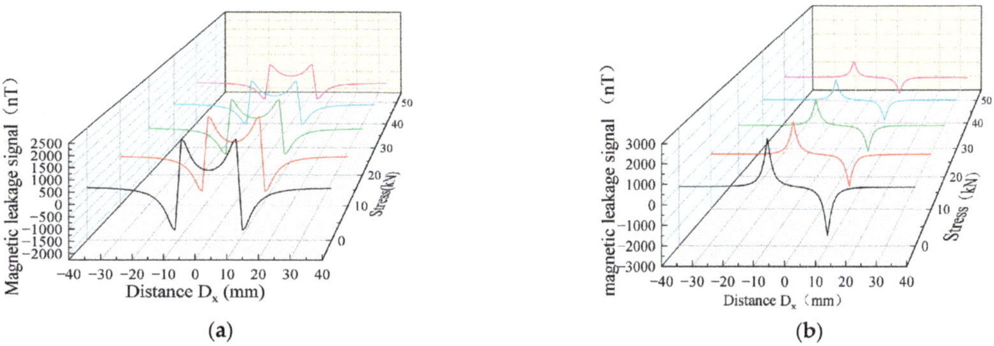

Figure 6. Changes of MFL signals under different stresses. (**a**) Axial components of MFL signals under different stresses. (**b**) Radial components of MFL signals under different stresses.

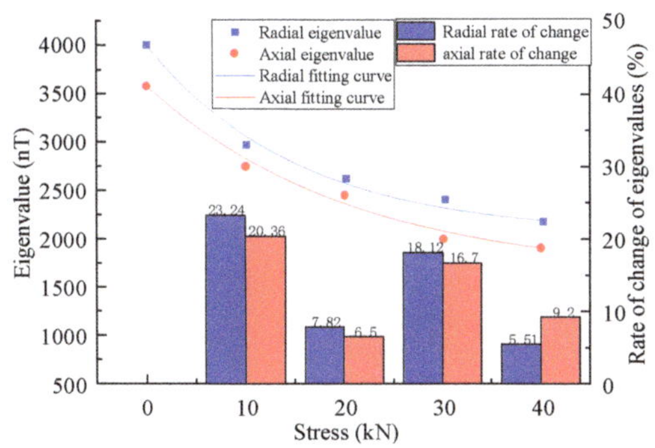

Figure 7. Relationship between stress and eigenvalue at the defect.

3.4. The Influences of Plastic Deformation on MFL Signals

In the transportation process of long-distance oil and gas pipelines, the wall of the pipeline is in continuous stress under the action of long-time stress, and the plastic deformation of the defects is noteworthy. In order to study the variation trend of the MFL signals in the plastic stage, the MFL signals were analyzed under the plastic deformation values of 0%, 5%, 10%, 15%, 20%, 25%, 30%, 35%, 40%, 45%, and 50% (Figure 8).

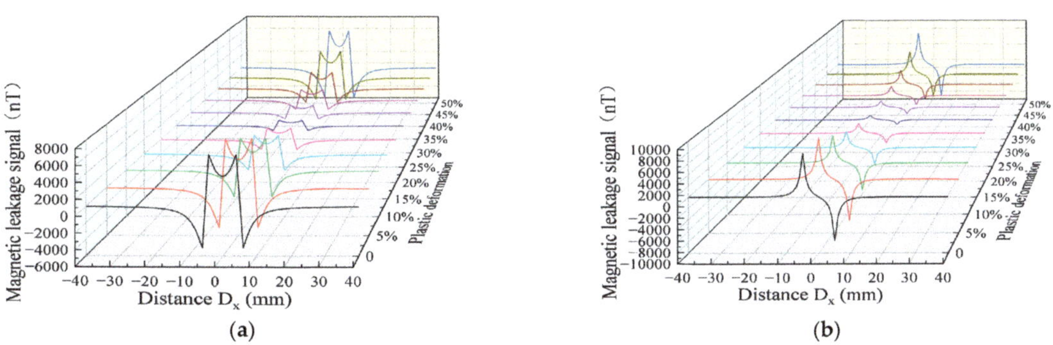

Figure 8. MFL signals with different plastic deformation values. (**a**) Axial component of MFL signals with different plastic deformation values. (**b**) Radial component of MFL signals with different plastic deformation values.

With the increase in the plastic deformation, the MFL signals presented an upward trend, followed by a downward trend. The eigenvalues of the axial and radial components in the plastic deformation were extracted (Figure 9).

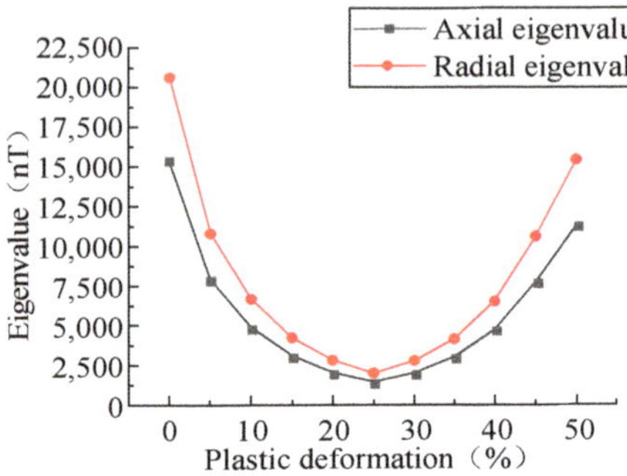

Figure 9. Relationship between plastic deformation and characteristic values.

The MFL signals had an inflection point, a very small value, and the trend could be divided into two stages: in the first stage, a non-linear exponential reduction in the characteristics of the MLF signals was found with the increase in the plastic deformation; in the second stage, the characteristic values of the MLF signals non-linearly increased with the elevation of the plastic deformation.

The effects of plastic deformation on the MFL signals under different stresses and the influences of the stress on the inflection point were analyzed. As shown in Figure 10, the effects of plastic deformation on the MFL signals at stresses of 10, 20, 30, 40, and 50 kN were analyzed.

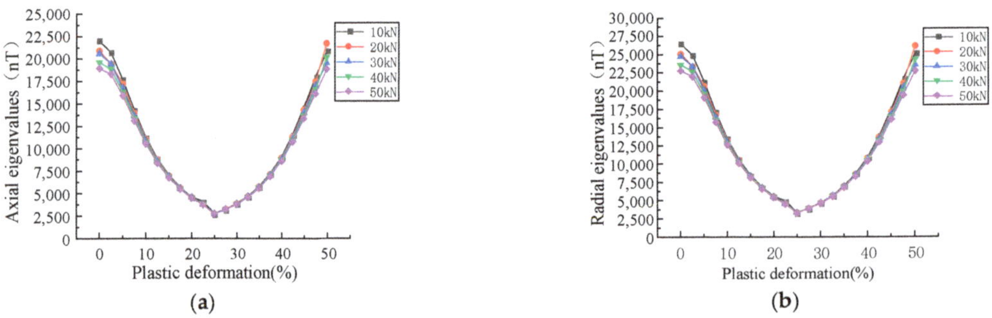

Figure 10. Relationship between plastic deformation and eigenvalues under different stresses. (**a**) The relationship between plastic deformation under different stresses and the eigenvalues of axial component of MFL signals. (**b**) The relationship between plastic deformation under different stresses and the eigenvalues of radial component of MFL signals.

In the plastic stage, the intensity of the MFL signals decreased with the increase in the stress, initially decreased and then increased with the elevation of the plastic deformation, and an inflection point was found in the plastic deformation of 0.2–0.3 (Figure 10). As the magnetostriction coefficient λs is material-dependent, λs was analyzed and a tensile force of 10 kN was applied with magnetostriction coefficients λs of 1, 2, 3, and 4 ppm (Figure 11).

It was revealed that the magnetostriction coefficient λs affected the position of the inflection point, and with the increase in the magnetostriction coefficient λs, the inflection point shifted to the right. Therefore, it could be concluded that the inflection point was associated with the properties of ferromagnetic materials (Figure 11).

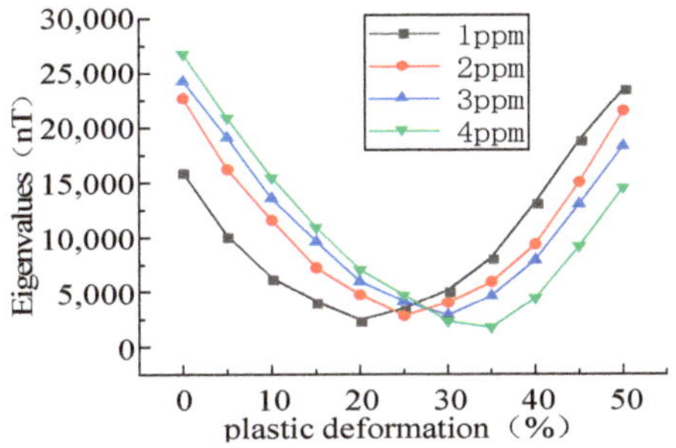

Figure 11. Positional relationship between magnetostriction and inflection point.

3.5. Comparative Analysis of Improved Magnetic Charge Model and Classical Magnetic Charge Model with Different Defect Sizes

3.5.1. Comparative Analysis of Improved Magnetic Charge Model and Classical Magnetic Charge Model under Different Defect Lengths

In order to study the variation rules of the magnetic leakage signals of the improved model and the classical model under different defect lengths, the width D_y = 1 mm, depth D_z = 1 mm, and length D_x were set as 1–10 mm to simulate the variation rules of the magnetic leakage signals with different defect lengths. (Figure 12).

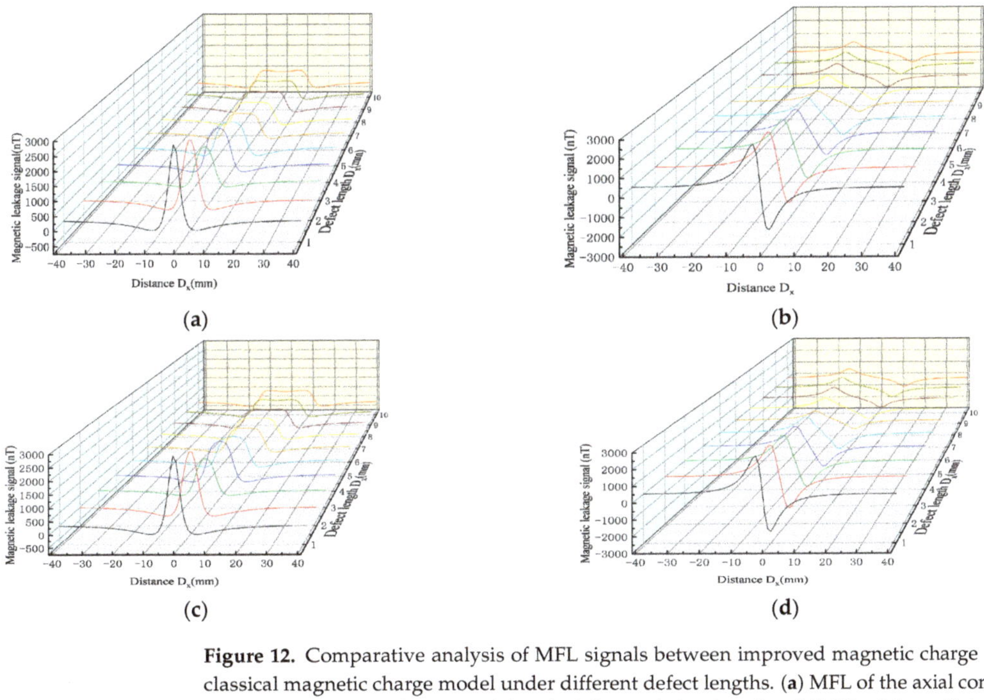

Figure 12. Comparative analysis of MFL signals between improved magnetic charge model and classical magnetic charge model under different defect lengths. (**a**) MFL of the axial component of improved magnetic charge model with different defect lengths. (**b**)MFL of the radial component of improved magnetic charge model with different defect lengths. (**c**) MFL of axial components of classical magnetic charge model with different defect lengths. (**d**) MFL of radial components of classical magnetic charge model with different defect lengths.

As shown in Figure 12, the improved magnetic charge model and the classical magnetic charge model decrease with the increase in the defect length, the axial component has a double peak value, and the distance between the peak and valley values of the radial component becomes larger, and the change trend is consistent. As shown in Figure 13, the variation rule of the eigenvalues is studied.

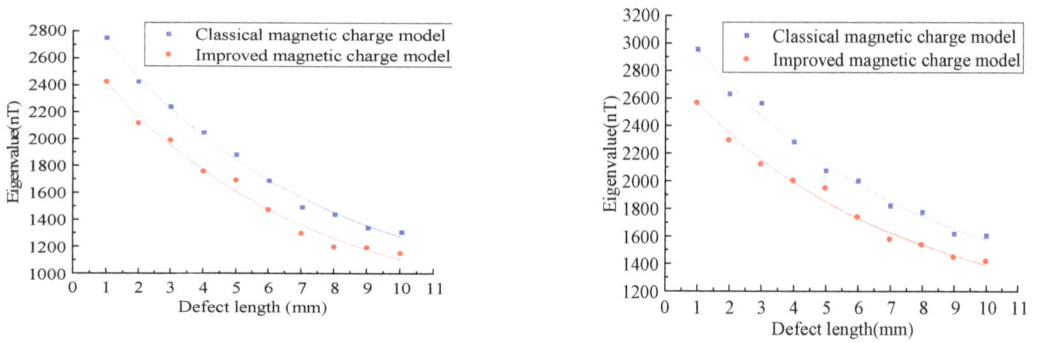

(a) (b)

Figure 13. Analysis of characteristic values under different defect lengths. (**a**) Analysis of axial eigenvalues under different defect lengths. (**b**) Analysis of radial eigenvalues under different defect lengths.

As shown in Figure 13, it can be seen that with the increase in the defect length, the characteristic value of the improved model decreases nonlinearly, and the characteristic value of the improved magnetic charge model is smaller than that of the classical magnetic charge model.

3.5.2. Comparative Analysis between the Improved Magnetic Charge Model and the Classical Magnetic Charge under Different Defect Depths

As can be seen from Equation (7), the influence results of the depth and width are consistent. Therefore, the variation rules of the magnetic leakage signals of the improved model and the classical model under different defect depths were studied. The width $D_y = 1$ mm, length $D_x = 1$ mm, and depth D_z were set as 1–10 mm to simulate the variation rules of the magnetic leakage signals with different defect lengths. (Figure 14).

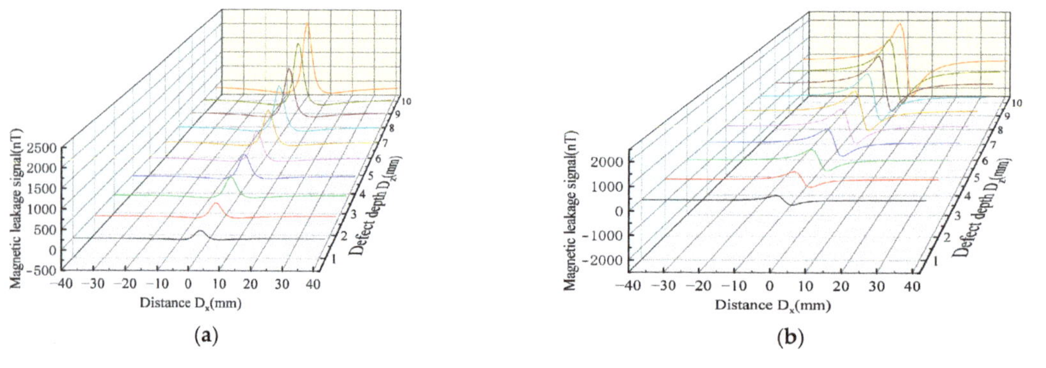

(a) (b)

Figure 14. *Cont.*

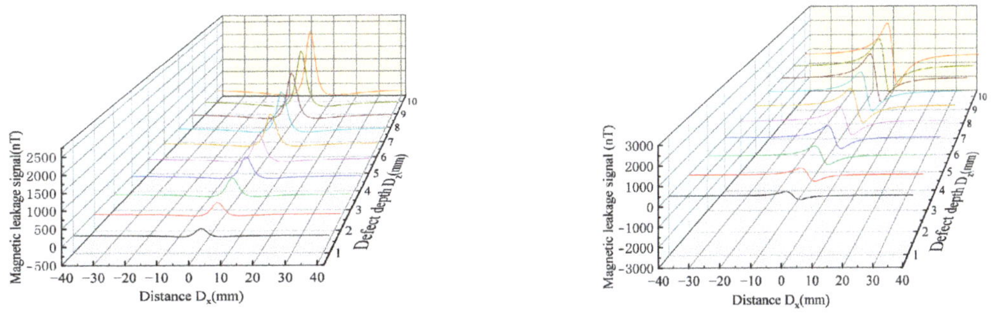

(c) (d)

Figure 14. Comparative analysis of MFL signals between the improved magnetic charge model and the classical magnetic charge model under different defect depth. (**a**) MFL of the axial component of the improved magnetic charge model with different defect depth. (**b**) MFL of the radial component of the improved magnetic charge model with different defect depth. (**c**) MFL of axial components of classical magnetic charge model with different defect depth. (**d**) MFL of radial components of classical magnetic charge model with different defect depth.

As shown in Figure 14, the MFL signals of the improved magnetic charge model and the classical magnetic charge model increase with the increase in the defect depth, and the peak-valley value spacing of the radial component does not change. Figure 15 shows the variation rules of the characteristic values under different defect depths.

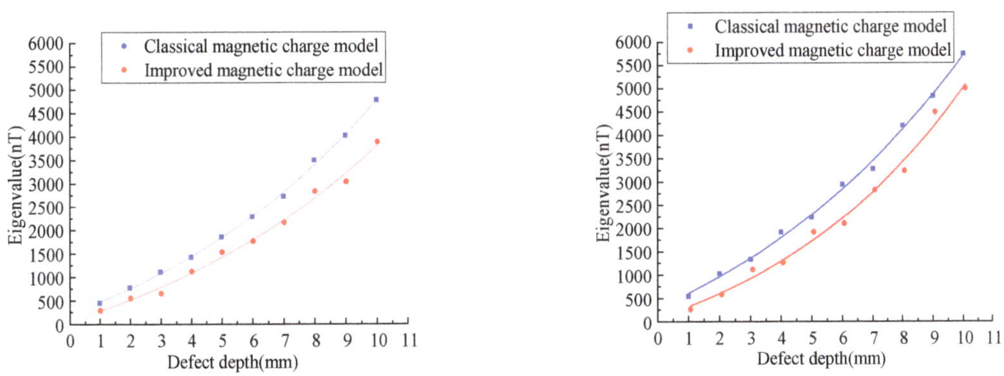

(a) (b)

Figure 15. Analysis of characteristic values under different defect depth. (**a**) Analysis of axial eigenvalues under different defect depth. (**b**) Analysis of radial eigenvalues under different defect depth.

As shown in Figure 15, it can be seen that with the increase in the defect depth, the eigenvalues increase nonlinearly, and the eigenvalues of the improved magnetic charge model are smaller than the classical magnetic charge model.

4. Experiment and Analysis

In order to verify the influences of the elastic-plastic stage on the MFL signals under different stresses, the trend of the change in the MFL signals was analyzed by tensile strip testing of an X70 steel pipe. The machining defects in the X70 steel pipe were analyzed using EDM technology. The chemical composition and mechanical properties of X70 steel are shown in Table 1.

Table 1. Chemical composition and mechanical properties of X70 steel pipe.

Chemical Component (%)					Mechanical Property	
C	Si	Mn	P	S	Strength of Extension (Mpa)	Yield Strength (Mpa)
0.12	0.45	1.07	0.025	0.015	570	485–638

4.1. Experimental Equipment and Materials

As shown in Figure 16, the experimental device was mainly composed of a tensile testing machine, X70 ferrimagnet, software control system, and signal acquisition system.

Figure 16. X70 steel strip tensile test device. (**a**) Tensile testing machine. (**b**) Software control systems and signal acquisition systems.

The size of the ferromagnetic material type used was 800 × 60 × 16 mm^3, the cross-sectional area was 60 × 16 mm^2, and the defect size was 16 × 1 × 2 mm^3. The tensile tester model SHT-4106, with clamps on the top and bottom sides, was utilized to tighten and apply tension to the ferromagnetic materials. The probe in the signal acquisition system was a three-axis Mlx90393 magnetic induction sensor with a 16-bit magnetic field resolution, the lifting distance of which is 2 mm. (Figure 16).

4.2. Experimental Analysis and Verification of Elastic Deformation

Uniform tensile forces of 0, 10, 20, 30, and 40 kN were applied to the steel pipe, and the magnitudes of the MFL signals were recorded. Figure 17 shows the MFL signals of the axial and radial components of the X70 specimen in the elastic stage under different tensile forces.

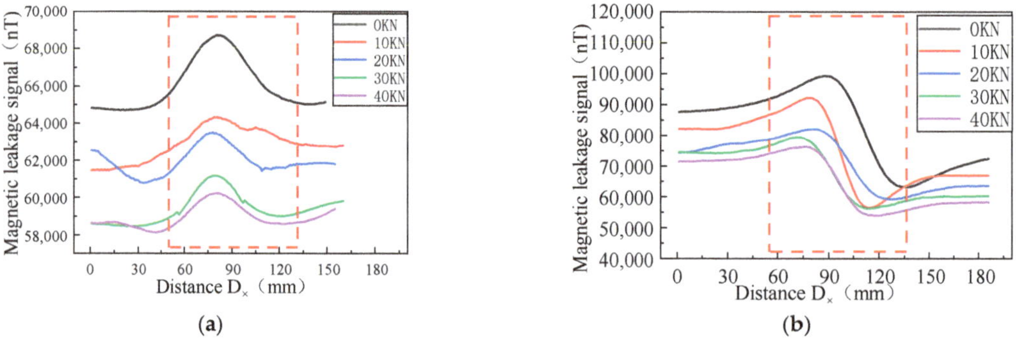

Figure 17. MFL signals under different stresses. (**a**) Axial component of MFL signals. (**b**) Radial component of MFL signals.

The MFL signals recorded by the experimental equipment are illustrated in Figure 17. With the increase in the pressure, the intensity of the MFL signals gradually decreased, and the peak values of the axial and radial components were gradually reduced. The peak value of the axial component and the peak value of the radial component were extracted and studied. The MFL signals under different stresses are shown in Figure 18.

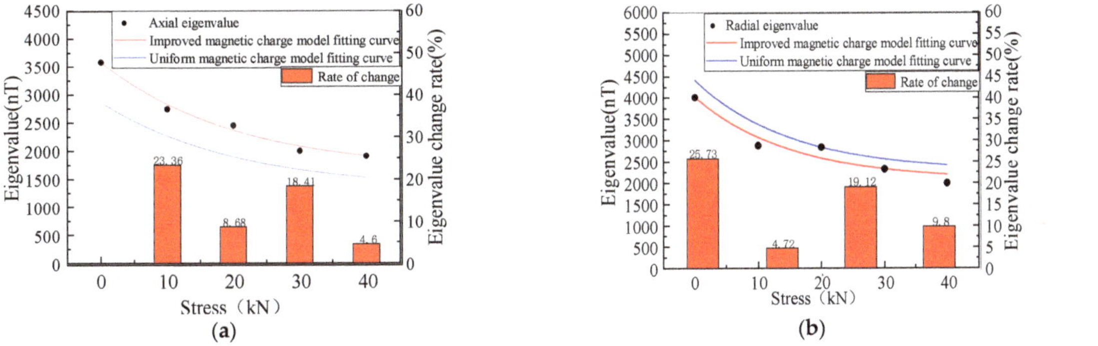

Figure 18. Eigenvalues under different stresses. (**a**) Axial component fitting and eigenvalue analysis. (**b**) Radial component fitting and eigenvalue analysis.

It was revealed that in the elastic stage, with the increase in the tension, the characteristic values of the axial and radial components nonlinearly decreased. The variation trend of the experimental data was consistent with that of the calculated model, and the characteristic values of the radial component were larger than those of the axial component. Compared with the uniform magnetic charge model, the accuracy of the characteristic values of the axial and radial components of the modified magnetic charge model was 87% and 88%, and the accuracy of the characteristic values of the axial and radial components of the uniform magnetic charge model was 70% and 72%, which increased by 17% and 16%, respectively.

4.3. Experimental Analysis and Verification of Plastic Deformation

As shown in Figure 15, the plastic deformation of the ferromagnetic signal material is 10%, 20%, 30%, 40% and 50% under different pulling forces.

It was revealed that the variation patterns of the characteristic values under different pulling forces were similar (Figure 19). With the increase in the pulling forces, the characteristic values of the MFL signals were reduced. With the increase in the plastic deformation, the eigenvalue initially decreased, and then increased gradually. The improved magnetic charge model and the uniform magnetic charge model were compared and analyzed (Table 2).

Table 2. Precision analysis of the improved magnetic charge model and uniform magnetic charge model.

	Accuracy of Improved Magnetic Charge Model (%)		Accuracy of Uniform Magnetic Charge Model (%)	
Stress (kN)	**Axial Component**	**Radial Component**	**Axial Component**	**Radial Component**
10	85.2	87.3	75.8	76.5
20	84.4	85.1	73.5	74.3
30	85.2	85.6	73.7	78.2
40	83.4	86.4	78.6	79.4

It can be concluded from Table 2 that the average accuracy of the eigenvalue of the axial component of the improved magnetic charge model was 84.55%, and that of the radial component was 86.1%. The average accuracy of the eigenvalues of the axial component and the radial component of the uniform magnetic charge model was 75.4% and 77.1%, which increased by 9.15% and 9%, respectively. Therefore, the model can be used to predict the variation pattern of the characteristic values of MFL signals in the plastic deformation stage under different stresses.

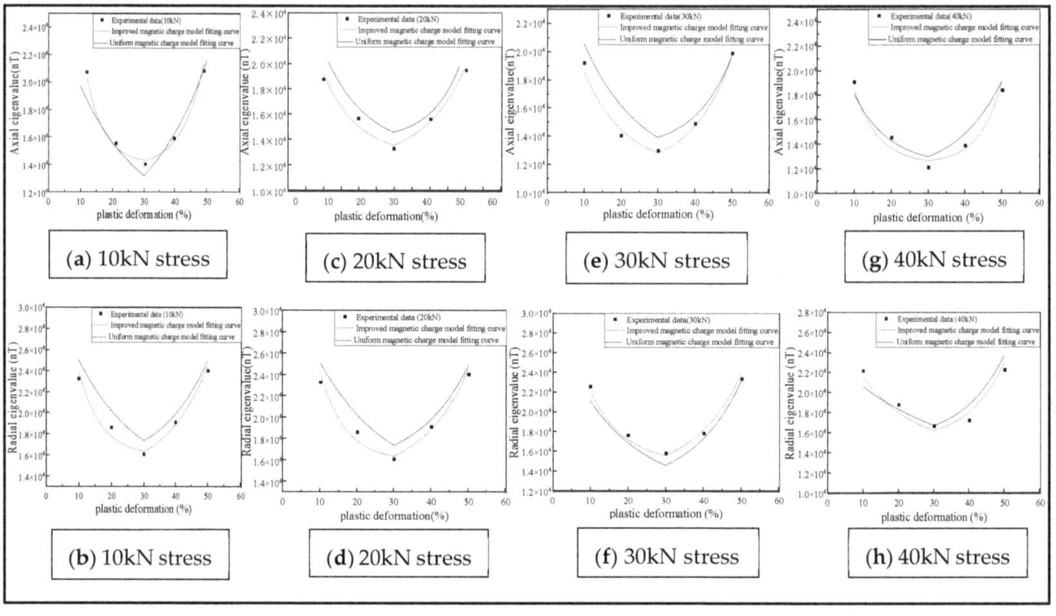

Figure 19. Relationship between plastic deformation and eigenvalue under different stresses.

5. Conclusions

This study presented the force-magnetic coupling relationship for the non-uniform magnetic charge model, established a theoretical model of compound magnetic charge, and analyzed the variation pattern of characteristic values of MFL signals under the influences of elastic stress and plastic deformation. The research results can be summarized as follows:

(1) The characteristic values of the MFL signals under elastic stress gradually decreased with the increase in the stress, and the characteristic values of axial and radial components conformed to the nonlinear decline.

(2) At the plastic stage, the characteristic values of the MFL signals initially increased, and then decreased, with the rise of the deformation, showing an inflection point. This may be attributed to the material properties. The radial component of the MFL signals exhibited sensitivity to the change in the plastic deformation.

(3) The MFL signals in the elastic and plastic stages were compared with the uniform magnetic charge model. The accuracy of the axial component of the MFL signals in the elastic stage rose by 16%, and the accuracy of the radial component of the MFL signals was elevated by 17%. The accuracy of the axial component of the MFL signals and the radial component of the MFL signals increased by 9.15% and 9%, respectively, in the plastic stage. It can effectively predict the magnitude of MFL signals under stress.

Author Contributions: Conceptualization, Q.G., Z.W. and Z.L.; Methodology, Q.G. and H.G.; Resources, L.Y.; Writing—original draft, Q.G.; Writing—review & editing, B.L. All authors have read and agreed to the published version of the manuscript.

Funding: This research was funded by the National Natural Science Foundation of China (Grant No. 62101356).

Institutional Review Board Statement: Not applicable.

Informed Consent Statement: Not applicable.

Data Availability Statement: Not applicable.

Conflicts of Interest: The authors declare no conflict of interest.

References

1. Chen, Y.; Zhang, L.; Hu, J.; Liu, Z.; Xu, K. Emergency response recommendation for long-distance oil and gas pipeline based on an accident case representation model. *J. Loss Prev. Process Ind.* **2022**, *77*, 104779. [CrossRef]
2. Li, M.; Lin, S.; Xu, J.; Jia, S. Implementation of PetroChina Pipeline Company on Pipeline Integrity Management to Improve the Safety and Efficient Operation of Long-distance Pipeline. In Proceedings of the 8th International Symposium on Project Management, Bejing, China, 4–5 July 2020; pp. 573–585. [CrossRef]
3. Zhou, M.; Huang, X.; Chen, Q.; Yan, Y.; Li, J.; Zhang, Y. Research on steel stress measurement based on weak AC magnetization. *J. Electron. Meas. Instrum.* **2021**, *35*, 23–31. [CrossRef]
4. Ong, Z.C.; Eng, H.C.; Noroozi, S. Non-destructive testing and assessment of a piping system with excessive vibration and recurrence crack issue: An industrial case study. *Eng. Fail. Anal.* **2016**, *82*, 280–297. [CrossRef]
5. Yang, L.; Geng, H.; Gao, S. Magnetic flux leakage internal detection technology of the long distance oil pipeline. *Chin. J. Sci. Instrum.* **2016**, *37*, 1736–1746. [CrossRef]
6. Zhang, D.; Huang, W.; Zhang, J.; Jin, W.; Dong, Y. Prediction of fatigue damage in ribbed steel bars under cyclic loading with a magneto-mechanical coupling model. *J. Magn. Magn. Mater.* **2021**, *530*, 167943. [CrossRef]
7. Xu, L.; Liu, F.; Ding, Y.; Li, Z.; Xie, Y. Residual thickness detection of pipeline base on electromagnetic uitrasonic shear wave. *J. Beijing Univ. Aeronaut. Astronaut.* **2022**, *48*, 1767–1773. [CrossRef]
8. Peng, L.; Huang, S.; Wang, S.; Zhao, W. A Simplified Lift-Off Correction for Three Components of the Magnetic Flux Leakage Signal for Defect Detection. *IEEE Trans. Instrum. Meas.* **2021**, *70*, 1–9. [CrossRef]
9. Long, Y.; Huang, S.; Peng, L.; Wang, S.; Zhao, W. A Characteristic Approximation Approach to Defect Opening Profile Recognition in Magnetic Flux Leakage Detection. *IEEE Trans. Instrum. Meas.* **2021**, *70*, 6003912. [CrossRef]
10. Hu, J.; Jiao, X.; Zheng, L.; Gang, B.; Liu, S. An improved support vector regression method for quantifying magnetic leakage defects in three-axis pipeline. *Nondestruct. Test.* **2021**, *43*, 62–68.
11. Clausse, B.; Lhémery, A.; Walaszek, H. A model to predict the ultrasonic field radiated by magnetostrictive effects induced by EMAT in ferromagnetic parts. *J. Phys. Conf. Ser.* **2017**, *797*, 012004. [CrossRef]
12. Wei, M.; Tu, F.; Zhang, P.; Jiang, L.; Jiang, P.; Jing, Y. Composite excitation multi-extension direction defect magnetic flux leakage detection technology. *J. Meas. Sci. Instrum.* **2022**, *13*, 156–165.
13. Geng, H.; Xia, H.; Wang, G. Study on defect location method of inner and outer wall of pipeline during high-speed magnetic flux leakage testing. *Chin. J. Sci. Instrum.* **2022**, *43*, 70–78. [CrossRef]
14. Lang, X.; Li, P.; Hu, Z.; Ren, H.; Li, Y. Leak Detection and Location of Pipelines Based on LMD and Least Squares Twin Support Vector Machine. *IEEE Access* **2017**, *5*, 8659–8668. [CrossRef]
15. Zhu, H.; Liu, H.; Li, H.; Huang, S.; Su, Z. Quantitative Analysis Method for Pipeline Defects Based on Optimized RBF Neural Network. *Instrum. Tech. Sens.* **2016**, 83–86. [CrossRef]
16. Xu, P.; Fang, Z.; Wang, P.; Geng, M.; Xu, Y. Quantitative Evaluation Method of Defects Based on Transient Characteristics of PMFL. *China Mech. Eng.* **2021**, *32*, 860–866, 881.
17. Liu, B.; Luo, N.; Feng, G. Quantitative Study on MFL Signal of Pipeline Composite Defect Based on Improved Magnetic Charge Model. *Sensors* **2021**, *21*, 3412. [CrossRef]
18. Wang, Y.; Melikhov, Y.; Meydan, T.; Yang, Z.; Wu, D.; Wu, B.; He, C.; Liu, X. Stress-Dependent Magnetic Flux Leakage: Finite Element Modelling Simulations Versus Experiments. *J. Nondestruct. Eval.* **2020**, *39*, 1–9. [CrossRef]
19. Yang, L.; Zheng, F.; Gao, S.; Huang, P.; Bai, S. An anaiytical model of electormagnetic stress detection for pipeline based on magneto-mechanical coupling model. *Chin. J. Sci. Instrum.* **2021**, *41*, 249–258. [CrossRef]
20. Liu, B.; Cao, Y.; Wang, D.; He, L.; Yang, L. Quantitative analsis of the magnetic memory yielding singal characteristics based on the LMTO algorithm. *Chin. J. Sci. Instrum.* **2017**, *38*, 1413–1420. [CrossRef]
21. Dutta, S.M.; Ghorbel, F.H.; Stanley, R.K. Dipole Modeling of Magnetic Flux Leakage. *IEEE Trans. Magn.* **2009**, *45*, 1959–1965. [CrossRef]
22. Wu, D.; Liu, Z.; Wang, X.; Su, L. Mechanism analysis of influence of surface-breaking orientation on magnetic leakage field distribution. *Acta Phys. Sin.* **2017**, *66*, 293–304.
23. Hosseingholizadeh, S.; Filleter, T.; Sinclair, A.N. Evaluation of a Magnetic Dipole Model in a DC Magnetic Flux Leakage System. *IEEE Trans. Magn.* **2019**, *55*, 6200407. [CrossRef]
24. Liu, T.; Liu, B.; Feng, G.; Lian, Z.; Yang, L. Quantization of pipeline magnetic flux leakage detection signal under load. *Chin. J. Sci. Instrum.* **2022**, *43*, 262–273. [CrossRef]
25. Huang, X.; Wu, J.; Sun, Y.; Kang, Y. 3D magnetic dipole models of magnetic flux leakage for 'concave' and 'bump' defects. *Int. J. Appl. Electromagn. Mech.* **2018**, *59*, 1305–1312. [CrossRef]
26. Huang, R.; Li, H.; Jiang, M.; Wang, Y.; Zhao, C.; Qiu, J.; Xiong, K.; Ji, H. Magnetic charge model of defect in magnetic flux leakage testing. *Int. J. Appl. Electromagn. Mech.* **2020**, *64*, 1315–1323. [CrossRef]

27. Xu, Z.S.; Xu, Y.; Wang, J.B. *Quantitative Detection Principle and Application of Crack in MFL Method*; National Defend Industry Press: Beijing, China, 2005; Volume 1, p. 124. (in Chinese)
28. Zhong, W. The Watersheds of Magnetic Forces and Magneitc Charges on the End Surfaces of a Rectangular Steel Component Magnetized Longitudinally. *Nondestruct. Test.* **2005**, 349–351, +363. Available online: https://kns.cnki.net/kcms2/article/abstract?v=3uoqIhG8C44YLTlOAiTRKgchrJ08w1e7F1IFNsBV5UuooOozlxTTWC9DEgN391DFYL7aUUMM4a0FhFFJ_TLq80NuvpjIN1Ck&uniplatform=NZKPT (accessed on 1 February 2023).
29. Zhong, W. The Repulsive Interaction Between Uniformly Distributed aREAL mAGENTIC Charges in a Round Olane. *Nondestruct. Test.* **2010**, *32*, 160–162.
30. Zhong, W. On the Proportional Coefficient α of Areal Density of Magnetic Charges σs toLinear Density of Magnetic Charges σl. *Nondestruct. Test.* **2009**, *31*, 2.
31. Zhong, W. Distribution of Magnetic Charges in Arbitrarily-shaped Region on the Plane. *Nondestruct. Test.* **2008**, 23–25.
32. Liu, M.; Xu, Z.; Wang, J. Inverse Imaging for Defects Based on Energy Level of the Magnetic Dipole. *China Mech. Eng.* **2005**, *16*, 959. [CrossRef]
33. Shi, P.; Bai, P.; Chen, H.; Su, S.; Chen, Z. The magneto-elastoplastic coupling effect on the magnetic flux leakage signal. *J. Magn. Magn. Mater.* **2020**, *504*, 166669. [CrossRef]
34. Bratov, V.; Borodin, E. Comparison of dislocation density based approaches for prediction of defect structure evolution in aluminium and copper processed by ECAP. *Mater. Sci. Eng. A* **2015**, *631*, 10–17. [CrossRef]
35. Shi, P.; Zhang, P.; Jin, K.; Chen, Z.; Zheng, X. Thermo-magneto-elastoplastic coupling model of metal magnetic memory testing method for ferromagnetic materials. *J. Appl. Phys.* **2018**, *123*, 145102. [CrossRef]
36. Liu, B.; Ma, Z.; Liu, Z.; Luo, N.; Xu, X. Research on internal detection technology for axial crack of long-distance oil and gas pipeline based on micromagnetic method. *Struct. Health Monit.* **2020**, *19*, 1123–1136. [CrossRef]
37. Shi, P.; Jin, K.; Zheng, X. A general nonlinear magnetomechanical model for ferromagnetic materials under a constant weak magnetic field. *J. Appl. Phys.* **2016**, *119*, 145103. [CrossRef]
38. Liu, B.; Ma, Z.; He, L.; Wang, D.; Zhang, H.; Ren, J. Quantitative study on the propagation characteristics of MMM signal for stress internal detection of long distance oil and gas pipeline. *NDT E Int.* **2018**, *100*, 40–47. [CrossRef]
39. Shi, P.; Zheng, X. Magnetic charge model for 3D MMM signals. *Nondestruct. Test. Eval.* **2016**, *31*, 45–60.
40. Li, Q.; Zhao, M.; Ren, X.; Wang, L.; Feng, X.; Niu, Y. Construction Status and Development Trend of Chinese Oil & Gas Pipeline. *Oil-Gas Field Surf. Eng.* **2019**, *38*, 14–17.
41. Zahoor, A. Closed Form Expressions for Fracture Mechanics Analysis of Cracked Pipes. *J. Press. Vessel. Technol.* **1985**, *107*, 203–205. [CrossRef]
42. Liu, B.; Liu, Z.; Luo, N.; He, L.; Ren, J.; Zhang, H. Research on Features of Pipeline Crack Signal Based on Weak Magnetic Method. *Sensors* **2020**, *20*, 810. [CrossRef]

materials

MDPI

Article

Historical Silk: A Novel Method to Evaluate Degumming with Non-Invasive Infrared Spectroscopy and Spectral Deconvolution

Ludovico Geminiani [1,2,*], Francesco Paolo Campione [2,3,4], Carmen Canevali [5,6], Cristina Corti [2,3], Barbara Giussani [1], Giulia Gorla [1], Moira Luraschi [2,4], Sandro Recchia [1] and Laura Rampazzi [2,3,6]

1 Dipartimento di Scienza e Alta Tecnologia, Università degli Studi dell'Insubria, Via Valleggio 11, 22100 Como, Italy
2 Centro Speciale di Scienze e Simbolica dei Beni Culturali, Università degli Studi dell'Insubria, Via Sant'Abbondio 12, 22100 Como, Italy
3 Dipartimento di Scienze Umane e dell'Innovazione per il Territorio, Università degli Studi dell'Insubria, Via Sant'Abbondio 12, 22100 Como, Italy
4 Museo delle Culture, Villa Malpensata, Riva Antonio Caccia 5, 6900 Lugano, Switzerland
5 Dipartimento di Scienza dei Materiali, Università di Milano-Bicocca, Via Roberto Cozzi 55, 20125 Milan, Italy
6 Istituto per le Scienze del Patrimonio Culturale, Consiglio Nazionale delle Ricerche (ISPC-CNR), Via Cozzi 53, 20125 Milano, Italy
* Correspondence: lgeminiani@uninsubria.it; Tel.: +39-031-238-6475

Abstract: To correctly manage a collection of historical silks, it is important to detect if the yarn has been originally subjected to degumming. This process is generally applied to eliminate sericin; the obtained fiber is named soft silk, in contrast with hard silk which is unprocessed. The distinction between hard and soft silk gives both historical information and useful indications for informed conservation. With this aim, 32 samples of silk textiles from traditional Japanese samurai armors (15th–20th century) were characterized in a non-invasive way. ATR-FTIR spectroscopy has been previously used to detect hard silk, but data interpretation is challenging. To overcome this difficulty, an innovative analytical protocol based on external reflection FTIR (ER-FTIR) spectroscopy was employed, coupled with spectral deconvolution and multivariate data analysis. The ER-FTIR technique is rapid, portable, and widely employed in the cultural heritage field, but rarely applied to the study of textiles. The ER-FTIR band assignment for silk was discussed for the first time. Then, the evaluation of the OH stretching signals allowed for a reliable distinction between hard and soft silk. Such an innovative point of view, which exploits a "weakness" of FTIR spectroscopy—the strong absorption from water molecules—to indirectly obtain the results, can have industrial applications too.

Keywords: samurai armor; silk; sericin; FTIR spectroscopy; spectral deconvolution; PCA

Citation: Geminiani, L.; Campione, F.P.; Canevali, C.; Corti, C.; Giussani, B.; Gorla, G.; Luraschi, M.; Recchia, S.; Rampazzi, L. Historical Silk: A Novel Method to Evaluate Degumming with Non-Invasive Infrared Spectroscopy and Spectral Deconvolution. *Materials* **2023**, *16*, 1819. https://doi.org/10.3390/ma16051819

Academic Editors: Bogusław Łazarz, Grzegorz Peruń and Tangbin Xia

Received: 31 December 2022
Revised: 2 February 2023
Accepted: 20 February 2023
Published: 22 February 2023

1. Introduction

Preventive conservation in museums is essential to carry cultural heritage from the past to future generations. To correctly manage different materials, an assessment about their nature should be scheduled, by means of micro-invasive and non-invasive analytical techniques [1,2]. Moreover, chemical analyses for conservation purposes often offer the chance to deepen the knowledge about a collection, thus enhancing its historical value. Silk textiles are a good example of the issue, as few scientific works deeply investigated silk variability and decay despite the wide geographical and historical diffusion of this precious yarn. Silk textiles have been appreciated for their strength, luster, and vivid color which is obtained through the dyeing process.

Silk is obtained from the *Bombyx mori* silkworm. The silk cocoon is treated with hot water to obtain the single filaments which are successively wound together to give a raw silk thread or *grège* [3]. Then, the raw silk generally undergoes the degumming process, where hot water and other substances are employed to remove the gum covering [3–6]. Gum is mainly composed of sericin, which is soluble in hot water, unlike the two core brins of fibroin. More details are given in the Appendix A. According to the degree of degumming of the silk fiber, three different yarns could be distinguished.

(i) Raw or hard silk. It is obtained after twisting some single filaments together in order to obtain the thread. The sericin coating is still present, which makes dyeing difficult [3].
(ii) Partially degummed or supple silk [3].
(iii) Degummed or soft silk. It is the most diffused material, which possesses the typical lustrousness and smoothness. The fiber can be dyed easily [3,7].

The practice of partially or totally degumming silk is related to the geographical, historical, and cultural context which the silk was intended for [3]. There are few recent examples of hard silks in collections, such as the United States first ladies' gowns at the Smithsonian Institution [8], but it is probable that many other collections of silk textiles could be made of supple silk. On the contrary, hard silk can easily be found in the archaeological contexts of East-Asian countries, on condition that they are found in arid burial environments that are essential to preserving ancient fibers with an intact sericin coating [6]. In China [3,6], during the first millennium C.E., silk with gum was often found as the ground cloth for paintings and writings, due to its stiffness. Instead, soft silk was present in a profusion of garments and banners, often resist-dyed or embroidered in order to create a polychrome design. For a long time, in Japan [9], only samurai and rich merchants could afford silk (*kinu*), which had been imported already degummed from China and Korea since the third century C.E. Since the Edo period, it started to be produced locally, but it remained for a long time as a luxury product, which was reserved for the well-off, as sumptuary laws prevented peasants from using this yarn. They could use only raw silk and production waste, which were spun into a coarse and matt yarn. High-quality silk was extensively used in samurai armors to make the *odoshi-ge* (lacing among metal plates) and for linings brocades. It is reported that hard silk was chosen for the samurai armor, as the stronger fiber improved armor quality in battle [10]. Soft silk has been diffused in armor making since the Edo period for the sake of aesthetics to obtain more vivid colors and a more lustrous effect. In that period, the armor had in fact lost its practical value and became a ceremonial dress.

The presence of the remnant sericin gum may have a consequence on the preservation of silk artefacts. In 1995, Becker [8,11] first proposed that even if sericin yellows, it can protect silk fibroin against light-induced damage. The role of sericin as a free-radical-scavenging antioxidant was confirmed recently [12]. In 2011, Zhang and co-workers [6,13] tested mock-up samples under heat and moisture and found that sericin can provide some protection against fibroin deterioration, but in high humidity environments, soft and hard silk ages at the same rate because of the leaching of sericin. For this reason, the most ancient samples, which could be made of hard silk, are hardly likely to retain their sericin coating. Since heat and moisture are important factors which may be relevant in promoting the decay of silk within historic collections, any wet treatment with aqueous solvents is detrimental to the preservation of the sericin coatings [6,8], as well as too high humidity conditions during exposure or conservation could be harmful [6,14]. Specific studies devoted to raw silk suggest that it should be stored below 50% RH. As a matter of fact, silk is known to be a very hygroscopic textile fiber, with hard silk being even more hygroscopic [14]. The reason for this behavior could be related to silk's amino acidic composition (more details in the Appendix A). Hard silk contains both fibroin and sericin, while soft silk is composed of fibroin only [15]. Sericin contains three times the polar side groups of fibroin. The higher content of polar side groups makes sericin more prone to bind water molecules and soluble in hot water [7]. As an example, under standard atmospheric conditions, i.e., 27 °C and 65% RH (relative humidity), mulberry raw silk fiber has a moisture regain of 11% w/w, which reduces to about 9% w/w after degumming [16]. Zhang et al. report that both hard and soft silk show a reduction of water sorption related to ageing [6]. Another issue which affects water uptake is crystallinity. While sericin has an amorphous structure [17], fibroin is characterized to show both crystalline and amorphous domains [15]. The superior mechanical properties in terms of toughness which are attributed to silk fiber depend on this particular mixed structure so any variation in the size of crystallites and in the crystallinity degree could affect the physical properties of the yarn. The amorphous region is responsible for the elasticity, while highly ordered regions play a major role in determining the strength and stiffness. As happens with other natural polymers, water acts as a plasticizer in fibroin, penetrating only the amorphous regions' fiber [7]. Therefore, when silk is stored in an environment with a relative humidity below 40% or at high temperatures, it can desiccate, becoming rigid, brittle and less soft.

Most of the scientific investigations about silk focus on fibroin, and only a few studies deal with sericin. Many of them investigate sericin as a biomaterial and deal with its extraction [4–6,17,18]. The ageing behavior of hard silk has been theoretically studied using mock-up samples [6,8,11,13,19]. To the best of our knowledge, only a few authors were interested in detecting sericin on historical silk [8,11,19,20]. Initially, amino acid analysis was used to detect sericin [8,11,19]. Zhang et al. first proposed the use of FTIR spectroscopy [13,18], even if they admitted that amino acid analysis showed the best performance [6]. Traditional transmission FTIR spectroscopy has proved to be a sensitive technique for the identification of natural fibers and in particular of silk fibroin, since the first studies in the Fifties [21,22]. However, ATR and reflection modes are definitely easier and more rapid to use, even if some modification of spectra can appear [23,24]. ATR-FTIR spectroscopy is commonly used to characterize fibroin [25–28] and especially its secondary structure [2,13,25,29], but pure sericin has also been investigated [30,31]. The possibility to use microsamples and no need for pretreatment makes the technique widely used for the study of cultural heritage materials [32–36]. External reflection FTIR (ER-FTIR) spectroscopy is a reflection technique as well and uses an extended MIR (medium infrared) region, collecting signals from 7500 to 375 cm^{-1}. It has demonstrated great possibilities in the last decade, thanks to its portability and non-invasiveness; it has been tested mainly in the study of mortars [37] and pigments [38–40], but it has also been confirmed to be a sensitive technique to detect silk fibroin and sericin [18,41,42]. Nevertheless, some problems with the interpretations of spectra can occur, in the form of bands' distortions and variations of their intensity ratio, due to the influence of both the physical and optical properties of the surface that is being investigated [24]. Thus, it is often difficult to make direct comparisons between peaks in ATR and external reflection modes, requiring the construction of dedicated databases [41]. An advantage of the extended spectral range is that a part of the NIR (near infrared) region is available for interpretation. Pure NIR spectroscopy in reflection mode has also been used to study silk, using chemometrics for the interpretation. Firstly, NIR spectroscopy can assess the nature of the yarn [43–45]. Secondly, some theoretical studies used the technique to indirectly estimate silk decay by measuring the loss of tensile strength [46] and the different moisture sorption of aged silk [47]. However, the last method cannot be applied to museum collections, due to the need to conditionate the textile under different fixed RHs, which is not possible in museum spaces and is problematic for the conservation of artefacts. On the contrary, NIR spectroscopy is a well-known method to measure dampness in the textile industry [45]. Finally, Mossotti et al. [48] first used NIR data associated with chemometrics to detect sericin on textiles, considering water as an interferent for the analyses.

Aim of the Work

External reflection FTIR spectroscopy is proposed in our work as a simple and non-invasive technique to discriminate hard and soft silk samples from a collection of historical silk. The present work arises as a part of a challenging work of characterization of the ancient materials of Japanese traditional armors belonging to Museo delle Culture (Lugano, Switzerland). The project is of particular interest, as this kind of applied art has never been tested so thoroughly and extensively before. ER-FTIR spectroscopy was chosen to analyze historical samples mainly due to its non-invasiveness. This permitted us to investigate a great number of samples, chosen to represent the majority of colors and textures. As ER-FTIR spectroscopy is a practically new technique for silk recognition, the band assignment has been discussed and compared with the well-studied ATR mode, focusing also on part of the NIR spectral range, which is not commonly considered. Studies about water uptake in silk have been evaluated and reviewed in accordance with hard silk features and with data from water absorption tests. Instead of considering water as an interferent for the analysis, the signals arising from water in the O-H stretching band were studied to detect differences in the affinity to the water of hard and soft silk. Peak fitting analysis was used to confirm and quantify the differences in the spectra. Finally, principal component analysis was applied to propose visual discrimination of the two kinds of silk.

This work demonstrates for the first time that ER-FTIR spectroscopy is a successful tool to differentiate hard and soft silk, in historical samples too. The recognition of hard silk textiles provides doubly valuable information. Besides its historical significance, the detection of sericin is essential for preventive conservation and for targeted restoration works.

2. Materials and Methods

2.1. Reference Silk

Modern samples of hard and soft *Bombyx mori* silk were used as reference materials. Some of them were obtained from Museo della Seta (Como, Italy), while others were borrowed from Centro Tessile Serico Sostenibile (Como, Italy). The reference samples are the following:

- HS1 and HS2: Hard silk cloth
- HS3: Hard silk yarn
- SS1, SS2, and SS3: Soft silk cloth

Water absorption by the reference silk samples was controlled using two different strategies. High humidity conditions were reproduced by storing the samples for 65 h in a desiccator over K_2SO_4 saturated aqueous solution. The relative humidity level was constantly monitored using a data logger and set at approximately 97%. The dry condition was reproduced by leaving the samples for 65 h at 40 °C in a ventilated laboratory oven. The samples were generally analyzed under ambient laboratory conditions, except where explicitly stated in the text.

2.2. The Morigi Collection of Traditional Japanese Armors

The analyses discussed in this paper were carried out on a representative core of the collection of Japanese samurai armors which Museo delle Culture in Lugano received from collector Paolo Morigi in 2017. After a limited exhibit in 2018, the works of art became part of the museum's permanent collections. The armors cover a wide range of styles and historical periods, from the Azuchi–Momoyama era (second half of the 16th century) to the Showa era (1926 to 1989) through to the Edo period (1603–1868), as reported in Tables 1 and S1.

Table 1. List of samples from the Morigi collection. A photograph of the full armor is reported with its inventory number. On the right, the samples are listed and the presumed dating for each analyzed part is reported.

Armor	Sample	Dating	Armor	Sample	Dating
2017.Mor.1 	1_2	Late 16th c.	**2017.Mor.2** 	2_2	Late 19th c.
	1_6	Late 16th c.		2_3	Late 19th c.
	1_8	Late 16th c.			
	1_9	Late 16th c.			
	1_26	Late 16th c.			
2017.Mor.3 	3_3	17th c.	**2017.Mor.4** 	4_2	18th c.
	3_4	17th c.		4_3	18th c.
	3_8	17th c.		4_4	18th c.
	3_10	17th c.		4_11	18th c.
	3_11	17th c.		4_13	18th c.
	3_12	17th c.			
	3_18	17th c.			
	3_23	Late 16th c.			
2017.Mor.5 	5_1	After 1926	**2017.Mor.7** 	7_4	17th c.
	5_2	After 1926		7_8	17th c.
	5_9	After 1926			
2017.Mor.8 	8_4	17th c.	**2017.Mor.9** 	9_5	Late 19th c.
	8_7	17th c.		9_6	Late 19th c.
	8_12	Early 16th c.		9_17	Late 19th c.
	8_18	Early 16th c.			

c. is the abbreviation for a century.

Some armors were actually used in battle, as can be inferred from the fit that was comfortable and protected; others were exhibited only for celebrations and parades. Further details on the eras to which they belong have been described elsewhere [32].

The armors which are the object of this study and the sampling points are shown in Table 1. The armors silk textiles showed great variability in their colors and typology. Details for each analyzed sample are reported in Table S1. The parts of the traditional Japanese armor are shown in Figure S1. Non-invasive analyses with ER-FTIR spectroscopy were performed on different areas which were chosen to represent the majority of the colors and textures.

2.3. *Attenuated Total Reflectance Fourier Transform Infrared Spectroscopy*

ATR-FTIR spectra were acquired with a Thermo Scientific Nicolet iS10 instrument equipped with a fast recovery deuterated triglycine sulphate (DTGS) detector. The parameters used were 32 scans, 4 cm^{-1} resolution, and a range between 4000 and 600 cm^{-1}. A background spectrum was acquired periodically to allow the software to automatically subtract the atmospheric air spectrum from that of the sample. The spectra obtained are generally presented omitting the 2400–1800 cm^{-1} region, which is not very informative because it shows, at around 2100 cm^{-1}, the typical absorption of the crystal of analysis consisting of a diamond. The ATR-FTIR analysis was conducted on the reference samples.

2.4. *External Reflection Fourier Transform Infrared Spectroscopy*

In situ analyses on the Morigi collection were performed in ER-FTIR mode, using a portable Alpha Bruker FTIR spectrophotometer equipped with an external reflection module for contactless measurements and a DTGS detector. The analysis parameters used are 200 scans, 4 cm^{-1} resolution, and a range of 7500–375 cm^{-1}. Periodically, a background spectrum was acquired using a flat gold mirror. The measurement area was approximately 6 mm in diameter and the instrument was placed in a frontal position relative to the analysis point, at a working distance of approximately 1–1.5 cm. Fine-tuning of the optimal distance was then achieved by searching for the maximum signal directly in the interferogram using the software. The acquired spectra were processed using pseudo-absorbance [log (1/R); R = reflectance] as the intensity unit.

2.5. *Data Treatment and Elaboration*

The spectra of the samples were interpreted by comparing them with the reference samples and the literature. The optical spectroscopy software Spectragryph, version 1.2.16.1, was used to visualize and manipulate the ATR-FTIR and ER-FTIR spectra [49]. The same software was used to convert native ER-FTIR spectra into .jdx files in order to manipulate them with chemometrics.

Baseline correction was applied to all spectra. FTIR spectra are also commonly pre-processed to remove the effects of light scattering phenomena. For this purpose, the SNV (standard normal variate) method is often used to effectively remove multiplicative interference of scattering and particle size [50]. This pre-processing method eliminates the information about the absolute intensity of the signals but enhances the subtle differences in the band shape of the different superimposed spectra. In this study, SNV pre-processing was applied, when necessary, to the entire spectrum or only to a region of it.

The application of SNV is based on the following mathematical operation:

$$y_{ij\ (SNV)} = \frac{y_{ij} - \overline{y}}{\sqrt{\frac{\Sigma(y_i - \overline{y})^2}{n-1}}} \qquad (1)$$

That is subtracting the mean spectra $\overline{y}$ to each intensity value y_i of the original spectrum and then dividing for the standard deviation value.

2.6. Spectral Deconvolution/Curve-Fitting Analysis

Based on the literature and our previous work, the OH stretching band was analyzed using a band fitting method [28,32,51,52]. The selected spectra were cut to the range of 3800–2400 cm^{-1}; then, a baseline correction was applied using a linear function passing through the ordinates at the endpoints of this interval, and SNV correction was performed too. The FitPeaks Pro function of the peak analyzer package of Origin Pro 2018 software (OriginLab Corporation) was used for band fitting as follows.

As a first step, the second derivative of the convoluted spectra was calculated and smoothed using the adjacency averaging method (measurement smoothing window 20). This made it possible to identify the position of the bands, which were then compared with the literature. Next, the spectra were deconvoluted using Gaussian curves and a constant baseline (constrained to zero absorbance). Some bands were allowed to shift from their initial position, within a specific range, while the full width at half height (FWH) of the bands was fit to a specific range based on the theoretical width of the band [53]. Table 2 shows the bounds setting. The fitting was iterated until convergence and a Chi-Sqr tolerance value of 10^{-6} was obtained.

Table 2. Bounds setting for the curve fitting of the OH-stretching band.

Peak	1	2	3	4	5	6	7	8	9
Centre (cm^{-1})	3630	3560	3400	3320	3220	3060	2977	2931	2875
Centre bounds (cm^{-1})	±15	±15	±30	±15	±30	±5	±5	±5	±5
FWHH bounds (cm^{-1})	0–200	0–200	0–200	0–80	0–350	0–80	0–50	0–50	0–50

A similar method was developed in order to deconvolute the water band at 5170 cm^{-1} in the NIR region. The spectral region between 5400 and 5000 cm^{-1} was selected, smoothed (Savitsky–Golay method, interval = 21, polynomial order = 2), and baselined. The position of the bands was found by means of the second derivative, accordingly to the literature [47]. The bands were assigned as follows: non-freezing water, 5050 cm^{-1}; freezing bound water, 5140 cm^{-1}; bulk water, 5220 cm^{-1}.

2.7. Principal Component Analysis

ER-FTIR datasets were subjected to principal component analysis (PCA). All data were centered before further analysis. Prior to model calculation, different preprocessing techniques were tested and evaluated in order to correct unwanted data modification such as, as an example, different scattering. The preprocessing step was optimized by assessing the suitable mathematical transformation to remove the unwanted artefacts from the spectra. The Savitsky–Golay derivative, SNV, baseline correction, range reduction, and a combination of them were tested. The software used for the chemometric calculations was R version 3.6.3 (Rstudio version 1.4.1106).

3. Results and Discussion

Firstly, the ATR-FTIR and ER-FTIR spectra of soft silk are reported and compared, as a complete band assignation for silk fibroin with ER-FTIR has never been discussed before. The spectral differences arising from water uptake are then evaluated for both ATR-FTIR and ER-FTIR spectra, by means of water absorption tests. Finally, reference hard silk is investigated, and compared with soft silk in order to find a key for discrimination. Peak fitting analysis is used to validate our supposition. At the end, the proposed method is tested on a case study, by applying it on historical silk samples. PCA is applied to visually detect samples made of hard silk.

3.1. ATR-FTIR and ER-FTIR Spectra of Soft Silk

The band assignment for the ATR-FTIR spectra has been discussed and published extensively for fibroin [18,25,29]. The main absorption bands are due to the absorptions by amides A, B, I, II, and III, which are typical for the protein backbone [53,54]; alongside these absorptions which are shared by all proteins with little variations, other signals arise from the amino acids' side chains, such as ν(CC) and δ(CH) in tyrosine, ν(C=O) in aspartic acid, and ν(CO) in serine [53–56]. Figure 1 compares the spectra of the same reference of soft silk taken with ATR-FTIR and ER-FTIR spectroscopy.

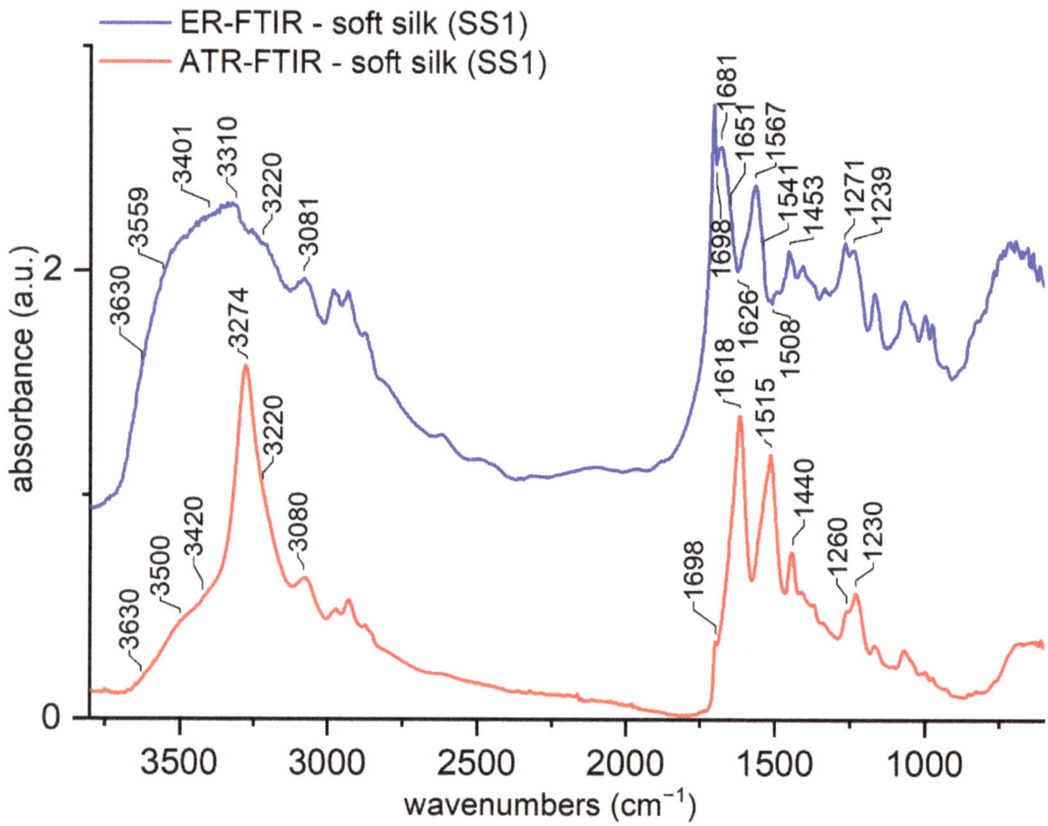

Figure 1. Comparison of ATR-FTIR and ER-FTIR spectra of soft silk. The region 7500–4000 cm^{-1} of the ER-FTIR spectrum is not shown.

The spectra appear very different; in particular, some shifts appeared mainly in amides A, I and II's peaks. At first sight, the peaks at 1706 cm^{-1} and 1680 cm^{-1} appear extremely enhanced by external reflection, while below 1450 cm^{-1}, no sensitive differences are noticed. Amides I and II'd peaks apparently show a great shift. In our opinion, their intensities were probably enhanced to the point that they appear as inverted peaks. This is a common problem with the ER-FTIR mode [40], but also with diffuse reflectance infrared Fourier transform spectroscopy (DRIFT) [27]. Thus, some peaks should be considered as inverted and their maxima could be identified at 1698, 1628–1618, and 1508 cm^{-1}. The assignment of ER-FTIR bands is proposed in Table 3, making a tentative comparison with ATR-FTIR spectral features which are reported in the literature.

Table 3. Comparison between the spectral features of the ER-FTIR and ATR-FTIR modes. The most evident shifts are in bold.

ER-FTIR (cm^{-1})	ATR-FTIR (cm^{-1})	ATR-FTIR Band Assignment
~3630	3630	Free OH (water dimers) [54]
~3560	**3500 (sh)**	Non-freezing water (O-H—O=C) [57,58]
~3400	**3420 (sh)**	Freezing bound water (O-H—polar groups) [57–59]
3320	**3274**	Amide A, N-H stretching [18,42]
3220	3220	O-H stretching, bulk water (—OH—OH—OH—) [59–61]
3080	3080	Amide B, N-H stretching [54]
1698 (inv)	1698 (sh)	ν(C=O) amide I bond, β-sheets [25,29]
1682	**1675 (sh)**	Amide I, β-turn [25,29]
1650	1650	Amide I, α-helix/random coil [25,29]
1627–17 (inv)	1628–1621	Amide I, intermolecular β-sheets [25,29]
	1621–1616	Aggregated β-strand/intermolecular β-sheet
1567	**1555**	Amide II, β-sheets [25,29]
1541	1545	Amide II, α-helix/random coil [25,29]
1508 (inv)	1515	Amide II, β-sheets [25,29]
1453	1440	CH_2, CH_3 bending in alanine [25,29]
1270	1260	Amide III, β-sheets [62,63]
1238	1230	Amide III, α-helix/random coil [62,63]

(sh): shoulder; (inv): inverted peak.

3.2. *Water Uptake Behavior of Soft Silk*

Another characteristic of the ER-FTIR spectrum (Figure 1) is the broadening of the band at 3330 cm^{-1}, which creates a single band together with the water OH stretching band between 3400 and 3600 cm^{-1}, as previously reported [18]. The enhancement of the -OH signal with respect to the ATR spectra is typical of the ER-FTIR mode [41].

The spectral region between 3600 and 3000 cm^{-1} is generally associated with intramolecular and intermolecular hydrogen bonding and free hydroxyls in polar macromolecules, such as cellulose, but also with the free or the bound water linked to the substrate. Water FTIR signals are strongly influenced by their state of aggregation [59]. In particular, water molecules bind in different forms when they are adsorbed on a protein film [54,64] or a biocompatible polymer [57,58], such as silk [7]. On the interface, water and the C=O and N-H groups of the protein backbone form hydrogen bonds, some of which can replace direct N-H···O=C hydrogen bonds which are typical of crystalline domains of fibroin. This water is the so-called non-freezing water, as it never crystallizes due to the tight bond to carbonyl groups. Freezing bound water, which instead crystallizes below 0 °C, interacts with the other polar groups in the side chain. Finally, bulk or freezing water, which crystallizes at ~0 °C, has a bulk-water-like structure with an O-H—O-H hydrogen bond network and it weakly adsorbs to the surface. As the degree of hydrogen bonding between water and protein increases, the FTIR peaks are shifted to higher wavenumbers [54]. Assignments for each type of O-H stretching are reported in Table 3 and compared with the same vibrations in the ATR mode. Some shifts between the ER and ATR modes were experienced. Some soft silk samples were conditioned under different RH conditions to evaluate spectral differences which arise from the water uptake of silk. The spectra were taken both with ATR-FTIR mode and the ER-FTIR mode, and they are shown in Figure 2.

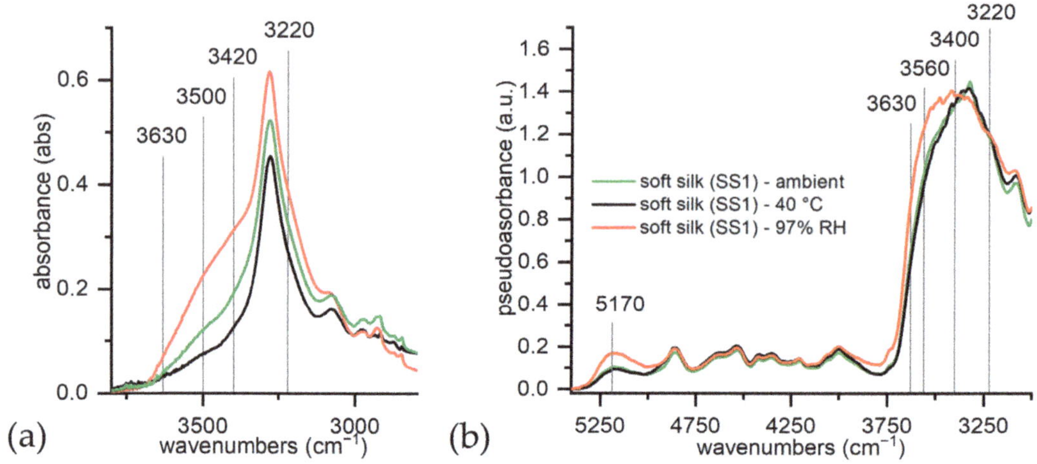

Figure 2. (**a**) ATR-FTIR spectra of soft silk under different relative humidity conditions; (**b**) ER-FTIR spectra of soft silk under different relative humidity conditions.

As expected, soft silk shows very hygroscopic behavior. The bands which are associated with water uptake are highlighted in Figure 2a,b. Under increasing RH conditions, the overall intensity of the spectra increases, thus suggesting the enhancement of the broad bulk water absorption at around 3220 cm^{-1}. Similarly, both signals associated respectively to freezing bound (3420–3400 cm^{-1}) and non-freezing water (3560–3500 cm^{-1}) are strongly influenced by both low and high humidity conditions. Their intensity is enhanced, and the spectral shape is changed. No shifts appear yet. When humidity conditions are changed, the ATR-FTIR and ER-FTIR modes show different responses. The ATR-FTIR spectrum at low humidity appears very different from the spectra under ambient and high humidity conditions, while the ER-FTIR spectra under low and ambient conditions are similar. In the ER-FTIR spectra, it is interesting to also note the OH combination band at 5170 cm^{-1}, whose intensity is strongly enhanced only under high humidity conditions. In Figure 2b, the different contributions to the overall band are indicated and assigned to the different types of water. These contributions are discussed in Section 3.3. As the conditioned samples were analyzed with ATR-FTIR and ER-FTIR paying particular attention to maintaining the correct conditioning, the results suggest that ER-FTIR is not so strongly influenced by low humidity conditions with respect to the ATR mode.

3.3. Characterization of Hard and Soft Silk

The possibility to discriminate hard and soft silk with ATR-FTIR spectroscopy was the theoretical basis of this research project. Band assignment for the ATR-FTIR spectra has been published extensively for fibroin [18,25,29]. Sericin shows similar signals [18,30,31] and the main source of the slight differences lies in their distinct conformation of the secondary structure, in addition to differences in amino acid composition [25,65]. The literature [17,18,30,31] reports that slight shifts in amides I and II's peaks are the main signs of sericin's presence, together with a broader amide A band at 3270 cm^{-1} and a water sorption band centered at 3400 cm^{-1}, which are signals for raised hydroxyl content. Generally, the authors report that peaks at about 1400 cm^{-1} (C-H and O-H bending [17,25,31]) and 1075–52 cm^{-1} (C-OH stretching [25,31] or C-C bending [54,65]) are distinctive of serine [18,30,31,54] and as a consequence are useful to distinguish sericin (mainly composed of serine) from fibroin [20,65]. Moreover, Zhang et al. [18] suggest that a decrease in intensity of the 1000 and 975 cm^{-1} peaks, which are typical of fibroin, could infer the presence of a sericin coating. Generally speaking, it appears that the distinction between hard and soft silk is challenging, as there are no evident peak shifts or spectral features belonging to hard silk only. It is important to also consider the possibility of sericin leaching due to high humidity conditions [18].

The reference samples of hard and soft silk were analyzed with ATR-FTIR spectroscopy to evaluate the best markers for differentiating hard and soft silk. The spectra of reference samples of hard and soft silk (Figure 3) confirm what is reported in the literature. In particular, the decreased intensity of peaks at 3270 and 1440 cm^{-1} and the increased intensity at 2920, 2850, and 1071 cm^{-1} could be markers for hard silk detection. The decrease in the peaks at 1000 and 975 cm^{-1} is quite difficult to notice. In our opinion, the best indicator is the broadening of the bands at 3500–3420 cm^{-1}. Such broad bands can be attributed to hydrogen-bonded water, whose absorption can be found between 3600 and 2900 cm^{-1} according to the strength of hydrogen bonding [54,66]. The shoulder at 3500 cm^{-1} is ascribable to H-bonded OH to C=O of the amide [67]. Another interesting difference is the increase in the band at 3220 cm^{-1}. Both silk types have a high capability to adsorb moisture, but hard silk is more prone to bind water due to its composition and amorphous structure, as discussed in the Introduction.

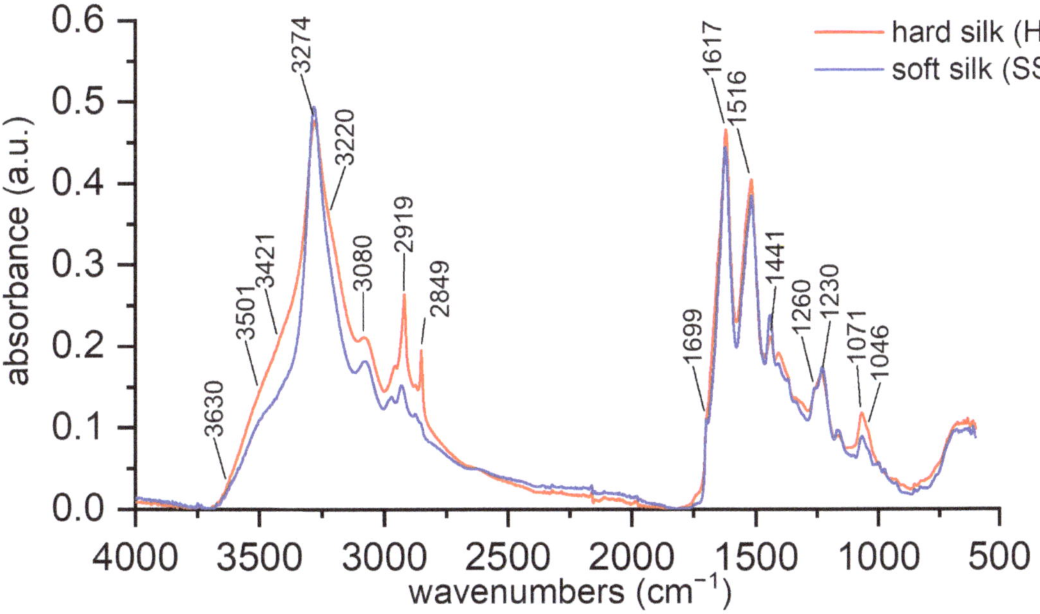

Figure 3. ATR-FTIR spectra of the reference samples of hard and soft silk.

External reflection infrared spectroscopy was applied on the same reference materials to test if the same spectral features which characterize hard silk in the ATR-FTIR spectra could be recognized. Figures S2 and S3 show the ER-FTIR spectra for all of the references of hard and soft silk. The instrumental spectral range is split into two spectra (range 6100–3800 cm^{-1} and range 3800–400 cm^{-1}). The region 7500–6100 cm^{-1} of the ER-FTIR spectrum is not shown. For clarity, only two references are shown in Figure 4, which represent the range 3800–900 cm^{-1}.

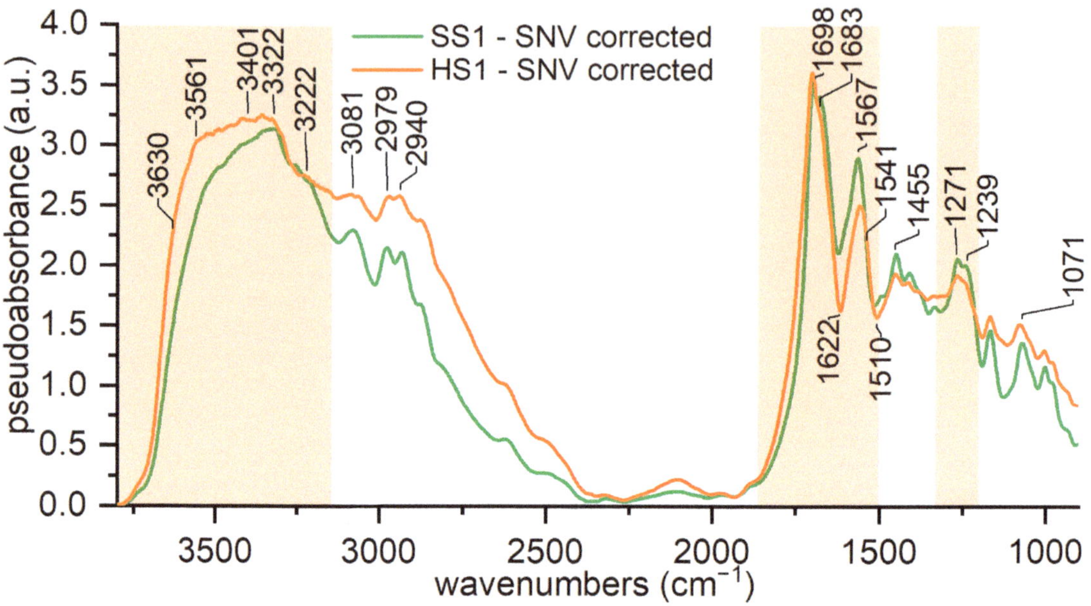

Figure 4. ER-FTIR spectra of hard and soft silk. The region of 3800–900 cm^{-1} is shown.

In this region, the main differences are located in the band of the hydrogen-bonded water and in the intensities of some of amides I and II's peaks. Assuming that the maxima of amides I and II are inverted peaks, we point out the shift of the absorption from 1618 (hard silk) to 1626 cm^{-1} (soft silk) and the differences between the samples and references at around 1510 cm^{-1}. The inverted peak at 1618 cm^{-1} attests the higher content of β-strands for hard silk, while the shoulder at 1650 cm^{-1} is a sign of the high content of random coil conformation. Both findings agree with the description of hard silk. Similarly, a decrease in intensity is observed at 1680 cm^{-1}, probably due to β-turn content which is lower in hard silk. The amide II peak at 1560 cm^{-1} shows a shift and a decrease in intensity, too. At around 1510 cm^{-1}, the inverted peaks are due to C-N and N-H from amide II. Another characteristic of the hard silk spectrum is the impressive broadening of the band between 3400 and 3600 cm^{-1}, which is more enhanced with respect to soft silk and centered around the new peak at 3560 cm^{-1}. The variations in intensities which are distinctive for hard silk in the ATR mode are not present, so other markers for sericin should be identified.

We think that the different water uptake values under the same environmental conditions are the key to discrimination. The amorphous fraction of silk (sericin and the disordered fraction of fibroin) is mainly affected by the absorption of non-freezing water, as crystalline regions are hardly accessible to water [47,68]. As a matter of fact, the amorphous phase is responsible for 70% of the uptake of environmental water [69]. The amino acid composition could have some influence too, as the polar amino acid serine is the main component of sericin with 30% w/w in contrast to the serine content in fibroin which is nearly 15% w/w [18]. In particular, where the structure presents grooves, as in the case of hard silk, the nature of the side chain has greater importance in the interactions with water molecules [70]. Therefore, the combined influence of the morphology and the amino acid composition could make hard silk more prone to bind water with respect to soft silk under the same environmental conditions. These considerations are visualized in Figure 5.

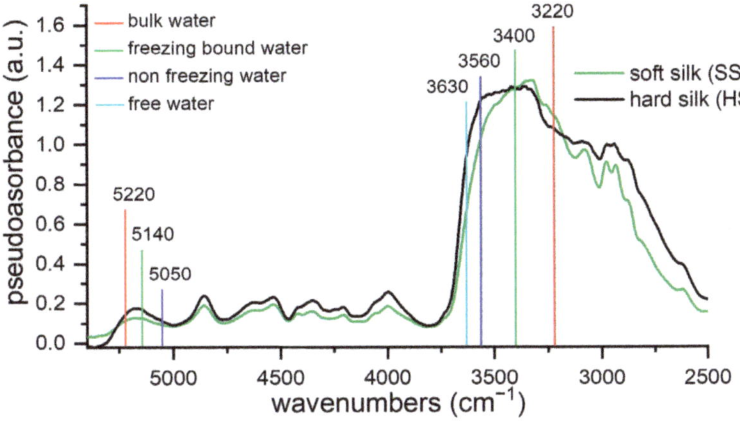

Figure 5. Details of the ER-FTIR spectra of hard and soft silk reference (region 5500–2250 cm^{-1}). Absorptions due to different types of water are highlighted.

In order to study the different contributions of water to the overall band, a peak fitting analysis was carried out. Figure 6a,b shows the deconvolution models of the region of 3800–2400 cm^{-1}. Some areas under the Gaussian curve forming the overall curve are reported in form of percentages to highlight the difference between the two kinds of silk (Figure 6c).

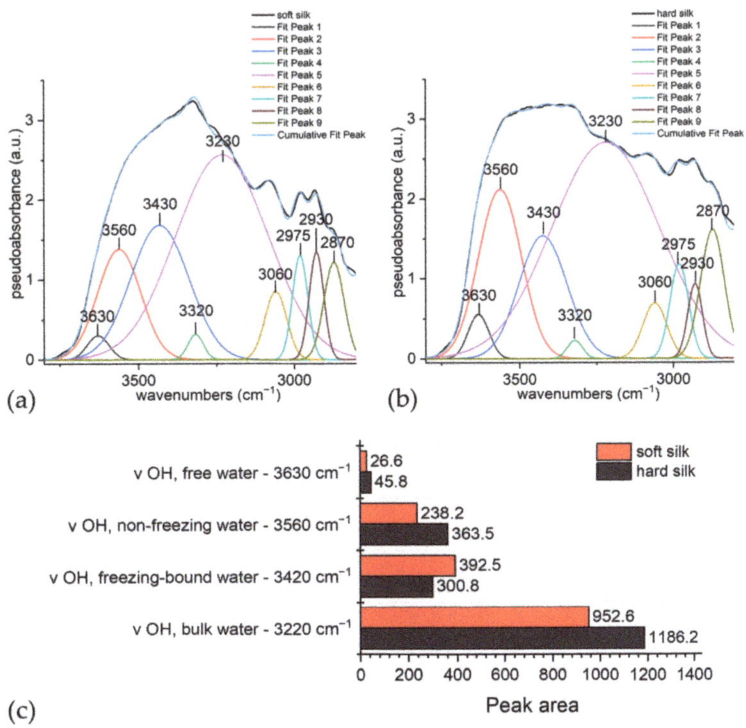

Figure 6. Deconvolution of the O-H stretching band (region 3715–2800 cm^{-1}): (**a**) soft silk; (**b**) hard silk; (**c**) comparison of the peak areas for each contribution in soft and hard silk.

In hard silk, the area of the water band centered at 3560 cm^{-1} is definitely higher than in soft silk. This band is related to non-freezing water, which is evidently associated to the amorphous structure which is accessible to water. The band at around 3630 cm^{-1}, which is assigned to non-hydrogen bonded water, is higher in hard silk than in soft silk, while the signal of freezing bound water (around 3420 cm^{-1}) is unexpectedly higher in soft silk, but this band can also be influenced by the contribution of the amide A peak. In the ATR mode, the signal is located at 3270 cm^{-1}, but the literature [54] reports that H-bonding could be responsible for a blueshift towards 3310 cm^{-1}. Actually, both are probably present, as N-H groups exist in two forms, both C=O-bonded (NH—OC) and water-bonded (NH—OH_2), giving rise to two different signals [71]. At 3270 cm^{-1}, the absorption is due to intermolecular bonded N-H stretching, as confirmed by theoretical calculations [72]; at 3313 cm^{-1}, the signal could be ascribable to water-bonded N-H stretching, as it is experimentally seen in the polyamides' spectra [61,73]. Manas et al. [74] confirmed that amides show different signals at the same time, when solvent-exposed domains differ from the bulk which has little interaction with water. The blueshift is due to the NH—OH_2 hydrogen bond, which tends to increase the force constant of the NH in-plane bending motion. Actually, it is important to recall that the region 3400–3200 cm^{-1} is overlapped with O-H stretching signals, so that it is difficult to clearly discriminate the contributions. Similarly, the signal of bulk water at 3220 cm^{-1} can be partially overlapped with N-H stretching. Anyway, it is definitely higher in hard silk than in soft silk, as we can expect as it is reported that hard silk absorbs much more water than soft silk [16].

The NIR range (7000–4000 cm^{-1}) of the spectrum of hard and soft silk references HS1 and SS1 is shown in Figure 7, while Figure S3 shows the same range for all of the references. The literature [47,48,75–77] about silk reports overtone and combination bands arising from OH (water and serine), NH (peptides), and CH (peptides, alanine, serine, etc.). The main assignments are described in Table 4. Significant changes in intensities occur in this region; while the water band at 5170 cm^{-1} (O-H combination) increases in hard silk, amides bands at 4850, 4620, and 4520 cm^{-1} (hydrogen bonded N-H and C=O vibration) are more prominent in soft silk, as experienced by Mossotti et al. [48]. Mo et al.'s study [76] showed that NIR bands are sensitive to conformational changes, so higher intensities in soft silk are probably due to the dominant conformation of fibroin, which is β-sheet. Hard silk is covered by amorphous sericin, so the signal from crystalline fibroin is lower. On the other hand, sericin makes hard silk more prone to adsorb water, probably causing enhancement of the band.

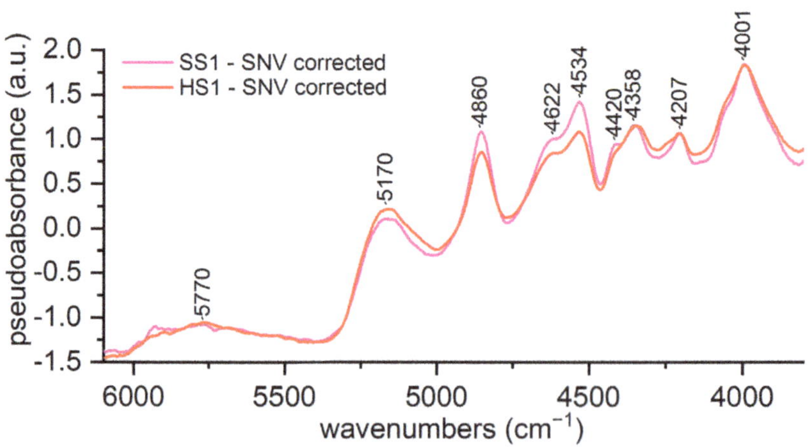

Figure 7. ER-FTIR spectra of hard and soft silk references. The region 6100–3800 cm^{-1} is shown.

Table 4. List of peaks and their assignations in the NIR range of the spectrum (7000–4000 cm^{-1}).

Wavenumber (cm^{-1})	Assignment [47,48,75,76]
7500–6000	Water, first overtone
5900–5700	ν(CH), first overtone
5170	O-H combination
4860	Amide A ν(NH) + amide I/amide II
4620	Amide A ν(NH) + amide III/amide B + amide II
4534	Amide A ν(NH) + amide III/amide B + amide II
4420	ν + δ(CH)
4358	ν + δ(CH)
4205	ν + δ(CH)

If the band at 5170 cm^{-1} is deconvoluted (Figure 8), it reveals other important information about the water degree of association with silk [47]. Due to its asymmetric aspect, it can be deconvolved into at least three components (5219, 5139, and 5046 cm^{-1}). As the degree of C=O and N-H groups' hydrogen bonding to water increases, peaks are shifted to lower wavenumbers. There are three types of signals, which could be attributed to non-freezing, freezing bound, and bulk water [76]. In all cases, water is more prone to bind hard silk, which contains an amorphous sericin covering and less β-sheet structured fibroin (the degumming process increases the crystallinity index of silk), hence the higher areas of hard silk shown in Figure 8c, obtained through the peak fitting analysis. The two higher wavenumber bands offer a distinction of bound water with different degrees of hydrogen bonding. They are located at 5222 and 5141 cm^{-1}, assigned to bulk and bound-freezing water, respectively. The band at 5050 cm^{-1} is assigned to strongly hydrogen-bonded structural water, which is non-freezing water. Values associated with this band account in hard silk for more than twice as much as for soft silk. This demonstrates that the capacity to strongly bond water is higher in hard silk than in soft silk due to structural differences. The percentages for bound-freezing water are significant, too; as higher values are associated with aminoacidic composition richer in polar side chains, hard silk is confirmed to be more prone to bond freezing bound water. As for bulk water, differences in absorption are still present, even less evidently, because hydrated proteins offer a similar surface to bind bulk water. The different contributions of the combination band of water are shown in Figure 5.

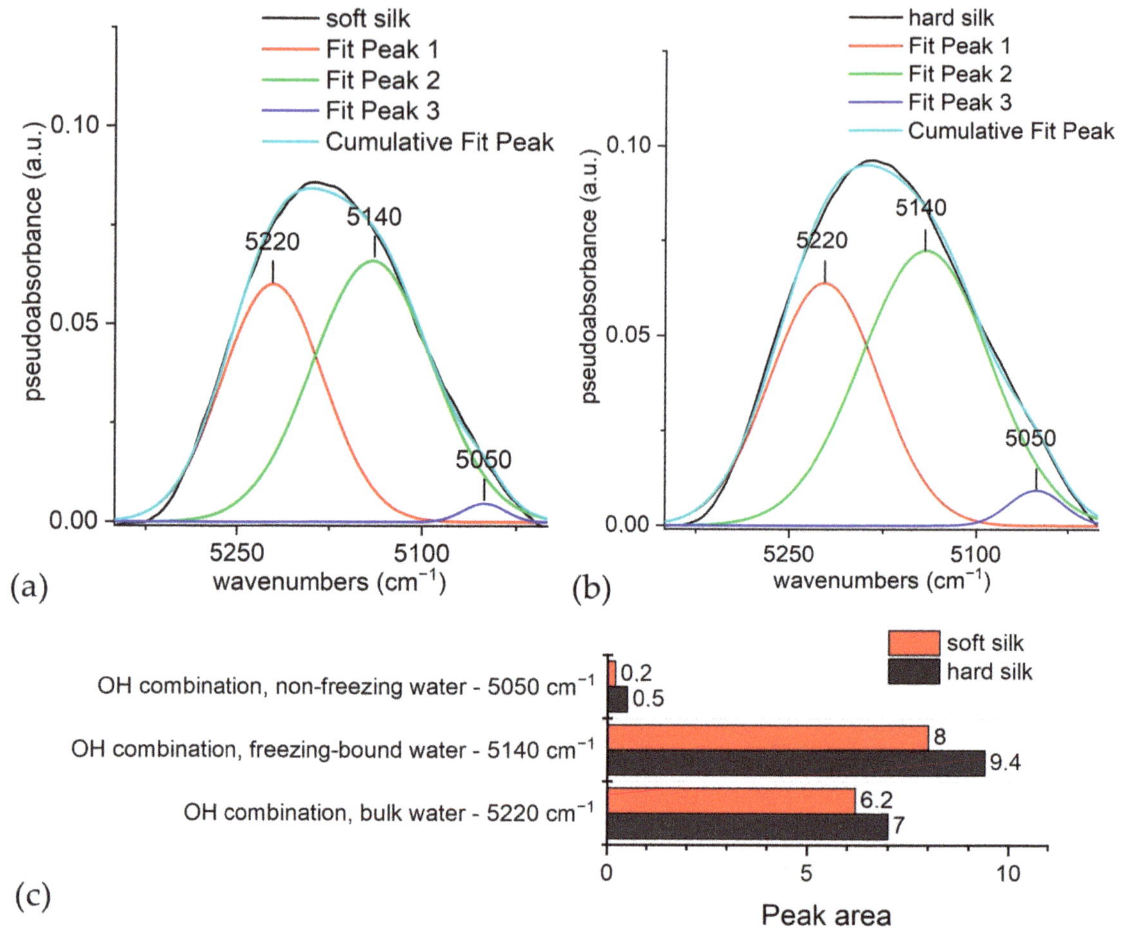

Figure 8. Deconvolution of the band at 5170 cm^{-1}. Deconvolution of the O-H combination band (region 5350–5000 cm^{-1}): (**a**) soft silk; (**b**) hard silk; (**c**) comparison of the peak areas for each contribution in soft and hard silk.

3.4. Characterization of Silk in Traditional Japanese Armors

External reflection FTIR spectroscopy was applied to a wide selection of textiles from armors in order to test extensively the variability in the appearance and color of silk yarns. The same main peaks (1706, 1566, 1454, 1265, and 1071 cm^{-1}) are found in all the spectra, so we can infer that they are all made of pure silk, without peculiar samples showing other recognizable signals. However, samples 4_2 and 4_11, belonging to the Mor.004 armor (17th century), appear different from the others, especially in their water sorption band (3200–3600 cm^{-1}), which is shown in Figure 9a, and in the amides region (1800–1500 cm^{-1}), which is not shown. In order to obtain Figure 9a, the spectra were truncated between 3800–2600 cm^{-1}, linearly baselined, and SNV corrected. It is clear that samples 4_2 and 4_11 are different from all of the other samples, mainly due to the shape of the OH stretching band. Through comparison of this feature, the spectra of hard and soft silk samples could be visually differentiated. For the sake of clarity, in Figure 9b, just two representative spectra of the samples are shown together with the references. The spectrum of sample 4_2 strictly resembles the spectrum of the hard silk reference, and the same happens for sample 5_1 and the soft silk reference.

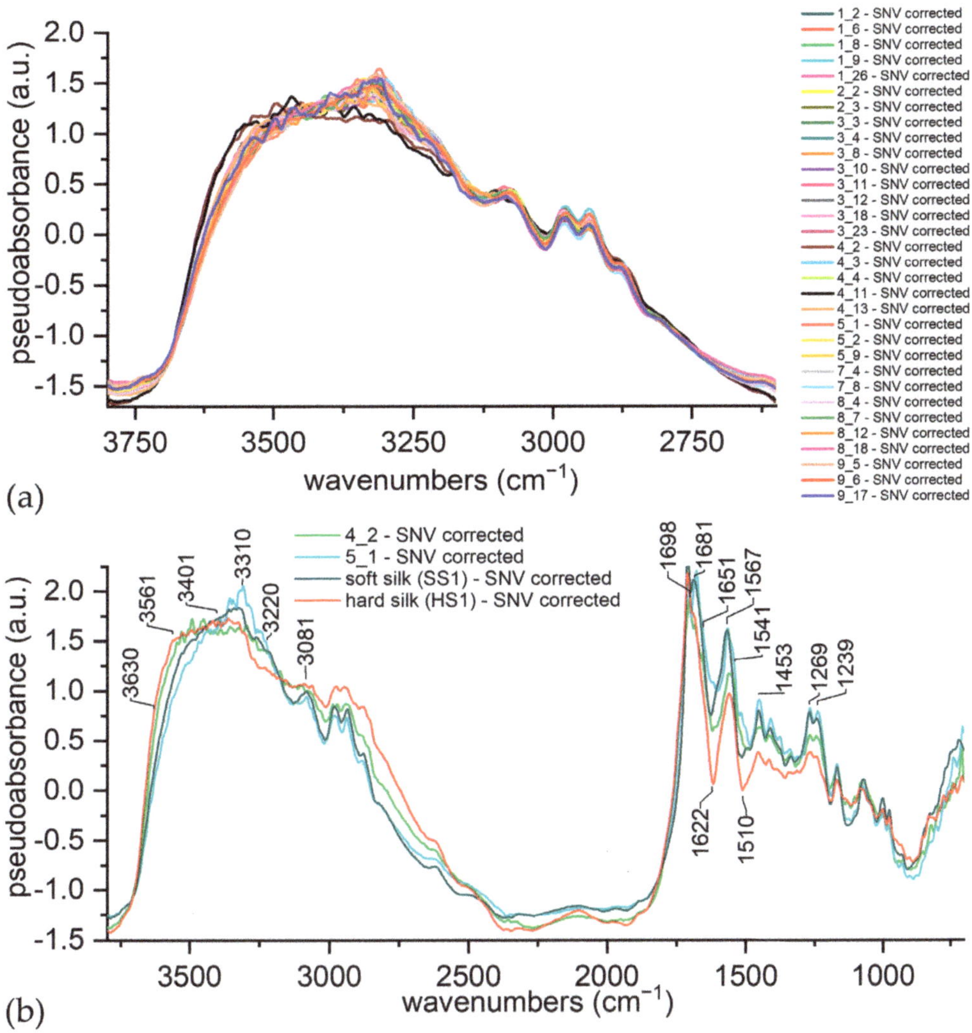

Figure 9. (**a**) ER-FTIR spectra of samples. (**b**) ER-FTIR spectra of samples 4_2 and 5_1 together with the spectra of hard and soft silk references.

In Figure 2a,b, we showed how OH stretching bands in soft silk spectra are influenced under different RH conditions. It could be argued that it is not possible to distinguish between the enhanced absorption due to high humidity conditions from that caused by the presence of the sericin coating. Actually, under the same temperature and RH, hard silk is doomed to show higher absorptions in the OH stretching band with respect to soft silk. Thus, the comparison method works provided that the textiles to be compared belong to the same collection or are stored under the same RH conditions. Within the analysis of a part of a collection, it could be useful to conditionate two reference clothes—made of hard and soft silk—under the same temperature and RH conditions of the collection. Recording their spectra together with other samples would make it easier to visually evaluate the spectra. As shown in Figures S2, S3 and 9b, the reference materials do not differ significantly among them and from historical hard silk samples. With respect to an earlier proposed method [48], there is no need to conditionate the textiles of the collection under fixed RH before the analysis, which would be difficult to obtain and potentially dangerous for their preservation.

Finally, it is worth pointing out that the recognition of hard silk was achieved on samples dating back to 17th century. It is a high-value consideration as it shows that the proposed method works on both modern mock-up samples and historical samples. It is proof that the decay condition does not generally prevent the identification of hard silk, provided that sericin has not been degraded. Indeed, the chance to detect sericin decreases as decay takes place. Thus, the most ancient samples, which have a higher probability of being made of hard silk for historical reasons, can hardly show spectral features of sericin. In this case, more sophisticated techniques should be used. Moreover, the proposed method cannot falsely indicate the presence of sericin in a degummed textile. Indeed, the rise of water sorption cannot be related to the aging of the samples, as both hard and soft silk show a reduction of water sorption related to aging [6].

Instead of visual comparison, principal component analysis (PCA) can be applied to the ER-FTIR spectra. The method is more rigorous and objective to evaluate differences among sets of samples and to search groups among them. Similar samples locate themselves in the same region of the scores plot, while samples belonging to different groups are far apart. PCA is an unsupervised learning algorithm, with it being able to find some patterns and regularities without direct supervision of an operator and thus objectively. The scores plot permits obtaining visual recognition of such differences. According to the purpose of this research, hard and soft silk samples should create two different groups. Samples appearing as disturbing or unusual are named as outliers, and care of them must be taken to obtain reliable models [78]. In general, the spectrum of the sample could be intended as an outlier if it lies outside the distribution obtained from those of the other samples, and it should be corrected or removed from the model. Outliers' evaluation can be carried out according to the Hoteling T^2 statistic and the Q statistic, and their presence is due to a gross error producing an anomalous acquisition or the peculiar features of the sample in respect to the others (e.g., strongly IR-absorbing substances adhering to the textile).

In this work, different spectral ranges were tested to choose the most significant. The MIR region and NIR region were initially considered independently in order to evaluate the specific preprocessing method for each part. A successively extended NIR region (7400–2400 cm^{-1}) was tested in order to enhance the information from the most important bands. Finally, the whole spectrum was considered and preprocessed with Savitsky–Golay smoothing (derivative order zero, second polynomial order, window width 71) and MSC. This appeared as the best choice as the sum of the explained variance for PC1 and PC2 was higher, and by observing these spectra, the artefacts were removed in an effective way. The relative scores plot of PC1 vs. PC2 is shown in Figure 10. These two components describe 69.9% and 7.4% of the total variation respectively.

In the scores plot, the samples are black colored. Most of them are located at low positive and low negative values on PC1 and PC2, thus forming a group in the middle of the graph. With respect to the main group, samples 4_2 and 4_11 appear at higher values on both PC1 and PC2. In order to have a visual evaluation of their nature, some hard and soft silk references were projected into the PCA. They are blue and red colored, respectively. As a matter of fact, the soft silk reference correctly joined the group in the middle of the graph, while hard silk references locate themselves near to samples 4_2 and 4_11. The hard silk reference HS3 is not shown as it was recognized as an outlier. Probably their spectra show similarities. In order to study the spectral features which contribute to such a scores plot, the analysis of the loadings plot of the first two PCs is needed.

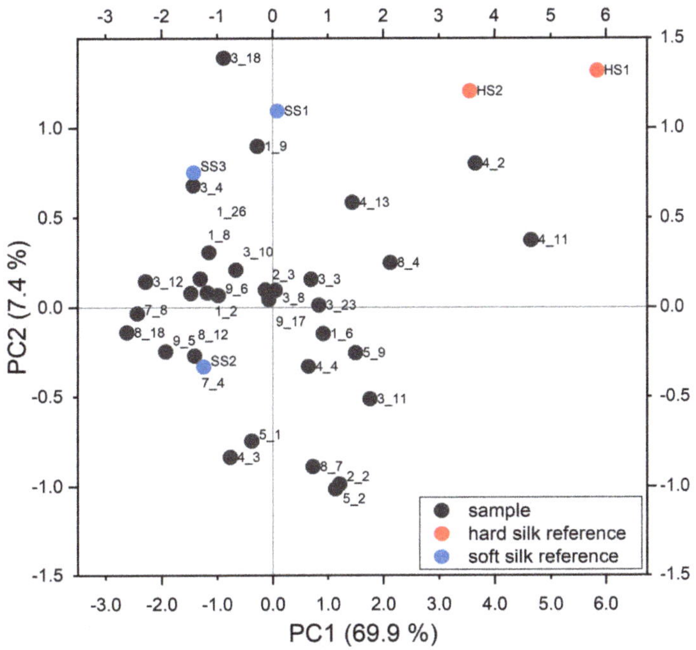

Figure 10. Scores plot for the first two principal components. Red and blue points are the hard and soft silk references, respectively, which were projected into the PCA.

The loadings for PC1 and PC2 are depicted as lines in Figure 11, together with the spectra of references for hard and soft silk. It is worthwhile to recall that high loading values (both positive and negative values) indicate important variables and thus, in this case, important peaks. The regions of the spectra where the loading values are higher are more significant, so they are highlighted.

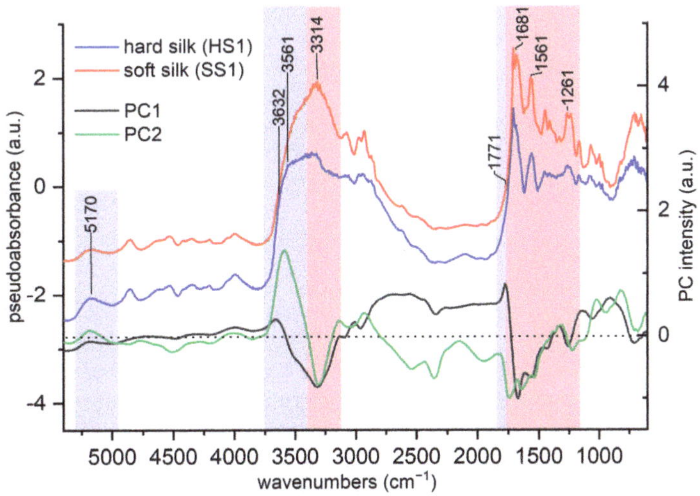

Figure 11. Loadings plot of PC1 and PC2, shown together with the ER-FTIR spectra of the references for hard and soft silk. The region 5400–700 cm^{-1} is shown. The blue and red regions represent the most important regions for the differentiation between hard and soft silk, respectively.

The PC1 loadings showed a shape that resembles the one of the hard silk spectrum. This is not surprising as the first PC mainly describes the differences between the hard silk and the other silk samples. Negative high loadings values are strongly associated with peaks which are typical of soft silk (red colored bands in Figure 11), such as the amide A peak (3313 cm^{-1}), β-turn (1681 cm^{-1}), amide II (1560 cm^{-1}), and amide III (1260 cm^{-1}). The positive values at around 3650–3520 cm^{-1} could be attributed to an enhancement of the contribution due to non-bonded water and to non-freezing water (blue colored bands in Figure 11). Similarly, positive values at around 1770 cm^{-1} are associated with hard silk (blue colored band in Figure 11), as they appear in the band broadening in Figures 4 and S2.

PC2 shows high values for variables (peaks) which can be associated with water uptake in the region 3580–3400 cm^{-1} (blue colored bands in Figure 11). A positive value at around 3580 cm^{-1} is associated with the vibration of non-freezing water, whose content is much higher in hard silk than in soft silk. On the contrary, the negative value at around 3305 cm^{-1} is associated with NH stretching (red colored band in Figure 11), which appears more evidently in soft silk. As for the NIR region, the positive value at around 5200 cm^{-1} could be associated with the raised water content of hard silk (blue colored band in Figure 11).

4. Conclusions

The proposed method to discriminate hard and soft silk is based on the different water uptake during the storage at the same temperature and RH, followed by ER-FTIR spectroscopy. This is an innovative point of view, which exploits a "weakness" of FTIR spectroscopy—the strong absorption from water molecules—in order to indirectly obtain the results. Indeed, OH stretching bands are generally considered "forbidden" regions, since the analytical information about the molecule under analysis is covered by environmental water, which is very difficult to remove from silk textiles too. Actually, we showed that these bands are useful to study adsorbed water, by means of peak fitting analysis which appeared as an interesting tool to evaluate the different contributions of the OH stretching band. In particular, we found that the contribution of non-freezing water is decisive for differentiating hard and soft silk.

When samples are stored under the same conditions, the higher water content, which causes the broad absorption at around 3600 cm^{-1}, is linked to the presence of sericin and, indirectly, to the detection of hard silk. Thus, through the analysis of the shape of the OH stretching band, it is possible to differentiate hard and soft silk textiles, using a rapid and completely non-invasive technique. With respect to previous methods developed with ATR-FTIR spectroscopy, the proposed method is also easier, as only a single broad band had to be taken into consideration. The recognition of hard silk is obtained by visual comparison of spectra—two conditioned reference clothes made of hard and soft silk can be used to make identification easier. Alternatively, principal component analysis allows a more rigorous and objective comparison of the spectra, paying particular attention to the presence of outliers.

The proposed approach could infer very useful information about the nature and authenticity of historical silk textiles, thus suggesting the best conservation conditions and leading to targeted restoration works. Great campaigns of analyses can be managed, as the technique is rapid and non-invasive. The method could be useful within the industrial refining of silk too. Quality control analyses are fundamental to assure that the product achieves standard levels, but the measurement of the degumming extent of raw silk is difficult with traditional protocols. Instead, reflection infrared spectroscopy could be applied by manufacturers for continuous process control, as the working parameters are controlled and constant. The indirect measurement of degumming extent in the industrial context could be an interesting future outlook, even if further studies are needed to obtain quantitative data and chemometrics would be fundamental to manage them.

Supplementary Materials: The following supporting information can be downloaded at: https://www.mdpi.com/article/10.3390/ma16051819/s1. Table S1: List of samples from the Morigi collection. For each sample the presumed dating, color, the armor part which the sample comes from, and the kind of textile are reported; Figure S1: Scheme of the main parts of the traditional Japanese armor; Figure S2: ER-FTIR spectra of hard and soft silk references (region 3800–400 cm^{-1}); Figure S3: ER-FTIR spectra of hard and soft silk references (region 6100–3800 cm^{-1}).

Author Contributions: Conceptualization, L.G., M.L. and L.R.; investigation, C.C. (Cristina Corti), L.G., G.G. and L.R.; writing—original draft preparation, C.C. (Cristina Corti), L.G., B.G., G.G., M.L. and L.R.; writing—review and editing, F.P.C., C.C. (Carmen Canevali), C.C. (Cristina Corti), L.G., B.G., G.G., M.L., L.R. and S.R. All authors have read and agreed to the published version of the manuscript.

Funding: This research received no specific grant from any funding agency in the public, commercial, or not-for-profit sectors.

Institutional Review Board Statement: Not applicable.

Informed Consent Statement: Not applicable.

Data Availability Statement: The data presented in this study are available on request from the corresponding author.

Conflicts of Interest: The authors declare no conflict of interest.

Appendix A

Fibroin and sericin are the main proteins that form *Bombyx mori* silk fiber [15]. Sericin, which accounts for 25–30% of the weight, is a coating for the two core brins of fibroin, so that the raw silk fiber has about a 20 µm width. In particular, the *B. mori* silk fibroin is composed of two protein chains, a heavy-chain (H-fibroin) and a light chain (L-fibroin) covalently linked by a disulfide bond at the carboxy terminus of the two subunits. The complete amino acid sequence of the *B. mori* fibroin heavy chain is composed of a highly repetitive (Gly-Ala)$_n$ sequence motif and tyrosine-rich domains, together with some other amino acids whose sum does not exceed 20% (molar mass). On the contrary, in sericin, the majority of the amino acids are polar including serine and aspartic acid, with non-polar amino acids accounting for only 22% of the protein [5].

Silk fibers in cocoons [5], before any treatment, exhibit a layered structure, where slightly different types of sericin are found moving closer to the inner fibroin core. The outermost layer of sericin is the richest in polar amino acids, so it is easily solubilized in hot or boiling water. It is obtained through raw silk yarn or *grège*. This procedure is named *silk reeling*. The subsequent process of removing sericin, or "gum", from the hard silk is named *degumming* or *refining*. Generally, hot water was used [4], adding soap in Europe [3,5], while alkaline substances and/or enzymes derived from porcine pancreas were used together with water in China [3,6]. This process makes the final silk yarn lustrous and smooth, also making the fiber more able to bind to dyes. After degumming, silk loses up to 25% of its weight, so it generally undergoes the weighting process. Inorganic salts or other substances are employed to replace some of the lost weight, or even to exceed the original weight. These substances could be also added to act as mordanting agents, to make fibers more prone to bind dyes [3,7]. Instead of removing sericin directly on the *grège*, which has lost only a minor part of sericin [3], the degumming can be performed after weaving the yarn with its gum into a cloth that can be degummed and dyed in piece; alternatively, the raw yarn can be twisted to obtain a stronger thread and finally degummed and dyed in yarn. Actually, textiles may be also used in a partially degummed form, named *supple silk* [3], or used with their gum, although dyeing gives poor results and is difficult to obtain.

References

1. Garside, P. The Role of Fibre Identification in Textile Conservation. In *Identification of Textile Fibers*; Houck, M.M., Ed.; Cambridge-Woodhead Publishing: Boca Raton, FL, USA, 2009; pp. 335–365. ISBN 9781845692667.
2. Garside, P.; Lahlil, S.; Wyeth, P. Characterization of Historic Silk by Polarized Attenuated Total Reflectance Fourier Transform Infrared Spectroscopy for Informed Conservation. *Appl. Spectrosc.* **2005**, *59*, 1242–1247. [CrossRef]
3. Desrosiers, S. Scrutinizing Raw Material between China and Italy: The Various Processing Sequences of Bombyx Mori Silk. Available online: https://journals.openedition.org/acrh/10323 (accessed on 14 July 2022).
4. Silva, A.S.; Costa, E.C.; Reis, S.; Spencer, C.; Calhelha, R.C.; Miguel, S.P.; Ribeiro, M.P.; Barros, L.; Vaz, J.A.; Coutinho, P. Silk Sericin: A Promising Sustainable Biomaterial for Biomedical and Pharmaceutical Applications. *Polymers* **2022**, *14*, 4931. [CrossRef] [PubMed]
5. Cao, T.T.; Zhang, Y.Q. Processing and Characterization of Silk Sericin from Bombyx Mori and Its Application in Biomaterials and Biomedicines. *Mater. Sci. Eng. C* **2016**, *61*, 940–952. [CrossRef] [PubMed]
6. Zhang, X.; van den Berghe, I.; Wyeth, P. Heat and Moisture Promoted Deterioration of Raw Silk Estimated by Amino Acid Analysis. *J. Cult. Herit.* **2011**, *12*, 408–411. [CrossRef]
7. Timar-Balazsy, A.; Eastop, D. Materials. In *Chemical Principles in Textile Conservation*; Routledge: New York, NY, USA, 1998; pp. 3–66.
8. Becker, M.A.; Willman, P.; Tuross, N.C. The U.S. First Ladies Gowns: A Biochemical Study of Silk Preservation. *J. Am. Inst. Conserv.* **1995**, *34*, 141–152. [CrossRef]
9. van Assche, A. *Avvolti Nel Mito, Tessuti e Costumi Tra Settecento e Novecento. Dalla Collezione Montgomery*; Ideart: Milano, Italy, 2005.
10. Civita, F. Le Sete e Le Lacche Vanno in Battaglia: Le Armature Giapponesi. In *Il Samurai. Da Guerriero a Icona*; Luraschi, M., Ed.; Silvana Editoriale: Cinisello Balsamo, Italy, 2018; pp. 80–87.
11. Becker, M.A.; Magoshi, Y.; Sakai, T.; Tuross, N.C. Chemical and Physical Properties of Old Silk Fabrics. *Stud. Conserv.* **1997**, *42*, 27–37.
12. Kaur, J.; Rajkhowa, R.; Tsuzuki, T.; Millington, K.; Zhang, J.; Wang, X. Photoprotection by Silk Cocoons. *Biomacromolecules* **2013**, *14*, 3660–3667. [CrossRef]
13. Zhang, X.; Wyeth, P. Performance Measurement of Sericin-Coated Silks during Aging. *Sci. China Chem.* **2011**, *54*, 1011–1016. [CrossRef]
14. Luxford, N.; Thickett, D.; Wyeth, P. Applying Preventive Conservation Recommendations for Silk in Historic Houses. In Proceedings of the Multidisciplinary Conservation: A Holistic View for Historic Interiors. Joint Interim Meeting of Five ICOM-CC Working Groups; ICOM-CC: Rome, Italy, 2010.
15. Liu, X.; Zhang, K.-Q. Silk Fiber—Molecular Formation Mechanism, Structure- Property Relationship and Advanced Applications. In *Oligomerization of Chemical and Biological Compounds*; Lesieur, C., Ed.; IntechOpen: London, UK, 2014; pp. 69–102.
16. Padaki, N.V.; Das, B.; Basu, A. *Advances in Understanding the Properties of Silk*; Elsevier Ltd.: Amsterdam, The Netherlands, 2015; ISBN 9781782423249.
17. Gupta, D.; Agrawal, A.; Rangi, A. Extraction and Characterization of Silk Sericin. *Indian J. Fibre. Text Res.* **2014**, *39*, 364–372.
18. Zhang, X.M.; Wyeth, P. Using FTIR Spectroscopy to Detect Sericin on Historic Silk. *Sci. China Chem.* **2010**, *53*, 626–631. [CrossRef]
19. Vilaplana, F.; Nilsson, J.; Sommer, D.V.P.; Karlsson, S. Analytical Markers for Silk Degradation: Comparing Historic Silk and Silk Artificially Aged in Different Environments. *Anal. Bioanal. Chem.* **2015**, *407*, 1433–1449. [CrossRef]
20. Akyuz, S.; Akyuz, T.; Cakan, B.; Basaran, S. Investigations of the Historic Textiles Excavated from Ancient Ainos (Enez-Turkey) by Multiple Analytical Techniques Dedicated to Professor Simion Simon. *J. Mol. Struct.* **2014**, *1073*, 37–43. [CrossRef]
21. Asai, M.; Tsuboi, M.; Shimanouchi, T. Infrared Spectra of Polypeptides and Related Compounds. *J. Phys. Chem.* **1964**, *59*, 322–325. [CrossRef]
22. Miyazawa, T.; Blout, E.R. The Infrared Spectra of Polypeptides in Various Conformations: Amide I and II Bands. *J. Am. Chem. Soc.* **1961**, *83*, 712–719. [CrossRef]
23. Grdadolnik, J. Atr-ftir spectroscopy: Its advantages and limitations. *Acta Chim. Slov.* **2002**, *49*, 631–642.
24. Nodari, L.; Ricciardi, P. Non-Invasive Identification of Paint Binders in Illuminated Manuscripts by ER-FTIR Spectroscopy: A Systematic Study of the Influence of Different Pigments on the Binders' Characteristic Spectral Features. *Herit. Sci.* **2019**, *7*, 7. [CrossRef]
25. Boulet-Audet, M.; Vollrath, F.; Holland, C. Identification and Classification of Silks Using Infrared Spectroscopy. *J. Exp. Biol.* **2015**, *218*, 3138–3149. [CrossRef] [PubMed]
26. Shao, J.; Zheng, J.; Liu, J.; Carr, C.M. Fourier Transform Raman and Fourier Transform Infrared Spectroscopy Studies of Silk Fibroin. *J. Appl. Polym. Sci.* **2005**, *96*, 1999–2004. [CrossRef]
27. van Nimmen, E.; de Clerck, K.; Verschuren, J.; Gellynck, K.; Gheysens, T.; Mertens, J.; van Langenhove, L. FT-IR Spectroscopy of Spider and Silkworm Silks. Part I. Different Sampling Techniques. *Vib. Spectrosc.* **2008**, *46*, 63–68. [CrossRef]
28. Carissimi, G.; Baronio, C.M.; Montalbán, M.G.; Víllora, G.; Barth, A. On the Secondary Structure of Silk Fibroin Nanoparticles Obtained Using Ionic Liquids: An Infrared Spectroscopy Study. *Polymers* **2020**, *12*, 1294. [CrossRef] [PubMed]
29. Koperska, M.A.; Pawcenis, D.; Bagniuk, J.; Zaitz, M.M.; Missori, M.; Łojewski, T.; Łojewska, J. Degradation Markers of Fibroin in Silk through Infrared Spectroscopy. *Polym. Degrad. Stab.* **2014**, *105*, 185–196. [CrossRef]

30. Teramoto, H.; Miyazawa, M. Molecular Orientation Behavior of Silk Sericin Film as Revealed by ATR Infrared Spectroscopy. *Biomacromolecules* **2005**, *6*, 2049–2057. [CrossRef]
31. Teramoto, H.; Miyazawa, M. Analysis of Structural Properties and Formation of Sericin Fiber by Infrared Spectroscopy. *J. Insect. Biotechnol. Sericology* **2003**, *72*, 157–162.
32. Geminiani, L.; Campione, F.P.; Corti, C.; Luraschi, M.; Mo, S.; Recchia, S.; Rampazzi, L. Differentiating Natural and Modified Cellulosic Fibres by ATR- FTIR Spectroscopy. *Heritage* **2022**, *5*, 4114–4139. [CrossRef]
33. Akyuz, S. Investigation of the Degradation Stages of Archaeological and Historical Silk Textiles: An ATR-FTIR Spectroscopic Study. *Archaeol. Anthropol. Sci.* **2019**, *3*, AAOA.000573. [CrossRef]
34. Gorassini, A.; Adami, G.; Calvini, P.; Giacomello, A. ATR-FTIR Characterization of Old Pressure Sensitive Adhesive Tapes in Historic Papers. *J. Cult. Herit.* **2016**, *21*, 775–785. [CrossRef]
35. Vahur, S.; Teearu, A.; Peets, P.; Joosu, L.; Leito, I. ATR-FT-IR Spectral Collection of Conservation Materials in the Extended Region of 4000-80 Cm-1. *Anal. Bioanal. Chem.* **2016**, *408*, 3373–3379. [CrossRef] [PubMed]
36. Coletti, F.; Romani, M.; Ceres, G.; Zammit, U.; Guidi, M.C. Evaluation of Microscopy Techniques and ATR-FTIR Spectroscopy on Textile Fibers from the Vesuvian Area: A Pilot Study on Degradation Processes That Prevent the Characterization of Bast Fibers. *J. Archaeol. Sci. Rep.* **2021**, *36*, 102794. [CrossRef]
37. Brunello, V.; Corti, C.; Sansonetti, A.; Tedeschi, C.; Rampazzi, L. Non-Invasive FTIR Study of Mortar Model Samples: Comparison among Innovative and Traditional Techniques. *Eur. Phys. J. Plus* **2019**, *134*, 270. [CrossRef]
38. Rampazzi, L.; Brunello, V.; Corti, C.; Lissoni, E. Non-Invasive Techniques for Revealing the Palette of the Romantic Painter Francesco Hayez. *Spectrochim. Acta A Mol. Biomol. Spectrosc.* **2017**, *176*, 142–154. [CrossRef]
39. Rampazzi, L.; Brunello, V.; Campione, F.P.; Corti, C.; Geminiani, L.; Recchia, S.; Luraschi, M. Non-Invasive Identification of Pigments in Japanese Coloured Photographs. *Microchem. J.* **2020**, *157*, 36–42. [CrossRef]
40. Miliani, C.; Rosi, F.; Daveri, A.; Brunetti, B.G. Reflection Infrared Spectroscopy for the Non-Invasive in Situ Study of Artists' Pigments. *Appl. Phys. A Mater. Sci. Process.* **2012**, *106*, 295–307. [CrossRef]
41. Peets, P.; Kaupmees, K.; Vahur, S.; Leito, I. Reflectance FT-IR Spectroscopy as a Viable Option for Textile Fiber Identification. *Herit. Sci.* **2019**, *7*, 93. [CrossRef]
42. Margariti, C. The Application of FTIR Microspectroscopy in a Non-Invasive and Non-Destructive Way to the Study and Conservation of Mineralised Excavated Textiles. *Herit. Sci.* **2019**, *7*, 63. [CrossRef]
43. Delaney, J.K.; Ricciardi, P.; Glinsman, L.; Palmer, M.; Burke, J. Use of near Infrared Reflectance Imaging Spectroscopy to Map Wool and Silk Fibres in Historic Tapestries. *Anal. Methods* **2016**, *8*, 7886–7890. [CrossRef]
44. Zhou, C.; Han, G.; Via, B.K.; Song, Y.; Gao, S.; Jiang, W. Rapid Identification of Fibers from Different Waste Fabrics Using the Near-Infrared Spectroscopy Technique. *Text. Res. J.* **2019**, *89*, 3610–3616. [CrossRef]
45. Cleve, E.; Bach, E.; Schollmeyer, E. Using Chemometric Methods and NIR Spectrophotometry in the Textile Industry. *Anal. Chim. Acta* **2000**, *420*, 163–167. [CrossRef]
46. Richardson, E.; Garside, P. The Use of near Infrared Spectroscopy as a Diagnostic Tool for Historic Silk Artefacts. *E-Preserv. Sci.* **2009**, *6*, 68–74.
47. Zhang, X.; Wyeth, P. Moisture Sorption as a Potential Condition Marker for Historic Silks: Noninvasive Determination by near-Infrared Spectroscopy. *Appl. Spectrosc.* **2007**, *61*, 218–222. [CrossRef] [PubMed]
48. Mossotti, R.; Innocenti, R.; Zoccola, M.; Anghileri, A.; Freddi, G. The Degumming of Silk Fabrics: A Preliminary near Infrared Spectroscopy Study. *J. Near Infrared Spectrosc.* **2006**, *14*, 201–208. [CrossRef]
49. Menges, F. Spectragryph—Optical Spectroscopy Software, Version 1.2.16. 2021. Available online: http://www.effemm2.de/spectragryph/ (accessed on 30 December 2022).
50. Barnes, R.J.; Dhanoa, M.S.; Lister, S.J. Standard Normal Variate Transformation and De-Trending of near-Infrared Diffuse Reflectance Spectra. *Appl. Spectrosc.* **1989**, *43*, 772–777. [CrossRef]
51. Bramanti, E.; Benedetti, E. *Determination of the Secondary Structure of Isomeric Forms of Human Serum Albumin by a Particular Frequency Deconvolution Procedure Applied to Fourier Transform*; John Wiley & Sons, Inc: Hoboken, NJ, USA, 1996; Volume 38.
52. Badillo-Sanchez, D.; Chelazzi, D.; Giorgi, R.; Cincinelli, A.; Baglioni, P. Characterization of the Secondary Structure of Degummed Bombyx Mori Silk in Modern and Historical Samples. *Polym. Degrad. Stab.* **2018**, *157*, 53–62. [CrossRef]
53. Barth, A.; Zscherp, C. What Vibrations Tell Us about Proteins. *Q. Rev. Biophys.* **2002**, *35*, 369–430. [CrossRef] [PubMed]
54. Barth, A. Infrared Spectroscopy of Proteins. *Biochim. Biophys. Acta Bioenerg.* **2007**, *1767*, 1073–1101. [CrossRef] [PubMed]
55. de Meutter, J.; Goormaghtigh, E. Amino Acid Side Chain Contribution to Protein FTIR Spectra: Impact on Secondary Structure Evaluation. *Eur. Biophys. J.* **2021**, *50*, 641–651. [CrossRef] [PubMed]
56. Hernández, B.; Pflüger, F.; Adenier, A.; Kruglik, S.G.; Ghomi, M. Vibrational Analysis of Amino Acids and Short Peptides in Hydrated Media. VIII. Amino Acids with Aromatic Side Chains: L-Phenylalanine, l-Tyrosine, and l-Tryptophan. *J. Phys. Chem. B* **2010**, *114*, 15319–15330. [CrossRef] [PubMed]
57. Tanaka, M.; Hayashi, T.; Morita, S. The Roles of Water Molecules at the Biointerface of Medical Polymers. *Polym. J.* **2013**, *45*, 701–710. [CrossRef]
58. Morita, S.; Tanaka, M.; Ozaki, Y. Time-Resolved in Situ ATR-IR Observations of the Process of Sorption of Water into a Poly(2-Methoxyethyl Acrylate) Film. *Langmuir* **2007**, *23*, 3750–3761. [CrossRef] [PubMed]

59. Vasylieva, A.; Doroshenko, I.; Vaskivskyi, Y.; Chernolevska, Y.; Pogorelov, V. FTIR Study of Condensed Water Structure. *J. Mol. Struct.* **2018**, *1167*, 232–238. [CrossRef]
60. Maeda, Y.; Ide, M.; Kitano, H. Vibrational Spectroscopic Study on the Structure of Water in Polymer Systems. *J. Mol. Liq.* **1999**, *80*, 149–163. [CrossRef]
61. Jin, Y.; Wang, W.; Su, Z. Spectroscopic Study on Water Diffusion in Aromatic Polyamide Thin Film. *J. Memb. Sci.* **2011**, *379*, 121–130. [CrossRef]
62. Ling, S.; Qi, Z.; Knight, D.P.; Shao, Z.; Chen, X. Synchrotron FTIR Microspectroscopy of Single Natural Silk Fibers. *Biomacromolecules* **2011**, *12*, 3344–3349. [CrossRef]
63. Yoshimizu, H.; Asakura, T. The Structure of Bombyx Mori Silk Fibroin Membrane Swollen by Water Studied with ESR, 13C-NMR, and FT-IR Spectroscopies. *J. Appl. Polym. Sci.* **1990**, *40*, 1745–1756. [CrossRef]
64. Chong, S.H.; Ham, S. Dynamics of Hydration Water Plays a Key Role in Determining the Binding Thermodynamics of Protein Complexes. *Sci. Rep.* **2017**, *7*, 8744. [CrossRef] [PubMed]
65. Chen, F.; Porter, D.; Vollrath, F. Silk Cocoon (Bombyx Mori): Multi-Layer Structure and Mechanical Properties. *Acta Biomater.* **2012**, *8*, 2620–2627. [CrossRef]
66. Célino, A.; Gonçalves, O.; Jacquemin, F.; Fréour, S. Qualitative and Quantitative Assessment of Water Sorption in Natural Fibres Using ATR-FTIR Spectroscopy. *Carbohydr. Polym.* **2014**, *101*, 163–170. [CrossRef]
67. Morita, S.; Kitagawa, K.; Ozaki, Y. Hydrogen-Bond Structures in Poly(2-Hydroxyethyl Methacrylate): Infrared Spectroscopy and Quantum Chemical Calculations with Model Compounds. *Vib. Spectrosc.* **2009**, *51*, 28–33. [CrossRef]
68. Cheng, Y.; Koh, L.D.; Li, D.; Ji, B.; Han, M.Y.; Zhang, Y.W. On the Strength of β-Sheet Crystallites of Bombyx Mori Silk Fibroin. *J. R. Soc. Interface* **2014**, *11*, 20140305. [CrossRef] [PubMed]
69. Lucas, F.; Smith, S.G. 43—The Moisture Sorption of the Silk of Bombyx Mori in Relation to the Proportion and Chemical Composition of the Crystalline and Amorphous Phases. *J. Text. Inst. Trans.* **1959**, *50*, T695–T700. [CrossRef]
70. Kuhn, L.A.; Siani, M.A.; Pique, M.E.; Fisher, C.L.; Getzoff, E.D.; Tainer, J.A. The Interdependence of Protein Surface Topography and Bound Water Molecules Revealed by Surface Accessibility and Fractal Density Measures. *J. Mol. Biol.* **1992**, *228*, 13–22. [CrossRef]
71. Myshakina, N.S.; Ahmed, Z.; Asher, S.A. Dependence of Amide Vibrations on Hydrogen Bonding. *J. Phys. Chem. B* **2008**, *112*, 11873–11877. [CrossRef]
72. Moore, W.H.; Krimm, S. Vibrational Analysis of Peptides, Polypeptides, and Proteins. II. B-Poly(L-alanine) and B-poly(L-alanylglycine). *Biopolymers* **1976**, *15*, 2465–2483. [CrossRef] [PubMed]
73. Zhang, S.; Ma, J. Study on the Unsaturated Hydrogen Bond Behavior Abstract: Let F Denote a Field and Let V Denote a Vector Space over F with Finite Positi of Bio-Based Polyamide 56. *e-Polymers* **2019**, *19*, 23–31. [CrossRef]
74. Manas, E.S.; Getahun, Z.; Wright, W.W.; DeGrado, W.F.; Vanderkooi, J.M. Infrared Spectra of Amide Groups in α-Helical Proteins: Evidence for Hydrogen Bonding between Helices and Water. *J. Am. Chem. Soc.* **2000**, *122*, 9883–9890. [CrossRef]
75. Wang, J.; Sowa, M.G.; Ahmed, M.K.; Mantsch, H.H. Photoacoustic near-infrared investigation of homo-polypeptides. *J. Phys. Chem.* **1994**, *98*, 4748–4755. [CrossRef]
76. Mo, C.; Wu, P.; Chen, X.; Shao, Z. The Effect of Water on the Conformation Transition of Bombyx Mori Silk Fibroin. *Vib. Spectrosc.* **2009**, *51*, 105–109. [CrossRef]
77. Fukuda, M.; Miyagawa, M. Fundamental Studies on the Interactions between Moisture and Textiles V. FT-IR Study on the Moisture Sorption Isotherm of Nylon 6. *Polym. J.* **1987**, *19*, 785–804. [CrossRef]
78. Bro, R.; Smilde, A.K. Principal Component Analysis. *Anal. Methods* **2014**, *6*, 2812–2831. [CrossRef]

applied sciences

MDPI

Article

Possibilities of Detecting Damage Due to Osmosis of GFRP Composites Used in Marine Applications

Waldemar Swiderski * and Martyna Strag

Military Institute of Armament Technology, Prymasa Stefana Wyszynskiego 7 St., 05-220 Zielonka, Poland
* Correspondence: waldemar.swiderski@wp.pl; Tel.: +48-509-420-225

Abstract: The marine composites market is driven by the increasing demand for lightweight, corrosion-resistant, and impact-resistant boats. Polymer matrix composites are currently the most popular composite material in marine applications. Fiberglass composites are practically the main type of fiber composites that are used extensively in marine applications. Due to the aggressive sea environment, composite structural elements of ships are exposed to damage due to the phenomenon of osmosis. This damage is also favored by defects that result from impacts and technological errors during the production of these elements. Non-destructive testing methods are necessary to detect damage in the internal structure of the composite. The paper presents a numerical analysis of the possibility of using vibrothermography in the detection of defects in glass–fiber reinforced laminates in marine applications. Numerical simulations have shown that the most favorable method for detecting defects will be acoustic waves. This is an unusual application because, as a rule, the range of ultrasonic waves is used in vibrothermography. In our further works, it is planned to verify numerical calculations through experimental research. The applicability of the terahertz technique was also assessed. During the experimental testing, all defects in the test sample of the glass–fiber reinforced composite were detected using this technique. The presented results indicate the applicability of the presented methods for the detection of defects in composites used in marine applications.

Keywords: composites; vibrothermography; terahertz radiation; non-destructive testing

Citation: Swiderski, W.; Strag, M. Possibilities of Detecting Damage Due to Osmosis of GFRP Composites Used in Marine Applications. *Appl. Sci.* **2023**, *13*, 4171. https://doi.org/10.3390/app13074171

Academic Editors: Bogusław Łazarz, Grzegorz Peruń and Tangbin Xia

Received: 17 January 2023
Revised: 16 March 2023
Accepted: 23 March 2023
Published: 24 March 2023

1. Introduction

Interest in composite materials in many industries results from the possibility of designing their functional properties required in a specific application. This creates the possibility of new design solutions that have been previously unavailable with the use of traditional materials, e.g., metals. Fiber-reinforced composites are particularly attractive materials. The use of fibers in the composite makes the material stronger and stiffer [1]. It may have physical parameters comparable to metals but have a lower specific weight. Most often, composites employ textile materials joined together with plastics, such as a binder, to create multi-layered composite materials used, for example, in marine applications. Glass–fiber reinforced composites are most often used in the construction of yachts. More and more often there are constructions where there is a layer of rigid polyvinyl chloride foam between two layers of glass–fiber reinforced composite X.

Damage to the composite structure may occur both as a result of technological errors in the production phase and during operation as material fatigue. The main type of damage to watercraft is damage to the composite structure caused by impacts, which is one of the most critical failures. The effects of impact damage can be considered as the internal fracture surface is characterized

(1) impact damage resistance which is related to the response and damage caused by impact [2],
(2) impact damage tolerance, associated with the reduced stability and strength of the structure due to the damage [3,4].

Another factor that contributes to the damage to the composite hulls of watercraft is the phenomenon of osmosis (Figure 1). Osmosis in hulls is the penetration of water through the gelcoat layer into the laminate. Even the best-made laminate is not a homogeneous structure. There are micro-air bubbles and microcracks, both in the resin itself and at the junction of resin and glass. In these voids of the laminate, water collects, which forms acetic and hydrochloric acids and glycol as a result of hydrolysis. Since glycol is a strongly hygroscopic substance, the amount of water absorbed into the hull increases and the degradation process accelerates. The molecules of water contaminated with compounds formed as a result of hydrolysis strive to equalize the degree of their contamination with the environment. As a result of diffusion, pure water molecules penetrate the laminate unhindered, accelerating the osmosis phenomenon. Hydrolysis and diffusion are more intense in seawater, which is alkaline (pH 8–8.5). As a result of these phenomena, the connections between the fibers of the glass reinforcement and resins were destroyed in the first place. Water spreads along the reinforcement fibers, separating them from each other. The effect of this phenomenon is the gradual degradation of the laminate manifested by the formation of blisters filled with a specifically fragrant, acidic liquid with a pH of 0 to 6.5. Osmotic blisters can vary in size—from the size of a pinhead to a diameter of 10 cm. As a result of osmosis, not only does the laminate structure gradually deteriorate but the water content in the laminate increases, and the weight of the hull increases. Changing the water content in the laminate has a direct impact on the depth of immersion, maneuverability, speed, and fuel consumption [5].

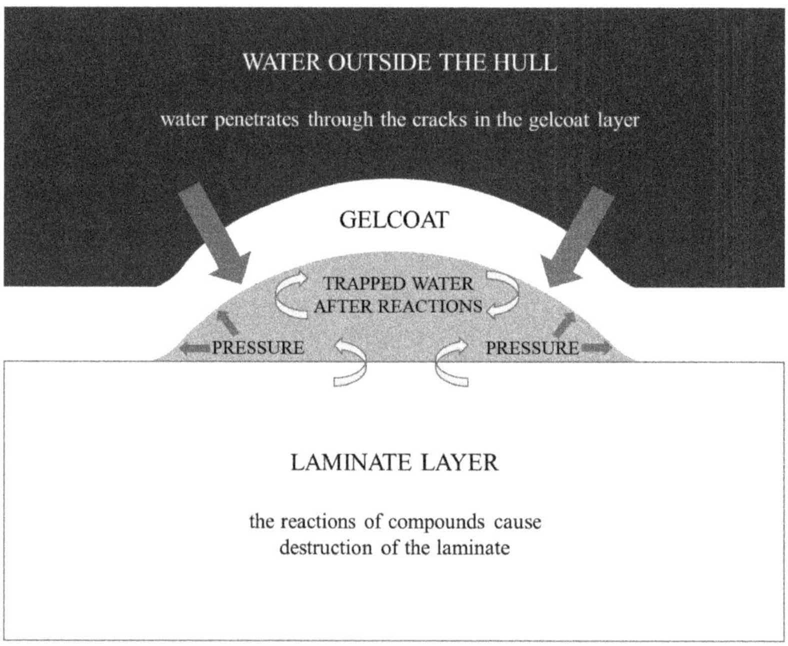

Figure 1. The process of the osmosis phenomenon [5].

Damage resulting from impacts or technological errors in the production of composite hulls of floating units accelerate the formation of damage caused by the phenomenon of osmosis. Figure 2 shows osmosis foci on the surface of the yacht's hull after removing the gelcoat layer. In this picture you can see what damage can be caused by the phenomenon of osmosis.

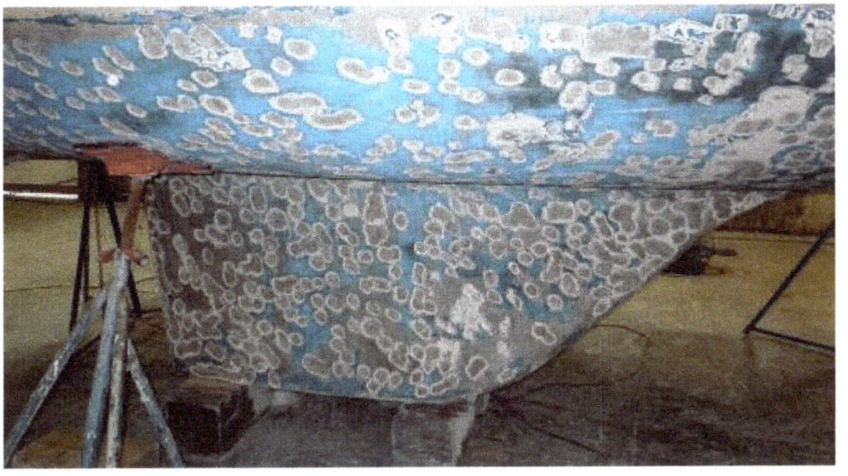

Figure 2. Osmosis foci on the yacht's hull [5].

So far, a commonly used method of detecting the effects of osmosis has been tapping, for example, with a hammer on the surface of the yacht's hull. The characteristic sound caused by tapping is information about the existence of damage due to osmosis. However, such a damage detection technique requires the repairperson to have a lot of experience, especially when the damage surface is very small. This method, in many cases, does not allow us to detect the places where the osmosis process begins.

Compared to the literature on non-destructive testing of composites, publications on testing composites used in marine applications are a small part. More often, it is only mentioned that composites used in marine applications can be tested with the presented NDT methods [6]. In the study of composites in marine applications, in most cases, the methods used in similar composites for other purposes are used [7]. These are methods such as ultrasonic tests, optical methods (holography and sherography [8]), microwaves, thermography, and radiography [9]. A certain exception is the acoustic method, which is often used in the study of composites used in marine applications [10], and occasionally in other applications. In recent years, results from the use of terahertz radiation have been presented, most often from the use of THz SWT (Stationary Wavelet Transform) [11] and THz TDS (Time–Domain Spectroscopy) [12] methods.

Non-destructive testing methods make it possible to detect the defects described above, formed in composite hulls of watercraft. The purpose of our work was to analyze the possibility of determining the parameters of non-destructive testing method, which would be both effective in detecting defects resulting from the phenomenon of osmosis and would allow relatively fast testing of the composite hull. Our many years of experience in non-destructive composites showed that one of these such methods is vibrothermography, and the possibilities of its use for detecting defects in composite hulls will be analyzed in this paper. The possibility of using the terahertz transmission method to detect defects caused by osmosis is also analyzed.

2. Vibrothermography

In infrared thermography, the distribution of the temperature field on the surface of the test object is used to assess the condition of the test object. Modern thermal imaging cameras can record changes in electromagnetic energy in the infrared band radiated from the surface of the tested object with high frequency and temperature resolution (approx. 15 mK). The whole process, after transforming the radiated energy into an electronic signal, is recorded in the form of a sequence of images [13].

Non-destructive testing with infrared thermography can be performed with both passive and active techniques [14–22]. When using an active technique, it is necessary to use the source of thermal excitation of the tested object. For this purpose, either the heating or the cooling source of the test object can be used. Material defects which, before the test begins, have a homogeneous temperature equal to the ambient temperature do not generate "useful" temperature signals and, for this purpose, require heating or cooling the whole object or its part. During such a test a dynamic temperature field is created, and the results of the test of temperature distribution depending on the observation time.

The name vibrothermography itself can be slightly misleading because this method does not rely on a direct combination of the physical phenomena used in the vibration method and thermography. The vibration method uses information on the change of direction and time of passage of waves inside the tested structure, which are affected by the elasticity and homogeneity of this structure. The detection of defects is influenced by, among other factors, the length of the mechanical wavelength and the ratio of the wavelength to the size of the defect. Only the source-generating mechanical waves were adopted from the vibration method to vibrothermography.

In vibrothermography, the evaluation of latent structural heterogeneities of materials is based on changes in the surface temperature field under cyclic mechanical loads. The reason for the increase in temperature is the internal friction of the cavity walls when they are stimulated by mechanical waves. If the cyclic loads do not exceed the elasticity of the material and the rate of their changes is high, then heat losses due to thermal conductivity are small. After removing the load, the tested object returns to its original shape and temperature. Thermographic methods are usually non-contact methods. Vibrothermography differs from this principle, and the generator of mechanical waves must be in contact with the surface of the tested object.

In order to effectively detect defects using vibrotermography, the most favorable values of the frequency and amplitude of the mechanical waves for thermal excitation of the tested object should be determined. The ThermoSon software was developed for this purpose.

In vibrothermography, ultrasounds are usually used to excite the material under test, the frequency range of which is from 20 kHz to 1 GHz. Occasionally, sound waves with a frequency range of 20 Hz to 20 kHz are used.

3. Modelling Vibrothermography NDT

3.1. Matematical Problem

The principles of vibrothermography are illustrated in Figure 3.

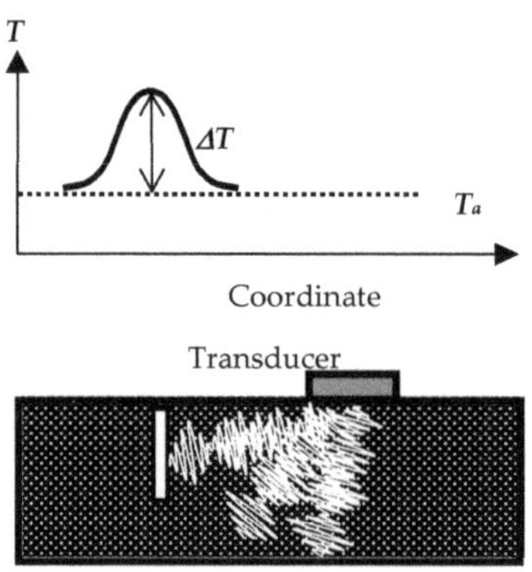

Figure 3. Vibrothermography principle [21].

Theoretical numerical calculations can be divided into two steps with 3D solutions to the following problems:

- propagation of mechanical waves in the object and the thermal power generated by the defect,
- propagation of heat generated by the defects based on the results of the first stage.

Below, a description of the steps for numerical calculations is presented. The presented algorithms were developed on the basis of the theoretical approach described in the publications [23,24].

The thermomechanical model implements parallel numerical solutions for the propagation of thermal energy and mechanical waves. Figure 3 shows the mechanical stresses (normal and tangential) that act on each face of the elementary volume of the tested object. Along the two coordinates, the tangential stresses have two components. By adding up one normal component and two tangent components there are three stress components on each surface of the parallelepiped. For example, on the surface perpendicular to X we have one normal component, σ_x, and two tangent components, τ_{xy} and τ_{xz} (the first subscript shows the coordinate axis which is parallel to the normal vector of the parallelepiped surface, and the second subscript indicates the coordinate axis which is parallel to the tangent stress component).

As described in Figure 4, the mechanical stresses acting on each plane of a solid parallelepiped act as three components of stress. For example, on the plane perpendicular to the x coordinate, these are σ_x, τ_{xy}, and τ_{xz} stresses. The balance of forces along x can be described by the formula:

$$(\sigma_{x+0} - \sigma_{x-0})\Delta y \Delta z + (\tau_{y+0,x} - \tau_{y-0,z})\Delta x \Delta z + (\tau_{z+0,x} - \tau_{z-0,x})\Delta x \Delta y + X \Delta x \Delta y \Delta z = 0 \quad (1)$$

where Δx, Δy, and Δz are the parallelepiped dimensions and X is the projection of volume forces (if they are present) on the x axis. The subscripts $x + 0$ and $x - 0$ correspond to the maximum and minimum x coordinates.

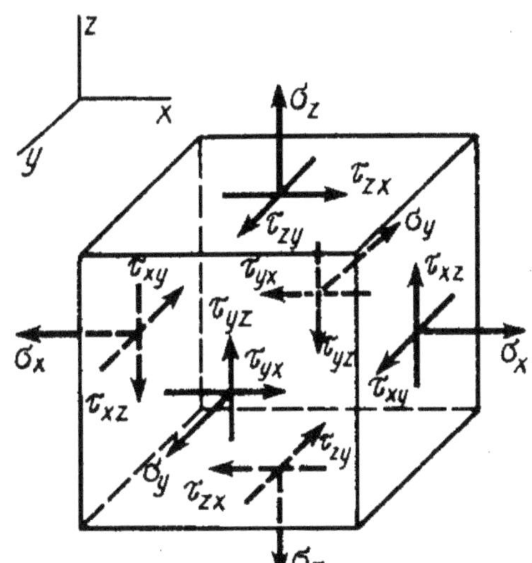

Figure 4. Distribution of stresses acting on the tested sample with an elemental volume [25].

The differential form of Equation (1) will be as follows:

$$\frac{\partial \sigma_x}{\partial x} + \frac{\partial \sigma_{yz}}{\partial y} + \frac{\partial \sigma_{zx}}{\partial z} + X = 0 \quad (2)$$

The numerical solution of the obtained set of basic equations was made using ThermoSon software:

$$(\lambda+2G)\frac{\partial^2 U}{\partial x^2}+G\left(\frac{\partial^2 U}{\partial y^2}+\frac{\partial^2 U}{\partial z^2}\right)+(\lambda+G)\frac{\partial^2 V}{\partial x\partial y}+(\lambda+G)\frac{\partial^2 W}{\partial x\partial z}=0 \quad (3)$$

where λ is the Lamė constant, G is the shear modulus, and U, V, and W are the displacement projections on the x, y, z coordinates.

The right element of the Lame Equation (3) was replaced with the corresponding components of the inertia forces F_{in}. This is because we are considering elastic waves in a solid.

F_{in} in a chosen elementary volume. For instance, in the x direction:

$$F_{in,x}=\rho\frac{\partial^2 U}{\partial \tau^2} \quad (4)$$

where ρ is the material density.

The following model boundary conditions were adopted:

(1) normal and tangential stresses on the surface of the upper and lower plate do not occur
(2) zero vertical displacements on the lower surface and forced harmonic vibrations at the point of stimulation.

The internal fracture surface is characterized by nonlinear boundary conditions. At the crack compression stage, no tangential stresses are assumed, and at the crack deformation stage, the condition of zero normal stresses is added. Assuming that the appropriate elements of Equation (3) are equal to zero, the above conditions will be achieved. As a starting condition, the displacements are assumed to be zero at $\tau = 0$. By solving Equation (3), it is possible to calculate the displacement for the x, y and z coordinates at each time step. The vertical displacement of the surface at the stimulation point is described by the equation:

$$W = A[1\Delta cos(2\pi f\tau)]/2 \quad (5)$$

where A and f are the vibration amplitude and frequency, respectively.

The heating power in a crack perpendicular to the x direction is calculated by:

$$P=\frac{k_{fr}\sigma_x S_{fr}}{\tau^*}\int_0^{\tau^*}\left|\frac{\partial U}{\partial \tau}\right|d\tau \quad (6)$$

where k_{fr} is the crack wall friction coefficient, σ_x is the stress normal to the crack surface, and S_{fr} is the crack surface.

For the second step, a 3D heat conduction equation was numerically solved to calculate the temperature distribution in a sample [16].

3.2. Model

The multi-layered structure to be tested consists of layers of glass fabric joined with an epoxy resin glue. The analysed model composite has a thickness of 10 mm and lateral size 250 mm. Three air-filled defects with dimensions D1—50 × 100 mm, D2—20 × 50 mm, and D3—20 × 50 mm (Figure 5) were placed at different depths (D1—0.25 mm, D2—0.5 mm, and D3—0.25 mm) below the surface of the tested composite sample model. All defects are very thin and have a thickness of 0.1 mm.

The model takes into account differences in the propagation of heat generated by the defect as a result of simulation with mechanical waves, which are when heat propagates along the fibers and perpendicular to them. The structure of the model corresponds to the structure of the real composite used for the hulls of the boats.

Table 1 contains the thermal and strength parameters that were determined experimentally. The thermal properties of air were adopted on the basis of the literature (Table 2) [14].

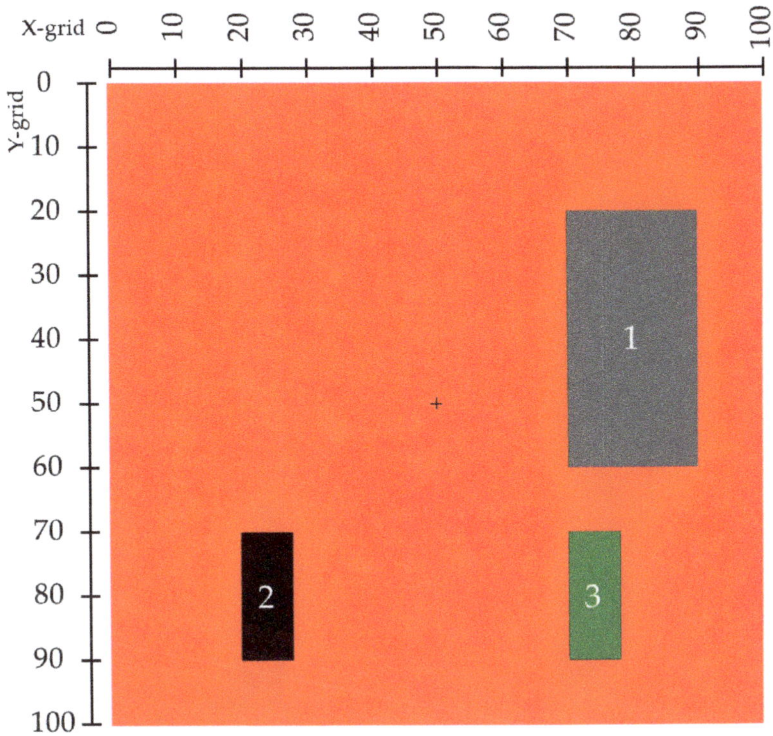

Figure 5. Model of composite where 1, 2, and 3 correspond to the defects 1, 2, and 3, respectively.

Table 1. Thermal and strength properties of the composite [26].

Parameter	Value	Unit
Specific heat	1.62	$J \cdot kg^{-1} \cdot K^{-1}$
Density	2000	$kg \cdot m^{-3}$
Thermal conductivity	0.38	$W \cdot m^{-1} \cdot K^{-1}$
Poisson's coefficient	0.24	-
Young's modulus	36	GPa

Table 2. Thermal properties of air.

Parameter	Value	Unit
Specific heat	1005	$J \cdot kg^{-1} \cdot K^{-1}$
Density	1.2	$kg \cdot m^{-3}$
Thermal conductivity	0.07	$W \cdot m^{-1} \cdot K^{-1}$

4. Results and Discussion

The performed numerical calculations showed that the best imaging of all defects was obtained at the frequency f = 1 kHz (this is the frequency of the sound wave) and the amplitude A = 0.1 mm. Figure 6 shows the graphs of temperature changes over time on the sample surface over the defects. The simulation time of generating vibrations (heating) was t_h = 3 s, and the total simulation time t = 5 s. The point of application of the center of the vibration transducer (with an area of 1 cm^2) on the sample surface was: x = 50, y = 50 (Figure 5).

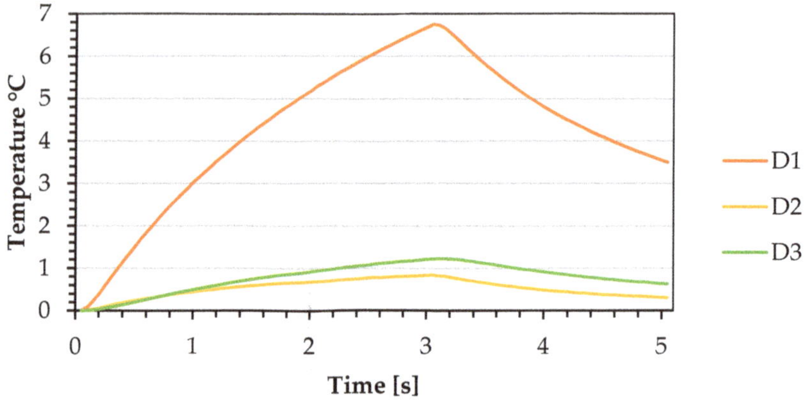

Figure 6. The course of temperature changes on the sample surface over the defects (location of the center of the vibration transducer x = 50, y = 50).

Figure 7 shows the temperature distribution after 3 s from the start of vibration generation. Then, as shown in Figure 5, the increases in temperature values above the defects were the highest.

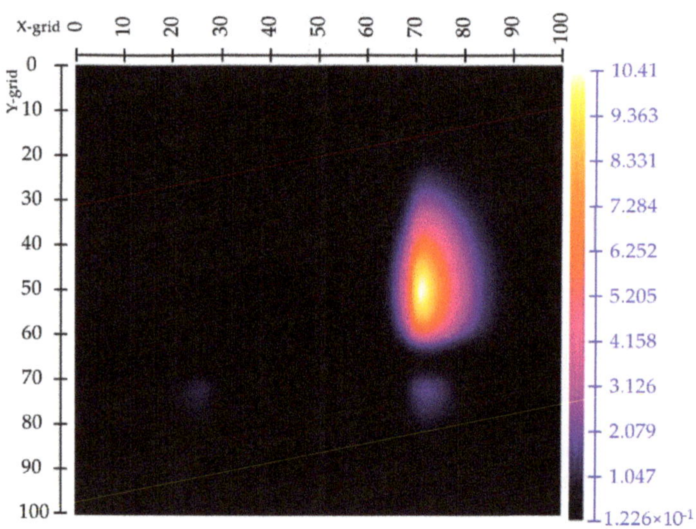

Figure 7. The temperature distribution after 3 s from the start of vibration generation (location of the center of the vibration transducer x = 50, y = 50).

The shape of defects cannot be determined in Figure 7, which shows the thermogram simulation. The Fourier transform was used to improve this visualization (Figure 8). Such an operation is possible thanks to the ThermoSon software used in the simulation. The principle of processing thermograms using the Fourier transform is described in the paper [27]. As you can see in Figure 8, the shape of the defects is very visible and consistent with the model shown in Figure 5. The image shown in Figure 5 was obtained for the 14th harmonic.

Figure 8. Image of the thermogram from Figure 6 after analysis using the Fourier transform for the 14th harmonic.

Determining the shape and dimensions of the defect, especially when it is not visually visible, is important to repair the damaged part of the hull of the watercraft. The example of using the Fourier transform presented in the article shows how helpful the thermogram processing methods (described in [27]) can be in detecting defects using infrared thermography. Figures 9 and 10 show how the temperature rise changes with time during the heating and cooling phase over the defects when the position of the center of the vibration transducer changes.

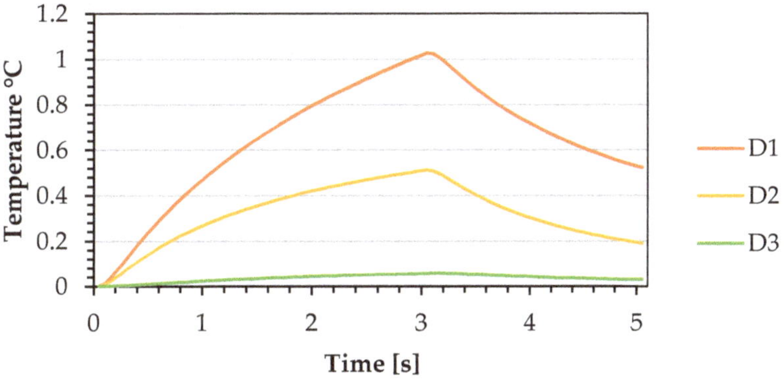

Figure 9. The course of temperature changes on the sample surface over the defects (location of the center of the vibration transducer x = 40, y = 50).

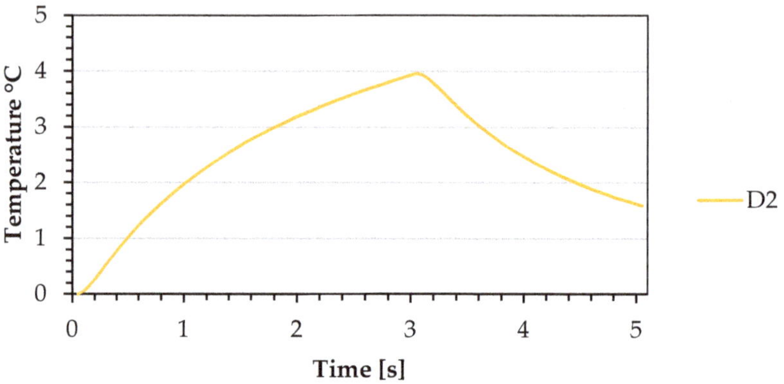

Figure 10. The course of temperature changes on the sample surface over the defects (location of the center of the vibration transducer x = 25, y = 55).

The presented simulation results clearly show that the impact on the detection of the defect (increase in the temperature value on the surface of the sample above the defect), with the same thermal forcing parameters (vibration transducer operating parameters), is significantly influenced by three factors: the size of the defect area, the distance between the defect, the transducer application point, and the depth under the surface where the defect is located. The defect D1, which has the largest surface area, generates the highest increase in temperature, as it can be seen in the graphs in Figures 7 and 9. Figure 7 shows the effect of the depth at which the defect is located under the sample surface. The transducer application point is at a similar distance from the D2 and D3 defects. The defect D2 is located deeper under the surface of the sample and the value of the temperature rise on this defect is lower than that on the defect D3, even though the surfaces of these defects are the same. The shift of the transducer application points towards the D2 defect (Figure 10), which resulted in a significant increase in the value of the temperature signal over this defect compared to the graph shown in Figure 7. However, there is no increase in the temperature signal over the D1 and D3 defects. This proves that the defect detection is possible with the distance of the transducer application not more than a few centimeters from the defect. No experimental tests have been carried out that would confirm the results of numerical calculations due to the lack of access to a source generating vibrations with parameters specified in the numerical calculations.

In order to compare the results with a different frequency of generated longitudinal waves but with the same other parameters, simulations were carried out for the frequencies of 20 kHz and 30 kHz. These are frequencies from the range of ultrasonic waves most often used for thermal excitation in ultrasonic thermography. The obtained results are presented in Table 3.

Table 3. The temperature increase over defects by thermal stimulation with ultrasounds.

Defect Number	20 kHz	30 kHz
D1	4.55	2.2
D2	0.11	0.25
D3	0.47	0.05

From the results presented in Table 3, it can be seen that the temperature rises over the defects when using a sound wave (Figure 3) is higher than when using ultrasound. This shows that thermal stimulation of the composite used to build boat hulls with the use of sound waves increases the probability of detecting a defect caused by the phenomenon of osmosis compared to ultrasonic waves.

5. Terahertz Test

As an alternative and future-proof method of non-destructive testing that can be used for damage testing of boat hulls, we present the terahertz transmission method.

As part of the work on the use of terahertz radiation in non-destructive testing of composites, experimental tests of the composite reinforced with glass fiber were also carried out. A sample of the composite approximately 10 mm thick with circular incomplete holes milled to various depths and diameters of different sizes was made at the Air Force Institute of Technology.

The transmission technique was chosen as the test method. The source of terahertz radiation and the scanner were stationary, and the test sample moved at a constant speed. Figure 11 shows a diagram of the stand.

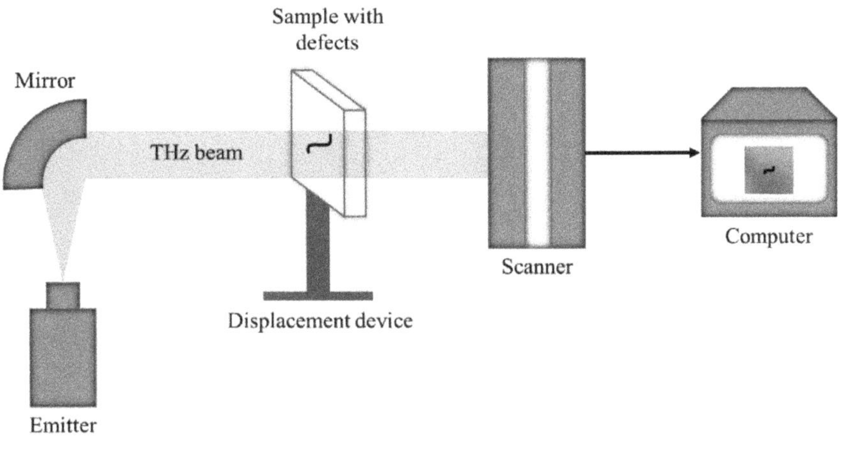

Figure 11. Schematic representation of experimental setup operated via Terahertz method in transmission mode [28].

Terasense Development Labs equipment was used in the study. A line scanner with the following parameters was used as a detector of terahertz radiation: image resolution 512 × 1 pixel with a pixel pitch of 0.5 mm and a frequency of ~300 GHz. The IMPATT THz generator, with the parameters of frequency 292 GHz ± 5 GHz and power ~10 mW, was the source of terahertz radiation. This source incorporates innovative THz reflective optics based on a specially configured high-gain horn antenna in combination with a metallic mirror. The distance between the radiation source antenna and the center of the mirror was 280 mm, and the distance between the center of the mirror and the scanner was 310 mm. The scanner was located about 5 mm from the sample surface.

Due to the adopted solution, the amount of power reaching the linear sensor matrix increases significantly. This has a decisive influence on the improvement of THz imaging. The data were recorded using dedicated software.

As can be seen in Figure 12, the dimensions and shapes of the incomplete holes are clearly visible. The outlines of the circles visible outside the periphery of the holes may be the result of terahertz radiation scattering through the walls of the holes or the detected damage resulting from hole milling. This phenomenon will be explained in the course of our further research.

In the THz image (Figure 12), disturbances (darker spots) are visible on the surface of the sample, apart from defects. These disturbances can be reduced by carefully selecting the appropriate sample speed.

Figure 12. THz image of a glass–fiber reinforced composite with defects in the form of incomplete holes.

6. Conclusions

The computer simulations carried out have shown that vibration thermography can be an effective method in detecting defects in the hulls of vessels made of fiberglass-based composite. Both damages caused by the phenomenon of osmosis and impacts can be effectively detected by this method. The selection of the vibration source parameters is important.

In our simulations, we used sound waves that are unusual and very rarely used in vibrothermography, which (as our simulations have shown) can be more effective in detecting defects in boat hulls than the use of the ultrasound. The highest temperature increase over the defects was achieved at the frequency of 1 kHz (acoustic wave). This shows that at this frequency it will be possible to detect defects of much smaller dimensions that are located much deeper in the structure of the composite than at other frequencies.

In further work, we plan to conduct experimental research in order to verify our numerical calculations regarding the use of acoustic waves in the detection of defects caused by the phenomenon of osmosis. We intend to determine what the limitations are of the method resulting from the dimensions of the defects and at what stage of osmosis we can detect its destructive effects.

As results from the test carried out with the terahertz method show, it has great potential for its application to the detection of defects in composites. This technique is constantly evolving and has great potential in the future for non-destructive testing.

In further research, we will test this method in the reflection configuration, which is more practical due to the research object, which is the hull of the vessel. Both the radiation source and the scanner are located on the same side of the tested object. In this case, the hull will be stationary and the source with the scanner will move along the hull.

Author Contributions: Conceptualization, W.S.; methodology, W.S.; software, W.S.; formal analysis, M.S.; investigation, W.S.; resources, M.S.; data curation, M.S.; writing—original draft preparation, W.S.; writing—review and editing, M.S.; visualization, M.S.; supervision, W.S.; project administration, W.S.; funding acquisition, W.S. All authors have read and agreed to the published version of the manuscript.

Funding: This research was funded by The National Centre for Research and Development, grant number DOB-SZAFIR/02/A/001/01/2020".

Institutional Review Board Statement: Not applicable.

Informed Consent Statement: Not applicable.

Data Availability Statement: Not applicable.

Acknowledgments: We would like to thank the employees of the Air Force Institute of Technology for making a sample of the glass-fiber reinforced composite used in our research.

Conflicts of Interest: The authors declare no conflict of interest.

References

1. Chawla, K.K. *Composite Materials, Science and Engineering*; Springer: New York, NY, USA, 2012.
2. Sarasini, F.; Tirillo, J.; D'Altilia, S.; Valente, T.; Santulli, C.; Touchard, F.; Chocinski-Arnault, L.; Melier, D.; Lampani, L.; Gaudenzi, P. Damage tolerance of carbon/flax hybrid composites subjected to low velocity impact. *Compos. Part B Eng.* **2016**, *91*, 144–153. [CrossRef]
3. Koo, J.-M.; Choi, J.-H.; Seok, C.-S. Prediction of residual strength after impact of CFRP composite structures. *Int. J. Precis. Eng. Manuf.* **2014**, *15*, 1323–1329. [CrossRef]
4. Li, B.; Hoo Fatt, M.S. Impact damage and residual strength predictions of 2D woven SiC/SiC composites. *Finite Elem. Anal. Des.* **2016**, *113*, 30–42. [CrossRef]
5. krzysztofkluza.pl. Available online: https://krzysztofkluza.pl/osmoza/ (accessed on 6 October 2022).
6. Gupta, R.; Mitchell, D.; Blanche, J.; Harper, S.; Tang, W.; Pancholi, K.; Baines, L.; Bucknall, D.G.; Flynn, D. A Review of Sensing Technologies for Non-Destructive Evaluation of Structural Composite Materials. *J. Compos. Sci.* **2021**, *5*, 319. [CrossRef]
7. Wang, B.; Zhong, S.; Lee, T.L.; Fancey, K.; Mi, J. Non-destructive testing and evaluation of composite materials/structures: A state-of-the-art review. *Adv. Mech. Eng.* **2020**, *12*, 1687814020913761. [CrossRef]
8. Tao, N.; Anisimov, A.G.; Groves, R.M. Shearography non-destructive testing thick GFRP laminates: Numerical and experimental study on defect detection with thermal loading. *Compos. Struct.* **2022**, *282*, 115008. [CrossRef]
9. Battley, M.; Skeates, A.; Simpkin, R.; Holmqvist, A. Non-destructive inspection of marine composite structures. In Proceedings of the High Performance Yacht Design Conference, Auckland, New Zealand, 4–6 December 2002.
10. Franstsev, M.E. Nondestructive Testing of Ship Hulls Made of Composite Materials Using Acoustic Methods. *Russ. J. Nodestruct. Test.* **2013**, *49*, 1–7. [CrossRef]
11. Tu, W.; Zhong, S.; Shen, Y.; Incecik, A. Nondestructive Testing of Marine Protective Coatings Using Terahertz Waves with Stationary Wavelet Transform. *Ocean Eng.* **2016**, *111*, 582–592. [CrossRef]
12. Ibrahim, M.E.; Headland, D.; Withayachumnankul, W.; Wang, C.H. Nondestructive Testing of Defects in Polymer-Matrix Composite Materials for Marine Applications Using Terahertz Waves. *J. Nondestruct. Eval.* **2021**, *40*, 37. [CrossRef]
13. Carlomagno, G.M.; Luca, L. *Infrared Thermography in Heat Transfer*, 2nd ed.; Taylor&Francis: New York, NY, USA, 2001.
14. Maldague, X.P. *Theory and Practice of Infrared Technology for Nondestructive Testing*; John Wiley&Sons, Inc.: New York, NY, USA, 2001.
15. Dragan, K.; Świderski, W. Multimode NDE approach for structure health assessment of composite elements in aerospace applications. *Acta Phys. Pol.* **2010**, *117*, 877–882.
16. Świderski, W. Metody i techniki termografii w podczerwieni w badaniach nieniszczących materiałów kompozytowych. *Probl. Tech. Uzbroj.* **2009**, *4*, 75–92.
17. Sethy, D.; Makireddi, S.; Varghese, F.V.; Balasubramaniam, K. Passive thermo-elastic behaviour of surface enhanced glass-epoxy composites with spray coated graphene nanoplatelets (GNP). *Mater. Res. Express* **2019**, *6*, 125044. [CrossRef]
18. Wiecek, B. Review on thermal image processing for passive and active thermography. In Proceedings of the 2005 IEEE Engineering in Medicine and Biology 27th Annual Conference, Shanghai, China, 17–18 January 2006.
19. Mulaveesala, R.; Vaddi, J.S.; Singh, P. Pulse compression approach to infrared nondestructive characterization. *Rev. Sci. Instrum.* **2008**, *79*, 94901. [CrossRef] [PubMed]
20. Zhang, H.; Yang, R.; He, Y.; Foudazi, A.; Cheng, L.; Tian, G. A Review of Microwave Thermography Nondestructive Testing and Evaluation. *Sensors* **2017**, *17*, 1123. [CrossRef] [PubMed]
21. Vavilov, V. Thermal NDT: Historical milestones, state-of-the-art and trends. *Quant. Infrared Thermogr. J.* **2014**, *11*, 66–83. [CrossRef]

22. Swiderski, W. Detection of very thin defects in multi-layer composite made of carbon fibre with IR thermography methods. In *Nondestructive Testing of Materials and Structures*; Güneş, O., Akkaya, Y., Eds.; Springer: Dordrecht, The Netherlands, 2013; pp. 633–639.
23. Thomson, D.O.; Chimenti, D.E. Review of Progress in Quantitative Nondestructive Evaluation. *AIP Conf. Proc.* **2012**, *1430*, 3.
24. Świderski, W.; Pracht, M.; Vavilov, V. Analiza możliwości zastosowania ultradźwiękowej termografii w podczerwieni w badaniach nieniszczących wielowarstwowych kompozytów aramidowych. *Probl. Tech. Uzbroj.* **2015**, *134*, 7–21.
25. Świderski, W.; Hłosta, P.; Chuda-Kowalska, M. Non-destructive evaluation of sandwich plates by an ultrasonic IR thermographic method. *Meas. Autom. Monit.* **2016**, *62*, 406–410.
26. Guangfa, G.; Yongchi, L.; Zheng, J.; Shujie, Y. Experimental investigation on thermal properties of an advanced glass fiber composite material. *Phys. Procedia* **2012**, *25*, 339–344. [CrossRef]
27. Świderski, W.; Vavilov, V. Thermograms and data processing in nondestructive testing by IR thermography methods. *Probl. Tech. Uzbroj.* **2009**, *3*, 57–81. (In Polish)
28. Strąg, M.; Świderski, W. The Terahertz-based non-destructive evaluation of military designated materials. *Proc. SPIE* **2022**, *12274*, 173–178.

materials

MDPI

Article

An Approach to the Automated Characterization of Out-of-Plane and In-Plane Local Defect Resonances

Paweł Zdziebko [1], Mateusz Krzemiński [1], Maciej Okoń [1], Gabriela Loi [2], Francesco Aymerich [2], Łukasz Pieczonka [1] and Andrzej Klepka [1,*]

[1] Department of Robotics and Mechatronics, AGH University of Krakow, Al. Mickiewicza 30, 30-059 Krakow, Poland; zdziebko@agh.edu.pl (P.Z.); mkrzemin@agh.edu.pl (M.K.); lukasz.pieczonka@agh.edu.pl (Ł.P.)

[2] Department of Mechanical, Chemical and Materials Engineering, University of Cagliari, Piazza d'Armi, 09123 Cagliari, Italy; gabriela.loi@unica.it (G.L.); francesco.aymerich@unica.it (F.A.)

* Correspondence: klepka@agh.edu.pl

Abstract: The paper presents an approach to efficiently detect local defect resonances (LDRs) in solids with localized defects. The 3D scanning laser Doppler vibrometry (3D SLDV) technique is applied to acquire vibration responses on the surface of a test sample due to a broadband vibration excitation applied by a piezoceramic transducer and modal shaker. Based on the response signals and known excitation, the frequency characteristics for individual response points are determined. The proposed algorithm then processes these characteristics to extract both out-of-plane and in-plane LDRs. Identification is based on calculating the ratio between local vibration levels and the mean vibration level of the structure as a background. The proposed procedure is verified on simulated data obtained from finite element (FE) simulations and validated experimentally for an equivalent test scenario. The obtained results confirmed the effectiveness of the method in identifying in-plane and out-of-plane LDRs for both numerical and experimental data. The results of this study are important for damage detection techniques utilizing LDRs to enhance the efficiency of detection.

Keywords: local defect resonance; damage detection; structural dynamics; non-destructive testing; laser vibrometry; structural health monitoring

Citation: Zdziebko, P.; Krzemiński, M.; Okoń, M.; Loi, G.; Aymerich, F.; Pieczonka, Ł.; Klepka, A. An Approach to the Automated Characterization of Out-of-Plane and In-Plane Local Defect Resonances. *Materials* **2023**, *16*, 3084. https://doi.org/10.3390/ma16083084

Academic Editors: Bogusław Łazarz, Grzegorz Peruń and Tangbin Xia

Received: 23 March 2023
Revised: 7 April 2023
Accepted: 11 April 2023
Published: 13 April 2023

1. Introduction

The increasing use of advanced materials in modern engineering structures demands new inspection strategies offering reliable and accurate data. In recent years, a wide range of non-destructive testing (NDT) methods and structural health monitoring (SHM) techniques have been developed for assessing the structural integrity of engineering materials [1–3]. Among them, methods based on the analysis of non-linear vibration/acoustic phenomena, such as higher and subharmonics generation and elastic waves modulation, have been gaining special attention [2–8]. This is mainly due to their better sensitivity to small damage severities than their linear counterparts. In this context, a novel non-invasive procedure that uses sonic or ultrasonic frequency excitation tuned to the local dynamic properties of defect to activate a resonant response was recently proposed by Solodov et al. [9]. The principle behind this is that the presence of embedded defects results in a local loss of stiffness that gives rise to characteristic resonant frequencies of the defect itself, known as Local Defect Resonances (LDRs). As for the classical modal testing approach, the match between excitation and LDR frequency corresponds to the maximum wave–defect interaction. The energy delivered by the impinging wave is selectively trapped within the damaged area, leading to a significant increase in the defect response amplitude. This increase is strongly localized in the defect area, providing an excellent contrast between the damage and the intact specimen. Solodov et al. further applied the concepts of LDR to a Flat Bottom Hole (FBH) [10]. Experimental results were validated through numerical modeling, and an analytical formulation was proposed to determine the LDR frequencies of defects such as FBH, delaminations in composite materials, and laminar defects in rolled sheet metals. The effects of local defects on the non-linear ultrasonic response were also investigated in the literature, revealing that the defect excited at its LDR frequency exhibited a transition to a non-linear regime. Consequently, the input energy was efficiently converted into non-linear frequencies components, such as sub- and higher harmonics inside the damaged region [11,12]. LDR frequencies have also been used to enhance the thermal response of a damaged structure [13]. Solodov et al. [14] developed an analytical solution for different planar defect shapes, which was validated through a series of experimental tests. In the experiments, a wide-bans chirp signal was used to excite the specimens, and the laser vibrometer was used for response measurements. Laser Doppler vibrometry has been used by other researchers to extract LDR frequencies of aluminum and PMMA plates with FBHs, and delaminations in glass-fiber reinforced polymer (GFRP) and carbon-fiber reinforced polymer (CFRP) [15,16]. Moreover, it has been demonstrated that LDR behavior does not limit itself to out-of-plane direction but can be extended towards in-plane characteristics [17,18]. Experiments on different types of defects, i.e., FBHs, surface cracks and BVIDs (Barely Visible Impact Damage), showed a clear in-plane LDR at an elevated frequency range due to the high in-plane bending stiffness.

Even though many papers deal with the LDR frequencies for imaging damage, only a few attempts were made to develop a robust algorithm to identify the LDR frequencies among the system's natural frequencies, making the procedure cumbersome and time-consuming, starting from the analytical formulation proposed in [14], which can be applied only for a few classes of idealized defects and only when the geometry and position of the fault are known. More recently, in [19,20], the authors proposed an approach based on the bicoherence analysis to obtain the LDR of FBH in an aluminum plate and of delamination in a GFRP composite plate.

The current study proposes a novel algorithm for the efficient detection of LDRs. The main goal of the algorithm is to automate the process of identifying LDR frequencies, which is a particularly tedious task for structures with a considerable number of resonances. It is designed to exhibit unique features, including:

- Possibility of investigating structures with multiple defects;
- Detection of multiple frequencies (and their mode shapes) for the same defect, which is needed for in-plane LDRs extraction associated with higher-order modes;
- Possibility of examining structures without a priori knowledge of defects' locations or limits regarding their size.

The above-listed features make the proposed approach novel with respect to the existing approaches, such as the algorithm proposed in [12], which is based on a similar index to that presented in this paper. However, the processing procedure of the proposed index is modified to improve the algorithm's features.

In this paper, we present the work carried out to develop, validate and verify the effectiveness of the method of searching the LDRs (both out-of-plane and in-plane) in the broad spectrum of vibration modes. Firstly, we describe the algorithm in Section 2. The test sample, experimental setup, and finite element models are presented in Section 3. Damage detection and localization results are presented in Section 4. Finally, the paper is concluded in Section 5.

2. Materials and Methods

As a result of the local loss of stiffness, the LDRs are specific for the selected area of a structure. Hence, vibrations in the corresponding frequencies are increased in damaged regions, and a negligibly small background response is observed simultaneously. It indicates that for the local modes, the vibration amplitudes in the areas of damage are significantly higher than the mean value of the amplitudes of the whole structure. Based on this observation, we introduce the Average Ratio (AR):

$$AR_{i,f} = \frac{A_{i,f}}{E\left[A_f\right]}, \quad (1)$$

where $A_{i,f}$ is the velocity amplitude of vibration of the single point i in the single frequency f, and $E\left[A_f\right]$ is the mean value of amplitudes of all points in the single frequency. Having frequency response functions (FRF) for all the points of a structure, we can create the $M \times N$ matrix of absolute FRF values, where M is the number of points and N is the number of frequencies. This allows for the calculation of the AR matrix. Then, selecting the maximum AR value in each column gives the $ARmax$ characteristic (see Figure 1).

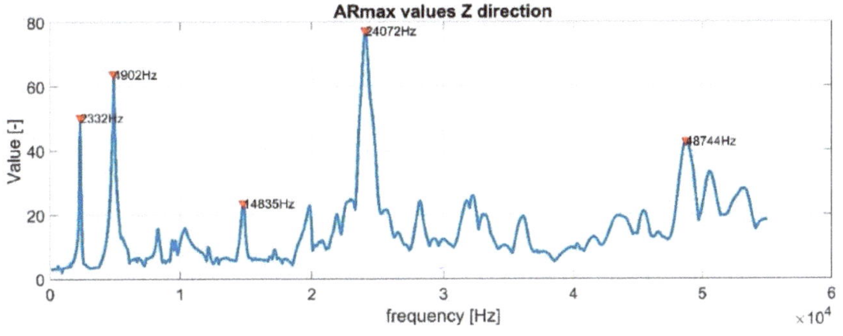

Figure 1. Example of the ARmax characteristic.

The $ARmax$ characteristic contains information about the relative differences between points with the highest amplitudes and mean values for a whole structure. The higher the value is, the more distinct LDR is detected. Two parameters determine the threshold for LDR frequency selection. The first one is the sum of the mean and standard deviation of prominence to define the minimum peak prominence. The second one is the standard deviation of $ARmax$ values, respectively, for the whole characteristic to define minimum peak height. The algorithm's sensitivity is controlled by the maximum number of peaks to be detected. Usually, it is set to 9–24. The concept and main steps of the calculations are schematically summarized in Figure 2.

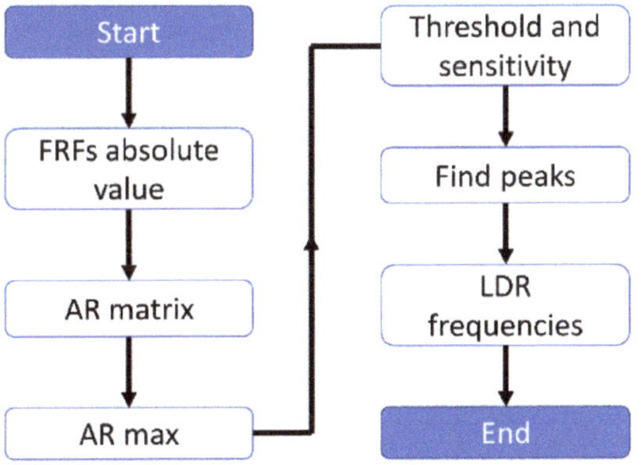

Figure 2. The flowchart of the automated LDR extraction algorithm.

To sum up, the algorithm works as follows: the FRF computed for measured or simulated data is the input for the algorithm. Firstly, the absolute value of FRF is computed. Next, the *AR* matrix is calculated according to Equation (1) and the *ARmax* index is calculated as introduced above. Then, the peaks of the *ARmax* are found, taking into account threshold and sensitivity presets. Frequencies corresponding to those peaks are interpreted as LDR frequencies.

3. Examination Setup

3.1. Test Sample with FBHs

The test sample made of poly(methyl methacrylate) (PMMA) was manufactured for the experimental testing of the algorithm's operation. The dimensions of the plate are as follows: 300 × 300 × 18 mm. Damages of FBH-type were introduced in the plate to represent three sizes of deep defects with nominal diameters of 58, 40, and 18 mm. The designed depth of all FBHs was 17 mm, which corresponds to the 1 mm thickness of the residual material in the damaged area. The test specimen's dimensions are presented in Figure 3A. Please note that the sample is presented from the intact-side view, while in Figure 3B the manufactured sample is presented from the damaged-side perspective.

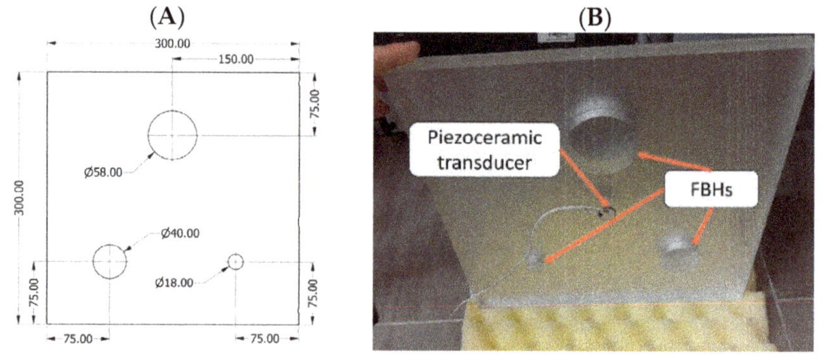

Figure 3. PMMA test sample with FBHs: nominal dimensions in [mm] (**A**) and the sample prepared for experiments (**B**).

A preliminary study revealed that non-negligible differences were observed between the simulated and experimentally measured natural frequencies, which was not expected for a relatively simple structure under consideration. Therefore, we decided to verify the geometrical dimensions of the sample with respect to their desired nominal values in order to tune the numerical model accordingly. The dimensions of FBHs of the test sample are summarized in Table 1.

Table 1. Design and measured dimensions of the test sample.

Dimension Type	Nominal Value	Identified Values
Thickness of plate	18 mm	18.455 mm
FBH "Ø58 mm" diameter	58 mm	57.646 mm
FBH "Ø58 mm" residual thickness	1 mm	0.686 mm
FBH "Ø40 mm" diameter	40 mm	39.64 mm
FBH "Ø40 mm" residual thickness	1 mm	0.707 mm
FBH "Ø18 mm" diameter	18 mm	17.566 mm
FBH "Ø18 mm" residual thickness	1 mm	0.715 mm

It has been confirmed that the differences observed between the designed and measured dimensions of FHBs were the reason for the inconsistencies between the preliminary results of the numerical calculations and the experiments. The following formula allows for the prediction of the LDR frequencies for FBHs, and it was used to determine the influence of the variation of FBHs' dimensions on LDR frequencies:

$$f_{LDR} = \frac{6.4t}{a^2}\sqrt{\frac{E}{12\rho\left(1-v^2\right)}} \tag{2}$$

where t is the defect's residual material thickness, a is the diameter of the FBH, E is the Young's Modulus, ρ is the density, and v is the Poisson ration of the material.

The data shown in Figure 4 prove that a slight change in the residual thickness or the diameter of the damage can significantly shift the LDR frequency.

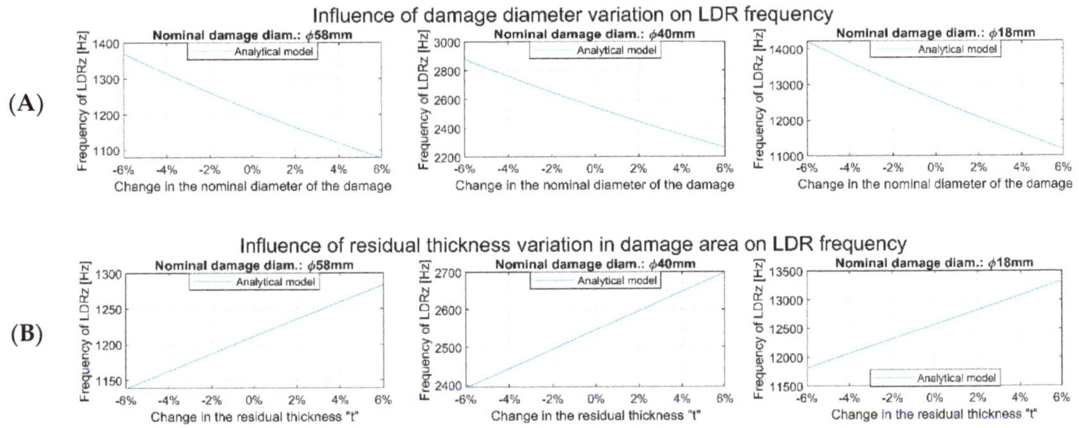

Figure 4. Influence of damage diameter (**A**) and residual thickness (**B**) variation on theoretical LDR frequencies calculations made for the design values of FBH dimensions.

3.2. Experimental Modal Analysis

In experiments, the sample was freely suspended to avoid boundaries' nonlinearities. Preliminary studies have shown that a model shaker is required to excite the largest FBH's normal modes effectively. At the same time, the piezoceramic transducer was needed to excite higher-frequency modes of LDRs. A frequency sweep signal was used to excite the sample in the following frequency ranges: 0.5–3 kHz using a modal shaker and 0.5–20 kHz in the case of using a piezoceramic transducer. The excitation amplitudes were set as 0.2 V and 8 V for the shaker and piezoceramic transducer, respectively. An external signal generator generated the signal, and next the power amplifier amplified it ten times and passed it on to the piezoceramic transducer. In the case of using a modal shaker, the generated signal was amplified using a built-in amplifier. In both cases, the test sample's response was measured using 3D scanning laser Doppler vibrometry (3D SLDV). The sample surface was mapped with 362 equally spaced measurement grid points. The *Polytec PSV400 3D* laser vibrometer was used for non-contact measurements of vibration responses. The sampling frequency and single point measurement duration were 5 kHz and 1.6 s in the shaker measurement and 125 kHz and 2.048 s in the piezoceramic transducer measurement. For both acquisitions, we used three averages per point. The Frequency Response Functions (FRFs) were calculated from the experimental input and output data using the *Polytec PSV* software v9.0. The experimental arrangement for measurements is presented graphically in Figure 5.

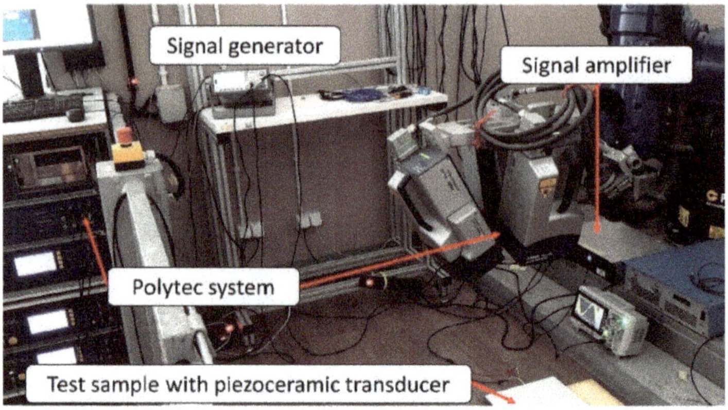

Figure 5. Measurement system used in experiments—configuration with piezoceramic transducer.

3.3. *Numerical Models*

Numerical simulations can be used to predict the dynamic properties of a structure. In the context of this work, a computational modal analysis and FRF synthesis were used to generate input data for the identification algorithm. Two numerical models employing the Finite Element Method (FEM) to simulate the test specimen with FBHs were formulated. The first model was based on 3D 8-node brick Finite Elements (FEs) and is assumed to accurately reflect the real structure's response. The second proposed model, based on 2D 4-node shell FEs, is supposed to allow for a rough determination of LDRs in a shorter computing time. Both models were formulated using the *Altair HyperMesh* v2020 software and computed with the *MSC.Nastran* v2020 solver.

Figure 6 presents the FE mesh used in the 2D and 3D numerical models. The geometric dimensions of the model correspond to the measured ones. The thickness of shell 2D FEs corresponds to the thickness of the material in a given area of the sample. The thicknesses of shell FEs used in the 2D model are described in the legend in Figure 6A. The influence of the piezoceramic transducer was modeled as constraints in X and Y directions on the transducer's montage side. The division of elements in the XY plane is the same in both models.

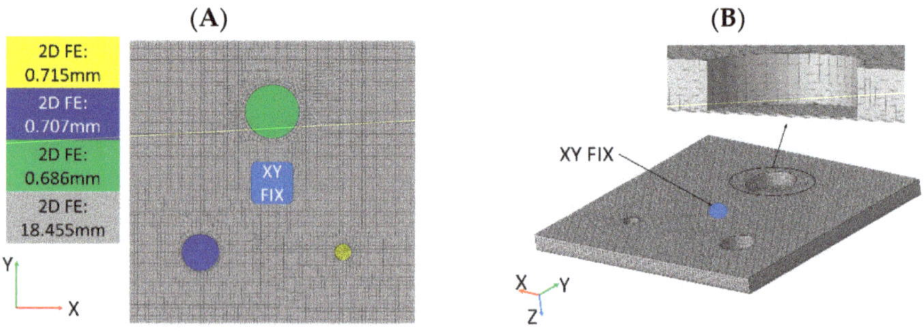

Figure 6. Formulated numerical models based on 2D FEs (**A**) and 3D FEs (**B**).

A mesh convergence analysis was carried out, comparing the first LDR frequency of the damage. According to the outcomes of this analysis, the recommended number of FEs per FBH diameter was determined as 16. Three FEs were modeled on the thickness of the residual material in the damaged area in the case of the 3D model and eleven FEs in the intact area of the plate. The material parameters adopted in the simulations are summarized in Table 2.

Table 2. Material properties of the material used in numerical models.

Parameter	Value
Young's modulus	4919 MPa
Poisson ratio	0.4
Density	1.204 g/cm^3

A comparison of selected normal models was carried out to check the correctness of the material model parameters. The numerical results of the natural frequencies of the test sample were compared with the measured ones. One of the analyzed normal modes is presented in Figure 7—hereinafter referred to as global mode 'B'.

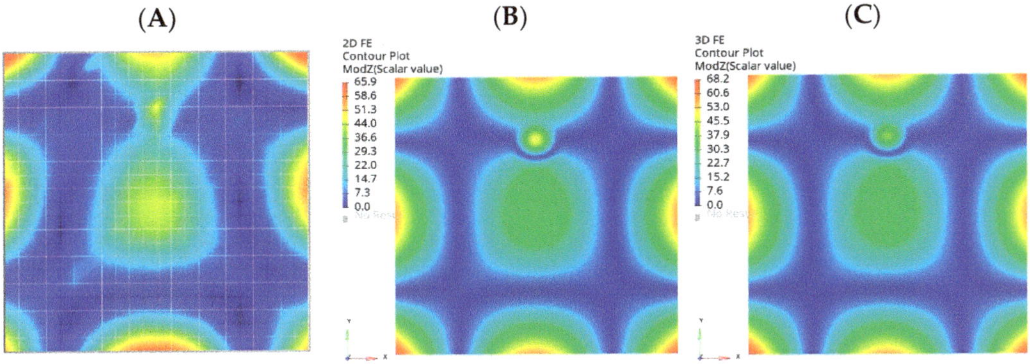

Figure 7. One of the global normal modes selected for validation: experiment (**A**), 2D FE model (**B**), 3D FE model (**C**).

Analogous comparisons were made for two other global normal modes, called 'A' and 'C'. The results of the validation are presented in Table 3.

Table 3. Natural frequencies of the test sample.

		Model with 2D FEs		Model with 3D FEs	
Global Mode	**Experiment**	**Simulation**	**Error**	**Simulation**	**Error**
'A'	478 Hz	476 Hz	0.4%	480 Hz	0.4%
'B'	1207 Hz	1121 Hz	7.1%	1151 Hz	4.6%
'C'	2360 Hz	2181 Hz	7.6%	2296 Hz	2.7%

The obtained results confirm a good agreement of the numerical model formulated with 3D FEs with the experimental results. The maximum error was observed for the normal mode marked as 'B', which was less than 5% between the model and experiment. This error is more significant for the model made of 2D FEs, and amounts to a maximum of 7.6% for the 'C' normal mode. The model made of 3D FEs can be considered to be more accurate. It should be noted that the thickness of the test sample in numerical models was assumed to be constant, but in reality it most likely varied over the sample, which may be the reason for the discrepancy between numerical results and experimental ones in global normal modes.

4. Results and Discussion

4.1. Out-of-Plane LDRs

Four datasets for the detection of out-of-plane LDRs were examined. The results were derived from experiments with a piezoceramic transducer, a modal shaker (described in Section 3.2) and the numerical results (2D and 3D FE models defined in Section 3.3). The determined *ARmax* characteristics for datasets were computed and analyzed, and finally LDR frequencies were found by the proposed algorithm. Exemplary results of the *ARmax* index for experimental data and the 3D numerical model are presented in Figure 8. Similar results were computed for the 2D numerical model, but are not presented here to limit the manuscript's size. The maximum number of peaks to be detected in the algorithm was limited to nine. The LDR frequencies determined by the algorithm are marked with red triangles.

(A)

(B)

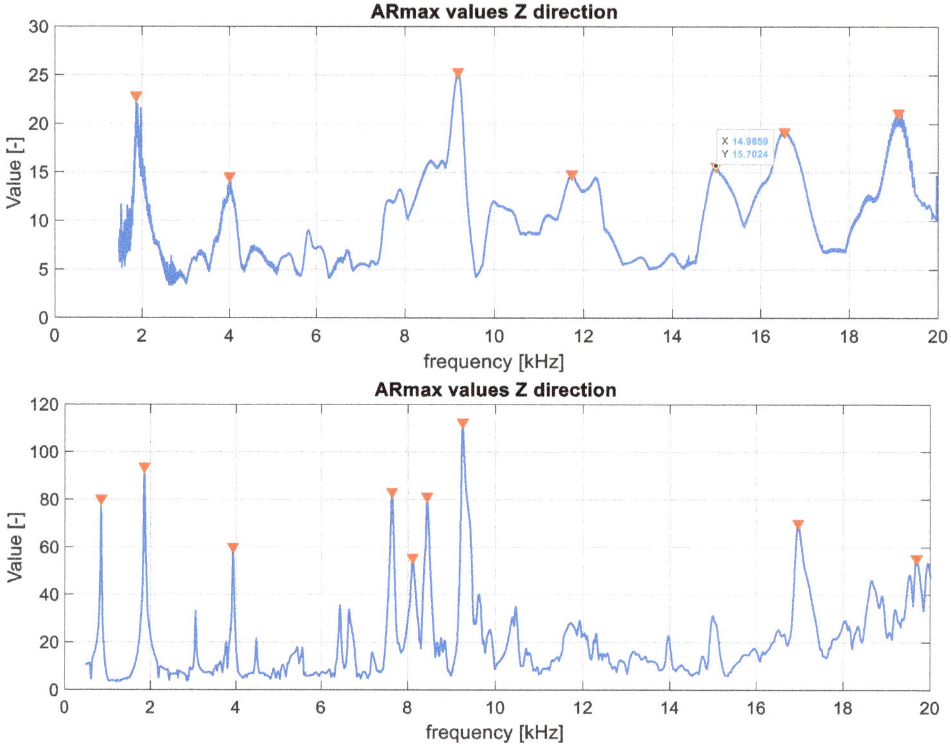

Figure 8. Computed ARmax characteristics and LDR frequencies selected by the algorithm for experimental data (**A**) and the numerical results (**B**).

When comparing numerical and experimental results, it can be noticed that the damping in the actual structure is higher than in the numerical model, as the ratio of amplitudes shown in the *ARmax* characteristic is lower. The higher frequencies in the real system are damped more. The numerical models consider a 0.3% modal damping, which is the same for all modes. Normal mode shapes corresponding to the first and second LDRs were determined by the algorithm and are visualized in Figures 9–11 for subsequent datasets. The results are presented for experimental data and numerical ones using the 3D FE model. Similar results were obtained for the 2D FE model.

The visualizations of LDRs of the FBH Ø58 mm allow for determining its location and shape. The algorithm identified similar frequencies to LDRs for numerical and experimental data. The first out-of-plane LDR turned out to occur below 1 kHz, so it was necessary to use a modal shaker instead of a piezoceramic transducer in the experimental setup. Piezo-stack was used to excite the structure at higher frequencies, but it was ineffective in the low-frequency range. It resulted in heavily noised signals, so using the detection algorithm on low-quality data did not make sense. For this reason, the tests for the low-frequency range were repeated using a modal shaker. It allowed for correct excitation of the first mode of out-of-plane LDR for the largest FBH. Unfortunately, the excitation of the second out-of-plane mode of LDR of FBH Ø58 mm was accompanied by the first out-of-plane mode of LDR of FBH Ø40 mm. As a result, the average vibration level was high, and thus the *ARmax* index was low. The same observations were noted in experimental and numerical results (both 2D and 3D models)—see Figure 9B,D. This caused a false negative indication for the second out-of-plane LDR mode of FBH Ø58 mm. It must also be noted that the algorithm also found higher LDRs, but the presentation is limited to the first two ones (first and second out-of-plane modes).

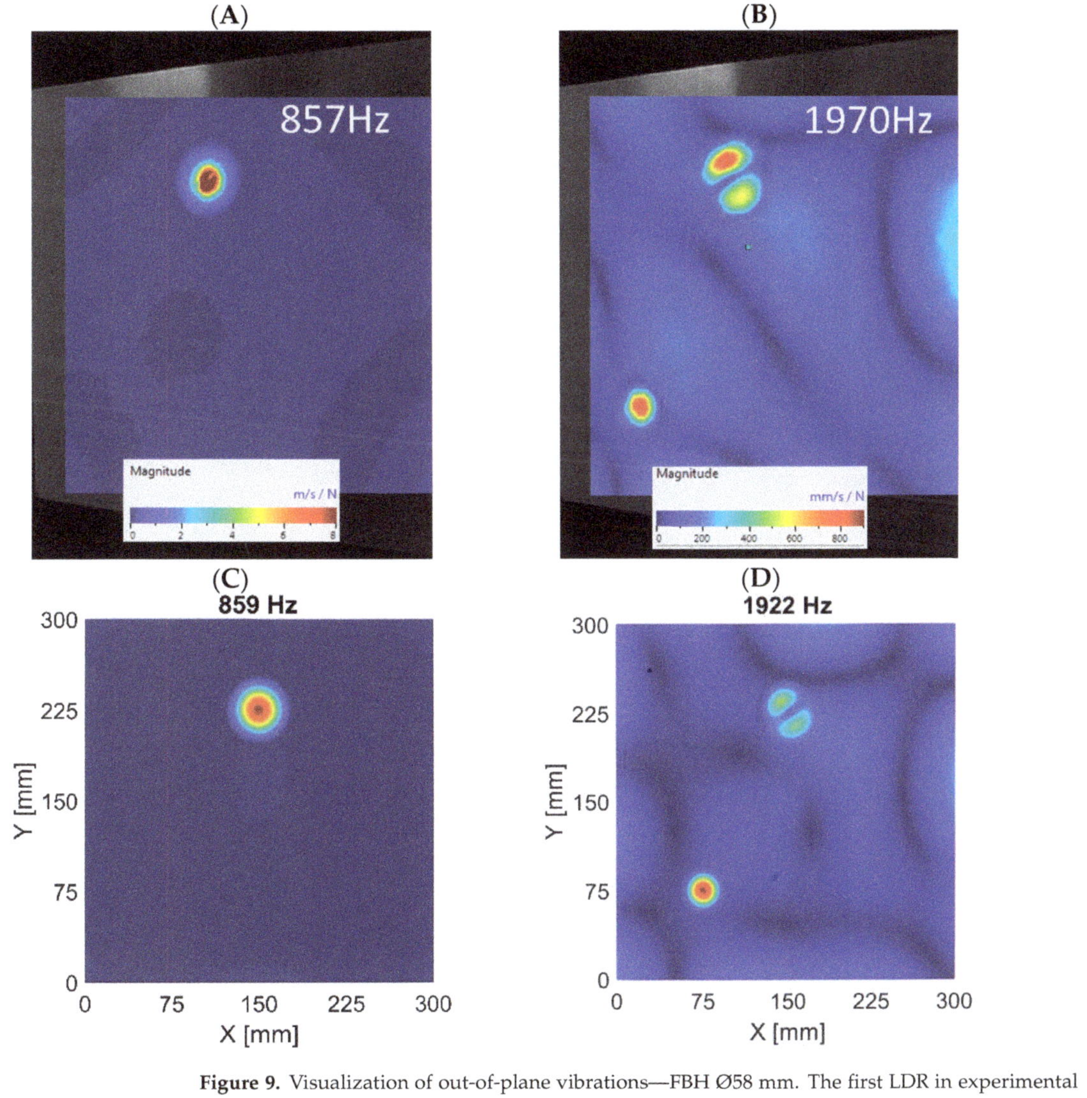

Figure 9. Visualization of out-of-plane vibrations—FBH Ø58 mm. The first LDR in experimental data (**A**) and numerical results based on the 3D FE model (**C**). The second LDR in experimental data (**B**) and numerical results based on the 3D FE model (**D**).

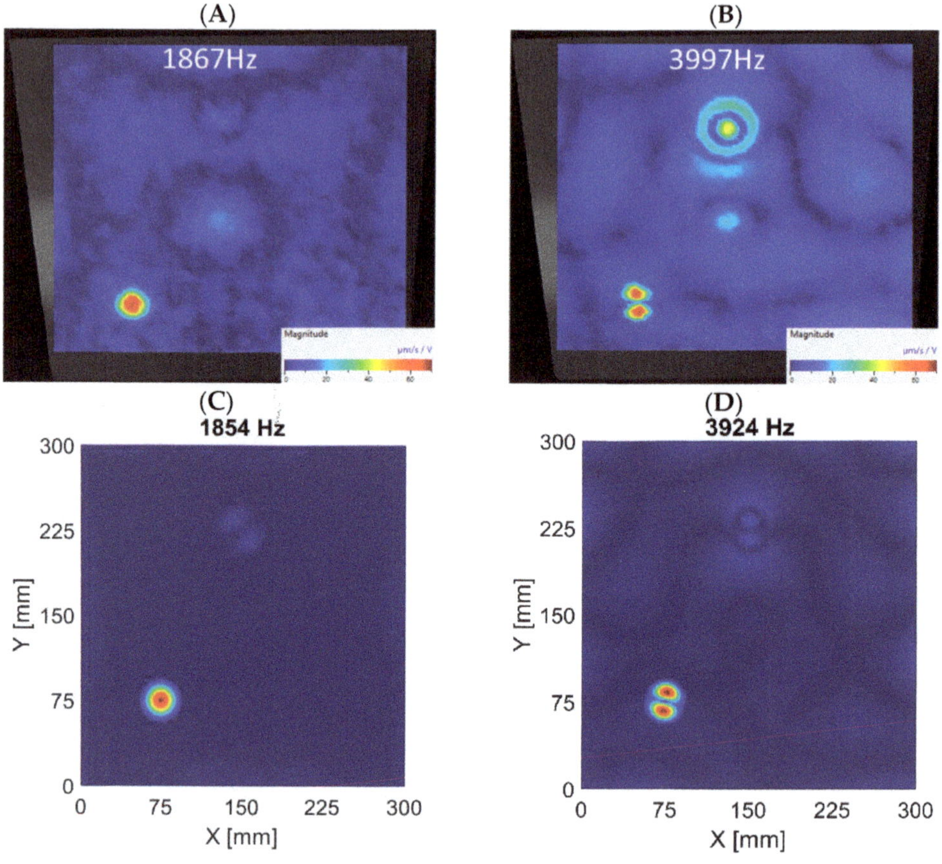

Figure 10. Visualization of out-of-plane vibrations—FBH Ø40 mm. The first LDR in experimental data (**A**) and numerical results based on the 3D FE model (**C**). The second LDR in experimental data (**B**) and numerical results based on the 3D FE model (**D**).

In the FBH Ø40 mm case, the algorithm correctly identified the first two modes of out-of-plane LDRs. The discrepancies between the experimental data and the results of the 3D model are minor, as presented in Figure 10. However, it should be added that these conclusions apply only to the model based on 3D FEs. The algorithm working on the numerical results from the 2D FEs model did not correctly indicate the second out-of-plane mode of LDR of FBH Ø40 mm (false negative indication). On the other hand, the first out-of-plane mode was correctly extracted.

The FBH Ø18 mm is the only defect for which the algorithm's indications are only true positive. It applies to all datasets, i.e., experimental data and 2D and 3D FE models. The visualizations of first and second out-of-plane LDRs of FBH Ø18 mm extracted by the algorithm for various datasets are depicted in Figure 11.

The summary of detected modes for three FBHs is presented in Table 4. The LDR frequencies determined by the algorithm were compared for different datasets to assess the accuracy of the numerical models. The summary is given in Table 4, and is limited to the first and second detected LDR frequencies for given defects and presents relative errors for the two proposed FE models (based on 2D and 3D FEs). It can be observed that the results obtained from numerical models correspond very closely with the experimental results, which prove the successful validation of the numerical models. As expected, when focusing on the first out-of-plane mode of the LDR, the 3D model turned out to be more accurate—the maximum error was observed for FBH Ø40 mm, and it amounted to 0.7% compared to the experimental data.

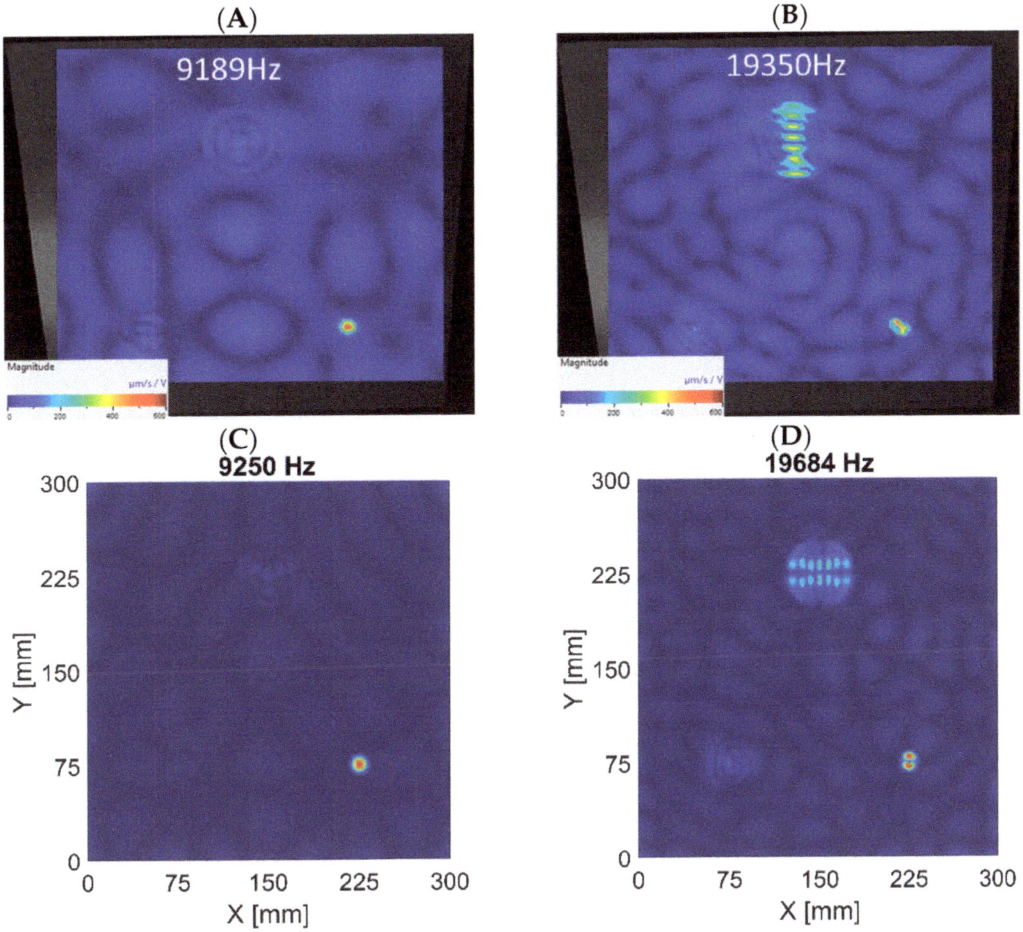

Figure 11. Visualization of out-of-plane vibrations—FBH Ø18 mm. The first LDR in experimental data (**A**) and numerical results based on the 3D FE model (**C**). The second LDR in experimental data (**B**) and numerical results based on the 3D FE model (**D**).

Table 4. Comparison of out-of-plane LDRs frequencies detected by the algorithm for numerical models and experiments.

			The Model with 2D FEs		The Model with 3D FEs	
	Defect Type	Experiment	Simulation	Error	Simulation	Error
First LDR (out-of-plane)	FBH Ø58 mm	857 Hz	855 Hz	0.23%	859 Hz	0.23%
	FBH Ø40 mm	1867 Hz	1859 Hz	0.43%	1854 Hz	0.70%
	FBH Ø18 mm	9189 Hz	9455 Hz	2.90%	9250 Hz	0.67%
Second LDR (out-of-plane)	FBH Ø58 mm	Not found	Not found	—	Not found	—
	FBH Ø40 mm	3997 Hz	Not found	—	3924 Hz	1.83%
	FBH Ø18 mm	19,350 Hz	19,771 Hz	2.18%	19,684 Hz	1.73%

The maximal error in the 2D model was observed for the FBH Ø18 mm, which amounted to 2.9% compared to the experimental data, but it was still an acceptable result in most applications. Similarly, when comparing the second out-of-plane mode of LDRs, the 3D FE model proved slightly more accurate. The maximum error was observed for FBH Ø18 mm, but the difference was not as evident as for the first out-of-plane mode of LDRs, i.e., 1.73% and 2.18% for 3D and 2D FE models, respectively. It must also be noted that for any datasets, the algorithm found the second out-of-plane mode of FBH Ø58 mm. This problem was addressed earlier in the text. Unfortunately, the second out-of-plane mode of FBH Ø40 mm was also not recognized by the algorithm working with 2D FE model results.

It can therefore be summed up that 3D modeling gives more realistic results. In the case of 2D modeling, the results showed a more significant maximum divergence compared to the experiment than the results of the 3D FEs model. The algorithm working on the experimental data and the results from the 3D model correctly indicated the second out-of-plane mode of LDR (FBH Ø40 mm), which was not shown in the 2D data. On the other hand, the 2D model deserves attention, as it is much faster to formulate and is characterized by a three times lower computing time than for the 3D model. It can be successfully employed for a rough determination of the first out-of-plane LDR modes. Moreover, the 2D FE model allows the easy manipulation of the material's thickness, saving a lot of time in the model-tuning process. The simulations were performed on a workstation with an AMD Ryzen® 9 5950X 16-core processor.

4.2. Damage Size Assessment

The higher order out-of-plane LDR modes allowed for a more accurate identification of the FBH damage size. Figure 12 presents a visualization of the detected out-of-plane modes of FBH Ø40 mm using the discussed algorithm. The view was limited only to the damaged area. The algorithm's sensitivity was increased to 24 peaks across the entire frequency spectrum. It allowed for the identification of higher out-of-plane LDRs of the FBH Ø40 mm damage. Horizontal lines were drawn to allow the quantification of the damage size based on the shape of the highest noted mode (23,062 Hz). The determined size of the damage was Ø37 mm. This indication was therefore underestimated in relation to the nominal damage diameter by 3 mm. Nevertheless, the lower modes underestimated the damage diameter to a much greater extent. The first out-of-plane mode, especially, underestimated the damage diameter by about 15 mm, which resulted in a damage size of Ø25 mm.

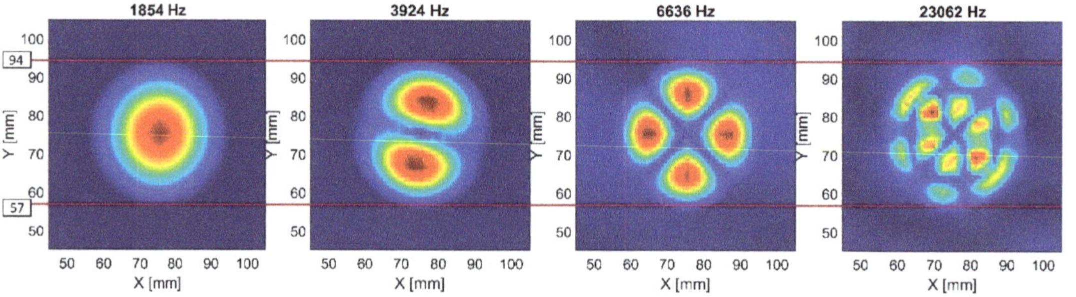

Figure 12. Comparison of the diameter of FBH Ø40 mm based on different LDR modes.

4.3. In-Plane LDRs

The proposed algorithm successfully detects out-of-plane modes of LDRs. Nonetheless, another examination was performed to check the possibility of identifying the in-plane LDRs using the presented algorithm. In this case, the *ARmax* characteristic was computed based on the velocity amplitude of vibrations in the XY plane instead of the velocity Z component. In the case of experimental data, visualization was provided with the Y (vertical) component of vibration velocities. The algorithm's sensitivity was set to 24 because in-plane LDRs are usually related to higher mode shapes.

Although the analysis concerned in-plane modes, most detected resonance frequencies corresponded to out-of-plane modes of LDRs. Further investigation showed that out-of-plane modes are, in fact, always accompanied by increased vibrations in the in-plane directions, as presented in Figure 13, where for the same eigenvalue (3925 Hz) the out-of-plane (Figure 13A) and in-plane (Figure 13B) vibration components are depicted.

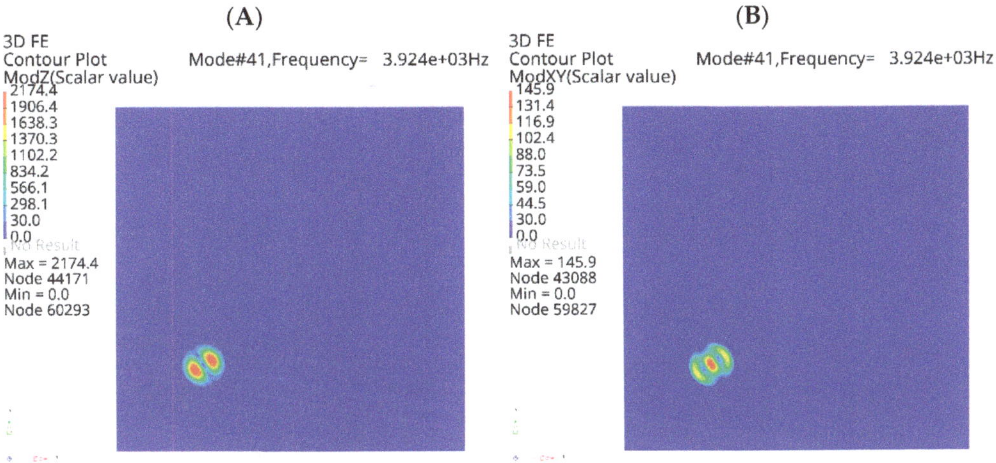

Figure 13. One of the mode shapes of FBH Ø40 mm LDR: out-of-plane component (**A**) and in-plane component (**B**).

The assessment of the normal modes determined by the algorithm showed that all indications referred to the FBHs introduced into the plate. None of the indications was a false-positive, proving the validity of the proposed approach. As mentioned earlier, most indications referred to the out-of-plane modes, for which significant in-plane vibrations were also observed. For this reason, these indications cannot be considered incorrect even though they do not correspond to 'clear' in-plane modes. In the determined results, two modes could be recognized as 'clear' in-plane modes. Those modes refer to the Ø58 mm and Ø40 mm FBHs, and are visualized in Figure 14. No 'clear' in-plane LDRs of FBH Ø18 mm were found, which is likely to be out of range of the analysis. The summary of detected in-plane LDRs is presented in Table 5.

Table 5. Comparison of in-plane LDRs detected by the algorithm for numerical models and experiments.

		The Model with 2D FEs		The Model with 3D FEs	
Defect Type	**Experiment**	**Simulation**	**Error**	**Simulation**	**Error**
FBH Ø58 mm	19,895 Hz	23,164 Hz	16.43 %	20,966 Hz	5.38%
FBH Ø40 mm	29,402 Hz	33,443 Hz	13.74 %	29,623 Hz	0.75%
FBH Ø18 mm	Not found	Not found	—	Not found	—

In the case of using the algorithm to identify in-plane LDR modes, the developed approach also proved to be very effective. It can be noticed that the 3D numerical model allows the representation of in-plane modes of FBHs more accurately than the model based on 2D FEs. The errors reported for the 2D FE model (more than 13%) questioned the correctness of in-plane LDR frequency representation for this modeling technique. The observed differences with the experimental results were more significant than for the out-of-plane modes for both FE models. It was associated with a worse representation of real structures by numerical models in the higher frequency range.

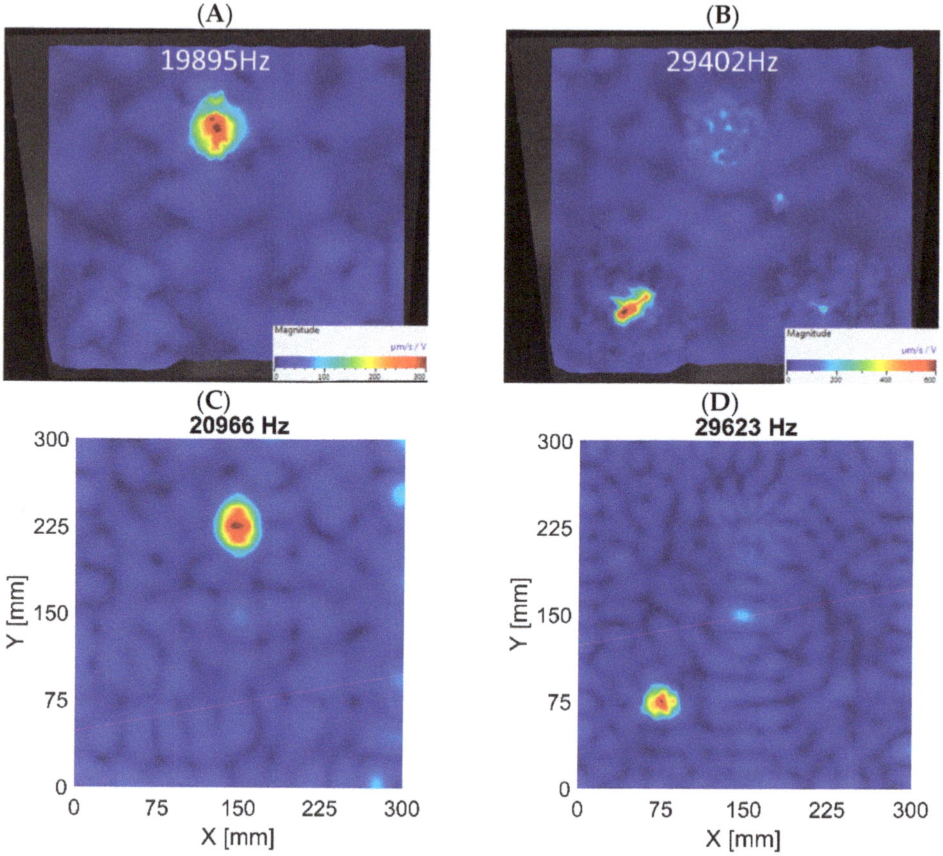

Figure 14. 'Clear' in-plane LDRs of FBH Ø58 mm based on experimental data (**A**) and 3D FE model results (**C**); 'Clear' in-plane LDRs of FBH Ø40 mm based on experimental data (**B**) and 3D FE model results (**D**).

5. Conclusions

This paper presents a method for determining the frequencies of local defect resonances. The algorithm was based on the observation that the vibration amplitudes were significantly higher for the local defect's mode shape than the mean value of the whole structure. Based on that observation, the *ARmax* was computed as a ratio between the maximum amplitude of vibrations for a given frequency and the mean vibrations level. The algorithm found the *ARmax* index's peaks and identified them as LDR frequencies.

To sum up, the most significant achievements reported in this paper are as follows:

- A novel algorithm for the automatic extraction of LDRs was developed. The algorithm was successfully implemented to identify out-of-plane and in-plane modes of LDRs. Various datasets, including experimental and numerical data of a test specimen with FBH of different diameters, were examined;
- The algorithm allowed the determination of higher-order LDR modes, which usually better represented FBH shape and size. Finding higher-order LDRs was also essential to correctly extract in-plane modes. This feature was a significant improvement compared to the current state-of-the-art methods;

- Two numerical models based on the FEM were developed. One employed 3D FEs, and the other was based on 2D FEs. The algorithm worked successfully with the results provided by both models;
- The indicated LDRs proved that, in most cases, the 2D model was sufficient for determining FBH-type out-of-plane LDRs. The recorded error between corresponding LDRs in experiments and results of the 2D FE model was less than 3%. On the other hand, a false negative indication was noted for this dataset. The results provided by the 3D FE model were better fitted with an actual structure, which corresponded with the relative error of less than 2% in out-of-plane modes;
- The 3D FE numerical model allowed the capturing of in-plane modes of FBHs more accurately than the 2D FE model. The errors of in-plane LDRs called into question the applicability of the 2D modeling technique to represent in-plane modes.

It should be emphasized that the presented work is the basis for further research. Future works will concern the use of the developed algorithm for LDRs detection for other types of damage, such as delaminations in composites or fatigue cracks.

Author Contributions: Conceptualization, A.K., M.K., G.L. and P.Z.; methodology, M.O., M.K.; software, M.O., M.K. and P.Z.; validation, P.Z., M.K. and G.L.; formal analysis, P.Z. and M.K.; investigation, M.K. and P.Z.; and resources, Ł.P. and F.A.; data curation, M.K. and G.L.; writing—original draft preparation, P.Z., G.L. and M.K.; writing—review and editing, Ł.P.; visualization, P.Z.; supervision, A.K. and F.A.; project administration, A.K. and Ł.P.; funding acquisition, A.K. All authors have read and agreed to the published version of the manuscript.

Funding: This research was funded by the National Science Center, Poland, within the project no. 2018/29/B/ST8/02207.

Institutional Review Board Statement: Not applicable.

Informed Consent Statement: Not applicable.

Data Availability Statement: Data available upon request.

Conflicts of Interest: The authors declare no conflict of interest.

References

1. Moore, P.O. *Nondestructive Testing Overview*; The American Society for Nondestructive Testing: Columbus, OH, USA, 2012; ISBN 978-1-57117-187-0.
2. Stepinski, T.; Uhl, T.; Staszewski, W. *Advanced Structural Damage Detection: From Theory to Engineering Applications*; John Wiley & Sons: Hoboken, NJ, USA, 2013; ISBN 978-1-118-53612-4.
3. Kundu, T. *Nonlinear Ultrasonic and Vibro-Acoustical Techniques for Nondestructive Evaluation*; Springer International Publishing AG: Cham, Switzerland, 2019; ISBN 978-3-319-94474-6.
4. Delsanto, P.P. *Universality of Nonclassical Nonlinearity: Applications to Non-Destructive Evaluations and Ultrasonics*; Springer: New York, NY, USA, 2006; ISBN 978-0-387-33860-6.
5. Jhang, K. Nonlinear ultrasonic techniques for nondestructive assessment of micro damage in material: A review. *Int. J. Precis. Eng. Manuf.* **2009**, *10*, 123–135. [CrossRef]
6. Pieczonka, L.; Klepka, A.; Martowicz, A.; Staszewski, W.J. Nonlinear vibroacoustic wave modulations for structural damage detection: An overview. *Opt. Eng.* **2015**, *55*, 011005. [CrossRef]
7. Broda, D.; Mendrok, K.; Silberschmidt, V.V.; Pieczonka, L.; Staszewski, W.J. The Study of Localized Crack-Induced Effects of Nonlinear Vibro-Acoustic Modulation. *Materials* **2023**, *16*, 1653. [CrossRef] [PubMed]
8. Kim, Y.; Choi, S.; Jhang, K.Y.; Kim, T. Experimental Verification of Contact Acoustic Nonlinearity at Rough Contact Interfaces. *Materials* **2021**, *14*, 2988. [CrossRef] [PubMed]
9. Solodov, I.; Bai, J.; Bekgulyan, S.; Busse, G. A Local Defect Resonance to Enhance Acoustic Wave-Defect Interaction in Ultrasonic Nondestructive Evaluation. *Appl. Phys. Lett.* **2011**, *99*, 211911. [CrossRef]
10. Solodov, I.; Bai, J.; Busse, G. Resonant Ultrasound Spectroscopy of Defects: Case Study of Flat-Bottomed Holes. *J. Appl. Phys.* **2013**, *113*, 223512. [CrossRef]
11. Solodov, I. Resonant Acoustic Nonlinearity of Defects for Highly-Efficient Nonlinear NDE. *J. Nondestruct. Eval.* **2014**, *33*, 252–262. [CrossRef]
12. Pieczonka, L.; Zietek, L.; Klepka, A.; Staszewski, W.J.; Aymerich, F.; Uhl, T. Damage Imaging in Composites Using Nonlinear Vibro-Acoustic Wave Modulations. *Struct. Control Health Monit.* **2018**, *25*, e2063. [CrossRef]

13. Hedayatrasa, S.; Segers, J.; Poelman, G.; Paepegem, W.V.; Kersemans, M. Vibro-Thermal Wave Radar: Application of Barker Coded Amplitude Modulation for Enhanced Low-Power Vibrothermographic Inspection of Composites. *Materials* **2021**, *14*, 2436. [CrossRef] [PubMed]
14. Solodov, I.; Rahammer, M.; Kreutzbruck, M. Analytical Evaluation of Resonance Frequencies for Planar Defects: Effect of a Defect Shape. *NDT E Int.* **2019**, *102*, 274–280. [CrossRef]
15. Derusova, D.A.; Vavilov, V.P.; Druzhinin, N.V.; Shpil'noi, V.Y.; Pestryakov, A.N. Detecting Defects in Composite Polymers by Using 3D Scanning Laser Doppler Vibrometry. *Materials* **2022**, *15*, 7176. [CrossRef] [PubMed]
16. Hettler, J.; Tabatabaeipour, M.; Delrue, S.; van den Abeele, K. Detection and Characterization of Local Defect Resonances Arising from Delaminations and Flat Bottom Holes. *J. Nondestruct. Eval.* **2017**, *36*, 2. [CrossRef]
17. Segers, J.; Kersemans, M.; Hedayatrasa, S.; Calderon, J.; van Paepegem, W. Towards in-Plane Local Defect Resonance for Non-Destructive Testing of Polymers and Composites. *NDT E Int.* **2018**, *98*, 130–133. [CrossRef]
18. Segers, J.; Hedayatrasa, S.; Verboven, E.; Poelman, G.; van Paepegem, W.; Kersemans, M. In-Plane Local Defect Resonances for Efficient Vibrothermography of Impacted Carbon Fiber-Reinforced Polymers (CFRP). *NDT E Int.* **2019**, *102*, 218–225. [CrossRef]
19. Roy, S.; Bose, T.; Debnath, K. Detection of Local Defect Resonance Frequencies Using Bicoherence Analysis. *J. Sound Vib.* **2019**, *443*, 703–716. [CrossRef]
20. Roy, S.; Bose, T. Efficient Determination of Local Defect Resonance Frequencies from Bicoherence Plots Using Double Excitations. *Mech. Syst. Signal Process.* **2019**, *127*, 595–609. [CrossRef]

applied sciences

MDPI

Article

A Technique for Multi-Parameter Signal Processing of an Eddy-Current Probe for Measuring the Thickness of Non-Conductive Coatings on Non-Magnetic Electrically Conductive Base Metals

Michael Syasko [1], Pavel Solomenchuk [2,*], Igor' Soloviev [1] and Natalia Ampilova [1]

[1] Faculty of Mathematics and Computer Science, St Petersburg University, 199034 St. Petersburg, Russia
[2] JSC "CONSTANTA", 198095 St. Petersburg, Russia
* Correspondence: pavel257@mail.ru

Featured Application: The presented results can be used in devices for measuring the thickness of a non-conductive coating on non-magnetic electrically conductive base metals and non-metals.

Abstract: The known amplitude-sensitive eddy-current method for measuring the thickness of non-conductive coatings on conductive non-magnetic base metals does not satisfy the accuracy requirements. A primary consideration is the significant influence of a change in the specific electrical conductivity of the base metals on results of measurements. In this study, we developed a technique for measuring the thickness of non-conductive coatings on non-magnetic conductive base metals by using the eddy-current amplitude-phase method and implemented algorithms to process obtained information. Our method considered the influence of the specific electrical conductivity of the base metals by forming a two-dimensional graduation characteristic of the thickness gauge by using several base metals with different specific electrical conductivity. The algorithm for point-in-polygon determination was applied, which allowed us to measure the thickness of the coatings and the specific electrical conductivity of the base metals as independent values. The equipment necessary to construct the two-dimensional graduation characteristic and the algorithm for calculation of the thickness are described in detail.

Keywords: eddy-current; thickness gauge; amplitude-phase algorithm; two-dimensional graduation

Citation: Syasko, M.; Solomenchuk, P.; Soloviev, I.; Ampilova, N. A Technique for Multi-Parameter Signal Processing of an Eddy-Current Probe for Measuring the Thickness of Non-Conductive Coatings on Non-Magnetic Electrically Conductive Base Metals. *Appl. Sci.* **2023**, *13*, 5144. https://doi.org/10.3390/app13085144

Academic Editors: Bogusław Łazarz, Grzegorz Peruń and Tangbin Xia

Received: 26 February 2023
Revised: 30 March 2023
Accepted: 17 April 2023
Published: 20 April 2023

1. Introduction

1.1. Amplitude-Sensitive Eddy-Current Method for Measuring the Thickness of Coating

To measure the thickness d of a non-conductive coating on non-magnetic electrically conductive base metals, the amplitude-sensitive eddy-current method of non-destructive testing is provided [1,2]. The operation principle of eddy-current devices is based on the fact that a high-frequency electromagnetic field generated by an eddy-current probe (ECP) produces eddy currents in a conductive base metal beneath the coating on which the ECP is placed (Figure 1). The density of eddy currents depends on the distance between the excitation coil and the surface of the conductive base metal. These eddy currents cause a change in the electromagnetic field surrounding the coil of the ECP and affect the impedance of the coil.

The change in the impedance permits obtaining a set of calibration values by using reference standards for thickness of coating, and then, to calculate the thickness of a coating on an electrically conductive base metal.

In most practical applications, the single ECP excitation coil is a part of a self-oscillating circuit, as shown in Figure 2.

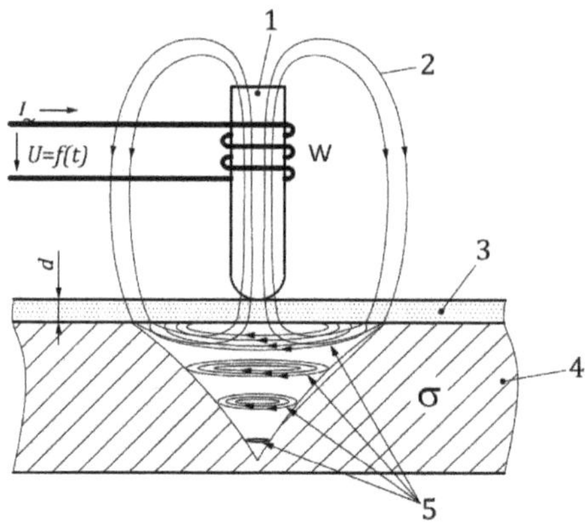

Figure 1. Amplitude-sensitive eddy-current method of measurement of the thickness of coating. 1—probe core with coil W, 2—high frequency magnetic field, 3—non-conductive coating, 4—non-magnetic electrically conductive base metals, 5—induced eddy current, *I*—exciting current, *U*—signal of eddy-current converter.

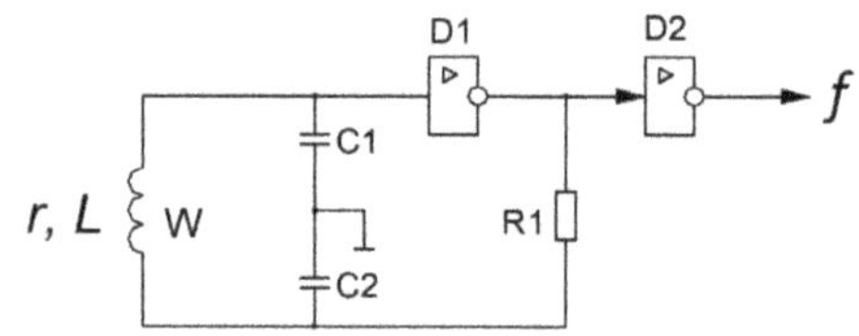

Figure 2. Structure of eddy-current probe with self-oscillating circuit.

The output signal of the self-oscillating circuit is the resonance frequency *f*, determined by the formulas:

$$f = \frac{1}{2\pi\sqrt{LC}} \qquad C = \frac{C1 \cdot C2}{C1 + C2} \tag{1}$$

ECP of this type is called parametric, the parameter determining the output signal being the inductance L of the ECP coil, depending on the thickness of a coating d. The ohmic resistance r of the ECP coil does not affect the output signal of the self-oscillating circuit directly. The value f is a measure of d, and it is converted to d by means of a graduation curve.

1.2. Stray Parameters When Measuring d by Using the Amplitude-Sensitive Eddy-Current Method

Stray parameters affecting the value of measurement uncertainty can be divided into the following groups:

- Geometric: curvature and roughness of a base metal and coating surfaces; the tilt of ECP; contamination of the coating and ECP surfaces resulting in a gap between the ECP and the test object; edge effect.
- Technological: the pressure of ECP on the coating; intermediate coatings; the influence of temperature and external electromagnetic fields.
- Electromagnetic: the specific electrical conductivity σ of the base metal.

Practice shows that the value of σ and its changing along the surface of the base metal and/or from one object to another make the greatest contribution to the measurement uncertainty [3].

To study the influence of the test and stray parameters, one should use the analysis of the ECP parameters on the complex plane.

The coil impedance $\overline{Z}$ of a parametric ECP is a complex value determined by the formula:

$$\begin{aligned} \overline{Z} &= r + jXl \\ Xl &= 2\pi f L, \end{aligned} \tag{2}$$

where r is ohmic resistance and Xl is inductive resistance.

Figure 3 (on the left) shows the dependence of the coil impedance $\overline{Z}$ of the parametric ECP on d on the complex plane, Figure 3 (on the right) shows the local hodograph diagrams of the coil impedance of the parametric ECP when changing d from 0 to ∞ at σ = 5 MS/m, 17 MS/m and 58 MS/m and σ from 0 to 58 MS/m.

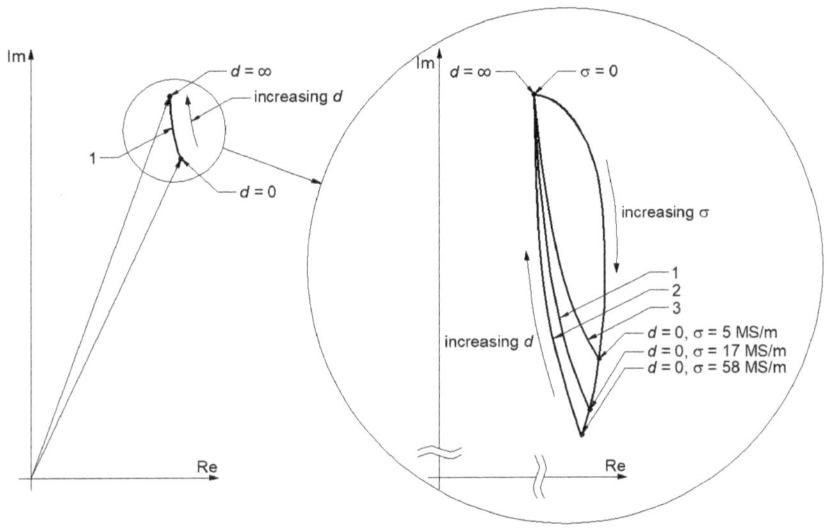

Figure 3. Dependence of the coil impedance of the parametric ECP on the thickness of coating and specific electrical conductivity.

On placing an ECP on an uncoated base metal (d = 0) with nominal value σ = 17 MS/m, the impedance on the complex plane corresponds to the point d = 0, σ = 17 MS/m. With an increase in d, the hodograph diagram of the impedance change describes line 1. When an ECP is placed on an uncoated base metal (d = 0) with a different specific electrical conductivity; for example, σ = 58 MS/m or 5 MS/m, the impedance on the complex plane corresponds to the point d = 0, σ = 58 MS/m or d = 0, σ = 5 MS/m. As d increases, the hodograph diagram of impedance change is line 2 or 3. In this case, the dependence f on d at σ = 5 MS/m, σ = 17 MS/m and at σ = 58 MS/m is described by the curves shown in Figure 4a, and the dependence the deviation of d on σ on a logarithmic scale is shown in Figure 4b.

When measuring d on base metals with different values of σ, a multiplicative change in f occurs, which can be compensated by adjusting (calibrating) the ECP on an uncoated base metal or a coated base metal with a known thickness d. Such a compensation will be observed in a small range of d near the calibration point. However, sometimes, we cannot perform a compensation; for example, in the following cases:

- The thickness is measured on a fully painted product, and removing the coating from a part of the surface (for ECP calibration) is not possible;
- σ changes considerably on the surface of base metal;
- The test object has different temperatures on the surface. In this case, a deviation of σ is observed.

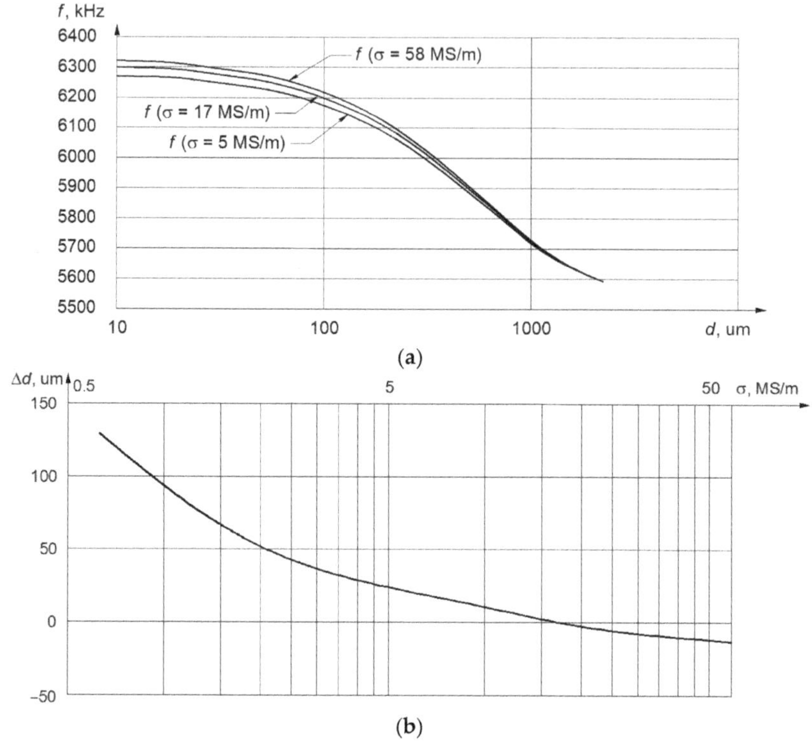

Figure 4. (**a**) Dependence f on d at σ = 5 MS/m, σ = 17 MS/m and at σ = 58 MS/m, (**b**), dependence the deviation of d on σ.

In these cases, it is necessary to provide the suppression the influence of σ by implementing the adaptive structure of the measuring probe, which has functional redundancy [4,5].

In indirect measurement methods using the ECP graduation curve, the interpolation error of the graduation curve is an additional source of measurement uncertainty. The curve is constructed for a limited number of graduation points, but at the same time, it should cover the entire measurement range, for the interpolation of the curve power or logarithmic functions are usually applied. Large distances between graduation and interpolation points may result in significant deviations of obtained values from the real transmission characteristic of the measuring probe. This is due to the uncertainty of the points of graduation curve and the choice of interpolating functions.

1.3. A Review of the Methods for a Suppression of the Influence of σ Deviation

In the patent [6], the problem of reducing the influence of σ is solved by turning the hodographs of signals on the complex plane. This method allows the suppressing the influence of σ for the base metal, but it does not take into account a change in the sensitivity of the ECP to the thickness of the coating when σ of the base metal changes. In general, the proposed method is effective for insignificant changes in σ when different objects are made of metal of the same brand.

The paper [7] describes the algorithm for measuring the thickness of an electrically conductive plate, which is not sensitive to σ. This algorithm assumes the use of a sweep-frequency eddy current testing. It may be applied to measure the thickness of the coating on an electrically conductive base. At the same time, the variable-frequency method necessitates equipping the thickness gauge with a complex electronic circuit, which is acceptable for stationary devices and unacceptable for portable ones.

In the paper [8], the authors proposed the algorithm for measuring the thickness of an electrically conductive film. It was not sensitive to the gap between the VTP and the film. This algorithm also assumed the application of a sweep-frequency eddy current testing. Along with it, a simplified variant based on a small set of excitation frequencies was also considered. In this case, the selection of suitable pre-calculated sets of film thicknesses and gap parameters satisfying the measured signals of the ECP at several frequencies was carried out.

In periodicals, besides those mentioned, a large number of articles describing the technology of measuring the thickness of the coating by the multi-frequency method, variable frequency method and pulse method were published. Unfortunately, all these methods are not applicable for implementation in portable devices. The works concerning ECP with two or more measuring coils (for example: [9]) are also of little use for implementation in miniature ECP.

One of obvious solutions of the problem discussed is to develop a method using a single fixed excitation frequency for an absolute transformer compensated ECP.

2. Materials and Methods

2.1. Description of an Amplitude-Phase Probe

It should be noted that the solving the discussed problem by implementing a self-oscillating ECP is practically impossible because the output signal f is one-parameter, but it is influenced by both d and σ, and their influences cannot be separated.

We propose to use a three-coil transformer compensated amplitude-phase eddy-current probe (AP ECP), the block diagram of which is shown in Figure 5 [10].

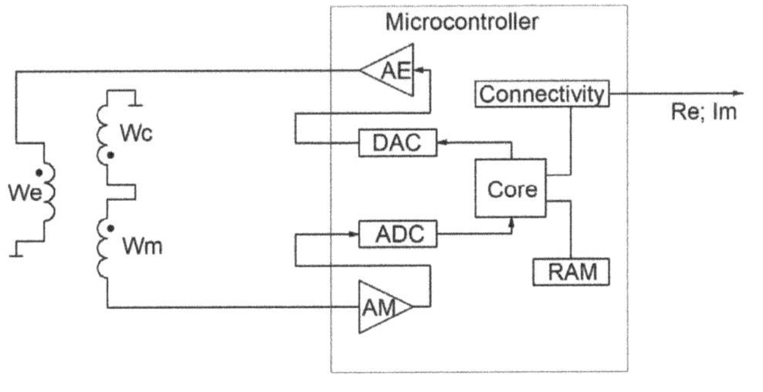

Figure 5. Block diagram of three-coil transformer compensated amplitude-phase eddy-current probe. Here, We—excitation coil, Wc—compensating coil, Wm—measuring coil, AE—excitation signal amplifier, AM—added voltage amplifier.

Using the built-in DAC, the microcontroller synthesizes a harmonic excitation signal of a given frequency. The excitation signal amplifier (AE) is the source of the required excitation voltage. The excitation voltage applied to the excitation coil (We) creates a magnetic field that interacts with the base metal of the test object. The resulting magnetic field, derived from the interaction of the magnetic field of the excitation coil and the eddy currents field in the base metal of the test object, is picked by the measuring coil (Wm). The compensating coil (Wc) senses the magnetic field We but does not pick the magnetic field of the eddy currents. Wm and Wc are connected in-series in opposite directions, which results in compensating the constant component of the Wm signal. The signal of the measuring coil output corresponds to the added voltage of the AP ECP. This voltage is applied to the added voltage amplifier (AM), and then, the signal arrives to the ADC. The ADC signal is transmitted through the communication unit for further processing. The output signal of the AP ECP is a complex value containing two independent signal parameters (Re; Im)—the real and imaginary components of the added voltage of the AP ECP [11].

2.2. Description of the Finite Element Model of AP ECP

To calculate and optimize the parameters of AP ECP and obtain model signals for debugging the mathematical processing algorithms, a finite element model of the primary measuring probe was developed.

To study the influence of test and stray parameters on the signal of a transformer-type AP ECP, the analysis of the AP ECP added voltage on the complex plane should be performed. Despite a number of obvious differences, the hodograph diagrams of transformer AP ECP signals were similar to the diagrams of the parametric ECP ones.

The graphical representation of the finite element model of AP ECP coils placed over a base metal is shown in Figure 6a and the hodograph diagrams of the signals are shown in Figure 6b.

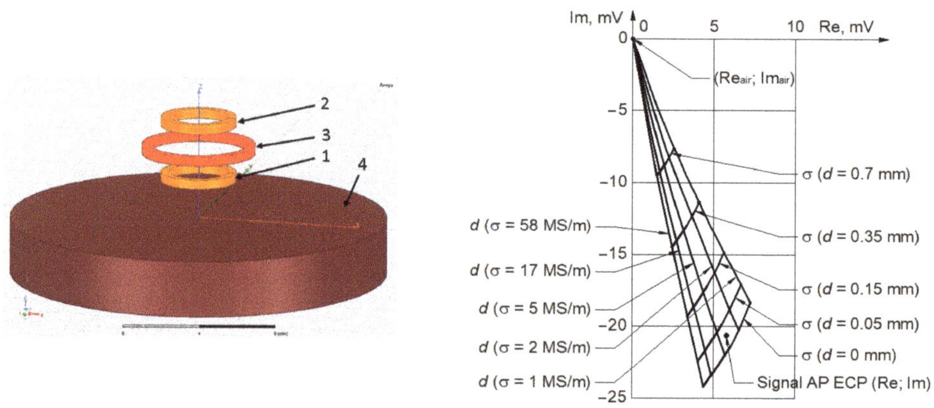

(a) (b)

Figure 6. (**a**) The representation of the finite element model of AP ECP coils over a base metal, (**b**) hodograph diagrams of the signals of the AP ECP finite element model. 1—measuring coil, 2—compensating coil, 3—excitation coil, 4—base metal.

2.3. Overview of Two-Parameter Measurement Methods

Among the methods for multidimensional analysis of ECP signals, one should mention the method of principal components [12]. Its application shows good results if the influences of both parameters are unidirectional when switching to an alternative coordinate system. The joint application of the amplitude-sensitive and phase-sensitive eddy-current methods was described in [13,14]. It allows correcting the results of measurement of the parameter obtained by the amplitude-sensitive method according to the results of measuring the second parameter obtained by the phase-sensitive method. It is assumed that a change in the parameter measured by the amplitude-sensitive method does not affect the results of measurements for the second parameter. Obviously, the achieved measurement accuracy does not meet the modern requirements of consumers of the equipment being developed. Methods of two-parameter measurements in eddy-current testing using excitation of eddy currents at different frequencies [15] cannot be applied for miniature ECPs and for small-sized test objects.

2.4. Description of the Proposed Graduation Technique

In existing devices, each point of the graduation curve corresponds to one or more values of the thickness of coating on the base metal.

The proposed technique involves a two-dimensional graduation characteristic. Instead of graduating the AP ECP on a single base metal with σ corresponding to a certain measurement average [16], it is necessary to graduate the AP ECP on a set of reference conductive non-magnetic base metals, for which values of σ are different and cover the entire range of actual σ values. The graduation characteristic is similar to hodograph diagrams in Figure 6b.

A two-dimensional graduation characteristic involving several base metals complicates the graduation procedure considerably. To eliminate this problem, we propose to use a simulating system for the thickness of a non-conductive coating (CTSS). It consists of a precision linear actuator, a servo drive, a basic supporting structure and software providing an automated ECP graduation procedure (Figure 7).

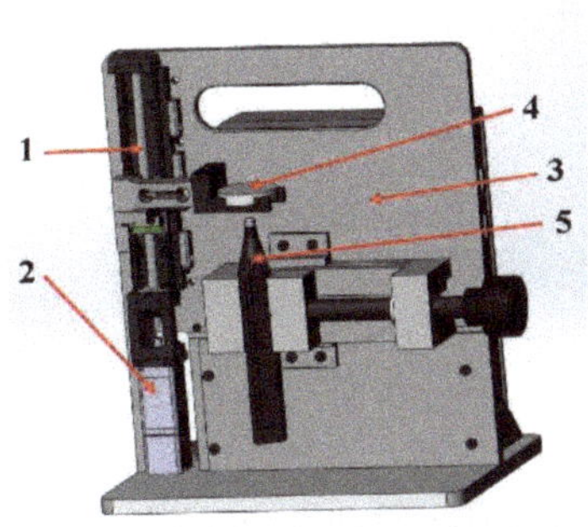

Figure 7. Non-conductive coating thickness simulating system. 1—precision linear actuator, 2—servo drive, 3—basic supporting structure, 4—reference base metal, 5—ECP.

During the automated graduation process, CTSS software generates a graduation curve containing a large number of graduation points. This number is comparable to the resolution of the thickness gauge implemented by AP EC. It enables one to omit inappropriate interpolating functions. Reducing the complexity of construction of the curve for one base metal allows graduating the AP ECP on several reference base metals with known σ without any significant complication in the graduation procedure.

2.5. Description of the Proposed Measurement Technique

The test parameter is d, while σ is the stray parameter. The value σ is not displayed on the indicator of the thickness gauge, as it is not relevant for the user.

The graduation characteristic is represented as a fan-shaped grid on the plane. The grid thickens in d direction and thinning out in σ direction. In these conditions, the algorithm for processing the measurement information has only one task—to determine the section of this grid that corresponds to the point of the signal obtained during the measurement. Obviously, for the effective operation of the proposed technique in conditions of a large set of graduation points (approximately several thousands) and relatively low-performance of single-core ARM architecture of microcontrollers, it is necessary to optimize the computational algorithm.

The proposed two-dimensional algorithm to calculate σ and d is based on the checking whether a point lies in a polygon [17] (pp. 142–146).

2.5.1. General Description of Two-Dimensional Algorithm for the Calculation σ and d

A generalized graphical interpretation of the two-dimensional algorithm for calculation of σ and d is shown in Figure 8. In Figure 8, for brevity, we denote graduation table as GT; graphical interpretation of the graduation characteristic in the form of hodograph diagrams of the signals of AP ECP finite element model is shown in Figure 6b.

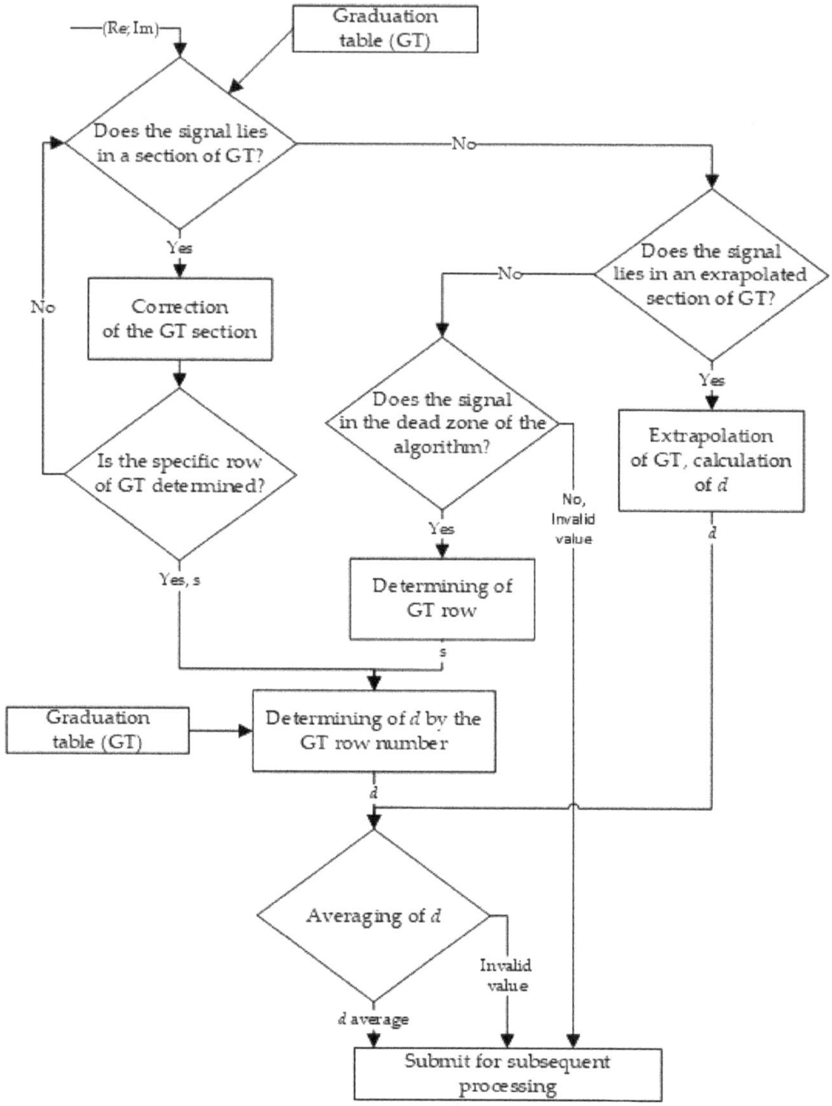

Figure 8. Two-dimensional algorithm of the calculation σ and *d*.

The lines of the graduation characteristic obtained on different reference base metals are presented as lines $d(\sigma)$. These lines converge fanwise at a single point—an air point with coordinates (Re_{Air}; Im_{Air}). For clarity, the points with the same *d* value on different lines of the graduation characteristic are connected by lines which are denoted as $\sigma(d)$. During a measurement process, the signal which has coordinates (Re; Im) is calculated. In the figure, it is denoted as the point "(Re; Im)".

The task of the algorithm for calculation σ and *d* is to determine whether the signal belongs to one of the sections bounded by four lines of the grid and to calculate its value relative to these boundaries.

Due to the large number of sections of the graduation characteristic, the calculation is carried out by using the "half section" method [18].

An example of the graduation characteristic of AP ECP in tabular form is presented in Table 1.

Table 1. Graduation characteristic of AP ECP in tabular form.

d	Σ			
	$\sigma 1$	$\sigma 2$	-//-	σ_{column}
$d0$	$(\mathrm{Re}_{d0,\sigma1}; \mathrm{Im}_{d0,\sigma1})$	$(\mathrm{Re}_{d0,\sigma2}; \mathrm{Im}_{d0,\sigma2})$	-//-	$(\mathrm{Re}_{d0,\sigma column}; \mathrm{Im}_{d0,\sigma column})$
$d1$	$(\mathrm{Re}_{d1,\sigma1}; \mathrm{Im}_{d1,\sigma1})$	$(\mathrm{Re}_{d1,\sigma2}; \mathrm{Im}_{d1,\sigma2})$	-//-	$(\mathrm{Re}_{d1,\sigma column}; \mathrm{Im}_{d1,\sigma column})$
$d2$	$(\mathrm{Re}_{d2,\sigma1}; \mathrm{Im}_{d2,\sigma1})$	$(\mathrm{Re}_{d2,\sigma2}; \mathrm{Im}_{d2,\sigma2})$	-//-	$(\mathrm{Re}_{d2,\sigma column}; \mathrm{Im}_{d2,\sigma column})$
-//-	-//-	-//-	-//-	-//-
d_{row}	$(\mathrm{Re}_{d\,row,\sigma1}; \mathrm{Im}_{d\,row,\sigma1})$	$(\mathrm{Re}_{d\,row,\sigma2}; \mathrm{Im}_{d\,row,\sigma2})$	-//-	$(\mathrm{Re}_{d\,row,\sigma column}; \mathrm{Im}_{d\,row,\sigma column})$

2.5.2. "Signal-in-Section" Check

The graduation table is divided in half and the belonging of the signal to a section of the upper or lower half of the graduation table is determined.

Only 3 rows from the table are selected: $d = d0$, $d = d_{\text{row}/2}$ and $d = d_{\text{row}}$.

The belonging the signal to the sections separated by the polyline $\sigma(d_{\text{row}/2})$ is determined by condition 2:

$$\begin{cases} n = 0 \\ (starting\ conditions) \\ if\ (\mathrm{Im}_{d,\sigma} \le \mathrm{Im} < \mathrm{Im}_{d+1,\sigma}\ or\ \mathrm{Im}_{j,k} \ge \mathrm{Im} > \mathrm{Im}_{d+1,\sigma})\ and\ \left(\mathrm{Re} \ge \frac{(\mathrm{Re}_{d,\sigma} - \mathrm{Re}_{d+1,\sigma})\cdot(\mathrm{Im} - \mathrm{Im}_{d,\sigma})}{\mathrm{Im}_{d,\sigma} - \mathrm{Im}_{d+1,\sigma}} + \mathrm{Re}_{d,\sigma+1}\right),\ then\ n = n+1 \\ (check\ of\ lower\ boundary) \\ if\ (\mathrm{Im}_{d,\sigma+1} \le \mathrm{Im} < \mathrm{Im}_{d+1,\sigma+1}\ or\ \mathrm{Im}_{d,\sigma+1} \ge \mathrm{Im} > \mathrm{Im}_{d+1,\sigma+1})\ and\ \left(\mathrm{Re} > \frac{(\mathrm{Re}_{d,\sigma+1} - \mathrm{Re}_{d+1,\sigma+1})\cdot(\mathrm{Im} - \mathrm{Im}_{d,\sigma+1})}{\mathrm{Im}_{d,\sigma+1} - \mathrm{Im}_{d+1,\sigma+1}} + \mathrm{Re}_{d,\sigma+1}\right),\ then\ n = n+1 \\ (check\ of\ upper\ boundary) \\ if\ (\mathrm{Im}_{d+1,\sigma} \le \mathrm{Im} < \mathrm{Im}_{d+1,\sigma+1}\ or\ \mathrm{Im}_{d+1,\sigma} \ge \mathrm{Im} > \mathrm{Im}_{d+1,\sigma+1})\ and\ \left(\mathrm{Re} > \frac{(\mathrm{Re}_{d+1,\sigma} - \mathrm{Re}_{d+1,\sigma+1})\cdot(\mathrm{Im} - \mathrm{Im}_{d+1,\sigma})}{\mathrm{Im}_{d,\sigma+1} - \mathrm{Im}_{d+1,\sigma+1}} + \mathrm{Re}_{d+1,\sigma}\right),\ then\ n = n+1 \\ (check\ of\ right\ boundary) \\ if\ (\mathrm{Im}_{d,\sigma} \le \mathrm{Im} < \mathrm{Im}_{d,\sigma+1}\ or\ \mathrm{Im}_{d,\sigma} \ge \mathrm{Im} > \mathrm{Im}_{d,\sigma+1})\ and\ \left(\mathrm{Re} \ge \frac{(\mathrm{Re}_{d,\sigma} - \mathrm{Re}_{d,\sigma+1})\cdot(\mathrm{Im} - \mathrm{Im}_{d,\sigma})}{\mathrm{Im}_{d,\sigma} - \mathrm{Im}_{d,\sigma+1}} + \mathrm{Re}_{d,\sigma}\right),\ then\ n = n+1 \\ (check\ of\ left\ boundary\ ) \\ n = 1\ or\ n = 3 \\ (check\ of\ the\ number\ of\ intersections\ ) \end{cases} \quad (3)$$

where d—selected line number of the graduation table ($d = 0$ or $row/2$), σ—column number of the graduation table ($\sigma = 1 \ldots$ column).

- If the condition is met for $d = 0$, then the signal belongs to the upper half of the table. In this case, for the next iteration, points from the upper half of the table ($d = 0$, $d = row/4$ and $d = row/2$) are selected;
- If the condition is met for $d = row/2$, then the signal belongs to the lower half of the table. In this case, for the next iteration, points from the lower half of the table ($d = 0$, $d = row3/4$ and $d = row$) are selected;
- If the condition is not met neither for $d = 0$, nor for $d = row/2$, then either the signal belongs to one of the extrapolated sections, or the signal is in one of the dead zones of the algorithm. In this case, the signal is checked for belonging to one of the extrapolated sections of the table;
- If the condition is met for none of the interpolated sections of the graduation characteristic, then the signal belongs to one of "dead zones" of the algorithm; in this case, the signal is checked for belonging to "dead zones";
- If the condition is met for none of "dead zones" of the algorithm, then the signal is an error and is not subject to further processing.

The signal can belong only to one section of the table; when determining the section, it is necessary to proceed to the next iteration.

In the next iteration, the involving half of the graduation table is divided in half, the belonging the signal to one of the sections of a quarter of the table is checked by analogy with previous iteration. It should be marked that the case when the signal does not belong to any quarter of the graduation table cannot occur. The result of this calculation in the described branch of the two-dimensional graduation algorithm is the determination of the number of the row and column of the graduation table to which the signal (d; σ) belongs. In this case, we can perform an additional optimization of the algorithm, namely not check the columns σ which differ significantly from the value determined at the previous stages.

The thickness of coating is determined by the line of the graduation characteristic. In this case, the number of lines of the graduation characteristic can be several times fewer than the number of possible values of d in the measurement range of the thickness gauge. To ensure the specified resolution of the thickness gauge measurement, it is necessary to use the accumulation and averaging of the results of calculation of d.

2.5.3. Extrapolation of the Graduation Characteristic and "Dead Zones" of the Algorithm

A two-dimensional graduation characteristic is a field bound by values of d from 0 to ∞ and σ from σmin to σmax. It should be taken into consideration that in practice measurements outside the specified boundaries are not excluded:

- Example 1. The measure σmax is taken for copper, $\sigma \approx 58$ MS/m. If we measure d on a copper base at low temperature of the test object, the value of σ may exceed the σmax.
- Example 2. As was mentioned before, in practice, the curvature of the surface of the test object is a non-excluded stray parameter. When graduating, we use only measures for flat surfaces. If measurements are taken on an object with a concave surface, the d readings decrease. With a small d value, the readings may be negative.

However, such situations should not cause failures in the operation of the algorithm for mathematical processing of AP ECP signals. For this purpose, the graduation characteristic is extrapolated in the directions $\sigma > \sigma max$, $\sigma < \sigma min$ and $d < 0$ by using empirically selected polynomial extrapolation functions.

Descriptions of methods for extrapolating the graduation characteristic and mathematical processing of the signal belonging to the "dead zone" of the algorithm are beyond the scope of this article.

3. Results

Tests of the described algorithm were carried out by using the finite element model described above. The tests had two stages:

At the first stage, the graduation of the ECP model was carried out. To carry it out, we used a base metal model with values of σ corresponding to the real base metal measures used for the graduation of ECP: 0.63, 0.94, 2.12, 3.77, 5.33, 9.81, 14.4, 16.9, 26.3, 35.0, 40.5, 50.6, 58.8 MS/m. The values of σ for the graduation base metals were selected so that they had an approximately monotonous arrangement on a logarithmic scale. The values of d were set in a vicinity of the test values. Due to the fact that, in general, the algorithm has a high accuracy of calculation, the graduation of the model was carried out with a resolution of 0.1 microns, which is significantly higher than necessary for practical applications. The data obtained during the calibration of the ECP model were loaded into the program implementing the algorithm described above.

At the second stage, the ECP model was tested. For this, we used a base metal model with a value of σ that changes uniformly on a logarithmic scale with a small step. The following test values d were set: 0, 10, 100, 1000 microns. The test results in the form of a dependence of the deviation of the calculated value of d on the value of σ on a logarithmic scale are shown in Figure 9.

From the data shown in Figure 9, it can be seen that the deviation of d value was expected to be minimal in a vicinity of the graduation points. The largest deviation d was observed at $\sigma \approx 1.4$ MS/m. This value corresponds to a group of low-alloyed titanium alloys. Despite the fact that the practical significance of measuring the thickness of the coating on such base metals is not obvious, including an additional graduation base metal with $\sigma \approx 1.4$ MS/m in the table significantly improved the quality of the graduation characteristics for this type of ECP. The test results of the proposed method in comparison with the amplitude-sensitive eddy-current method showed a decrease in the deviation d

under the influence of the change σ by about two orders of magnitude, which was obvious when comparing the data given in Figures 4b and 9.

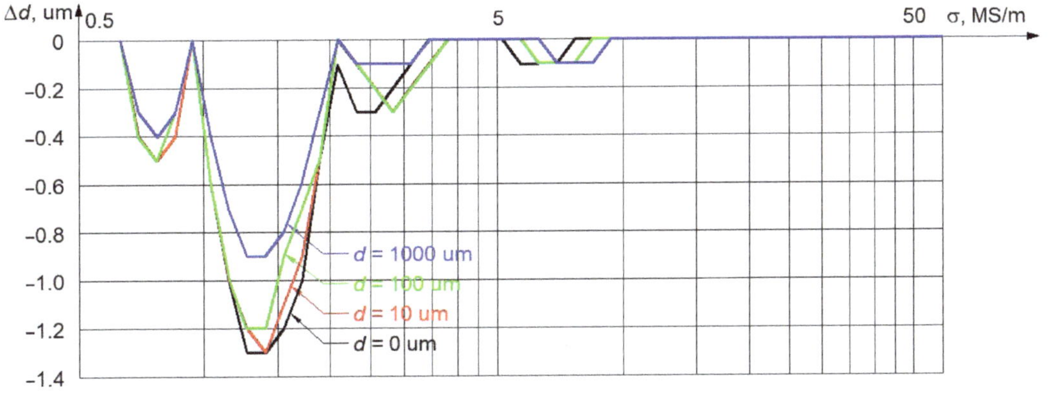

Figure 9. Dependence of the deviation of the measured value *d* on the value σ.

The tests carried out showed the expected component of the measurement uncertainty conditioned by the operation of the algorithm for calculating *d* and the correctness of the selection of the values of the graduation base metals.

The study of the influence of a measuring device, actions of the operator and external factors on the measurement uncertainty are beyond the scope of this article.

4. Discussion

The proposed algorithm, in addition to the tasks of measuring the thickness of a non-conductive coating on a non-magnetic conductive base metal, can be used in the following tasks:

- Measurement of the thickness of a non-conductive coating on a magnetic and non-magnetic conductive base metal and non-metal base;
- Simultaneous measurement of the thickness of layers of a non-conductive coating with a sublayer of a conductive non-magnetic coating on a conductive magnetic base metal, for example, a paint coating applied to a steel base with a zinc underlayer;
- Measurement of the thickness of non-magnetic conductive sheet with suppression of its specific electrical conductivity;
- Measurement of the specific electrical conductivity of the base metal with suppression of the influence of the lift-off or the thickness of the paint coating in a wide range of the base metal specific electrical conductivity values.

Author Contributions: Conceptualization, P.S.; methodology, M.S.; software, M.S.; validation, I.S. and N.A.; formal analysis, N.A.; investigation, P.S.; resources, M.S.; data curation, P.S.; writing—original draft preparation, P.S.; writing—review and editing, N.A.; visualization, P.S.; supervision, I.S.; project administration, I.S.; funding acquisition, M.S. All authors have read and agreed to the published version of the manuscript.

Funding: This research received no external funding.

Institutional Review Board Statement: Not applicable.

Informed Consent Statement: Not applicable.

Data Availability Statement: Data sharing not applicable.

Conflicts of Interest: The authors declare no conflict of interest.

References

1. *ISO 2360:2017*; Non-Conductive Coatings on Non-Magnetic Electrically Conductive Base Metals—Measurement of Coating Thickness—Amplitude-Sensitive Eddy-Current Method. ISO: Geneva, Switzerland, 2017.
2. *GOST R ISO 2360-2021*; Non-Conductive Coatings on Non-Magnetic Electrically Conductive Base Metals. Measurement of Coating Thickness. Amplitude-Sensitive Eddy-Current Method. Russian Technical Standard: Moscow, Russia, 2021.
3. Syasko, V.A. Measuring the thicknesses of nonferromagnetic metal coatings on nonferrous metal products using the eddy-current frequency method. *Russ. J. Nondestruct. Test.* **2010**, *46*, 898–905. [CrossRef]
4. *GOST R 8.673-2009*; State System for Ensuring the Uniformity of Measurements. Intelligent Sensors and Intelligent Measuring Systems. Basic Terms and Definitions. Federal Agency for Technical Regulation and Metrology: Moscow, Russian, 2009.
5. *GOST R 8.734-2011*; State System for Ensuring the Uniformity of Measurements. Intelligent Sensors and Intelligent Measuring Systems. Methods of Metrological Self-Checking. Federal Agency for Technical Regulation and Metrology: Moscow, Russian, 2011.
6. Tian, Y.; Rebinsky, D.A.; Kinney, C.A.; Luick, K. Eddy Current Based Method for Coating Thickness Measurement. U.S. Patent 9,377,287 B2, 28 January 2016.
7. Cheng, W. Thickness Measurement of Metal Plates Using Swept-Frequency Eddy Current Testing and Impedance Normalization. *IEEE Sens. J.* **2017**, *17*, 4558–4569. [CrossRef]
8. Yin, W.; Peyton, A.; Dickinson, S. Simultaneous Measurement of Distance and Thickness of a Thin Metal Plate With an Electromagnetic Sensor Using a Simplified Model. *IEEE Trans. Instrum. Meas.* **2004**, *53*, 1335–1338. [CrossRef]
9. Meng, X.; Lu, M.; Yin, W.; Bennecer, A.; Kirk, K.J. Evaluation of Coating Thickness Using Lift-Off Insensitivity of Eddy Current Sensor. *Sensors* **2021**, *21*, 419. [CrossRef] [PubMed]
10. Syasko, V. Optimization of structure and operation algorithms for electromagnetic plated coatings thickness meters with the use of digital technologies. In Proceedings of the BINDT/45th Annual British Conference on Non-Destructive Testing 2006, Shanghai, China, 25–28 October 2008.
11. Potapov, A.I.; Syasko, V.A. *Non-Destructive Methods and Means of Thickness Control of Coatings and Products/SCIENTIFIC, Methodological and Reference Manual*; Humanistic Studies: Saint Petersburg, Russia, 2009; pp. 203–213.
12. Egorov, A.V.; Polyakov, V.V. Eddy Current Testing of Metallic Materials Using Projection Methods. *Russ. J. Nondestruct. Test.* **2018**, *54*, 377–384. [CrossRef]
13. Nerush, M.N. Eddy current method of measuring the thickness of an electrically conductive plate and the thickness of a non-conductive coating. In Proceedings of the Collection of Scientific Papers of the II International Conference of Schoolchildren, Students, Post-Graduates, Young Scientists "Resource-Efficient Systems in Management and Control: A look into the Future", Tomsk, Russia, 18–19 September 2013; pp. 164–168.
14. Gromyka, D.S.; Gogolinskiy, K.V. Method of State and Residual Resource Assessment of Excavator Bucket Tooth Caps. *Russ. J. Nondestruct. Test.* **2022**, *58*, 381–390. [CrossRef]
15. Goldstein, A.E.; Belyankov, V.Y. An eddy-current gauge for measuring the wall thickness of light-alloy drill pipes. *Russ. J. Nondestruct. Test.* **2017**, *53*, 588–595. [CrossRef]
16. Syasko, V.; Polacek, A. Electromagnetic thickness measurement of coatings. Situation and prospects. In Proceedings of the 11 European Conference on Non-Destructive Testing (ECNDT 2014), Prague, Czech Republic, 6–11 October 2014.
17. Laszlo, M. *Computational Geometry and Computer Graphics in C++ BINOM*; Prentice Hall: Moscow, Russia, 1997; p. 304.
18. Litovka, Y.V. *Obtaining Optimal Design Solutions and Their Analysis Using Mathematical Models*; Tambov State Technical University: Tambov, Russia, 2012; p. 161.

applied sciences

MDPI

Article

Learning from Projection to Reconstruction: A Deep Learning Reconstruction Framework for Sparse-View Phase Contrast Computed Tomography via Dual-Domain Enhancement

Changsheng Zhang [1], Jian Fu [1,2,3,*] and Gang Zhao [1]

[1] School of Mechanical Engineering and Automation, Beihang University, Beijing 100190, China; by1807136@buaa.edu.cn (C.Z.); zhaog@buaa.edu.cn (G.Z.)
[2] Jiangxi Research Institute, Beihang University, Nanchang 330224, China
[3] Ningbo Institute of Technology, Beihang University, Ningbo 315000, China
* Correspondence: fujian706@buaa.edu.cn

Abstract: Phase contrast computed tomography (PCCT) provides an effective non-destructive testing tool for weak absorption objects. Limited by the phase stepping principle and radiation dose requirement, sparse-view sampling is usually performed in PCCT, introducing severe artifacts in reconstruction. In this paper, we report a dual-domain (i.e., the projection sinogram domain and image domain) enhancement framework based on deep learning (DL) for PCCT with sparse-view projections. It consists of two convolutional neural networks (CNN) in dual domains and the phase contrast Radon inversion layer (PCRIL) to connect them. PCRIL can achieve PCCT reconstruction, and it allows the gradients to backpropagate from the image domain to the projection sinogram domain while training. Therefore, parameters of CNNs in dual domains are updated simultaneously. It could overcome the limitations that the enhancement in the image domain causes blurred images and the enhancement in the projection sinogram domain introduces unpredictable artifacts. Considering the grating-based PCCT as an example, the proposed framework is validated and demonstrated with experiments of the simulated datasets and experimental datasets. This work can generate high-quality PCCT images with given incomplete projections and has the potential to push the applications of PCCT techniques in the field of composite imaging and biomedical imaging.

Keywords: phase contrast computed tomography; sparse-view sampling; dual domain; convolutional neural network; radon inversion layer

Citation: Zhang, C.; Fu, J.; Zhao, G. Learning from Projection to Reconstruction: A Deep Learning Reconstruction Framework for Sparse-View Phase Contrast Computed Tomography via Dual-Domain Enhancement. *Appl. Sci.* **2023**, *13*, 6051. https://doi.org/10.3390/app13106051

Academic Editor: Jan Egger

Received: 16 April 2023
Revised: 8 May 2023
Accepted: 12 May 2023
Published: 15 May 2023

1. Introduction

Attenuation and refraction occur when X-rays penetrate objects, which correspond to the absorption and phase contrast. Conventional absorption-based X-ray computed tomography (CT) is widely used in clinical diagnosis [1–4] and industrial testing [5–8]. It plays a crucial role in imaging strong absorption objects, while it performs poorly when encountering weak absorption objects such as soft tissue, rare Earth materials and composite materials.

Phase contrast computed tomography (PCCT) provides better image contrast for weak absorption objects than absorption-based CT [9–14]. Several PCCT techniques have been developed in the past years, and the results have indicated that PCCT can greatly improve image quality for weak absorption objects [15–18]. Grating-based PCCT is the most sensitive and universal approach since a coherent X-ray tube is not required during imaging. It is based on the Talbot effect. However, limited by the phase stepping principle, grating-based PCCT usually requires several samplings at each view to extract the contrast signals, which results in a high radiation dose. Sparse-view sampling is usually performed to reduce imaging radiation [19,20], while it introduces artifacts and noise in reconstruction.

In recent years, deep learning (DL) has been popular in image processing [21–26]. DL has also been applied to the field of CT [27–29], which generally is grouped into two categories. The first category can be classified as enhancement in the projection sinogram domain. By using the residual network (ResNet) for better convergence and patch-wise training to reduce memory, Lee et al. proposed a DL framework to in-paint the missing data in the sparse-view projection sinogram [30]. It significantly outperformed conventional linear interpolation algorithms. Moreover, their subsequent work that utilized UNet [31] and residual learning [32] outperformed the existing interpolation methods and IR approaches [33]. Different from using UNet to correct sparse-view sinograms, Fu et al. proposed a deep learning filtered back-projection (DLFBP) framework to use differential forward projection of the image reconstructed with incomplete data as input and a dense connection net to output a complete sinogram [34]. The results showed that this framework can generate high-quality reconstructed images with given incomplete data. However, these approaches may introduce unpredictable artifacts, since the reconstruction process is extremely susceptible to the inherent consistency of the sinogram.

The second category can be classified as enhancement in the image domain. Chen et al. developed a deep convolutional neural network (CNN) to map low-dose CT reconstructed images to their corresponding normal-dose images in a patch-by-patch fashion [35]. The results demonstrated the great potential of the proposed method for artifact reduction. By using a directional wavelet transform to extract the directional component of artifacts and to exploit the intra- and inter-band correlations, Min et al. proposed a DL method that utilized the wavelet transform coefficients of low-dose images [36]. It could effectively suppress CT-specific noise. Zhang et al. used Dense Net and deconvolution to remove streaking artifacts from sparse-view CT images [37]. The results showed that it can effectively suppress artifacts in reconstructed images. These approaches offer a significant advantage in reducing artifacts and noise in reconstruction, while they may oversmooth the images.

Several methods working in dual domains (i.e., the projection sinogram domain and image domain) have been developed [38,39]. They are grouped into two categories, and each of them has its own limitations: (i) using fully connected layers to connect dual domains, which incurs a huge computational overhead; (ii) training networks in dual domains separately, which superimposes the degradation of dual domains. In addition, most of the studies focus on conventional absorption-based CT, while there is currently a scarcity of studies on applying DL to low-dose PCCT, and the development of related techniques is still in great demand. In this paper, we propose an end-to-end DL framework for PCCT with sparse-view projections. Different from these mentioned methods, the CNNs of dual domains are trained together, allowing network parameters of both CNNs to be updated simultaneously for further removal of artifacts. Therefore, the network in this framework consists of an enhanced network in the projection sinogram domain to restore the projection structure, an enhanced network in the image domain to reduce artifacts in reconstruction and a phase contrast Radon inversion layer (PCRIL) to connect them. PCRIL can achieve PCCT reconstruction, and it allows for backpropagation of the gradients from the image domain to the projection sinogram domain, which enables CNNs in dual domains to be trained simultaneously. In addition, the differential forward projections of the images reconstructed with sparse-view projections are used as input of the network, and the reconstructed images with complete-view projections are used as the targets. Once trained, the network is fixed and can reduce artifacts in the reconstructed images. The experiments with the simulated datasets and experimental datasets are performed to validate the effect of this framework. The results show that the proposed framework can output high-quality reconstructed images with incomplete PCCT projections.

2. Materials and Methods

2.1. Framework Overview

Figure 1 shows the end-to-end DL reconstruction framework for PCCT with sparse-view projections. The network in the framework can update parameters of the CNNs in dual domains synchronously, which is indicated with the green dotted rectangle. This framework is referred to as DDPC-Net. The differential forward projection operator combined with the PCCT filtered back-projection (FBP) algorithm is required to transform the size of the sparse-view sinogram to be the same as the complete sinogram. In addition, a PCCT reconstruction layer allowing the gradients to backpropagate from the image domain to the projection sinogram domain is needed to achieve the mapping between dual domains and to output the reconstructed image while taking the projection sinogram as input. Therefore, the proposed framework has five components: (i) the FBP reconstruction for PCCT, (ii) the differential forward projection, (iii) the enhanced network in the projection sinogram domain, (iv) the PCCT reconstruction layer allowing for the backpropagation of gradients, and (v) the enhanced network in the image domain.

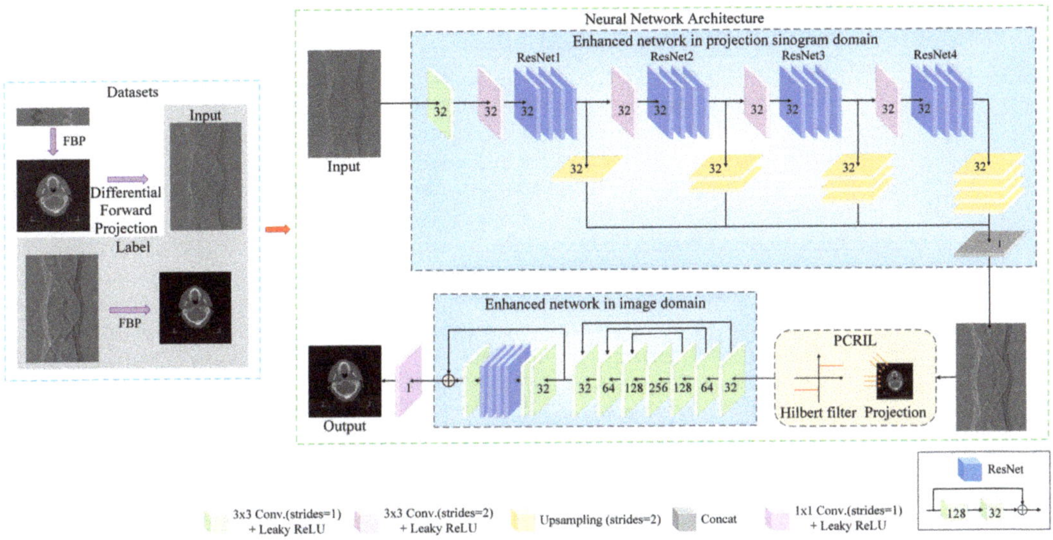

Figure 1. The architecture of the proposed DL framework for PCCT. It consists of the FBP algorithm for PCCT, the differential forward projection operation, and the neural network that allows the network parameters in dual domains to be synchronously updated.

Equations (1) and (2) present the fan-beam FBP algorithm for PCCT, where $\delta(x, y)$ represents the reconstructed image, U represents the geometrical weight factor, $\alpha_\theta(s)$ represents the sinogram, s represents the sinogram index, h represents the Hilbert filter, v represents the frequency, and θ represents the rotation angle.

$$\delta(x,y) = \frac{1}{2}\int_0^{2\pi} U \cdot \alpha_\theta(s) * h(v) d\theta \tag{1}$$

$$h(v) = \frac{1}{2\pi} isgn(v) \tag{2}$$

Equations (3) and (4) present the three-point differential forward projection operator in the proposed framework to generate PCCT sinograms, where $P(s, \theta)$ represents the forward projection, and l represents the forward projection path.

$$\alpha_\theta(s) \approx \frac{P(s+1,\theta) - P(s-1,\theta)}{2} \tag{3}$$

$$P(s,\theta) = \int_l \delta(x,y) dl \tag{4}$$

Equation (5) presents the end-to-end neural network. The information-missing sinogram is transmitted into the network, and then, the corresponding high quality reconstructed image $\overline{rec}$ is output.

$$\overline{rec} = DDPC(\alpha_\theta(s)) \tag{5}$$

2.2. Neural Network Architecture

2.2.1. The Enhanced Network in the Projection Sinogram Domain

As shown in Figure 1, the enhanced network in the projection sinogram domain is indicated with the larger blue solid rectangle, which adopts a multi-scale feature extraction network. PCCT is commonly used in medical diagnosis, where medical images often consist of tissue, organs, and structures of different scales. After projection, these different scales of information are distributed in the projected sinogram. Therefore, the network can effectively capture information at different scales from the projected sinogram, improving the accuracy of image feature extraction. Here, initialization is performed as the first step. Then, four downsamplings of different scales are performed for multi-scale feature extraction. Finally, the multi-scale features are fused using the concatenate block represented by the gray rectangle to output the restored sinogram.

Initialization: Initialization is performed with the convolution filter to convert the corrupted PCCT projection sinogram into its feature image, which is represented with the green cuboid. Increasing the size of the convolution kernel could improve the effect of feature extraction, while it exponentially increases the learning parameters and even causes overfitting. Studies have shown that multi-layer convolutional filters with smaller-sized convolution kernels could enlarge the receptive field and decrease the parameters. Therefore, the convolution filter with a size of 3×3 is used as the feature extractor. The stride is set to 1 to ensure that the sinogram has the same size as its feature. Rectified linear units (ReLU) and batch normalization (BN) techniques are integrated into initialization, so as to overcome the problem of vanishing gradients and to greatly speed up training.

Multi-scale feature extraction: Multi-scale feature extraction is performed by four downsampling branches, where each branch contains a different number of downsampling blocks and the subsequent ResNets. Each downsampling block has a convolution kernel size of 3×3 and a stride of 2, as represented with the pink cuboid in Figure 1. The downsampling convolution intersects the conventional downsampling methods in DL, such as max-pooling or mean-pooling operations, to achieve higher learning accuracy and efficiency.

However, the multi-scale feature extraction may cause degradation problems due to the network depth. ResNet provides an effective solution to the degradation problem of deep neural networks and accelerates convergence. Therefore, ResNets are introduced for the multi-scale feature extraction to enable the convergence and the acceleration of network training. As shown in Figure 1, four ResNets labeled "ResNet1", "ResNet2", "ResNet3" and "ResNet4" are used, and each of them is connected after the previous downsampling blocks.

ResNet consists of four layers of convolutions with the linear rectification function (ReLU) and batch normalization (BN), where each layer has the structure as shown in the lower right corner of Figure 1. It adopts the highway network architecture for introducing an additional identity mapping transmission, which is performed by directly transmitting each layer's input to its subsequent layer's outputs. ResNet keeps the integrity of information to a certain extent, ensuring that the performance of the deep network is at least the same as the performance of the shallow one, not worse. Moreover, it only requires learning the difference between the input and output to speed up the learning process by simplifying its objectives and difficulty.

Feature restoration: Upsampling is required to restore low-resolution features of the downsampling branches, since the previous step yields features with four proportionally decreased sizes. As represented with the yellow cuboid, upsampling is performed with the deconvolution operation referred to as the transpose convolution (ConvTranspose), which is the reverse operation of convolution. In addition, features of different scales match upsampling of different multiples. Finally, the concatenation layer is used to merge features of these ConvTransposes.

2.2.2. Phase Contrast Radon Inversion Layer

The PCCT reconstruction is required since it can achieve mapping from the projection sinogram domain to the image domain. However, conventional reconstruction algorithms do not allow for the backpropagation of gradients, resulting in that only parameters of the enhanced networt in the image domain are updated. The Radon inversion layer (RIL) proposed by Lin [40], acting as an efficient and differentiable variant of FBP, allows for the backpropagation of gradients. It is adopted in the absorption-based CT and obtains excellent performance on reducing metal artifacts. Based on RIL, the PCRIL is derived in this work, which consists of the phase contrast filter, the back-projection derivation and the gradients of the backpropagation. The fan-beam back-projection is required since the grating-based PCCT allows for the use of the laboratory source.

Hilbert Transform Filter: The phase contrast filter is performed with the Hilbert transform in the PCCT reconstruction. It can provide a phase shift of 90° without affecting the amplitude. Therefore, the Hilbert transform is equivalent to the quadrature phase shift of the signal, making them quadrature pairs [41]. As presented in Equations (6) and (7), the sinogram is filtered with the Hilbert transform filter, where H represents the filter, ω represents the frequency, x represents the initial sinogram, X represents the filtered sinogram, F and F^{-1} represent the discrete Fourier transform (DFT) and inverse discrete Fourier transform (iDFT), respectively.

$$H(\omega) = -i \cdot sgn(\omega) = \begin{cases} -i, & \omega \geq 0 \\ i, & \omega < 0 \end{cases} \tag{6}$$

$$X = F^{-1}[-i \cdot sgn(\omega) \cdot F(x)] \tag{7}$$

Back-projection Module: Back-projection is when the value of each pixel in the reconstructed image is regarded as the sum of all projections passing through it. Equations (8) and (9) present the back-projection process, where Y represents the reconstructed image with a size of $row \times col$, θ represents the rotation angle, D and D_0, respectively, represent the distance between the source and detector and that between the source and object, $offset$ represents the offset between the rotation center and the detector center, i represents the sinogram index, and $\lceil\rceil$ and $\lfloor\rfloor$ represent the round up and round down operators. Moreover, the computation can be highly parallel since the back-projection at each view is independent.

$$\begin{aligned} Y &= \int_0^{2\pi} X(\theta, \frac{D_0 * (row \cdot \cos\theta + col \cdot \sin\theta)}{D - row \cdot \sin\theta + col \cdot \cos\theta} + offset) d\theta \\ &\approx \Delta\theta \sum\nolimits_i X(\theta_i, \frac{D_0 * (row \cdot \cos\theta + col \cdot \sin\theta)}{D - row \cdot \sin\theta + col \cdot \cos\theta} + offset) \\ &\approx \Delta\theta \sum\nolimits_i (\lceil t_i \rceil - t_i) X(\theta_i, \lfloor t_i \rfloor) + (t_i - \lfloor t_i \rfloor) X(\theta_i, \lceil t_i \rceil) \end{aligned} \tag{8}$$

$$t_i = \frac{D_0 * (row \cdot \cos\theta + col \cdot \sin\theta)}{D - row \cdot \sin\theta + col \cdot \cos\theta} + offset \tag{9}$$

Backpropagation gradients: While backpropagating, the gradients from the image domain to the projection sinogram domain are presented as Equation (10), where the symbols have the same representation as those of Equation (8).

$$\frac{\partial Y}{\partial X} = \begin{cases} \Delta\theta(\lceil t_i \rceil - t_i), & t = \lfloor t_i \rfloor \\ \Delta\theta(t_i - \lfloor t_i \rfloor), & t = \lceil t_i \rceil \\ 0, & otherwide \end{cases} \quad (10)$$

2.2.3. The Enhanced Network in the Image Domain

The enhanced network in the image domain adopts an improved UNet. UNet is a classic CNN that is particularly suitable for image processing tasks due to its special symmetric downsampling and upsampling structure. In addition, skip connections are used to connect the downsampling module and the symmetric upsampling module, allowing UNet to simultaneously utilize features at different levels. As shown in Figure 1, the enhanced network in the image domain is indicated with the smaller blue solid rectangle. By cascading a ResNet, advanced feature extraction can be performed while reducing the depth of the UNet.

Primary feature extraction: The primary features of the reconstructed images optimized in the sinogram domain are extracted by a UNet. The architecture of the UNet refers to [31].

Advanced feature extraction: Advanced features of the reconstructed images optimized in the sinogram domain are extracted by a series of convolution layers. High-quality reconstructed images are then generated as output. It consists of two convolutional layers with a size of 3×3, a stride of 1 and a filter of 32, four residual blocks, one convolutional layer with a size of 3×3, a stride of 1, and a filter of 32. The output is obtained by adding the result to the primary feature. The enhanced network in the image domain aims to eliminate artifacts while preserving the image structure as much as possible.

3. Experiments

3.1. Data Preparation

3.1.1. Simulation

The simulated datasets are generated by performing the differential fan-beam forward projection operation to images in the head and neck CT image database of The Cancer Imaging Archive (TCIA) [42,43], as shown in Figure 2. TCIA is a large-scale open-access database that contains medical images of common tumors and the corresponding clinical information, such as magnetic resonance imaging (MRI), positron emission computed tomography (PET), and CT. While performing the differential forward projection operation, the sampling step is set to 0.5, 2, 3, 4, and 6 with complete scanning of 360°, corresponding to 720, 180, 120, 90, and 60 views, respectively. Projection sinograms with 720 views are considered as complete and others as sparse-view. The distance between the source and detector and that between the source and object are set to 20,000 and 18,000 pixels. The offset is set to 0.600 CT images from 30 patients, and a size of 368×368 pixels is used to generate the simulated datasets, where 400 CT images are used to train the network and 200 CT images to test the network. Each patient provides 20 CT images.

Specifically, the differential forward projection as expressed in Equations (3) and (4) of the mentioned sampling factors are performed on the phantoms, where complete projection sinograms are used as the labels of the enhanced network in the projection sinogram domain. The PCCT FBP reconstruction as expressed in Equations (1) and (2) is executed on these sinograms to obtain the reconstructed images, where the images reconstructed with complete projection sinograms are used as the labels of the enhanced network in the image domain. Finally, the differential forward projection of 720 views is performed on the degraded images reconstructed with sparse-view projection sinograms, and the results are used as the input of the network. In addition, the projection sinograms used in this network have a size of 720×368 pixels and the CT images 368×368 pixels.

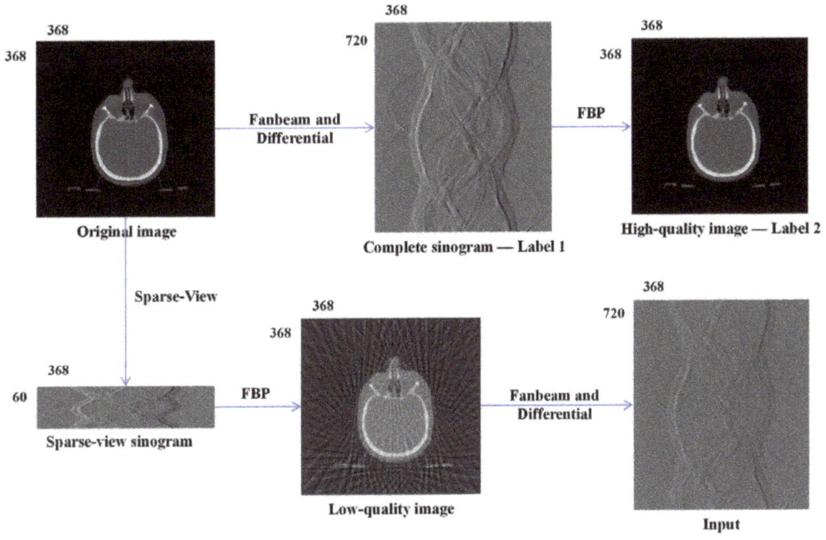

Figure 2. The data preparation process of the simulated datasets.

3.1.2. Experimental

The experimental datasets were generated by performing the fan-beam PCCT experiments on the mouse paw, which was provided by Institute of High Energy Physics, Chinese Academy of Sciences. The used mouse was kept in a pathogen-free environment and was fed ad lib. The procedures for care and use of this mouse were conducted in accordance with the "Guiding Principles in the Care and Use of Animals" [44] and were approved by the Ethics Committee of the Institute of High Energy Physics, Chinese Academy of Sciences. While scanning, 720 views were acquired within 360 degrees using the laboratory fan-beam X-ray source to obtain complete projections. Four phase steppings occurred at each sampling view. Then, sparse sampling was carried out on the complete projections to obtain sparse-view projections with 180, 120, 90, and 60 views. The distances between the source and detector and that between the source and object were 22,400 and 20,200 pixels. The offset was four pixels. The acquired projection images had a size of 512×512 pixels, and the corresponding sinograms and reconstructed images had sizes of 720×512 and 512×512 pixels, respectively. In the experiments, 600 tomographic images were obtained, where the first 400 images from top to bottom were chosen for training and the remaining 200 images for testing.

3.2. *Implementation*

The proposed DDPC-Net was implemented by Python 3.5.2 and Tensorflow 1.8, and the Adam [45] optimizer with a mini-batch size of 2 was applied to train this framework. All the models were trained for 100 epochs on Nvidia GTX 1080Ti graphics processing unit (GPU).

Equations (11)–(13) present the loss function of this framework containing the penalties on the dual domains, where the subscripts 1 and 2 represent the projection sinogram domain and the image domain, and α and $\hat{\alpha}$ represent the learning result and the ground truth. The loss function in each domain is the same, composed of the weighted sum of the mean square error (MSE) and the multi-scale structure similarity (MS-SSIM). MSE helps to reduce the difference in pixel values, and MS-SSIM is closer to subjective quality evaluation methods. The learning rate gradually decreased from 1×10^{-4} to 1×10^{-6} while training.

$$loss = loss_1 + loss_2 \tag{11}$$

$$loss_1 = MSE(\alpha_1, \widehat{\alpha_1}) + 0.2 \cdot (1 - MS\_SSIM(\alpha_1, \widehat{\alpha_1})) \quad (12)$$

$$loss_2 = MSE(\alpha_2, \widehat{\alpha_2}) + 0.2 \cdot (1 - MS\_SSIM(\alpha_2, \widehat{\alpha_2})) \quad (13)$$

3.3. Comparison Methods

Several existing DL-based CT approaches are used as comparisons for DDPC-Net, including the denseness-deconvolution network (DD-Net) [37], the DLFBP framework [34], and the hybrid domain neural network (HD-Net) [38], which respectively represent the enhanced network in the projection sinogram domain, the image domain, and the dual domains.

3.4. Image Evaluation

Image evaluation consisted of qualitative and quantitative evaluation. Qualitative evaluation was achieved by observing the reconstructed images and the regions of interest (ROI). The feature similarity (FSIM) and the information weighted SSIM (IW-SSIM) were used for quantitative evaluation, which outperforms other evaluation methods on accuracy [46].

In addition, the relative improvement ratios ($relI$) for the above two evaluation indexes are defined in Equation (14), where M_{FBP} and M represent the image evaluation indexes of the results from FBP and other methods.

$$relI = \frac{M - M_{FBP}}{M_{FBP}} \quad (14)$$

3.5. Efficiency

The efficiency of the used deep learning methods was evaluated based on the number of parameters included in each framework and the runtime with the same epochs. The number of parameters was calculated by adding one of each layer in the network, as presented in Equation (16), where N_lp represents the number of parameters in each layer, N_i represents the number of input feature images, N_o represents the number of output feature images, and f_h and f_w respectively represent the height and width of the convolutional filter. The runtime was obtained by subtracting the end time and start time.

$$N_{lp} = (N_i \times f_h \times f_w + 1) \times N_o \quad (15)$$

3.6. Results

3.6.1. Simulation

Figure 3 presents the results of the simulated testing datasets with 60 views. The ROI is indicated with the dashed square, which is enlarged and shown for better visualization. The image profiles along the blue line in Figure 3 are drawn and shown in Figure 4.

As expected, severe streak artifacts introduced by sparse-view sampling exist in FBP reconstruction and much less in the results of other methods. However, for DD-Net, the image is blurred, and some image structure still vanishes. For DLFBP, great unpredictable artifacts exist, which affect the visual observation of the image structure. HD-Net and DDPC-Net efficiently suppress artifacts and restore the vanished structure, while the result of HD-NET is a little more blurred compared with DDPC-Net. As presented in Figure 4, the intensity curves in the images from DLFBP and DDPC-Net are noticeably closer to the ground truth, while the intensity curve of DLFBP is relatively more undulating. Table 1 lists the FSIM and IW-SSIM values of the images in Figure 3. DDPC-Net achieves at least 5% higher values in terms of FSIM and IW-SSIM, which support the conclusion of the visual observation.

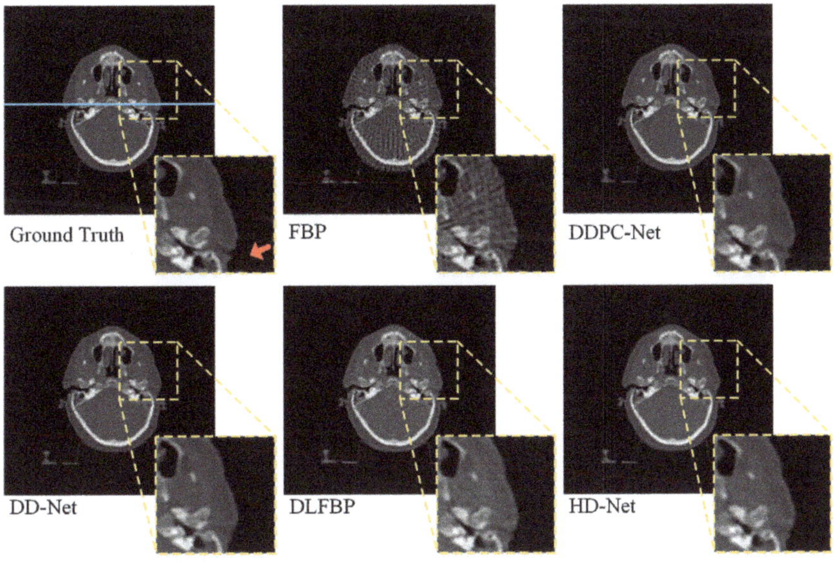

Figure 3. The reconstructed images of one of the results of the simulated testing datasets. This sinogram has 60 sampling views, and the reconstructed images were obtained by five methods.

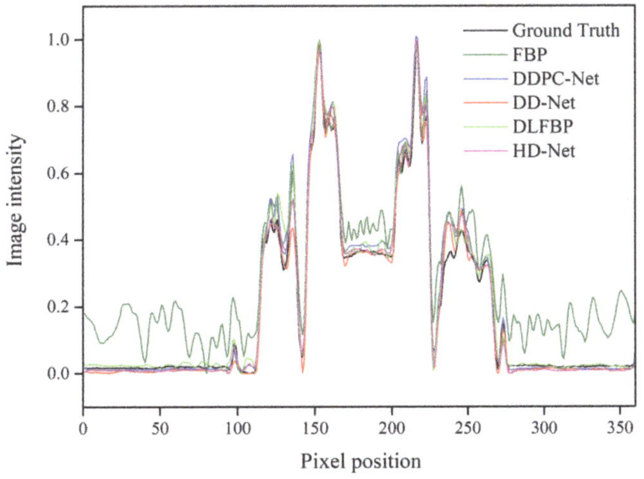

Figure 4. The profiles along the blue solid line in Figure 3.

Table 1. The FSIM and IW-SSIM values of the images in Figure 3.

Evaluation	FBP	DD-Net	DLFBP	HD-Net	DDPC-Net
FSIM	0.5632	0.8922	0.8780	0.9153	0.9652
IW-SSIM	0.5673	0.9278	0.8922	0.9300	0.9793

Table 2 lists the average FSIM and IW-SSIM values of the results of the mentioned five methods. It can be observed that as the number of sampling views increases, the average FSIM and IW-SSIM values increase, and the methods except for FBP obtain values higher than 90%. In addition, DDPC-Net achieves slightly better values than the other methods. The *relI* of the average values of the average FSIM and IW-SSIM values are drawn in Figure 5. The same conclusion can be drawn that DDPC-Net outperforms other methods.

Moreover, Figure 5 shows that the image quality of the results decreases drastically with the decrease in number of the sampling views. Table 3 lists the efficiency of the four deep learning methods. As expected, the efficiency of the dual-domain reconstruction frameworks is slightly lower than that of the single-domain reconstruction framework, both in terms of the number of parameters and runtime. However, compared to HD-Net, which trains enhancement networks in the projection sinogram domain and image domain separately and cascades them, DDPC-Net is more efficient. This indicates that the proposed method can balance image quality and efficiency.

Table 2. The average values of FSIM and IW-SSIM for the results of the simulated testing datasets.

Evaluation	Methods	Cases			
		60	90	120	180
FSIM	FBP	0.5851	0.6387	0.6671	0.7495
	DD-Net	0.9168	0.9234	0.9333	0.9475
	DLFBP	0.9037	0.9139	0.9221	0.9432
	HD-Net	0.9234	0.9357	0.9552	0.9604
	DDPC-Net	0.9722	0.9772	0.9874	0.9884
IW-SSIM	FBP	0.5879	0.6831	0.7961	0.9074
	DD-Net	0.9174	0.9254	0.9439	0.9589
	DLFBP	0.8955	0.9109	0.9360	0.9528
	HD-Net	0.9233	0.9318	0.9470	0.9616
	DDPC-Net	0.9795	0.9876	0.9957	0.9978

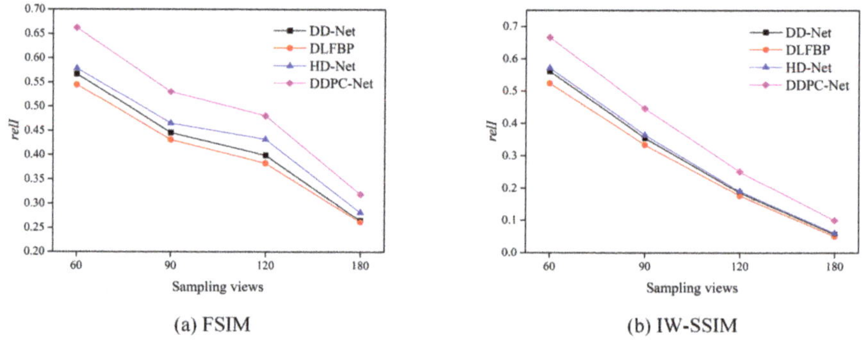

Figure 5. The *rell* curves of the average values of FSIM and IW-SSIM of the simulated testing datasets.

Table 3. The efficiency of the used methods with simulated datasets.

Efficiency	DD-Net	DLFBP	HD-Net	DDPC-Net
Parameters (million)	1.06	1.36	4.14	3.28
Runtime (s)	0.17	0.21	0.47	0.41

3.6.2. Experimental

Figure 6 shows the results of the experimental testing datasets with 60 views. The ROI is indicated with the dashed square, which is enlarged and shown to obtain better visualization. The analysis of the experimental datasets was performed as the same as that of the simulation datasets. The corresponding curves and index values are presented in Figures 7 and 8 and Tables 4–6. The same conclusion can be drawn as that of the simulation datasets. The images of DD-Net and HD-Net are blurred and lose some structure. There are severe artifacts existing in the images of DLFBP. Furthermore, DDPC-Net outperforms the comparison methods. In addition, the FSIM and IW-SSIM values of the experimental datasets are significantly worse than those of the simulation datasets, since noise introduced during the experiment degrades the experimental datasets.

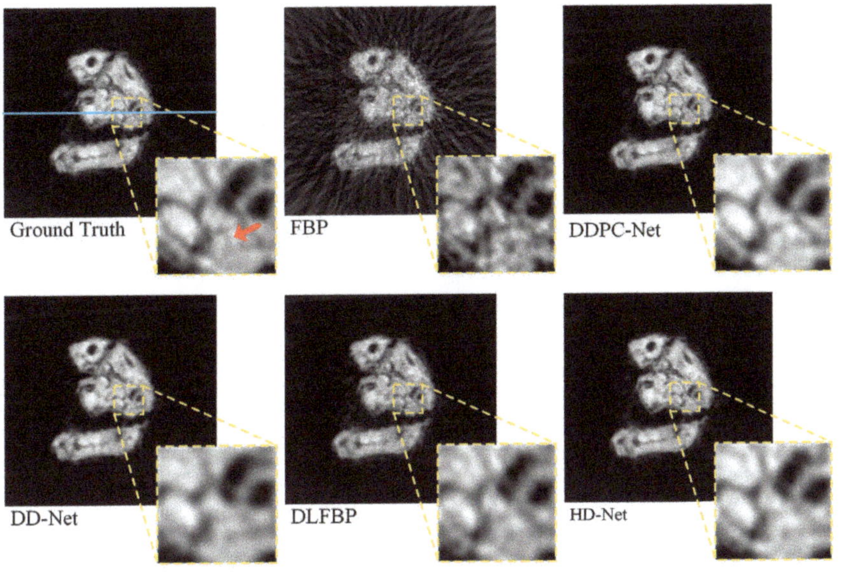

Figure 6. The reconstructed images of the results of the experimental testing datasets. This sinogram has 60 sampling views, and the reconstructed images were obtained by five methods.

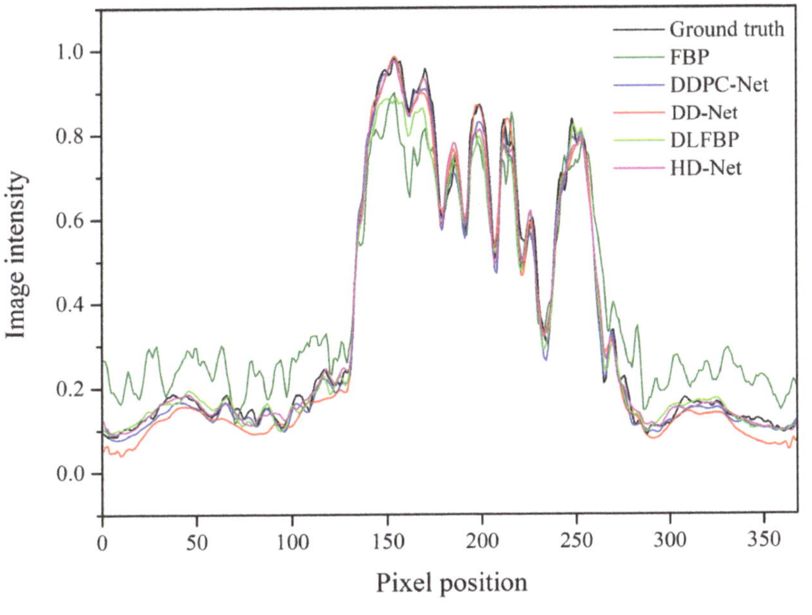

Figure 7. The profiles along the blue solid line in Figure 5.

Table 4. The FSIM and IW-SSIM values of the images in Figure 6.

Evaluation	FBP	DD-Net	DLFBP	HD-Net	DDPC-Net
FSIM	0.7272	0.8957	0.8751	0.9026	0.9584
IW-SSIM	0.5957	0.8844	0.8750	0.9168	0.9690

Table 5. The average values of FSIM and IW-SSIM of the results of the experimental testing datasets.

Evaluation	Methods	Cases			
		60	90	120	180
FSIM	FBP	0.7279	0.7756	0.8172	0.8858
	DD-Net	0.8922	0.8979	0.9006	0.9265
	DLFBP	0.8707	0.8841	0.8901	0.9138
	HD-Net	0.8959	0.9063	0.9145	0.9350
	DDPC-Net	0.9453	0.9564	0.9642	0.9791
IW-SSIM	FBP	0.5812	0.7052	0.7994	0.8753
	DD-Net	0.8758	0.8820	0.8988	0.9299
	DLFBP	0.8626	0.8784	0.8917	0.9287
	HD-Net	0.9047	0.9196	0.9294	0.9482
	DDPC-Net	0.9574	0.9697	0.9779	0.9915

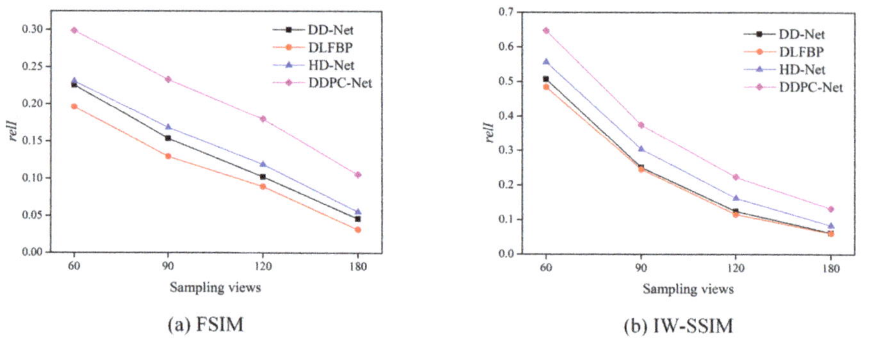

Figure 8. The *rell* curves of the average values of FSIM and IW-SSIM of the experimental testing datasets.

Table 6. The efficiency of the used methods with experimental datasets.

Efficiency	DD-Net	DLFBP	HD-Net	DDPC-Net
Parameters (million)	1.46	2.04	5.91	4.48
Runtime (s)	0.22	0.29	0.61	0.52

4. Discussion

After the network architecture is determined, the loss function has a great effect on the results. In this work, the weighted sums of MSE and MS-SSIM are adopted as the loss function, as shown in Equation (16), where ω_1 and ω_2 represent the weight of MSE and MS-SSIM. ω_1 of 1 and ω_2 of 0.2 are adopted in the experiments. To discuss the influence of the weight values on the image quality and to validate that the best weight values are adopted, the experiments are repeated with different ω_1 and ω_2. Considering the experimental datasets as examples, the network is trained with several commonly used loss functions (i.e., Loss1, Loss2, Loss3, Loss4, Loss5, and Loss6), as presented in Table 7.

$$loss = \omega_1 MSE(\alpha, \hat{\alpha}) + \omega_2(1 - MS\_SSIM(\alpha, \hat{\alpha})) \quad (16)$$

Table 7. The weight values of loss functions used in this work.

Weight	Loss1	Loss2	Loss3	Loss4	Loss5	Loss6
ω_1	1	1	1	1	1	0
ω_2	0	0.1	0.2	0.5	1	1

Figure 9 shows one of the results of the experimental testing datasets with 60 views, and the ROI is indicated with a dashed square, which is enlarged and shown for better visualization. It can be observed that Loss2 and Loss3 help to obtain high-quality results, and the result with Loss3 has a relatively clearer structure. Table 8 lists the FSIM and IW-SSIM values of the images in Figure 9. These values provide evidence that the network trained with Loss3 outperforms those trained with other loss functions mentioned.

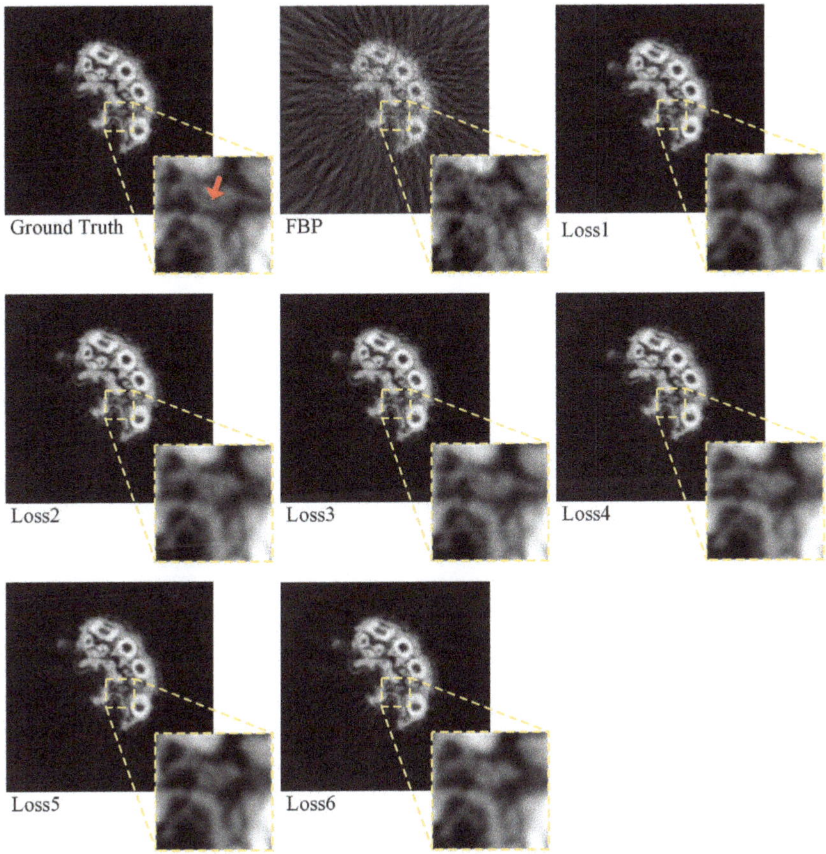

Figure 9. The reconstructed images of one of the results of the experimental testing datasets with different loss functions. This sinogram has 60 sampling views, and the reconstructed images are obtained by DDPC-Net with loss functions as presented in Table 7.

Table 8. The weight values of loss functions used in this work.

Evaluation	Loss1	Loss2	Loss3	Loss4	Loss5	Loss6
FSIM	0.9301	0.9348	0.9509	0.9340	0.9339	0.9270
IW-SSIM	0.9531	0.9573	0.9676	0.9558	0.9535	0.9458

Table 9 lists the average FSIM and MS-SSIM values of the results with different loss functions. Figure 10 presents the *relI* of the average values of FSIM and IW-SIIM. Furthermore, regarding the number of the sampling view, DDPC-Net with Loss3 enables the best performance in imaging. It also indicates that the CNNs with a combination of several losses may outperform that with a single loss of the applications in the field of CT.

Table 9. The average values of FSIM and IW-SSIM of the results of the experimental testing datasets.

Evaluation	Methods	Cases 60	90	120	180
FSIM	Loss1	0.9262	0.9359	0.9448	0.9662
	Loss2	0.9300	0.9374	0.9520	0.9666
	Loss3	0.9403	0.9514	0.9582	0.9741
	Loss4	0.9293	0.9430	0.9522	0.9684
	Loss5	0.9275	0.9427	0.9491	0.9642
	Loss6	0.9207	0.9369	0.9462	0.9554
IW-SSIM	Loss1	0.9353	0.9492	0.9606	0.9789
	Loss2	0.9402	0.9501	0.9662	0.9797
	Loss3	0.9524	0.9647	0.9729	0.9866
	Loss4	0.9400	0.9560	0.9667	0.9806
	Loss5	0.9382	0.9561	0.9632	0.9796
	Loss6	0.9295	0.9502	0.9615	0.9782

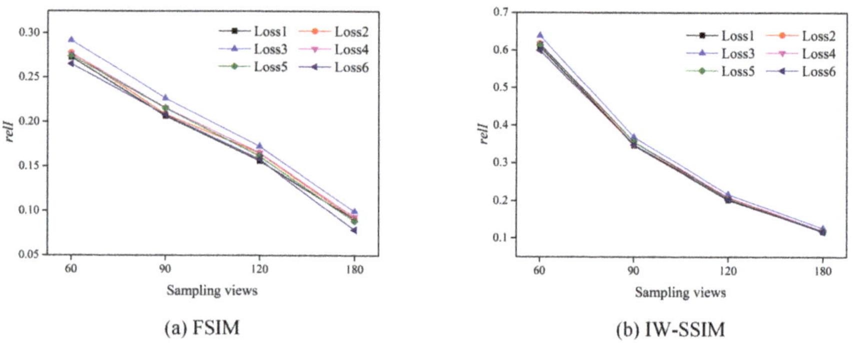

Figure 10. The *rell* curves of the average values of FSIM and IW-SSIM of the experimental testing datasets with loss functions as presented in Table 7.

5. Conclusions

In this paper, we reported a DL reconstruction framework for PCCT with sparse-view projections and validated it with experiments of the simulation datasets and experimental datasets. The proposed framework consists of CNNs in dual domains and PCRIL as the connection between them. PCRIL can achieve PCCT reconstruction, and it allows for the backpropagation of gradients from the image domain to the projection sinogram domain. Therefore, this framework enables the CNNs in dual domains to be trained simultaneously for further reduction of artifacts and to restore the missing structure introduced by sparse-view sampling. In addition, the differential forward projection of the image reconstructed with the sparse-view projection sinogram is adopted as the input of the network, instead of the interpolation of the sparse-view projection sinogram. It efficiently improves the image quality of the images reconstructed with given sparse-view PCCT projections. This work has the potential to push PCCT techniques to applications in the field of composite imaging and biomedical imaging.

Author Contributions: Conceptualization, C.Z. and J.F.; methodology, C.Z. and J.F.; software, C.Z.; validation, C.Z. and J.F.; formal analysis, C.Z.; investigation, C.Z.; resources, J.F. and G.Z.; data curation, C.Z.; writing—original draft preparation, C.Z.; writing—review and editing, J.F. and G.Z.; visualization, C.Z.; supervision, J.F. and G.Z.; project administration, J.F. and G.Z.; funding acquisition, J.F. All authors have read and agreed to the published version of the manuscript.

Funding: This research was funded by the Ningbo Major Projects of Science and Technology Innovation 2025 (2020Z074), the National Natural Science Foundation of China (51975026), the Joint Fund of Research Utilizing Large-scale Scientific Facilities by the National Natural Science Foundation of China and Chinese Academy of Science (U1932111), the Innovation Leading Talent Short Term Projects of Natural Science by Jiangxi Double Thousand Plan (S2020DQKJ0355), the Jiangxi Provincial Science and Technology Innovation Base Plan—Introduction of workpiece research and development institutions (20203CCH45003), and the Jiangxi Provincial Science and Technology Innovation Base Plan—Introduction of workpiece research and development institutions (20212CCH45001).

Institutional Review Board Statement: The procedures for care and use of the used mouse were conducted in accordance with the "Guiding Principles in the Care and Use of Animals" and were approved by the Ethics Committee of the Institute of High Energy Physics, Chinese Academy of Sciences.

Informed Consent Statement: Not applicable.

Data Availability Statement: Data was obtained from the Institute of High Energy Physics, Chinese Academy of Sciences and are available from J.F. with the permission of the Institute of High Energy Physics, Chinese Academy of Sciences.

Acknowledgments: The authors are grateful to Peiping Zhu (Institute of High Energy Physics, Chinese Academy of Sciences) for providing us the experimental datasets.

Conflicts of Interest: The authors declare no conflict of interest.

References

1. Venkatesh, E.; Elluru, S.V. Cone beam computed tomography: Basics and applications in dentistry. *J. Istanb. Univ. Fac. Dent.* **2017**, *51*, S102. [CrossRef]
2. Ardila, D.; Kiraly, A.P.; Bharadwaj, S.; Choi, B.; Reicher, J.J.; Peng, L.; Tse, D.; Etemadi, M.; Ye, W.; Corrado, G.; et al. End-to-end lung cancer screening with three-dimensional deep learning on low-dose chest computed tomography. *Nat. Med.* **2019**, *25*, 954–961. [CrossRef]
3. Zhu, T.; Wang, Y.; Zhou, S.; Zhang, N.; Xia, L. A comparative study of chest computed tomography features in young and older adults with corona virus disease (COVID-19). *J. Thorac. Imaging* **2020**, *35*, W97. [CrossRef] [PubMed]
4. Li, W.; Cui, H.; Li, K.; Fang, Y.; Li, S. Chest computed tomography in children with COVID-19 respiratory infection. *Pediatr. Radiol.* **2020**, *50*, 796–799. [CrossRef] [PubMed]
5. Du Plessis, A.; Rossouw, P. X-ray computed tomography of a titanium aerospace investment casting. *Case Stud. Nondestruct. Test. Eval.* **2015**, *3*, 21–26. [CrossRef]
6. Thompson, A.; Maskery, I.; Leach, R.K. X-ray computed tomography for additive manufacturing: A review. *Meas. Sci. Technol.* **2016**, *27*, 072001. [CrossRef]
7. Asadizanjani, N.; Tehranipoor, M.; Forte, D. PCB reverse engineering using nondestructive X-ray tomography and advanced image processing. *IEEE Trans. Components Packag. Manuf. Technol.* **2017**, *7*, 292–299. [CrossRef]
8. Townsend, A.; Pagani, L.; Scott, P.; Blunt, L. Areal surface texture data extraction from X-ray computed tomography reconstructions of metal additively manufactured parts. *Precis. Eng.* **2017**, *48*, 254–264. [CrossRef]
9. Bonse, U.; Hart, M. An X-ray interferometer with Bragg case beam splitting and beam recombination. *Z. Phys.* **1966**, *194*, 1–17. [CrossRef]
10. Ingal, V.; Beliaevskaya, E. X-ray plane-wave topography observation of the phase contrast from a non-crystalline object. *J. Phys. D Appl. Phys.* **1995**, *28*, 2314. [CrossRef]
11. Wilkins, S.; Gureyev, T.E.; Gao, D.; Pogany, A.; Stevenson, A. Phase-contrast imaging using polychromatic hard X-rays. *Nature* **1996**, *384*, 335–338. [CrossRef]
12. Nugent, K.; Gureyev, T.; Cookson, D.; Paganin, D.; Barnea, Z. Quantitative phase imaging using hard X-rays. *Phys. Rev. Lett.* **1996**, *77*, 2961. [CrossRef] [PubMed]
13. Weitkamp, T.; Diaz, A.; David, C.; Pfeiffer, F.; Stampanoni, M.; Cloetens, P.; Ziegler, E. X-ray phase imaging with a grating interferometer. *Opt. Express* **2005**, *13*, 6296–6304. [CrossRef]
14. Pfeiffer, F.; Weitkamp, T.; Bunk, O.; David, C. Phase retrieval and differential phase-contrast imaging with low-brilliance X-ray sources. *Nat. Phys.* **2006**, *2*, 258–261. [CrossRef]
15. Cloetens, P.; Ludwig, W.; Baruchel, J.; Van Dyck, D.; Van Landuyt, J.; Guigay, J.; Schlenker, M. Holotomography: Quantitative phase tomography with micrometer resolution using hard synchrotron radiation X-rays. *Appl. Phys. Lett.* **1999**, *75*, 2912–2914. [CrossRef]
16. Bech, M.; Jensen, T.H.; Feidenhans, R.; Bunk, O.; David, C.; Pfeiffer, F. Soft-tissue phase-contrast tomography with an X-ray tube source. *Phys. Med. Biol.* **2009**, *54*, 2747. [CrossRef]
17. Donath, T.; Pfeiffer, F.; Bunk, O.; Grünzweig, C.; Hempel, E.; Popescu, S.; Vock, P.; David, C. Toward clinical X-ray phase-contrast CT: Demonstration of enhanced soft-tissue contrast in human specimen. *Investig. Radiol.* **2010**, *45*, 445–452. [CrossRef] [PubMed]
18. Momose, A.; Takeda, T.; Itai, Y.; Hirano, K. Phase–contrast X–ray computed tomography for observing biological soft tissues. *Nat. Med.* **1996**, *2*, 473–475. [CrossRef] [PubMed]

19. Zhu, P.; Zhang, K.; Wang, Z.; Liu, Y.; Liu, X.; Wu, Z.; McDonald, S.A.; Marone, F.; Stampanoni, M. Low-dose, simple, and fast grating-based X-ray phase-contrast imaging. *Proc. Natl. Acad. Sci. USA* **2010**, *107*, 13576–13581. [CrossRef] [PubMed]
20. Ge, Y.; Li, K.; Garrett, J.; Chen, G.H. Grating based X-ray differential phase contrast imaging without mechanical phase stepping. *Opt. Express* **2014**, *22*, 14246–14252. [CrossRef] [PubMed]
21. Quan, Y.; Chen, Y.; Shao, Y.; Teng, H.; Xu, Y.; Ji, H. Image denoising using complex-valued deep CNN. *Pattern Recognit.* **2021**, *111*, 107639. [CrossRef]
22. Zhu, H.; Xie, C.; Fei, Y.; Tao, H. Attention mechanisms in CNN-based single image super-resolution: A brief review and a new perspective. *Electronics* **2021**, *10*, 1187. [CrossRef]
23. Nguyen, Q.H.; Nguyen, B.P.; Nguyen, T.B.; Do, T.T.; Mbinta, J.F.; Simpson, C.R. Stacking segment-based CNN with SVM for recognition of atrial fibrillation from single-lead ECG recordings. *Biomed. Signal Process. Control.* **2021**, *68*, 102672. [CrossRef]
24. Zhou, W.; Liu, M.; Xu, Z. The dual-fuzzy convolutional neural network to deal with handwritten image recognition. *IEEE Trans. Fuzzy Syst.* **2022**, *30*, 5225–5236. [CrossRef]
25. Lu, J.; Tan, L.; Jiang, H. Review on convolutional neural network (CNN) applied to plant leaf disease classification. *Agriculture* **2021**, *11*, 707. [CrossRef]
26. Yu, J.; Fan, Y.; Yang, J.; Xu, N.; Wang, Z.; Wang, X.; Huang, T. Wide activation for efficient and accurate image super-resolution. *arXiv* **2018**, arXiv:1808.08718.
27. Wang, J.; Liang, J.; Cheng, J.; Guo, Y.; Zeng, L. Deep learning based image reconstruction algorithm for limited-angle translational computed tomography. *PLoS ONE* **2020**, *15*, e0226963.
28. Han, Y.; Wu, D.; Kim, K.; Li, Q. End-to-end deep learning for interior tomography with low-dose X-ray CT. *Phys. Med. Biol.* **2022**, *67*, 115001. [CrossRef]
29. Liu, Y.; Kang, J.; Li, Z.; Zhang, Q.; Gui, Z. Low-dose CT noise reduction based on local total variation and improved wavelet residual CNN. *J. X-ray Sci. Technol.* **2022**, *30*, 1229–1242. [CrossRef] [PubMed]
30. Lee, H.; Lee, J.; Cho, S. View-interpolation of sparsely sampled sinogram using convolutional neural network. In *Proceedings of the Medical Imaging 2017: Image Processing*; International Society for Optics and Photonics: San Diego, CA, USA, 2017; Volume 10133, p. 1013328.
31. Ronneberger, O.; Fischer, P.; Brox, T. U-net: Convolutional networks for biomedical image segmentation. In Proceedings of the International Conference on Medical Image Computing and Computer-Assisted Intervention, Munich, Germany, 5–9 October 2015; pp. 234–241.
32. He, K.; Zhang, X.; Ren, S.; Sun, J. Deep residual learning for image recognition. In Proceedings of the IEEE Conference on Computer Vision and Pattern Recognition, Las Vegas, NV, USA, 27–30 June 2016; pp. 770–778.
33. Lee, H.; Lee, J.; Kim, H.; Cho, B.; Cho, S. Deep-neural-network-based sinogram synthesis for sparse-view CT image reconstruction. *IEEE Trans. Radiat. Plasma Med. Sci.* **2018**, *3*, 109–119. [CrossRef]
34. Fu, J.; Dong, J.; Zhao, F. A deep learning reconstruction framework for differential phase-contrast computed tomography with incomplete data. *IEEE Trans. Image Process.* **2019**, *29*, 2190–2202. [CrossRef] [PubMed]
35. Chen, H.; Zhang, Y.; Kalra, M.K.; Lin, F.; Chen, Y.; Liao, P.; Zhou, J.; Wang, G. Low-dose CT with a residual encoder-decoder convolutional neural network. *IEEE Trans. Med. Imaging* **2017**, *36*, 2524–2535. [CrossRef] [PubMed]
36. Kang, E.; Min, J.; Ye, J.C. A deep convolutional neural network using directional wavelets for low-dose X-ray CT reconstruction. *Med. Phys.* **2017**, *44*, e360–e375. [CrossRef] [PubMed]
37. Zhang, Z.; Liang, X.; Dong, X.; Xie, Y.; Cao, G. A sparse-view CT reconstruction method based on combination of DenseNet and deconvolution. *IEEE Trans. Med. Imaging* **2018**, *37*, 1407–1417. [CrossRef] [PubMed]
38. Hu, D.; Liu, J.; Lv, T.; Zhao, Q.; Zhang, Y.; Quan, G.; Feng, J.; Chen, Y.; Luo, L. Hybrid-Domain Neural Network Processing for Sparse-View CT Reconstruction. *IEEE Trans. Radiat. Plasma Med. Sci.* **2020**, *5*, 88–98. [CrossRef]
39. Lee, D.; Choi, S.; Kim, H.J. High quality imaging from sparsely sampled computed tomography data with deep learning and wavelet transform in various domains. *Med. Phys.* **2019**, *46*, 104–115. [CrossRef]
40. Lin, W.A.; Liao, H.; Peng, C.; Sun, X.; Zhang, J.; Luo, J.; Chellappa, R.; Zhou, S.K. Dudonet: Dual domain network for ct metal artifact reduction. In Proceedings of the IEEE/CVF Conference on Computer Vision and Pattern Recognition, Long Beach, CA, USA, 15–20 June 2019; pp. 10512–10521.
41. Pfeiffer, F.; Kottler, C.; Bunk, O.; David, C. Hard X-ray phase tomography with low-brilliance sources. *Phys. Rev. Lett.* **2007**, *98*, 108105. [CrossRef] [PubMed]
42. Lai, H.; Chen, W.; Fu, H. A new double-sampling method for mediastinal lymph nodes detection by deep conventional neural network. In Proceedings of the 2018 Chinese Control And Decision Conference (CCDC), Shenyang, China, 9–11 June 2018; pp. 6286–6290.
43. Seff, A.; Lu, L.; Cherry, K.M.; Roth, H.R.; Liu, J.; Wang, S.; Hoffman, J.; Turkbey, E.B.; Summers, R.M. 2D view aggregation for lymph node detection using a shallow hierarchy of linear classifiers. In Proceedings of the International Conference on Medical image Computing and Computer-Assisted Intervention, Boston, MA, USA, 14–18 September 2014; pp. 544–552.
44. Guidelines for the Ethical Review of Laboratory Animal Welfare (GB/T 35892-2018). Standardization Administration of China Beijing ICP 09001239. Available online: http://www.gb688.cn/bzgk/gb/newGbInfo?hcno=9BA619057D5C13103622A10FF4BA5D14 (accessed on 20 March 2022).

45. Aerts, H.J.; Velazquez, E.R.; Leijenaar, R.T.; Parmar, C.; Grossmann, P.; Carvalho, S.; Bussink, J.; Monshouwer, R.; Haibe-Kains, B.; Rietveld, D.; et al. Decoding tumour phenotype by noninvasive imaging using a quantitative radiomics approach. *Nat. Commun.* **2014**, *5*, 4006. [CrossRef] [PubMed]
46. Zhang, L.; Zhang, L.; Mou, X.; Zhang, D. A comprehensive evaluation of full reference image quality assessment algorithms. In Proceedings of the 2012 19th IEEE International Conference on Image Processing, Orlando, FL, USA, 30 September–3 October 2012; pp. 1477–1480.

applied sciences

Article

Design and Experimental Research of Intelligent Inspection and Classification System for Yuba Skin Quality

Yanhu Tao [1], Yinjie Shen [1], Liangyuan Xu [1,2,*], Qiansheng Tang [1,2] and Haibo Yang [1]

[1] College of Engineering, Anhui Agricultural University, Hefei 230036, China; tyh17344059942@163.com (Y.T.); 17755066969@163.com (Y.S.); qiansh_tang@ahau.edu.cn (Q.T.)
[2] Anhui Intelligent Agricultural Machinery Equipment Engineering Laboratory, Hefei 230036, China
* Correspondence: xlyjwh@ahau.edu.cn; Tel.: +86-138-5691-9745

Abstract: At present, the surface quality of Yuba skin is determined by sensory methods. In order to realize the intelligent classification detection of Yuba skin quality, this study designed a system that automatically determines the quality of Yuba skin surfaces based on image processing and support vector machine (SVM) approaches. Specifically, the system uses image preprocessing to extract the grayscale eigenvalues, gray level co-occurrence matrix (GLCM) eigenvalues, and gray level run length matrix (GLRLM) eigenvalues of the sample image and uses them as input values for a quality grading system. Through model evaluation of three classification models, the SVM classification model was selected according to the evaluation results, and different kernel functions were used in the model for sample training. Based on Matlab, the quality grading software of Yuba skin was developed and designed. Intelligent detection and grading were realized through the radial basis kernel function support vector machine (RBF-SVM) grading model. The best penalty factor (c = 3.50) and kernel parameter value (g = 0.98) were obtained through cross-validation. The accuracy of the model was 95.31% and 94.16% for the training and test sets, respectively. The grading accuracy of the RBF-SVM grading system was 93.56%, and the error was less than 5% compared with the traditional sensory method of grading; thus, the quality classification method based on the SVM classification system for Yuba skin is feasible and can be used for quality detection.

Keywords: Yuba; multiple eigenvalues; RBF-SVM hierarchical model; quality detection

Citation: Tao, Y.; Shen, Y.; Xu, L.; Tang, Q.; Yang, H. Design and Experimental Research of Intelligent Inspection and Classification System for Yuba Skin Quality. *Appl. Sci.* **2023**, *13*, 7070. https://doi.org/10.3390/app13127070

Academic Editors: Shu Taira, Grzegorz Peruń, Tangbin Xia and Bogusław Łazarz

Received: 28 March 2023
Revised: 12 May 2023
Accepted: 9 June 2023
Published: 13 June 2023

1. Introduction

Yuba skin, a common nonfermented soybean product, is a film formed on the surface of boiled soybean milk maintained at a constant temperature [1,2]. Yuba skin has a fine texture and contains protein, oil, carbohydrates, zinc, magnesium, iron, calcium, and other nutrients; it is highly sought after because it is nutritious, tasty, and easy to digest (digestibility rate: 93%). Many studies have evaluated the correlations between soybean variety, slurry concentration during processing, heating temperature and method, and the perceptible quality of Yuba skin [3]. Nonetheless, few studies have formulated automated detection methods for judging Yuba skin quality [4]. Determinations of Yuba skin quality are based on the color of undried Yuba skin. Thus, machine vision can be used to obtain assessments of Yuba skin quality based on its color and shape [5]. The principal component analysis for Yuka skin involves several factors and is conducted according to the "DB 52/519-2007 Yuba Quality Standard". The results indicate that the surface color of Yuba skin reflects its freshness and contributes the most to its surface quality.

In recent years, researchers have analyzed the acoustic properties [6,7], optical properties [8,9], and acoustic vibration [10,11] for nondestructive testing. With the development of machine vision technology [12–15], extensive research has been con-ducted on sorting operations for numerous agricultural products. Su et al. analyzed the appearance and size characteristics of potatoes using a convolutional neural network. Non-destructive grading of potatoes was achieved [16]. Tian et al. analyzed the image features of apples using the YOLOv3 model. Non-destructive grading of apples was achieved [17]. Liu et al. analyzed the color characteristics of tomatoes with different ripeness using a color HSV model. Nondestructive grading of tomatoes was achieved [18]. Kumar et al. analyzed the color, shape, and texture features of tomatoes using an SVM model. Non-destructive grading of tomatoes was achieved [19]. Olaniyi et al. analyzed the GLCM texture features of bananas using a neural network model. Non-destructive grading of bananas was achieved [20]. Huang et al. analyzed the image features of spinach samples using a BPNN model with multisensory data fusion. Non-destructive grading of spinach was achieved [21]. At present, the quality inspection of Yuba skin relies on human experience to determine the color and texture characteristics of the Yuba skin surface through human eyes, and then grade the Yuba skin, which is not accurate and cannot achieve intelligent grading of the Yuba skin quality. Thus, machine vision holds promise as an objective and nondestructive means for improving Yuba skin testing and accuracy, thereby adding value.

This study formulated an image processing method for automated Yuba skin quality assessment. The Yuba image is captured at the terminal of the conjunctival trough, and the image is preprocessed [22]. The average gray value is used to extract the eigenvalues of the degree of darkness on the surface of the Yuba skin, and the grayscale run length matrix (GLRLM) and the gray-level co-occurrence matrix (GLCM) are used to extract the surface texture eigenvalues of the conjunctival terminal Yuba skin. Finally, the nine extracted eigenvalue factors are used as the support vector machine (SVM) input values and Yuba skin quality is output as a classification result [23,24]. Thus, the SVM quality classification training model was established to detect the quality of the Yuba skin.

2. Materials and Methods

2.1. Sensory Classification Experiment Design

The test material was made from Heilongjiang Hairui No. 2 soybean. Through the analysis of the bean curd quality questionnaire produced by different soybeans, it can be seen that the quality of the bean curd skin produced from Heilongjiang Hairui No. 2 soybean meets the requirements of this test. The surface quality of Yuba skin has an important influence on human appetite. According to the "DB 52/519-2007 Yuba Quality Standard", a principal component analysis was conducted on several influencing factors, and the results showed that the surface color of Yuba skin had the highest contribution to the surface quality of Yuba skin. The color of Yuba skin is a general term for color and luster, which is a mixture of the color of light (light not absorbed and transmitted by the food) reflected by the Yuba skin when it is illuminated by light. The color and luster of the Yuba skin can reflect the freshness of the Yuba skin. For soybean products, the formulation and the conjugation process of Yuba skin can affect the color of the surface. A suitable processing process can make the Yuba skin neatly shaped, light yellow in color, uniform and consistent in color, and fine in texture, without short strips and broken pieces. The traditional method of judging the shade of the surface of the Yuba skin is to rely on the human eyes to qualitatively discriminate, such as dark color, light color, etc., but the qualitative analysis of similar colors is generally difficult to distinguish or the results of the judgment will vary from person to person. Based on the national standard for the quality of Yuba skin in this paper, a color sensory scale was developed to provide a judgment basis for the subsequent sensory method, and color scoring was performed by six reviewers on five groups of test samples through the sensory method.

2.1.1. Design of Scoring Criteria

The scores for each grade are shown in Table 1. The region of interest (ROI) selection of the sample images was performed before the quality grading, and the sensory scoring table of the Yuba skin was developed based on the criteria.

Table 1. Criteria for sensory grading of Yuba skin.

Sensory Level	Color	Color Uniformity	Surface Texture	ROI Region Extraction
9–10 (Master)	Light yellow	Uniform and consistent	Fine texture and no impurities	
7–9 (Medium)	Light yellow	Basically uniform	Short strips of broken pieces without obvious impurities	
0–7 (Inferior)	Dark yellow or deep red	Large differences	Short strips of debris and obvious impurities	

2.1.2. Sensory Method Grading Test

Five concentrations of bean slurry (6.5%, 7%, 7.5%, 8%, and 8.5%) were used to carry out the Yuba skin production test. Each concentration of Yuba skin was collected at the end of the production line, and a total of 5 groups of test samples were collected. The collected Yuba skin was stored in a drying oven and labeled. In this evaluation experiment, six evaluation experts were selected, aged 20–30, including three males and three females, all of whom were in good health and had normal vision. Relevant evaluation training was provided to the reviewers before the evaluation to ensure accurate evaluation results. The test site was a room with normal light, and the six judges scored the color of the five groups of test samples, and the subjective evaluation value of the color of the fermented soybean peel was obtained; the results are shown in Table 2. As can be seen from Table 2, the surface color scores of the five groups of Yuba skin samples at different concentrations ranged from 6.28 to 9.05, and the Yuba skin sample with the highest color score was group 5. When the original image of the sample was visually inspected, the color change of the Yuba skin sample corresponds to different slurry concentrations, which represents that an increase in the pre-pulp concentration which can moderately deepen the Yuba skin color.

Table 2. Sensory evaluation from six experts.

Concentration (%)	Sensory Score of The Luster						Average
	Expert A	Expert B	Expert C	Expert D	Expert E	Expert F	
6.5	6.2	6.1	6.5	6.3	6.2	6.4	6.28
7	6.8	7.1	6.3	6.5	6.9	6.1	6.62
7.5	7.5	7.6	7.8	7.8	7.1	7.9	7.62
8	8.2	8.5	8.1	8.6	8.9	8.5	8.47
8.5	8.6	9.2	9.3	8.9	8.8	9.5	9.05

The process of sensory scoring at different concentrations is shown in Figure 1, and the color score of the Yuba skin samples showed an increasing trend with the increase in concentration. This is because, as the slurry concentration increases, the protein content increases, and the color of the Yuba skin also deepens. However, when the concentration increases to a certain value, with the evaporation of the water on the surface of the slurry, the slurry is further concentrated, the protein content increases significantly, and the color of the Yuba skin becomes light yellow and shiny, which improves the color score. It can also be observed through the raw images that the compounds formed by the increased concentration of the slurry cause a visible color change in the Yuba skin, and this color change in the Yuba skin samples can be considered as a suitable method to estimate the surface color of the Yuba skin through image processing.

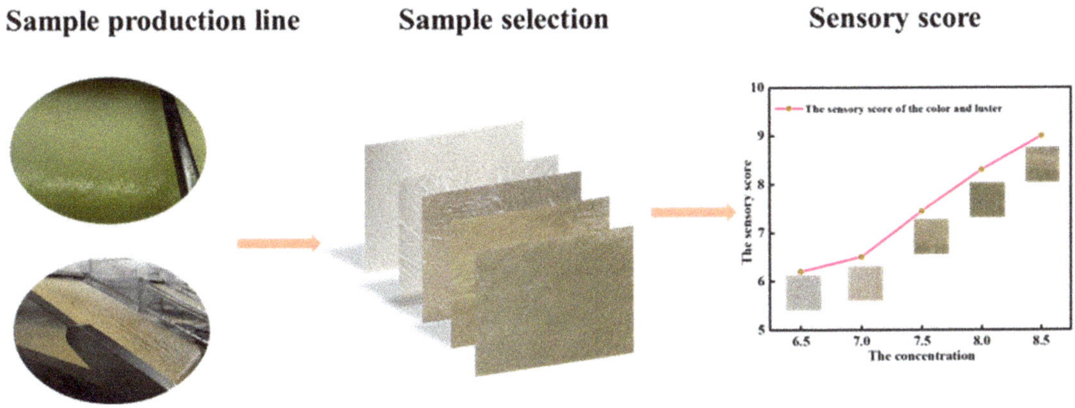

Figure 1. Sensory scoring processes at different concentrations.

2.2. *RBF-SVM Intelligent Grading System Design*

In the processing of the conjunctiva of Yuba skin, different grades of Yuba skin have some differences in texture and morphology. Therefore, the texture features on the surface of Yuba skin can be extracted to distinguish the surface image features of Yuba skin and use them to design an intelligent grading system for Yuba skin.

2.2.1. System Solution Design

The specific process of the grading system is shown in Figure 2. Firstly, the ROI region extraction and pre-processing are performed on the fermented skin images captured by the industrial camera. The gray histogram distribution of sample images are analyzed through MATLAB, and histogram equalization was carried out to reduce errors. The grayscale eigenvalues, gray level co-occurrence matrix (GLCM) eigenvalues, and gray level run length matrix (GLRLM) eigenvalues of the equalized image are extracted. The gray characteristic value contains only one factor of the degree of darkness (DOD). GLCM contains four factors of Angular Second Moment (ASM), Contrast (CON), Correlation (COR), and Inverse Different Moment (IDM). GLRLM contains four factors of Short Run Emphasis (SRE), Long Run Emphasis (LRE), Gray Level Different Uniformity (GLD), and Run Length Different Uniformity (RLD). The SVM quality detection model was established based on the data of the three groups of characteristic values, and the 9 factors with different combination methods are detected by the model. To select the appropriate kernel function, we used different kernel functions to test the quality detection model and conduct a comparative analysis of the test results. Additionally, we determined the optimal quality grading model by comparing and analyzing the evaluation results and grading accuracy of 3 different grading models. After the design of the grading system was completed, a quality grading test was carried out to verify the rationality of the grading system.

2.2.2. Image Pre-Processing and Feature Value Extraction

The initial acquisition of the terminal Yuba skin conjunctiva image is completed by the image acquisition system, and then the target extraction of the region of interest (ROI) in the original image is performed, and the unwanted information in the image is eliminated by noise reduction pre-processing. Finally, the image is converted into a clear black-and-white effect image, which in turn provides the necessary basis for the extraction of multi-texture feature values in the next step. Through comparison, a Gaussian filter and Sobel operator edge detection were finally selected to achieve noise reduction of the image, and a clear black-and-white effect map is obtained by segmentation of the Big Law method in binarization. The results after pre-processing are shown in Figure 3.

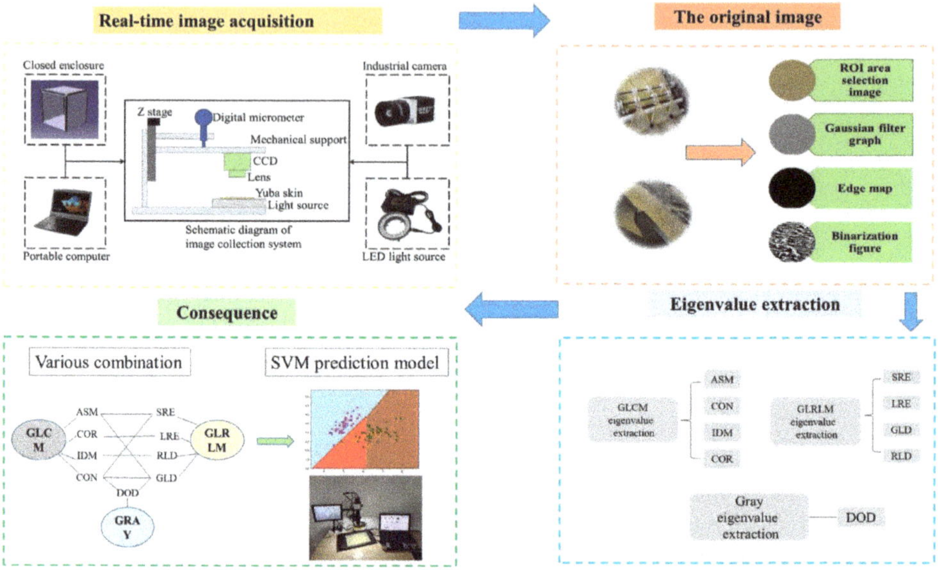

Figure 2. Software design flow chart.

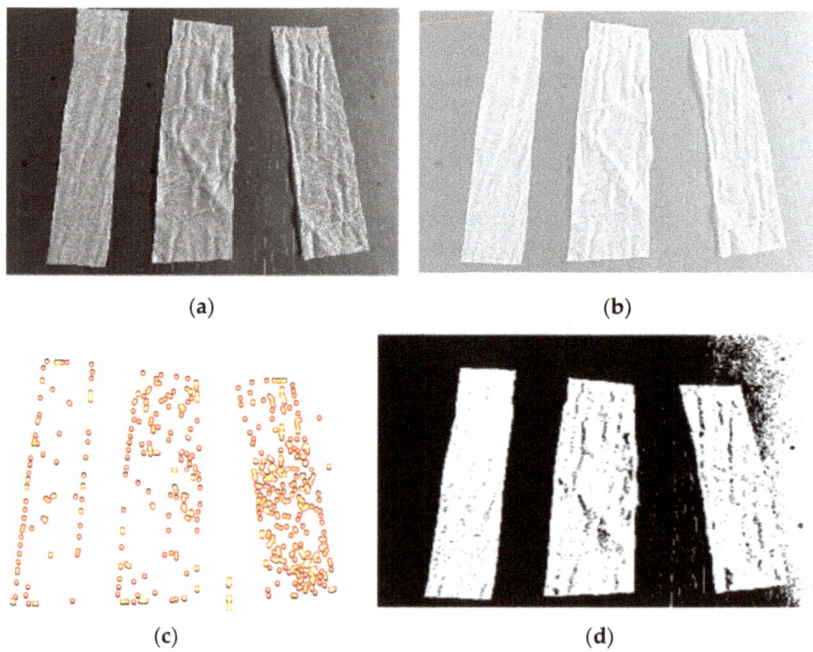

Figure 3. Image preprocessing. (**a**) Original image; (**b**) Gaussian filter graph; (**c**) Sobel edge detection; (**d**) Big Law division.

The factors affecting the quality score of Yuba skin mainly include the processing equipment and technological process. In the technological process, if the concentration of slurry increases, the protein content becomes higher, and the color of Yuba skin deepens which the main reasons for the quality change. Therefore, we need to further explore the average gray level to determine whether there is a certain correlation between the value and the quality of Yuba skin. The quality scores of Yuba skin samples produced at different concentrations were determined by the sensory method, and the average gray value of the surface of Yuba skin samples at different concentrations was determined by the image processing method. When the image is an 8-bit pixel image, the brightest pixel has a value of 255 (white), the darkest pixel has a value of 0 (black), and the other pixels have different grayscale intensity values between 0 and 255. According to this, the average gray intensity value was normalized using Equation (1), and the obtained normalized value was defined as the degree of darkness (DOD) of the Yuba skin. The three groups of image processing results (prediction data) and sensory scoring results (experimental data) were linearly fitted, as shown in Figure 4a. The data fitting results show that there is a strong relationship between the average gray value and the quality of the Yuba skin. Correlation (R^2 of the three groups are 0.99453, 0.99247, 0.99077, respectively), therefore, of the grayscale eigenvalues can be used as an alternative method for the quality score of the Yuba skin.

$$\text{Degree of Darkness} = \frac{255 - \text{Mean Gray Value}}{255} \quad (1)$$

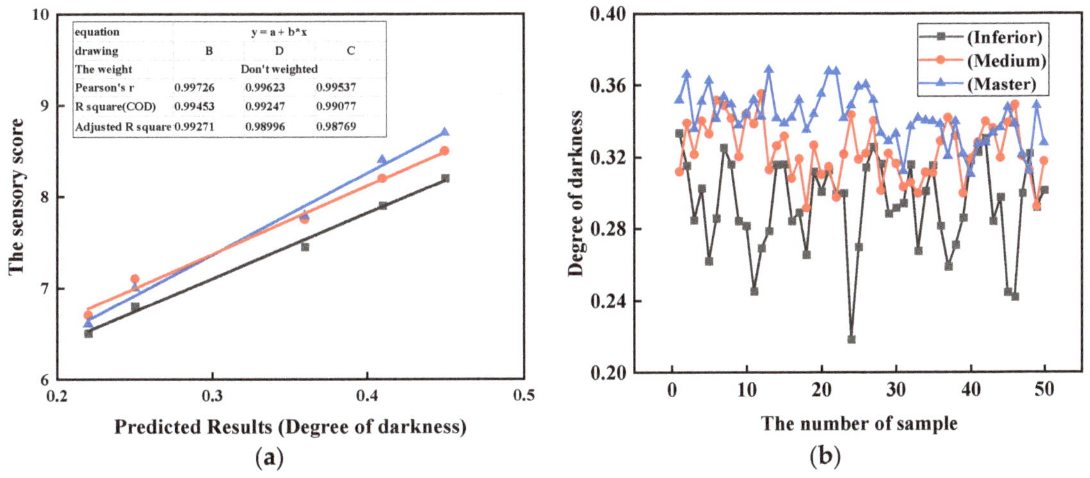

Figure 4. Comparison of sensory scores with image processing data fitting and DOD values. (**a**) Sensory scores fit to image processing data; (**b**) DOD numerical comparison.

Master, medium, and inferior Yuba skin sample images were randomly selected, with 50 for each grade. First, the gray histogram distribution of the Yuba skin sample image was analyzed by MATLAB, the histogram equalization is carried out, and then the gray characteristic value was extracted from the image after the equalization. The results shown in Figure 4b show that the grayscale darkness value of the superior and medium Yuba skins changed relatively smoothly, and their distribution were concentrated in the ranges of 0.34–0.38 and 0.29–0.33, respectively. There is still quite a difference between these two grades of Yuba skin and the dark degree value range of the inferior Yuba skin which was mainly concentrated in the range of 0.22–0.31. Therefore, the average gray characteristic value is suitable to be used as the quality classification standard for Yuba skin.

The greycomatrix function in MATLAB was used to generate the GLCM and calculate the required parameters. A total of 150 samples were randomly selected at the terminal of the yuba skin and conjunctiva. The GLCM was obtained by calling the "Gray-Comatrix" function, and then GLCM characteristic values were extracted. The characteristic value factors were calculated, including angular second moment (ASM), contrast (CON), Correlation (COR), and inverse different moment (IDM). According to the analysis of the GLCM characteristic results, the GLCM characteristic values of the three different grades of Yuba skin have obvious differences; therefore, the GLCM characteristic values are suitable for the quality detection standard of Yuba skin. The GLCM eigenvalues are shown in Figure 5.

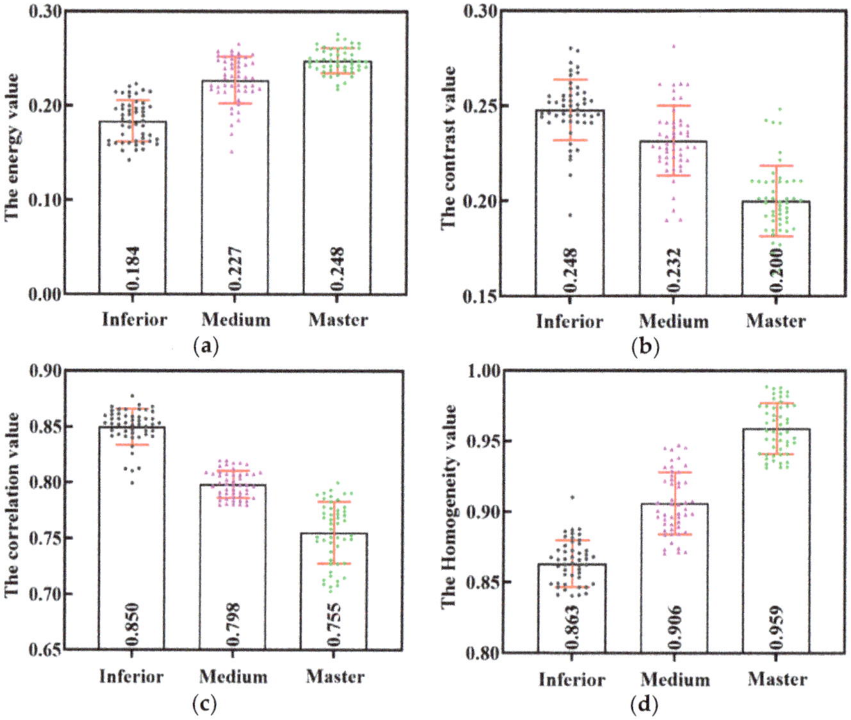

Figure 5. GLCM eigenvalue comparisons. (**a**) ASM numerical comparison; (**b**) CON numerical comparison; (**c**) COR numerical comparison; (**d**) IDM numerical comparison. In the figure, black represents inferior Yuba skin, red represents medium Yuba skin, green represents master Yuba skin.

The getGraylevelRumatrix function in MATLAB was used to obtain the GLRLM and calculate the parameters to be collected. The same 150 Yuba skin samples were selected, the images were subjected to grayscale processing and quantification, and the GLRLM eigenvalues were extracted. The chosen parameter directions were 0° and 45°. We calculated the eigenvalues of the GLRLM, and the mean values of the data were collected from the two angles. The GLRLM eigenvalues of SRE, LRE, GLD and, RLD were calculated. Finally, the mean values of the data collected from the two angles were calculated. The comparative analysis of the GLRLM eigenvalue parameters indicated notable differences in the eigenvalues of SRE, GLD, LRE, and RLD among the three different grades of Yuba skin. Our results suggest that the mean value of GLRLM eigenvalue is suitable as a classification standard for Yuba skin quality. The GLRLM eigenvalue results are shown in Figure 6.

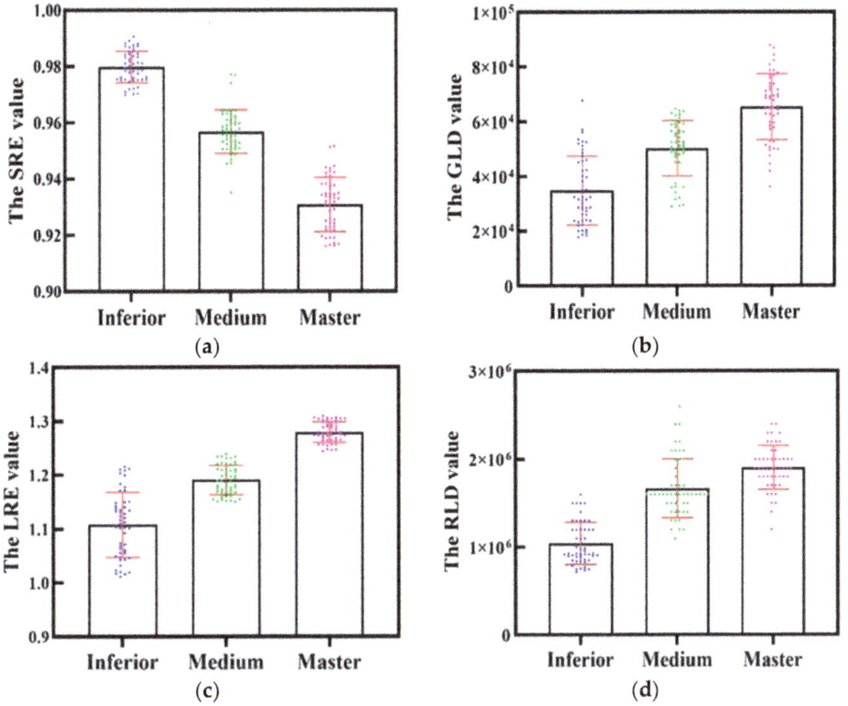

Figure 6. GLRLM eigenvalue comparisons. (**a**) SRE numerical comparison; (**b**) GLD numerical comparison; (**c**) LRE numerical comparison; (**d**) RLD numerical comparison. In the figure, blue represents inferior Yuba skin, green represents medium Yuba skin, red represents master Yuba skin.

2.2.3. RBF-SVM Hierarchical Model Construction

SVM is a binary classification model that uses a linear classifier and determines the largest interval defined on the feature space [25]. This largest interval function distinguishes it from a perceptron. SVM contains different kernel functions, which can effectively solve linear or nonlinear classification and prediction problems. We set the Yuba skin quality classification training sample to the following:

$$D = \{(x_i, y_i), i = 1, 2, \cdots, l\} x_i \in R^d \quad (2)$$

The following is the hyperplane equation of the SVM classifier:

$$\omega x_i + b = 0 \quad (3)$$

where:

x_i—Training vectors;
ω—normal vector;
b—Offset of the hyperplane.

Because the quality of Yuba is indicated by one of three grades, a three-classifier module was used. When the system automatically carries out the quality classification test of Yuba skin, it first loads the sample set for treatment classification, and obtains the optimal value of the penalty factor c and kernel parameter g using the characteristic parameters of the sample set. The training and learning module in the system carries out repeated training and learning through the treatment of the test sample set, and can finally automatically form a quality grade classifier model. The test module in the system carries out grade classification through the treatment of the test sample set. When the kernel function of the classification model is the radial basis function kernel (RBF) or the polynomial kernel function, the features of GLCM and GLRLM perform better than those of other kernel functions. Overall, when the kernel function of the classification model was the RBF radial basis kernel function, the model was the most accurate on the training and testing sets than when the other kernel functions were used. When the feature category was TOTAL, the accuracy rates for the model on the training and testing sets were 95.31% and 94.16%, respectively, and the optimal parameters of the model were c = 3.50 and g = 0.98. Additionally, the robustness of the SVM classification model was the best under these conditions.

2.2.4. Establishment and Analysis of Three Classification Model Evaluation Criteria

To verify the performance of the RBF-SVM algorithm proposed in this paper, two classification algorithms, the GBDT algorithm and BP neural network (Back Propagation), were used to compare with the RBF-SVM algorithm. The maximum error (Max error), mean absolute error (*MAE*), mean absolute percentage error (*MAPE*), root mean square error (*RMSE*), and coefficient of determination (R^2) of the model were calculated. The maximum error describes the extreme case of the prediction result, and the mean absolute error and the average absolute percentage error are based on the absolute error, which can be used together to describe the model's prediction of the limit value, and the root mean square error and coefficient of determination describe the overall regression of the model. The results of the evaluation indicators are shown in Table 3.

$$MAE(y,\hat{y}) = \frac{1}{n}\sum_{i=1}^{n}|y_i - \hat{y}_i| \tag{4}$$

$$MAPE(y,\hat{y}) = \frac{1}{n}\sum_{i=1}^{n}\frac{|y_i - \hat{y}_i|}{\hat{y}_i} \tag{5}$$

$$RMSE(y,\hat{y}) = \sqrt{\frac{1}{n}(y_i - \hat{y}_i)^2} \tag{6}$$

$$R^2(y,\hat{y}) = 1 - \frac{\sum_{i=1}^{n}(y_i - \hat{y}_i)^2}{\sum_{i=1}^{n}(y_i - \overline{y}_i)^2} \tag{7}$$

Table 3. Comparison of three models' regression evaluation index data.

	SVM	BPNN	GBDT
Max error	7.69%	19.3%	9.68%
MAE	0.38	0.72	0.42
MAPE	5.38%	10.35%	6.04%
RMSE	0.392	0.82	0.43
R^2	0.9846	0.9528	0.9786

The comparison diagram and absolute percentage absolute errors of the prediction results of the SVM, BPNN, and GBDT quality prediction models are shown in Figure 7. The bias of the BPNN prediction was large, all its regression indicators were the worst, and its comprehensive prediction accuracy was the lowest. The prediction deviations of GBDT and SVM were similar, and the average absolute error and average absolute percentage error indicators were also relatively close; however, the maximum error, root mean square error, and coefficient of determination index were different, with the GBDT model having extreme errors. The subjective evaluation results of humans are linear, and the use of non-sexual models can cause bias during the training process of the sample, resulting in a decrease in the stability of the accuracy of the prediction results.

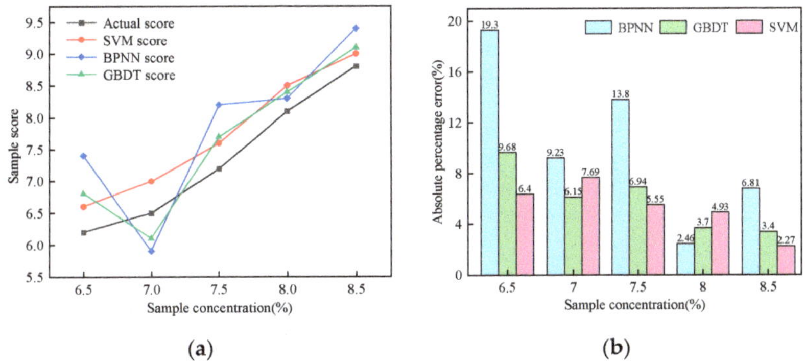

Figure 7. Comparison of prediction results and absolute percentage errors of three models. (**a**) Model prediction comparison; (**b**) comparison of model prediction errors.

3. Results and Analysis

3.1. Quality Grading Software Interface

SVM-based software for grading the quality of Yuba skin was developed using a graphical user interface library in MATLAB. The grading software interface consisted of a preprocessing image module, feature value extraction module, and quality result module (which has a final kernel function). The software preprocesses and displays the image of the Yuba skin to be evaluated. The corresponding eigenvalue in the feature value extraction module is then calculated, and the user clicks the RBF-SVM grading button in the final quality result module to display the corresponding quality grading results. The software interface is presented in Figure 8.

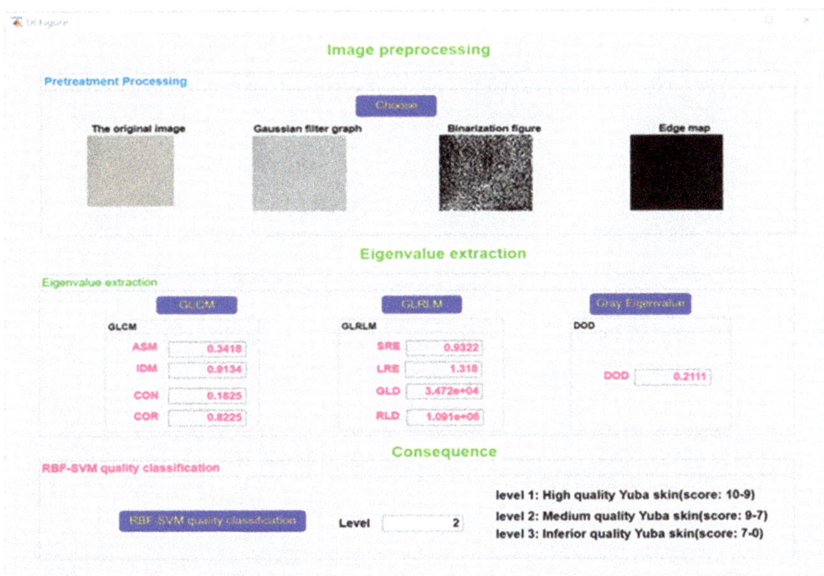

Figure 8. Yuba quality grading software interface.

3.2. RBF-SVM Intelligent Grading Test Analysis

The software was run on a test platform (Figure 9). The platform positions the charge-coupled device (CCD) industrial camera to a position where the best pictures can be taken, and the light emissivity controller is used to set the brightness of the light source. When a stable brightness is reached, the image is captured, and the photos are uploaded to the host computer through a USB connection. After opening the Yuba skin quality classification software, the Yuba skin images that need to be classified are automatically selected by clicking the image selection button, and the images are preprocessed and displayed. Then, in the feature value extraction module, the corresponding feature values are displayed in the software by clicking the corresponding button. Finally, the corresponding results of the quality classification can be attained by clicking the RBF-SVM classification button in the final quality result module.

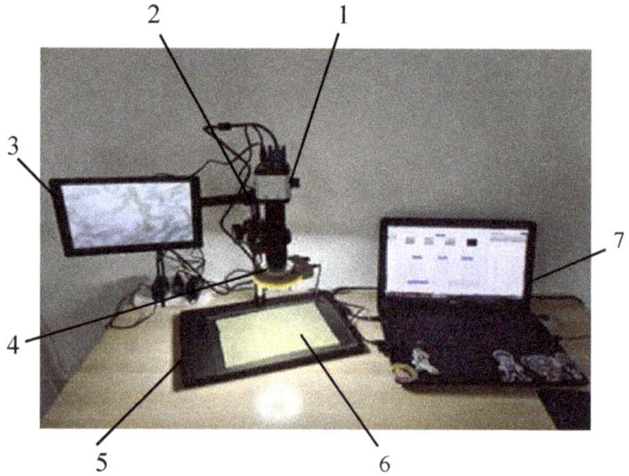

Figure 9. Quality grading system test platform. (**1**) CCD; (**2**) Z stage; (**3**) LCD; (**4**) LED light source; (**5**) plane light source; (**6**) Yuba skin; (**7**) portable computer.

The collected multi-eigenvalue factors were input into the pre-established SVM grading model as input values, and the SVM grading models with different kernel functions were used to determine Yuba skin quality. Different grading results were obtained with different sum function models, and the specific grading test results are shown in Table 4.

Table 4. Test results of different SVM kernel functions.

Eigenvalue Combination Method	**SVM Kernel Function**		
	LKF Linear Kernel	**PKF Polynomial Kernel**	**RBF Radial Basis Core**
ASM	62.36%	67.62%	68.89%
CON	68.38%	72.06%	70.32%
COR	72.55%	74.65%	60.24%
IDM	78.54%	75.36%	67.75%
SRE	64.32%	73.68%	75.66%
LRE	73.59%	79.69%	69.98%
GLD	69.72%	82.69%	70.63%
RLD	72.28%	81.63%	72.39%
DOD	83.57%	82.77%	74.68%
ASM+CON	77.68%	83.64%	79.68%
ASM+CON+COR	82.32%	84.66%	85.55%
ASM+CON+COR+IDM	83.54%	85.78%	87.07%
ASM+CON+COR+IDM+SRE	84.36%	86.69%	88.46%
ASM+CON+COR+IDM+SRE+LRE	86.65%	87.68%	89.97%
ASM+CON+COR+IDM+SRE+LRE+GLD	86.64%	88.55%	90.54%
ASM+CON+COR+IDM+SRE+LRE+GLD+RLD	87.32%	88.87%	91.35%
ASM+CON+COR+IDM+SRE+LRE+GLD+RLD+DOD	88.87%	89.75%	93.56%

The test results indicate that the SVM grading system's accuracy increased gradually with longer combinations of eigenvalues. The RBF radial basis kernel function had a 93.56% accuracy rate for the eigenvalue combination of ASM+CON+COR+IDM+SRE+LRE+GLD+RLD+DOD. Despite this promising result, further improvements can be made, possibly with a larger or more varied image dataset for training.

3.3. Sensory Method and RBF-SVM Grading Systematic Error Analysis

Two methods were used to score the Yuba skins with different concentrations, and 10 groups of Yuba skin samples were produced using the same concentration, and the scoring results were shown in Figure 10. Both methods showed that, in the concentration range of 6.5% to 8.5%, the score of the Yuba skin gradually increased with the increase in concentration. Moreover, the average score difference obtained by the two methods was small, which verifies that the results of the RBF-SVM grading system are in good agreement with the results of the actual sensory method. The average score data of each concentration using the two methods were calculated and analyzed, and the relative error results of the actual sensory method score value and the RBF-SVM grading system score value are shown in Figure 11. As shown in Figure 11, the relative error of the two methods reached a maximum of 3.31% when the concentration was 8.0%, and the minimum error of the two methods was 0.06% when the concentration was 7.0%. The relative error of the different concentration test groups was within 5%, and the test results were in line with expectations, which further proves the applicability of the RBF-SVM grading system in the quality detection of Yuba skin.

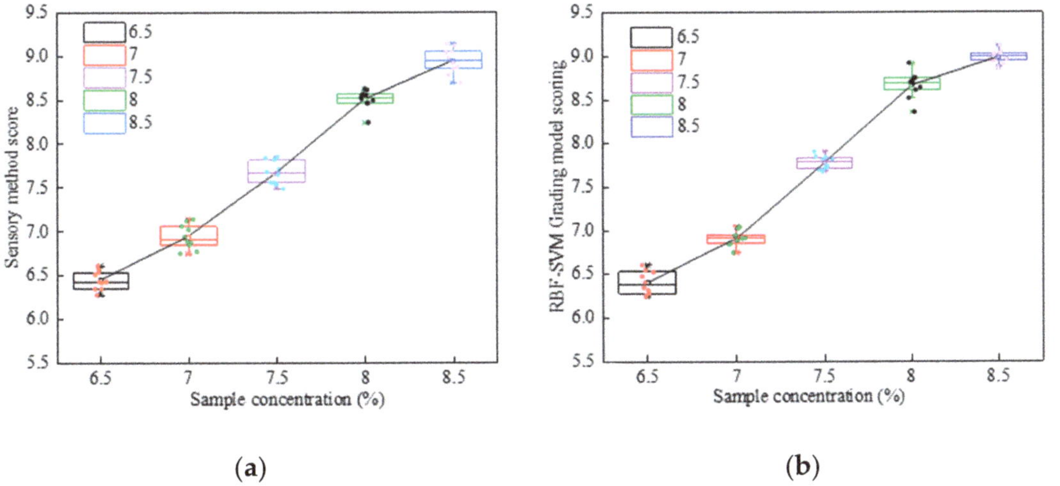

Figure 10. Comparison of sensory method and RBF-SVM grading system scores. (**a**) Sensory method scores; (**b**) RBF-SVM grading system scores.

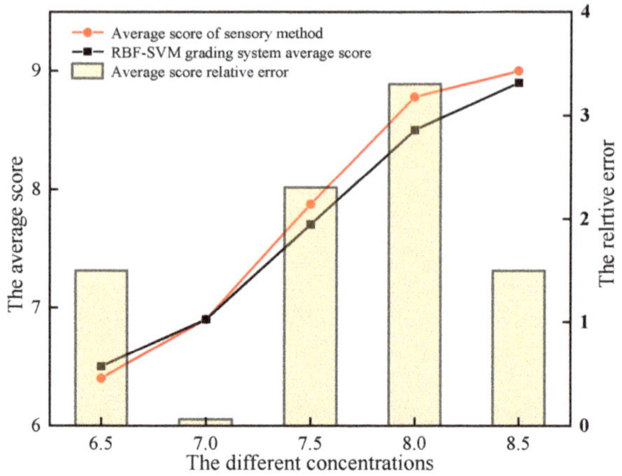

Figure 11. Average score relative errors of sensory method and RBF-SVM grading system.

4. Discussion

In this study, an RBF-SVM classification system was designed based on image processing. The initial image is preprocessed, and then the required eigenvalues are extracted. Finally, the data of each eigenvalue are processed, and the quality grade of Yuba skin is obtained by RBF-SVM classification recognition system. However, only 150 sample images were collected for training in the establishment of the RBF-SVM model. On this basis, the number of sample images can be further increased to improve the accuracy of the RBF-SVM model.

At present, there are few studies in the field of quality classification of Yuba skin. In this paper, industrial cameras were used to shoot images of bean curd skin. In the actual test process, the accuracy of the characteristic values extracted from Yuba skin images may be affected due to changes in lighting conditions, thus affecting the accuracy of the classification. Therefore, in future research, we can carry out detailed research on image feature extraction of Yuba skin and design a classification system with a higher accuracy.

5. Conclusions

Our study collected 150 sample images to train and construct an SVM hierarchical model that evaluates Yuba skin quality. Different kernel functions were used in the model for sample training. The best penalty factor (c = 3.50) and kernel parameter value (g = 0.98) were obtained through cross-validation. The accuracy of the model was 95.31% and 94.16% for the training and test sets, respectively.

In this study, a quality grading software of Yuba skin based on the SVM model was designed. By preprocessing the initial image, then extracting the required eigenvalues, and finally processing the data of each eigenvalue, the quality of Yuba skin can be obtained through the SVM classification and recognition system. In terms of quality level, the accuracy rate reached 93.56%.

The traditional sensory method and RBF-SVM grading system were used to grade samples of Yuba skin made from different concentrations of bean curd slurry. The results show that the sensory method has good consistency with the results of the RBF-SVM grading system, and the detection error of the RBF-SVM grading system was less than 5%, which can meet the needs of automated production of Yuba skin.

Author Contributions: Conceptualization, Y.T. and L.X.; methodology, Y.T. and L.X.; software, Y.T. and Y.S,; valida-tion, Y.T., L.X. and Q.T.; formal analysis, Y.T.; investigation, H.Y. and Y.S.; resources, L.X.; data curation, Y.T. and L.X; writing—original draft preparation, Y.T.; writing—review and editing, Y.T., Q.T. and L.X.; visualization, Y.T.; supervision, L.X.; project administration, L.X. and Q.T. All authors have read and agreed to the published version of the manuscript.

Funding: This research was supported by Anhui Province Key Research and Development Program Project (grant NO. 202204c06020057).

Institutional Review Board Statement: Not applicable.

Informed Consent Statement: Not applicable.

Data Availability Statement: Not applicable.

Acknowledgments: The financial support by Anhui Province Key Research and Development Program Project is gratefully acknowledged.

Conflicts of Interest: The authors declare no conflict of interest.

References

1. Su, J.-F.; Huang, Z.; Yuan, X.-Y.; Wang, X.-Y.; Li, M. Structure and properties of carboxymethyl cellulose/soy protein isolate blend edible films crosslinked by Maillard reactions. *Carbohydr. Polym.* **2010**, *79*, 145–153. [CrossRef]
2. Li, S.; Liu, X.; Zhang, Y. Development of SFZCX125 automatic forming machine for curd bamboo. *Grain Oil Process.* **2006**, *12*, 72–74. (In Chinese)
3. Zhang, J.; Peng, X.; Guo, S. Protein-lipid film (fuzhu) prepared from soymilk: Effects of soymilk convection on its formation, composition, and quality. *LWT-Food Sci. Technol.* **2021**, *141*, 110909. [CrossRef]
4. Zhang, S.; Lee, J.; Kim, Y. Chemical composition, water vapor permeability, and mechanical properties of yuba film influenced by soymilk depth and concentration. *J. Sci. Food Agric.* **2018**, *98*, 1751–1756. [CrossRef] [PubMed]
5. Zhao, T. Development and application of machine vision technology in food inspection. *Food Res. Dev.* **2021**, *42*, 233–234. (In Chinese)
6. Qin, J.; Liu, X.; Van Den Abeele, K.; Cui, G. The study of wood knots using acoustic nondestructive testing methods. *Ultrasonics* **2018**, *88*, 43–50. [CrossRef]
7. Donskoy, D.; Liu, D. Vibro-acoustic modulation baseline-free non-destructive testing. *J. Sound Vib.* **2021**, *492*, 115808. [CrossRef]
8. Yin, X. Driven by machine learning to intelligent damage recognition of terminal optical components. *Neural Comput. Appl.* **2021**, *33*, 789–804. [CrossRef]
9. Zhu, Y.-K.; Tian, G.-Y.; Lu, R.-S.; Zhang, H. A Review of Optical NDT Technologies. *Sensors* **2011**, *11*, 7773–7798. [CrossRef] [PubMed]
10. Fathizadeh, Z.; Aboonajmi, M.; Beygi, S.R.H. Nondestructive firmness prediction of apple fruit using acoustic vibration response. *Sci. Hortic.* **2020**, *262*, 109073. [CrossRef]
11. Zhang, H.; Wu, J.; Zhao, Z.; Wang, Z. Nondestructive firmness measurement of differently shaped pears with a dual-frequency index based on acoustic vibration. *Postharvest Biol. Technol.* **2018**, *138*, 11–18. [CrossRef]
12. Shu, Y.; Xiong, C.; Fan, S. Interactive design of intelligent machine vision based on human-computer interaction mode. *Microprocess. Microsyst.* **2020**, *75*, 103059. [CrossRef]
13. Zhou, L.; Chalana, V.; Kim, Y. PC-based machine vision system for real-time computer-aided potato inspection. *Int. J. Imaging Syst. Technol.* **2015**, *9*, 423–433. [CrossRef]
14. Hou, B.; Zhang, C.; Yang, S. Computer Vision Tool-Setting System of Numerical Control Machine Tool. *Sensors* **2020**, *20*, 5302. [CrossRef] [PubMed]
15. Wu, Y.; Lu, Y. An intelligent machine vision system for detecting surface defects on packing boxes based on support vector machine. *Meas. Control.* **2019**, *52*, 1102–1110. [CrossRef]
16. Su, Q.; Kondo, N.; Riza, D.F.A.; Habaragamuwa, H. Potato Quality Grading Based on Depth Imaging and Convolutional Neural Network. *J. Food Qual.* **2020**, *2020*, 8815896. [CrossRef]
17. Tian, Y.; Yang, G.; Wang, Z.; Wang, H.; Li, E.; Liang, Z. Apple detection during different growth stages in orchards using the improved YOLO-V3 model. *Comput. Electron. Agric.* **2019**, *157*, 417–426. [CrossRef]
18. Liu, L.; Li, Z.; Lan, Y.; Shi, Y.; Cui, Y. Design of a tomato classifier based on machine vision. *PLoS ONE* **2019**, *14*, e0219803. [CrossRef]
19. Kumar, S.D.; Esakkirajan, S.; Bama, S.; Keerthiveena, B. A microcontroller based machine vision approach for tomato grading and sorting using SVM classifier. *Microprocess. Microsyst.* **2020**, *76*, 103090. [CrossRef]
20. Olaniyi, E.O.; Adekunle, A.A.; Odekuoye, T.; Khashman, A. Automatic system for grading banana using GLCM texture feature extraction and neural network arbitrations. *J. Food Process Eng.* **2017**, *40*, e12575. [CrossRef]
21. Huang, X.; Yu, S.; Xu, H.; Aheto, J.H.; Bonah, E.; Ma, M.; Wu, M.; Zhang, X. Rapid and nondestructive detection of freshness quality of postharvest spinaches based on machine vision and electronic nose. *J. Food Saf.* **2019**, *39*, e12708. [CrossRef]
22. Shen, Y.; Tao, Y.; Tang, Q.; Xu, L. Study on Multifactor Control Experiment and Design of Quality Detection System for Yuba Skin Production. *Mob. Inf. Syst.* **2022**, *2022*, 3067830. [CrossRef]
23. Manavalan, B.; Shin, T.H.; Lee, G. PVP-SVM: Sequence-based prediction of phage virion proteins using a support vector machine. *Front. Microbiol.* **2018**, *9*, 476. [CrossRef] [PubMed]

24. Suyoto, R.; Komarudin, M.; Nama, G.F.; Yulianti, T. Classification of Civet and Canephora coffee using Support-Vector Machines (SVM) algorithm based on order-1 feature extraction. *IOP Conf. Ser. Mater. Sci. Eng.* **2021**, *1173*, 012006. [CrossRef]
25. Otchere, D.A.; Ganat TO, A.; Gholami, R.; Ridha, S. Application of supervised machine learning paradigms in the prediction of petroleum reservoir properties: Comparative analysis of ANN and SVM models. *J. Pet. Sci. Eng.* **2021**, *200*, 108182. [CrossRef]

sensors

MDPI

Article

Exploration of Damage Identification Method for a Large-Span Timber Lattice Shell Structure in Taiyuan Botanical Garden based on Structural Health Monitoring

Guoqing Wang, Chenjia Xu, Shujia Zhang, Zichun Zhou, Liang Zhang, Bin Qiu, Jia Wan and Honggang Lei *

College of Civil Engineering, Taiyuan University of Technology, Taiyuan 030024, China;
wangguoqing0066@link.tyut.edu.cn (G.W.); xuchenjia0072@link.tyut.edu.cn (C.X.);
zhangshujia0076@link.tyut.edu.cn (S.Z.); zhouzichun0049@link.tyut.edu.cn (Z.Z.);
zhangliang0073@link.tyut.edu.cn (L.Z.); qiubin@tyut.edu.cn (B.Q.); wanjia@tyut.edu.cn (J.W.)
* Correspondence: leihonggang@tyut.edu.cn

Abstract: Large-span spatial lattice structures generally have characteristics such as incomplete modal information, high modal density, and high degrees of freedom. To address the problem of misjudgment in the damage detection of large-span spatial structures caused by these characteristics, this paper proposed a damage identification method based on time series models. Firstly, the order of the autoregressive moving average (ARMA) model was selected based on the Akaike information criterion (AIC). Then, the long autoregressive method was used to estimate the parameters of the ARMA model and extract the residual sequence of the autocorrelation part of the model. Furthermore, principal component analysis (PCA) was introduced to reduce the dimensionality of the model while retaining the characteristic values. Finally, the Mahalanobis distance (MD) was used to construct the damage sensitive feature (DSF). The dome of Taiyuan Botanical Garden in China is one of the largest non-triangular timber lattice shells worldwide. Relying on the structural health monitoring (SHM) project of this structure, this paper verified the effectiveness of the damage identification model through numerical simulation and determined the damage degree of the dome structure through SHM measurement data. The results demonstrated that the proposed damage identification method can effectively identify the damage of large-span timber lattice structures, locate the damage position, and estimate the degree of damage. The constructed DSF had relatively strong robustness to small damage and environmental noise and has practical application value for SHM in engineering.

Keywords: damage identification; damage sensitive feature; spatial lattice structure; structural health monitoring; timber structure; time series model

Citation: Wang, G.; Xu, C.; Zhang, S.; Zhou, Z.; Zhang, L.; Qiu, B.; Wan, J.; Lei, H. Exploration of Damage Identification Method for a Large-Span Timber Lattice Shell Structure in Taiyuan Botanical Garden based on Structural Health Monitoring. *Sensors* **2023**, *23*, 6710. https://doi.org/10.3390/s23156710

Academic Editors: Bogusław Łazarz, Grzegorz Peruń and Tangbin Xia

Received: 11 June 2023
Revised: 21 July 2023
Accepted: 24 July 2023
Published: 27 July 2023

1. Introduction

During long-term service, structures may suffer damage due to material aging, corrosion, prolonged loading, and natural disasters such as fire and earthquakes [1]. For major engineering structures, such as large-span bridges, high-rise buildings, large-span spatial lattice structures, large-scale water conservancy structures, and large-scale offshore platforms, failure can cause significant economic damage and human casualties. Therefore, SHM is of great significance in civil engineering and related fields [2].

Damage identification and localization are crucial processes in SHM since damage alters the stiffness, mass, or damping of structures. Consequently, stiffness reduction caused by cracking or loosening of connections will lead to changes in structural vibration modes, resulting in alterations in its dynamic response [3,4]. However, obtaining complete and accurate high-order modal information through theoretical analysis for actual structures, particularly for large-span spatial structures with numerous degrees of freedom, is challenging due to the complex vibration data involved. In addition, in practical engineering, it is almost impossible to obtain input and output signals by artificially exciting the structure and accurately identifying the structural modal information.

The damage identification method based on time series does not involve modal parameter identification. This method is based on statistical theory and can represent a large amount of effective information contained in structural response data with fewer parameters. By identifying the change pattern of the system through the interrelationships within the data, the changes in the model parameters can be used as the basis for detecting the existence and location of damage [5]. The time series model is sensitive to the identification of small structural damage, with strong noise reduction ability and operability, and has great potential in theoretical research and engineering applications. In recent years, it has received widespread attention and development from scholars [6,7].

The ARMA model, as an important component of time domain methods, is often used to fit regression structures to acceleration response data and extract DSF from it. Zuo and Guo [8] proposed a nonlinear damage identification method based on the autoregressive (AR) model and Kullback–Leibler distance, which has high sensitivity to minor damage. Razavi et al. [4]. suggested a damage feature identification method based on ARMA model and residual sample power spectral density, introducing Jeffrey distance and Smith distance for damage localization, and verifying the effectiveness of the damage identification model through vibration response experiments. Zhu et al. [9] studied the correlation between AR coefficients and structural stiffness reduction, proposing a method to identify structural damage using underdetermined equations established by AR coefficients, and solving underdetermined equations using the sparse regularization method. The effectiveness of the method was verified through experiments. Chen et al. [10] proposed a method for identifying structural nonlinear damage based on ARMA model and vector space cosine similarity (VSCS), and verified the feasibility of the method through experimental studies, solving the nonlinear problems that traditional methods cannot effectively handle caused by structural damage. Zeng et al. [11] advanced a time series model based on fuzzy c-means clustering algorithm, characterizing the degree of structural damage by the change of model coefficients, and verifying the feasibility and accuracy of the method through experimental and numerical studies. The method has low computational cost and is suitable for real-time monitoring of civil engineering. Diao et al. [12] combined the AR model with cointegration in econometrics to cointegrate variables at different nodes of the offshore platform. The method employs cointegration residuals as damage indicators and utilizes X-bar control charts for structural damage identification. The effectiveness of this approach was validated through both numerical simulation and experimental studies. Hu et al. [13] discussed a nonlinear ARMA(n,m) time series structural damage assessment method based on residual algorithm, proposing that higher-order models are more sensitive to disguised outliers, and verifying the method through dynamic monitoring tests. Liu et al. [7] explored the implementation method of damage feature extraction and damage warning for structures based on the ARMA model, and verified the significant changes of DSF index before and after damage through a t-test.

After obtaining the DSF, structural damage identification is reduced to a pattern recognition problem. Machine learning, as the main method of pattern recognition, includes Gaussian processes, support vector machines, neural networks, etc. Tang et al. [14] presented a damage identification method based on the AR model and Gaussian process, introducing parameters that characterize damage location and state information, and achieving a probability output of multi-damage localization and damage severity. Sui et al. [15] obtained the feature vector by arranging the damage indicators obtained by the AR model in order, and inputting it into the support vector machine optimized by a Bayesian algorithm for structural damage identification. The effectiveness of the method in identifying structural damage location was verified through experiments. Xu et al. [16] proposed a damage identification method based on the AR model and a BP neural network for lattice shell structures, which does not require modal parameter identification and excitation information, avoids the dependence of damage identification results on the accuracy of structural finite element (FE) models, and has high damage identification accuracy.

This article introduced PCA on the basis of existing research to extract features from the residual sequence. The advantage of incorporating PCA lies in its ability to reduce the dimensionality of the data while retaining the most informative features [17]. This not only simplifies the subsequent analysis but also helps to identify the most significant damage indicators. In addition, the MD was used in constructing DSF because it takes into account the differences and correlations in the variability of each observed variable. In addition, it can effectively eliminate the influence of scale on different measurement values.

Furthermore, existing studies were limited to theoretical research and simple experimental verification, lacking engineering applications and practices for large and complex structures. Based on the SHM project for a timber lattice shell dome in Taiyuan Botanical Garden, this paper proposes a damage identification method based on time series analysis. The effectiveness of the damage identification model was verified using numerical simulation, and the degree of damage to the lattice shell structure was identified using measured data from SHM. This paper combined theoretical research with engineering applications to verify the effectiveness and feasibility of applying the time-series-analysis-based damage identification method for practical engineering.

2. Research Background

This study took the timber lattice shell of Taiyuan Botanical Garden as the research background, which is located in Taiyuan, Shanxi, China. It is one of the largest non-triangular timber lattice structures worldwide. It was designed by Delugan Meissl Associated Architects (DMMA) from Austria and is now a landmark building in Taiyuan. The dome, from above, is shaped like a shell and glazed with double-curved panes of glass. The span of the dome is 89.5 m, with a span-to-height ratio of 3.0 and a projected area of approximately 6000 m^2. The dome consists of double-curved laminated timber beams, which are arranged in two or three intersecting layers. The beams in the intersection joint area are connected by steel pins and self-tapping screws. The splicing connection of the beam adopts the half-lap joint form and is connected by self-tapping screws. To improve the overall rigidity of the structure, a bidirectional prestressed steel cable net was installed inside it, arranged diagonally to the timber beams. The steel support of the dome is connected to the timber beam using anti-shear steel plates and self-tapping screws and connected to the concrete foundation using chemical anchor bolts. The appearance and structural details of the dome are shown in Figure 1. For more details of the dome, please refer to reference [18].

(a)

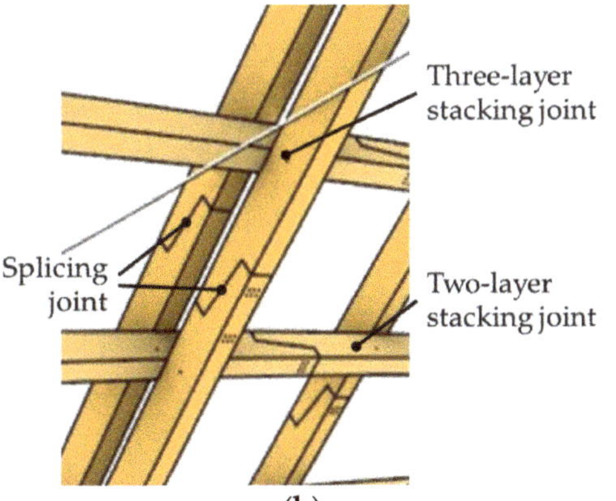

(b)

Figure 1. *Cont.*

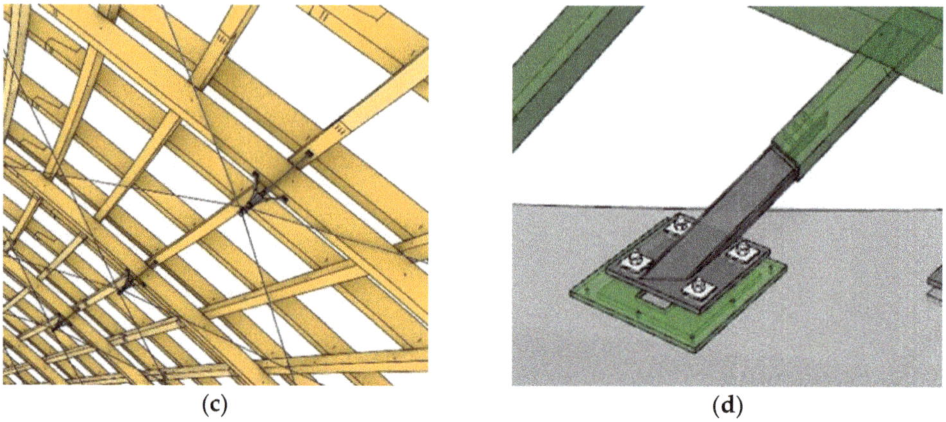

Figure 1. Timber lattice shell dome of Taiyuan Botanical Garden. (**a**) Appearance; (**b**) joint and its spatial relationship; (**c**) bidirectional cross steel cable; (**d**) root support.

3. Methodology

This section provides an overview of the fundamental principles of utilizing the ARMA model for time series analysis. It also elucidates the techniques for determining the order and estimating the parameters of the ARMA model using AIC and LAR. Furthermore, this section offers a theoretical exposition of the principal component analysis method introduced and defines a DSF construction method based on MD.

3.1. Time Series Analysis Modeling

Time series analysis is a statistical method for dynamic data processing, which is based on the theory of stochastic processes and mathematical statistics. It uses parameter models to process ordered random sampling data and studies the statistical laws followed by data sequences, in order to perform system identification [19,20]. Any structure can be regarded as a mechanical system composed of stiffness, mass, and damping matrices, which can be described by a motion differential equation and can be transformed into a difference equation in the discrete time domain. This difference equation has the same expression form as the time series model. Therefore, the equivalent relationship between the time series model and the structural motion differential equation can be used to determine the state of the structure and perform damage identification of the system.

3.1.1. Basic Theory of ARMA Model

The ARMA model is a statistical model commonly used in time series analysis. It takes past observations of the time series as independent variables and current observations as the dependent variables to describe the randomness and autocorrelation of the time series data. The ARMA model consists of two parts: the AR model and the moving average (MA) model, and has the characteristics of both models. The AR or MA model is a special case of the ARMA model, so the ARMA model is more general for system response and is the most commonly used model in time series analysis.

For a multi-degree-of-freedom system, its vibration differential equation is:

$$M\ddot{x}(t) + C\dot{x}(t) + Kx(t) = F(t), \tag{1}$$

where M, C, and K represent the mass matrix, damping matrix, and stiffness matrix of the structure, respectively. $\ddot{x}(t)$, $\dot{x}(t)$, and $x(t)$ represent the acceleration vector, velocity vector, and displacement vector of the structure, respectively. $F(t)$ is the excitation force applied to the system.

Equation (1) is equivalent to a $2n$-order non-homogeneous differential equation system. Under the action of a single excitation $f(t)$, the vibration differential equation is expressed as follows:

$$\begin{aligned}&\varphi_{2n}x^{(2n)} + \varphi_{2n-1}x^{(2n-1)} + \cdots + \varphi_1\dot{x} + \varphi_0 x \\ &= \theta_{2n-2}f_t^{(2n-2)} + \theta_{2n-1}f_t^{(2n-1)} + \cdots + \theta_1\dot{f}_t + \theta_0 f_t\end{aligned} \tag{2}$$

Discretizing Equation (2) and using a white noise sequence $a_t \sim ND(0, \sigma_a^2)$ as the system input yields a single degree of freedom ARMA (p,q) model [21]:

$$x_t - \sum_{i=1}^{p} \varphi_i x_{t-i} = \theta_0 a_t - \sum_{j=1}^{q} \theta_j a_{t-j} \tag{3}$$

In the equation, φ_i represents the i-th order AR coefficient, θ_j represents the j-th order MA coefficient, and p and q are the AR and MA orders, respectively. By introducing a backward shift operator B, defined as $B^k x_t = x_{t-k}$ [5], Equation (3) can be transformed into:

$$x_t = \frac{1 - \sum_{j=1}^{q} \theta_j B^j}{1 - \sum_{i=1}^{p} \varphi_i B^i} a_t = \frac{\theta(B)}{\varphi(B)} a_t \tag{4}$$

According to Equation (4), the ARMA model describes a system with a transfer function of $\theta(B)/\varphi(B)$, where $\varphi(B)$ represents the inherent characteristics of the system and $\theta(B)$ represents the relationship between the system and the external environment. As the input signals of large civil engineering structures are often difficult to test or measure, the performance and behavior of the structure can only be inferred through the analysis of output signals. The ARMA model can be established by analyzing the output signal of the structure, without considering the specific information of the input signals, and only using the white noise sequence $\{a_t\}$ as input to establish an analysis model. This makes the ARMA model increasingly widely used in the field of civil engineering. In the analysis of time series models, the AR, MA, and ARMA models have their own characteristics, which can be used to preliminarily determine the appropriate time series model. The characteristics of the models are shown in Table 1.

Table 1. Basic characteristics of time series models.

Model Type	Autocorrelation Function (ACF)	Partial Autocorrelation Function (PACF)
AR	Tailing	Truncation
MA	Truncation	Tailing
ARMA	Tailing	Tailing

3.1.2. Determination of ARMA Order

When applying the ARMA model for time series analysis, it is necessary to determine the order of the ARMA model, and the quality of the order setting has a significant impact on the parameter identification results. If the order is too high, the model will be too complex, resulting in overfitting and unnecessary calculations; if the order is too low, the model will be too simple to capture important features of the time series, leading to poor model fitting. Currently, one of the mostly used methods for determining the order of the ARMA model is AIC [22]. The function of AIC is defined as follows:

$$\text{AIC}(p+q) = \ln\left(\sigma_a^2\right) + 2(p+q)/N \tag{5}$$

In Equation (5), N represents the length of the time series x_t, σ_a^2 represents the variance of the ARMA model residuals, and p and q represent the AR and MA order, respectively. As the sum of model orders $(p+q)$ increases, $\ln(\sigma_a^2)$ will decrease and $2(p+q)/N$ will increase. Therefore, under the given parameter estimation method, the order with the minimum $\text{AIC}(p+q)$ value should be chosen.

3.1.3. Estimation of ARMA Parameter

This article used the long autoregressive (LAR) method to estimate the parameters of the ARMA model. Compared with other methods for parameter estimation, the LAR method can transform nonlinear regression problems into linear regression problems, which can simplify the calculation process and improve efficiency. According to the theory of LAR method, the ARMA model and the AR model are equivalent mathematical models. Therefore, the parameters of the AR model can be estimated first, and then the parameters of the ARMA model can be estimated based on the relationship of the transfer functions. The equivalent relationship between the transfer functions of AR and ARMA models is as follows:

$$\frac{1}{1-\sum_{i=1}^{p} I_i B^i} = \frac{1-\sum_{j=1}^{m} \theta_j B^j}{1-\sum_{k=1}^{n} \varphi_i B^k} \tag{6}$$

In Equation (6), the left side represents the transfer function of an AR(p) model, and the inverse function I_i is equal to the parameter φ_i of the AR model. The right side represents the transfer function of an ARMA(n,m) model. By moving the same power coefficients of the shift operator B in Equation (6), it can be deduced that

$$\begin{aligned}
&\varphi_1 = \theta_1 + I_1 \\
&\varphi_2 = \theta_2 - \theta_1 I_1 + I_2 \\
&\cdots\cdots \\
&\varphi_n = \theta_m I_{n-m} - \ldots - \theta_2 I_{n-2} - \theta_1 I_{n-1} + I_n \\
&0 = \theta_m I_{k-m} - \ldots - \theta_2 I_{k-2} - \theta_1 I_{k-1} + I_k
\end{aligned} \tag{7}$$

According to Equation (7), the solution for $\varphi_i (i = 1, 2, \ldots, n)$ can be obtained as follows:

$$\begin{bmatrix} \varphi_1 \\ \varphi_2 \\ \varphi_3 \\ \vdots \\ \varphi_n \end{bmatrix} = \begin{bmatrix} \theta_1 \\ \theta_2 \\ \theta_3 \\ \vdots \\ \theta_n \end{bmatrix} + \begin{bmatrix} 1 & 0 & 0 & \ldots & 0 \\ -\theta_1 & 1 & 0 & \ldots & 0 \\ -\theta_2 & -\theta_1 & 1 & \ldots & 0 \\ \vdots & \vdots & \vdots & \ddots & \vdots \\ -\theta_{n-1} & -\theta_1 & -\theta_1 & \ldots & 1 \end{bmatrix} \begin{bmatrix} I_1 \\ I_2 \\ I_3 \\ \vdots \\ I_n \end{bmatrix} \tag{8}$$
$$(\text{when } j > 0,\ \theta_j = 0)$$

For the last term in Equation (7), let $k = n+1, n+2, \ldots, n+m$, where $n + m = p$. Then, $I_i (i = n+1, n+2, \ldots, n+m)$ can be expressed as

$$\begin{bmatrix} I_{n+1} \\ I_{n+2} \\ I_{n+3} \\ \vdots \\ I_{n+m} \end{bmatrix} = \begin{bmatrix} I_n & I_{n-1} & I_{n-2} & \ldots & I_{n+1-m} \\ I_{n+1} & I_n & I_{n-1} & \ldots & I_{n+2-m} \\ I_{n+2} & I_{n+1} & I_n & \ldots & I_{n+3-m} \\ \vdots & \vdots & \vdots & \ddots & \vdots \\ I_{n+m-1} & I_{n+m-2} & I_{n+m-3} & \ldots & I_n \end{bmatrix} \begin{bmatrix} \theta_1 \\ \theta_2 \\ \theta_3 \\ \vdots \\ \theta_n \end{bmatrix} \tag{9}$$

Both Equations (8) and (9) are linear equation sets about θ_j, so θ_j can be solved first according to Equation (9), and then φ_i can be solved according to Equation (8).

Based on the theoretical research of Ljung [23], this paper adopted the order scheme of $p = 2m$, $q = 2m - 1$ to establish the ARMA model of a multi-degree-of-freedom system, which is also a widely used model scheme in engineering [24].

3.1.4. Principal Component Analysis

There is a large amount of environmental noise in the on-site measurement data of a structure, which greatly reduces the efficiency and reliability of damage monitoring. Therefore, it is necessary to conduct feature analysis on the data to remove the noise influence and amplify the impact of structural damage on data changes. In this paper, PCA was adopted for feature extraction, which can filter out noise and redundancy in the original high-dimensional feature space data and transform it into interpretable data in a low-dimensional feature space. It can eliminate confusing data while preserving the main information [25].

For a dataset $X \in R^{n\times p}$ with p variables and n samples,

$$X = \begin{bmatrix} x_{11} & x_{12} & \cdots & x_{1p} \\ x_{21} & x_{22} & \cdots & x_{2p} \\ \vdots & \vdots & \ddots & \vdots \\ x_{n1} & x_{n2} & \cdots & x_{np} \end{bmatrix} = \begin{bmatrix} X_1 & X_2 & \ldots & X_p \end{bmatrix} \tag{10}$$

the i-th principal component of X can be represented as

$$\begin{gathered} Y_i = u_{1i}X_1 + u_{2i}X_2 + \cdots + u_{pi}X_p \\ (i = 1,\ 2,\ \cdots,\ p) \end{gathered} \tag{11}$$

where u_{pi} represents the characteristic vector. After standardizing the original data matrix X, the resulting matrix X^* has a covariance matrix S that is equal to the correlation coefficient matrix:

$$S = \frac{1}{n-1} X^* X^{*T} \tag{12}$$

The eigenvalues of the covariance matrix of X^* are denoted by $\lambda_1,\ \lambda_2,\ \ldots,\lambda_p$, and $\lambda_1 \geq \lambda_2 \geq \ldots \geq \lambda_p > 0$. The number of principal components n is determined by the variance accumulative contribution rate (ACR) $\sum_{i=1}^{n} \lambda_i / \sum_{i=1}^{p} \lambda_i$. If the ACR of the first n principal components is not less than 85%, it indicates that the first n principal components have incorporated most of the original data information [26].

3.2. *Damage Sensitive Feature*

DSF refers to the characteristic or indicators used to identify and locate potential structural damage, which can reflect the internal state changes of the structure or system. It requires the establishment of an ARMA model to analyze the signal, and the parameters of the model include the inherent characteristics of the system. By extracting the carrier containing structural features, it can be determined whether the structure is damaged. Because the AR coefficients reflect the dynamic characteristics of the system, they can be used to construct the DSF.

3.2.1. Mahalanobis Distance

The principle of the structural damage detection method based on statistical pattern recognition is to compare two sets of model parameters, one set is the model feature parameters of the structure in the healthy state, and the other set is the feature parameters in the unknown (to be identified) state. In the damage identification method based on the ARMA model, the MD is usually used to determine the distance between the sample data and the reference data after feature extraction, and then evaluate the health status of the structure. MD calculates the distance between two multivariate sets by considering the correlation between them, which is an effective method for computing the similarity between two unknown-state sample sets. The expression of the MD between the sample set z and the population G is:

$$d_M(z, G) = \left[\left(z - v_G \right)^T \sum\nolimits^{-1} (z - v_G) \right]^{\frac{1}{2}} \tag{13}$$

where v_G represents the mean vector of G, and Σ represents the covariance matrix of G.

3.2.2. Construction of DSF

ARMA models were established for the unknown-state sample set, reference sample set, and training sample set. The DSF was constructed by measuring the differences between the data parameters of the ARMA model under the unknown state and the healthy state. The formula is as follows:

$$\mathrm{DSF}(u,r) = \frac{d_M^2(u,T)}{d_M^2(r,T)} \tag{14}$$

where u represents the unknown-state sample set, r represents the reference sample set of the healthy state, and T represents the training sample set of the healthy state. A DSF value close to 1 indicates that the structure is in a healthy state, while a DSF value greater than 1 indicates damage. Moreover, the severity of the damage is positively correlated with the magnitude of the DSF.

4. Results and Discussions

This section provides an explanation and discussion of the damage identification results using the FE models and SHM measured data. Firstly, a concise introduction is provided for the FE model, which includes the setup of five simulated DCs. The establishment of a time series model and the detailed results of damage identification are then elaborated. Finally, the obtained results are thoroughly explained and discussed. Furthermore, this section is based on two years of SHM data collected from the timber lattice shell dome of Taiyuan Botanical Garden. The proposed damage identification method from this article is applied to identify the damage in the dome, and the identification results are subsequently analyzed and discussed.

4.1. Damage Identification Based on Numerical Models

Structural damage in practical engineering exhibits characteristics such as randomness and finiteness, making it difficult to determine the effectiveness and availability of time series models through measured data. However, FE methods can be used to simulate the health and damage status of the structure, in order to establish and validate the ARMA model.

To verify the feasibility and applicability of the ARMA model for large-span timber lattice structures, this paper used Midas Gen, a general structural analysis software, to perform FE modeling and analysis for the timber lattice dome of Taiyuan Botanical Garden. The stiffness of individual or multiple components was reduced to simulate the structural damage conditions (DC). Gaussian white noise was used to excite the structure, and the response results of multiple measuring points (MP) were extracted to evaluate the damage identification effect of the ARMA model.

4.1.1. Finite Element Model

This paper used 3D line elements to analyze the spatial structure of the dome. The timber beams were modeled with beam elements, while the cable net was modeled with cable (tensile only) elements. In this study, the maximum main stress and maximum vertical deformation were employed as the convergence indicators. The analysis findings demonstrated that as the number of elements increased from 7732 to 334,817, both the maximum stress and maximum deformation exhibited a gradual convergence. The mesh convergence plot is shown in Figure 2. Notably, the maximum stress was more sensitive to the element number, thus serving as the primary controlling factor. Taking into account both computational accuracy and efficiency, the 134,372-element model was ultimately selected, which achieved a maximum stress convergence within a 5% range and satisfied the required computational accuracy.

The FE model took into account the fact that the three-layer timber beams are not coplanar, as shown in Figure 3. The timber beams are made of European spruce glued laminated timber GL28h, with a main beam section of 200 mm × 400 mm (width × height) and a secondary beam section of 200 mm × 300 mm (width × height). The components of the lattice structure are mainly subjected to axial forces, and glued laminated timber has relatively ideal material properties. According to material tests [27] and related studies [28–30], the FE material model adopted an orthotropic bilinear elastic-plastic model. The material test data and FE material model were shown in Figure 4, and the material properties are detailed in Table 2. The cable has a diameter of 26 mm, and the material used is austenitic steel 06Cr17Ni12Mo2 (316 stainless steel), with an elastic modulus of 160,000 MPa and a design pre-tension value of 40 kN.

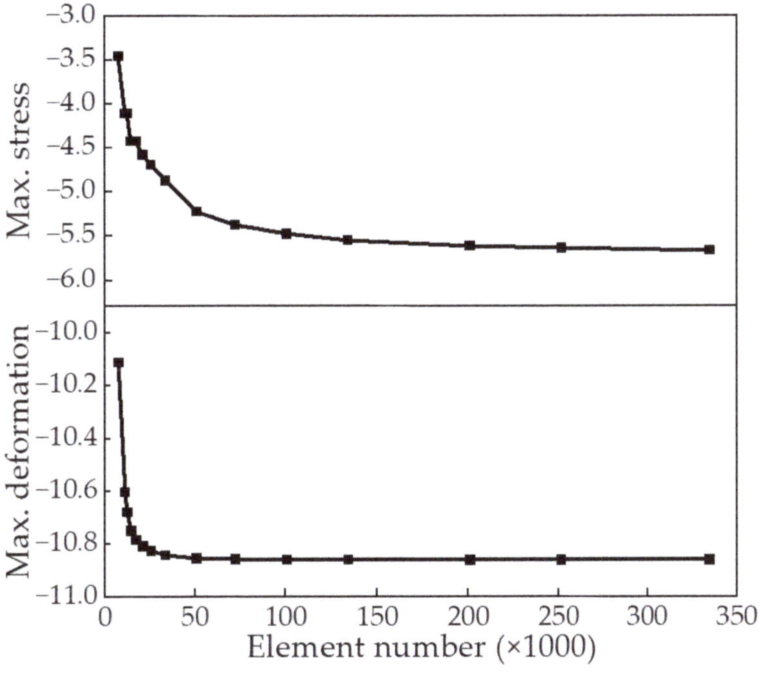

Figure 2. Mesh convergence analysis.

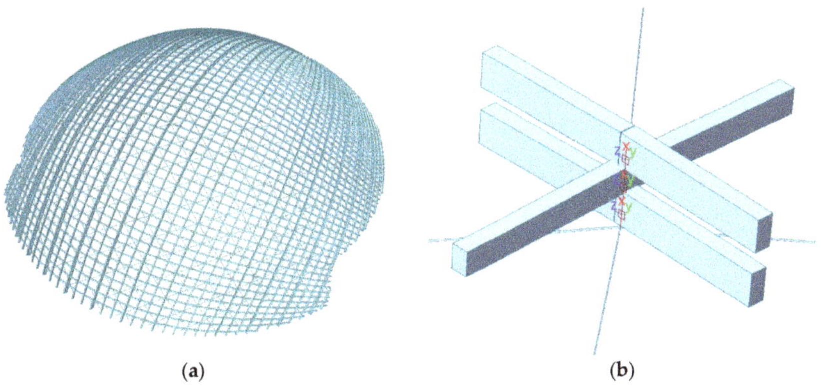

Figure 3. FE model. (**a**) Overall structure; (**b**) timber beam's stacking joint.

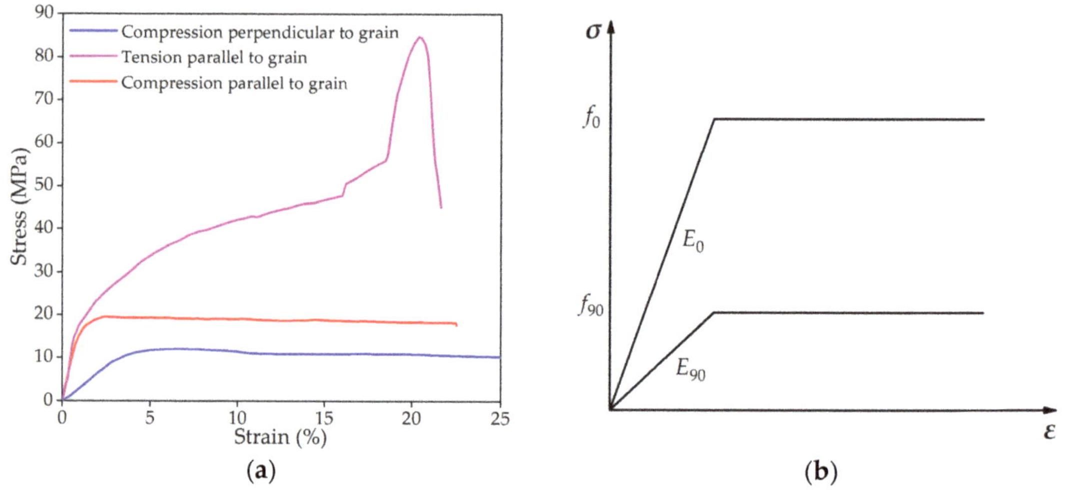

Figure 4. Material properties. (**a**) Material test data (GL28h), reprinted with permission from Ref. [27]. 2022, Shuizhong Jia; (**b**) bilinear material model of glued laminated timber [31].

Table 2. Material property of GL28h.

$E_{c,0}$ (MPa)	$E_{t,0}$ (MPa)	$E_{c,90}$ (MPa)	$f_{c,0}$ (MPa)	$f_{t,0}$ (MPa)	$f_{c,90}$ (MPa)
13,901	7438	185	29.6	77.5	5.0

Based on the measured data of SHM, simple modifications were attempted on the FE model. Initially, the acceleration signal during the early stage of structural service was processed and analyzed. The time-domain data were then transformed into frequency-domain data using Fourier transform. To eliminate high-frequency noise, a low-pass filter of 50 Hz was applied. The power spectral density (PSD) of the acceleration data was calculated, and the dominant frequency with significant amplitude in the signal was extracted through peak detection. This allowed for the determination of the natural frequency of the structure.

Subsequently, modifications were made to the FE model parameters. The elastic modulus and density of the material were adjusted, along with the magnitude of the additional constant load (converted into mass) on the roof, and the connection stiffness of the joints. These modifications aimed to align the natural vibration characteristics of the FE model with the analysis results obtained from the measured data.

4.1.2. Damage Conditions

When establishing a time series model, the first step is to determine the sampling time interval Δt and the sample length L for the continuous signal. When sampling a continuous signal, a reasonable sampling time interval is $\Delta t < 1/(2f_{max})$, and a reasonable sample length is $L > 1/\Delta f_{min}$, where f_{max} is the highest frequency in the interested vibration modes and Δf_{min} is the minimum difference between adjacent frequencies [32].

Given the complex natural vibration characteristics of the lattice shell structure, a 100-order natural vibration modal analysis was conducted for the FE model. The results of the modal analysis are presented in Figure 5 and Table 3. The analysis revealed that the natural frequencies of the lower order modes are closely spaced, with a difference of only 8.92 Hz between the first and 70th modes. By the 94th mode, 90% of the cumulative participation mass of the vibration modes was reached, indicating that the first 94 modes could capture the majority of the natural vibration characteristics of the structure. Therefore, the appropriate sampling frequency and sample length were calculated based on the first 94 modes. The maximum frequency in the first 94 modes was approximately 49.85 Hz, and the minimum difference between adjacent frequencies was 0.0109 Hz. According to the sampling theories [32], the sampling time interval of this structure should be less than 0.0101 s, and the sample length should be greater than 91.75 s. Therefore, a sampling frequency of 100 Hz and a sample length of 120 s were adopted to establish a time series model. According to the basic principle of time series analysis, the signal input needs to meet the characteristics of stationary, normal, and zero-mean, and Gaussian white noise can just meet the above characteristics. Thus, Gaussian white noise generated by MATLAB was used as the environmental excitation to conduct the time history analysis on the FE model.

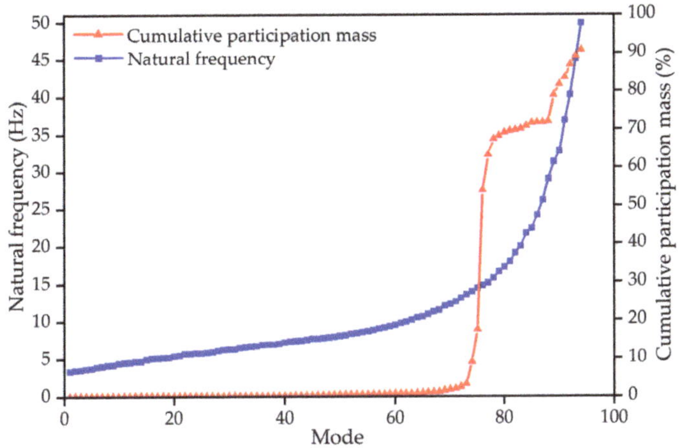

Figure 5. 100-order Z-axis modal analysis.

Table 3. Natural frequencies of the FE model.

Mode Order	Frequency (Hz)	Mode Order	Frequency (Hz)	Mode Order	Frequency (Hz)	Mode Order	Frequency (Hz)
1	3.3537	26	5.8696	51	8.2039	76	14.8373
2	3.5002	27	5.9719	52	8.3625	77	15.1701
3	3.5525	28	6.134	53	8.4575	78	15.8307
4	3.6825	29	6.2517	54	8.5635	79	16.6591
5	3.7451	30	6.2953	55	8.7213	80	17.2487
6	3.9603	31	6.3233	56	8.847	81	17.9974
7	4.0067	32	6.5079	57	9.0585	82	19.1284
8	4.1308	33	6.6175	58	9.1913	83	20.0353
9	4.2118	34	6.6899	59	9.3714	84	21.7652
10	4.431	35	6.7379	60	9.5392	85	22.4354
11	4.4916	36	6.9291	61	9.7873	86	24.1665
12	4.5736	37	6.9573	62	9.9823	87	26.2336
13	4.6526	38	6.9682	63	10.2407	88	29.05
14	4.6659	39	7.0649	64	10.5235	89	31.3453
15	4.9154	40	7.2344	65	10.6672	90	32.7382
16	5.0791	41	7.3614	66	10.9764	91	36.9209
17	5.1145	42	7.3811	67	11.3259	92	40.3238
18	5.1646	43	7.4171	68	11.5494	93	45.1509
19	5.2276	44	7.5568	69	12.0646	94	49.8536
20	5.3595	45	7.6886	70	12.2717	95	59.3389
21	5.4691	46	7.7193	71	12.6162	96	72.188
22	5.6584	47	7.7773	72	13.0992	97	88.9345
23	5.6804	48	7.8668	73	13.6149	98	121.4361
24	5.7494	49	7.9797	74	13.9864	99	180.6925
25	5.7668	50	8.1043	75	14.4609	100	349.0827

In practical engineering, vibration sensors were used for SHM of the dome structure. Specifically, nine vibration MPs were set up at the joint positions, taking into account the shape and characteristics of the structure. From above, the MPs were distributed at the central point of the dome and in the directions of east, west, south, and north. In the vertical direction, the MPs were evenly distributed at the top, middle, and bottom of the structure. During the time history analysis on the FE models, acceleration response data were extracted from the nine MPs. The layout of the MPs was shown in Figure 6. In the FE analysis process, stiffness reduction was applied to local timber beam and steel cable elements to simulate damage or stiffness degradation of the components. This paper developed five sets of DCs, including damage to single and multiple components. The five simulated DCs are detailed in Table 4.

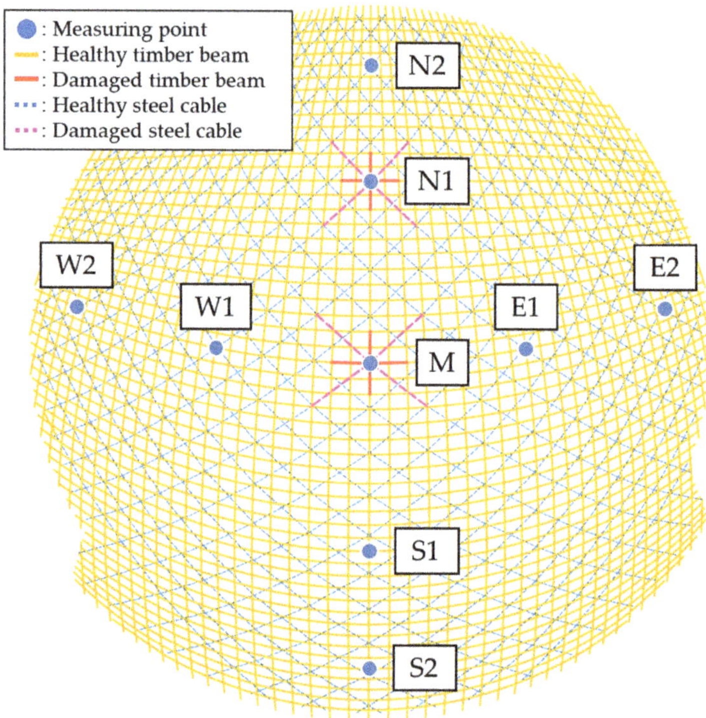

Figure 6. Distribution of MPs and damaged components.

Table 4. Damage conditions.

Condition No.	MP M		MP N1	
	Beam	Cable	Beam	Cable
1	●	●	●	●
2	● (yellow)	●	●	●
3	● (orange)	●	●	●
4	● (orange)	● (orange)	●	●
5	● (orange)	● (orange)	● (orange)	● (orange)

Note: ● (green): in a healthy state; ● (yellow): damaged by 40%; ● (orange): damaged by 80%.

4.1.3. Establishment of ARMA Model

This study established time series models for all nine MPs under various DCs as shown in Table 4. In this paper, the process of establishing a time series model was demonstrated using the acceleration time history data of MP M, which is located at the top of the dome structure.

Before establishing the ARMA model, it is necessary to standardize the sample dataset. The normal distribution probability plot of the processed data, which was shown in Figure 7c, indicated that the processed sample is a stationary, normal, and zero-mean time series, satisfying the prerequisite for establishing a time series model. After preprocessing the data, the ACF and PACF of the data sample were calculated, as shown in Figure 7a,b. The ACF and PACF gradually decreased to within the 95% confidence interval, and both exhibited tailing characteristics. According to Table 1, an ARMA model could be established for the data sample.

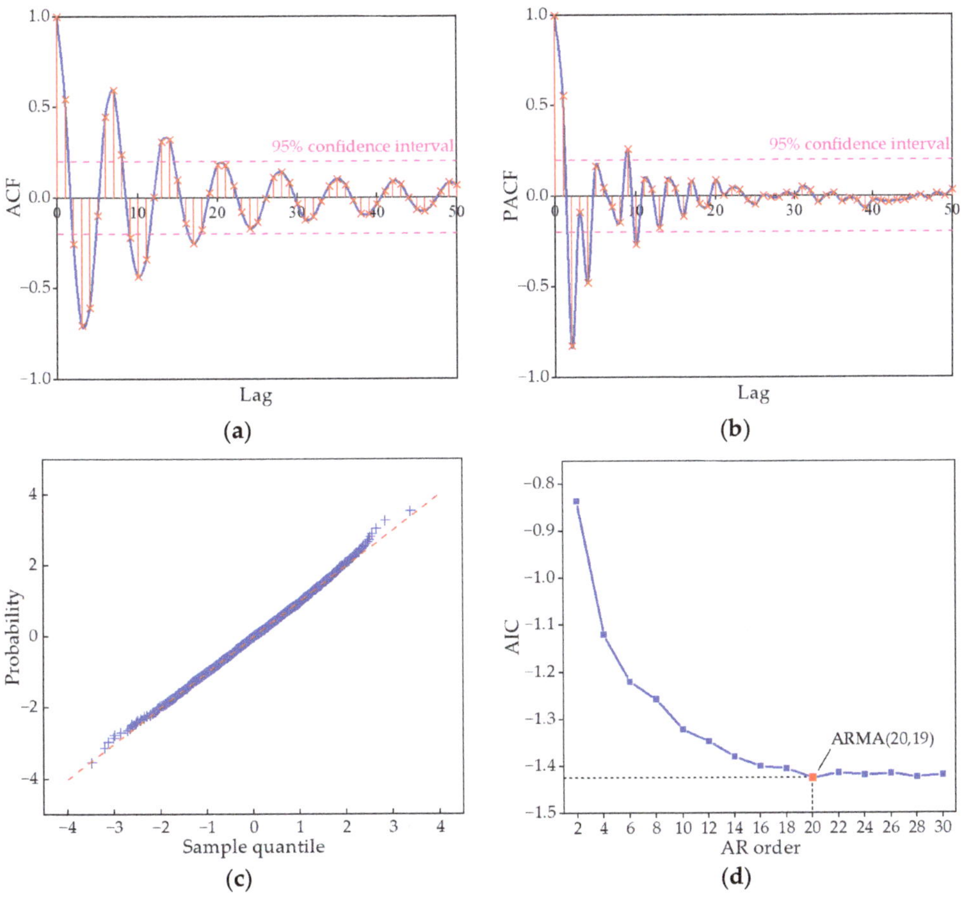

Figure 7. ARMA order selection. (**a**) ACF; (**b**) PACF; (**c**) normal distribution probability; (**d**) AIC of each order combination.

Subsequently, the AIC and LAR method were used to determine the order and estimate the parameters of the model. Figure 7d shows the AIC results of the ARMA($2m$,$2m-1$) model for the first 30 orders. Based on the AIC results, an ARMA(20,19) model was established for the MP M, and its estimated parameters are shown in Table 5.

Table 5. Estimated ARMA model parameters.

Order	AR Coef.	MA Coef.	Order	AR Coef.	MA Coef.
1	0.1956	0.9294	11	0.0487	−0.3235
2	−0.3641	0.1359	12	−0.0731	−0.1088
3	−0.0754	−0.3965	13	−0.0082	0.3625
4	−0.2056	−0.4014	14	−0.4730	0.8166
5	−0.1359	−0.0875	15	0.1008	0.4343
6	−0.1749	0.1776	16	−0.2690	−0.0291
7	0.2484	0.4023	17	−0.1583	−0.1125
8	−0.1414	0.1495	18	−0.3139	0.0210
9	0.0141	−0.2229	19	0.0208	0.1027
10	0.0242	−0.2021	20	0.0635	-

4.1.4. Damage Identification of FE Models

ARMA(p,q) models were established for the responses obtained from the FE models under the five DCs shown in Table 4. To alleviate the computational burden arising from the high AR order, dimensionality reduction could be achieved by conducting PCA on its parameters. Based on the principle of PCA, the AR parameters of each MP were analyzed for five distinct DCs.

Table 6 presents the eigenvalue and ACR obtained through PCA for the data of each MP. The eigenvalue quantifies the extent to which each principal component explains the variation in the data, while the ACR represents the proportion of total variance explained by the first n principal components. An ACR exceeding 85% indicates that the primary components already encompass most of the feature information [26]. Based on the PCA results, the eigenvalues of the first and second order AR coefficients for each MP were nearly 10 and 4, respectively. Furthermore, the ACRs for the first three orders at each MP all exceeded 90%, surpassing the 85% threshold. In some cases, the ACR even reached approximately 99%. Consequently, it can be inferred that employing the first three order AR coefficients to construct DSF can yield sufficiently accurate outcomes in theory.

Table 6. PCA results.

Principal Component	M		E1		E2	
	Eigenvalue	ACR	Eigenvalue	ACR	Eigenvalue	ACR
1	10.3245	51.62%	14.4691	72.35%	13.7320	68.66%
2	4.6304	74.77%	3.2225	88.46%	3.1710	84.52%
3	3.8133	93.84%	1.2549	94.73%	1.9219	94.12%
4	0.8551	98.12%	1.0002	99.73%	1.0343	99.30%
5	0.3768	100.00%	0.0532	100.00%	0.1407	100.00%
Principal Component	**W1**		**W2**		**S1**	
	Eigenvalue	**ACR**	**Eigenvalue**	**ACR**	**Eigenvalue**	**ACR**
1	15.0565	75.28%	10.4912	52.46%	12.8791	64.40%
2	4.0753	95.66%	6.4063	84.49%	4.0294	84.54%
3	0.7505	99.41%	1.8273	93.62%	2.8699	98.89%
4	0.1117	99.97%	1.0842	99.05%	0.2155	99.97%
5	0.0060	100.00%	0.1909	100.00%	0.0061	100.00%
Principal Component	**S2**		**N1**		**N2**	
	Eigenvalue	**ACR**	**Eigenvalue**	**ACR**	**Eigenvalue**	**ACR**
1	10.8759	54.38%	8.3965	41.98%	11.1385	55.69%
2	4.3566	76.16%	7.4009	78.99%	7.1796	91.59%
3	2.7723	90.02%	2.4547	91.26%	1.1040	97.11%
4	1.9425	99.74%	1.2340	97.43%	0.4086	99.15%
5	0.0527	100.00%	0.5139	100.00%	0.1693	100.00%

The acceleration time series data from each MP under the healthy state (DC 1) was subjected to windowing processing, resulting in two equally divided sections. The first 60 s of the resulting time series data, comprising 6000 sample data points, was utilized as training sample data, while the last 60 s, also comprising 6000 sample data points, was used as reference sample data. The remaining data obtained from other DCs were unknown-state sample data. The first three order AR coefficients of the ARMA model were extracted for each sample data. A principal component matrix was established and the MD was calculated between the principal component matrices of a single MP under different DCs. Corresponding DSF for damage identification was constructed according to Equation (14). The DSF for each MP is presented in Figure 8.

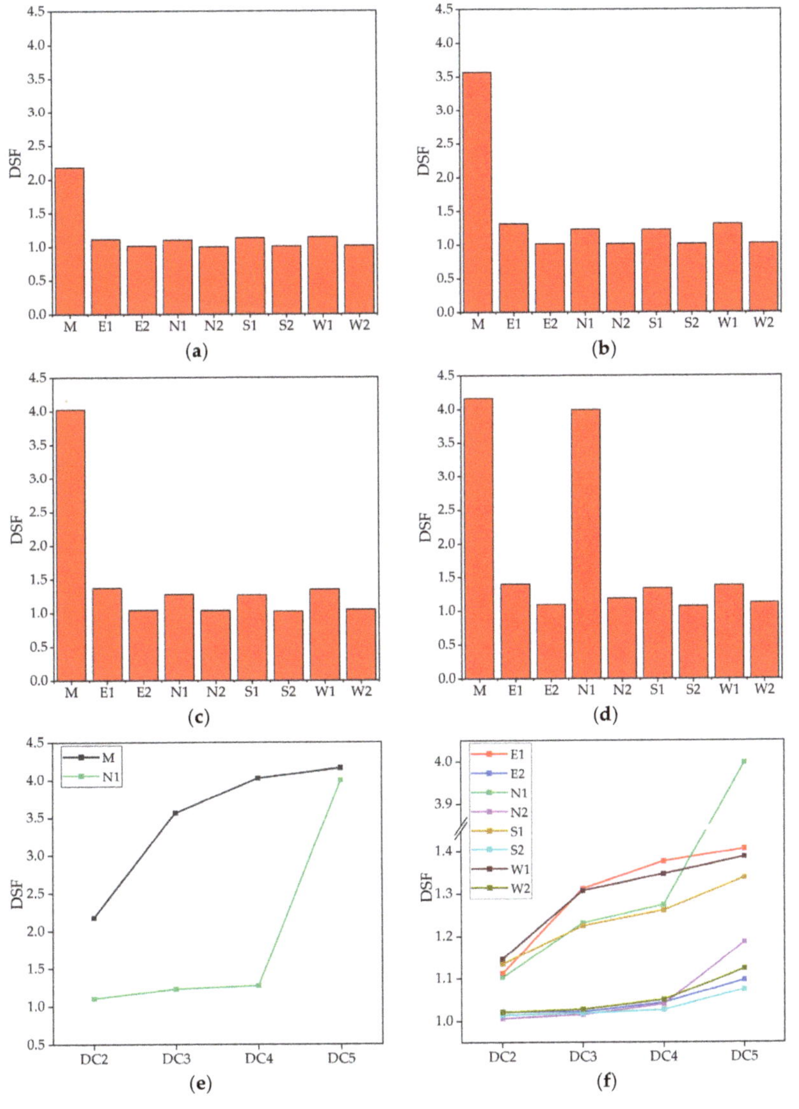

Figure 8. Damage identification results of FE models. (**a**) Results under DC 2; (**b**) results under DC 3; (**c**) results under DC 4; (**d**) results under DC 5; (**e**) results of M and N1; (**f**) results of undamaged MPs.

Based on the results of damage identification, the following observations could be made:

(1) When the damage of the timber beam at MP M was 40% (DC 2), the DSF at M was 2.1799, which was significantly higher than the DSF at other MPs, not exceeding 1.15. As the damage increased to 80% (DC 3), the DSF at M increased to 3.5720, while the average DSF at other MPs was 1.1450. This represented a 63.9% increase in DSF compared to the 40% damage level. When both the timber beam and steel cable at M were damaged simultaneously (DC 4), the DSF further increased to 4.0281, which was approximately 12.77% higher than the previous DC. The average DSF value at other MPs was 1.1773. These results demonstrate that the DSF constructed using PCA and MD is highly sensitive to component damage, capable of identifying and locating damage to timber beams and steel cables, and positively correlates with the degree of damage. Therefore, it could reflect the degree of damage to the damaged components to a certain extent.

(2) When the timber beam and steel cable at MPs M and N1 were simultaneously damaged by 80% (DC 5), the DSF of M and N1 were 4.1645 and 3.9983, respectively, which were significantly higher than the DSF values of other MPs, ranging from 1.07 to 1.38. This demonstrates that the proposed damage identification model can effectively identify conditions where multiple components are damaged simultaneously.

(3) As the damage at MP M intensified from DC 2 to 4, the DSF of M significantly increased, while the DSF of adjacent MPs N1, E1, S1, and W1 also slightly increased, with an average increase of approximately 0.19. However, the DSF of MPs N2, E2, S2, and W2 did not increase significantly, with an average increase of no more than 0.1. Under DC 5, further damage occurred at MP N1. During this process, the DSF of N2 and M, which were closest to N1, increased the most, with values of 0.2070 and 0.1364, respectively. However, the DSF of S2, W2, and E2, which were farthest from N1, remained almost unchanged, with values of 0.049, 0.074, and 0.054, respectively. This suggests that structural damage at a specific location can cause an increase in the DSF of surrounding MPs, and the degree of increase is inversely proportional to the distance. This phenomenon may be attributed to the changes in the local natural vibration characteristics caused by component damage, which were not significantly far from the damage location.

(4) Assuming that no MPs are set at M and N1, meaning that the DSF of M and N1 were not taken into account, under DC 4, the DSF values of E1, W1, and S1 experienced a significant increase, ranging from 1.25 to 1.35, while the DSF of the remaining MPs showed no noticeable changes. This suggests that the damage occurred in areas near E1, W1, and S1. It can be inferred that the damage may have occurred around M, which aligns with the severe damage observed at M under DC 4. From DC 4 to 5, the DSF of N2 increased significantly, while the DSF of the other MPs increased slightly and to a similar extent. This implies that the damage may have occurred in a location close to N2. As determined, N1 sustained damage during the transition from DC 4 to 5, which is consistent with this characteristic. The aforementioned statement indicates that even if the damage does not occur at MPs, the approximate location of the damage can be determined based on the DSF of the existing MPs, thereby achieving the objective of damage monitoring. This is advantageous for optimizing the number of MPs.

4.2. Damage Identification Based on SHM Data

This paper presented a damage identification study of the dome structure at Taiyuan Botanical Garden, utilizing measured data from vibration sensors. The structure employed glued laminated timber beam as its primary load-bearing component, and exhibited a large span, complex natural vibration characteristics, and susceptibility to wind vibration, earthquakes, and other impacts. To ensure the structure's safety, long-term SHM has been conducted since its service. Based on the structure's characteristics and FE analysis results, nine vibration MPs shown in Figure 6 were selected to monitor the structure's vibration behavior. These MPs were equipped with the 2D001 magnetoelectric vibration sensors produce by Donghua Testing Technology Co. Ltd. (Taizhou, China), with a maximum sampling frequency of 100 Hz. The data acquisition device is shown in Figure 9, and the technical parameters of the 2D001 sensor are shown in Table 7.

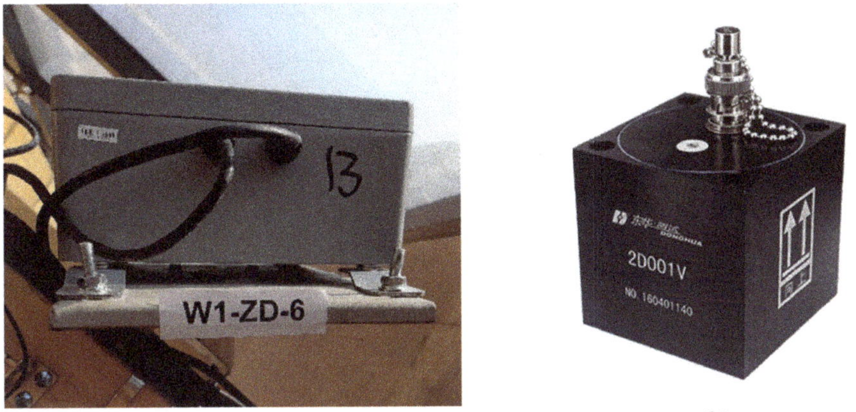

(**a**) (**b**)

Figure 9. Data acquisition device. (**a**) Tri-dictional vibration monitoring module; (**b**) 2D001 magnetoelectric vibration sensor.

Table 7. Technical parameters of the 2D001 magnetoelectric vibration sensor.

Mode	**0**	**1**	**2**	**3**
Parameter	**Acceleration**	**Low Velocity**	**Medium Velocity**	**High Velocity**
Sensitivity	0.3 V/m·s^{-2}	20 V/m·s^{-1}	5 V/m·s^{-1}	0.3 V/m·s^{-1}
Capacity	20 m·s^{-2}	0.125 m·s^{-1}	0.3 m·s^{-1}	0.6 m·s^{-1}
Bandwidth (−3–+1 dB)	(0.25–100) Hz	(1–100) Hz	(0.5–100) Hz	(0.17–80) Hz
Output Impedance	50 kΩ			
Working Temperature	(−20–80) °C			
Dimensions	63 mm × 63 mm × 63 mm			
Weight	0.6 kg			

Under normal working conditions, the only dynamic load on this structure was wind load. However, due to the fact that the dome is covered with glass, the structure was exposed to sunlight radiation during the day, resulting in temperature gradients on the surface that may have an uncertain impact on the sampling data and final analysis results. Therefore, this study selected measured data at night as the sample data. Specifically, acceleration data from the initial stage of structural service (October 2020) were chosen as the healthy-state sample data, while data after two years of service (October 2022) were selected as the unknown-state sample data, with a sample length of 120 s. Hence, ARMA modeling was performed and the health status of the structure after two years of service was evaluated. Figure 10 displays partial data for the initial stage and after two years of service.

The results of the calculation indicated that the ACF and PACF of the model are tailing, which satisfies the prerequisite for ARMA modeling. The AIC was employed to determine the order of the ARMA($2m$,$2m-1$) model for the measured data. The findings reveal that the ARMA(28,27) model had the lowest AIC value of -1.4235 among the first

30 orders, indicating that this order scheme can achieve a high degree of fitting. The vertical acceleration data of nine vibration sensors at each MP were preprocessed, and ARMA(28,27) models were established. PCA was performed on the AR parameters using a set of measured data from each MP under both healthy and unknown states. The analysis results showed that the first 5 order AR parameters of any ARMA model contained 90% of the model's damage characteristic values. Therefore, the first 5 order AR coefficients were utilized to construct subsequent DSF.

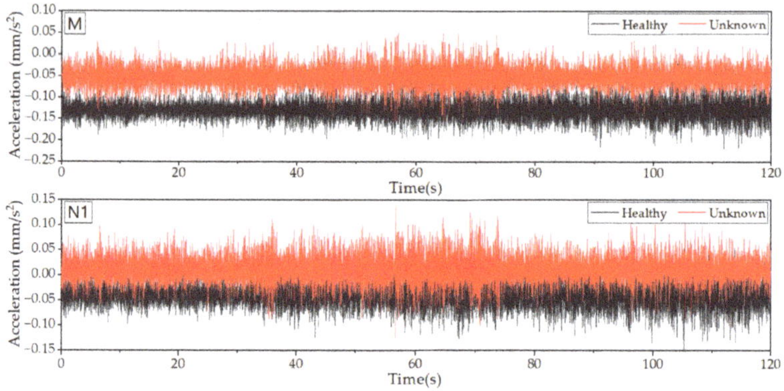

Figure 10. Original SHM data (M and N1).

The acceleration time series of each MP in a healthy state was extracted and partitioned into two equal segments by windowing. The first 60 seconds' 6000 sample data points were designated as training sample data, while the last 60 seconds' 6000 sample data points were designated as reference sample data. The remaining data were used as unknown-state sample data. The MD was computed to generate DSF for damage identification.

According to the analysis results, the MDs of each sample were not significantly different, and the DSF of all MPs were close to 1.0, indicating that the structure was almost not damaged after two years of service. The results of MD and DSF of each MP were shown in Figure 11. In addition, it could be found that the DSF of MP S2 was 1.097, while the maximum value of DSF of other MPs was 1.062, and the average value was 1.046, indicating that the damage degree of the component at S2 was slightly higher than that of other MPs. Figure 12 shows the static FE analysis results of the dome. Under the combined action of dead load and roof live load, the stress level of the timber components around S2 was higher than that in other areas. Moreover, real-time dynamic loads such as wind loads and temperature effects might have adverse effects on local areas. These adverse effects might be the potential factors leading to the increase of DSF at S2. Therefore, it is necessary to pay attention to the SHM data of this dome structure, and focus on the acceleration data of MP S2, so as to timely warn when the structure suffers serious damage.

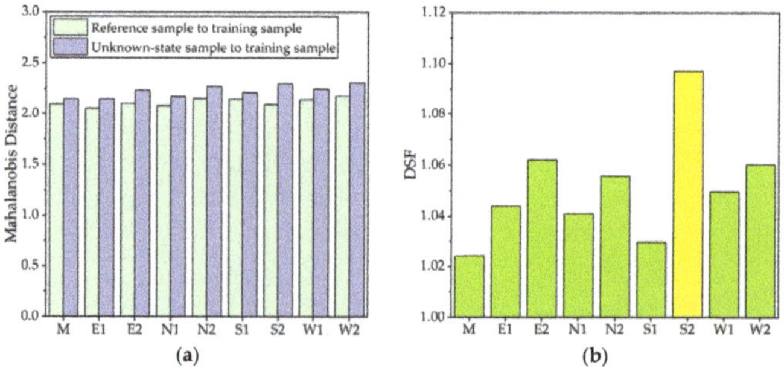

Figure 11. Damage identification results. (**a**) MD results; (**b**) DSF results.

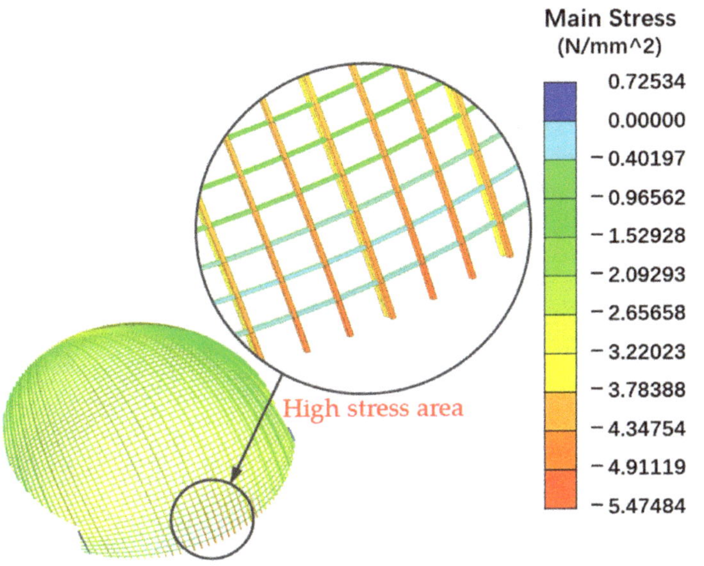

Figure 12. High stress area exhibited in static FE analysis result.

5. Conclusions

Structural damage identification is a crucial analysis method in SHM, with significant scientific and engineering applications. To address the challenges of incomplete modal information, high modal density, and large degrees of freedom in large-span spatial lattice structures, this paper proposed a structural damage identification method based on the ARMA model and PCA method for structures of this kind. Based on the above research, the following conclusions can be drawn:

(1) This article was based on a time series model and introduced PCA for extracting principal components. In this study, only the first three AR coefficients were required to achieve damage identification and it demonstrated high accuracy.
(2) The MD is advantageous compared to other distance methods in constructing DSF because it considers the differences and correlations in the variability of variables for each observation. Additionally, it can effectively calculate the influence of scale on different measurement values.
(3) Verified by the FE model, the damage identification method proposed in this article was found to be sensitive to structural damage. It could, to a certain extent, accurately locate the site of structural damage and reflect the extent of the damage. Therefore, it can be effectively utilized for practical structural damage identification.
(4) Based on SHM data, the paper identified the damage of the structure after two years of service. The results indicated that the structure is in a relatively healthy state, but the DSF of MP S2 is slightly higher than those of other MPs. In the future, it is necessary to pay attention and give timely warnings if necessary.

However, there are still some shortcomings in this article. The next step should be to conduct more in-depth research on the rationality of the number and distribution of measurement points, as well as the refinement of the FE model correction.

Author Contributions: Conceptualization, G.W. and H.L.; methodology, G.W., C.X. and H.L.; software, G.W. and C.X.; validation, G.W. and S.Z.; formal analysis, G.W., C.X. and S.Z.; investigation, G.W., C.X., S.Z. and Z.Z.; data curation, Z.Z. and L.Z.; writing—original draft preparation, G.W.; writing—review and editing, H.L.; visualization, J.W. and B.Q.; supervision, H.L.; project administration, L.Z. and B.Q.; funding acquisition, C.X. and L.Z. All authors have read and agreed to the published version of the manuscript.

Funding: This research was funded by National Natural Science Foundation of China, grant number 52278198; and by the Structural Health Monitoring Project for Taiyuan Botanical Garden, grant number 0632-1920FW3L2061-01.

Institutional Review Board Statement: Not applicable.

Informed Consent Statement: Not applicable.

Data Availability Statement: The data presented in this study are available from the first author, (wangguoqing0066@link.tyut.edu.cn), upon reasonable request.

Conflicts of Interest: The authors declare no conflict of interest.

References

1. Zhao, Y.; Gong, M.; Yang, Y. A review of structural damage identification methods. *World Earthq. Eng.* **2020**, *36*, 73–84.
2. Yan, G.; Duan, Z.; Ou, J. Review on structural damage detection based on vibration data. *J. Earthq. Eng. Eng. Vib.* **2007**, *27*, 95–103. [CrossRef]
3. Lakshmi, K.; Rao, A.R.M. Damage identification technique based on time series models for LANL and ASCE benchmark structures. *Insight-Non-Destr. Test. Cond. Monit.* **2015**, *57*, 580–587. [CrossRef]
4. Razavi, B.S.; Mahmoudkelayeh, M.R.; Razavi, S.S. Damage Identification under Ambient Vibration and Unpredictable Signal Nature. *J Civ. Struct Health Monit* **2021**, *11*, 1253–1273. [CrossRef]
5. Box, G.E.P.; Jenkins, G.M. *Time Series Analysis, Forecasting and Control*; John Wiley & Sons: Hoboken, NJ, USA, 1971; Volume 134.
6. Bao, C.; Hao, H.; Li, Z.-X. Integrated ARMA Model Method for Damage Detection of Subsea Pipeline System. *Eng. Struct.* **2013**, *48*, 176–192. [CrossRef]
7. Liu, Y.; Li, A.; Ding, Y.; Fei, Q. Time Series Analysis with Structural Damage Feature Extraction and Alarming Method. *Chin. J. Appl. Mech.* **2008**, *25*, 253–257+357.
8. Zuo, H.; Guo, H. Structural Nonlinear Damage Identification Method Based on the Kullback-Leibler Distance of Time Domain Model Residuals. *Remote Sens.* **2023**, *15*, 1135. [CrossRef]
9. Zhu, H.; Yu, H.; Gao, F.; Weng, S.; Sun, Y.; Hu, Q. Damage Identification Using Time Series Analysis and Sparse Regularization. *Struct. Control Health Monit.* **2020**, *27*, e2554. [CrossRef]
10. Chen, L.; Yu, L.; Fu, J.; Ng, C.-T. Nonlinear Damage Detection Using Linear ARMA Models with Classification Algorithms. *Smart Struct. Syst.* **2020**, *26*, 23–33. [CrossRef]
11. Zeng, Y.; Yan, Y.; Weng, S.; Sun, Y.; Tian, W.; Yu, H. Fuzzy Clustering of Time-Series Model to Damage Identification of Structures. *Adv. Struct. Eng.* **2019**, *22*, 868–881. [CrossRef]
12. Diao, Y.; Cao, Y.; Sun, Y. Structural Damage Identification based on AR Model Coefficients and Cointegration for Offshore Platform Under Environmental Variations. *Eng. Mech.* **2017**, *34*, 179–188.
13. Hu, M.-H.; Tu, S.-T.; Xuan, F.-Z. Statistical Moments of ARMA(n,m) Model Residuals for Damage Detection. *Procedia Eng.* **2015**, *130*, 1622–1641. [CrossRef]
14. Tang, Q.; Xin, J.; Zhou, J.; Fu, L.; Zhou, B. Structural damage identification method based on AR-GP model. *J. Vib. Shock* **2021**, *40*, 102–109. [CrossRef]
15. Sui, Z.; Wang, Q.; Jia, D.; Yang, Y. Structural Damage Identification based on AR Model and Bayesian Optimization SVM. *Low Temp. Archit. Technol.* **2020**, *42*, 74–76. [CrossRef]
16. Xu, J.; Guo, W.; Wang, X.; Cui, S. Study on the damage detection method for latticed shell structures based on AR model and BP neural network. *Spat. Struct.* **2015**, *21*, 66–71. [CrossRef]
17. Wang, C.; Guan, W.; Gou, J.; Hou, F.; Bai, J.; Yan, G. Principal Component Analysis Based Three-Dimensional Operational Modal Analysis. *Int. J. Appl. Electromagn. Mech.* **2014**, *45*, 137–144. [CrossRef]
18. Jia, S. Research on key technologies of large-span glued laminated timber reticulated shell structure design in Taiyuan Botanical Garden. *Build. Struct.* **2022**, *52*, 1–5+62. [CrossRef]
19. Liu, Z. Spatial Steel Structural Damage Alarming Based on Statistical Pattern Recognition. Master's Thesis, Harbin Institute of Technology, Harbin, China, 2010.
20. Zhou, X. Research on the Damage Identification of Girder Bridge Structures Based on Auto-Regressive Moving Average Model. Master's Thesis, Southwest Jiaotong University, Chengdu, China, 2008.
21. Du, Y.; Li, W.; Li, H.; Liu, D. Structural damage identification based on time series analysis. *J. Vib. Shock* **2012**, *31*, 108–111. [CrossRef]
22. Burnham, K.; Anderson, D. *Model Selection and Multimodel Inference: A Practical Information-Theoretic Approach*, 2nd ed.; Springer: Berlin/Heidelberg, Germany, 2002; Volume 67, ISBN 978-0-387-95364-9.
23. Ljung, L. System Identification: Theory for the User. In *System Identification: Theory for the User*; Prentice-Hall: Hoboken, NJ, USA, 1999.
24. Yang, S. *Time Series Analysis in Engineering Application*, 2nd ed.; Huazhong University of Science & Technology Press: Wuhan, China, 1991.

25. Zhou, Y.; Liu, Z.; Zhang, G.; Zhou, L.; Liu, Y.; Tang, L.; Jiang, Z.; Yang, B. Structural Damage Identification Method Based on Moving Principal Component Analysis and Ensemble Learning. *J. Univ. Jinan Sci. Technol.* **2023**, *37*, 116–126. [CrossRef]
26. Kumar, K.; Biswas, P.K.; Dhang, N. Time Series-Based SHM Using PCA with Application to ASCE Benchmark Structure. *J. Civ. Struct. Health Monit.* **2020**, *10*, 899–911. [CrossRef]
27. Jia, S.; Li, R.; Li, Y. Research on material properties test and application of imported glued laminated timber in Taiyuan Botanical Garden. *Build. Struct.* **2022**, *52*, 17–21. [CrossRef]
28. Zhang, L.; Xie, Q.; Wu, Y.; Zhang, B. Research on the mechanical properties and constitutive model of wood under parallel-to-grain cyclic loading. *China Civ. Eng. J.* **2023**, *44*, 286–304.
29. Wang, M.; Song, X.; Gu, X. Nonlinear analysis of wood based on three-dimensional combined elastic-plastic and damage model. *China Civ. Eng. J.* **2018**, *51*, 22–28+49. [CrossRef]
30. Wang, M.; Gu, X.; Song, X.; Tang, J. State-of-the-art for description methods of nonlinear mechanical behavior of wood. *J. Build. Struct.* **2021**, *42*, 76–86. [CrossRef]
31. Xu, B.-H.; Bouchaïr, A.; Racher, P. Appropriate Wood Constitutive Law for Simulation of Nonlinear Behavior of Timber Joints. *J. Mater. Civ. Eng.* **2014**, *26*, 04014004. [CrossRef]
32. Shannon, C.E. Communication in the Presence of Noise. *Proc. IEEE* **1998**, *86*, 447–457. [CrossRef]

sensors

MDPI

Article

A Transformer-Optimized Deep Learning Network for Road Damage Detection and Tracking

Niannian Wang [1], Lihang Shang [1] and Xiaotian Song [2,*]

[1] School of Water Conservancy and Transportation, Zhengzhou University, Zhengzhou 450001, China
[2] School of Engineering and Technology, China University of Geosciences (Beijing), Beijing 100083, China
* Correspondence: slh240054@163.com

Abstract: To solve the problems of low accuracy and false counts of existing models in road damage object detection and tracking, in this paper, we propose Road-TransTrack, a tracking model based on transformer optimization. First, using the classification network based on YOLOv5, the collected road damage images are classified into two categories, potholes and cracks, and made into a road damage dataset. Then, the proposed tracking model is improved with a transformer and a self-attention mechanism. Finally, the trained model is used to detect actual road videos to verify its effectiveness. The proposed tracking network shows a good detection performance with an accuracy of 91.60% and 98.59% for road cracks and potholes, respectively, and an F1 score of 0.9417 and 0.9847. The experimental results show that Road-TransTrack outperforms current conventional convolutional neural networks in terms of the detection accuracy and counting accuracy in road damage object detection and tracking tasks.

Keywords: road damage detection; object tracking; self-attention mechanism; transformer

Citation: Wang, N.; Shang, L.; Song, X. A Transformer-Optimized Deep Learning Network for Road Damage Detection and Tracking. *Sensors* **2023**, *23*, 7395. https://doi.org/10.3390/s23177395

Academic Editor: Biswanath Samanta

Received: 20 July 2023
Revised: 11 August 2023
Accepted: 23 August 2023
Published: 24 August 2023

1. Introduction

For economic development and social benefits, the health of roads is crucial. In daily life, repeated crushing by vehicles can cause damage to the structural layer of the road, which in turn produces cracks, potholes and other damage. The road performance and load carrying capacity will suffer as a result of pavement degradation [1,2]. If pavement damage is not repaired in a timely manner, rain and snow, as well as vehicle loads, will deepen the degree of pavement damage, which will seriously affect people's travel and safety and thus have an impact on social benefits. Therefore, regular maintenance of roads is very important. For road maintenance, one of the main aspects lies in efficient and accurate road damage detection. Currently, manual inspection and analysis is the main method of detecting pavement damage in China; however, manual inspection is often tedious and inefficient [3]. Although manual inspection has obvious operational advantages, when the inspector is inexperienced, the assessment of the degree of damage can be inaccurate, thus adversely affecting the pavement evaluation process [4–6]. The drawbacks of these manual inspections mean that this method no longer meets the increasing requirements of modern society for road damage detection.

1.1. Related Works

1.1.1. Conventional Methods

In addition to the above manual detection methods, conventional methods of road damage detection include automatic detection and image processing techniques. With the development of research and technological support, the usage of automatic road damage detection is constantly expanding, with conventional equipment such as infrared or sensor-equipped road inspection vehicles [7,8]. However, due to the complexity of the actual environment in the road detection process, automated detection equipment is often unable to meet the actual needs in terms of recognition accuracy and speed, and this type of equipment often incurs higher hardware costs, corresponding to an increase in detection costs. For example, some vibration-based detection methods are suitable for real-time assessment of pavement conditions [9], but they cannot measure pavement damage in areas outside the vehicle wheel path or identify the size of pavement damage. Laser-measurement-based inspection methods use special equipment, such as a laser scanner, mounted on a separate inspection vehicle [10–13] to convert the pavement into a three-dimensional object in a coordinate system, and this method allows for the direct calculation of various metrics for an accurate evaluation of pavement condition. However, real-time processing at high speeds is difficult and relatively expensive due to the increased amount of computation required. Compared with the high cost of automatic detection, the benefits of image processing technology include a great effectiveness and low cost. As technology advances, its recognition accuracy also gradually improves. Therefore, numerous researchers have chosen to use image processing methods for the detection of pavement damage [14–16]. Traditional image processing methods use manually chosen features, such as color, texture and geometric features, to first segment pavement faults, and then machine learning algorithms are used to classify and match them for pavement damage detection purposes. For instance, Fernaldez et al. [17] began by preprocessing cracked photos of a road in order to highlight the major aspects of the cracks, and then chose a decision tree heuristic algorithm and finally achieved classification of the images. Rong G et al. [18] performed entropy and image dynamic threshold segmentation of pavement crack pixels based on thresholds obtained from image histograms as a way to classify cracked and non-cracked pixels. Bitelli G et al. [19] proposed another application of image processing to crack recognition, focusing and obtaining additional noise-free images of specific cracks. Li Q et al. [20] presented an image processing algorithm for accuracy and efficiency, which was specifically used for fast evaluation of pavement surface cracks. Song E P et al. [21] proposed an innovative optimized two-phase calculation method for primary surface profiles to detect pavement crack damage. Traditional image processing techniques cannot, however, meet the requirements of model generalization capability and resilience in real-world engineering through manually planned feature extraction due to the complexity of the road environment. For example, it is often impossible to segment an image effectively when it contains conditions such as uneven illumination.

1.1.2. Deep Learning Methods

The issues with the aforementioned conventional methods can be successfully resolved thanks to the recent rapid advancements in artificial intelligence and deep learning technology. Deep learning has advantages over the aforementioned techniques, including the absence of manual feature extraction and good noise robustness. With their strong feature extraction capabilities, deep-learning-based models are used more and more, for example, convolutional neural networks [22] are commonly employed in image classification [23], object detection [24] and semantic segmentation [25]. Nu et al. [26] proposed a unique method for detecting tunneling defects based on a masked region convolutional neural network (RCNN) and optimized the network with a path-enhanced feature pyramid network (PAFPN) and an edge detection branch to increase the detection accuracy. Wang et al. [27] used an improved network model based on Faster RCNN to detect and classify damaged roads, and used data augmentation techniques before training to address the imbalance in the number of different damage datasets to achieve better network training results. In the same vein, for crack detection, Kaige Zhang et al. [28] suggested a depth-generating adversarial network (GAN), which successfully addressed the issue of data imbalance, thus achieving a better training effect and a higher detection accuracy. Yi-zhou Lin et al. [29] suggested a cross-domain structural damage detection method based on transfer learning, which enhanced the performance of damage identification. Wang Zifeng et al. [30] used DeepLabV3 + model to achieve precise segmentation of certain building site objects and three-dimensional object reconstruction. With the help of fully convolutional networks (FCN), Yang et al. [31] were able to successfully identify cracks at the pixel level in pavement and wall images, but there was still a shortcoming of poor detection of small cracks. Jeong et al. [32] improved a model based on You Only Look Once (YOLO)v5x with Test-Time Augmentation (TTA), which could generate new images for data enhancement then combine the original photographs with the improved images in the trained u-YOLO. Although this method achieved a high detection accuracy, the detection speed was not good. Many other researchers have worked to test lightweight models. Shim et al. [33] developed a semantic segmentation network with a small volume. They improved the network's parameters, but at the same time affected the detection speed of the model. Sheta et al. [34] developed a lightweight convolutional neural network model, which had a good crack detection effect. However, this model still had the problem of a single application scenario and could not deal with multiple road damage detection. Guo et al. [35] improved the model based on YOLOv5s to achieve the purpose of detecting a variety of road damage, and achieved a high accuracy in damage detection. However, the improved model was somewhat higher in weight, and meeting the criteria of embedded devices proved difficult. In addition, Ma D. et al. [36] proposed an algorithm called YOLO-MF that combines an acceleration algorithm and median flow for intelligent recognition of pavement cracks, achieving high recognition accuracy and a good PR curve. All of the above researchers have made reasonable contributions to road damage detection, but there are some deficiencies. For example, the models only detect crack damage, they cannot find a reasonable balance between detection efficiency and accuracy, they cannot effectively detect damage in road videos, etc. These are problems that still need to be studied and solved.

YOLOv5 is a single-stage target detection algorithm. Four versions of the YOLOv5 single-stage target detection model exist: YOLOv5s, YOLOv5m, YOLOv5l and YOLOv5x. For this study, the fastest and smallest model, YOLOv5s, with parameters of 7.0 M and weights of 13.7 M, was selected. YOLOv5 makes the following improvements compared to YOLOv4: For input side, the model training phase makes use of mosaic data augmentation, adaptive anchor frame computation and adaptive picture scaling. The benchmark network makes use of the FOCUS structure and the Cross Stage Partial (CSP) structure. In the Neck network, between the Backbone and the final Head output layer, the Feature Pyramid Network (FPN)_Path Aggregation Network (PAN) structure is added. The loss function named Generalized Intersection over Union Loss (GIOU_Loss) is added to the Head output layer during training and predicts the Distance-IOU_nns of the screening frame.

As shown in Figure 1, the algorithm framework is split into three major sections: the backbone network (Backbone), the bottleneck network (Neck) and the detection layer (Output). The Backbone consists of a focus module (focus), a standard convolution module (Conv), a C3 module and a spatial pyramid pooling module (SPP). In YOLOv5, the network architecture is the same for all four versions, and two variables determine the network structure's size: depth_multiple and width_multiple. For instance, the C3 operation of YOLOv5s is performed just once, while YOLOv5l is three times as deep as v5s and three C3 surgeries will therefore be carried out. Since the one-stage network YOLOv5s technique leverages multilayer feature map prediction, it produces improved outcomes in terms of detecting speed and accuracy.

1.1.3. Aircraft-Based Evaluation Methods

Manual and automated detection methods are ground-based evaluation methods. Another method of assessing the pavement surface is through aerial observation. Aircraft-based evaluation methods are more efficient, cost-effective and safer than labor-intensive evaluation methods. Su Zhang et al. [37] explored the utility of the aerial triangulation (AT) technique and HSR-AP acquired from a low-altitude and low-cost small-unmanned aircraft system (S-UAS), and the results revealed that S-UAS-based hyper-spatial resolution imaging and AT techniques can provide detailed and reliable primary observations suitable for characterizing detailed pavement surface distress conditions. Susan M. Bogus et al. [38]

evaluated the potential of using HSR multispectral digital aerial photographs to estimate overall pavement deterioration using principal component analysis and linear least squares regression models. The images obtained from aerial photography can also be used to train models for pavement damage recognition. Ahmet Bahaddin Ersoz et al. [39] processed a UAV-based pavement crack recognition system by processing UAV-based images for support vector machine (SVM) model training. Ammar Alzarrad et al. [40] demonstrated the effectiveness of combining AI and UAVs by combining high-resolution imagery with deep learning to detect disease on roofs. Long Ngo Hoang, T et al. [41] presented a methodology based on the mask regions with a convolutional neural network model, which was coupled with the new object detection framework Detectron2 to train a model that utilizes roadway imagery acquired from an unmanned aerial system (UAS).

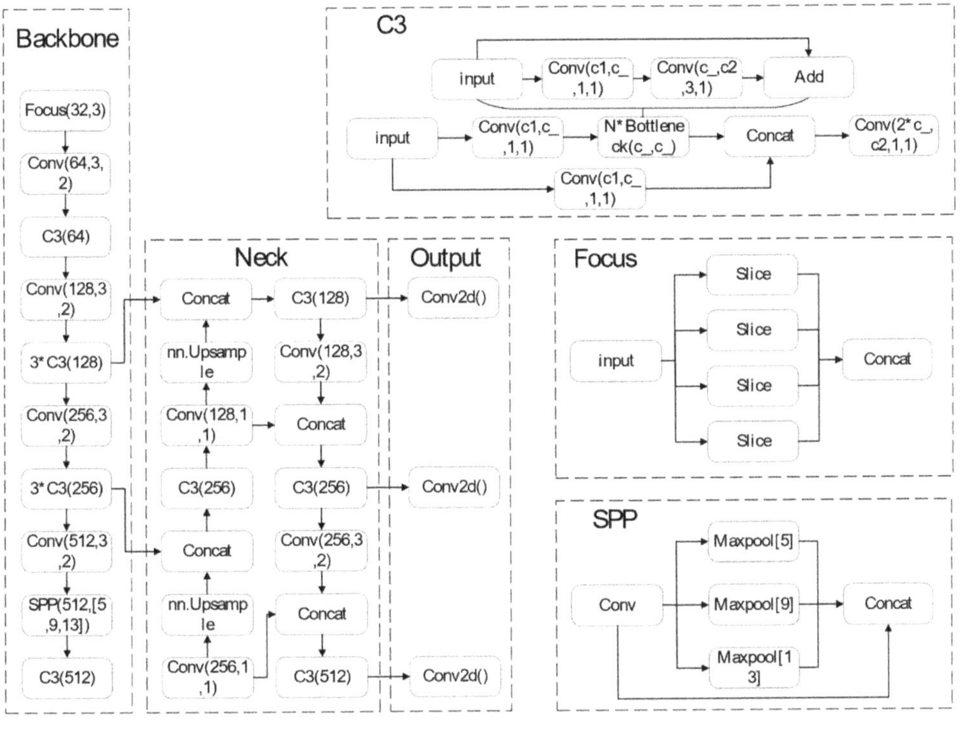

Figure 1. The detail of the network of YOLOv5s.

1.2. Contribution

Aiming at the problem of repeated missed detections due to a low detection accuracy during video detection of pavement damage, the paper's primary contribution is to propose and train a tracking and counting model named Road-TransTrack and improve the tracking model by using a transformer and a self-attention mechanism, which increases the detection precision of pavement damage when the model is tracking and achieves accurate counting of damage without damaging the detection speed, making it more appropriate for work detecting pavement damage.

2. Methodology

2.1. Transformer and Self-Attention

The transformer model is an attention-based neural network architecture that learns interdependencies between sequences through the self-attention mechanism. In a one-dimensional signal classification task, the signal can be considered as a sequence, and the transformer model can be employed to study the interdependence of various points in the sequence. Then, the signal is classified based on the learned information. As shown in Figure 2, based on the correlation between the input samples, a self-attention network is built. Initially, the input sequence x, as shown in Formulas (1)–(3), is multiplied by the weight matrices (W_k, W_v, W_q) to obtain the key vector k_i, the value vector v_i and the query vector q_i, respectively.

$$k_i = W_k x_i \tag{1}$$

$$v_i = W_v x_i \tag{2}$$

$$q_i = W_q x_i \tag{3}$$

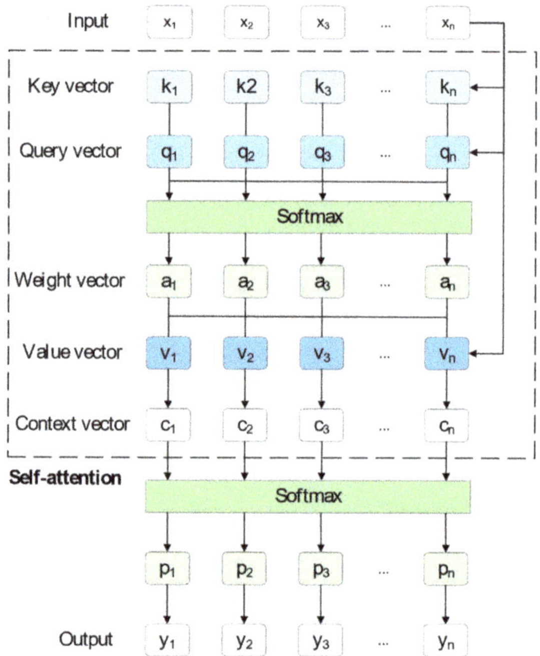

Figure 2. The self-attention mechanism's structural diagram.

Secondly, the key vector is multiplied by the query vector, as shown in Formula (4), and the weight vector a_i can be obtained under the softmax function processing. The weight vector represents the correlation of xi with the sequence of [x_1, x_2, ..., x_n] of the sequence, i.e., the degree of attention of x_i.

$$a_i = Softmax\left(\left[k_1^T, k_2^T, k_3^T, \cdots, k_n^T\right] q_i\right) \tag{4}$$

After that, as demonstrated by Formula (5), the product of the value vector and the weight vector is the semantic vector c_i, where the value vector represents the value of each input x_i.

$$c_i = \left[v_1^T, v_{21}^T, v_3^T, \cdots, v_n^T\right] a_i \tag{5}$$

In the end, the distribution of probabilities can be obtained by softtmax function processing and the corresponding output can be obtained by label coding.

The transformer is a pile of self-attention networks, which, in contrast to typical models, uses only self-attention mechanisms as a way to reduce computational effort and not corrupt the final experimental results [42]. As shown in Figure 3, the transformer model has two main parts: an encoder and a decoder. The input patches are fed into the multi-headed self-attention network, which is a type of self-attention network. The multi-headed self-attention network divides the result into eight subspaces and more relevant information can be learned in different subspaces [43]. To improve the deep network, residual connectivity and layer normalization are employed in the full network. As demonstrated by Formula (6), the multilayer perceptron (MLP) consists of two fully connected layers and a nonlinear activation function.

$$MLP(x) = max(0, xW_1 + b_1)W_2 + b_2 \quad (6)$$

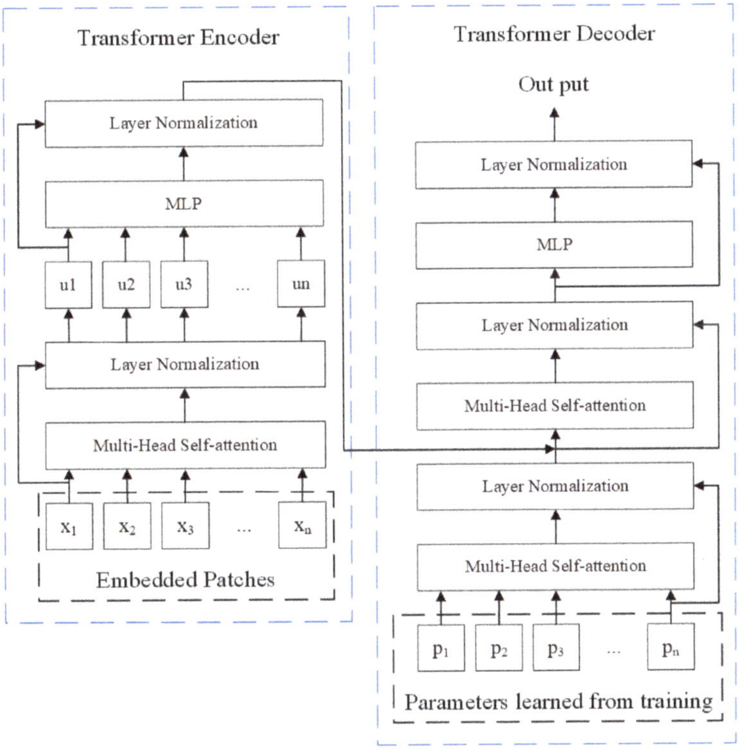

Figure 3. The structure diagram of the transformer.

2.2. *Road-TransTrack Detection and Tracking Model*

Traditional deep-learning-based pavement damage detection algorithms are often effective in obtaining the class and location of damage. However, for sequences of consecutive frames, conventional detection algorithms cannot effectively identify the same impairment and cannot accurately count multiple impairments. In this study, the proposed detection tracking model called Road-TransTrack can solve the above problem. Detection is a static task that generally finds regions of interest based on a priori knowledge or salient features. Tracking, however, is a fluid job, finding the same thing in a series of successive frames by means of characteristics carried over from the earlier frame. The tracking task checks the picture similarity of the previous and current frames to find the best matching position to find the target's dynamic path.

As illustrated in Figure 4, successive frames of the pavement video are first fed into the model, defects are detected when they first appear in frame Ft and the amount of defects is increased by one. The frames F_t and F_{t+1} are then fed into the tracking model. This damage continues to be tracked till it vanishes from the video, and IOU (Intersection over Union) matching is performed between the tracked and detected frames to obtain the tracking result. The detection and counting of the next damage continue. Finally, the overall number of discovered defects is determined. Meanwhile, the network is improved with the transformer to enhance the performance of the network.

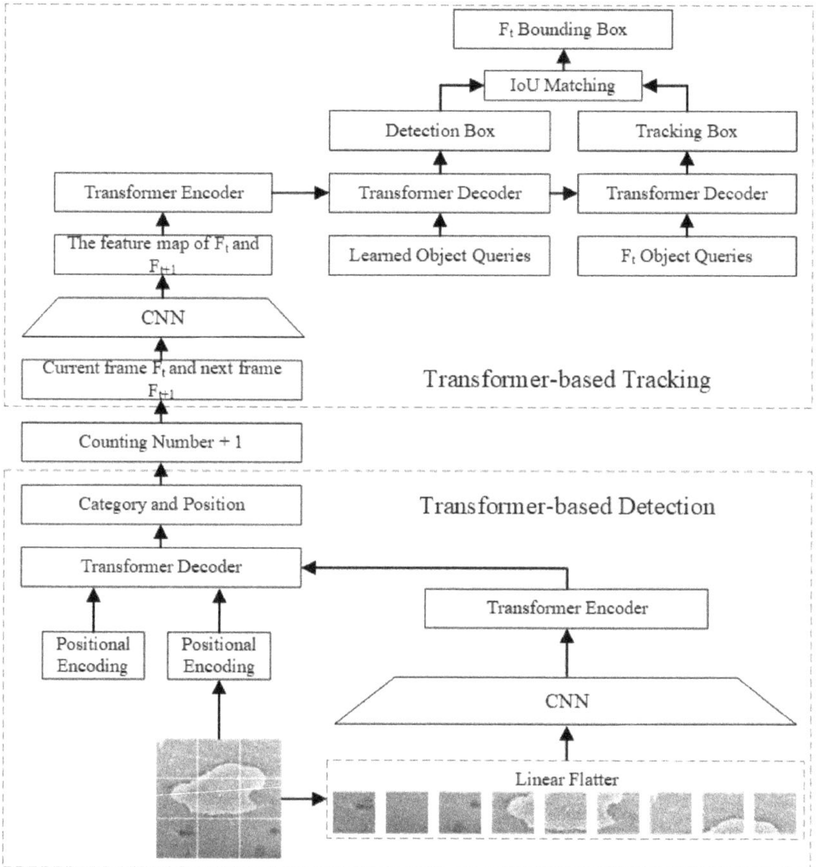

Figure 4. The detailed network of Road-TransTrack.

3. Dataset Construction

3.1. Data Collection

Like the deep convolutional neural network model, the transformer-improved network model also necessitates a lot of image data as the dataset. Images in today's road damage datasets have problems like erratic resolution, inconsistent picture data capturing equipment and extrinsic influences such as lighting and shadows. These have a significant impact on the criteria for the datasets used to train the models. Therefore, this study used a pavement damage dataset that was collected and produced by us. The initial image acquisition device is an integrated vehicle used for pavement detection, as shown in Figure 5. The parameters of the on-board camera are shown in Table 1. Combined with the actual acquisition needs, the shooting height was set between 40 and 80 cm to ensure the right size of damage in the images. Images were captured under normal lighting for several asphalt as well as concrete roads, and then images with high clarity and a balanced amount of damage were manually retained for the next step of processing.

Figure 5. Integrated vehicle used for pavement detection.

Table 1. Camera parameters.

Sensor Pictures	Equipment Parameters	HIKVISION U68
	Highest resolution	4 K
	Highest resolution video output	3840 × 2160 30/25 FPS
	Maximum Field of View	83° × 91°
	Digital zoom	fourfold
	Autofocus	support
	TOF Sensing	support

3.2. Data Processing

YOLOv5-Based Classification Network

Since this study focuses on the two most common types of road damage, potholes and cracks, the original data images collected need to be extracted and classified, i.e., two types of images with pothole and crack damage were selected to build the dataset. In order to achieve efficient and high accuracy classification, we adopted the YOLOv5 network with better performance for image classification. Four versions of the YOLOv5 model exist: YOLOv5s, YOLOv5m, YOLOv5l and YOLOv5x. After testing the four models, the smallest and quickest model, YOLOv5s, was used in this study under the condition of guaranteed accuracy. The acquired images were normalized and scaled down to 640 × 640 size prior to model training to ensure that the YOLO model performs optimally for training. After standardization, the data were manually annotated using annotation according to different types of road damage, where the annotation file format was txt. For data preparation, a total of 1000 images of potholes and fractures were prepared. During training, the learning rate was 0.01 and the mini-batch number and momentum coefficient were set to 2 and 0.937.

For the classification model, the true and predicted classification permutations are as follows: True Positives (*TP*): the number of true positive classes predicted as positive classes; False Positives (*FP*): the number of true negative classes predicted as positive classes; False Negatives (*FN*): the number of true positive classes predicted as negative classes; and True Negatives (*TN*): the number of true negative classes predicted as negative classes. The following indicators can be defined based on the values of the above four categories.

Accuracy is calculated as:

$$Accuracy = \frac{TP + TN}{TP + FP + TN + FN} \tag{7}$$

Precision is calculated as:

$$Precision = \frac{TP}{TP + FP} \tag{8}$$

Recall is calculated as:

$$Recall = \frac{TP}{TP + FN} \tag{9}$$

When the number of classification targets is unbalanced, the *F*1 *score* is used as the numerical evaluation index, and the *F*1 *score* is calculated as in Formula (10):

$$F_1 score = \frac{2 \times Precision \times Recall}{Precision + Recall} \tag{10}$$

After the training of the model, in the testing phase, the IOU threshold was set to 0.5 and the confidence threshold was set to 0.4. The final results were calculated according to the above formula and shown in Table 2. The classification accuracy of cracks and potholes reached 85.10% and 92.47%, and the F1 scores were 0.8512 and 0.9259, respectively. Most of the images of cracks and potholes can be correctly selected.

Table 2. The output of the classification network based on YOLOV5.

Class	Accuracy	Precision	Recall	F1 Score
Crack	0.8510	0.8407	0.862	0.8512
Pothole	0.9247	0.9076	0.945	0.9259

The images of potholes and cracks filtered by the classification network are shown in Figure 6. Data annotation was performed on these images to construct the dataset required for training. In total, there are 310 potholes and 300 cracks in the training set, 104 potholes and 101 cracks in the validation set and 103 potholes and 100 cracks in the test set.

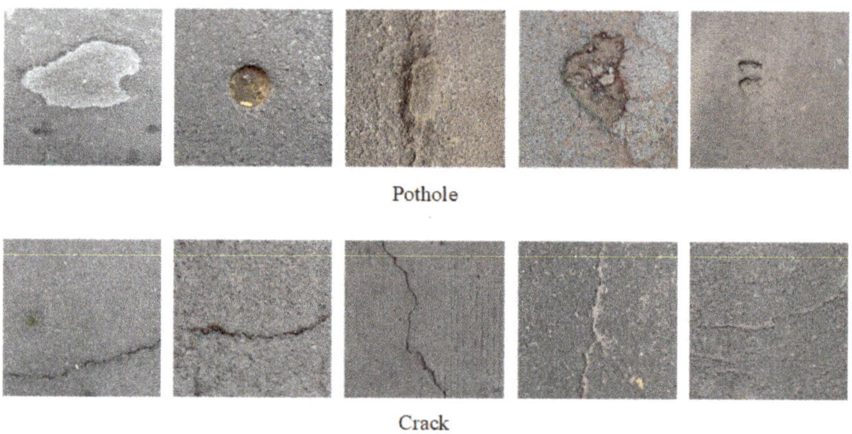

Figure 6. Damage images obtained from classification networks.

4. Road-TransTrack-Based Road Damage Detection and Tracking

4.1. Model Initialization

Computer vision algorithms based on deep learning require an abundance of labeled images as datasets. Similarly, the transformer relies on a large amount of data. Studies and tests have shown that as the size of the dataset increases, the CNN model is eventually surpassed by the transformer model in terms of detection performance [44]. Thus, in order to improve the performance of the model, migration learning can be used to improve the model detection performance before training with the prepared dataset. Transfer learning is a way of improving learning effectiveness by transferring the knowledge structure of a related domain to the target domain [45]. In this study, a model that has been trained on the Microsoft COCO dataset was chosen for the transformer-based detection network setup.

4.2. Hyperparameter Tuning

In the case of deep learning networks, the model parameters include common parameters and hyperparameters. The common parameters are the weight parameters of each network layer. Back propagation and ongoing training can be used to find the best public parameters. Unlike public parameters, the values of the hyperparameters, which are generally set artificially through experience, were set before the start of training. In general, to enhance learning performance and effectiveness, manually optimizing the hyperparameters and selecting an ideal set of hyperparameters is required for model training. The hyperparameters that have an important effect on the model performance primarily comprise the learning rate, the weight decay coefficient and the mini-batch size. In this experiment, the learning rate and weight decay coefficients were adjusted, and six combinations were trained; the outcomes are shown in Table 3.

Table 3. Hyperparameter tuning.

Case	Learning Rate	Weight Decay	Accuracy
1	10^{-5}	5×10^{-4}	90.73%
2	10^{-5}	10^{-5}	89.38%
3	10^{-5}	10^{-3}	89.75%
4	5×10^{-5}	10^{-4}	90.88%
5	2×10^{-4}	10^{-4}	91.59%
6	10^{-4}	10^{-4}	89.06%

As shown in Table 3, the model obtained the highest accuracy when the learning rate was 2×10^{-4} and the weight decay coefficient was 10^{-4}. As shown in Figure 7, the loss of the model gradually decreases as the training proceeds. As shown in Figure 8, the model accuracy gradually increases as the training proceeds. After 68 epochs of training, the model reached a maximum accuracy of 91.59% and the loss curve became flat. This model was saved and the hyperparameters set during the training of this model were selected for the next step of the study.

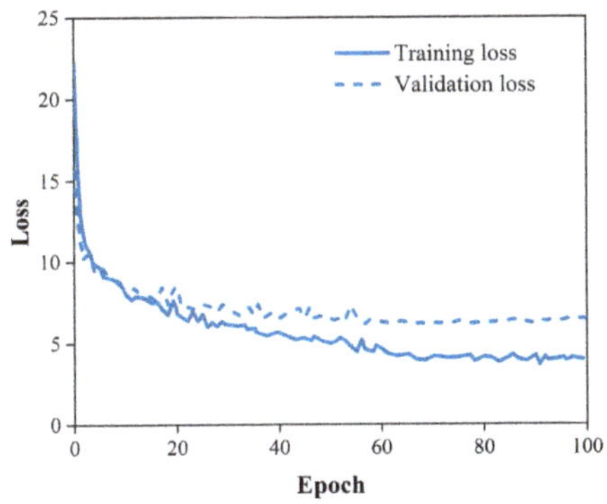

Figure 7. The decline curve of loss.

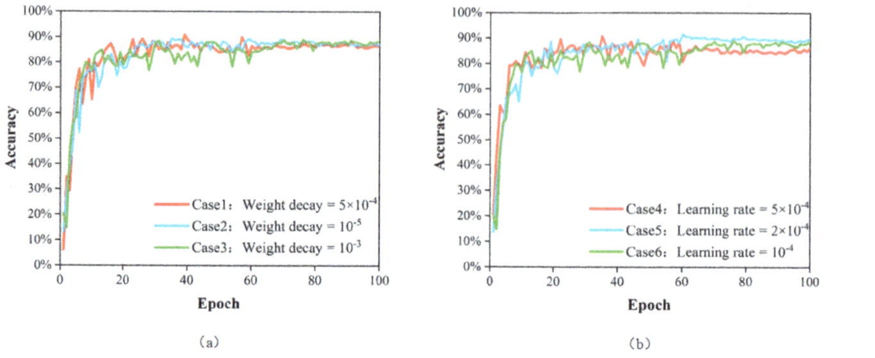

Figure 8. The upcurve of accuracy (**a**) with different weight decays; (**b**) with different learning rates.

4.3. Transformer-Based Detection and Tracking Network

Some traditional CNN-based networks have achieved good performance in pavement image damage detection. However, for the same damage present in consecutive frames of video, these networks often either fail to detect it or perform duplicate counts without achieving good detection results. To address the above issues, the improved detection tracking network with a transformer was trained and tested on the dataset. Since the data are static images, adjacent frames are simulated during training by randomly scaling and transforming the static images. The optimal combination of hyperparameters derived in the above model initialization was selected for model training. As shown in Figures 9 and 10, as the epoch increases, the loss value decreases, the accuracy increases and the optimal model is saved.

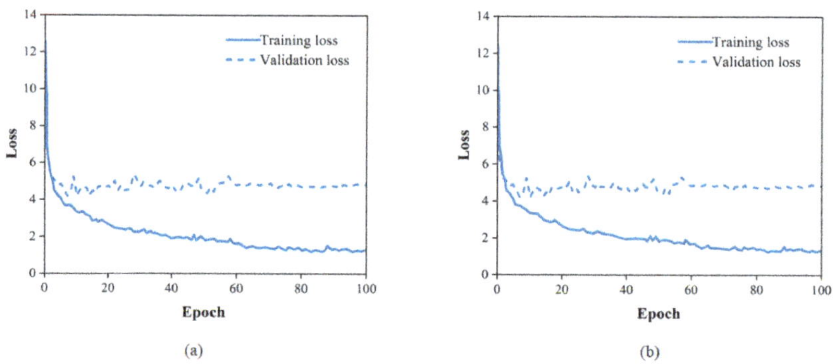

Figure 9. The decline curve of loss: (**a**) cracks; (**b**) potholes.

The upgraded tracking network was put to the test using the test set. Table 4 demonstrates the tracking network results; for the detection of pavement damage, the average accuracy score was 95.09% and the average F1 score value was 0.9646. As shown in Figure 11, the PR curve is close to the upper right corner of the coordinate system. This implies that the trained network performs well for tracking pavement diseases.

Table 4. The results of the transformer-based tracking network.

Class	Accuracy	Precision	Recall	F1 Score
Crack	91.60%	91.6%	96.9%	0.9417
Pothole	98.59%	98.6%	98.9%	0.9874
Mean	95.095	95.1%	97.9%	0.9646

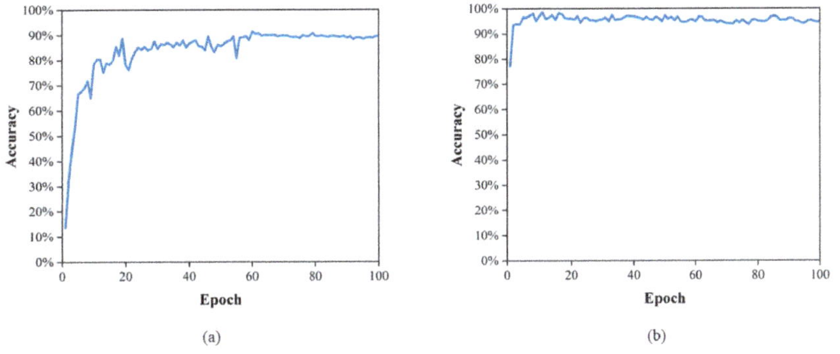

Figure 10. The upcurve of accuracy: (**a**) cracks; (**b**) potholes.

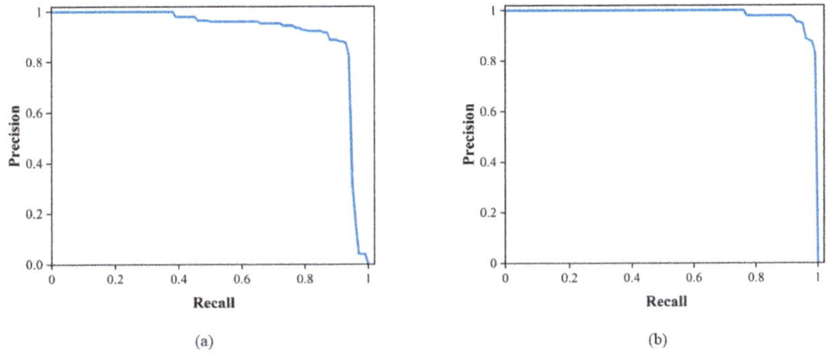

Figure 11. The PR curves of tracking network: (**a**) cracks; (**b**) potholes.

During the whole tracking process, the frame sequence is first detected. If the detection network detects the presence of damage in frame F_t, the frame image is fed to the tracking network and the number of damages is increased by 1. Next, defects are detected in the next frame based on the features in frame F_{t+1} and are tracked based on the features in frame F_t. Finally, IOU (Intersection over Union) matching is performed between the tracked and detected frames to obtain the tracking results.

To visualize the effects of model training more intuitively, two videos of pavement damage were selected to test the trained network.

As shown in Figure 12, two crack damages appear sequentially in the first video. The figure shows the tracking process from the appearance of the first crack to the appearance of the second crack and the simultaneous presence of both cracks, with the serial numbers of the two cracks in the upper left corner of the detection box in that order. The detection and counting results are consistent with the results of manual identification in the field.

As shown in Figure 13, three pothole damages appear in sequence in the second video. The diagram shows the tracking process from the appearance to disappearance of the first damage, the appearance to disappearance of the second damage, the coexistence of the first two damages and the appearance of the third damage, with the serial numbers of the three damages in the upper left corner of the detection box in that order. The detection and counting results are the same as those of manual identification in the field.

Both the above illustrations and results show that the trained model has good results for the detection tracking and counting of pavement potholes and crack damage and can basically meet the actual detection requirements.

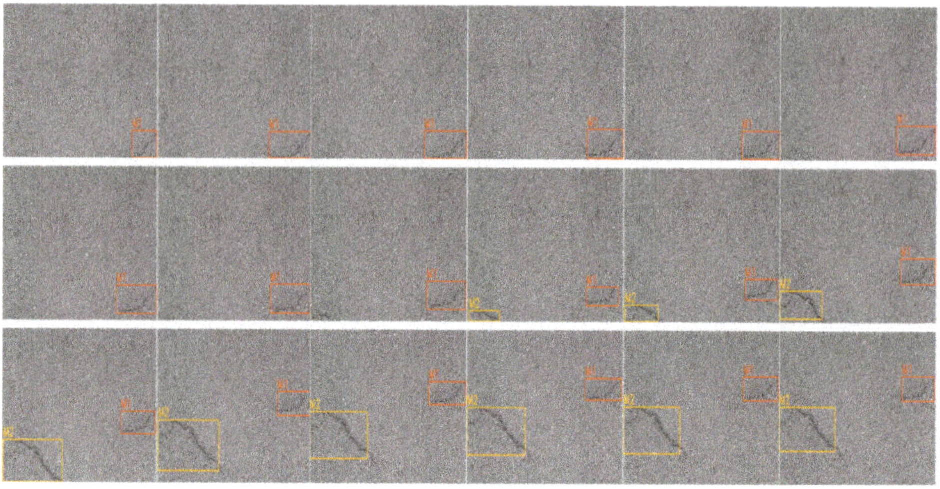

Figure 12. Tracking results for video 1.

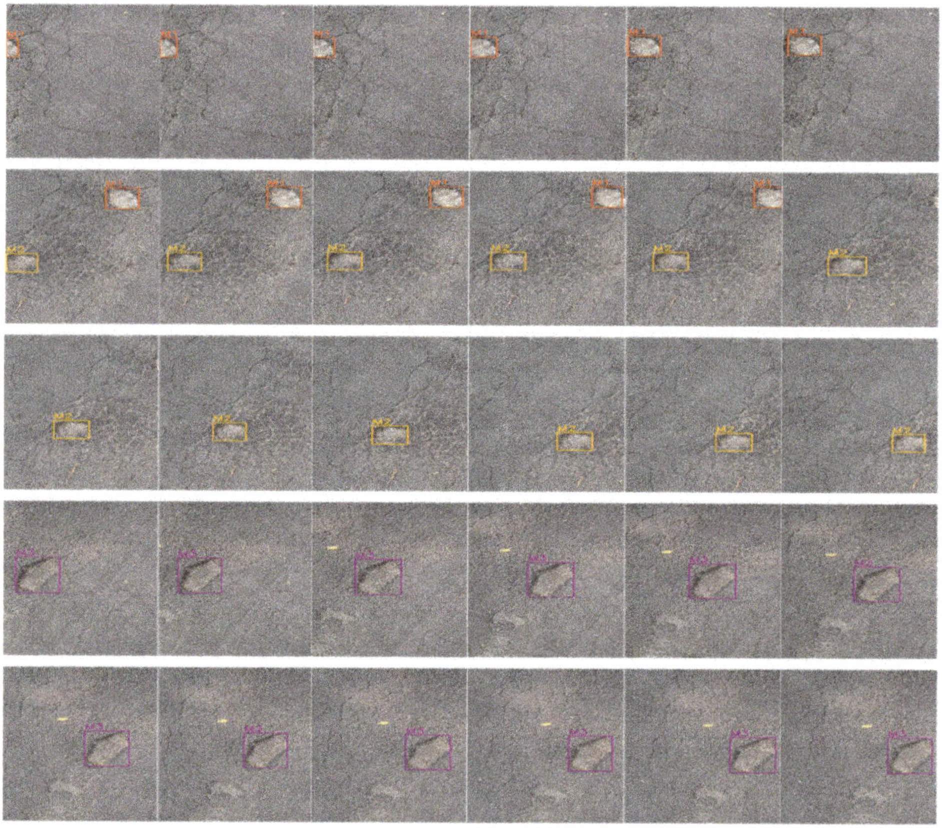

Figure 13. Tracking results for video 2.

5. Results and Discussion

In this study, the trained models were tested on a test set. To validate the performance of the model, the improved algorithm was compared with the CNN-based algorithm using evaluation metrics such as accuracy, precision, recall and F1 score. The algorithms YOLOv3, Single Shot MultiBox Detector (SSD) and Faster RCNN, which are commonly used for pavement damage detection, were selected for comparison [46,47]. For detection algorithms, the PR curve is an intuitive comparison graph; the closer the curve is to the upper right, the better the performance of the algorithm. As shown in Table 5, the classical CNN-based network was used to test the pavement damage dataset and the results were obtained. As shown in Table 6, compared with the classical CNN network, the F1 score of the detection network optimized by a transformer is the best, at 96.64, and the accuracy is 95.10%, which are 12.49% and 2.74% higher than the optimal CNN model, respectively. As illustrated in Figure 14a, for the class of cracks, when compared to various CNN models, the PR curve of the transformer optimized detection network is closest to the upper right corner. As illustrated in Figure 14b, for the class of potholes, the red curve of our network encloses the curves of YOLOv3, SSD and Faster RCNN. These comparative results effectively demonstrate that the transformer has a superior performance over CNN-based networks for the classification and detection of pavement damage.

Table 5. The results of the classical CNN-based detection network.

Class	Accuracy	Precision	Recall	F1 Score
(YOLOv3)				
Crack	75.26%	75.06%	80.60%	77.73
Pothole	90.74%	89.56%	91.60%	90.57
(SSD)				
Crack	92.72%	91.49%	71.67%	80.37
Pothole	92.00%	80.39%	91.11%	85.41
(Faster RCNN)				
Crack	87.29%	35.37%	95.08%	51.55
Pothole	93.46%	75.00%	95.33%	83.95

Table 6. A comparison of different detection networks.

Network	Mean Precision	Mean Recall	Mean F1 Score	Mean Accuracy
Our Network	95.09%	97.90%	96.46	95.10%
YOLOv3	82.31%	86.10%	84.15	83.00%
SSD	85.94%	81.39%	82.98	92.36%
Faster RCNN	55.18%	95.21%	67.75	90.38%

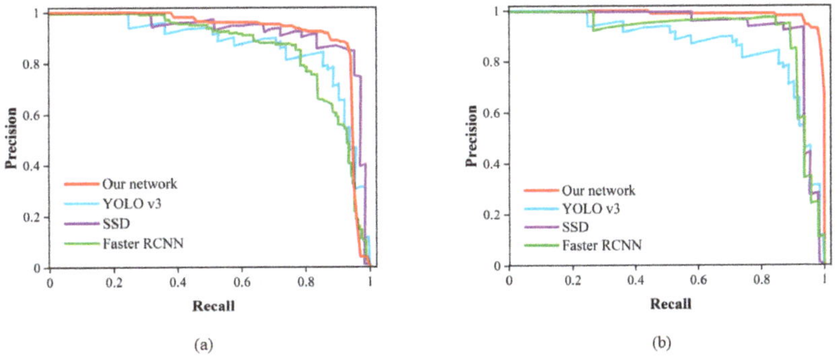

Figure 14. A comparison of PR curves: (**a**) cracks; (**b**) potholes.

In order to show the comparison results more intuitively, the same frame was detected with our network and the traditional CNN network, respectively, and the results were compared. As shown in Figure 15, for crack images, (a) the set of detection images demonstrates that the four networks detect approximately the same effect when there is only one crack in the figure and (b) the group detection images show that when multiple cracks appear in the figure, our network shows a better detection effect, without missing or wrong detections, and counts are carried out. For potholes, (a) the set of detection images shows that each network can accurately detect the two potholes present in the figure when the pothole size feature is obvious, and also our network counts the potholes and (b) the group detection images show that our network detects a pothole, while all other networks produce false detections, i.e., parts of the ground that are similar in shape to potholes are detected as potholes.

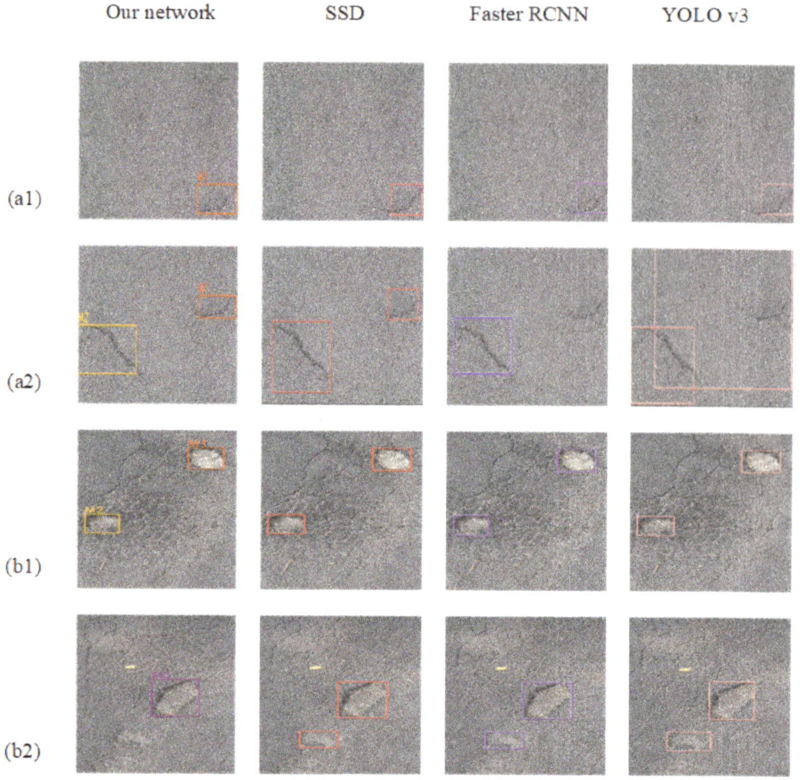

Figure 15. Comparison of different network detection results: (**a1**) comparison of individual crack detection results; (**a2**) comparison of multiple crack detection results; (**b1**) comparison of individual pothole detection results; (**b2**) comparison of multiple pothole detection results.

The above comparative tests show that the proposed model has good performance in terms of detection accuracy and accuracy of damage statistics. However, the detection speed of the current model does not meet the requirement of real-time execution. From the establishment of the dataset to the subsequent part of model testing, the current study uses pavement images and videos taken on the ground, so the generalization degree of the model needs to be further investigated. For example, the collection of pavement damage images can be carried out using the UAS technique to enrich the dataset required for model training; the model can then be used for the detection of images and videos captured by the UAS for better and efficient assessment of pavement damage.

6. Conclusions

For pavement damage video inspections, the detection accuracy is not high, resulting in the problem of repeated counting of missed detections. The main contribution of this study is the proposed tracking counting network called Road-TransTrack. When damage first appears in a video, it is detected and tracked until the defect disappears and the number of damages increases by 1. The tracking and counting model is improved with a transformer and a self-attention mechanism to improve the accuracy of damage detection and counting in road videos. Compared to the classic CNN network, the F1 score of the transformer-optimized detection network is 96.64, with an average accuracy of 95.10%, which are 12.49% and 2.74% higher than the optimal CNN model, respectively. A comparison of actual frame image detections shows that compared to other classical CNN networks, the model does not have the phenomena of missing and wrong detections. Additionally, the detection results of two road videos show that the model can track and count potholes and cracks correctly. All the above results indicate that the model in this study possesses better performance in video detection and tracking of road damage. In the future, we will consider training and testing models for more types of road damage.

Author Contributions: Writing—original draft, L.S.; Writing—review and editing, N.W.; Investigation, X.S. All authors have read and agreed to the published version of the manuscript.

Funding: This research was funded by the National Key Research and Development Program of China (No. 2022YFC3801000), the National Natural Science Foundation of China (No. 51978630), the Program for Innovative Research Team (in Science and Technology) in University of Henan Province (No. 23IRTSTHN004), the National Natural Science Foundation of China (No. 52108289), the Program for Science & Technology Innovation Talents in Universities of Henan Province (No. 23HASTIT006), the Postdoctoral Science Foundation of China (No. 2022TQ0306), the Key Scientific Research Projects of Higher Education in Henan Province (No. 21A560013) and the Open Fund of Changjiang Institute of Survey, Lanning, Design and Research (No. CX2020K10).

Institutional Review Board Statement: Not applicable.

Informed Consent Statement: Not applicable.

Data Availability Statement: Data are contained within the article.

Conflicts of Interest: The authors declare that there are no conflicts of interest regarding the publication of this paper.

References

1. Yao, Y.; Tung, S.-T.E.; Glisic, B. Crack detection and characterization techniques-An overview. *Struct. Control. Health Monit.* **2014**, *21*, 1387–1413. [CrossRef]
2. Jahanshahi, M.R.; Masri, S.F. A new methodology for non-contact accurate crack width measurement through photogrammetry for automated structural safety evaluation. *Smart Mater. Struct.* **2013**, *22*, 035019. [CrossRef]
3. Barreira, E.; de Freitas, V.P. Evaluation of building materials using infrared thermography. *Constr. Build. Mater.* **2007**, *21*, 218–224. [CrossRef]
4. Wang, N.; Zhao, X.; Zhao, P.; Zhang, Y.; Zou, Z.; Ou, J. Automatic damage detection of historic masonry buildings based on mobile deep learning. *Autom. Constr.* **2019**, *103*, 53–66. [CrossRef]
5. Gattulli, V.; Chiaramonte, L. Condition assessment by visual inspection for a bridge management system. *Comput.-Aided Civ. Infrastruct. Eng.* **2005**, *20*, 95–107. [CrossRef]
6. O'Byrne, M.; Schoefs, F.; Ghosh, B.; Pakrashi, V. Texture Analysis Based Damage Detection of Ageing Infrastructural Elements. *Comput.-Aided Civ. Infrastruct. Eng.* **2013**, *28*, 162–177. [CrossRef]
7. Torbaghan, M.E.; Li, W.; Metje, N.; Burrow, M.; Chapman, D.N.; Rogers, C.D.F. Automated detection of cracks in roads using ground penetrating radar. *J. Appl. Geophys.* **2020**, *179*, 104118. [CrossRef]
8. Hadjidemetriou, G.M.; Vela, P.A.; Christodoulou, S.E. Automated Pavement Patch Detection and Quantification Using Support Vector Machines. *J. Comput. Civ. Eng.* **2018**, *32*, 04017073. [CrossRef]
9. Chang, K.T.; Chang, J.R.; Liu, J.K. Detection of Pavement Distresses Using 3D Laser Scanning Technology. In Proceedings of the International Conference on Computing in Civil Engineering, Cancun, Mexico, 12–15 July 2005.
10. Li, S.; Yuan, C.; Liu, D.; Cai, H. Integrated Processing of Image and GPR Data for Automated Pothole Detection. *J. Comput. Civ. Eng.* **2016**, *30*. [CrossRef]

11. Huang, Y.; Xu, B. Automatic inspection of pavement cracking distress. *J. Electron. Imaging* **2006**, *15*, 013017. [CrossRef]
12. Zou, Q.; Cao, Y.; Li, Q.; Mao, Q.; Wang, S. Crack Tree: Automatic crack detection from pavement images. *Pattern Recogn. Lett.* **2012**, *33*, 227–238. [CrossRef]
13. Oliveira, H.; Correia, P.L. Automatic Road Crack Segmentation Using Entropy and Image Dynamic Thresholding. In Proceedings of the 2009 17th European Signal Processing Conference, Glasgow, UK, 24–28 August 2009.
14. Nguyen, T.S.; Bégot, S.; Duculty, F.; Avila, M. Free-form anisotropy: A new method for crack detection on pavement surface images. In Proceedings of the IEEE International Conference on Image Processing, Brussels, Belgium, 11–14 September 2011.
15. Nguyen, H.T.; Nguyen, L.T.; Sidorov, D.N. A robust approach for road pavement defects detection and classification. *Irkutsk. Natl. Res. Tech. Univ.* **2016**, *3*, 40–52. [CrossRef]
16. Safaei, N.; Smadi, O.; Masoud, A.; Safaei, B. An Automatic Image Processing Algorithm Based on Crack Pixel Density for Pavement Crack Detection and Classification. *Int. J. Pavement Res. Technol.* **2021**, *15*, 159–172. [CrossRef]
17. Cubero-Fernandez, A.; Rodriguez-Lozano, F.J.; Villatoro, R.; Olivares, J.; Palomares, J.M. Efficient pavement crack detection and classification. *Eurasip J. Image Video Process.* **2017**, *2017*, 1. [CrossRef]
18. Rong, G.; Xin, X.; Dejin, Z.; Hong, L.; Fangling, P.; Li, H.; Min, C. A Component Decomposition Model for 3D Laser Scanning Pavement Data Based on High-Pass Filtering and Sparse Analysis. *Sensors* **2018**, *18*, 2294.
19. Bitelli, G.; Simone, A.; Girardi, F.; Lantieri, C. Laser Scanning on Road Pavements: A New Approach for Characterizing Surface Texture. *Sensors* **2012**, *12*, 9110–9128. [CrossRef] [PubMed]
20. Li, Q.; Yao, M.; Yao, X.; Xu, B. A real-time 3D scanning system for pavement distortion inspection. *Meas. Sci. Technol.* **2010**, *21*, 015702. [CrossRef]
21. Park, S.E.; Eem, S.H.; Jeon, H. Concrete crack detection and quantification using deep learning and structured light. *Constr. Build. Mater.* **2020**, *252*, 119096. [CrossRef]
22. Aloysius, N.; Geetha, M. A review on deep convolutional neural networks. In Proceedings of the 2017 International Conference on Communication and Signal Processing (ICCSP), Chennai, India, 6–8 April 2017.
23. Wang, F.; Jiang, M.; Qian, C.; Yang, S.; Tang, X. Residual Attention Network for Image Classification. In Proceedings of the 2017 IEEE Conference on Computer Vision and Pattern Recognition (CVPR), Honolulu, HI, USA, 21–26 July 2017.
24. Zhao, Z.Q.; Zheng, P.; Xu, S.T.; Wu, X. Object Detection with Deep Learning: A Review. *arXiv* **2018**. [CrossRef]
25. Wang, P.; Chen, P.; Yuan, Y.; Liu, D.; Huang, Z.; Hou, X.; Cottrell, G. Understanding Convolution for Semantic Segmentation. In Proceedings of the 2018 IEEE Winter Conference on Applications of Computer Vision (WACV), Lake Tahoe, NV, USA, 12–15 March 2018.
26. Xu, Y.; Li, D.; Xie, Q.; Wu, Q.; Wang, J. Automatic defect detection and segmentation of tunnel surface using modified Mask R-CNN. *Measurement* **2021**, *178*, 109316. [CrossRef]
27. Wang, W.; Wu, B.; Yang, S.; Wang, Z. Road Damage Detection and Classification with Faster R-CNN. In Proceedings of the 2018 IEEE International Conference on Big Data (Big Data), Seattle, WA, USA, 10–13 December 2018.
28. Zhang, K.; Zhang, Y.; Cheng, H.D. CrackGAN: Pavement Crack Detection Using Partially Accurate Ground Truths Based on Generative Adversarial Learning. *IEEE Trans. Intell. Transp. Syst.* **2020**, *22*, 1306–1319. [CrossRef]
29. Lin, Y.; Nie, Z.; Ma, H. Dynamics-based cross-domain structural damage detection through deep transfer learning. *Comput.-Aided Civ. Infrastruct. Eng.* **2022**, *37*, 24–54. [CrossRef]
30. Wang, Z.; Zhang, Y.; Mosalam, K.M.; Gao, Y.; Huang, S. Deep semantic segmentation for visual understanding on construction sites. *Comput.-Aided Civ. Infrastruct. Eng.* **2022**, *37*, 145–162. [CrossRef]
31. Yang, X.; Li, H.; Yu, Y.; Luo, X.; Huang, T.; Yang, X. Automatic Pixel-Level Crack Detection and Measurement Using Fully Convolutional Network. *Comput.-Aided Civ. Infrastruct. Eng.* **2018**, *33*, 1090–1109. [CrossRef]
32. Hegde, V.; Trivedi, D.; Alfarrarjeh, A.; Deepak, A.; Shahabi, C. Yet Another Deep Learning Approach for Road Damage Detection using Ensemble Learning. In Proceedings of the 2020 IEEE International Conference on Big Data (Big Data), Atlanta, GA, USA, 10–13 December 2020.
33. Shim, S.; Kim, J.; Lee, S.-W.; Cho, G.-C. Road surface damage detection based on hierarchical architecture using lightweight auto-encoder network. *Autom. Constr.* **2020**, *130*, 103833. [CrossRef]
34. Sheta, A.F.; Turabieh, H.; Aljahdali, S.; Alangari, A. Pavement Crack Detection Using Convolutional Neural Network. In Proceedings of the Computers and Their Applications, San Francisco, CA, USA, 23-25 March 2020.
35. Guo, K.; He, C.; Yang, M.; Wang, S. A pavement distresses identification method optimized for YOLOv5s. *Sci. Rep.* **2022**, *12*, 3542. [CrossRef]
36. Ma, D.; Fang, H.; Wang, N.; Zhang, C.; Dong, J.; Hu, H. Automatic Detection and Counting System for Pavement Cracks Based on PCGAN and YOLO-MF. *IEEE Trans. Intell. Transp. Syst.* **2022**, *23*, 22166–22178. [CrossRef]
37. Zhang, S.; Lippitt, C.D.; Bogus, S.M.; Neville, P.R.H. Characterizing Pavement Surface Distress Conditions with Hyper-Spatial Resolution Natural Color Aerial Photography. *Remote Sens.* **2016**, *8*, 392. [CrossRef]
38. Zhang, S.; Bogus, S.M.; Lippitt, C.D.; Neville, P.R.H.; Zhang, G.; Chen, C.; Valentin, V. Extracting Pavement Surface Distress Conditions Based on High Spatial Resolution Multispectral Digital Aerial Photography. *Photogramm. Eng. Remote Sens.* **2015**, *81*, 709–720. [CrossRef]

39. Ersoz, A.B.; Pekcan, O.; Teke, T. Crack identification for rigid pavements using unmanned aerial vehicles. In Proceedings of the International Conference on Building up Efficient and Sustainable Transport Infrastructure (BESTInfra), Prague, Czech Republic, 21–22 September 2017.
40. Alzarrad, A.; Awolusi, I.; Hatamleh, M.T.; Terreno, S. Automatic assessment of roofs conditions using artificial intelligence (AI) and unmanned aerial vehicles (UAVs). *Front. Built Environ.* **2022**, *8*, 1026225. [CrossRef]
41. Long Ngo Hoang, T.; Mora, O.E.; Cheng, W.; Tang, H.; Singh, M. Deep Learning to Detect Road Distress from Unmanned Aerial System Imagery. *Transp. Res. Rec.* **2021**, *2675*, 776–788.
42. Cheng, J.; Dong, L.; Lapata, M. Long Short-Term Memory-Networks for Machine Reading. *arXiv* **2016**, arXiv:1601.06733.
43. Vaswani, A.; Shazeer, N.; Parmar, N.; Uszkoreit, J.; Jones, L.; Gomez, A.N.; Kaiser, L.; Polosukhin, I. Attention Is All You Need. *arXiv* **2017**, arXiv:1706.03762.
44. Dosovitskiy, A.; Beyer, L.; Kolesnikov, A.; Weissenborn, D.; Houlsby, N. An Image is Worth 16x16 Words: Transformers for Image Recognition at Scale. *arXiv* **2020**, arXiv:2010.11929.
45. Arya, D.; Maeda, H.; Ghosh, S.K.; Toshniwal, D.; Mraz, A.; Kashiyama, T.; Sekimoto, Y. Transfer Learning-based Road Damage Detection for Multiple Countries. *arXiv* **2020**, arXiv:2008.13101.
46. Zhong, Q.; Li, C.; Zhang, Y.; Xie, D.; Yang, S.; Pu, S. Cascade Region Proposal and Global Context for Deep Object Detection. *Neurocomputing* **2017**, *395*, 170–177. [CrossRef]
47. Tao, X.; Gong, Y.; Shi, W.; Cheng, D. Object Detection with Class Aware Region Proposal Network and Focused Attention Objective. *Pattern Recognit. Lett.* **2018**, *130*, 353–361. [CrossRef]

MDPI
St. Alban-Anlage 66
4052 Basel
Switzerland
www.mdpi.com

MDPI Books Editorial Office
E-mail: books@mdpi.com
www.mdpi.com/books

www.ingramcontent.com/pod-product-compliance
Lightning Source LLC
Chambersburg PA
CBHW061507220926
51850CB00004B/191
9783725810819